# CALCULUS

OF

# VARIATIONS

# CALCULUS

## OF

# VARIATIONS

BY

## A. R. FORSYTH,

Sc.D., LL.D., Math.D., F.R.S.

EMERITUS PROFESSOR,
IMPERIAL COLLEGE OF SCIENCE AND TECHNOLOGY, LONDON:
FORMERLY SADLERIAN PROFESSOR
IN THE UNIVERSITY OF CAMBRIDGE.

## CAMBRIDGE
### AT THE UNIVERSITY PRESS
### 1927

CAMBRIDGE UNIVERSITY PRESS
Cambridge, New York, Melbourne, Madrid, Cape Town,
Singapore, São Paulo, Delhi, Tokyo, Mexico City

Cambridge University Press
The Edinburgh Building, Cambridge CB2 8RU, UK

Published in the United States of America by Cambridge University Press, New York

www.cambridge.org
Information on this title: www.cambridge.org/9781107640832

First published 1927
First paperback edition 2011

A catalogue record for this publication is available from the British Library

ISBN 978-1-107-64083-2 Paperback

TO

MARION

IN

REMEMBRANCE

# PREFACE.

The subject, commonly called the Calculus of Variations, has attracted a rather fickle attention at more or less isolated intervals in its growth. Its progress has been neither steady nor consecutive. From some cause, in its nature, or in its incompleteness, or in its presentation, it has not secured an abiding interest.

Not infrequently, investigators have been concerned with applications of the Calculus and, for their purpose, have been known to use fragmentary results.

Thus, in the theory of the potential, Dirichlet's Principle has been invoked. In instances when regard has been paid to the establishment of the Principle beyond an assumption of its intuitive truth, only the initial test belonging to weak conditions has been imposed ; and a general inference has been drawn, which was not justified by that test alone.

Again, the Principle of Least Action has been made the support, and sometimes the occasional basis, of theoretical explanations of the physics of the universe : though it should be added that the introduction of kinetic foci in dynamics is the equivalent of another necessary canonical test. Even so, all the recognised tests have assumed that variations in natural phenomena must be gently regular. Variations which, remaining small and continuous in their magnitude, change in a violently regular or irregular manner within a very restricted range, have usually been ignored . yet the theory of small vibrations wields a far-flung domination.

In Newton's problem of the Solid of Least Resistance, the formal solution satisfies all the customary tests which arise through variations of the gently regular type. Still, more than a century ago, Legendre proved that the solution is mathematically unsatisfactory, though its neglect by engineers is not due solely to mathematical deficiencies.

The significance of the investigations, due to Weierstrass, is not always recognised; but their importance need not be emphasised, as though complete finality has been attained. The results, usually associated with his name, relate to only the simplest class among the problems which present themselves and which require no more than the simplest form of his specially devised analysis. There is ample scope for further research by his method, in extension of the range of its application.

The present volume attempts a systematic exposition of the subject by what, in the main, is a uniform composite process. Though it does not purport to be a history, the gradual historical growth of the successive tests has governed the arrangement. A fundamental (yet quite elementary) simplification, derived from the Weierstrass method, has been used from the beginning, even to obtain the results originally due to the founders of the subject. These limited results maintain their standing, because they provide tests which must be satisfied in simple forms of enquiry, and because they remain significant even when they are merged in the wider results obtained by the more general method of Weierstrass.

Moreover, the volume has no pretensions to an encyclopædic range. Processes and investigations, however useful in the exploration of other regions, are omitted unless they fall into the course of exposition adopted. So far as I am aware, much of its material is novel. Two sources, more than others, have been useful to me. The first of them is the Moigno-Lindelöf volume *Calcul des Variations*, published in 1861; except for the Sarrus formalities, it seems to me an admirable exposition of the older range of investigation. The other source is to be found in such access to the work of Weierstrass as has been possible. Before the year 1895, I had read a manuscript copy of notes of lectures by

Weierstrass on his treatment of single integrals of the first order, including the associated isoperimetrical problems; for the loan of the volume from their College Library, I remain indebted to the authorities of St John's College, Cambridge. Since that date, Professor Harris Hancock has published (1903) his volume, based on similar notes and on lectures by Schwarz. Unfortunately, a general expectation, that an authoritative edition of the Weierstrass lectures would be published, has not yet been realised.

Beyond the sources just mentioned and such other sources as are quoted in the text, my work is independent. Some mathematicians may wish that the exposition had been differently balanced. Some will feel regret, and may award blame, for the omission of the work of writers such as Clebsch and Hilbert—an omission not due to lack of appreciation of their researches. Whatever its merits or its demerits, the presentation is that which has appealed to me, as leading most directly to a comprehension of the subject.

An abstract of the contents of the book may be useful, as an indication of its scope.

In the first chapter, the simplest form of integral is discussed. It involves only one dependent variable, together with the first derivative. The method adopted is, in substance, the older method for restricted variations; and the results obtained, including Jacobi's test which limits the extent of the range of the integration, are typical of those that persist in all subsequent investigations, though they do not constitute the aggregate of tests of a general character. The second chapter deals with the same type of integral by the method of Weierstrass, which makes both the dependent variable and the independent variable in the older process to be functions of a new independent variable, usually selected so as not to be intrinsic to the problem; thus simultaneous independent variations can then be imposed from the beginning upon both the variables which occur. It is found that, for gently regular variations, no

new tests emerge from the use of the Weierstrass method,—a conclusion not unimportant in itself—though the formal expression of
the tests is modified. In the third chapter, both methods are
applied to integrals, which still involve only a single original dependent variable and now include derivatives of the second order as
well as those of the first order. Of the analytical material in
these three chapters, convenient geometrical illustration is provided by plane curves.

The next three chapters are devoted to the discussion, by both
methods, of single integrals which involve two dependent variables
and one independent variable in their initial form, together with
derivatives of the first order, and (less generally) of the second
order, though the analytical development in the latter case is not
carried so far as in the former. The increase in the number of
variables does not lead to an increase in the number of significant
tests, though (as is almost to be expected) the expression of the
several tests tends to become more complicated. For the material
in these chapters, convenient geometrical illustration is provided
by skew curves.

The seventh chapter introduces the essential advance made by
the Weierstrass method, through the emergence of a new additional
test. The advance comes through the consideration of variations
which are not restricted to be of a gently regular type. The variations are naturally required to be continuous and, as maxima and
minima are being considered, they are required to be small in
magnitude; but, within that small range, they are permitted to
vary even abruptly, as violently as continuous curves representing
rapid small oscillations or even as continuous serrated curves.
Many such variations can be compounded from rudimentary variations of a selected type; and the use of the latter variation leads
to the construction of a new test which, necessarily satisfied for
the most elementary form, is cumulative in its effect for the composite form. This Weierstrass test is applied to single integrals

which, of course, involve only ordinary derivatives. In the case of
the Solid of Least Resistance, it is shewn that the solution, satisfactory under the tests associated with the gently regular type of
variation, does not obey the further test associated with the strong
variation, and therefore does not supply a minimum. It appears
also that the Principle of Least Action does not supply a minimum : the demands of the tests, arising out of gentle variations,
are satisfied ; but the demand of the Weierstrass test, arising out
of strong variations, is not satisfied.

The eighth chapter is devoted to the consideration of simpler
problems of relative maxima and minima—the isoperimetrical
problems of even ancient interest. In particular, those problems
are discussed, in which the requirement of a maximum or of a
minimum is obliged to fulfil the condition of allowing a coexistent
related integral to maintain an assigned value. Other types of
relative problems—in which, for example, persistent relations hold
among the variables—are considered, though only briefly, partly
because the first stage in their treatment is to be found in treatises
and memoirs easily accessible.

The ninth chapter deals with double integrals which, in their
initial postulation, involve one dependent variable and its two first
derivatives. The concurrent geometrical illustration is, of course,
provided by surfaces in ordinary space. Both the older method and
the later method are used for the discussion. The treatment of the
most interesting of all problems of this kind—minimal surfaces—
is simplified when the Weierstrass method is used from the beginning. Schwarz's theorem, which secures the determination of
a minimal surface by initially assigned conditions, has been extended so as to obtain an analytical expression of the Jacobi test
in limitation of the range. The tenth chapter is devoted to two
issues : one, the construction of the Weierstrass test for double
integrals and a proof that it is satisfied by minimal surfaces: the
other, the simplest type of isoperimetrical problem. The eleventh

chapter is concerned with double integrals which involve the partial derivatives of the second order; but there is no attempt at a full discussion, mainly because, after the application of even the simpler tests, the analysis becomes unwieldy and the developments demand the differential geometry of the curvature of surfaces.

A final chapter is devoted to triple integrals, involving the first derivatives of a single dependent variable. The convenient geometrical illustration is provided by the consideration of volumes in quadruple space. Only a slight use is made of the mathematical notions of such space; and, because the geometrical considerations are mainly concerned with volumes, a three-fold amplitude finds, for most purposes, a working representation in the ordinary space of experience. The analysis, which is requisite for the full application of the Weierstrass method to triple integrals, soon becomes laboured; it is here developed only so far as to construct the necessary tests which shew that, owing to failure under the Weierstrass test, Dirichlet's Principle is not valid.

Before parting from the volume, I would thank Professor H. F. Baker, for his kindness in reading the earliest sheets of the volume. Above all, I must mention the Staff of the University Press, Cambridge. Their steady and unfailing co-operation has been my mainstay during the printing of the book. Now that my task is ended, I tender my grateful thanks to all of them who have shared our joint labour.

A. R. FORSYTH.

31 *December* 1926

# CONTENTS.

## INTRODUCTION.

## CHAPTER I.

### INTEGRALS OF THE FIRST ORDER : MAXIMA AND MINIMA FOR SPECIAL WEAK VARIATIONS : EULER TEST, LEGENDRE TEST, JACOBI TEST.

# CHAPTER II.

## INTEGRALS OF THE FIRST ORDER: GENERAL WEAK VARIATIONS: THE METHOD OF WEIERSTRASS.

# CHAPTER III.

### INTEGRALS INVOLVING DERIVATIVES OF THE SECOND ORDER : SPECIAL WEAK VARIATIONS, BY THE METHOD OF JACOBI ; GENERAL WEAK VARIATIONS, BY THE METHOD OF WEIERSTRASS.

# CHAPTER IV.

### INTEGRALS INVOLVING TWO DEPENDENT VARIABLES AND THEIR FIRST DERIVATIVES: SPECIAL WEAK VARIATIONS.

# CHAPTER V.

### INTEGRALS INVOLVING TWO DEPENDENT VARIABLES AND THEIR FIRST DERIVATIVES: GENERAL WEAK VARIATIONS.

# CHAPTER VI.

### INTEGRALS WITH TWO DEPENDENT VARIABLES AND DERIVATIVES
### OF THE SECOND ORDER : MAINLY SPECIAL WEAK VARIATIONS.

# CHAPTER VII.

### ORDINARY INTEGRALS UNDER STRONG VARIATIONS, AND THE WEIERSTRASS
### TEST : SOLID OF LEAST RESISTANCE : ACTION.

# CHAPTER VIII.

### RELATIVE MAXIMA AND MINIMA OF SINGLE INTEGRALS: ISOPERIMETRICAL PROBLEMS.

# CHAPTER IX.

### DOUBLE INTEGRALS WITH DERIVATIVES OF THE FIRST ORDER : WEAK VARIATIONS : MINIMAL SURFACES.

# CHAPTER X.

## STRONG VARIATIONS AND THE WEIERSTRASS TEST, FOR DOUBLE INTEGRALS INVOLVING FIRST DERIVATIVES: ISOPERIMETRICAL PROBLEMS.

# CHAPTER XI.

### DOUBLE INTEGRALS, WITH DERIVATIVES OF THE SECOND ORDER: WEAK VARIATIONS.

# CHAPTER XII.

### TRIPLE INTEGRALS WITH FIRST DERIVATIVES.

# INTRODUCTION.

### *General range of the subject.*

1. The range of Mathematical Analysis, usually known as the Calculus of Variations, deals with one of the earliest problems of ordinary experience. The requirement was, and is, to obtain the most profitable result from imperfectly postulated data; and the data may possibly be subjected to conditions, which likewise are imperfectly postulated. When data and conditions are expressed in analytical form, the necessary mathematical calculations cannot be effected directly, because of some essential deficiency in the information. The gap has to be filled before the resolution of the problem is attained; and the process of supplying the lacking information is indirect, as compared with the regular methods of calculation. It consists of the construction of tests, which are the mathematical expression of general conditions; it is composed of various gradual stages, sometimes independent of one another; and the ensuing requirements are combined into an aggregate which is adequate for the purpose. Usually, the predominating interest lies in the qualitative results that are constructed. Not infrequently, the subsequent quantitative calculations are ignored; they involve processes which belong to an elementary range, that is unconcerned with the mode or modes of obtaining the information lacking in the initial stage of the original statement.

### *Early beginnings.*

2. To the ancient Greeks, with their wonderful geometry, the problem came not infrequently in a practical form: how to secure the greatest amount of land, which could be enclosed within a boundary capable of being ploughed as a contour furrow in a given time: or, geometrically expressed, how to find the shape of the closed curve, which shall enclose the largest area within a perimeter of assigned length. The Greeks—perhaps not the earliest race, in spite of their scientific achievements, and certainly not the only race, desirous of saving the most and wasting the least out of opportunities subject to assigned limitations—obtained a number of results, partly by what may be called inspired guessing or intuition; partly by assumptions as to harmonies, and perfection, in curves; partly by experiments. Though the results would not be regarded by later mathematical precisians as having been established with complete rigour, they often were sufficiently satisfactory for working purposes.

Little substantial progress beyond the old attainments was made down through the middle ages. In the East, and among the Arabs in the South West of Europe, the old mathematical subjects were not merely maintained,

but even flourished, within the range of the old methods; which came to be
processes of record and description, mostly of an arithmetical and a geometrical
type. Indeed, algebra was the main subject where progress was noted and
notable, though often attained in geometrical guise: even the beginning of
Descartes' *Geometry* is occupied with the geometrical constructions devised for
the solution of quadratic equations. Real progress began with the new world
of science, which was created in the seventeenth century in Western Europe—
not least with the new mechanics and new astronomy, under Galileo, Kepler,
and Newton. It soon appeared that the old geometrical methods, even in
their new amplifications, must merge into an entirely new method which, in
character, was analytic and not synthetic. Thenceforward, what now is called
Analysis came to be the main instrument in vast fields of mathematical
research. The infinitesimal calculus was developed: its historical origins are
marked by the morbid hostility between the partisans of Newton and the
partisans of Leibnitz, which raged for many a year to the detriment of achievement in recognised knowledge. Old problems were re-stated; and they were
solved once more, not infrequently with new additions and unsuspected limitations. New investigations, hitherto belonging to the region of fancy, were
brought within the range of practice. In particular, during the late seventeenth century and through the eighteenth century, problems were propounded
by individuals as public challenges—a habit that has survived among Aca-
demies in the present day when proceeding to the award of their prizes.
The actual propounding of the problem was frequently an intimation that
the challenger had achieved the solution. Specially in mathematics, this was
a form of initial publication. The Bernoulli family, in its successive generations, was conspicuous for contributions of this type. Others, less publicly
aggressive in tone, were content to achieve a result and to be satisfied with
the knowledge thus attained, without challenge or publication in any similar
form. Of such men, perhaps Newton is the most conspicuous instance.

### *Newton, John Bernoulli.*

**3.** The gradual development of the infinitesimal calculus led to the
formulation and the solution of new problems; and the new inquiries were
not less frequent in problems, the aspect of which was mathematical rather
than astronomical or mechanical or physical. Particularly within the range
of problems connected with maximum attainment or minimum reduction, the
new calculus proved effective; usually, the data were sufficient to allow a
direct attack to be made upon the problem. But, soon, the problems sometimes became of a subtler indirect nature: the very character, not merely the
magnitude, of the unknown quantities was the essence of the problem; and
even behind this difficulty lay processes that could not be effected.

Such difficulties arose most directly when the maximum or the minimum
was to be possessed by a quantity which, however veiled in expression at the

time, can be formulated as an integral, definite or indefinite. The quadrature —the actual evaluation—of the integral could not be made, because it would contain unknown quantities. On the one hand, the quest of these quantities was the main problem; on the other hand, progress towards solution of the problem was barred, so far as the old methods were concerned, precisely because the quantities were unknown. Among such preliminary and isolated problems, two (among others) survive in interest to the present day.

One is Newton's problem of the determination of the shape of the solid of revolution which shall meet with the least resistance to its motion through a fluid, on the assumption of a law of resistance conforming roughly to obser-vation. Newton's solution was not satisfactory so far as practice is concerned, even if the law of resistance be regarded as adequate; and Weierstrass's analysis provided, more than half a century ago, a reason why the Newtonian solution is not satisfactory, even from a theoretical point of view. This general problem has come into more importance in recent times, owing to the develop-ments of submarines and air-craft; and the necessary association of new knowledge, derived from successive physical and mechanical experiments, has added grave complications to the mathematics of the general problem.

A second problem, of historical importance and still of interest, is that of the Brachistochrone, associated (1696) with the name of John Bernoulli. In its simplest form, it requires the determination of the curve joining two given points such that the time of passage of a given mass from one point to the other along the curve (as in a smooth groove), under the influence of gravity alone and subject to no retarding force, is a minimum, that is, less than the time of passage in like circumstances along any other curve joining the two points. The original solution, which requires the curve to be a cycloidal arc, is accurate as regards the quality of the characteristic curve; later investiga-tions have added limiting specifications on the range of the curve. And, naturally, analogous problems have been propounded and solved, when the motion is due to external forces other than gravity, and when retarding forces can come into play.

Without multiplying the citation of instances unduly, it may suffice to mention the mathematical-physical conception denoted by the word Action. Philosophers have been fain to deduce the mechanical movements of a dyna-mical configuration, even much of the physics of the universe, from the single property, that the Action between any two states of the configuration in continuous change is a minimum; and the property has been elevated in postulation to the Principle of Least Action.

### Euler and Lagrange.

4. The mathematical development of that section of Analysis, which has to be considered here, has been fitful, sporadic, slow; and, as so often is the case, not entirely free from controversy. Amid many names that now

belong to the history of mathematical science, four stand out, because of the significance of their contributions towards the attainment of the first stage in the systematic construction of the calculus.

The earliest name is that of Euler, fruitful in that branch (as in all branches) of Analysis in his day. He discovered* the characteristic differential equation which, as a necessary requisite, must be satisfied. It was obtained by considering the increment of the integral which arose through the variation of a rudimentary arc. In the first stage, Euler stayed his investigations at this mathematical result; it secured a stationary quality, as a preliminary common to a maximum or a minimum property of the magnitude under discussion.

Lagrange† discussed the problem by regarding variations as active through the range of integration, not restricted to any rudimentary portion of that range. Indeed, he made a new foundation of the subject, entirely analytical in character as was his mathematical wont. To him is due the introduction of the symbol $\delta$, which discharged useful duty through successive generations of mathematicians: though, as can happen from lack of definite and precise explanation, initially unexpressed, inferences by later investigators were sometimes drawn, more comprehensive in statement than could be justified. He also extended his analysis so that it could be applied to problems involving two independent variables; and, in particular, his researches constitute a foundation of the theory of minimal surfaces‡.

Subsequently, Euler resumed his consideration of the subject, on the basis of Lagrange's work. His completed account§ is based upon the geometrical representation of the analysis, in which he regarded $\delta y$ as a change in the position of a point on an original curve, along the ordinate, to a contiguous point on the varied curve.

There, for a period, advance in the progress of the subject was arrested. Detailed amplifications within the range were forthcoming, but always in the field of work already achieved by Euler and Lagrange, whose names dominate the first stage in the systematic calculus.

## Legendre.

5. The results obtained by Euler and Lagrange were of the nature of necessary qualifying tests. The discrimination among merely stationary values, as between a maximum or a minimum, was left to intuitive considerations, foreign to the analysis.

* *Methodus inveniendi lineas curvas maximi minimive proprietate gaudentes, sive solutio problematis isoperimetrici latissimo sensu accepti* (Lausanne et Geneva, 1744).

† *Miscellanea Taurinensia*, t. ii (1760–1), pp. 173—195; *ib.*, t. iv (1766–9), pp. 163—187: *Œuvres*, I, pp. 335—362, II, pp. 37—63.

‡ For reference, see the preliminary note to Chapter IX.

§ *Institutiones calculi integralis*, t. iii (2nd ed., 1793), pp. 381—475.

The next achievement in progress is due to Legendre. Thus far, all the investigations had concentrated on the stationary requirement—the Euler test, when there is one independent variable; and the Lagrange test, when there are two independent variables. All of them were concerned with the terms of the first degree in the small quantities in the increment of the integral (the so-called 'first' variation), consequent upon a small increment of the independent variable. The customary requirements for a maximum or a minimum of an ordinary function, when the function is given explicitly, demand the consideration of the aggregate of terms of the second degree in those small quantities in its increment (the so-called 'second' variation). Legendre was the first to discuss this aggregate. To him is due the idea of modifying its initial expression, by adding suitable arbitrary variations within the range, and subtracting their corporate effect in the balance at the boundary, always subject to unaltered conditions at the boundary. This device enabled him to express the second variation in a compact form, the sign of which became of essential significance for the main purpose.

Legendre thus added * a further test, discriminating as to character, which must be satisfied by integrals in one independent variable if they are to possess a maximum or a minimum. The new requirement, like the Euler test and the Lagrange test, is a necessary qualifying test.

### *Jacobi.*

**6.** The next name to be mentioned is that of Jacobi. Discussing the second variation as formulated by Legendre, he proved that a quantity, of critical importance in Legendre's expression, could always (without any inverse process) be derived from the knowledge presumed as attained in the Legendre stage. Making this advance in the purely mathematical range of the problem, he obtained extended forms for the tests already established; he constructed an essential modification of Legendre's form; and from this modified form, he deduced † a further test, limiting the range of the integral.

A full proof, and an extension (also with full proof), of Jacobi's results were given in a memoir by Hesse‡; and further investigations connected with the transformation of the second variation are due to Clebsch§, and to Mayer‖, though their inferences for the most part are left without a statement of the geometrical significance of the form of the second variation, such as was made by Jacobi.

The character of Jacobi's test admits of simple illustration. The shortest arc on the surface of a sphere, between two points on that surface, is the

* *Mémoires de l'Académie Royale des Sciences* (1786), pp. 7—37.
† *Crelle's Journal*, t. xvii (1837), pp. 68—82.
‡ *Crelle's Journal*, t. liv (1857), pp. 227—273.
§ *Crelle's Journal*, t. lv (1858), pp. 254—273, 335—355.
‖ *Beiträge zur Theorie der Maxima und Minima einfacher Integrale* (Leipzig, 1866); *Crelle's Journal*, t. lxix (1868), pp. 238—263.

portion of the great circle passing through them and lying between them: so far, the tests of Euler and Legendre are satisfied. But the portion of the great circle must be less than half the circle; if it is a half-circle, the length (minimal in this case) is not unique; if it is greater than a half-circle, the length is certainly not a minimum between its extreme points, because (among other reasons) it can always be reduced by small variations. The requirement in this instance (and the cognate requirement in any corresponding problem) comes into Jacobi's work: the limit of the integral, as ranging between 'conjugate' points on the curve obtained, is bounded by a point on the circle and the diametrically opposite point.

The general test devised by Jacobi indicates the limit (if there be a limit) to the range along the characteristic curve within which the magnitude of the integral under consideration is a maximum or a minimum.

With the achievement of these results, essential progress in the solution of the problem again came to an end for a long period. Mathematical amplifications, and generalisations wider in range but cognate in character, were made from time to time; but, in real effect, there were only the three types of test, to be associated initially with the names of Euler and Lagrange, Legendre, and Jacobi.

### General notion of variations.

**7.** In all the analysis hitherto applied, whatever its form, the general notion behind the mathematical expression was the same. A hypothetical maximum or a hypothetical minimum was postulated; if it were a maximum, every small change (called a *variation*) must lead to a decrease in the hypothetical maximum value; if it were a minimum, every such variation must lead to an increase in the hypothetical minimum value. Much explicit virtue lay in the assumption of the smallness of the variation. Not a little implicit difficulty was involved (though not recognised) in the lack of adequate specification of the properties constituting that smallness. Latent limitations (also not recognised) were imposed by the use of the symbol $\delta$, which was initially employed to indicate the actual small variation of the unknown magnitude and which, by argument presumably deemed too obvious for statement, was allowed to exercise a supposed and unchallenged parsimonious influence solely by its literal character, wherever and whenever it was made to appear. When once the range, within which the tacit assumptions are legitimate, is actually examined, the result is to shew that the small variations are themselves gravely limited in character. Other small variations can be postulated, initially the same as regards type of magnitude, but with action gravely different from the restricted influence exercised by the other magnitudes, which are derived from them by some of the operations that were freely used. One consequence is that the old tests, when satisfied, lead to inferences which are legitimate only within the range of the small variations that remain small, not only in their own influence, but through the influence to be exercised by these derivatives. Another consequence is that the old tests can be trusted solely as connected with such range of variation : and that, precisely because they cease to be applicable when the hypotheses leading to their establishment are no longer valid, these old tests can furnish no information for small variations of the other kind indicated. They will remain, as necessary tests, because the maximum or the minimum property must be possessed for all small variations and there-fore for variations of the limited character. They are inadequate for the discussion of the magnitude, when it is subjected to small variations not of this limited character. In fact, the three tests are necessary for the solution of the problem; they are not sufficient to secure the completeness of the solution.

8. Further, another limitation was imposed upon the small variations adopted. It was imposed explicitly; no attempt was made to shew that it is immaterial; and any significance, that the limitation might have borne, was so far ignored as not even to come within the range of examination. The limitation consisted in restricting the variations so that they should affect the dependent variable alone and should not apply to the independent variable, either independently or concurrently with the changes made in the dependent variable. If we turn to the graphical representation, whereby the relation between the two variables is illustrated by a plane curve, the small variations adopted consisted of a small displacement along the ordinate, coupled with a small displacement of the direction of the tangent; but small displacements, made sideways off the ordinate, were never entertained. Occasionally, such a displacement had to be considered for an isolated point or for isolated points; thus the extreme point might be required to lie on a given curve and be not merely a fixed point. In such instances, a special detailed argument was applied to take account of the effect caused by such a displacement of the particular point, while the possible effect of such a displacement throughout the range was ignored.

When the corresponding investigation, necessary to take these small variations into consideration (subject to the old limitations as to the character of the variation of their derivatives), has been completed, it is found that no new test emerges. The sole change is of a formal character; but such a result requires definite establishment. The net conclusion, however, is that, for small variations which are subject to the limitations indicated, three necessary critical tests exist.

*Weak variations: strong variations.*

9. Now these small variations, even when imposed upon both the depen-dent variable and the independent variable, are only the simplest type of all small variations. Thus, to return to the graphical representation, we could

have small variations of a curve such as are given by displacements corresponding to very rapid small oscillations. In these, while the change of position is small, the change of direction may be finite though not small; the change of curvature may be very large; and so on. For all such variations, the analysis that has been effective in the other stages no longer applies; some new process must be devised that will cause these less simple variations to be taken into account.

Small variations of the earlier type, viz. *those in which the derivatives of the variation are of the same order of smallness as the variation itself*, are often called *weak* variations. Among weak variations, those which are restricted so that they affect the dependent variable alone, will be called *special* variations (the quality of weakness being tacitly included in the name). Small variations of the later type, viz. those in which the·derivatives are not limited to be small, though the actual displacement is definitely small,—are called *strong* variations. Manifestly strong variations will be much more versatile in character than weak variations; and progress will initially be secured by considering selected types of strong variations.

### Weierstrass.

**10.** An adequate method, new in form and (as regards later developments) new in substance, was initiated by the work of Weierstrass. Although he died as long ago as 1897, no fully authoritative exposition of his researches has yet appeared. He gave lectures on the subject in 1872 and 1879, perhaps earlier, in Berlin. Notes of his lectures have circulated, though without indications as to whether their range is complete or only partial. The work of Schwarz on minimal surfaces, as on branches of the theory of analytical functions, may be not unfairly presumed to have had its foundations in the teaching and the work of his master, whom he succeeded at Berlin. The results of the earlier stages of Weierstrass's work have found a partial record in some books* published since his death. Renewed interest was stirred in the subject owing to his influence; and writers, in varying degrees of independence, have been stimulated to further researches. Besides Schwarz, the names of Hilbert, Goursat, Hadamard may fitly be mentioned, as well as those of Kobb, Bolza, Kneser, Veblen, Hedrick, and Tonelli†. The work of Clebsch also was important, though it belongs to the older range; and Todhunter's two volumes, dealing with the history of that range, may be consulted with advantage.

Two contributions, different in kind yet connected, and both of funda-

---

* Hancock's *Calculus of Variations* may be consulted in this connection.

† A considerable amount of research has appeared in scattered articles in the *Bulletin of the American Mathematical Society*. An article, by Love, in the supplement to the *Encyclopædia Britannica*, gives a concise account of the earlier stages of the older theory; and a fuller account of the subject will be found in two articles, by Kneser, and by Zermelo and Hahn, in the *Encyclopädie d. math. Wissenschaft*, vol. ii, A. 1, pp. 571—625, 626—641.

mental importance, are made by the work of Weierstrass. In one of these, he ignores the preferential selection of the dependent variable as the subject of variation; with him, the dependent variable and the independent variable are alike subjected to arbitrary small variations, independently of one another. The old tests, in modified amplification and with added elucidation, are obtained, in connection with weak variations. The discussion of strong variations, and the deduction of one new test—the so-called Excess-function— derived from the consideration of one simple class of strong variations, are entirely due to his initial researches in the matter. But he dealt with only the simplest class of strong variations, the particular class being sufficient for the purpose of the particular type of problem which he had under con- sideration. It may be that similar researches, involving less restricted strong variations for less particular types of problem, will only lead to generalisa- tions, however important, of his Excess-function; it would appear as if such researches still remained for investigation.

**11.** As regards the weak variations, a further remark may be made to confirm the desirability of examining variations other than those which have been entitled 'special.' The more immediate geometrical representation, of the relation between a dependent variable and an independent variable by means of a plane curve, would imply one kind of special variation, were the analytical equivalent expressed by Cartesian coordinates; it would imply a different kind of special variation, were that equivalent represented by the usual polar coordinates; for homogeneous coordinates, and for bipolar co- ordinates, other different kinds of special variation would be used. All possible kinds of special variations are included when the most general kind of weak variation is adopted, because each special variation is only one particular case of the general type.

Moreover, the general variation includes, as a particular case, that apparent variation which consists of displacement merely along the length of a curve without any alteration of its shape. Such an apparent variation is not a variation of the value of the integral because, when the integral is regarded as a sum of infinitesimal elements between its upper limit and its lower limit, the value of the integral is independent of the precise detailed mode of selection of its constituent elements.

Lastly, the use of general variations makes it possible to consider (with- out the supplementary discussions needed for special variations) the cases, when the limits of the integral are liable to partial (though not quite arbitrary) mobility instead of being definitely fixed.

*Assumed limitations on functionality.*

**12.** Throughout all the investigations, we shall suppose, and now explicitly assume, that the subject of integration either (i) is a regular function* of its

---

* That is, a function which (with its derivatives) is uniform, finite and continuous.

arguments within the range of integration: or (ii), when not a uniform function for all possible values of its arguments within the whole field of variation, remains a uniform branch of some function, so that it behaves as a regular function in each part of the range of integration.

If, at any place or places, an argument or arguments should suffer discontinuity in value, so that the function or any derivative of the function potentially may cease to be regular, explicit note of the fact or of the possibility will be taken. Again, during the analysis and unless a warning to a contrary effect is given, the solutions of ancillary differential equations will be regarded as providing regular functions, when these are determined by regular data. In particular instances, it may happen that the conditions, which are imposed, cannot be satisfied by regular functions; it may even happen that, of a postulated problem, no solution exists of a type which is regular: the impossibility will be noted. But we shall proceed on an initial assumption that a regular solution may exist and shall make its discovery our quest. The hypothetical solution of that type (perhaps with the assistance of existence-theorems, established with similar initial assumptions) will be subjected to all the conditions, and will be made to conform to all the data, that are imposed and provided. When the conditions are satisfied and the data are incorporated, the result will be regarded as the solution. In the contrary event, note will be taken of the fact, which thence will have emerged, that the imposed conditions and the assigned data forbid a solution of the foregoing type. It may be that conditions could be modified and that data could be changed; such alterations could be considered in their turn, as raising a new specific problem.

Moreover, unless some contrary demand occurs requiring satisfaction, we shall expect that the functions concerned are of an ordinary simple character, using these words as in common parlance. Unless expressly forbidden, we shall suppose that our functions have derivatives which are unique, save at isolated singularities or in the immediate vicinity of singularities; e.g., we shall not be concerned with uniform continuous functions, which nowhere have a unique differential coefficient. Equally we shall suppose that all the functions, with which we have to deal, are capable of unconditional integration, e.g., that they are not beset with singularities. The present purpose is the development of a calculus, without delaying over the foundations of Analysis, which will be regarded as having been duly laid and recognised in any bearing that has significance. It may happen that, when strong variations are considered in a full detail which would be essential in a complete discussion of their influence, corresponding refinements will prove necessary. For the present, its more ordinary and less critical commonplace range will be held sufficient; and, of course, the functions admitted will be subject to the corresponding limitations as regards regularity, continuity, and variation.

# CHAPTER I.

The chapter is devoted to the discussion of the conditions for the possession of maxima and minima by integrals which involve a single dependent variable and its first derivative. The conditions emerge from the application of weak variations, which are imposed solely upon the dependent variable and do not affect the independent variable. The problem thus is the simplest of all the problems in the calculus; and the analysis, though modified and amplified from the original form, is substantially in accord with the analysis that occurs in the researches of Euler, Lagrange, Legendre, and Jacobi.

The Weierstrass method, in which variations of both the dependent variable and the independent variable are considered, is reserved for the succeeding chapter.

### *Types of problems arising from integrals maxima and minima.*

**13.** The calculus of variations deals with those problems which, enunciated in mathematical phraseology, require the determination of the form of an unknown quantity or unknown quantities as a function or functions of a variable (being a dependent variable or dependent variables), so that some integral may assume a maximum or a minimum value. The dependent variable or variables will be supposed to occur in the expression of the integral: a derivative or derivatives of the unknown variable may occur in the integral; and it is supposed that the integral cannot be evaluated for unspecified forms of that variable, so as to become free from quadratures. Sometimes the demand is that the maximum or the minimum is unconditional. Sometimes there is a demand that the required maximum or minimum is to be possessed subject to assigned conditions, which may have a variety of forms: thus one form of condition requires that another integral of similar form may possess a fixed value.

Even when there is only a single dependent variable, various cases may occur. Thus there may be only one independent variable; advantage then accrues, if only by way of illustration, from the association of the problem with the geometry of a plane curve. There may be two independent variables; the corresponding advantage comes from an association with the geometry of a surface.

There may be two dependent variables and one independent variable: the corresponding advantage comes from an association with the geometry of a twisted curve. Or there may be any number of dependent variables and a single independent variable; and then a natural association, for purposes or needs of illustration, is provided by some dynamical system.

The central demand is for the possession of a maximum or of a minimum. The maximum is to be a true and a full maximum in the mathematical sense of the term, as distinct from the phrase 'largest value' in common parlance: that is to say, the change in the value of the integral, which is caused by every small variation, however general or however particular, or by any number of small variations of every type, whether taken together or acting individually, must be an increase, uniformly definite in character as an increase, though its magnitude will be small because the variation or variations effected from continuous finite quantities, are small. Similarly, a minimum is to be a true and a full mathematical minimum; that is to say, every small variation (of the preceding range in all its extent) must lead to a decrease in the value of the integral, uniformly definite in character as a decrease. Should the integral provide a decrease for only some variations but not for all, its value cannot be declared a maximum; and, similarly, if it should happen to provide an increase for some variations but not for all, its value cannot be declared a minimum. Partial maxima and partial minima are not recognised as maxima or as minima. Thus a mountain peak is a place of maximum elevation; the bottom of a tarn is a place of minimum elevation; the top of a mountain pass is not a place of either true maximum or true minimum in elevation.

In discussing questions of maxima and minima, it is an advantage when the independent variable increases continuously throughout the range; if it decreased continuously, a mere change to the opposite direction of the range would allow the desirable continuous increase. Sometimes care has to be exercised, at least initially, in the selection of an appropriate independent variable; thus, if there were a question of determining a curve of shortest length so as to satisfy some permanent condition, it is preferable not to choose the length of the arc for the purpose.

### Integral of the first order: its variation.

**14.** In order to give an indication of the method of analysis, we shall begin with the simplest general problem, and shall proceed to obtain necessary tests for the possession of a maximum or a minimum by an integral

$$\int f(x, y, y')\, dx,$$

where $y'$ denotes $dy/dx$. The integral* is supposed taken between fixed constant limits; and no external conditions are imposed. The unknown element, to be obtained for the purpose, is the form of $y$ as a function of the variable $x$; and, in the first instance, we shall use only special variations—that is, we shall assume that $y$ alone is subjected to weak variations. In the earlier

---

* Such an integral, involving only the first derivative, will be called an integral of the *first order*.

historical development of the subject, such a variation was represented by
the mathematical expression

$$\delta y,$$

where the symbol $\delta$ (perhaps in imitation of the $d$ in the infinitesimal
element $dy$: it was introduced by Lagrange) is assumed to impose the
restriction to smallness, and where the properties of the quantity $\delta y$ are not
always explicitly stated, sometimes not even hinted. A varied value of $y$ can
be taken in a more definite form

$$y + \kappa v,$$

where $\kappa$ is an arbitrary constant and is so small in magnitude that any positive integral power is unimportant relative to every lower positive integral
power. Also, $v$ is any regular uniform function of $x$ within the range of the
integral, and all its derivatives also are regular uniform functions of $x$
within that range. Moreover, $v$ otherwise is an arbitrary function; and we
further postulate that $v$ is independent of $\kappa$. Denoting the original value and
the varied value of the integral by $I$ and $J$ respectively, we have

$$J - I = \int \{f(x,\, y + \kappa v,\, y' + \kappa v') - f(x,\, y,\, y')\}\, dx,$$

the integration extending between the fixed constant limits. We further
assume the function $f$ to be of such a form that $f(x,\, y + \kappa v,\, y' + \kappa v')$ can be
expanded in a uniformly converging series, which proceeds in powers of $\kappa$.
Thus we have, after this expansion,

$$J - I = \kappa I_1 + \tfrac{1}{2} \kappa^2 I_2 + R_3;$$

where

$$I_1 = \int \left( v \frac{\partial f}{\partial y} + v' \frac{\partial f}{\partial y'} \right) dx,$$

$$I_2 = \int \left( v^2 \frac{\partial^2 f}{\partial y^2} + 2vv' \frac{\partial^2 f}{\partial y \partial y'} + v'^2 \frac{\partial^2 f}{\partial y'^2} \right) dx,$$

with the same range of integration as for $I$; and where $R_3$ denotes the aggregate of terms that involve third and higher powers of $\kappa$. Manifestly the
quantity $\kappa I_1$ (called the 'first' variation), if it does not vanish, will dominate *
the value of the right-hand side; and $R_3$ is unimportant in comparison with
$\kappa^2 I_2$, if $I_2$ does not vanish.

### Modification of the First Variation.

**15.** The expression for $I_1$ is to be modified. We have

$$\int v' \frac{\partial f}{\partial y'}\, dx = \left[ v \frac{\partial f}{\partial y'} \right] - \int v \frac{d}{dx} \left( \frac{\partial f}{\partial y'} \right) dx,$$

where the quantity in square brackets is to be taken at the limits. But when
the limits are fixed, no variation of $y$ is admissible for those values of $x$;

---

\* See foot-note on p. 17.

consequently, $v$ vanishes at the upper limit and at the lower limit. Hence

$$I_1 = \int v \left\{ \frac{\partial f}{\partial y} - \frac{d}{dx} \left( \frac{\partial f}{\partial y'} \right) \right\} dx.$$

Now the quantity $\kappa I_1$, when it does not vanish, dominates the value of $J - I$. In that event, a change of sign for $\kappa$ changes the sign of $\kappa I_1$, that is, changes the sign of the value of $J - I$: in other words, one variation would lead to an increase, and another variation to a decrease. As a maximum must be characterised by a decrease for all variations, and a minimum by an increase for all variations, the preceding possibility must be excluded: consequently the quantity $\kappa I_1$, and therefore the integral $I_1$, must vanish.

Let $Y$ denote the quantity $\dfrac{\partial f}{\partial y} - \dfrac{d}{dx} \left( \dfrac{\partial f}{\partial y'} \right)$. Then the integral $I_1$, which is equal to $\int v Y dx$, must vanish for all arbitrary values and forms assigned to the regular function $v$. This result cannot be attained unless $Y$ vanishes throughout the range. If $Y$ were positive in any portion of the range, the integral would be positive, for an assigned positive value of $v$ through the portion and a zero value elsewhere; and the integral would be negative, for an assigned negative value of $v$ through the portion and a zero value elsewhere. Similar definite values, positive or negative at will, could be secured for the integral, if $Y$ were negative in any portion of the range. Hence we must have

$$Y, = \frac{\partial f}{\partial y} - \frac{d}{dx} \left( \frac{\partial f}{\partial y'} \right) = 0.$$

**16.** This relation $Y = 0$ is either an identity, or an equation not identically satisfied.

If $Y = 0$ is an identity, then the term $\dfrac{\partial^2 f}{\partial y'^2} y''$ which would arise from the second term in $Y$ cannot exist, for there is no other term involving $y''$ to cancel it in the supposed identity. Thus $\dfrac{\partial^2 f}{\partial y'^2}$ is zero, and therefore $f$ is of the form

$$f = y' U + V,$$

where $U$ and $V$ are functions of $x$ and $y$ only. Then

$$Y = y' \frac{\partial U}{\partial y} + \frac{\partial V}{\partial y} - \frac{dU}{dx}$$

$$= y' \frac{\partial U}{\partial y} + \frac{\partial V}{\partial y} - \frac{\partial U}{\partial x} - y' \frac{\partial U}{\partial y},$$

an expression which can vanish identically only if

$$\frac{\partial V}{\partial y} - \frac{\partial U}{\partial x} = 0,$$

that is, if some quantity $M(x, y)$ exists such that

$$V = \frac{\partial M}{\partial x}, \quad U = \frac{\partial M}{\partial y}.$$

In that case, we have

$$I = \int f(x, y, y')\, dx$$

$$= \int \left( y' \frac{\partial M}{\partial y} + \frac{\partial M}{\partial x} \right) dx$$

$$= M(x, y);$$

and so the integral can be evaluated explicitly, contrary to the initial requirement (§ 13) that the initial integral is not capable of explicit evaluation *.

This possibility is excluded, as being irrelevant to our purpose; and we therefore assume that $Y = 0$ is an equation not identically satisfied.

Further, if $f$ has the same linear form $f = y'U + V$, and if the equation

$$\frac{\partial V}{\partial y} - \frac{\partial U}{\partial x} = 0$$

(which now is the form of $Y = 0$) is not identically satisfied, $Y = 0$ becomes an equation involving $x$ and $y$ alone: it provides $y$ as a function of $x$ which, when substituted, renders the integral $I$ merely dependent upon known constants. Such a result is insignificant for our purpose; no element is left which can be used for the construction of tests concerning maximum or minimum. This second possibility is therefore excluded: that is, we reject from further consideration the case when $f$ is a mere linear function of $y'$.

### Characteristic equation, and curve; the Euler test.

**17.** In all other cases, $\frac{\partial^2 f}{\partial y'^2}$ is not zero: the equation $Y = 0$ is a differential equation, manifestly of the second order. In passing, the form of the equation is to be noted; the second derivative $y''$ occurs only through the term $\frac{\partial^2 f}{\partial y'^2} y''$, to which the other terms are additive; and a later test, which will be imposed upon $\frac{\partial^2 f}{\partial y'^2}$ if a maximum or a minimum is to exist, gives to this differential equation $Y = 0$ a character that is of organic importance for its primitive.

This equation $Y = 0$ is often called the *characteristic equation*. It determines $y$ as a function of $x$; and the curve, which is the geometrical repre-

---

* When $I$ can be thus evaluated, the problem does not fall within the range of the calculus of variations; the value of $I$ then depends solely upon the values of $x$ and of $y$ at the two limits of the range.

sentation of the functional relation, is called the *characteristic curve*. The equation was first obtained by Euler, in 1744, and indeed was obtained for the more general integral which involves derivatives of $y$ up to order $n$. Sometimes it is called Euler's equation, hardly a distinctive name; it is the first of the essential tests to be satisfied if the integral is to have a maximum or a minimum.

*Note.* The integral of the equation, or properties of the integral when the actual integration of the equation is not feasible, will arise later (§§ 22–25) for detailed consideration. But two simple cases may be noted where one stage in the integration can be attained at once.

(i) Let the function $f(x, y, y')$ not involve $y$ explicitly, so that $\dfrac{\partial f}{\partial y} = 0$; then the characteristic equation is

$$\frac{d}{dx}\left(\frac{\partial f}{\partial y'}\right) = 0,$$

so that we have

$$\frac{\partial f}{\partial y'} = A,$$

where $A$ is an arbitrary constant.

(ii) Let the function $f(x, y, y')$ not involve $x$ explicitly. Then

$$\frac{df}{dx} = \frac{\partial f}{\partial y}y' + \frac{\partial f}{\partial y'}y'',$$

$$= y'\frac{d}{dx}\left(\frac{\partial f}{\partial y'}\right) + \frac{\partial f}{\partial y'}y'',$$

on using the equation $Y = 0$: integrating, we have

$$f = y'\frac{\partial f}{\partial y'} + B,$$

where $B$ is an arbitrary constant.

In each of these simple cases, the equation remaining to be integrated is only of the first order.

*Ex.* 1. *Required the curve constituting the shortest distance between two points, or a point and a curve, or two curves, in the same plane.*

We shall deal only with the characteristic equation; terminal conditions, and other possible conditions or tests, either intrinsic to all the problems or accidental to any particular problem, belong to a different (and later) range of investigation. In the present case,

$$I = \int (1 + y'^2)^{\frac{1}{2}}\, dx,$$

so that $f(x, y, y')$, which is $(1 + y'^2)^{\frac{1}{2}}$, is independent of $x$ and of $y$; and the example belongs to each of the simple cases adduced. The first gives

$$y'(1 + y'^2)^{-\frac{1}{2}} = A,$$

and the second, on reduction, gives

$$(1+y'^2)^{-\frac{1}{2}}=B.$$

From either, we have

$$y'=m,$$

and therefore

$$y=mx+c,$$

where $m$ and $c$ are constants to be determined by other considerations.

Here $A^2+B^2=1$. Generally, if the function $f$ is free from explicit occurrence of $x$ and of $y$, the arbitrary constants $A$ and $B$ in (i) and (ii) are not independent.

*Ex.* 2. A unit mass moves in a plane field of force so that its velocity depends only upon its distance from a fixed point, and $\int v\,ds$ along the path, between any two points, is a maximum or a minimum. Prove that the force is directed to that fixed point; and determine the orbit.

### *Quadratic terms in the small variation: the 'second' variation.*

**18.** We shall assume that the characteristic curve is adopted, so that the Euler test is satisfied. The term $\kappa I_1$ in the expression for $J-I$ now disappears; and $\kappa^2 I_2$ (called the *second* variation) becomes the governing* term. As $\kappa$ is constant, the sign of this term depends upon the sign of $I_2$. If the sign is positive for all variations of the type considered, then $J-I$ is positive for all such variations; and $I$ is (so far) a minimum. If the sign is negative, then $J-I$ is negative for all the variations; and $I$ is (so far) a maximum. It therefore becomes necessary to make a detailed examination of $I_2$, which denotes the quantity

$$\int\left(v^2\frac{\partial^2 f}{\partial y^2}+2vv'\frac{\partial^2 f}{\partial y\,\partial y'}+v'^2\frac{\partial^2 f}{\partial y'^2}\right)dx.$$

For brevity, we shall denote $\dfrac{\partial^2 f}{\partial y^2}$, $\dfrac{\partial^2 f}{\partial y\partial y'}$, $\dfrac{\partial^2 f}{\partial y'^2}$, by $f_{00}, f_{01}, f_{11}$, respectively.

Because $\dfrac{d}{dx}(v^2\lambda)=v^2\lambda'+2vv'\lambda$ for any quantity $\lambda$, we have

$$I_2+[v^2\lambda]=\int\{v^2(f_{00}+\lambda')+2vv'(f_{01}+\lambda)+v'^2 f_{11}\}\,dx,$$

where $[v^2\lambda]$ denotes the term $v^2\lambda$ taken at the limits, and it is implicitly assumed that $v^2\lambda$ is continuous through the range of the integral. As the limits are fixed, $v$ vanishes at both of them; hence, if $\lambda$ is assumed to remain finite at each limit, the quantity $[v^2\lambda]$ vanishes. Bearing this last assumption in mind, as one of which account must be taken, we have

$$I_2=\int\{v^2(f_{00}+\lambda')+2vv'(f_{01}+\lambda)+v'^2 f_{11}\}\,dx.$$

Thus far, no quantitative condition has been imposed upon $\lambda$; now let it be chosen so that

$$f_{11}(f_{00}+\lambda')=(f_{01}+\lambda)^2.$$

---

* In a series $a_0+\kappa a_1+\kappa^2 a_2+...$, of finite sum $S$, the sign of $S$ is governed by that of $a_0$ for sufficiently small values of $\kappa$. Let $S_1$, necessarily finite, denote the largest numerical value of the sum $a_1+\mu a_2+\mu^2 a_3+...$, for $-\lambda<\mu<\lambda$ where $\lambda$ is a small positive quantity; and choose $\kappa$ less than $\lambda$ and small enough to secure that $\kappa S_1=\theta a_0$, where $-1<\theta<1$. Then $S,=a_0(1+\theta)$, is of the same sign as $a_0$.

In this equation of condition, let the value of $y$ derived from the characteristic equation be substituted, so that $f_{00}, f_{01}, f_{11}$ become functions of $x$ alone. To determine $\lambda$, let a quantity $z$ be chosen so that

$$f_{01} + \lambda = -\frac{f_{11}}{z}\frac{dz}{dx};$$

then $z$ satisfies the equation

$$f_{11}\frac{d^2z}{dx^2} + f_{11}'\frac{dz}{dx} - (f_{00} - f_{01}')z = 0,$$

a linear equation of the second order, of which the coefficients are known functions of $x$. And then, with such a value of $\lambda$, it follows that

$$I_2 = \int f_{11}\left(v' + \frac{f_{01} + \lambda}{f_{11}}v\right)^2 dx$$

$$= \int f_{11}\left(v' - \frac{v}{z}\frac{dz}{dx}\right)^2 dx.$$

In order to take account of the most general admissible variation $\kappa v$ for $y$, it is desirable to have the most general value of $z$ possible : that is, we must have the primitive of the linear equation in $z$. But subject to restrictions actually imposed, $\lambda$ is at our disposal, because the main purpose of its introduction is merely a modification of the expression of $I_2$; therefore the general elements in $z$, which survive after the restrictions are obeyed, can be used for other aims that may be in view.

If the value of $y$, satisfying the characteristic equation, is only a particular solution of that equation, then the foregoing equation for $z$ must be completely integrated. If the primitive of the characteristic equation has been obtained, the primitive of the equation in $z$ can be deduced by direct processes alone, as follows.

### The subsidiary characteristic equation.

**19.** To obtain the primitive of the equation in $z$, we return to the forms of $I_1$ and $I_2$. Two expressions for $I_1$ have been obtained, viz. the original form

$$I_1 = \int\left(v\frac{\partial f}{\partial y} + v'\frac{\partial f}{\partial y'}\right)dx,$$

and a form

$$I_1 = \int v\left\{\frac{\partial f}{\partial y} - \frac{d}{dx}\left(\frac{\partial f}{\partial y'}\right)\right\}dx, = \int vY dx,$$

deduced from the first in association with the terminal conditions; and the two expressions are equal to one another, because terms at the fixed limits vanish. To both forms of $I_1$ let the same variation be applied as was applied to the original form of the integral $I$; let $J_1$ be the deduced value of $I_1$, and let $Y_1$ be the deduced form of $Y$. From the earlier form of $I_1$, we have

$$J_1 - I_1 = \kappa \int \left( v^2 \frac{\partial^2 f}{\partial y^2} + 2vv' \frac{\partial^2 f}{\partial y \partial y'} + v'^2 \frac{\partial^2 f}{\partial y'^2} \right) dx + K_2,$$

where $K_2$ represents the aggregate of terms in $\kappa^2$ and higher powers of $\kappa$; that is, neglecting $K_2$ in comparison with quantities involving only the first power of $\kappa$, we have

$$J_1 - I_1 = \kappa I_2.$$

From the later form of $I_1$, we have

$$J_1 - I_1 = \int v(Y_1 - Y) \, dx,$$

there being no variation of $v$, because it is assumed to be a function of $x$ which is not subjected to small variations. Now, as

$$Y = \frac{\partial f}{\partial y} - \frac{d}{dx} \left( \frac{\partial f}{\partial y'} \right),$$

we have

$$Y_1 - Y = \kappa \left( f_{00} v + f_{01} v' \right) - \kappa \frac{d}{dx} \left( f_{01} v + f_{11} v' \right)$$

$$= \kappa \left( f_{00} - f_{01}' \right) v - \kappa \frac{d}{dx} \left( f_{11} v' \right).$$

Our immediate concern is with the sign of $I_2$, when the equation $Y = 0$ is satisfied so as to make $I_1$ vanish; so, as before, we imagine the value of $y$, determined by that equation, to be now substituted in the coefficients. As $y$ is a solution of $Y = 0$, let $y + \kappa \eta$ be a solution of $Y_1 = 0$: that is, $y + \kappa \eta$ is a solution of what still is the characteristic equation. Hence when the value $\eta$ is assigned to $v$, we have $Y_1 = 0$ as well as $Y = 0$, and so

$$\left( f_{00} - f_{01}' \right) \eta - \frac{d}{dx} \left( f_{11} \eta' \right) = 0,$$

manifestly a linear equation of the second order in $\eta$, the coefficients in which are functions of $x$.

The quantity $y + \kappa \eta$ has been introduced as a solution of the characteristic equation when small variations are imposed, so that $\eta$ can be derived from the general value of $y$ when this latter is known. But, if the value of $y$ satisfying $Y = 0$ be only special and particular, the quantity $\eta$ must be derived as the primitive of the new equation which, owing to its source, may be called the *subsidiary characteristic equation*.

We now have

$$\kappa I_2 = J_1 - I_1,$$

$$= \int v(Y_1 - Y) \, dx;$$

and therefore

$$
\begin{aligned}
I_2 &= \int v \left\{ (f_{00} - f_{01}{}') \, v - \frac{d}{dx} (f_{11} v') \right\} dx \\
&= \int \frac{v}{\eta} \left\{ v \frac{d}{dx} (f_{11} \eta') - \eta \frac{d}{dx} (f_{11} v') \right\} dx \\
&= \int \frac{v}{\eta} \frac{d}{dx} \{ f_{11} (v \eta' - v' \eta) \} \, dx \\
&= \left[ \frac{v}{\eta} (v \eta' - v' \eta) f_{11} \right] + \int f_{11} \left( v' - v \frac{\eta'}{\eta} \right)^2 dx.
\end{aligned}
$$

Thus

$$
I_2 + \left[ v \left( v' - v \frac{\eta'}{\eta} \right) f_{11} \right] = \int f_{11} \left( v' - v \frac{\eta'}{\eta} \right)^2 dx,
$$

where the terms in square brackets are to be taken at the fixed limits of the integral, and where (as before) an implicit assumption has been made as to the finiteness of the subject of integration through the range. With this assumption, and remembering that $v$ vanishes at each of the fixed limits, we make these limit-terms zero, and finally obtain

$$
I_2 = \int f_{11} \left( v' - v \frac{\eta'}{\eta} \right)^2 dx.
$$

The equation for $\eta$ is the same as the equation for $z$: we therefore can take $\eta$ as the value of $z$, and the two expressions for $I_2$ are now the same. Whether $\eta$ be deduced from the value of $y$, when the primitive of the characteristic equation is known, or be obtained by the independent integration of the subsidiary characteristic equation, the magnitude $\eta$ can be taken as any form of the primitive of the latter equation that may have been found. The function $v$ must be allowed all arbitrary values, subject to the requirements of regularity and of vanishing at the limits of the integral.

*Ex.* Prove that the intermediate forms of $I_2$ are equivalent to one another, in virtue of the variation of the quantity the evanescence of which makes the two forms of $I_1$ equivalent.

### The Legendre test.

**20.** The sign of the quantity $I_2$ is to be a steady positive, or a steady negative, for all possible variations of $y$ represented by $\kappa v$, where $v$ is limited to be a regular function of $x$, which must vanish at the extremities of the range of the integral but otherwise can be arbitrarily chosen. As regards the sign of $I_2$, we note that $\left( v' - v \frac{\eta'}{\eta} \right)^2$ is always positive unless it is zero; and, later, we shall assign the condition that it shall not be zero unless $v$ itself is zero, in which event there is no variation. Then as a first condition for the permanence of sign of $I_2$, *it is necessary that $f_{11}$ shall keep the same sign throughout the range.*

The requirement is established as follows. Suppose, for a part of the range, say from a lower limit $a$ of the integral up to a value $x_1$ which is less than the upper limit $c$ of the integral, that $f_{11}$ has one sign, say the positive sign, and for the rest of the range that $f_{11}$ has the other (the negative) sign; and consider the effect of the two following variations. In the first portion of the range, let $v = (x-a)^2(x_1-x)^2 V$, where $V$ is a regular function of $x$, so that $v$ and $v'$ obviously vanish at $x = a$ and $x = x_1$; and everywhere in the second portion of the range beginning at $x = x_1$, let $v$ be zero. For this composite small variation, the sign of $I_2$ is positive. Next, everywhere in the first portion of the same range, let $v$ be zero; and in the second portion of the range, let $v = (x-x_1)^2(c-x)^2 U$, where $U$ is a regular function of $x$, so that $v$ and $v'$ obviously vanish at $x = x_1$ and $x = c$. For this composite small variation, the sign of $I_2$ is negative. Hence, on the hypothesis assumed as to the change in the sign of $f_{11}$ within the range, the sign of $I_2$ can be made positive by the choice of one kind of variation, and can be made negative by the choice of another kind of variation. The same alternative sign can be obtained for $I_2$ by appropriate selected variations, whenever changes of sign* of $f_{11}$ occur otherwise in the range of integration. We therefore infer that, for the possession of a maximum or a minimum by the integral $I$, *the quantity $f_{11}$ cannot change its sign in the course of the range of integration.* If the sign is regularly positive, the value obtained for the integral is (so far) a minimum; if the sign is regularly negative, the value obtained for the integral is (so far) a maximum. But the new condition has only been proved to be necessary; it has not proved to be sufficient, and it is not in fact sufficient, to secure a maximum or a minimum.

This second test was first obtained (in 1786) by Legendre†; and it is not infrequently called after him, as the first test is called after Euler. The Legendre criterion manifestly is distinct in character from the Euler criterion. When the Legendre criterion is satisfied, the property of $f_{11}$ thus possessed is important, as regards both the nature of the integral of the Euler characteristic equation, and the nature of the primitive of the subsidiary characteristic equation.

**21.** The quantity $v' - v\dfrac{\eta'}{\eta}$ remains for consideration, in relation to $I_2$. As it occurs squared, there is no question of sign; the collapse of the significance of $I$ will ensue if, through the appropriate selection of $v$, this quantity can be chosen so as to vanish over a continuous portion of the range—a zero value at merely isolated places is not of material importance. Such an eventuality would occur if we could choose $v$ as a constant multiple of $\eta$, the only relation (except the irrelevant persistent zero for $v$) which can secure

---

* The character, postulated for the function $f$, implicitly requires that, in any finite range, the number of changes of sign, for any derivative such as $f_{11}$, shall be finite.

† *l.c.*, § 5.

a continuous zero for the magnitude in question. Now $v$ is subject to the conditions that it shall vanish at each extremity of the range of the integral. If, therefore, the suggested relation is a possibility, the quantity $\eta$ would be such that it could be chosen so as to vanish at each of these extremities. (The relation would be *a fortiori* possible if $\eta$ could be chosen so as to vanish at two places within the range: for we then could take $v$ a constant multiple of $\eta$ within that portion of the range, and a steady zero outside that portion.) As $\eta$ is the primitive of a linear equation of the second order, its general form is

$$\eta = c_1\eta_1 + c_2\eta_2,$$

no matter how the primitive is derived; the arbitrary constants $c_1$ and $c_2$ are at our disposal, so that they can be used for any desired purpose. Manifestly a relation can be established which would make $\eta$ vanish at one extremity—say, at the lower extremity of the range. If the functional character of $\eta$ is then such that it does not again vanish within the range, or (what is organically the same restriction) if the range of the integral does not extend as far as the first place, where $\eta$ vanishes next after vanishing at the lower limit: the contemplated relation is excluded from the possibility of occurrence. The quantity $v' - v\dfrac{\eta'}{\eta}$ could not be made to vanish for admissible variations, as these are bound to vanish at the limits; and, when the Legendre test is satisfied, $I_2$ would have a permanent sign for all such variations.

### *The primitive of the subsidiary characteristic equation.*

**22.** The function $\eta$ thus becomes of critical importance; we therefore proceed to its discussion, and we shall deal with its construction by the most general process, viz. by derivation from a supposed knowledge of the primitive of the characteristic equation.

This characteristic equation is

$$Y = \frac{\partial f}{\partial y} - \frac{d}{dx}\left(\frac{\partial f}{\partial y'}\right)$$
$$= \frac{\partial f}{\partial y} - \frac{\partial^2 f}{\partial x \partial y'} - \frac{\partial^2 f}{\partial y \partial y'}y' - \frac{\partial^2 f}{\partial y'^2}y'' = 0,$$

being of the second order and (usually) not linear. The only term, which involves $y''$, is $\dfrac{\partial^2 f}{\partial y'^2}y''$; and, when the Legendre test is completely satisfied, the coefficient $\dfrac{\partial^2 f}{\partial y'^2}$ does not vanish for any value of $x$ in the range of $x$ that is considered. Then the general existence-theorem* for ordinary differential equations gives us the result:—if values $x = a$, $y = C_1$, $y' = C_2$, be chosen

---

* See my *Theory of Differential Equations*, vol. iii, § 209.

arbitrarily, subject to the conditions that $a$ lies within the range of integration and that these values constitute a set of ordinary (that is, non-singular) values for the derivatives of $f$ which occur; a uniform integral of the equation exists, uniquely determined by the requirement that $y$ and $y'$ acquire the respective values $C_1$ and $C_2$ when $x = a$. This integral is of the form

$$y = C_1 + C_2 (x - a) + (x - a)^2 R (x - a, C_1, C_2),$$

where $R$ is a regular function; we may denote it more briefly by

$$y = \phi (x, C_1, C_2),$$

where $\phi$ is a regular function, and where $C_1$ and $C_2$ may be regarded as arbitrary constants, the presence of which constitutes a primitive. The first form is not the unique expression of a primitive for all initial requirements. But the quoted theorem establishes the existence of a primitive, containing a couple of arbitrary constants; the later form, taking no special account of the initial value of $x$, will be adopted as the analytical basis for further developments.

Every primitive $y = \phi (x, C_1, C_2)$ satisfies the characteristic equation $Y = 0$, for all positive arbitrary values assigned to the constants $C_1$ and $C_2$. It therefore will provide a primitive, when variations are applied to $C_1$ and $C_2$, such as a choice of $C_1 + \kappa c_1$ and $C_2 + \kappa c_2$ in place of $C_1$ and $C_2$ respectively, where $c_1$ and $c_2$ themselves are arbitrary, and where $\kappa$ is the current small constant. When the new value of $y$ is denoted by $y + \kappa \eta$ on the customary assumption as to the smallness of $\kappa$ in magnitude, then

$$\eta = c_1 \frac{\partial \phi}{\partial C_1} + c_2 \frac{\partial \phi}{\partial C_2};$$

or, if

$$\eta_1 = \frac{\partial \phi}{\partial C_1}, \quad \eta_2 = \frac{\partial \phi}{\partial C_2},$$

we have

$$\eta = c_1 \eta_1 + c_2 \eta_2.$$

Now the new value of $y$ satisfies the equation $Y_1 = 0$; hence $\eta$ satisfies the equation $Y_1 - Y = 0$, that is, it satisfies the subsidiary characteristic equation

$$(f_{00} - f_{01}') \eta - \frac{d}{dx} (f_{11} \eta') = 0.$$

As $\eta$ involves the two arbitrary independent constants, the value of $\eta$ provides the primitive of the equation, which may be written in the form

$$\mathbf{Y} = f_{11} \eta'' + f_{11}' \eta' - (f_{00} - f_{01}') \eta = 0.$$

*Properties of the primitive of the subsidiary equation* $\mathbf{Y} = 0$.

**23.** One or two useful properties of this integral may conveniently be established at this stage. The quantities $\eta_1$ and $\eta_2$ are clearly not in a constant ratio, because (when they are taken in the initial form)

$$\eta_1 = 1 + (x-a)^2 \frac{\partial R}{\partial C_1},$$

$$\eta_2 = x - a + (x-a)^2 \frac{\partial R}{\partial C_2},$$

and $R$ is a regular function of $x - a$: that is, $\eta_1$ and $\eta_2$ are linearly independent. Now

$$f_{11}\eta_1'' + f_{11}'\eta_1' - (f_{00} - f_{01}')\,\eta_1 = 0,$$

$$f_{11}\eta_2'' + f_{11}'\eta_2' - (f_{00} - f_{01}')\,\eta_2 = 0;$$

and therefore

$$f_{11}(\eta_1\eta_2'' - \eta_2\eta_1'') + f_{11}'(\eta_1\eta_2' - \eta_2\eta_1') = 0.$$

Thus

$$f_{11}(\eta_1\eta_2' - \eta_2\eta_1') = A,$$

where $A$ is a constant. As $\eta_1$ and $\eta_2$ are linearly independent, the quantity $\eta_1\eta_2' - \eta_2\eta_1'$ does not vanish. The Legendre test is supposed to be satisfied throughout the range, and therefore $f_{11}$ does not vanish; consequently $A$ does not vanish. But $f_{11}$ contains no arbitrary element in its first form, merely as derived from the subject of integration in the original integral; it usually will contain $C_1$, or $C_2$, or both $C_1$ and $C_2$, after substitution is made for $y$. Also $\eta_1$ and $\eta_2$ contain no arbitrary element other than $C_1$ and $C_2$. Hence $A$ may involve both $C_1$ and $C_2$, because it is equal to the value of the left-hand side in the last equation; but $A$, different from zero, is not a new independent arbitrary constant.

From the last equation, it follows that

$$\frac{d}{dx}\left(\frac{\eta_2}{\eta_1}\right) = \frac{A}{\eta_1^2}\, f_{11}.$$

On the right-hand side, the signs of $A$ and of $\eta_1^2$ do not change; and, by the Legendre test, the sign of $f_{11}$ does not change throughout the range; hence the sign of the right-hand side remains unchanged through the range. If the sign is always positive, then $\frac{\eta_2}{\eta_1}$ increases throughout the range. If the sign is always negative, then $\frac{\eta_2}{\eta_1}$ decreases throughout the range. This result, as to the persistence of sign, is (or may be regarded as) a consequence of the uniformity in the sign of $f_{11}$ through the range; therefore it is a consequence of the Legendre test.

**24.** There is a temptation to infer, as an immediate and obvious corollary, that the quantity $\frac{\eta_2}{\eta_1}$, which is always increasing or always decreasing throughout the range, can never acquire again, in the course of the range, the value which it possesses at the lower limit. The inference would be justified if the quotient were always a continuous function; it is not necessarily correct, if the range should include a zero of $\eta_1$, which would cause a discontinuity through an infinite value, always provided such a zero is not simultaneously a zero of $\eta_2$. For example, let $\eta_1 = \cos x$ and $\eta_2 = \sin x$, so that $\frac{d}{dx}\left(\frac{\eta_2}{\eta_1}\right)$ is always positive and never zero; we know that the function $\tan x$ increases as $x$ increases. But $\tan x$ can resume an initial value, say the value zero for $x = 0$, if the range is extended as far as $x = \pi$; the explanation is that, in spite of the steady increase in its value as $x$ increases, it suffers a sudden setback through its discontinuity as $x$ passes through the value $x = \frac{1}{2}\pi$. It is not therefore justifiable to assume, as a consequence of the persistence of sign of $\frac{\eta_2}{\eta_1}$, that this quantity cannot resume the value which it had at the beginning of the range.

*Note.* From the relation

$$f_{11}\,(\eta_1\eta_2' - \eta_2\eta_1') = A,$$

where the constant $A$ is not zero, it follows that there can be no value of $x$ within the range, for which $\eta_1$ and $\eta_2$ vanish simultaneously. (This result has been anticipated in the preceding argument.)

It equally follows that there can be no value of $x$ within the range, for which $\eta_1'$ and $\eta_2'$ vanish simultaneously.

It is unnecessary to consider a possibility of a value of $x$ for which $\eta_1$ and $\eta_1'$ (or, alternatively, $\eta_2$ and $\eta_2'$) vanish simultaneously; it would follow, from the linear differential equation $\mathbf{Y} = 0$ of which $\eta_1$ and $\eta_2$ are integrals, that $\eta_1$ (or $\eta_2$) would be zero over a continuous range.

**25.** The preceding construction of $\eta$ is based upon a supposed knowledge of the primitive $y$ of the characteristic equation. It shews that, if such primitive is known, then formal integration of the subsidiary characteristic equation is superfluous: the primitive of that equation can be obtained otherwise, by purely direct operations.

But if the primitive $y$ is not known—if, for example as an unfavourable case, only a very special value of $y$ can be obtained satisfying the characteristic equation—then, to complete the set of present operations, the formal integration of the subsidiary equation must be achieved. By some process, its primitive must be obtained.

*Note.* It might happen in this event, that the couple of fundamental integrals composing the primitive are different from $\eta_1$ and $\eta_2$. If so, let them be denoted by $\xi_1$ and $\xi_2$. Then, by the properties of an ordinary linear equation of the second order, we know that there are relations

$$\eta_1 = a_1 \xi_1 + a_2 \xi_2, \quad \eta_2 = b_1 \xi_1 + b_2 \xi_2,$$

where $a_1$, $a_2$, $b_1$, $b_2$ are constants such that $a_1 b_2 - a_2 b_1$ does not vanish; and then

$$\eta_1 \eta_2' - \eta_2 \eta_1' = (a_1 b_2 - a_2 b_1)(\xi_1 \xi_2' - \xi_2 \xi_1').$$

The corresponding equation

$$f_{11}(\xi_1 \xi_2' - \xi_2 \xi_1') = B,$$

where $B$ is a constant, agrees with the earlier equation

$$f_{11}(\eta_1 \eta_2' - \eta_2 \eta_1') = A,$$

provided the relation $A = (a_1 b_2 - a_2 b_1) B$ is established between the constants $A$ and $B$. All the argument and the conclusion are the same as before.

### *The Jacobi test: conjugate points on a characteristic.*

**26.** It has been seen that, if a maximum or a minimum is to exist for the integral, we must be precluded from the choice of a non-vanishing function $v$ which (i), could be a constant multiple of the magnitude $\eta$ throughout the range and (ii), could also be consistent with the condition that $v$ is to vanish at both the lower and the upper limit of the integral. If the choice were possible, then $\eta$ would have to vanish at the lower limit. Because

$$\eta = c_1 \eta_1 + c_2 \eta_2,$$

this requirement can always be satisfied by choosing one relation between the two constants which are at our disposal, so that $-c_1/c_2$ is equal to the value (say $m$) of $\dfrac{\eta_2}{\eta_1}$ at that lower limit. Now, owing to the other conditions that have been imposed, $\dfrac{\eta_2}{\eta_1}$ is always increasing through the range of the integral or is always decreasing through the range.

If, then, whatever be the range, the quantity $\dfrac{\eta_2}{\eta_1}$ never returns to the value which it acquires at the lower limit, then $\eta$ cannot again vanish. A non-vanishing function $v$ cannot be chosen (i), to be a constant multiple of $\eta$ through the range and (ii), to satisfy the terminal conditions. No new test emerges as an additional requirement to be satisfied in connection with the integral.

If, however, the quantity $\dfrac{\eta_2}{\eta_1}$ can return to the value which it acquires at the lower limit, and if the range is sufficiently extended so that it includes the value of $x$ where this return is made (and made for the first time, should the initial value recur more than once), then $\eta$ acquires the value zero at this place. We then can take $v = \mu\eta$, where $\mu$ is any constant (its magnitude is significant only as regards the scale of the small variation $\kappa\eta$ to which $y$ is subjected); and, for that variation, $I_2$ would vanish. In other words, some variation is possible under which a claim for a maximum or a minimum could not be regarded as established. Such a possibility must therefore be prevented; hence the range must not include the first place at which the monotonic function again assumes the value assumed at the lower limit. It therefore follows that the range of the integral must not extend so far as that first place, if such a place exist; while, if no such place exist, there is no corresponding restriction on the range of the integral when there is to be a maximum or a minimum.

The place, where the initial value of $\dfrac{\eta_2}{\eta_1}$ is next assumed, is customarily called the *conjugate* of the initial place. We thus have a new test—the third: viz., *the range of the integral must not extend as far along the characteristic curve as the conjugate (if any) of the lower limit.*

This test, imposing a limit upon the range of the integral, is due to Jacobi*, and is frequently called the Jacobi test. The earlier discussion as to the variation of $\dfrac{\eta_2}{\eta_1}$ shews that, as a test, it is independent of, and additional to, the Legendre test.

**27.** The Jacobi test can be expressed in a simple geometrical form. Let a characteristic curve $AP...$ be drawn, passing through an initial point $A$; the position of the point $A$, and the tangent at $A$ to the characteristic curve, are the geometrical representation of the arbitrarily assigned initial values determining the solution $y$ of the characteristic equation. The special small variation consists of a small displacement of a point $P$ of the curve along the ordinate, as to a 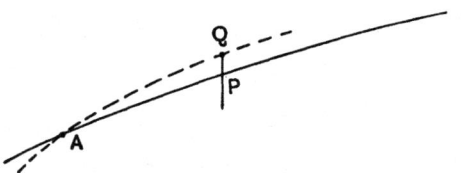 point $Q$; when this small variation is effected along the curve, then (for weak variations) the inclination of the tangent $Q$ to the displaced curve differs from the inclination of the tangent at $P$ by a correspondingly small amount.

The new ordinate $y + \kappa\eta$ is the ordinate of a new characteristic curve $AQ...$, passing through the initial point $A$, and having its tangential inclination at $A$ differing by the appropriate small quantity from the tangential

---

\* It was first stated by him in 1837 (*l.c.*, § 6); for the establishment, reference may be made to a memoir by Hesse, *Crelle*, t. liv (1857), pp. 227—273.

inclination at $A$ on the original characteristic. If this consecutive characteristic cuts the original characteristic in some point after $A$, we consider the first of such points of intersection; its limiting position, say $B$, as the initial deviation between the two curves at $A$ is incessantly diminished, is the conjugate of $A$.

The Jacobi test requires that the range of the integral, taken along the characteristic curve, and beginning at $A$, shall not extend so far as the point $B$, thus defined as the conjugate of the initial point $A$.

*Ex.* 1. We know that a straight line is the shortest distance between two points in a plane. Thus the characteristic curve (Ex. 1, § 17) is a straight line. Another characteristic curve (also a straight line) through the initial point, slightly inclined to the first, does not again meet the initial straight line. The initial point has no conjugate on the characteristic curve; there is no finite limit to the range of the integral.

*Ex.* 2. We know that the smaller arc of the great circle through two points on a sphere is the shortest distance on the surface of the sphere between the two points. Later (§ 75), as an illustration of the analysis, the great circle will be obtained as the characteristic curve for the spherical geodesic.

A consecutive characteristic, through a point $A$ on the great circle, is a consecutive great circle; the two circles intersect at the point $B$, which is diametrically opposite to $A$. The particular great circle between $A$ and $B$ cannot be declared a unique minimum between $A$ and $B$: in order to have an actual minimum, the great-circle range, which begins at $A$, must not extend beyond $B$.

*Ex.* 3. Shew that, if $B$ is the forward conjugate of $A$ on the characteristic curve, $A$ is the backward conjugate of $B$ on that curve.

## *Summary of the tests.*

**28.** It may be convenient to summarise the results thus far obtained in connection with special variations, as follows:

I. The *Euler test*: the characteristic equation

$$Y = \frac{\partial f}{\partial y} - \frac{d}{dx}\left(\frac{\partial f}{\partial y'}\right) = 0$$

must be satisfied; and (for subsequent purposes, but not as a part of the Euler test) the primitive of the subsidiary characteristic equation

$$\mathbf{Y} = (f_{00} - f_{01}')\,\eta - f_{11}'\,\eta' - f_{11}\,\eta'' = 0$$

must be obtained, in a form $\eta = c_1\eta_1 + c_2\eta_2$ where $c_1$ and $c_2$ are arbitrary constants.

II. The *Legendre test*: the quantity

$$\frac{\partial^2 f}{\partial y'^2}$$

(when, in it, the value of $y$ derived from the characteristic curve has been substituted) must keep its sign unchanged through the whole range of the integral.

III. The *Jacobi test*: the range of the integral must not extend so far, as to allow the monotonic magnitude $\eta_2/\eta_1$ to resume the value it assumes at the lower limit; or, in geometric phrase, the range of the characteristic curve must not extend so far as the conjugate of its initial point.

## Conjugate as actual limit of a range.

**29.** It may however happen that an assigned range extends so far as to include the conjugate of the initial point. In that event, a small variation, given by $v = \mu\eta$ where $\mu$ is a constant, will make the quantity $I_2$ vanish: that is, will make the term in $J - I$, depending on $\kappa^2$, to vanish; while, for all other variations, $I_2$ does not vanish. Except for this particular variation, the terms in $J - I$ involving powers of $\kappa$ higher than the second are negligible. But for this particular variation, those terms require consideration; and, among them, the most important initially is the term $\frac{1}{6}\kappa^3 I_3$, where

$$I_3 = \int \left( v^3 \frac{\partial^3 f}{\partial y^3} + 3v^2 v' \frac{\partial^3 f}{\partial y^2 \partial y'} + 3vv'^2 \frac{\partial^3 f}{\partial y \partial y'^2} + v'^3 \frac{\partial^3 f}{\partial y'^3} \right) dx.$$

In order that $J - I$ may keep an unvarying sign for all variations—so that a maximum or a minimum may be secured—it is necessary that $I_3$ should vanish for the variation $v = \mu\eta$ which makes $I_2$ vanish; otherwise a change of sign for the arbitrary constant $\kappa$ would alter the sign of $\kappa^3 I_3$. The requisite condition is that

$$\int \left( \eta^3 \frac{\partial^3 f}{\partial y^3} + 3\eta^2 \eta' \frac{\partial^3 f}{\partial y^2 \partial y'} + 3\eta\eta'^2 \frac{\partial^3 f}{\partial y \partial y'^2} + \eta'^3 \frac{\partial^3 f}{\partial y'^3} \right) dx$$

taken between the limits of the original integral, or taken between the range bounded by the conjugate points, shall vanish.

If this condition is not satisfied, then the integral does not provide a maximum or a minimum.

If this condition is satisfied, then the sign of $I_4$, where

$$I_4 = \int \left( \eta^4 \frac{\partial^4 f}{\partial y^4} + 4\eta^3 \eta' \frac{\partial^4 f}{\partial y^3 \partial y'} + 6\eta^2 \eta'^2 \frac{\partial^4 f}{\partial y^2 \partial y'^2} + 4\eta\eta'^3 \frac{\partial^4 f}{\partial y \partial y'^3} + \eta'^4 \frac{\partial^4 f}{\partial y'^4} \right) dx,$$

must be uniformly positive for a minimum and uniformly negative for a maximum.

A fuller discussion is given at a later stage (§§ 76—80).

*Note.* The same kind of consequence would follow, if an assigned range of integration extended beyond a conjugate. In that case, it is possible that $I_2$ might be made to vanish; then the term, involving $\kappa^3$ in the expression for $J - I$, would have to vanish under the conditions making $I_1$ and $I_2$ vanish, and the term involving $\kappa^4$ would have to keep one sign under those conditions. But it will be proved (§§ 76—80) that no maximum or minimum can be possessed by an integral over a range, which extends beyond the conjugate of its initial place.

## EXAMPLES.

**30.** We pass to some examples : in the first of these, the catenoid is discussed.

*Ex.* 1. *Find the plane curve joining two points such that, when the curve is rotated round a line in a plane which passes also through both the points, the surface of the generated solid is a minimum.*

(A) Take the line for the axis of $x$. The element of arc is $(1+y'^2)^{\frac{1}{2}} dx$, where the positive sign is taken for the radical. The area of the generated surface is

$$2\pi \int y\, ds, \;=2\pi \int y\,(1+y'^2)^{\frac{1}{2}}\, dx,$$

the integral being taken in the positive direction of the axis of $x$ between the values of $x$ for the two points. Here the subject of integration is $f$, where

$$f=y\,(1+y'^2)^{\frac{1}{2}},$$

which does not involve $x$ explicitly ; and therefore (§ 17, *Note*)

$$y\,(1+y'^2)^{\frac{1}{2}}=y'\,\frac{yy'}{(1+y'^2)^{\frac{1}{2}}}+c,$$

where $c$ is an arbitrary constant. Thus

$$y=c\,(1+y'^2)^{\frac{1}{2}};$$

and therefore

$$y=c\cosh\frac{x-b}{c},$$

where $b$ is another arbitrary constant independent of $c$. This equation, containing two arbitrary constants $b$ and $c$, is the primitive of the characteristic equation ; it represents a catenary, the directrix of which is the axis of revolution (the curve is the form of a uniform heavy string, when it hangs freely).

(B) Deferring temporarily the discussion of the possibility of drawing such a catenary through the two points, we proceed to the other tests.

For the Legendre test, we have

$$\frac{\partial^2 f}{\partial y'^2}=\frac{y}{(1+y'^2)^{\frac{3}{2}}}.$$

The sign of the radical is positive. The catenary does not meet the axis, so that $y$ is always positive and does not vanish. Hence $\frac{\partial^2 f}{\partial y'^2}$ is always positive ; and therefore, so far as the Legendre test is concerned, the catenary provides a minimum superficial area.

(C) To apply the Jacobi test, the quantities $\eta_1$ and $\eta_2$ must be constructed. As the arbitrary constants in the primitive are $b$ and $c$, we take

$$\eta_1=\frac{\partial y}{\partial b}=-\sinh\frac{x-b}{c},$$

$$\eta_2=\frac{\partial y}{\partial c}=\cosh\frac{x-b}{c}-\frac{x-b}{c}\sinh\frac{x-b}{c};$$

and therefore the critical quantity for the Jacobi test is

$$\frac{\eta_2}{\eta_1}=\frac{x-b}{c}-\coth\frac{x-b}{c}.$$

To trace the value of $\eta_2/\eta_1$, we use the geometrical representation. At any point $P$ of the curve, draw the tangent cutting the directrix in $T$; the vertex is $A$; and $AB$, $PN$ are ordinates: and $OB = b$, $AB = c$. Then

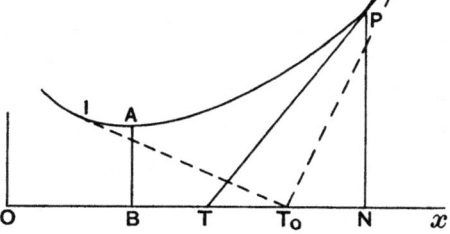

$$PN = c\cosh\frac{x-b}{c}, \quad \tan PTN = \sinh\frac{x-b}{c};$$

and so

$$TN = c\coth\frac{x-b}{c}.$$

Thus

$$c\frac{\eta_2}{\eta_1} = (x-b) - c\coth\frac{x-b}{c} = BN - TN = BT;$$

consequently, for successive points on the curve, the position of the point $T$ on the directrix gives the measure of the value of $\frac{\eta_2}{\eta_1}$. Also $\frac{d}{dx}\left(\frac{\eta_2}{\eta_1}\right)$ is positive, so that $\frac{\eta_2}{\eta_1}$ increases with $x$, corresponding to the motion of $T$ to the right as $P$ moves along the curve in the direction of increase of $x$.

Now let a moving point $Q$ pass from the initial point $I$ of the curve corresponding to the lower limit of the integral, $IT_0$ being the initial tangent. As $Q$ moves along the curve towards $A$, the intersection of the tangent at $Q$ and $Ox$ moves to the right; as $Q$ moves through $A$, that intersection suffers a violent discontinuity in position from the extreme right of $Ox$ to the extreme left of $Ox$; and then, as $Q$ still moves onward, the intersection still moves in the positive direction towards the right. When the intersection returns to its initial position, the position of $Q$ is the point of contact (say $J$) of the other tangent from $T_0$ to the catenary. This point $J$ is the conjugate of the initial point $I$.

Accordingly, the Jacobi test requires that a range of the integral, which begins at $I$, shall terminate before $J$, the conjugate of the point $I$. We shall assume that the Jacobi test is satisfied: the range of the integral is to be restricted within the indicated limits.

(D) The three critical tests are satisfied by the curve, as drawn. But there persists the fundamental question: can a catenary be drawn, passing through two arbitrarily given points and having the axis of $x$ for its directrix? There are subsidiary questions: if a catenary can be so drawn, is it unique? if it is not unique, which (if any) of the possible catenaries can be taken as satisfying the requirements?

In the first place, no such catenary can be drawn when the two points lie on opposite sides of the axis $Ox$; the analysis becomes illusory. If, in the first integral

$$y(1+y'^2)^{-\frac{1}{2}} = c$$

of the characteristic equation, a zero value is assigned to the arbitrary constant $c$, we could have $y = 0$, or $y'$ indefinitely large, or both; but such variations do not accord with the original assumption that $y(1+y'^2)^{\frac{1}{2}}$, the subject of integration, is a regular function for such arguments. Manifestly, some different analysis must be devised, in order to admit the hypothesis that the points lie on opposite sides of $Ox$—a hypothesis which still leaves a reasonable problem.

Next, when the two points are on the same side of the axis, we shall assume (in order, at this stage, merely to simplify the immediate algebra) that the two points are at the same distance $k$ from the axis; and we shall denote their distance apart by $2a$. When the origin is taken at the point $B$, we have $b = 0$; the equation of the catenary is

$$y = c\cosh\frac{x}{c},$$

and the quantity $c$ is to satisfy the equation

$$k = c \cosh \frac{a}{c}.$$

To determine whether this transcendental equation has any real positive root $c$, let

$$\frac{a}{c} = \frac{X}{a}, \qquad \frac{k}{c} = \frac{Y}{a};$$

then any real point common to the curves

$$Y = \frac{k}{a} X, \qquad Y = a \cosh \frac{X}{a},$$

will give a real value of $c$.

Let $\xi$ denote the abscissa of a point $K$ on the new catenary where the tangent to the curve passes through the origin $O$; as the equation of that tangent is

$$Y - a \cosh \frac{\xi}{a} = (X - \xi) \sinh \frac{\xi}{a},$$

we have

$$\coth \frac{\xi}{a} = \frac{\xi}{a}.$$

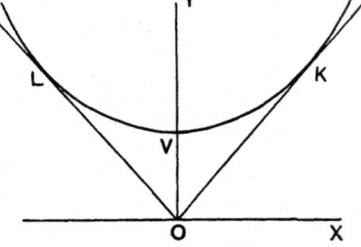

Simple geometrical considerations shew that this equation has only a single positive root*. From tables, we find that

$$\xi = a \sec 33° \, 32'$$

approximately. Thus, if the angle $XOK$ be denoted by $a$, then

$$a = 56° \, 28',$$

also approximately.

(i) If $k < a \tan a$, the line $Y = \frac{k}{a} X$ lies within the angle $XOK$, and consequently does not cut the catenary $LVK$ of the second figure in any real point. No real value of $c$ occurs as a root of the equation $k = c \cosh \frac{a}{c}$; and no catenary for our problem can be drawn through the two assigned points having $Ox$ for its directrix.

The analysis has hypothetically assumed the existence of a real catenary associated with a real finite value of $c$; and the assumption does not accord with present data. It may be that a zero value for $c$ has to be considered; as before, some different analysis must be devised.

(ii) If $k = a \tan a$, the line $Y = \frac{k}{a} X$ is the line $OK$, which touches the catenary $LVK$ of the second figure in the real point $K$. There is one catenary (and only one catenary) for the original problem: the pure constant $a$ (which is $< \frac{1}{2}\pi$) is determined by the equation

$$\coth (\operatorname{cosec} a) = \operatorname{cosec} a;$$

the value of $k$ is $a \tan a$; and the parameter $c$ of the catenary is equal to $a \sin a$.

The two limiting points are conjugate to one another. The case is of the character which sometimes arises as intermediate between the cases, where two real solutions exist, and the cases, where no real solutions exist. For the curve, thus bounded by the two

* There is also a single negative root $\xi$, equal in magnitude and opposite in sign to the single positive root; it leads, for our purpose, to the same result as the positive root.

conjugate points, small non-zero variations can occur, which cause the term $\kappa^2 I_2$ in $J - I$ to vanish because the range extends from a point to its conjugate. Taken in that sense, the Jacobi test is not satisfied: and the exact discrimination requires the investigation of terms of higher order in $\kappa$ in the expression for $J - I$ in powers of $\kappa$.

(iii) If $k > a \tan a$, the line $Y = \dfrac{k}{a} X$ lies within the angle $YOK$ and therefore cuts the catenary of the second figure in two real points. Each point of intersection gives a real

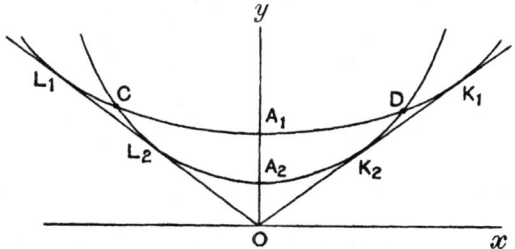

value of $c$; and therefore, in this case, two distinct catenaries can be drawn through the given points $C$ and $D$, so as to have $Ox$ for their directrix.

It is not difficult to prove that the envelope of the family of catenaries

$$y = c \cosh \frac{x}{c}$$

(with $c$ as the parameter of the family) is the two straight lines

$$y = \pm x \tan a,$$

with the former significance of $a$. Thus the two catenaries through $C$ and $D$ are as in the third figure. For both of these catenaries, the Euler test and the Legendre test are satisfied. The Jacobi test is not satisfied for $CL_2 A_2 K_2 D$, because $K_2$ and $L_2$ within its range are conjugate. The Jacobi test is satisfied for the catenary $CA_1 D$; the conjugate of $C$ on $CA_1 D$ lies beyond $D$, and the conjugate of $D$ lies behind $C$.

Thus, in the present case, there is a single catenary for our problem, satisfying the three critical tests. Its parameter is the larger of the two real values of the quantity $c$ satisfying the equation

$$\frac{k}{c} = \cosh \frac{a}{c},$$

when $k > a \tan 56° 28'$.

(iv) We have to examine the catenary in the case (ii); that is, we must consider the quantity $I_3$, where

$$I_3 = \int \left( \eta^3 \frac{\partial^3 f}{\partial y^3} + 3\eta^2 \eta' \frac{\partial^3 f}{\partial y^2 \partial y'} + 3\eta \eta'^2 \frac{\partial^3 f}{\partial y \partial y'^2} + \eta'^3 \frac{\partial^3 f}{\partial y'^3} \right) dx,$$

for the value of $v$, which makes $I_2$ vanish; it can be taken as equal to $\eta$. Now as

$$f = y (1 + y'^2)^{\frac{1}{2}},$$

we have

$$\frac{\partial^3 f}{\partial y^3} = 0, \qquad \frac{\partial^3 f}{\partial y^2 \partial y'} = 0,$$

$$\frac{\partial^3 f}{\partial y \partial y'^2} = \frac{1}{(1 + y'^2)^{\frac{3}{2}}}, \qquad \frac{\partial^3 f}{\partial y'^3} = -\frac{3yy'}{(1 + y'^2)^{\frac{5}{2}}};$$

and therefore

$$I_3 = \int \frac{3\eta'^2}{(1 + y'^2)^{\frac{3}{2}}} \left( \eta - \eta' \frac{yy'}{1 + y'^2} \right) dx.$$

Taking

$$y = c \cosh \frac{x-b}{c}$$

and one of the foregoing forms of $\eta$, viz. $\eta_2$, so that

$$\eta = \cosh \frac{x-b}{c} - \frac{x-b}{c} \sinh \frac{x-b}{c},$$

we find

$$\eta - \eta' \frac{yy'}{1+y'^2} = \cosh \frac{x-b}{c}.$$

The quantity

$$\int \frac{3\eta'^2}{(1+y'^2)^{\frac{3}{2}}} \cosh \frac{x-b}{c}\, dx$$

has the subject of integration everywhere finite and positive through the range; hence $I_3$ does not vanish. Consequently, in this case, when the Jacobi test is in the critical marginal stage between being satisfied and not satisfied, the integral does not provide a minimum or a maximum. Most small variations make $J - I$ positive; one set, as above, certainly can make the sign of $J - I$, which then is dominated by $\kappa^3 I_3$, positive or negative at will.

(v) Hence there is a unique minimum. It exists when $k > a \tan a$, and is given by the larger of the two real roots of the equation

$$k = c \cosh \frac{a}{c} :$$

if $c_1$ denote that larger root, the curve is the catenary

$$y = c_1 \cosh \frac{x}{c_1}.$$

(E) To illustrate the changes in the area of the surface of revolution when different curves are taken joining the two points, we may state the magnitudes of these areas in several cases as follows. (It will be noted that some cases are given when discontinuities of direction occur along the continuous curve.)

For simplicity, we take a square $ABCD$ of side $2a$; the side $AB$ is along the axis $Ox$: the origin is the middle point of $AB$: the terminal points of the curve are $C$ and $D$: and whatever the curve from $C$ to $D$, the area of the surface of revolution is

$$4\pi \int_0^a y\, (1+y'^2)^{\frac{1}{2}}\, dx.$$

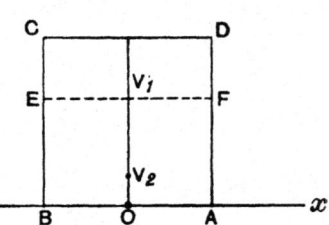

(a) The curve joining $C$ and $D$ is a *catenary*. As the angle $AOD$, $=\tan^{-1} 2$, is greater than $a$, there are two such catenaries having $BOAx$ for directrix. The parameter $c$ satisfies the equation

$$2a = c \cosh \frac{a}{c},$$

of which there are two real roots.

For the greater root $c_1$, we have

$$\frac{a}{c_1} = \cdot 5894,$$

and the vertex of the catenary is at $V_1$; for the smaller root $c_2$, we have

$$\frac{a}{c_2} = 2 \cdot 1276,$$

and the vertex of the catenary is at $V_2$. The area of the catenoidal surface is

$$2\pi c\left(a+c\cosh\frac{a}{c}\sinh\frac{a}{c}\right).$$

The area $S_1$ of the catenoidal surface through the catenary $CV_1D$, having the larger parameter $c_1$, is

$$S_1=2\pi a^2\times 3\cdot 814\ ;$$

this is the *unique minimum* area for the problem.

The area $S_2$ of the catenoidal surface through the catenary $CV_2D$, having the smaller parameter $c_2$, is

$$S_2=2\pi a^2\times 4\cdot 358\ ;$$

this is neither a maximum nor a minimum.

(b) The curve joining $C$ and $D$ is a *semicircle* on $CD$ as diameter, the vertex being turned downwards so that the arc is convex to $AB$. The area $K$ of the surface is

$$K=2\pi a^2\,(2\pi-2)$$
$$=2\pi a^2\times 4\cdot 283.$$

(c) The curve joining $C$ and $D$ is the *straight line $CD$*. The area $L_1$ of the surface is

$$L_1=8\pi a^2$$
$$=2\pi a^2\times 4.$$

(d) The curve joining $C$ and $D$ is the *broken line*, made up of $CO$ (straight) and $OD$ (straight). The area $L_3$ of the surface is

$$L_3=2\pi a^2\times 4\cdot 472.$$

(e) The curve joining $C$ and $D$ is the *broken line*, made up of $CB$ (straight), $BA$ (straight), and $AD$ (straight). The area $L_2$ of the surface is

$$L_2=2\pi a^2\times 4.$$

(f) The curve joining $C$ and $D$ is the *parabola*, having $O$ as its vertex and $OV_2V_1$ for its axis. The area $P$ of the surface is

$$P=2\pi a^2\times 4\cdot 236.$$

(g) The curve joining $C$ and $D$ is the *semi-ellipse* on $CD$ as major axis, with $V_1$ as an extremity of the minor axis. The area $E_1$ of the surface is

$$E_1=2\pi a^2\times 3\cdot 899.$$

(h) The curve joining $C$ and $D$ is the *broken line $CG$* along $CB$ (straight), any line (straight) $GH$ parallel to $BA$, the line $HD$ along $AD$ (straight). When $CG=k$, the area of the surface is

$$10\pi a^2-2\pi\,(k-a)^2,$$

the least values of which are $L_1$ and $L_2$ above, and the largest value of which is $2\pi a^2\times 5$, occurring when $GH$ is midway between $AB$ and $CD$.

(i) The curve joining $C$ and $D$ is the *semi-ellipse* with $CD$ as minor axis, with $O$ as an extremity of the major axis. The area $E_2$ of the surface is

$$E_2=2\pi a^2\times 5\cdot 028.$$

*Ex.* 2. Find the surface of the generated solid when $k=a\tan 56°\,28'$ ; and compare it with the surface $2\pi k^2$, when the generating curve is taken to be a broken line as in the foregoing case (e).

*Ex.* 3. A catenary $LVK$ has $OX$ for its directrix, $V$ is its vertex, and $VO$ is perpendicular to the directrix (fig., p. 32); the tangents from $O$ to the curve are $OL$ and $OK$. Any point $L'$ is taken in $OL$, and a point $K'$ is taken in $OK$, so that $OK' = OK$; and a catenary $L'V'K'$ is drawn, with $OX$ for directrix.

Prove that the surface generated by the revolution of $LVK$ about the directrix is equal to the surface generated by the revolution of the curve $LL'V'K'K$ about the directrix; and that each of these equal surfaces is equal to that of the surface generated by the revolution of the broken line $LL'OK'K$ about the directrix.

### *Mobile limits of an integral: terminal conditions.*

**31.** The preceding investigations have proceeded on the assumption that the extremities of the range of integration are definite: in analytical vocabulary, that the upper limit and the lower limit of the integral are constants: in geometrical vocabulary, that the curve, along which the point $(x, y)$ moves during the range of integration, is terminated by fixed points.

Frequently, however, the terminal conditions are not so simple and so precise. Thus the upper limit of the integral may be provided by the abscissa of a point, which is required to lie on a given curve, but which is not otherwise specified; and similarly for the lower limit. In such a case, the full solution of the problem requires the determination of the limits, in accordance with the assigned requirement, as well as the determination of the characteristic curve between these initially unknown limits. Thus further unknown magnitudes may arise for determination; and they can be obtained only in accordance with added requirements.

The admissible maxima and minima are to possess their distinctive property for all possible small variations that can be effected, alike individually and in all combinations. Now among such variations, there must occur the aggregate of those which are possible when the limits of the integral are not subjected to small variations, that is, when these limits are fixed. All the preceding conditions and tests must therefore apply; but now, instead of constituting the total of such tests, they constitute only a part of the total. There will be—at least, there may be—further conditions, although descriptive (rather than intrinsic) properties are sufficient and necessary to satisfy these conditions.

### *Variation of the integral with mobile limits.*

**32.** Accordingly, we resume the consideration of the integral

$$I = \int_{x_1}^{x_2} f(x, y, y')\, dx,$$

with the same function $f(x, y, y')$ as before. The limits $x_2$ and $x_1$ are now no longer necessarily invariable; they may be subjected to arbitrary small variations, independent of the small variations actually within the course of the range between $x_2$ and $x_1$, and restricted only by the need of being continuous with these at the extremities of the range. Within the range, we still take no variation for $x$; and we take the same variation for $y$ as before, namely, the value $y + \kappa v$ in place of $y$. But now we cannot use the requirement that $v$ must vanish at the limits; though, throughout, $v$ remains a regular function of $x$. As the limits are to admit possible small variations, we shall suppose an upper limit $x_2 + \kappa \xi_2$ in place of $x_2$, and a lower limit $x_1 + \kappa \xi_1$ in place of $x_1$, where $\xi_2$ and $\xi_1$ are finite quantities, which have an arbitrary quality independently of one another in the absence of further specified conditions. (It will be necessary to bring the value of $v$ at $x_2$ into relation with $\xi_2$, and the value of $v$ at $x_1$ into relation with $\xi_1$.) Denoting the value of the integral under this wider variation by $J'$, we have

$$J' = \int_{x_1+\kappa\xi_1}^{x_2+\kappa\xi_2} f(x, y + \kappa v, y' + \kappa v') \, dx$$

$$= \left( \int_{x_2}^{x_2+\kappa\xi_2} + \int_{x_1}^{x_2} - \int_{x_1}^{x_1+\kappa\xi_1} \right) f(x, y + \kappa v, y' + \kappa v') \, dx;$$

and therefore

$$J' - I = \int_{x_1}^{x_2} \{ f(x, y + \kappa v, y' + \kappa v') - f(x, y, y') \} \, dx$$

$$+ \int_{x_2}^{x_2+\kappa\xi_2} f(x, y + \kappa v, y' + \kappa v') \, dx$$

$$- \int_{x_1}^{x_1+\kappa\xi_1} f(x, y + \kappa v, y' + \kappa v') \, dx.$$

The first of these three integrals is the same in form as the integral which occurred in the initial expression for $J - I$; but $v$ is not now restricted to the value zero at each limit. We accordingly have

$$\int_{x_1}^{x_2} \{ f(x, y + \kappa v, y' + \kappa v') - f(x, y, y') \} \, dv = \kappa \int_{x_1}^{x_2} \left\{ \frac{\partial f}{\partial y} - \frac{d}{dx} \left( \frac{\partial f}{\partial y'} \right) \right\} v \, dx$$

$$+ \kappa \left[ v \frac{\partial f}{\partial y'} \right] + K_2,$$

where the term $[\ ]$ is taken at the limits $x_2$ and $x_1$, and where $K_2$ denotes the aggregate of terms of the second order and higher orders in powers of $\kappa$.

In the second of the three integrals, the whole range (from $x_2$ to $x_2 + \kappa\xi_2$) of the integral is very small. The subject of integration is finite, one-signed, and continuous through that range; and therefore, by one of the 'theorems of mean value,' there is some place within the range, such that the value of the integral is

$$\kappa \xi_2 F,$$

where $F$ denotes the value of $f(x, y + \kappa v, y' + \kappa v')$ at that place. But owing to the finiteness and continuity of $f$, the value of $F$ differs from the value of $f(x, y, y')$ at $x_2$, say from $f_2$, by small quantities of the first order and higher orders in powers of $\kappa$. When the aggregate of these quantities is denoted by $\kappa\Delta_2$, the value of the second integral is expressible in the form

$$\kappa \xi_2 f_2 + \kappa^2 \xi_2 \Delta_2.$$

Similarly, the value of the third integral in the expression for $J' - I$ is expressible in the form

$$\kappa \xi_1 f_1 + \kappa^2 \xi_1 \Delta_1,$$

where $f_1$ denotes the value of $f(x, y, y')$ at $x_1$, and $\kappa \Delta_1$ denotes the difference between $f_1$ and the value of $f(x, y + \kappa v, y' + \kappa v')$ at some place in the range mediate between $x_1$ and $x_1 + \kappa \xi_1$.

When these values of the respective integrals are substituted, the value of $J' - I$ takes the form

$$J' - I = \kappa \int_{x_1}^{x_2} \left\{ \frac{\partial f}{\partial y} - \frac{d}{dx} \left( \frac{\partial f}{\partial y'} \right) \right\} v\, dx + \kappa \left[ v \frac{\partial f}{\partial y'} \right] + \kappa (\xi_2 f_2 - \xi_1 f_1) + R_2,$$

where

$$R_2 = K_2 + \kappa^2 \xi_2 \Delta_2 - \kappa^2 \xi_1 \Delta_1,$$

so that $R_2$ is an aggregate of terms of the second and higher orders in $\kappa$. In this expression, the quantity $v$ is an arbitrary regular function of $x$ throughout the range; the quantities $\xi_1$ and $\xi_2$ are (or can be) independent of one another, and are independent of the variation of $v$ in the range; but $\xi_1$, $\xi_2$, and the values acquired by $v$ at $x_1$ and $x_2$, are subject to the relations imposed by the terminal conditions.

**33.** In order that $I$ may have a maximum or minimum, the aggregate of terms of the first order must vanish. If they do not vanish, they govern the value of $J' - I$; and they can be made positive or negative at choice—a result which, if attainable, excludes the possibility of a maximum or minimum. Further, this vanishing of the aggregate of terms of the first order must ensue under all possible small variations, in whatever form or forms they may arise; that is, for every form, and not merely for some individual form or for some group of forms.

Consider three particular forms: (i), a variation continuous through the range and vanishing at each extremity; (ii), a variation continuous through the range, vanishing at the lower limit, and subject to an assigned external condition at the upper limit; and (iii), a variation continuous through the range, vanishing at the upper limit and subject to an assigned external condition at the lower limit.

In the variation represented by (i), we make $v = 0$ at each limit, also $\xi_2 = 0$ and $\xi_1 = 0$; and then

$$\int_{x_1}^{x_2} \left\{ \frac{\partial f}{\partial y} - \frac{d}{dx} \left( \frac{\partial f}{\partial y'} \right) \right\} v\, dx = 0,$$

for all admissible arbitrary regular functions $v$, the limits $x_1$ and $x_2$ now not being subject to variation. These requirements are precisely the same as the requirements in the earlier discussion: the former argument is applicable and so the relation

$$Y, = \frac{\partial f}{\partial y} - \frac{d}{dx} \left( \frac{\partial f}{\partial y'} \right) = 0$$

must be satisfied throughout the whole of the range: that is, we have the Euler characteristic equation. Accordingly, the first term in the foregoing expression for $I$ vanishes.

In the variation represented by (ii), there is no variation at the lower limit; hence $v = 0$ when $x = x_1$, and $\xi_1 = 0$. Also the integral term in $J' - I$ has disappeared; there thus survives, for the test, the relation

$$\kappa \left( v \frac{\partial f}{\partial y'} \right)_{x=x_2} + \kappa \xi_2 f_2 = 0,$$

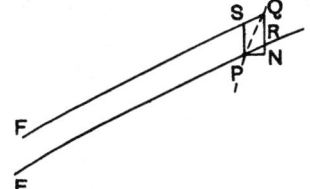

applying solely at the upper limit at $P$. Let the upper limit be required to lie on a given curve of which the equation is

$$g(x, y) = 0,$$

represented by the dotted curve $PQ$, while $EPR$ is the characteristic curve. Let $FSQ$ be the varied curve, obtained from $EPR$ by small displacements along the ordinates; $P$ is displaced by the variation $\kappa v$ along the ordinate to $S$; the upper limit of the integral is varied, so that $PN = \kappa \xi_2$; and $NQ$ is the ordinate through $N$, cutting in $R$ the characteristic curve, and in $Q$ the curve on which $P$ is bound to lie. Then

$$RN = y' \cdot \kappa \xi_2,$$

where $y'$ is the value at $P$ of the derivative of $y$ belonging to the characteristic curve. Because $P$ and $Q$ are consecutive points on the curve $g(x, y) = 0$, we have, at $P$,

$$\frac{\partial g}{\partial x} \kappa \xi_2 + \frac{\partial g}{\partial y} \cdot QN = 0.$$

But

$$QN = QR + RN = \kappa v + \kappa y' \xi_2;$$

consequently the relation becomes

$$\xi_2 \left[ \left\{ f_2 - \left( y' \frac{\partial f}{\partial y'} \right)_2 \right\} - \left( \frac{\partial f}{\partial y'} \frac{\frac{\partial g}{\partial x}}{\frac{\partial g}{\partial y}} \right)_2 \right] = 0.$$

If the upper limit were fixed, we should have $\xi_2 = 0$, and no new relation would ensue. When the upper limit is not fixed, $\xi_2$ can have a non-zero value; and then the relation becomes

$$\frac{\partial g}{\partial y} \left( f - y' \frac{\partial f}{\partial y'} \right) - \frac{\partial f}{\partial y'} \frac{\partial g}{\partial x} = 0,$$

which is to be satisfied at the mobile upper limit $x_2$ of the integral.

The same line of argument applies to the variation (iii); and the result is that, when the lower limit of the integral belongs to a point required to lie on a given curve of which the equation is

$$\psi\,(x,\,y) = 0,$$

the relation

$$f - y'\frac{\partial f}{\partial y'} - \frac{\partial f}{\partial y'}\frac{\dfrac{\partial \psi}{\partial x}}{\dfrac{\partial \psi}{\partial y}} = 0$$

must be satisfied at the lower limit $x_1$ of the integral, provided $x_1$ be not a point absolutely fixed by the limiting conditions. If the point were absolutely fixed, no new relation would ensue.

*Note.* The relation (p. 39) between $v$ and $\eta_2$ can be obtained simply as follows. When $P$ is taken as origin, $PN = \kappa\xi_2$, $QN = \kappa\eta_2$, the equation of the tangent at $Q$ to the curve $FSQ$ is

$$Y - \kappa\eta_2 = y'\,(X - \kappa\xi_2),$$

accurate up to first powers of $\kappa$. The quantity $SP$, equal to $\kappa v$, is the intercept on the ordinate $PS$ of that tangent: that is, we have $Y = \kappa v$ and $X = 0$ simultaneously, and therefore

$$\kappa v - \kappa\eta_2 = -y'\kappa\xi_2,$$

so that

$$v = \eta_2 - y'\xi_2,$$

which is the relation in question.

### Characteristic equation: terminal conditions.

**34.** When all these conditions are satisfied, then the part of $J' - I$ dependent upon the first power of $\kappa$ disappears for all small variations of the kinds contemplated. The portion, arising from small variations within the range, vanishes because of the relation

$$Y = 0,$$

which holds throughout the range. The portion, arising from small variations at the upper limit, vanishes because the relation

$$f - y'\frac{\partial f}{\partial y'} - \frac{\partial f}{\partial y'}\frac{\dfrac{\partial \phi}{\partial x}}{\dfrac{\partial \phi}{\partial y}} = 0$$

is satisfied at that upper limit; and the portion, arising from small variations at the lower limit, vanishes because the relation

$$f - y'\frac{\partial f}{\partial y'} - \frac{\partial f}{\partial y'}\frac{\dfrac{\partial \psi}{\partial x}}{\dfrac{\partial \psi}{\partial y}} = 0$$

is satisfied at that lower limit.

Thus the characteristic equation and the two terminal relations are sufficient, as well as necessary, to secure that the part of $I$ depending on the first power of $\kappa$ (the *first* variation of $I$) shall vanish. In that event,

$$J' - I = R_2:$$

the change in the value of $I$ is measured by quantities of the second order and of higher orders in powers of $\kappa$.

The characteristic equation $Y = 0$ determines, as before, the characteristic curve. The two new conditions at the limits, taken conjointly with $\phi(x, y) = 0$ and $\psi(x, y) = 0$, determine the final values and the initial values of $x$ and $y$: that is, potentially they suffice for the determination of the two limits of the integral and the two arbitrary constants in the primitive of the characteristic equation.

When these are known, the limits of the integral are fixed. The remaining tests are, therefore, the same as before.

*Ex.* 1.  Prove that, for any integral of the form $\int (1 + y_1{}^2)^{\frac{1}{2}} F(x)\, dx$, or of the form $\int (1 + y'^2)^{\frac{1}{2}} G(y)\, dy$—effectively the same integrals analytically—when the range of the integral is to terminate at a given curve, the characteristic curve and the given curve must cut orthogonally at that limit, if the integral is to provide a maximum or a minimum.

Complete the analytical solution of the problem in Ex. 1, § 17, when the straight line is to be the shortest distance between two circles external to one another.

*Ex.* 2.  Shew that, if the curve in Ex. 1, § 30, is to be drawn, not from the fixed points $(a, k)$, $(-a, k)$ as there required, but from one circle of radius $r$ to another circle of radius $r$ having these points for centres, the terminal inclination of the tangent to the catenary is given by the equation

$$\log \frac{1 + \sin \psi}{\cos \psi} = \frac{a \sec \psi - r}{k - r \sin \psi}.$$

*Ex.* 3.  *It is required to find the curve which shall provide the briefest time of passage*[*] *for a heavy particle, moving from rest at a given initial point $O$ to a given straight line the vertical plane through which contains $O$.*

(We may imagine the curve to be a smooth wire or hollow tube in a vertical plane, the particle being a bead.)

We take $O$ as origin, and the horizontal through $O$ towards the given line to be the axis of $x$; and we take the axis of $y$ vertically downwards. The velocity of the particle at a depth $y$ below the origin is $(2gy)^{\frac{1}{2}}$; and the time of passage is

$$\frac{1}{(2g)^{\frac{1}{2}}} \int_0 (1 + y'^2)^{\frac{1}{2}} y^{-\frac{1}{2}}\, dx,$$

the upper limit for the integral being the value of $x$ for some point on the given line yet to be determined.

(i) The subject of integration does not involve $x$ explicitly; hence (§ 17, *Note*) a first integral of the characteristic equation is

$$(1 + y'^2)^{\frac{1}{2}} y^{-\frac{1}{2}} = y' \frac{y'}{(1 + y'^2)^{\frac{1}{2}}} y^{-\frac{1}{2}} + \text{constant}.$$

---

[*] The brachistochrone problems were initiated by John Bernoulli (§ 3).

Let $y' = \tan \psi$, so that $\psi$ is the inclination to $Ox$ of the tangent to the characteristic curve, taken in the direction of $s$ increasing. Then

$$y = a\,(1 + \cos 2\psi),$$

where now $a$ is an arbitrary constant, so far as the equation is concerned. Also

$$dx = dy\,.\,\cot \psi = -a\,(2 + 2 \cos 2\psi)\,d\psi\,;$$

and therefore

$$x = k - a\,(2\psi + \sin 2\psi),$$

where $k$ is the second arbitrary constant for the primitive. The values of $x$ and $y$, taken together, shew that the characteristic curve is a cycloid having $Ox$ for directrix, while $a$ is the radius of the generating circle.

(ii) As $f$, the subject of integration, is $(1 + y'^2)^{\frac{1}{2}} y^{-\frac{1}{2}}$, we have

$$\frac{\partial^2 f}{\partial y'^2} = (1 + y'^2)^{-\frac{3}{2}} y^{-\frac{1}{2}}\,;$$

the radicals are positive, and therefore $\dfrac{\partial^2 f}{\partial y'^2}$ is positive. Thus the second test—the Legendre test—will allow the existence of a minimum.

(iii) Before proceeding to the third test—the Jacobi test—we deal with the terms at the limits. At $O$, we have $y = 0$ and $x = 0$. From $y = 0$, we have $\psi = \frac{1}{2}\pi$; and therefore the origin is a cusp of the cycloid. From $x = 0$, and using the initial value of $\psi$, we have $k = a\pi$.

The other extremity is to lie upon a given line. The equation of this line may be taken in the form

$$y = (x - c) \tan \beta,$$

where $c$ is positive owing to the positive direction chosen for $Ox$, and where $0 < \beta < \pi$. The terminal condition of § 33 becomes

$$(1 + y'^2)^{\frac{1}{2}} y^{-\frac{1}{2}} - y' \frac{y'}{(1 + y'^2)^{\frac{1}{2}}} y^{-\frac{1}{2}} - \frac{y'}{(1 + y'^2)^{\frac{1}{2}}} y^{-\frac{1}{2}} \left( \frac{-\tan \beta}{1} \right) = 0,$$

that is,

$$1 + y' \tan \beta = 0.$$

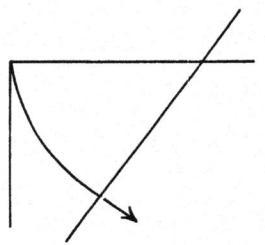 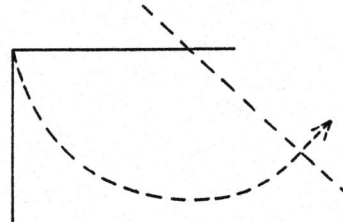

As $y' = \tan \psi$, we have $\psi = \beta - \frac{1}{2}\pi$ in the left-hand figure, and $-\psi = \frac{1}{2}\pi - \beta$ in the right-hand figure: that is, the terminal value of $\psi$ is always

$$\psi = \beta - \tfrac{1}{2}\pi.$$

But for that value of $\psi$,

$$y = a\,(1 - \cos 2\beta), \quad x = a\,(2\pi - 2\beta + \sin 2\beta),$$

and the point lies on $y = (x - c) \tan \beta$. Substituting, we find

$$2a\,(\pi - \beta) = c,$$

excluding the value $\sin \beta = 0$; and so the radius of the generating circle is known.

(iv) For the Jacobi test, we return to the unspecialised primitive of the characteristic equation. From the equation

$$k - a\,(2\psi + \sin 2\psi) = x,$$

it follows that $\psi$, and therefore any function of $\psi$, is a function of $\dfrac{k-x}{a}$; so we write

$$\cos 2\psi = \phi\left(\frac{k-x}{a}\right),$$

and the primitive becomes

$$y = a + a\phi\left(\frac{k-x}{a}\right).$$

From these, we easily find

$$\eta_1 = \frac{\partial y}{\partial k} = \phi'\left(\frac{k-x}{a}\right) = -\tan\psi,$$

$$\eta_2 = \frac{\partial y}{\partial a} = 1 + \phi\left(\frac{k-x}{a}\right) - \frac{k-x}{a}\,\phi'\left(\frac{k-x}{a}\right) = 2\,(1 + \psi\tan\psi),$$

so that

$$\frac{\eta_2}{\eta_1} = -2\,(\psi + \cot\psi).$$

It is also easy to shew that

$$\frac{d}{dx}\left(\frac{\eta_2}{\eta_1}\right) = -\frac{1}{2a\sin^2\psi},$$

so that $\dfrac{\eta_2}{\eta_1}$ is monotonic and is continually decreasing.

Now the initial value of $\psi$ is $\tfrac{1}{2}\pi$ and it decreases steadily along the arc of the cycloid. The initial value of $-\dfrac{\eta_2}{\eta_1}$ is $-\pi$. The value of $-\dfrac{\eta_2}{\eta_1}$ decreases towards $-\infty$ as $\psi$ decreases towards zero; it suffers a discontinuity as $\psi$ passes through zero from an infinitely large negative value to an infinitely large positive value; and as $\psi$ decreases to $-\tfrac{1}{2}\pi$ (that is, as $x$ increases), the value of $\dfrac{\eta_2}{\eta_1}$ decreases to $\pi$. At that place, $\psi$ suddenly changes from $-\tfrac{1}{2}\pi$ to $\tfrac{1}{2}\pi$; the value of $\dfrac{\eta_2}{\eta_1}$ again becomes $-\pi$, that is, the place is the conjugate of $O$. But the place is the first cusp of the cycloid next after the cusp at the origin.

Accordingly, the Jacobi test requires that the range of the cycloidal arc, which begins at $O$, shall not extend as far as the next cusp. The requirement is satisfied, because the values 0 and $\pi$ for $\beta$ have been excluded.

Hence the cycloidal arc, having its cusp at $O$ and cutting the given line orthogonally below the horizontal line $Ox$, and having no cusp in its range except at $O$, provides the curve of quickest descent from rest at $O$ to the line.

(v) If $T$ is the time of descent along this cycloidal arc,

$$T = \left\{\frac{2c}{g}\,(\pi - \beta)\right\}^{\frac{1}{2}};$$

while, if $T_1$ is the shortest time of descent along a straight line from $O$ to the given straight line,

$$T_1 = \left\{\frac{4c}{g}\cot\tfrac{1}{2}\beta\right\}^{\frac{1}{2}}.$$

As $\beta$ lies between 0 and $\pi$, we have $T_1 > T$.

*Note.* When dealing with the cycloid, advantage sometimes accrues in the discussion of the reality or the uniqueness of the curve by choosing the current variable to be the angle $\theta$ through which the generating circle of the cycloid has, in the usual generation rolled instead of the angle $\psi$ through which its tangent has turned. In particular, the difficulty at the cusp arising from the latter choice—the variable suddenly increases by $\pi$ at such a point—is avoided by the former choice. We have

$$\psi = \tfrac{1}{2}\pi - \tfrac{1}{2}\theta ;$$

the coordinates, referred to the cusp as origin and the directrix as the axis of $x$, are

$$X = a(\theta - \sin\theta), \quad Y = a(1 - \cos\theta).$$

In the preceding example, the initial value of $\theta$ is zero. The final value is $\theta = 2\pi - 2\beta$; for the final value the point $(X, Y)$ lies on the line

$$y = (x - c)\tan\beta,$$

and, as before, we have

$$2a(\pi - \beta) = c.$$

*Ex.* 4. When the particle starts downwards along the curve in Ex. 3 with an initial velocity $(2gh)^{\frac{1}{2}}$, and the curve is referred to the initial point as origin and the horizontal line as axis of $x$, obtain the coordinates of a point on the curve in the form

$$y + h = a(1 - \cos\theta), \quad x + k = a(\theta - \sin\theta),$$

where $a$ and $k$ are the two constants of integration. Shew that, if the time of quickest descent to a given point below the axis of $x$ is required, one (and only one) cycloid exists satisfying the requirement.

*Ex.* 5. Prove that an ellipse is a brachistochrone between two points on its range, under a central force to one focus varying inversely as the square of the distance from the other focus. Find the limit to the range between the two points, in order that the property may be secured.

*Ex.* 6. A plane curve is a free orbit under one central force, and is a brachistochrone under a different central force; and $v, \; v'$, are the velocities of the respective moving particles at any point. Prove that $vv'$ is constant.

*Ex.* 7. Prove that, in an ellipse under a central force to the focus varying inversely as the square of the distance, the quantity $\int v\,ds$ between two points on the curve ($v$ being the velocity at any point) is a minimum, provided the length of the arc is less than a semi-circumference by a vectorial angle greater than

$$2\tan^{-1}\frac{2e\sin\alpha}{1 + 2e\cos\alpha + e^2},$$

where $\alpha$ is the angle between the radii vectores to the nearer apse and the initial point of the arc, and $e$ is the excentricity of the ellipse.

*Ex. 8. Prove that a minimum of the integral*

$$\int \{1 + x^2 + 2xy\,y' + (1 + y^2)\,y'^2\}^{\frac{1}{2}}\,dx$$

*is provided by $y = 0$, if the upper limit and the lower limit of the integral have the same sign; and that, for a minimum when the lower limit of the integral is $-a$ and the upper limit is $c$, where $a$ and $c$ are positive, $c$ must be less than the root of the equation*

$$\sinh^{-1} u + \sinh^{-1} a = \frac{1}{u}(1 + u^2)^{\frac{1}{2}} + \frac{1}{a}(1 + a^2)^{\frac{1}{2}}.$$

A geometrical interpretation is immediate. The curve is a meridian on the paraboloid of revolution $2z = x^2 + y^2$; the meridian is a geodesic between any two points on its length, if the vertex does not lie between them; there is a limit to the range of the meridian as the geodesic between the points, when the vertex does lie between them. Also the example is adduced to indicate the analysis necessary (§ 25) when only a special integral—here, it is $y = 0$—of the characteristic equation is known.

The characteristic equation is

$$\frac{(x+yy')\,y'}{\Delta^{\frac{1}{2}}} - \frac{d}{dx}\left\{\frac{xy+(1+y^2)\,y'}{\Delta^{\frac{1}{2}}}\right\} = 0$$

where $\Delta$ denotes $1+x^2+2xyy'+(1+y^2)\,y'^2=0$: an equation which reduces to

$$(1+x^2+y^2)\,y'' = (xy'-y)\,(1+y'^2).$$

A first integral of the equation is

$$xy'-y = a\Delta^{\frac{1}{2}},$$

where $a$ is an arbitrary constant; the primitive is expressible by circular and logarithmic functions, and will be obtained later. Manifestly, $y=0$ satisfies the characteristic equation; it is a very special integral, and it cannot be used in the same way as the primitive for the construction of the critical function in Jacobi's test.

For the Legendre test, we have

$$\frac{\partial^2 f}{\partial y'^2} = \frac{1+x^2+y^2}{\Delta^{\frac{3}{2}}}$$

in general, and

$$= (1+x^2)^{-\frac{1}{2}}$$

for the particular solution. Thus $\dfrac{\partial^2 f}{\partial y'^2}$ is always positive. The stated integral, so far as concerns this test, is a minimum.

For the subsidiary characteristic equation $\mathbf{Y}=0$, we have

$$\frac{\partial^2 f}{\partial y^2}=0, \quad \frac{\partial^2 f}{\partial y\,\partial y'} = x\,(1+x^2)^{-\frac{1}{2}},$$

when the value $y=0$ is inserted in the general values; hence the equation for $\eta$ is

$$\frac{d}{dx}\{(1+x^2)^{-\frac{1}{2}}\,\eta'\} + \eta\,\frac{d}{dx}\{x\,(1+x^2)^{-\frac{1}{2}}\} = 0,$$

that is,

$$(1+x^2)\,\eta'' - x\eta' + \eta = 0.$$

The primitive of this equation is

$$\eta = Bx + A\,\{x\sinh^{-1}x - (1+x^2)^{\frac{1}{2}}\}.$$

Thus we can take

$$\eta_1 = x, \quad \eta_2 = x\sinh^{-1}x - (1+x^2)^{\frac{1}{2}},$$

and therefore

$$\frac{\eta_2}{\eta_1} = \sinh^{-1}x - \frac{(1+x^2)^{\frac{1}{2}}}{x}.$$

Hence

$$\frac{d}{dx}\left(\frac{\eta_2}{\eta_1}\right) = \frac{(1+x^2)^{\frac{1}{2}}}{x^2},$$

so that $\dfrac{\eta_2}{\eta_1}$ is always increasing.

If, in passing from the lower limit to the upper limit, $x$ does not acquire the value zero, then $\dfrac{\eta_2}{\eta_1}$ remains finite throughout; as this function increases through such a range, it cannot again acquire its initial value in the range.

Now let $x$ acquire the value zero in the range, supposed to begin with the value $-a$
For that value of $x$, the value of $\frac{\eta_2}{\eta_1}$ is

$$-\sinh^{-1}a + \frac{(1+a^2)^{\frac{1}{2}}}{a}.$$

As $x$ passes from negative to positive through the value zero, $\frac{\eta_2}{\eta_1}$ passes from a large positive value through infinity to a large negative value, and then again increases. If when $x=u$, where $u$ is positive, it again acquires the initial value, we have

$$\sinh^{-1}u - \frac{(1+u^2)^{\frac{1}{2}}}{u} = -\sinh^{-1}a + \frac{(1+a^2)^{\frac{1}{2}}}{a};$$

that is, $u$ is given by

$$\sinh^{-1}u + \sinh^{-1}a = \frac{1}{u}(1+u^2)^{\frac{1}{2}} + \frac{1}{a}(1+a^2)^{\frac{1}{2}}.$$

The possibility of acquiring the initial value must be excluded if a true minimum or a true maximum is to be acquired; hence the range of the integral beginning with the value $-a$ for $x$ must not extend as far as the positive root $u$ of this equation.

*Note.* The primitive of the characteristic equation for the general problem of minimising the given integral, viz.,

$$(1+x^2+y^2)\,y'' = (xy'-y)(1+y'^2),$$

can be obtained as follows. We have

$$\frac{y''}{(xy'-y)(1+y'^2)} = \frac{1}{1+x^2+y^2};$$

and so

$$\frac{xy''}{xy'-y} - \frac{y'y''}{1+y'^2} = \frac{x+yy'}{1+x^2+y^2}.$$

Thus

$$\frac{(xy'-y)^2}{(1+y'^2)(1+x^2+y^2)} = \text{constant} = \frac{a^2}{1+a^2},$$

where $a$ is an arbitrary constant. Let the variables be changed according to the relations

$$x = r\cos\theta, \qquad y = r\sin\theta.$$

The equation becomes

$$\frac{d\theta}{dr} = \frac{a}{r}\left(\frac{r^2+1}{r^2-a^2}\right)^{\frac{1}{2}};$$

and the primitive is

$$\theta - a = \tfrac{1}{2}a\log\left(\frac{u+1}{u-1}\right) - \tan^{-1}(au),$$

where

$$u^2 = \frac{r^2+1}{r^2-a^2} = \frac{x^2+y^2+1}{x^2+y^2-a^2}, \qquad \theta = \tan^{-1}\frac{y}{x}.$$

This equation gives the characteristic curve; interpreted geometrically in connection with the paraboloid of revolution, it is the projection of a surface geodesic upon a plane perpendicular to the axis.

For the Jacobi test, we have

$$\frac{\partial\theta}{\partial a} = 1,$$

$$\frac{\partial\theta}{\partial a} = \tfrac{1}{2}\log\frac{u+1}{u-1} - \frac{u}{1+a^2u^2};$$

so that the critical quantity is

$$\tfrac{1}{2}\log\frac{u+1}{u-1} - \frac{u}{1+a^2u^2},$$

or, if we write

$$\coth v = u = \left(\frac{x^2 + y^2 + 1}{x^2 + y^2 - a^2}\right)^{\frac{1}{2}},$$

this critical quantity is

$$V = v - \frac{\tanh v}{\tanh^2 v + a^2}.$$

Now $V$ always increases; hence the only way in which $V$ can resume its initial value is by passing on a positive increasing range for $v$ through the value infinity, then becoming negative and gradually increasing again. If then $v_1$ and $-v_2$ are the two limiting values, we have

$$v_1 - \frac{\tanh v_1}{\tanh^2 v_1 + a^2} = -v_2 + \frac{\tanh v_2}{\tanh^2 v_2 + a^2},$$

which limits the range.

In the form propounded*, the problem leads to a meridian geodesic; the projection is a straight line through the origin, and $a$ is zero. Then, for a limited range in a minimum, the projected range encloses the origin; if $d_1$ and $d_2$ are the distances of the limits of the range from the origin

$$\frac{1}{d_1}(1 + d_1{}^2)^{\frac{1}{2}} = \coth v_1, \qquad \frac{1}{d_2}(1 + d_2{}^2)^{\frac{1}{2}} = \coth v_2,$$

so that

$$\sinh v_1 = d_1, \qquad \sinh v_2 = d_2.$$

*Ex.* 9. *Discuss the integral* $\int (1 + y^2) y'^{-2} dx$ *between fixed limits, with a view to a maximum or minimum value.*

As $(1 + y^2) y'^{-2}$, the subject of integration, does not involve the independent variable, a first integral of the characteristic equation (§ 17) is

$$(1 + y^2) y'^{-2} - y' \{ -2 (1 + y^2) y'^{-3} \} = \text{constant},$$

that is,

$$1 + y^2 = c^2 y'^2,$$

where $c$ is an arbitrary constant. The primitive is

$$y = \sinh(cx + a),$$

where $a$ and $c$ are the arbitrary constants.

We have

$$\frac{\partial^2 f}{\partial y'^2} = 6 \frac{1 + y^2}{y'^4},$$

which does not change its positive sign for any range; hence the Legendre test will permit a minimum.

For the Jacobi test, we have

$$\eta_1 = \frac{\partial y}{\partial a} = \cosh(cx + a), \qquad \eta_2 = \frac{\partial y}{\partial c} = x \cosh(cx + a);$$

as $\frac{\eta_2}{\eta_1} = x$, which does not resume an initial value through any steadily increasing range, there is no conjugate of an initial point on the characteristic curve.

It thus appears that the three tests, which have been established in connection with special weak variations, would point to the existence of a minimum for the integral when $y$ is an ordinate of the characteristic curve $y = \sinh(cx + a)$.

* The question was set in the Mathematical Tripos, Part I, 1894.

But we must notice that this minimum, while provided by special weak variations, is not a true minimum. Its value can be decreased by other variations as follows. Let the curve be required to pass through two points $(x_1, y_1)$ and $(x_2, y_2)$ such that

$$x_2 > x_1 > 0, \quad y_2 > y_1 > 0;$$

then

$$c = \frac{\sinh^{-1} y_2 - \sinh^{-1} y_1}{x_2 - x_1},$$

so that $c$ is positive. Take a range for the integral such that, whatever be the value of $a$, the increasing argument $cx + a$ is positive through the range. Then $y$ is positive through the range; while $y$ and $y'$, along the curve, increase continuously with increase of $x$.

Now take successive points $P$, $Q$, $R$, $S$, ... along the curve, in the direction of $x$ increasing, these points being very close together. Draw tangents at $Q$, $R$, $S$, ... successively, meeting the ordinates through $P$, $Q$, $R$, ... respectively in $P_1$, $Q_1$, $R_1$, ..., so that the points $P_1$, $Q_1$, $R_1$, ... lie between the curve and the axis of $x$. Thus we have a broken line $PP_1QQ_1RR_1...$, which is a small variation of the curve $PQR...$ but is not a weak variation: it is a *strong* variation. Consider the integral along this broken line; and compare the portion along a section $PP_1Q$ with the portion along the corresponding section $PQ$ of the characteristic curve. Along $PP_1$, the quantity $y'$ is infinite; the contribution is zero. Along $P_1Q$, the value of $y'$ is greater than its value along $PQ$, while the value of $y$ is less than its value along $PQ$; on both grounds, $(1 + y^2) y'^{-2}$ is less for $P_1Q$ than for $PQ$. Thus the element of the integral for $PP_1Q$ is less than the element of the corresponding integral for the curve $PQ$. Similarly for every pair of associated elements. Thus the whole integral along $...PP_1QQ_1RR_1...$ is less than the integral along $...PQR....$

Consequently the minimum provided by the weak variations is not a true minimum; for there are other small variations which lessen that minimum (see § 215, Ex. 5).

*Note.* The systematic consideration of strong variations is deferred until Chapter VII.

### *First integral of the characteristic equation: Hilbert's theorem.*

**35.** The solution of any problem of the present type requires the integration of the characteristic equation

$$\frac{\partial f}{\partial y} - \frac{d}{dx}\left(\frac{\partial f}{\partial y'}\right) = 0,$$

that is, of

$$\frac{\partial f}{\partial y} - \frac{\partial^2 f}{\partial x \partial y'} - \frac{\partial^2 f}{\partial y \partial y'} y' - \frac{\partial^2 f}{\partial y'^2} y'' = 0.$$

The equation is of the second order. For a form of the function $f$, which is quite general and unspecialised, the explicit expression of a primitive or of a first integral (except as an infinite series, usually to be associated with an existence-theorem) is not possible.

When we postulate a first integral which, when resolved with regard to $y'$ may be expressed as

$$y' = p(x, y) = p,$$

we have

$$y'' = \frac{dp}{dx} = \frac{\partial p}{\partial x} + p \frac{\partial p}{\partial y}.$$

If, then, we regard $p$ as a new variable, to be determined as a function of $x$ and $y$, we can take it as satisfying the partial differential equation

$$\frac{\partial^2 F}{\partial p^2}\left(p\frac{\partial p}{\partial y}+\frac{\partial p}{\partial x}\right)=\frac{\partial F}{\partial y}-\frac{\partial^2 F}{\partial x\partial p}-p\frac{\partial^2 F}{\partial y\partial p},$$

where $F$ denotes $f(x, y, p)$, a known function of its arguments.

As this partial equation of the first order, when $p$ is the dependent variable and $x$ and $y$ are the independent variables, is of the Lagrange linear type—linear, that is to say, in the derivatives of $p$—we may be able to obtain an integral. Let some such integral*, necessarily involving $p$ in its expression and also involving an arbitrary constant, be obtained; it will arise as an integral of the subsidiary equations

$$\frac{dx}{1}=\frac{dy}{p}=\frac{dp}{\dfrac{1}{\dfrac{\partial^2 F}{\partial p^2}}\left(\dfrac{\partial F}{\partial y}-\dfrac{\partial^2 F}{\partial x\partial p}-\dfrac{\partial^2 F}{\partial y\partial p}p\right)}\ (=du),$$

and will be satisfactory, provided it contains the variable $p$. We denote this integral, resolved so as to express $p$ in terms of $x$ and $y$, by

$$p=p(x, y).$$

Now take

$$y'=p=p(x, y).$$

We have

$$\frac{\partial p}{\partial x}+p\frac{\partial p}{\partial y}=y'';$$

the function $F$ becomes $f(x, y, y')$; and the partial differential equation is transformed back into

$$\frac{\partial f}{\partial y}-\frac{\partial^2 f}{\partial x\partial y'}-\frac{\partial^2 f}{\partial y\partial y'}y'-\frac{\partial^2 f}{\partial y'^2}y''=0,$$

that is, into the characteristic equation. Hence *the relation*

$$y'=p(x, y)$$

*provides a first integrated equivalent of the characteristic equation, when $p$ satisfies the partial differential equation of the first order.*

The transformation is used by Hilbert†, in an investigation with the different aim of substituting, for the original integral $I$, an ultimately equivalent integral, chosen so as to become independent of the path between the lower limit and the upper limit.

Hilbert's investigation is based upon the requirement that an integral, allied to $\int f(x, y, y')\,dx$, and taken in the form

$$\int\{f(x, y, p)+(y'-p)\,A\}\,dx,$$

---

\* The primitive of the partial equation is not required for the purpose; any integral, of the specified type, will suffice.

† *Gött. Nachr.* (1900), p. 291.

where $A$ is some function of $x$, $y$, $p$, shall yield a maximum or minimum as regards the variations of $p$, and then shall be independent of the path, $p$ being some function of $x$ and $y$. For the first property, we must have

$$\frac{\partial f}{\partial p} + (y' - p)\frac{\partial A}{\partial p} - A = 0,$$

which can be satisfied by taking

$$A = \frac{\partial f}{\partial p},$$

as this relation leaves the result

$$y' - p = 0,$$

provided $\frac{\partial^2 f}{\partial p^2}$ is not zero.

With this value of $A$, the integral becomes

$$\int \left\{ y'\,\frac{\partial f}{\partial p} + \left( f - p\,\frac{\partial f}{\partial p} \right) \right\} dx.$$

If this is to be independent of the path, and therefore is to depend only upon the variables at the limits, the subject of integration must be the exact differential coefficient of some function $U$. Then

$$\frac{\partial U}{\partial y} = \frac{\partial f}{\partial p}, \quad \frac{\partial U}{\partial x} = f - p\,\frac{\partial f}{\partial p},$$

and therefore

$$\frac{\partial}{\partial y}\left( f - p\,\frac{\partial f}{\partial p} \right) = \frac{\partial}{\partial x}\left( \frac{\partial f}{\partial p} \right):$$

that is, as $p$ involves $x$ and $y$,

$$\left( \frac{\partial f}{\partial y} + \frac{\partial f}{\partial p}\frac{\partial p}{\partial y} \right) - \left\{ \frac{\partial p}{\partial y}\frac{\partial f}{\partial p} + p\left( \frac{\partial^2 f}{\partial p\,\partial y} + \frac{\partial^2 f}{\partial p^2}\frac{\partial p}{\partial y} \right) \right\} = \frac{\partial^2 f}{\partial x\,\partial p} + \frac{\partial^2 f}{\partial p^2}\frac{\partial p}{\partial x},$$

or

$$\frac{\partial f}{\partial y} - \frac{\partial^2 f}{\partial x\,\partial p} - p\,\frac{\partial^2 f}{\partial p\,\partial y} - \frac{\partial^2 f}{\partial p^2}\left( \frac{\partial p}{\partial x} + p\,\frac{\partial p}{\partial y} \right) = 0,$$

which is the characteristic equation.

For the use of the analytical property, reference may be made to Hilbert's memoir.

*Ex.* 1. When the integral is $\int y\,(1+y'^2)^{\frac{1}{2}}\,dx$, the integral for the catenoid (§ 30, Ex. 1), we have

$$F = y\,(1+y'^2)^{\frac{1}{2}}.$$

For this function,

$$\frac{\partial F}{\partial y} = (1+p^2)^{\frac{1}{2}}, \quad \frac{\partial^2 F}{\partial x\,\partial p} = 0, \quad \frac{\partial^2 F}{\partial y\,\partial p} = p\,(1+p^2)^{-\frac{1}{2}}, \quad \frac{\partial^2 F}{\partial p^2} = y\,(1+p^2)^{-\frac{3}{2}};$$

and so the partial equation for $p$ is

$$\frac{\partial p}{\partial x} + p\,\frac{\partial p}{\partial y} = \frac{1+p^2}{y}.$$

Of the subsidiary equations

$$\frac{dx}{1} = \frac{dy}{p} = \frac{y\,dp}{1+p^2},$$

an integral is

$$y = a\,(1+p^2)^{\frac{1}{2}}.$$

Hence a first integral of the characteristic equation is

$$y = a\,(1+y'^2)^{\frac{1}{2}}.$$

*Ex.* 2. When the integral is $\int \{1+x^2+2xyy'+(1+y^2)\,y'^2\}^{\frac{1}{2}}\,dx$, the integral for a geodesic on a paraboloid (§ 34, Ex. 8), we have

$$F = \{1+x^2+2xyp+(1+y^2)\,p^2\}^{\frac{1}{2}}.$$

After substitution and reduction, the partial equation for $p$ is

$$\frac{\partial p}{\partial x} + p\,\frac{\partial p}{\partial y} = \frac{(1+p^2)\,(px-y)}{1+x^2+y^2}.$$

From the subsidiary equations

$$\frac{dx}{1} = \frac{dy}{p} = \frac{1+x^2+y^2}{(1+p^2)\,(px-y)}\,dp = du,$$

we find

$$\frac{d}{du}\,(xp-y) = x\,\frac{(1+p^2)\,(xp-y)}{1+x^2+y^2},$$

$$\frac{d}{du}\,F^2 = 2x\,\frac{1+p^2}{1+x^2+y^2}\,F^2,$$

so that

$$\frac{1}{xp-y}\,\frac{d}{du}\,(xp-y) = \frac{1}{F}\,\frac{dF}{du},$$

and an integral is

$$xp - y = aF.$$

Hence a first integral of the characteristic equation is

$$xy' - y = a\,\{1+x^2+2xyy'+(1+y^2)\,y'^2\}^{\frac{1}{2}}.$$

# CHAPTER II.

## Integrals of the First Order; general weak variations; the method of Weierstrass.

The chapter is devoted to the discussion of maxima and minima of single integrals, involving one dependent variable and its first derivative. The method adopted is due to Weierstrass. It enables the application of variations, still limited to the weak type, to be made simultaneously to the independent variable and the dependent variable, instead of limiting the discussion to the effect of variations upon the dependent variable alone; and it thus allows the consideration of general weak variations.

The most direct account (as yet published) of the method seems to be that given in Harris Hancock's *Calculus of Variations* (*the Weierstrassian Theory*).

### *Weak variations in general.*

**36.** We now proceed to the consideration of weak variations, which are not restricted to the preceding type and have the general property of admitting simultaneous weak variations of both the dependent variable and the independent variable, each without regard to the other. The integrals, in the discussion that ensues immediately, will still be restricted to the first order, that is, they will involve derivatives of the first order and no derivatives of order higher than the first. The results to be obtained must include all the preceding results connected with the special weak variations. Moreover, they will have the added advantage of proving useful for the consideration of strong variations.

Accordingly, the variables $x$ and $y$ will now be regarded as subject to small variations, whereby the variable $x$ is changed to $x + \kappa u$ and the variable $y$ to $y + \kappa v$. As before, $\kappa$ is a small arbitrary constant quantity, so small that any positive integral power is negligible in comparison with every preceding positive integral power. Also, $u$ and $v$ are regular functions throughout the range; they are quite independent of one another; and they are completely at our arbitrary choice, subject (for the immediate purpose) solely to the condition of being regular. But in order to secure this general variation of $x$ and $y$, we make both of them functions of a new independent variable which can be chosen at will. It might, initially, seem natural to choose, as this new independent variable, a magnitude intrinsically connected with the characteristic curve to be obtained, such as the length of the arc of the curve measured from any assumable point. Yet any such intrinsic magnitude may itself be the subject of discussion; for instance, the problem may be limited by an arc of given length, or we may be required to find a geodesic arc on some surface. Thus there is an advantage in assuming an initial independent variable, unspecified by any particular intrinsic relation; when once the analytical results have been obtained, they may be subsequently modified or simplified by any specification that can be appropriate and convenient. We therefore take $x$ and $y$ to be functions of an independent variable denoted by $t$; and then $u$ and $v$ will be arbitrary independent regular functions of $t$. But it will be assumed (so far, the assumption is a partial limitation on arbitrary functions introduced) that the variable $t$ increases continually throughout any range of integration.

### Modification of the form of the integral.

**37.** Derivatives will be denoted by a single subscript number, so long as the derived quantity is a function of only a single argument; thus

$$\frac{dx}{dt} = x_1, \quad \frac{d^2x}{dt^2} = x_2, \quad \frac{dy}{dt} = y_1, \quad \frac{d^2y}{dt^2} = y_2,$$

and so on; modifications will be required for partial derivatives, when the derived quantity involves more than one argument. Thus

$$y' = \frac{dy}{dx} = \frac{y_1}{x_1},$$

$$y'' = \frac{d^2y}{dx^2} = \frac{x_1 y_2 - x_2 y_1}{x_1^3},$$

and so on. Making this change of variable in the integral, we have

$$\int f(x, y, y')\, dx = \int f\left(x, y, \frac{y_1}{x_1}\right) x_1\, dt = \int F(x, y, x_1, y_1)\, dt.$$

Thus our integral becomes

$$\int_{t_0}^{t_1} F(x, y, x_1, y_1)\, dt,$$

where the range of the real variable $t$ is from $t_0$ with continuous increase to $t_1$; and where $t_1$ and $t_0$ may be variable quantities, if occasion demands. Moreover, owing to its source, the function $F$ is such that

$$F(x, y, x_1, y_1) = x_1 f\left(x, y, \frac{y_1}{x_1}\right)$$

$$= x_1 f(x, y, y').$$

We at once have

$$\frac{\partial F}{\partial x_1} = f + x_1 \frac{\partial f}{\partial y'}\left(-\frac{y_1}{x_1^2}\right) = f - \frac{y_1}{x_1}\frac{\partial f}{\partial y'},$$

$$\frac{\partial F}{\partial y_1} = \frac{\partial f}{\partial y'},$$

and therefore

$$x_1 \frac{\partial F}{\partial x_1} + y_1 \frac{\partial F}{\partial y_1} = F,$$

a permanent identity satisfied by every function $F$ which arises in the manner indicated.

What is desired is the determination of $x$ and $y$ as functions of $t$ so that the integral, extended over an assigned range, shall acquire a maximum or a minimum for the weak variations indicated, of every type possible within the limitation.

**38.** Manifestly, the change to a variable $t$ is not unique; thus, if a variable $T$ were chosen, we should have

$$F\left(x, y, \frac{dx}{dT}, \frac{dy}{dT}\right) dT = f(x, y, y')\, dx$$

$$= F\left(x, y, \frac{dx}{dt}, \frac{dy}{dt}\right) dt.$$

In particular, let $t = \mu T$, where $\mu$ is taken to be a constant for our immediate purpose; then

$$F(x, y, \mu x_1, \mu y_1) = \mu F(x, y, x_1, y_1).$$

This relation holds for all values of $\mu$. Two important inferences can at once be made.

Firstly, let $\mu = 1 + \epsilon$, where $\epsilon$ is a small arbitrary quantity; then

$$F(x, y, x_1 + \epsilon x_1, y_1 + \epsilon y_1) = (1 + \epsilon) F(x, y, x_1, y_1).$$

Expanding the left-hand side in powers of $\epsilon$, and equating the coefficients of the first powers of $\epsilon$ on the two sides, we have

$$x_1 \frac{\partial F}{\partial x_1} + y_1 \frac{\partial F}{\partial y_1} = F,$$

the permanent identity satisfied by $F$; and from the other powers, relations such as

$$x_1^2 \frac{\partial^2 F}{\partial x_1^2} + 2x_1 y_1 \frac{\partial^2 F}{\partial x_1 \partial y_1} + y_1^2 \frac{\partial^2 F}{\partial y_1^2} = 0,$$

immediate consequences of the identity—being the customary property of an infinitesimal transformation in a continuous group.

Secondly, let $\mu = -1$; then

$$F(x, y, -x_1, -y_1) = -F(x, y, x_1, y_1).$$

It is an immediate corollary that, if

$$\frac{\partial F}{\partial x_1} = \Phi(x, y, x_1, y_1), \quad \frac{\partial F}{\partial y_1} = \Psi(x, y, x_1, y_1),$$

then

$$\Phi(x, y, -x_1, -y_1) = \Phi(x, y, x_1, y_1),$$
$$\Psi(x, y, -x_1, -y_1) = \Psi(x, y, x_1, y_1).$$

*General variation of the integral.*

**39.** We now consider general weak variations of the integral $I$, where

$$I = \int_{t_0}^{t_1} F(x, y, x_1, y_1)\, dt,$$

the function $F$ being subject to the permanent identity

$$F = x_1 \frac{\partial F}{\partial x_1} + y_1 \frac{\partial F}{\partial y_1}.$$

In the first place, we shall assume (though, later, the assumption will be modified) that all the arguments and all the functions of the arguments, which occur, are continuous.

We take a variation by which $x$ and $y$ are changed into $x + \kappa u$ and $y + \kappa v$ respectively, $\kappa$, $u$, $v$ being limited in the modes already stated. Such a variation changes a curve into an immediately neighbouring curve in such a way that, to each point of the original curve, there corresponds one (and only one) point of the modified curve: and that, as $\kappa$ is a small constant quantity at our arbitrary choice, the distance between two such corresponding points can be made less than any previously assigned small quantity. Initially, this is secured by the finiteness of $u$ and $v$ for all values of $x$ and $y$ within the range. But the wider restriction has been imposed that $u$ and $v$ are to be not merely finite, but also are to be regular functions. This restriction carries the property that, in these small variations, the inclination between the tangents to the original curve and the modified curve at the corresponding points also is small; for the trigonometrical tangent of this inclination is

$$\frac{\kappa (x_1 v_1 - y_1 u_1)}{x_1^2 + y_1^2 + \kappa (x_1 u_1 + y_1 v_1)},$$

a small quantity of the same order as $\kappa$.

We shall further assume that $u$ and $v$ are independent of $\kappa$, so that the magnitude of any derivative of $u$ or of $v$ remains unaffected by the magnitude of $\kappa$. Thus, for example, we shall not consider, at this stage, the possibility of a variation such as

$$u = a \sin \left( \frac{c}{\kappa^n} t \right),$$

for then

$$\kappa u_1 = a c \kappa^{1-n} \cos \left( \frac{c}{\kappa^n} t \right),$$

which, when $n$ differs from zero and is less than unity, is small but is not of the same order of small quantities as $\kappa u$; it is finite if $n = 1$; and it is large when $n$ is greater than unity.

**40.** Within these limitations, we take the most general small variation of the integral. Let the upper limit be varied to $t_1 + \kappa T_1$ and the lower to

$t_0 + \kappa T_0$, where $T_1$ and $T_0$ are finite quantities independent of one another, though $u$ and $v$ may possibly be subject to limitations at $t_1$, or at $t_0$, or at both $t_0$ and $t_1$, for such variations. Denoting the modified integral by $\mathbf{I}$, we have

$$\mathbf{I} - I = \int_{t_0 + \kappa T_0}^{t_1 + \kappa T_1} F\left(x + \kappa u, \ y + \kappa v, \ x_1 + \kappa u_1, \ y_1 + \kappa v_1\right) dt - \int_{t_0}^{t_1} F\left(x, y, x_1, y_1\right) dt.$$

Then, for a maximum or for a minimum, the magnitude $\mathbf{I} - I$ is to have a persistent sign, whether $\kappa$ be positive or negative, and whatever functional values be assigned to the functions $u$ and $v$, independently of one another. The right-hand side is

$$= \left\{ \int_{t_1}^{t_1 + \kappa T_1} + \int_{t_0}^{t_1} - \int_{t_0}^{t_0 + \kappa T_0} \right\} F\left(x + \kappa u, \ y + \kappa v, \ x_1 + \kappa u_1, \ y_1 + \kappa v_1\right) dt$$

$$- \int_{t_0}^{t_1} F\left(x, y, x_1, y_1\right) dt.$$

Now, as in § 32,

$$\int_{t_1}^{t_1 + \kappa T_1} F\left(x + \kappa u, \ y + \kappa v, \ x_1 + \kappa u_1, \ y_1 + \kappa v_1\right) dt = \kappa T_1 F^{(1)} + R^{(1)},$$

where $F^{(1)}$ denotes the value of $F\left(x, y, x_1, y_1\right)$ when $t = t_1$, and $R^{(1)}$ denotes an aggregate of terms of the second and higher orders in $\kappa$. Similarly

$$\int_{t_0}^{t_0 + \kappa T_0} F\left(x + \kappa u, \ y + \kappa v, \ x_1 + \kappa u_1, \ y_1 + \kappa v_1\right) dt = \kappa T_0 F^{(0)} + R^{(0)},$$

where $F^{(0)}$ denotes the value of $F\left(x, y, x_1, y_1\right)$ when $t = t_0$, and $R^{(0)}$ denotes another aggregate of terms of the second and higher orders in $\kappa$. Also

$$\int_{t_0}^{t_1} \left\{ F\left(x + \kappa u, \ y + \kappa v, \ x_1 + \kappa u_1, \ y_1 + \kappa v_1\right) - F\left(x, y, x_1, y_1\right) \right\} dt$$

$$= \kappa \int_{t_0}^{t_1} \left( u \frac{\partial F}{\partial x} + v \frac{\partial F}{\partial y} + u_1 \frac{\partial F}{\partial x_1} + v_1 \frac{\partial F}{\partial y_1} \right) dt + R,$$

where $R$ denotes an aggregate of terms of the second and higher orders in $\kappa$, arising out of the expansion of $F\left(x + \kappa u, \ y + \kappa v, \ x_1 + \kappa u_1, \ y_1 + \kappa v_1\right)$ in powers of $\kappa$. We thus have

$$\mathbf{I} - I = \kappa \left\{ U + T_1 F^{(1)} - T_0 F^{(0)} \right\} + K,$$

where

$$U = \int_{t_0}^{t_1} \left( u \frac{\partial F}{\partial x} + v \frac{\partial F}{\partial y} + u_1 \frac{\partial F}{\partial x_1} + v_1 \frac{\partial F}{\partial y_1} \right) dt,$$

and

$$K = R^{(1)} - R^{(0)} + R.$$

Here $U$, $F^{(1)}$, $F^{(0)}$, $T_1$, $T_0$ are independent of $\kappa$; and the quantity $K$ is the complete aggregate of terms of the second and higher orders in $\kappa$.

When regard is paid to the smallness of the arbitrary constant $\kappa$, so that $K$ becomes negligible in comparison with non-vanishing terms of the first order in $\kappa$, the value of $\mathbf{I} - I$ is governed * by the quantity

$$\kappa \left\{ U + T_1 F^{(1)} - T_0 F^{(0)} \right\}$$

when this quantity does not vanish. In the latter event, the value of $\mathbf{I} - I$ can be made to change its sign by a change in the sign of $\kappa$: that is, $\mathbf{I} - I$ does not have a persistent sign, the prime condition for either a maximum or a minimum.

It follows that, in order to secure our aim, the quantity

$$U + T_1 F^{(1)} - T_0 F^{(0)}$$

must vanish. It must vanish for all admissible variations, separately for each however arbitrarily chosen, and collectively in all combinations. For the present condition, the latter requirement is fulfilled if the former requirement is fulfilled, because of the linear character of the expression in the quantities $u$, $v$, $T_1$, $T_0$; and therefore it is sufficient to require that the quantity shall vanish for every admissible variation separately.

### Two critical equations.

**41.** The expression, denoted by $U$, can be modified through integration by parts, such integration being possible because the subject of integration is continuous through the range. We have

$$\int_{t_0}^{t_1} u_1 \frac{\partial F}{\partial x_1} \, dt = \left[ u \frac{\partial F}{\partial x_1} \right]_{t_0}^{t_1} - \int_{t_0}^{t_1} u \frac{d}{dt} \left( \frac{\partial F}{\partial x_1} \right) dt,$$

$$\int_{t_0}^{t_1} v_1 \frac{\partial F}{\partial y_1} \, dt = \left[ v \frac{\partial F}{\partial y_1} \right]_{t_0}^{t_1} - \int_{t_0}^{t_1} v \frac{d}{dt} \left( \frac{\partial F}{\partial y_1} \right) dt,$$

and therefore

$$U = \int_{t_0}^{t_1} \left[ u \left\{ \frac{\partial F}{\partial x} - \frac{d}{dt} \left( \frac{\partial F}{\partial x_1} \right) \right\} + v \left\{ \frac{\partial F}{\partial y} - \frac{d}{dt} \left( \frac{\partial F}{\partial y_1} \right) \right\} \right] dt + \left[ u \frac{\partial F}{\partial x_1} + v \frac{\partial F}{\partial y_1} \right]_{t_0}^{t_1}.$$

The whole expression for $U$ now consists (i), of an integral, which involves the arbitrary variations $u$ and $v$ but not their derivatives: (ii), of terms at the upper limit: (iii), of terms at the lower limit.

**42.** We write

$$X = \frac{\partial F}{\partial x} - \frac{d}{dt} \left( \frac{\partial F}{\partial x_1} \right) \bigg| $$
$$Y = \frac{\partial F}{\partial y} - \frac{d}{dt} \left( \frac{\partial F}{\partial y_1} \right) \bigg| $$

In the first place, take variations through the range of the curve such as to vanish at each extremity; thus we might have

$$u = (t_1 - t)(t - t_0) \, \phi(t), \quad v = (t_1 - t)(t - t_0) \, \psi(t),$$

* See foot-note on p. 17.

where $\phi$ and $\psi$ are regular through the range; and, simultaneously with variations through the range, keep the upper limit fixed and the lower limit fixed. In this case, we have

$$u\frac{\partial F}{\partial x_1} + v\frac{\partial F}{\partial y_1}$$

zero, both when $t = t_1$ and when $t = t_0$; also $T_1 = 0$ and $T_0 = 0$. Hence the requirement is that

$$\int_{t_1}^{t_0} (uX + vY)\, dt = 0.$$

First, take $v = 0$ throughout; then we must have $\int_{t_1}^{t_0} uX\, dt = 0$ for all values of $u$. If $X$, which is independent of $u$ and of $v$, does not vanish everywhere along the range, then $\int_{t_1}^{t_0} uX\, dt$ can be made positive, by taking $u$ positive when $X$ is positive, and $u$ negative when $X$ is negative; and it can be made negative, by taking $u$ negative when $X$ is positive, and $u$ positive when $X$ is negative, choices that are admissible. These possibilities must be excluded; and the exclusion is possible, only if

$$X = 0.$$

A similar argument leads to the necessary result that the equation

$$Y = 0$$

must be satisfied. Hence, from the foregoing requirement, we have

$$X = 0, \quad Y = 0,$$

as necessary equations; and when these are satisfied, the integral in the modified expression for $U$ vanishes, whatever admissible variations $u$ and $v$ are adopted.

It will appear that, if the limits of the integral are not given as definitely fixed by the data of the problem initially, they are definitely determined by other assigned conditions. When the limits are definitely fixed, we have $T_1 = 0$ and $T_0 = 0$. In the alternative when they have to be definitely determined, then, as soon as this determination is achieved, we have $T_1 = 0$ and $T_0 = 0$. Accordingly, in any event, we can take $T_1 = 0$ and $T_0 = 0$; and so no condition accrues from the terms involving $T_1$ and $T_0$.

**43.** It therefore remains to secure that the magnitude

$$u\frac{\partial F}{\partial x_1} + v\frac{\partial F}{\partial y_1}$$

shall vanish at the upper limit and at the lower limit, independently of one another.

Various inferences are drawn, according to the various terminal data that are given. If the range of the integral terminates, not merely in a fixed value $t_1$, but in a fixed point giving values of $x$ and $y$ that are not subject to variation at that point, then $u = 0$ and $v = 0$ at that extremity; and the terms, thus vanishing, provide no condition that is new. If the range of the integral terminates in a point that is required to lie upon a given curve represented by some equation such as

$$g(x, y) = 0,$$

then the varied extremity $x + \kappa u$, $y + \kappa v$ also lies on that curve; thus at the extremity, we have from the geometry

$$u \frac{\partial g}{\partial x} + v \frac{\partial g}{\partial y} = 0,$$

and therefore we have

$$\frac{\partial F}{\partial x_1} \frac{\partial g}{\partial y} - \frac{\partial F}{\partial y_1} \frac{\partial g}{\partial x} = 0,$$

thus providing a condition that is new. But it is only a terminal condition; and therefore it can be expected to prove of use for the determination of arbitrary constants which have arisen or may arise.

Similarly, the same type of equation is satisfied at the other extremity.

*Note.* This relation

$$\frac{\partial F}{\partial x_1} \frac{\partial g}{\partial y} - \frac{\partial F}{\partial y_1} \frac{\partial g}{\partial x} = 0,$$

to be satisfied at a mobile extremity (not determined by initially assigned data which require $u = 0$ and $v = 0$), is in actual agreement, though superficially not in formal agreement, with the relation obtained (§ 34) for the special variation. To harmonise the two forms, we note that

$$F = x_1 \frac{\partial F}{\partial x_1} + y_1 \frac{\partial F}{\partial y_1},$$

$$F = x_1 f(x, y, y'),$$

so that

$$- \frac{\partial F}{\partial y_1} = \frac{\partial f}{\partial y'},$$

and therefore

$$x_1 \frac{\partial F}{\partial x_1} = x_1 f(x) - y_1 \frac{\partial F}{\partial y_1},$$

that is,

$$\frac{\partial F}{\partial x_1} = f - y' \frac{\partial f}{\partial y'}.$$

The form of the relation, for the present type of weak variation, thus changes to

$$\left( f - y' \frac{\partial f}{\partial y'} \right) \frac{\partial g}{\partial y} - \frac{\partial f}{\partial y'} \frac{\partial g}{\partial x} = 0,$$

which is the form for the other type.

### The two equations are equivalent to one characteristic equation.

**44.** It follows that, in order to make the terms in $\mathbf{I} - I$ which involve the first power of $\kappa$ (the *first variation* of $I$) vanish, we have two kinds of conditions to be satisfied. The one set of conditions belongs to the limits we must have

$$u \frac{\partial F}{\partial x_1} + v \frac{\partial F}{\partial y_1} = 0$$

at each limit. The other set of conditions persists throughout the range of the integral; the two equations

$$X = 0, \quad Y = 0,$$

must be satisfied everywhere along the range.

But when the identity satisfied by $F$ is taken into account, these two equations are equivalent to a single equation, in virtue of which both are satisfied. This claim can be established as follows. We have

$$x_1 \frac{\partial F}{\partial x_1} + y_1 \frac{\partial F}{\partial y_1} = F.$$

Differentiating this identity with respect to $x_1$ and to $y_1$ separately, we find

$$x_1 \frac{\partial^2 F}{\partial x_1{}^2} + y_1 \frac{\partial^2 F}{\partial x_1 \partial y_1} = 0, \quad x_1 \frac{\partial^2 F}{\partial x_1 \partial y_1} + y_1 \frac{\partial^2 F}{\partial y_1{}^2} = 0.$$

Hence

$$\frac{1}{y_1{}^2} \frac{\partial^2 F}{\partial x_1{}^2} = -\frac{1}{x_1 y_1} \frac{\partial^2 F}{\partial x_1 \partial y_1} = \frac{1}{x_1{}^2} \frac{\partial^2 F}{\partial y_1{}^2} = P,$$

when we denote the common value by $P$, a critical quantity that recurs through the whole investigation.

We have

$$\frac{d}{dt}\left(\frac{\partial F}{\partial x_1}\right) = \frac{\partial^2 F}{\partial x \partial x_1} x_1 + \frac{\partial^2 F}{\partial y \partial x_1} y_1 + \frac{\partial^2 F}{\partial x_1{}^2} x_2 + \frac{\partial^2 F}{\partial x_1 \partial y_1} y_2$$

$$= \frac{\partial^2 F}{\partial x \partial x_1} x_1 + \frac{\partial^2 F}{\partial y \partial x_1} y_1 - y_1 (x_1 y_2 - x_2 y_1) P.$$

The same identity, when differentiated partially with respect to $x$, gives the relation

$$\frac{\partial F}{\partial x} = \frac{\partial^2 F}{\partial x \partial x_1} x_1 + \frac{\partial^2 F}{\partial x \partial y_1} y_1.$$

Hence

$$X = \frac{\partial F}{\partial x} - \frac{d}{dt}\left(\frac{\partial F}{\partial x_1}\right)$$

$$= y_1 \left\{ \frac{\partial^2 F}{\partial x \partial y_1} - \frac{\partial^2 F}{\partial y \partial x_1} + (x_1 y_2 - x_2 y_1) P \right\} = -y_1 \mathfrak{E},$$

where

$$\mathfrak{E} = \frac{\partial^2 F}{\partial y \partial x_1} - \frac{\partial^2 F}{\partial x \partial y_1} - (x_1 y_2 - x_2 y_1) P.$$

Similarly

$$Y = \frac{\partial F}{\partial y} - \frac{d}{dt}\left(\frac{\partial F}{\partial y_1}\right) = x_1 \mathfrak{E}.$$

It therefore follows that the two equations $X = 0$ and $Y = 0$ are satisfied in virtue of the single equation

$$\mathfrak{E} = 0;$$

and this is the *characteristic equation*.

**45.** Sometimes it is convenient to take the equation $\mathfrak{C} = 0$, sometimes the equation $X = 0$ or the equation $Y = 0$; and sometimes it is convenient to retain both equations. Moreover, the two relations

$$X = -y_1\mathfrak{C}, \quad Y = x_1\mathfrak{C},$$

have been deduced by means of the permanent identity affecting the form of the function, and without reference to the subsequent requirement that the first variation of $I$ shall vanish.

The integral in the former expression for $U$ is

$$\int_{t_0}^{t_1} (uX + vY)\,dt$$

$$= \int_{t_0}^{t_1} (x_1v - y_1u)\,\mathfrak{C}\,dt.$$

If then we write

$$W = \int_{t_0}^{t_1} (x_1v - y_1u)\,\mathfrak{C}\,dt,$$

and if we assume that the terms at the limits vanish—as, from the foregoing argument, they are bound to vanish—we have

$$\mathbf{I} - I = \kappa W + K_2,$$

where $K_2$ is the aggregate of terms of the second and higher orders in the small quantity $\kappa$.

## *Covariantive character of $P$ and $\mathfrak{C}$.*

**46.** We at once have the important properties that, *whatever independent variable be chosen, $P$ is a relative covariant, and $\mathfrak{C}$ is an absolute covariant.*

Let $t$ and $T$ be two distinct independent variables, so that (as in § 38)

$$F\left(x, y, \frac{dx}{dT}, \frac{dy}{dT}\right) dT = F(x, y, x_1, y_1)\,dt\,;$$

and write

$$F\left(x, y, \frac{dx}{dT}, \frac{dy}{dT}\right) = \mathbf{F}, \quad F(x, y, x_1, y_1) = F,$$

$$\frac{dx}{dT} = X_1, \quad \frac{dy}{dT} = Y_1, \quad \frac{d^2x}{dT^2} = X_2, \quad \frac{d^2y}{dT^2} = Y_2,$$

$$\frac{dt}{dT} = \mu,$$

so that $\mu$ is a variable quantity in general. Then

$$X_1 = \mu x_1, \quad Y_1 = \mu y_1, \quad \mathbf{F} = \mu F.$$

Also

$$\frac{\partial \mathbf{F}}{\partial X_1} = \frac{\partial F}{\partial x_1}, \quad \frac{\partial \mathbf{F}}{\partial Y_1} = \frac{\partial F}{\partial y_1},$$

$$\frac{\partial^2 \mathbf{F}}{\partial X_1 \partial Y_1}\,\mu = \frac{\partial^2 F}{\partial x_1 \partial y_1}\,;$$

and therefore

$$-\frac{1}{X_1 Y_1} \frac{\partial^2 \mathbf{F}}{\partial X_1 \partial Y_1} = \frac{1}{\mu^3}\left(-\frac{1}{x_1 y_1}\frac{\partial^2 F}{\partial x_1 \partial y_1}\right),$$

shewing that $P$ is a relative covariant.

Again, we have

$$X_2 = \mu\,(\mu x_2 + \mu_1 x_1), \quad Y_2 = \mu\,(\mu y_2 + \mu_1 y_1),$$

and therefore

$$X_1 Y_2 - X_2 Y_1 = \mu^3\,(x_1 y_2 - x_2 y_1).$$

Also

$$\frac{\partial^2 \mathbf{F}}{\partial x \partial Y_1} = \frac{\partial^2 F}{\partial x \partial y_1}, \quad \frac{\partial^2 \mathbf{F}}{\partial y \partial X_1} = \frac{\partial^2 F}{\partial y \partial x_1};$$

so that

$$\frac{\partial^2 \mathbf{F}}{\partial y \partial X_1} - \frac{\partial^2 \mathbf{F}}{\partial x \partial Y_1} - \frac{X_1 Y_2 - X_2 Y_1}{X_1 Y_1}\frac{\partial^2 \mathbf{F}}{\partial X_1 \partial Y_1} = \mathfrak{C},$$

shewing that $\mathfrak{C}$ is an absolute covariant.

In what follows, it will appear that the permanence of the sign of $P$ throughout a range is of fundamental importance as a test. The independent variable is assumed to increase continually throughout the range; hence, at any place, $dt$ and $dT$ have the same sign, and so $\mu$ is positive. As

$$\mathbf{P} = \frac{1}{\mu^3} P,$$

the signs of $P$ and $\mathbf{P}$ are the same; a test, as regards permanence of sign in $P$, is unaffected by the change.

Further, as the characteristic equation is thus essentially unaltered by change of the independent variable, we may change that variable after the equation has been constructed, should advantage accrue from any particular choice. Later, it will be found convenient (for some purposes) to choose the length of the arc measured from some fixed point as the variable, especially if intrinsic properties of the characteristic curve, or of variations from that curve, are under consideration; the change can be made after the characteristic equation has been obtained.

*Two quantities continuous in passage through a discontinuity.*

**47.** One important theorem can be established for the subject of integration in the case of a maximum or a minimum, when some of the assumptions already made as to continuity of arguments are not satisfied, viz.

*Should the characteristic curve suddenly change its direction at a free place, that is, a place unhampered by conditions as to fixity, the quantities $\dfrac{\partial F}{\partial x_1}$ and $\dfrac{\partial F}{\partial y_1}$ suffer no discontinuity in value.*

As there is a sudden change of direction, $y_1$ or $x_1$ or both $y_1$ and $x_1$ must suffer a sudden discontinuity in value there. Let such a place be $T$; and suppose that the curve possesses no singularities in the immediate vicinity. Take a point $t_3$, near $T$ between $T$ and $t_1$, and another point $t_2$ near $T$ but between $T$ and $t_0$; the direction is taken continuous from $t_2$ to $T$, taken continuous also from $T$ to $t_3$, and it undergoes a sudden change in passing through $T$.

Let the limiting values* of $\dfrac{\partial F}{\partial x_1}$ and

$\dfrac{\partial F}{\partial y_1}$ as $t_3$ approaches indefinitely near $T$

in the range $t_1 T$ be denoted by

$$\left(\frac{\partial F}{\partial x_1}\right)_+, \quad \left(\frac{\partial F}{\partial y_1}\right)_+;$$

and let their limiting values as $t_2$ approaches indefinitely near $T$ in the range $T t_0$ be denoted by

$$\left(\frac{\partial F}{\partial x_1}\right)_-, \quad \left(\frac{\partial F}{\partial y_1}\right)_-.$$

Take a variation of the characteristic curve such that $u = 0$ and $v = 0$ along $t_0 t_2$, and also along $t_3 t_1$; as the arc of the curve itself is continuous between $t_2$ and $t_3$, we can take

$$u = (t_3 - t)(t - t_2)\,\phi(t), \quad v = (t_3 - t)(t - t_2)\,\psi(t),$$

within that range, where $\phi(t)$ and $\psi(t)$ are arbitrary regular functions of $t$ which do not vanish at $T$. The first variation of the integral $I$ is to vanish; taking account of the fact that $u$ and $v$ vanish (for the present variation) between $t_0$ and $t_2$, and between $t_3$ and $t_1$, we have the analytical condition (as in §§ 41, 45)

$$\left\{\int_{t_2}^{T} + \int_{T}^{t_3}\right\} (x_1 v - y_1 u)\,\mathfrak{C}\,dt + \left[u\frac{\partial F}{\partial x_1} + v\frac{\partial F}{\partial y_1}\right]_{t_2}^{T} + \left[u\frac{\partial F}{\partial x_1} + v\frac{\partial F}{\partial y_1}\right]_{T}^{t_3} = 0.$$

Now $\mathfrak{C} = 0$ everywhere along the characteristic curve; thus the integral from $t_2$ to $T$, and the integral from $T$ to $t_3$, vanish. Also

$$\left[u\frac{\partial F}{\partial x_1} + v\frac{\partial F}{\partial y_1}\right]_{t_2}^{T} = u_T\left(\frac{\partial F}{\partial x_1}\right)_- + v_T\left(\frac{\partial F}{\partial y_1}\right)_-,$$

and

$$\left[u\frac{\partial F}{\partial x_1} + v\frac{\partial F}{\partial y_1}\right]_{T}^{t_3} = -u_T\left(\frac{\partial F}{\partial x_1}\right)_+ - v_T\left(\frac{\partial F}{\partial y_1}\right)_+.$$

The condition therefore becomes

$$u_T\left\{\left(\frac{\partial F}{\partial x_1}\right)_- - \left(\frac{\partial F}{\partial x_1}\right)_+\right\} + v_T\left\{\left(\frac{\partial F}{\partial y_1}\right)_- - \left(\frac{\partial F}{\partial y_1}\right)_+\right\} = 0.$$

* It is assumed that each part of the curve yields a limiting value for each of the two derivatives.

The quantities $u_T$ and $v_T$ are arbitrary; their values can be chosen, each independently of the other, and distinct from zero. Hence the condition can be satisfied for all such quantities, only if

$$\left(\frac{\partial F}{\partial x_1}\right)_- = \left(\frac{\partial F}{\partial x_1}\right)_+, \quad \left(\frac{\partial F}{\partial y_1}\right)_- = \left(\frac{\partial F}{\partial y_1}\right)_+.$$

Thus the proposition is established.

But the property that $\dfrac{\partial F}{\partial x_1}$ and $\dfrac{\partial F}{\partial y_1}$ are continuous in value, even if $x_1$ and $y_1$ are not continuous in value, in passing through any place, can be inferred solely if the place be free. When the place is fixed by external requirements, both $u_T$ and $v_T$ are zero: the condition is then satisfied without leaving a residuary relation.

*Ex.* 1. Let it be required to find the shortest curve passing through three fixed points $A$, $C$, $B$, not in a straight line.

The characteristic curve is always a straight line; so the shortest curve will consist of two sides of the triangle $ABC$. If the two sides are $AC$ and $CB$, the limiting conditions arise from

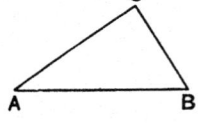

$$u_C\left[\frac{x_1}{(x_1{}^2+y_1{}^2)^{\frac{1}{2}}}\right]_{CA} - u_C\left[\frac{x_1}{(x_1{}^2+y_1{}^2)^{\frac{1}{2}}}\right]_C$$
$$+ v_C\left[\frac{y_1}{(x_1{}^2+y_1{}^2)^{\frac{1}{2}}}\right]_{CA} - v_C\left[\frac{y_1}{(x_1{}^2+y_1{}^2)^{\frac{1}{2}}}\right]_{CB} = 0.$$

But $u_C=0$ and $v_C=0$ in the present instance; the conditions at $C$ are satisfied without limitation.

*Ex.* 2. In the case of a curve, rotated about the axis of $x$ and providing a minimum area for the surface of revolution (§ 30), we have

$$F = x_1 \cdot y\,(1+y'^2)^{\frac{1}{2}} = y\,(x_1{}^2+y_1{}^2)^{\frac{1}{2}}.$$

Therefore, at a free point,

$$\frac{yx_1}{(x_1{}^2+y_1{}^2)^{\frac{1}{2}}}, \quad \frac{yy_1}{(x_1{}^2+y_1{}^2)^{\frac{1}{2}}},$$

suffer no discontinuity in value, in passing through the point. In particular, the curve is

$$y\,(1+y'^2)^{-\frac{1}{2}} = C;$$

that is, if there is a sudden discontinuity in direction, $C$ is unaltered when the point is passed.

An exceptional case was noted, when the two limiting points were on opposite sides of the axis; the catenary could not be drawn. But the possibility of the catenary has assumed a finite non-zero value for $C$. We must then consider $C=0$: that is,

$$\frac{y}{(1+y'^2)^{\frac{1}{2}}} = 0 = \frac{yx_1}{(x_1{}^2+y_1{}^2)^{\frac{1}{2}}}.$$

Before reaching the axis of $x$, we must have $x_1=0$ or $y'=\infty$; the curve is the straight line perpendicular to the axis of $x$. When it reaches the axis, the equation is satisfied by $y=0$ and $x_1$ not zero; the curve now lies along the axis. At the point of change, which is free from external conditions, there is a discontinuity of direction while the (zero) value of the constant is unchanged.

Similarly, in passing to the other extremity taken, in the present case, to lie on the other side of the axis, the remainder of the curve is the ordinate of that extremity.

*The number of discontinuities of direction in a finite range is limited.*

**48.** One important inference can be made. The result shews that a place may exist within the range of integration at which there can be discontinuity of direction and that, at such a place, $\dfrac{\partial F}{\partial x_1}$ and $\dfrac{\partial F}{\partial y_1}$ are continuous, if the place is not absolutely fixed. In order to establish the result, it is necessary that the place should be isolated as a place of discontinuity in direction; for it has been assumed that, on either side of the place as a position on a continuous curve, full small variations $\kappa u$ and $\kappa v$ are admissible, whether the place be absolutely fixed or not.

But there is nothing to restrict the number of such places in any range to a single unit; the general necessity for any such place is isolation. Accordingly, *there may exist a finite number of places within a finite range of integration at which discontinuity of direction may occur*; each such place is isolated. The number of such places within a finite range cannot be infinite; if it were unlimited, there would be concentration of a countless number of discontinuities at one or more of such places, and then a place of concentration would not admit solely continuous small variations $\kappa u$ and $\kappa v$ at each side along the curve.

*The primitive of the characteristic equation.*

**49.** The characteristic equation $\mathfrak{E} = 0$ is, effectively, the former (§ 17) characteristic equation $Y = 0$; the reduction can be made as follows. We have

$$F(x, y, x_1, y_1) = x_1 f\left(x, y, \frac{y_1}{x_1}\right) = x_1 f(x, y, y');$$

hence

$$\frac{\partial F}{\partial y_1} = \frac{\partial f}{\partial y'}, \quad \frac{\partial^2 F}{\partial x \partial y_1} = \frac{\partial^2 f}{\partial x \partial y'},$$

$$\frac{\partial F}{\partial x_1} = f - \frac{y_1}{x_1}\frac{\partial f}{\partial y'} = f - y'\frac{\partial f}{\partial y'},$$

$$\frac{\partial^2 F}{\partial y \partial x_1} = \frac{\partial f}{\partial y} - y'\frac{\partial^2 f}{\partial y \partial y'};$$

thus

$$\frac{\partial^2 F}{\partial y \partial x_1} - \frac{\partial^2 F}{\partial x \partial y_1} = \frac{\partial f}{\partial y} - \frac{\partial^2 f}{\partial x \partial y'} - y'\frac{\partial^2 f}{\partial y \partial y'}.$$

Also

$$\frac{\partial^2 F}{\partial y_1^2} = \frac{1}{x_1}\frac{\partial^2 f}{\partial y'^2},$$

so that

$$P = \frac{1}{x_1^2}\frac{\partial^2 F}{\partial y_1^2} = \frac{1}{x_1^3}\frac{\partial^2 f}{\partial y'^2}.$$

Again,

$$\frac{d}{dt}(y') = \frac{d}{dt}\left(\frac{y_1}{x_1}\right) = \frac{x_1 y_2 - x_2 y_1}{x_1^2};$$

hence, when $x$ is made the independent variable,

$$y'' = \frac{dy'}{dx} = \frac{1}{x_1}\frac{d}{dt}(y') = \frac{x_1 y_2 - x_2 y_1}{x_1^3};$$

and therefore

$$(x_1 y_2 - x_2 y_1)\, P = y'' \frac{\partial^2 f}{\partial y'^2}.$$

Consequently

$$\frac{\partial^2 F}{\partial y \partial x_1} - \frac{\partial^2 F}{\partial x \partial y_1} - (x_1 y_2 - x_2 y_1)\, P = \frac{\partial f}{\partial y} - \frac{\partial^2 f}{\partial x \partial y'} - y'\frac{\partial^2 f}{\partial y \partial y'} - y''\frac{\partial^2 f}{\partial y'^2}$$

$$= \frac{\partial f}{\partial y} - \frac{d}{dx}\left(\frac{\partial f}{\partial y'}\right),$$

establishing the exact equivalence of the old characteristic equation and of the new characteristic equation.

The old characteristic equation was a differential equation of the second order in $y$, with $x$ as the independent variable. Its primitive contained two arbitrary constants, which were to be determined by two assigned data such as an initial value of $y$ and an initial value of $y'$ for an initial value of $x$.

Initially there are two equations, in the new investigation, for the determination of $x$ and $y$, viz. $X = 0$, $Y = 0$, in each of which $x_2$ and $y_2$ occur raised only to the first power. When $x_2$ is eliminated, we have some relation such as

$$y_2 = \Phi\,(x,\, x_1,\, y,\, y_1).$$

Two derivatives of this, and subsequent elimination of $x$ and $x_1$, lead to an equation of the fourth order in $y$ alone; its primitive involves four arbitrary constants. When this value of $y$ is substituted in $y_2 = \Phi$ and in $y_3 = \dfrac{d\Phi}{dt}$, and $x_1$ is eliminated, we have $x$ expressed as a function of $t$, the expression involving those four arbitrary constants.

But the four constants are not essential for the equation of the characteristic curve in terms of $x$ and $y$ alone. Thus, as $t$ never occurs except in a differential coefficient, the quantity $t$ must occur in the primitive in a form $t - A$, where $A$ is an arbitrary constant; and $A$ will disappear with $t$, in the elimination of $t$ when the equation between $x$ and $y$ is obtained. Again, as initial conditions, we may have initial values of $x$, $x_1$, $y$, $y_1$ given for an initial value of $t$. But $y' = \dfrac{y_1}{x_1}$, and so these two constants for $y_1$ and $x_1$ merge into a single constant for $y'$; and the two initial constants for $x$ and $y$ become, for the equation in $x$ and $y$, an initial value of $y$ for an initial value of $x$. Thus when we pass to the final relation between $x$ and $y$, there will survive only two arbitrary independent constants, although in the initial form of the primitive of $X = 0$ and $Y = 0$, giving the most general values of $x$ and $y$ as functions of $t$, there are four arbitrary constants.

Moreover, because $X = -y_1\mathfrak{E}$, $Y = x_1\mathfrak{E}$, the primitive either of $X = 0$ or $Y = 0$ could lead to the primitive of $\mathfrak{E} = 0$; and a first integral of either would lead to a first integral of $\mathfrak{E} = 0$. It thus appears that, in practice, the integration of either $X = 0$ or $Y = 0$ would determine the curve: but, for the derivation of further results, it is convenient to retain both the equations.

Finally, unless there is express warning to the contrary, it is assumed (as stated in § 12) that the functions $x(t)$ and $y(t)$ are analytic functions of $t$, so that they are continuous in variable and in parameter, are differentiable, and are one-valued along a finite range or in each one of a finite number of portions of that range.

*The 'second' variation of the original integral.*

**50.** Now that the terms of the first order in $\kappa$, which occur in the expression for $\mathbf{I} - I$, have been made to vanish, we must consider terms of the second order in $\kappa$, so as to determine the necessary and sufficient conditions which shall secure a permanent sign for the aggregate of those terms for all admissible variations. We may write

$$\mathbf{I} - I = \kappa U + \tfrac{1}{2}\kappa^2 \int_{t_0}^{t_1} \Theta \, dt + K_3,$$

where, as before,

$$U = \int_{t_0}^{t_1} \left( u \frac{\partial F}{\partial x} + v \frac{\partial F}{\partial y} + u_1 \frac{\partial F}{\partial x_1} + v_1 \frac{\partial F}{\partial y_1} \right) dt;$$

here $K_3$ denotes the aggregate of terms in $\mathbf{I} - I$, which are of the third and higher orders in $\kappa$; and

$$\Theta = u_1^2 \frac{\partial^2 F}{\partial x_1^2} + 2u_1 v_1 \frac{\partial^2 F}{\partial x_1 \partial y_1} + v_1^2 \frac{\partial^2 F}{\partial y_1^2}$$

$$+ 2uu_1 \frac{\partial^2 F}{\partial x \partial x_1} + 2uv_1 \frac{\partial^2 F}{\partial x \partial y_1} + 2vu_1 \frac{\partial^2 F}{\partial y \partial x_1} + 2vv_1 \frac{\partial^2 F}{\partial y \partial y_1}$$

$$+ u^2 \frac{\partial^2 F}{\partial x^2} + 2uv \frac{\partial^2 F}{\partial x \partial y} + v^2 \frac{\partial^2 F}{\partial y^2}.$$

As the first variation has been made to vanish, when the characteristic curve is adopted and the boundary conditions are used, the terms of the second order now govern the sign of $\mathbf{I} - I$; they are often called the 'second' variation. We proceed to modify the expression for $\int_{t_0}^{t_1} \Theta \, dt$, on the assumption that the first variation vanishes; so we substitute the values of $x$ and $y$ in the coefficients $\frac{\partial^2 F}{\partial x_1^2}, \ldots, \frac{\partial^2 F}{\partial y^2}$, which accordingly may now be considered to be functions of $t$ alone.

Further, instead of using $u$ and $v$ separately, we find it convenient to introduce a quantity $w$, defined as the combination

$$w = x_1 v - y_1 u.$$

As the point $(x, y)$ is displaced to $(x + \kappa u, y + \kappa v)$ by the small variation, the quantity $\kappa (x_1^2 + y_1^2)^{-\frac{1}{2}} (x_1 v - y_1 u)$ is the distance of the displaced point from the tangent to the characteristic curve at the undisplaced point. Because $w$ thus measures that distance, we shall call $\kappa w$ the *deviation* due to the variation. The vanishing of $w$ means that the point is merely displaced along the curve; when this deviation vanishes everywhere, the ends being fixed, then there is no new curve obtained from the variation: in such an event, we should expect the second variation (and every other variation) to vanish.

### Relations among derivatives.

**51.** We have already obtained the relations

$$\frac{1}{y_1^2} \frac{\partial^2 F}{\partial x_1^2} = - \frac{1}{x_1 y_1} \frac{\partial^2 F}{\partial x_1 \partial y_1} = \frac{1}{x_1^2} \frac{\partial^2 F}{\partial y_1^2} = P.$$

Thus the first line in the preceding expression for $\Theta$ can be written

$$P (x_1 v_1 - y_1 u_1)^2.$$

From the characteristic equation $\mathfrak{C} = 0$, we have

$$\frac{\partial^2 F}{\partial x \partial y_1} + x_1 y_2 P = \frac{\partial^2 F}{\partial y \partial x_1} + y_1 x_2 P ;$$

and we introduce three quantities $L, M, N$, by the definitions

$$\left. \begin{aligned} \frac{\partial^2 F}{\partial x \partial x_1} &= L + y_1 y_2 P \\[1mm] \frac{\partial^2 F}{\partial x \partial y_1} &= M - x_1 y_2 P \\[1mm] \frac{\partial^2 F}{\partial y \partial x_1} &= M - y_1 x_2 P \\[1mm] \frac{\partial^2 F}{\partial y \partial y_1} &= N + x_1 x_2 P \end{aligned} \right\}$$

Thus the second line in the expression for $\Theta$ can be written

$$2 \{L u_1 u_2 + M (u_1 v_2 + v_1 u_2) + N v_1 v_2\} + 2P (x_1 v_1 - y_1 u_1) (x_2 v - y_2 u).$$

Next, we write temporarily

$$\frac{\partial^2 F}{\partial x^2} = A + y_2^2 P + \frac{dL}{dt} ,$$

$$\frac{\partial^2 F}{\partial x \partial y} = B - x_2 y_2 P + \frac{dM}{dt} ,$$

$$\frac{\partial^2 F}{\partial y^2} = C + x_2^2 P + \frac{dN}{dt} .$$

But, by partial differentiation of the identity

$$F = x_1 \frac{\partial F}{\partial x_1} + y_1 \frac{\partial F}{\partial y_1}$$

with regard to $x$, we have

$$\frac{\partial F}{\partial x} = x_1 L + y_1 M \,;$$

whence, differentiating completely with regard to $t$,

$$\frac{\partial^2 F}{\partial x^2} x_1 + \frac{\partial^2 F}{\partial x \partial y} y_1 + \frac{\partial^2 F}{\partial x \partial x_1} x_2 + \frac{\partial^2 F}{\partial x \partial y_1} y_2 = x_2 L + y_2 M + x_1 \frac{dL}{dt} + y_1 \frac{dM}{dt} \,,$$

or, on substituting for the second derivatives of $F$,

$$x_1 A + y_1 B = 0.$$

Differentiating the same identity partially with regard to $y$, and proceeding in the same manner, we find

$$x_1 B + y_1 C = 0.$$

Hence

$$\frac{A}{y_1{}^2} = \frac{B}{-x_1 y_1} = \frac{C}{x_1{}^2} = G,$$

introducing a new symbol $G$ to denote the common value* of the three fractions; and so we now take

$$
\left.
\begin{aligned}
\frac{\partial^2 F}{\partial x^2} &= \phantom{-} y_1{}^2 G + \phantom{-} y_2{}^2 P + \frac{dL}{dt} \\[4pt]
\frac{\partial^2 F}{\partial x \partial y} &= - x_1 y_1 G - x_2 y_2 P + \frac{dM}{dt} \\[4pt]
\frac{\partial^2 F}{\partial y^2} &= \phantom{-} x_1{}^2 G + \phantom{-} x_2{}^2 P + \frac{dN}{dt}
\end{aligned}
\right\}.
$$

Thus all the second derivatives of $F$, ten in number, are expressed in terms of five quantities $P, L, M, N, G$.

### Normal form of the second variation.

**52.** When these expressions for the derivatives of $F$ are substituted in $\Theta$, and the terms are suitably collected, we find

$$\Theta = P \{(x_1 v_1 - y_1 u_1) + (x_2 v - y_2 u)\}^2 + G (x_1 v - y_1 u)^2 + \frac{d}{dt} (Lu^2 + 2Muv + Nv^2)$$

$$= Pw_1{}^2 + Gw^2 + \frac{d}{dt} (Lu^2 + 2Muv + Nv^2).$$

Denoting by $H$, a quantity at our disposal and not involving $w$, we have

$$\Theta = Pw_1{}^2 - 2Hww_1 + (G - H_1) w^2 + \frac{d}{dt} (Lu^2 + 2Muv + Nv^2 + Hw^2).$$

Now let $H$ be so chosen, that $Pw_1{}^2 - 2Hww_1 + (G - H_1) w^2$ is a perfect square quâ function of $w$ and $w_1$; thus

$$H^2 = P (G - H_1).$$

---

* As $x_1$ and $y_1$ do not simultaneously vanish, this common value is finite.

In this relation, let

$$H = P\frac{z_1}{z},$$

where $z$ is a new variable in place of $H$. After reduction, the equation for $z$ becomes

$$Pz_2 + P_1 z_1 - Gz = 0,$$

which, as $P$ and $G$ do not involve $z$, is a linear equation in $z$ of the second order. To this equation we shall return later; meanwhile, for such a value of $z$,

$$Pw_1^2 - 2Hww_1 + (G - H_1)w^2 = P\left(w_1 - \frac{z_1}{z}w\right)^2,$$

and therefore

$$\Theta = P\left(w_1 - \frac{z_1}{z}w\right)^2 + \frac{d}{dt}\left(Lu^2 + 2Muv + Nv^2 + P\frac{z_1}{z}w^2\right).$$

Thus

$$\int_{t_0}^{t_1}\Theta\,dt = \int_{t_0}^{t_1}P\left(w_1 - \frac{z_1}{z}w\right)^2 dt + \left[Lu^2 + 2Muv + Nv^2 + P\frac{z_1}{z}w^2\right]_{t_0}^{t_1},$$

the last term being taken at the limits of the integral.

*Note.* It might appear as if, when taking solely the combination

$$w = x_1 v - y_1 u$$

in the construction of the modified expression for $\Theta$, a limited part of the small variation were taken into account; and that a combination

$$T = x_1 u + y_1 v$$

also ought to be considered. The quantity $w$ measures the normal deviation caused by the small variations $\kappa u$ and $\kappa v$; the quantity $T$ measures the tangential sliding due to the same source. Though $T$ will cause no change in the shape of the curve and therefore no apparent variation, it is not at first sight analytically clear that $T$ should disappear from the expressions.

That $T$ does so disappear is seen from the fact that $\Theta$ has been expressed in the form

$$P(x_1 v_1 - y_1 u_1 + x_2 v - y_2 u)^2 + G(x_1 v - y_1 u)^2 + \frac{d}{dt}(Lu^2 + 2Muv + Nv^2),$$

where $P, G, L, M, N$ do not involve $u$ and $v$; and this expression becomes

$$Pw_1^2 + Gw^2 + \frac{d}{dt}(Lu^2 + 2Muv + Nv^2),$$

so that the integral $\int_{t_0}^{t_1}\Theta\,dt$ becomes

$$\int_{t_0}^{t_1}(Pw_1^2 + Gw^2)\,dt + \left[Lu^2 + 2Muv + Nv^2\right]_{t_0}^{t_1}.$$

Now

$$u(x_1^2 + y_1^2) = -y_1 w + x_1 T, \quad v(x_1^2 + y_1^2) = x_1 w + y_1 T,$$

so that the quantity $T$ could only occur at the limits, if at all. At each of the limits, $u$ and $v$ (by given data, or after limits have been fixed from given conditions) vanish, so that $w$ and $T$ vanish at each of the limits. Thus $P$ disappears altogether from the expression of the second variation when the combination, denoted by $w$, is used.

**53.** The preceding expression for $\Theta$, which is of fundamental importance for the second variation of the integral $I$, can be established also as follows by somewhat different analysis.

Let, if possible, quantities $\alpha$, $\beta$, $\gamma$ be determined so that the magnitude

$$P\Theta - P\frac{d}{dt}(\alpha u^2 + 2\beta uv + \gamma v^2),$$

manifestly homogeneous and quadratic in $u$, $v$, $u_1$, $v_1$, is a perfect square, which $\left(\text{after the values of } \dfrac{\partial^2 F}{\partial x_1{}^2}, \ \dfrac{\partial^2 F}{\partial x_1 \partial y_1}, \ \dfrac{\partial^2 F}{\partial y_1{}^2}\right)$ must be

$$\{P(x_1 v_1 - y_1 u_1) + \mu v - \lambda u\}^2.$$

The necessary conditions are

$$
\left.
\begin{aligned}
\lambda y_1 &= L + y_1 y_2 P - \alpha \\
-\lambda x_1 &= M - x_1 y_2 P - \beta \\
-\mu y_1 &= M - y_1 x_2 P - \beta \\
\mu x_1 &= N + x_1 x_2 P - \gamma
\end{aligned}
\right\}, \qquad
\left.
\begin{aligned}
\lambda^2 &= P\frac{\partial^2 F}{\partial x^2} - P\alpha_1 \\
-\lambda\mu &= P\frac{\partial^2 F}{\partial x \partial y} - P\beta_1 \\
\mu^2 &= P\frac{\partial^2 F}{\partial y^2} - P\gamma_1
\end{aligned}
\right\}.
$$

From the second and third equations in the first set, we have

$$\lambda x_1 - \mu y_1 = (x_1 y_2 - x_2 y_1) P;$$

and therefore we may take

$$\lambda = y_2 P + y_1 \rho, \quad \mu = x_2 P + x_1 \rho,$$

so that

$$P(x_1 v_1 - y_1 u_1) + \mu v - \lambda u = Pw_1 + \rho w.$$

The first set of equations now reduces to

$$\alpha = L - y_1{}^2\rho, \quad \beta = M + x_1 y_1 \rho, \quad \gamma = N - x_1{}^2\rho,$$

so that

$$\alpha u^2 + 2\beta uv + \gamma v^2 = Lu^2 + 2Muv + Nv^2 + \rho w^2.$$

The second set of equations now gives

$$y_1{}^2\left(\frac{\rho^2}{P} - \rho_1\right) = \frac{\partial^2 F}{\partial x^2} - y_2{}^2 P - \frac{dL}{dt},$$

$$-x_1 y_1\left(\frac{\rho^2}{P} - \rho_1\right) = \frac{\partial^2 F}{\partial x \partial y} + x_2 y_2 P - \frac{dM}{dt},$$

$$x_1{}^2\left(\frac{\rho^2}{P} - \rho_1\right) = \frac{\partial^2 F}{\partial y^2} - x_2{}^2 P - \frac{dN}{dt}.$$

Hence there exists a quantity $G$, such that

$$\frac{\rho^2}{P} - \rho_1 = G \, ;$$

a result in exact accordance with the expressions for $\dfrac{\partial^2 F}{\partial x^2}, \dfrac{\partial^2 F}{\partial x \partial y}, \dfrac{\partial^2 F}{\partial y^2}$, deduced from the identity satisfied by the form of $F$ when account is taken of the characteristic equation. As regards this equation for $\rho$, let

$$\rho = - P \frac{z_1}{z} \, ;$$

then

$$Pz_2 + P_1 z_1 - Gz = 0,$$

the same equation as before. Gathering the results together, we obtain the expression for $\Theta$, given by

$$\Theta - \frac{d}{dt}\left( Lu^2 + 2Muv + Nv^2 + P\frac{z_1}{z}\, w^2 \right) = P\left( w_1 - \frac{z_1}{z}\, w \right)^2.$$

**54.** Before discussing the significance of the form obtained for the second variation, it is worth while to note assumptions which have been made, overtly or tacitly.

The variables $x$ and $y$, as functions of $t$ determined by the characteristic equation, and the quantities $u$ and $v$, vary continuously. The variables $x_1$ and $y_1$ vary continuously; but places may occur on the characteristic curve, where the ratio $y_1/x_1$ suffers a sudden discontinuity in value. Between such a place and each limit of the integral, and between every two consecutive such places (if there be more than one of them), these assumptions hold and the analysis is valid. The number of such places (if any) within a finite range of the integral must remain a finite integer.

The original integral involves no derivative higher than the first, and therefore is formally unaffected by possible discontinuities in $x_2$ and $y_2$. These quantities occur in the characteristic equation $\mathfrak{E} = 0$, as, of course, also in $X = 0$ and $Y = 0$; and their continuity is assumed, as also is a tacit assumption that $w_1$ is continuous. But the quantities $u_2$ and $v_2$ do not occur in the analysis, even by implication, although $u_1$ and $v_1$ are assumed continuous; yet such quantities arise, if the curvature of the varied curve has to be considered. Thus there might be a continuous varied curve, which is continuous also in direction, but which might have violent discontinuities of curvature at any number of successive points; for instance, any portion of the characteristic curve might be divided into very small arcs, and semicircles on these arcs in succession might be constructed on alternately different sides of the curve. Such possibilities must not be ignored; they belong, however, solely to the variations imposed upon the characteristic curve and not to the curve itself.

In order that the varied curve may be continuous, $u$ and $v$ must be continuous. But for such a curve, $u_1$ and $v_1$ are not necessarily continuous, merely because $u$ and $v$ are continuous; thus the varied curve might be sharply and rapidly serrated. For such variations, the preceding analysis is not valid; yet such variations must not be ignored: again, they belong not to the characteristic curve but to the imposed variations. For variations which thus (or in other respects) deviate from continuity, a different analytical calculation must be effected.

The more obvious occasions, where such deviations occur, arise at fixed points, where $u$ and $v$ are bound to vanish: terms

$$Lu^2 + 2Muv + Nv^2 + P\frac{z_1}{z}w^2$$

vanish at such places.

### Discussion of the normal form of the second variation.

**55.** We now return to the expression for the second variation, given by

$$\int \Theta dt = \int P\left(w_1 - \frac{z_1}{z}w\right)^2 dt + \left[Lu^2 + 2Muv + Nv^2 + P\frac{z_1}{z}w^2\right],$$

taken between limits.

If none of the preceding types of discontinuities occur in the whole range of integration, the integral extends from $t_0$ to $t_1$. In this event, we may assume (after the discussion of the first variation of $I$) that the limits are fixed, and that therefore $u$, $v$, $w$ vanish at each limit. The terms at the limits then vanish; and

$$\int_{t_0}^{t_1} P\left(w_1 - \frac{z_1}{z}w\right)^2 dt$$

is the quantity measuring—except as to the positive constant factor $\frac{1}{2}\kappa^2$—the second variation.

If some of the preceding types of discontinuities do occur and are subject to the prescribed limitations, we suppose the range to be made up of a finite number of sub-ranges, each sub-range extending from one such place to the immediately consecutive place. At each place, the limit terms vanish, and the whole second variation is the sum of the second variations

$$\int P\left(w_1 - \frac{z_1}{z}w\right)^2 dt$$

for the successive sub-ranges.

In the former case, the tests will apply to the whole range; in the latter case, they will apply to each sub-range separately and thus, cumulatively, to the whole range.

We thus have to deal with the magnitude

$$\Delta = \int P\left(w_1 - \frac{z_1}{z}w\right)^2 dt,$$

over the whole range, or over the aggregate of the sub-ranges taken separately. With the assumptions made, the second variation of $I$ is $\frac{1}{2}\kappa^2\Delta$; and therefore its sign is determined by the sign of $\Delta$, while it can vanish only if $\Delta$ can vanish.

The sign of $\Delta$, which is to be permanent for all variations $w$ that are admissible, depends upon the quantity $P$ and upon the quantity $\left(w_1 - \frac{z_1}{z}w\right)^2$. In the latter, $z$ is an integral of a definite linear differential equation, while $w$ is (within a broad range) at our arbitrary disposal; the quantity therefore is always positive, unless it can be made zero by a choice of $w$. If then $z$ is such that, consistently with the conditions, we can choose $w = mz$, where $m$ is a constant, and the choice is valid over the whole range or over part of the range, then the quantity is zero for the whole range or for that part. In the former case, we should have $\Delta = 0$; in the latter case, by choosing $w$ to be zero throughout all the range except that part, we again should have $\Delta = 0$. The second variation of $I$ then vanishes for that small variation or combination of small variations; we should no longer be in a position to assert a permanence of sign for $\mathbf{I} - I$ as derived from the coefficient of $\kappa^2$. It is therefore desirable, where possible, to exclude this chance: and such exclusion will depend upon the character of the quantity $z$, that is, upon the primitive of the equation

$$Pz_2 + P_1z_1 - Gz = 0.$$

### The Legendre test.

**56.** In the first place, we consider the effect upon $\Delta$ of the sign of $P$. If $P$ be sometimes negative and sometimes positive, then

(i) by choosing $w$ zero where $P$ is positive, and different from zero (or from $mz$) where $P$ is negative, we can make $\Delta$ negative: and

(ii) by choosing $w$ zero where $P$ is negative, and different from zero (or from $mz$) where $P$ is positive, we can make $\Delta$ positive:

that is, variations can be chosen at will which make $\Delta$ negative or $\Delta$ positive. This possibility must be excluded; as it depends upon having different signs for $P$ in different stretches of the range, this previous possibility must be excluded. Hence $P$ must not change its sign in the range: consequently $P$ must not pass through the value zero and change its sign in doing so, nor pass through an infinite value and change its sign in doing so. A zero value of $P$ at some place, without a change of sign, would not be excluded by this consideration; but owing to the differential equation satisfied by $z$, it is desirable to exclude even that possibility. Similarly as regards any infinite value of $P$ without change of sign in passing through such a value, quite independently of the assumptions concerning the continuity of the quantities that occur.

We therefore can summarise the results by the following requirement:

*In order that the integral I may acquire a maximum value or a minimum value, the quantity denoted by P must never vanish within the range.*

Manifestly this is the Legendre test.

### The subsidiary characteristic equation.

**57.** We now must deal with the equation satisfied by the quantity $z$, viz.

$$Pz_2 + P_1 z_1 - Gz = 0,$$

especially with reference to the foregoing requirements. We shall suppose that the Legendre test is satisfied, so that the quantity $P$ remains finite, continuous, and distinct from zero, for all values of $t$ within the range; thus $P$ can be regarded as a regular function of $t$ for all such values.

Further, suppose that $G$ does not become infinite within the range. As $G$ is the common value of the three equal fractions $\dfrac{A}{y_1{}^2}$, $\dfrac{B}{-x_1 y_1}$, $\dfrac{C}{x_1{}^2}$, and as $x_1$ and $y_1$ do not simultaneously vanish—otherwise $\dfrac{ds}{dt}$ would vanish—we then have the requirement that $A, B, C$ do not become infinite, a requirement usually satisfied in consequence of all the preceding assumptions as regards $F$, $x$, and $y$.

Thus the ordinary linear differential equation of the second order for $z$ has all its coefficients regular and finite for all the values of $t$ within the range, these coefficients being known functions after the (presumably) known values of $x$ and $y$ as functions of $t$ have been inserted. For such an equation, the customary existence-theorem [*] yields the following result:

An integral of the equation exists, which is a regular function of $t$, and which is uniquely determinate by the conditions that, for any assigned value of $t$, the variable $z$ acquires any arbitrarily assigned constant value and its first derivative $z_1$ acquires any other arbitrarily assigned constant value. Moreover, the two arbitrary constants occur linearly, so that $z$ is of the form

$$z = c\phi(t) + c'\psi(t),$$

where $c$ and $c'$ are arbitrary constants, while $\phi(t)$ and $\psi(t)$ are regular functions of $t$.

**58.** If this method of determining $z$ is adopted, we have to integrate a linear equation of the second order. We already have an equation, not linear, of the second order, $\mathfrak{C} = 0$ (or the equivalent equations $X = 0$ and $Y = 0$). We now proceed to prove that, *when the primitive of the characteristic equation is known, the primitive of the z-equation can be deduced by direct operations involving no quadrature.*

---

[*] See my *Theory of Differential Equations*, vol. iii, § 209.

As the second variation is to be a criterion for the possession of a maximum or a minimum by the characteristic curve, the values of $x$ and $y$ occurring in the coefficients in that variation are the values satisfying the characteristic equation, that is, satisfying the equations $X = 0$ and $Y = 0$. When $x$ and $y$ are submitted to small variations $\kappa u$ and $\kappa v$ in general, let $X$ become $\mathbf{X}$, $Y$ become $\mathbf{Y}$, and $\mathfrak{E}$ become $\mathbf{E}$. Then, as $X = -y_1\mathfrak{E}$, $Y = x_1\mathfrak{E}$, we have

$$-\mathbf{X} = (y_1 + \kappa v_1)\,\mathbf{E}, \quad \mathbf{Y} = (x_1 + \kappa u_1)\,\mathbf{E},$$

and therefore

$$-(\mathbf{X} - X) = y_1\,(\mathbf{E} - \mathfrak{E}) + \kappa v_1\mathbf{E}, \quad \mathbf{Y} - Y = x_1\,(\mathbf{E} - \mathfrak{E}) + \kappa u_1\mathbf{E},$$

so that

$$(x_1 v_1 - y_1 u_1)\,(\mathbf{E} - \mathfrak{E}) = u_1\,(\mathbf{X} - X) + v_1\,(\mathbf{Y} - Y).$$

Now

$$X = \frac{\partial F}{\partial x} - \frac{d}{dt}\left(\frac{\partial F}{\partial x_1}\right),$$

and therefore

$$\mathbf{X} = \frac{\partial F}{\partial x} + \kappa\left(u\,\frac{\partial^2 F}{\partial x^2} + v\,\frac{\partial^2 F}{\partial x \partial y} + u_1\,\frac{\partial^2 F}{\partial x \partial x_1} + v_1\,\frac{\partial^2 F}{\partial x \partial y_1}\right)$$
$$- \frac{d}{dt}\left\{\frac{\partial F}{\partial x_1} + \kappa\left(u\,\frac{\partial^2 F}{\partial x \partial x_1} + v\,\frac{\partial^2 F}{\partial y \partial x_1} + u_1\,\frac{\partial^2 F}{\partial x_1^2} + v_1\,\frac{\partial^2 F}{\partial x_1 \partial y_1}\right)\right\} + \dots,$$

the unexpressed terms being of the second and higher orders in $\kappa$. When these higher terms are neglected, because of their relative unimportance owing to the smallness of $\kappa$, we have

$$\mathbf{X} - X = \kappa\left(u\,\frac{\partial^2 F}{\partial x^2} + v\,\frac{\partial^2 F}{\partial x \partial y} + u_1\,\frac{\partial^2 F}{\partial x \partial x_1} + v_1\,\frac{\partial^2 F}{\partial x \partial y_1}\right)$$
$$- \kappa\,\frac{d}{dt}\left(u\,\frac{\partial^2 F}{\partial x \partial x_1} + v\,\frac{\partial^2 F}{\partial y \partial x_1} + u_1\,\frac{\partial^2 F}{\partial x_1^2} + v_1\,\frac{\partial^2 F}{\partial x_1 \partial y_1}\right).$$

Then

$$u_1\,\frac{\partial^2 F}{\partial x_1^2} + v_1\,\frac{\partial^2 F}{\partial x_1 \partial y_1} = - y_1 P\,(x_1 v_1 - y_1 u_1) = - y_1 P\,(w_1 - x_2 v + y_2 u),$$

so that

$$u\,\frac{\partial^2 F}{\partial x \partial x_1} + v\,\frac{\partial^2 F}{\partial y \partial x_1} + u_1\,\frac{\partial^2 F}{\partial x_1^2} + v_1\,\frac{\partial^2 F}{\partial x_1 \partial y_1} = uL + vM - y_1 P w_1.$$

Also

$$u_1\,\frac{\partial^2 F}{\partial x \partial x_1} + v_1\,\frac{\partial^2 F}{\partial x \partial y_1} = u_1 L + v_1 M + y_2 P\,(u_1 y_1 - v_1 x_1);$$

and therefore

$$\mathbf{X} - X = \kappa\left\{u\left(\frac{\partial^2 F}{\partial x^2} - \frac{dL}{dt} - y_2^2 P\right) + v\left(\frac{\partial^2 F}{\partial x \partial y} - \frac{dM}{dt} + x_2 y_2 P\right) + y_1\,\frac{d}{dt}\,(P w_1)\right\}$$
$$= \kappa y_1\left\{- Gw + \frac{d}{dt}\,(P w_1)\right\},$$

by the results obtained in § 51.

Proceeding in a similar manner, we find

$$\mathbf{Y} - Y = \kappa x_1 \left\{ Gw - \frac{d}{dt}(Pw_1) \right\}.$$

Hence

$$(x_1 v_1 - y_1 u_1)(\mathbf{E} - \mathfrak{E}) = \kappa (x_1 v_1 - y_1 u_1) \left\{ Gw - \frac{d}{dt}(Pw_1) \right\},$$

and therefore

$$\mathbf{E} - \mathfrak{E} = \kappa \left\{ Gw - \frac{d}{dt}(Pw_1) \right\}.$$

Now the quantities $x$ and $y$ satisfy the equation $\mathfrak{E} = 0$, that is, also the equations $X = 0$ and $Y = 0$; and they involve arbitrary constants which are completely independent of the characteristic equation. Let small arbitrary parametric variations be imposed upon these arbitrary constants, and let the consequently varied values of $x$ and $y$ be $x + \kappa\xi$ and $y + \kappa\eta$ respectively; also, let

$$\zeta = x_1\eta - y_1\xi,$$

so that $\zeta$ is the consequent value of $w$. Just as $x$ and $y$, when substituted, make $\mathfrak{E} = 0$, so $x + \kappa\xi$ and $y + \kappa\eta$, when substituted, make $\mathbf{E} = 0$. We therefore have

$$G\zeta - \frac{d}{dt}(P\zeta_1) = 0,$$

where $\kappa\xi$ and $\kappa\eta$ arise as the small variations of $x$ and $y$, when small parametric variations are imposed upon the arbitrary constants in the values of $x$ and $y$ that constitute the primitive of the characteristic equation. The equation for $\zeta$ and the equation for $z$ are identical; and therefore we can take

$$z = \zeta.$$

The quantity $\zeta$ is known, by direct operations upon $x$ and $y$ through these small parametric variations of the most general kind imposed upon the primitive of the characteristic equation; thus $z$, that is, the integral of the equation

$$Pz_2 + P_1 z_1 - Gz = 0,$$

can be regarded as known when the primitive of the characteristic equation is known.

On this account, the linear equation for $z$ (and $\zeta$) is called the *subsidiary characteristic equation*.

**59.** The expression for the second variation now becomes, on omitting the factor $\frac{1}{2}\kappa^2$,

$$\int \Theta dt = \int P \left( w_1 - \frac{\zeta_1}{\zeta} w \right)^2 dt,$$

with the usual limits for the integral.

This expression can also be derived as follows. The initial form of $U$ is

$$U = \int \left( u\frac{\partial F}{\partial x} + v\frac{\partial F}{\partial y} + u_1\frac{\partial F}{\partial x_1} + v_1\frac{\partial F}{\partial y_1} \right) dt.$$

Let $\mathbf{U}$ be the form of $u$ when $x$ and $y$ are subjected to the small variations $x + \kappa u$ and $y + \kappa v$; then

$$\mathbf{U} - U = \kappa \int \Theta\, dt,$$

in the original form of the function $\Theta$, because the varied value of $\dfrac{\partial F}{\partial x}$ is

$$\frac{\partial F}{\partial x} + \kappa \left( u\frac{\partial^2 F}{\partial x^2} + v\frac{\partial^2 F}{\partial x \partial y} + u_1\frac{\partial^2 F}{\partial x \partial x_1} + v_1\frac{\partial^2 F}{\partial x \partial y_1} \right),$$

and so for the other terms in $U$. The second form of $U$, when account is taken of the conditions at the limits, is (§ 41)

$$U = \int \left[ u\left\{ \frac{\partial F}{\partial x} - \frac{d}{dt}\left(\frac{\partial F}{\partial x_1}\right) \right\} + v\left\{ \frac{\partial F}{\partial y} - \frac{d}{dt}\left(\frac{\partial F}{\partial y_1}\right) \right\} \right] dt$$

$$= \int (uX + vY)\, dt;$$

and therefore

$$\mathbf{U} - U = \int \{ u\,(\mathbf{X} - X) + v\,(\mathbf{Y} - Y) \}\, dt$$

$$= \kappa \int w\left\{ Gw - \frac{d}{dt}(Pw_1) \right\} dt.$$

Hence

$$\int \Phi\, dt = \int w\left\{ Gw - \frac{d}{dt}(Pw_1) \right\} dt$$

$$= \int \frac{w}{\zeta}\left\{ w\frac{d}{dt}(P\zeta_1) - \zeta\frac{d}{dt}(Pw_1) \right\} dt$$

$$= \int \frac{w}{\zeta}\frac{d}{dt}\{ P(w\zeta_1 - \zeta w_1) \}\, dt.$$

When we integrate by parts, and in the portion outside the integral impose the condition that $u$ and $v$, and therefore also $w$, vanish at each extremity, we have

$$\int \Theta\, dt = -\int P(w\zeta_1 - \zeta w_1)\frac{1}{\zeta^2}(\zeta w_1 - w\zeta_1)\, dt$$

$$= \int P\left( w_1 - \frac{\zeta_1}{\zeta}w \right)^2 dt,$$

the former expression.

An integral of the subsidiary characteristic equation[*] derivable from the primitive of the characteristic equation, is necessary for this expression of the second variation. We proceed to consider this integral in closer detail.

---

[*] As already seen by an example (Ex. 8, § 34), the results for a particular case may be obtained more simply by solving the subsidiary equation than by first obtaining the primitive of the characteristic equation. The analysis, which now is to be given, relates to the general case when the primitive is known and applies, with unchanged inferences, to any particular case.

*The primitive of the subsidiary characteristic equation.*

**60.** The primitive of the equations $X = 0$ and $Y = 0$ leads to the $(x, y)$ equation of the characteristic curve, in which the variable $t$ does not appear. As $t$ does not appear explicitly in $X = 0$ or $Y = 0$, one of the arbitrary constants in the primitive is merely additive to $t$; it therefore is not an essential constant for the curve. Again, the initial values of $x_1$ and $y_1$ combine into a single constant (say $m$), as the initial value of $y'$, that is, of $y_1/x_1$: that is, these two arbitrary constants coalesce into a single constant for the curve. Thus, in effect, there are two essential arbitrary independent constants, which survive in the primitive whatever form be adopted. Let these be $A_1$ and $A_2$; and suppose that the expression of the primitive is

$$x = \phi(t, A_1, A_2), \qquad y = \psi(t, A_1, A_2).$$

At an initial point of the curve, let $x = a$, $y = b$, $t = t'$, where $t'$ may vary from one curve to another; then, if $\frac{\partial \phi}{\partial t}$ and $\frac{\partial \psi}{\partial t}$ be denoted by $\phi'(t)$ and $\psi'(t)$ respectively, we have

$$a = \phi(t', A_1, A_2), \qquad b = \psi(t', A_1, A_2),$$

$$m = \frac{\psi'(t')}{\phi'(t')}.$$

The quantity $a$, without loss of generality, can be regarded as non-parametric, because the origin can be chosen at will; hence, when $t'$ is eliminated, we have

$$b = h(A_1, A_2), \qquad m = k(A_1, A_2),$$

where the functions $h$ and $k$ must be independent of one another. Also, conversely, the independent constants $A_1$ and $A_2$ are functions of $b$ and $m$.

Moreover, we have

$$x = \phi(t, A_1, A_2), \qquad a = \phi(t', A_1, A_2),$$

and therefore

$$x - a = (t - t')\phi'(t', A_1, A_2) + \text{powers of } (t - t');$$

and similarly

$$y - b = (t - t')\psi'(t', A_1, A_2) + \text{powers of } (t - t').$$

Hence

$$y - b = m(x - a) + \text{powers of } x - a,$$

say

$$y = g(x, b, m),$$

the expansions having been taken in the vicinity of the place $(a, b)$.

Thus the primitive of the characteristic equation can be taken in the general form

$$x = \phi(t, A_1, A_2), \qquad y = \psi(t, A_1, A_2);$$

and also in the more particular form

$$y = b + m(x - a) + \dots$$
$$= g(x, b, m),$$

where $b$ and $m$ are functions of $A_1$ and $A_2$, as $A_1$ and $A_2$ are functions of $b$ and $m$, the two independent arbitrary constants in either pair being the essential arbitrary constants of the primitive.

These quantities $x$ and $y$ are such that the equation $\mathfrak{E} = 0$ is satisfied, whatever be the arbitrary constants $A_1$ and $A_2$. Let $A_1$ and $A_2$ suffer small arbitrary variations so that they become $A_1 + \kappa c_1$ and $A_2 + \kappa c_2$ respectively, where $c_1$ and $c_2$ themselves are arbitrary constants. The characteristic equation, now $\mathbf{E} = 0$, is satisfied. But the quantity $\kappa\xi$ is the term of the first order in $\kappa$ in the expansion of $\phi(t, A_1 + \kappa c_1, A_2 + \kappa c_2)$, and the quantity $\kappa\eta$ is the term of the first order in $\kappa$ in the expansion of $\psi(t, A_1 + \kappa c_1, A_2 + \kappa c_2)$; hence

$$\left. \begin{aligned} \xi &= c_1 \frac{\partial \phi}{\partial A_1} + c_2 \frac{\partial \phi}{\partial A_2} \\ \eta &= c_1 \frac{\partial \psi}{\partial A_1} + c_2 \frac{\partial \psi}{\partial A_2} \end{aligned} \right\} .$$

We write

$$\left. \begin{aligned} \chi(t) &= \phi'(t) \frac{\partial \psi}{\partial A_1} - \psi'(t) \frac{\partial \phi}{\partial A_1} \\ \omega(t) &= \phi'(t) \frac{\partial \psi}{\partial A_2} - \psi'(t) \frac{\partial \phi}{\partial A_2} \end{aligned} \right\} ;$$

and then

$$\zeta = x_1 \eta - y_1 \xi$$
$$= c_1 \chi(t) + c_2 \omega(t).$$

But this quantity $\zeta$, as was proved in § 58, satisfies the equation

$$\frac{d}{dt}(P\zeta_1) - G\zeta = 0;$$

and it contains linearly two independent arbitrary constants $c_1$ and $c_2$.

If then $\chi(t)$ and $\omega(t)$ are linearly independent functions of $t$, this quantity $\zeta$ provides the primitive of the subsidiary characteristic equation.

**61.** If $\chi(t)$ and $\omega(t)$ are not linearly independent, a relation

$$\alpha\chi(t) + \beta\omega(t) = 0$$

would subsist, $\alpha$ and $\beta$ being constants. Now

$$x = \phi(t, A_1, A_2), \qquad y = \psi(t, A_1, A_2),$$

and also

$$y = b + m(x - a) + (x - a)^2 R(x - a, b, m) = g(x, b, m),$$

where $R(x-a, b, m)$ is a regular function of $x-a$, and where $A_1$ and $A_2$ are two independent functions—any we please—of the two independent arbitrary constants $b$ and $m$. Now

$$\frac{\partial \psi}{\partial A_1} = \frac{\partial y}{\partial A_1} = \frac{\partial g}{\partial x}\frac{\partial x}{\partial A_1} + \frac{\partial g}{\partial b}\frac{\partial b}{\partial A_1} + \frac{\partial g}{\partial m}\frac{\partial m}{\partial A_1}$$

$$= \frac{\psi'(t)}{\phi'(t)}\frac{\partial \phi}{\partial A_1} + \frac{\partial g}{\partial b}\frac{\partial b}{\partial A_1} + \frac{\partial g}{\partial m}\frac{\partial m}{\partial A_1},$$

and therefore

$$\chi(t) = \phi'(t)\frac{\partial \psi}{\partial A_1} - \psi'(t)\frac{\partial \phi}{\partial A_1}$$

$$= \phi'(t)\left(\frac{\partial g}{\partial b}\frac{\partial b}{\partial A_1} + \frac{\partial g}{\partial m}\frac{\partial m}{\partial A_1}\right).$$

Similarly

$$\omega(t) = \phi'(t)\left(\frac{\partial g}{\partial b}\frac{\partial b}{\partial A_2} + \frac{\partial g}{\partial m}\frac{\partial m}{\partial A_2}\right).$$

If, then, the foregoing postulated relation could exist, we should have

$$\frac{\partial g}{\partial b}\left(\alpha\frac{\partial b}{\partial A_1} + \beta\frac{\partial b}{\partial A_2}\right) + \frac{\partial g}{\partial m}\left(\alpha\frac{\partial m}{\partial A_1} + \beta\frac{\partial m}{\partial A_2}\right) = 0,$$

for $\phi'(t)$ is not a permanent zero: that is, there would be a linear relation between the variable quantities $\frac{\partial g}{\partial b}$ and $\frac{\partial g}{\partial m}$, in which the coefficients are constant. But, as

$$\frac{\partial g}{\partial b} = 1 + (x-a)^2\frac{\partial R}{\partial b},$$

$$\frac{\partial g}{\partial m} = x - a + (x-a)^2\frac{\partial R}{\partial m},$$

it is clear that no such linear relation between $\frac{\partial g}{\partial b}$ and $\frac{\partial g}{\partial m}$ can exist; hence the foregoing coefficients must vanish, and therefore

$$\alpha\frac{\partial b}{\partial A_1} + \beta\frac{\partial b}{\partial A_2} = 0, \qquad \alpha\frac{\partial m}{\partial A_1} + \beta\frac{\partial m}{\partial A_2} = 0.$$

Now $b$ and $m$, as functions of $A_1$ and $A_2$, are independent of one another, so that

$$\frac{\partial b}{\partial A_1}\frac{\partial m}{\partial A_2} - \frac{\partial b}{\partial A_2}\frac{\partial m}{\partial A_1}$$

is not zero; we must therefore have $\alpha = 0$ and $\beta = 0$. Consequently, there is no linear relation between $\chi(t)$ and $\omega(t)$; and so

$$\zeta = c_1\chi(t) + c_2\omega(t),$$

where $c_1$ and $c_2$ are independent arbitrary constants, is the primitive of the subsidiary characteristic equation.

**62.** Thus $\chi(t)$ and $\omega(t)$ are individual integrals of this equation, and therefore

$$P\frac{d^2\chi}{dt^2} + \frac{dP}{dt}\frac{d\chi}{dt} - G\chi = 0,$$

$$P\frac{d^2\omega}{dt^2} + \frac{dP}{dt}\frac{d\omega}{dt} - G\omega = 0;$$

hence, by the well-known property of linear equations of the second order,

$$P\left(\chi\frac{d\omega}{dt} - \omega\frac{d\chi}{dt}\right) = \text{constant}.$$

Now $P$ does not vanish with the range of $t$, the Legendre test being supposed satisfied; and there is no relation of the form $\alpha\chi(t) + \beta\omega(t) = 0$, where $\alpha$ and $\beta$ are constants, so that the quantity multiplied by $P$ does not vanish. Thus the new constant is not zero and, because neither $P$ nor $\chi(t)$ nor $\omega(t)$ can involve the arbitrary constants in the primitive of the subsidiary equation, the new constant is determinate and not arbitrary. Denoting it by $C$, we have

$$P\left(\chi\frac{d\omega}{dt} - \omega\frac{d\chi}{dt}\right) = C.$$

We infer, from the fact that the constant $C$ is not zero: that

(i) the linearly independent integrals $\chi(t)$ and $\omega(t)$ cannot vanish simultaneously for any value of $t$ in its range;

(ii) the first derivatives of $\chi(t)$ and $\omega(t)$ cannot vanish simultaneously for any value of $t$ in its range; and

(iii) no root of $\chi(t)$, or of $\omega(t)$, or of any integral of the subsidiary equation, can be multiple.

*An integral $Z(t, t')$ of the subsidiary equation, affecting the second variation.*

**63.** We know that every integral of the equation can be expressed linearly in terms of $\chi(t)$ and $\omega(t)$, with coefficients that are independent of $t$; and any such linear combination is an integral. Thus, if $t'$ denote any particular value of $t$ in the range,

$$Z(t, t') = \omega(t')\chi(t) - \chi(t')\omega(t)$$

is such an integral. It will appear that this function $Z(t, t')$ is of critical importance, especially because of its zeros. We therefore note that, effectively, it is independent of any particular choice of fundamental integrals $\chi(t)$ and $\omega(t)$ of the subsidiary equation: for, if any two others be chosen, necessarily of the forms $k_1\chi(t) + k_2\omega(t)$ and $l_1\chi(t) + l_2\omega(t)$, where $k_1l_2 - k_2l_1$ is not zero, the new form of the function $Z(t, t')$ is

$$\begin{vmatrix} k_1\chi(t) + k_2\omega(t), & l_1\chi(t) + l_2\omega(t) \\ k_1\chi(t') + k_2\omega(t'), & l_1\chi(t') + l_2\omega(t') \end{vmatrix},$$

which is equal to $(k_1 l_2 - k_2 l_1) Z(t, t')$. The function $Z(t, t')$ is, in fact, covariantive amid all possible choices of fundamental integrals; and it cannot become evanescent, because $\chi(t')$ and $\omega(t')$ cannot vanish simultaneously for any selected value $t'$ in the range.

### *The Jacobi test.*

**64.** We return to the second variation of our integral $I$, and specially to the consideration of the permanence of sign of $\int \Theta \, dt$, where

$$\int \Theta \, dt = \int P \left( w_1 - \frac{\zeta_1}{\zeta} w \right)^2 dt,$$

where now the limits of the integral are regarded as fixed values. The quantity $P$ is to preserve a uniform sign along the range, in accordance with Legendre's test; the quantity $\left( w_1 - \frac{\zeta_1}{\zeta} w \right)^2$ is always positive when it does not vanish. The acquisition of a zero value, save for isolated values, must be avoided if possible; and such acquisition can be attained if, over some continuous stretch in the range, we could have $w = \gamma \zeta$, where $\gamma$ is a constant. Let the stretch be the whole range—the argument is *a fortiori* valid if it be only a portion of the range, because we could take $w = \gamma \zeta$ over that portion, $w = 0$ before that portion, and $w = 0$ after that portion. We have $w = 0$ at the lower limit $t_0$, and therefore at the lower limit $\zeta$ would have to vanish: that is, we should take

$$\zeta = Z(t, t_0).$$

We know (§ 62) that $\zeta$ cannot anywhere have a repeated zero; thus, in the immediate vicinity of $t_0$, we must have

$$\zeta = \lambda (t - t_0) + \text{higher powers of } t - t_0,$$

where $\lambda$ is a non-vanishing constant. Again, our quantity $w$ is to be a uniform function of the variable, which must vanish at $t_0$ because $w$ vanishes at the lower limit; hence any admissible value of $w$ is of the form

$$w = \mu (t - t_0)^n + \text{higher powers of } t - t_0,$$

where* $n \geqslant 1$. In the immediate vicinity of $t_0$, we have

$$w_1 - \frac{\zeta_1}{\zeta} w = \mu (n - 1)(t - t_0)^{n-1} + \text{higher powers of } t - t_0,$$

a quantity which vanishes at $t_0$ whether $n = 1$ or $n$ be greater than 1. Thus there is no infinity at the lower limit. For the rest of the range, the possibility of $w = \gamma \zeta$ makes

$$w_1 - \frac{\zeta_1}{\zeta} w = 0.$$

---

* In the text, $w$ is uniform and so $n$ is an integer; for the immediate argument, $w$ might be a radical. But $w_1$ is not to suffer discontinuity for values of $t$ that are considered; in that case, the positive quantity $n$ would be greater than unity.

If therefore at the other extremity of the range, or earlier than that other extremity, the selected quantity $\zeta$ can again vanish, the terminal condition as regards $w$ is satisfied by the assumption $w = \gamma\zeta$. All the other conditions are satisfied; and so the second variation would vanish for the particular small variation $\kappa w$. We should not then be in a position to assert the existence of a maximum or a minimum for the integral, because of the lack of a uniform sign, either negative or positive, for all small variations of the characteristic curve. To settle the doubt, we should (in the absence of other knowledge*) then have to consider the behaviour of the third and higher variations for the particular small variation $w = \kappa\zeta$ under which the second variation vanishes.

But if the quantity $\zeta$, $= Z(t, t_0)$, vanishing at the beginning of the range, does not again vanish in its course—or, in other words, if the range beginning at $t_0$ does not extend so far as the first zero of $Z(t, t_0)$ for a value of $t$ greater than $t_0$—then we cannot have a variation $\kappa w$, given by $w = \gamma Z(t, t_0)$, because $w$ would not vanish at the further extremity for such a variation. Over a range thus limited, $\left(w_1 - \dfrac{\zeta_1}{\zeta}w\right)^2$ is always positive for all non-zero small variations $\kappa w$; and then $\int \Theta\, dt$ remains permanent in sign for all small weak variations, that sign being the same as the sign of the quantity $P$ which does not vanish throughout the range. We thus find the extended form of what has been called (§ 26) the Jacobi test, viz.

*The range of the integral, beginning at a value $t_0$, must not extend so far as the first zero of the function $Z(t, t_0)$ which is greater than $t_0$.*

### Conjugate places in a range of integration.

**65.** Once again, attention must be called to an implicit elusive assumption. The range, beginning at $t_0$, must not extend so far as $T_0$, where $T_0$ is the first value of $t$ after $t_0$ for which $Z(t, t_0)$ vanishes. Let a value $t'$ be taken within that range, that is, such that $t_0 < t' < T_0$; and consider a function $Z(t, t')$, which of course vanishes at $t'$. We could have a small weak variation $\kappa w$, where $w$ is made zero from $t_0$ to $t'$, and is made equal to $\gamma Z(t, t')$ from $t'$ onwards; and such a variation could continue onwards, even if $Z(t, t')$ should again vanish within the range or at its limit, that is, if the range contains a value $T'$ of $t$ where $T'$ is the first zero of $Z(t, t')$ next after $t'$. Manifestly this possibility has been implicitly excluded in the preceding statement of the range; the implicit exclusion must be justified†.

For this purpose, and also specially in order to indicate a geometrical significance of the results, the characteristic curve may be considered with advantage. The expressions for $x$ and $y$ lead, when the variable $t$ is eliminated, to the equation of that curve. The quantities $\kappa\xi$ and $\kappa\eta$, leading to the forma-

---

* See § 29: and §§ 76—80, *post*.

† The analytical justification will be found in § 67, *post*.

tion of $\zeta$, *i.e.* of the function $Z(t, t')$, are small variations of $x$ and $y$ which still satisfy the characteristic equation and therefore give a consecutive characteristic curve. The latter, as represented by the variation $\kappa Z(t, t_0)$, meets the original characteristic curve in the point determined by the value $t_0$ of $t$; and the range of the integral, which is to possess a maximum or a minimum, is not to extend so far as the first value of $t$ greater than $t_0$, which corresponds to the next point of intersection of the two curves.

Accordingly we take a characteristic curve $...APB...$ and, on it, choose any ordinary point as the initial point for a range. In connection with the characteristic equation, the curve is made precise and unique, by the assign-

ment of the value of $y$ and the value of $\dfrac{dy}{dx}$ for the initial value of $x$—or, as

already explained, by the assignment of initial values of $x$, $x_1$, $y$, $y_1$ for an initial value of the variable $t$: that is, the curve is made precise by the assignment of an initial point and an initial direction. Now consider a characteristic curve $AP'...$ consecutive to $APB$, passing through the same initial point $A$; as the variation is small, the inclination of the tangents at $A$ is small. Let $P$ on the curve $...APB...$ be given by

$$x = \phi(t, A_1, A_2), \quad y = \psi(t, A_1, A_2),$$

so that, if $t_0$ be the value of $t$ at the initial point $A$, the coordinates of $A$ are $\phi(t_0, A_1, A_2)$, $\psi(t_0, A_1, A_2)$. For the consecutive curve, there is a small variation; let it be represented by $t_0 + \kappa T_0$, $A_1 + \kappa a_1$, $A_2 + \kappa a_2$ instead of $t_0$, $A_1$, $A_2$ respectively, and by $t + \kappa T$ instead of $t$, where $T$ is not necessarily the same as $T_0$. As the initial points are the same, we have

$$0 = \kappa \left\{ T_0 \phi'(t_0) + a_1 \frac{\partial \phi(t_0)}{\partial A_1} + a_2 \frac{\partial \phi(t_0)}{\partial A_2} \right\} + \text{terms of at least the second order in } \kappa,$$

$$0 = \kappa \left\{ T_0 \psi'(t_0) + a_1 \frac{\partial \psi(t_0)}{\partial A_1} + a_2 \frac{\partial \psi(t_0)}{\partial A_2} \right\} + \dots\dots\dots\dots\dots\dots\dots\dots ;$$

and therefore, on the elimination of $T_0$,

$$0 = \kappa \{a_1 \chi(t_0) + a_2 \omega(t_0)\} + \text{terms of the second order in } \kappa :$$

that is, in the limit as $\kappa$ diminishes indefinitely,

$$a_1 \chi(t_0) + a_2 \omega(t_0) = 0.$$

Let the point $P'$ be the displaced position of $P$ due to the small variation, so that the coordinates of $P'$ are $x + \kappa \xi$, $y + \kappa \eta$, where

$$\xi = T \phi'(t) + a_1 \frac{\partial \phi(t)}{\partial A_1} + a_2 \frac{\partial \phi(t)}{\partial A_2} + \text{terms of at least the first order in } \kappa,$$

$$\eta = T \psi'(t) + a_1 \frac{\partial \psi(t)}{\partial A_1} + a_2 \frac{\partial \psi(t)}{\partial A_2} + \dots\dots\dots\dots\dots\dots\dots\dots ;$$

that is,

$$\zeta = x_1\eta - y_1\xi$$
$$= a_1\chi(t) + a_2\omega(t) + \text{terms of at least the first order in } \kappa.$$

Let the curves, if they meet after $A$, meet again in $B$ for the first time after $A$; and let $\theta$ be the value of $t$ at $B$ for the characteristic curve. Then $\xi$ and $\eta$, and therefore $\zeta$, vanish when $t = \theta$; that is,

$$0 = a_1\chi(\theta) + a_2\omega(\theta) + \text{terms of at least the first order in } \kappa.$$

Hence, in the limit as $\kappa$ diminishes indefinitely,

$$0 = a_1\chi(\theta) + a_2\omega(\theta).$$

The arbitrary quantities $a_1$ and $a_2$ do not vanish; hence

$$Z(\theta, t_0) = \omega(t_0)\chi(\theta) - \chi(t_0)\omega(\theta) = 0,$$

and $\theta$ is the first value of $t$ greater than $t_0$, for which this equation gives another intersection of the consecutive characteristic with the original characteristic.

Then $B$ is called the *conjugate* of $A$ on the characteristic curve. On the one hand, it is the limiting position of the first point of intersection of the consecutive characteristic with the original after the initial point of intersection. On the other hand, it is given by the first value $\theta$ of $t$ greater than $t_0$ where $Z(t, t_0)$ vanishes. We may therefore re-enunciate the Jacobi test by requiring that *the range of integration must not extend along the characteristic curve as far as the conjugate of the initial point*.

**66.** We may take $\chi(t)$ and $\omega(t)$, more briefly denoted by $\chi$ and $\omega$ when there is no doubt as to the argument, as the fundamental integrals for the discussion of the covariantive function $Z(t, t_0)$; and the relation

$$P\left(\chi\frac{d\omega}{dt} - \omega\frac{d\chi}{dt}\right) = C$$

has been established, where $C$ is a specific constant. Also

$$Z(t, t_0) = \omega(t_0)\chi(t) - \chi(t_0)\omega(t),$$

so that

$$\left[\frac{d}{dt}Z(t, t_0)\right]_{t=t_0} = \omega(t_0)\chi'(t_0) - \chi(t_0)\omega'(t_0)$$
$$= -\frac{C}{P_0},$$

where $P_0$, not zero, is the value of $P$ at $t_0$; thus

$$Z(t, t_0) = -\frac{C}{P_0}(t - t_0) + \text{higher powers of } t - t_0,$$

verifying the former inference that $t_0$ is a simple zero of $Z(t, t_0)$. Then as $Z(\theta, t_0) = 0$, we can take

$$\chi(\theta) = \sigma\chi(t_0), \qquad \omega(\theta) = \sigma\omega(t_0),$$

where $\sigma$ is not zero, for otherwise $\chi(\theta)$ and $\omega(\theta)$ would vanish together; and so

$$\sigma Z(t, t_0) = \omega(\theta)\chi(t) - \chi(\theta)\omega(t)$$
$$= Z(t, \theta).$$

As before

$$Z(t, \theta) = -\frac{C}{P_\theta}(t - \theta) + \text{higher powers of } t - \theta,$$

where $P_\theta$ is the non-zero value of $P$ at $t = \theta$; and again, as is to be expected, $\theta$ is a simple zero of $Z(t, \theta)$, that is, it is a simple zero of $Z(t, t_0)$. Thus $Z(t, t_0)$ changes its sign as the variable $t$ passes through the zero $\theta$: and so for any zero.

*Note.* We have the corollary that, if we pass backwards along the curve, the limiting position of the first point of intersection of the original curve by a consecutive characteristic through $B$, is the point $A$. For

$$Z(t, \theta) = \sigma Z(t, t_0),$$

and the first value of $t$, less than $\theta$ in the range between $t_0$ and $\theta$, for which $Z(t, t_0)$ vanishes is $t_0$; thus $A$ is the backward conjugate of $B$.

### Exclusiveness of a range bounded by conjugates.

**67.** We can now establish the proposition that *when any point $A'$ is taken in an arc of the characteristic curve, limited by two reciprocally conjugate points $A$ and $B$, the conjugate of $A'$ lies outside the range $AB$.* (It will be sufficient if we establish the result for the forward conjugate of $A'$, on the assumption that $B$ is the forward conjugate of $A$.)

Let $A$ be the point determined by the value $t_0$ of $t$; and let $\theta$ be the value of $t$ at its conjugate $B$, so that

$$Z(\theta, t_0) = 0,$$

where $\theta$ is the first value of $t$ greater than $t_0$ for which this equation is satisfied. Let $t_1$ be the value of $t$ for $A'$, so that $t_0 < t_1 < \theta$. As $Z(\theta, t_0) = 0$, we have

$$\chi(\theta) = \sigma\chi(t_0), \quad \omega(\theta) = \sigma\omega(t_0),$$

where $\sigma$ is not zero; and therefore

$$Z(\theta, t_1) = \omega(t_1)\chi(\theta) - \chi(t_1)\omega(\theta)$$
$$= \sigma\{\omega(t_1)\chi(t_0) - \chi(t_1)\omega(t_0)\}$$
$$= -\sigma Z(t_1, t_0).$$

As $t_1$ lies between $t_0$ and $\theta$, we cannot have $Z(t_1, t_0)$ equal to zero: hence $Z(\theta, t_1)$ does not vanish. Consequently, the conjugate of $A'$ does not coincide with $B$; and $A'$ is any point within $AB$.

Now take a point $A_1$ within $AA'$ and near to $A$; and let the value of $t$ at that point be $t_0 + T_0$, where $T_0$ is small and positive. As $A_1$ is near to $A$, the conjugate $B_1$ of $A_1$ must be near to $B$. Let the value of $t$ for $B_1$ be $\theta + \Theta$; then $\Theta$ definitely is not zero when $T_0$ is not zero, because the conjugate of $A_1$ (a point in $AB$) cannot coincide with $B$. Also, when $T_0$ is zero, then $\Theta$ is zero; so that $\Theta$ and $T_0$ vanish together, and $\Theta$ must be small when $T_0$ is small. The sign of $\Theta$, as well as its magnitude, has to be determined. Because $\theta + \Theta$ gives the point conjugate to the point given by $t_0 + T_0$, we have

$$Z(\theta + \Theta, t_0 + T_0);$$

and therefore as $Z(\theta, t_0) = 0$, and $\Theta$ and $T_0$ are small,

$$\Theta \frac{\partial Z(\theta, t_0)}{\partial \theta} + T_0 \frac{\partial Z(\theta, t_0)}{\partial t_0} + \ldots = 0,$$

the unexpressed terms being of the second and higher orders in $\Theta$ and $T_0$ combined. But, as

$$Z(t, t_0) = \omega(t_0) \chi(t) - \chi(t_0) \omega(t),$$

we have

$$\frac{\partial Z(t, t_0)}{\partial t} = \omega(t_0) \chi'(t) - \chi(t_0) \omega'(t);$$

and therefore

$$\frac{\partial Z(\theta, t_0)}{\partial \theta} = \omega(t_0) \chi'(\theta) - \chi(t_0) \omega'(\theta).$$

Now

$$\frac{\chi(\theta)}{\chi(t_0)} = \frac{\omega(\theta)}{\omega(t_0)} = \sigma,$$

where $\sigma$ is not zero; and so

$$\frac{\partial Z(\theta, t_0)}{\partial \theta} = \frac{1}{\sigma} \{\omega(\theta) \chi'(\theta) - \chi(\theta) \omega'(\theta)\}$$

$$= -\frac{C}{\sigma P_\theta}.$$

Also

$$\frac{\partial Z(\theta, t_0)}{\partial t_0} = \omega'(t_0) \chi(\theta) - \chi'(t_0) \omega(\theta)$$

$$= \sigma \{\omega'(t_0) \chi(t_0) - \chi'(t_0) \omega(t_0)\}$$

$$= \sigma \frac{C}{P_0}.$$

Thus the equation connecting $\Theta$ and $T_0$ becomes

$$-\Theta \frac{C}{\sigma P_\theta} + T_0 \sigma \frac{C}{P_0} + \ldots = 0,$$

or, for sufficiently small values of $T_0$,

$$\Theta = \sigma^2 \frac{P_\theta}{P_0} T_0 + \text{higher powers of } T_0.$$

Now $P_0$ and $P_\theta$ have the same sign because the Legendre test is satisfied, and $\sigma$ is a non-zero constant; thus $\Theta$ is positive, because $T_0$ is positive. Consequently $\theta + \Theta$ is greater than $\theta$: and so the conjugate of $A_1$, which is near $A$ in the range $AA'$, lies near $B$ but beyond $B$. Let it be denoted by $B_1$; then $A_1B_1$ is another range bounded by two conjugate points, $BB_1$ being of the same order of magnitude as $AA_1$.

Now take a place $A_2$, in the range $A_1B_1$ and near $A_1$; its conjugate, say $B_2$, lies beyond $B_1$. And so on, with points $A$ in succession from $A$ towards $A'$; the successive conjugates lie beyond $B$ in succession, each beyond its predecessor: until, finally the conjugate of $A'$, say $B'$, is reached, and it thus lies in the stretch beyond $B$.

### *Lemma on consecutive characteristic curves.*

**68.** The function $Z(t, t_0)$ is of significance in a cognate investigation. Hitherto the small variations upon the characteristic curve have started from the curve itself; and the deduced characteristic curve has not been required to conform to any external condition. At a later stage, it will be found desirable to draw a characteristic curve, closely contiguous to a given characteristic curve in the sense that it is to pass through two assigned points very near the given curve: thus they may be taken near the initial point and the final point of some stretch in the range. The conditions, be it noted, differ from those hitherto imposed, viz. that the curve shall pass through an assigned point and shall have an assigned direction at the point; the new conditions require the curve to pass through two assigned points which lie very near a given characteristic curve. A question arises as to whether there are limitations—and, if so, what are the limitations—upon the data, in order that the new characteristic curve may be drawn.

We therefore postulate a characteristic curve, given by
$$x = \phi(t, A_1, A_2), \quad y = \psi(t, A_1, A_2).$$
We consider a stretch of the curve, beginning at a point $(x_0, y_0)$ with an initial value $t_0$, and ranging with increasing values of $t$ through a point $(x_1, y_1)$ where $t$ has the value $t_1$; thus
$$x_0 = \phi(t_0, A_1, A_2), \quad y_0 = \psi(t_0, A_1, A_2), \quad x_1 = \phi(t_1, A_1, A_2), \quad y_1 = \psi(t_1, A_1, A_2).$$
Let $(x_0 + X_0, y_0 + Y_0)$ be a point contiguous to $(x_0, y_0)$, so that $X_0$ and $Y_0$ are small: and $(x_1 + X_1, y_1 + Y_1)$ be a point contiguous to $(x_1, y_1)$, so that $X_1$ and $Y_1$ are small. If a consecutive characteristic curve can be drawn through these contiguous points, its arbitrary constants can be denoted by $A_1 + \kappa a_1$, $A_2 + \kappa a_2$; the value of $t$ for its initial point $(x_0 + X_0, y_0 + Y_0)$ by $t_0 + \kappa T_0$; and the value of $t$ for its final point by $t_1 + \kappa T_1$, where $T_0$ and $T_1$ are finite. Should this be possible for small values of $\kappa$, then, except for relatively

negligible quantities of the second and higher orders in $\kappa$, we must have

$$X_0 = \kappa \left\{ T_0 \phi'(t_0) + a_1 \frac{\partial \phi(t_0)}{\partial A_1} + a_2 \frac{\partial \phi(t_0)}{\partial A_2} \qquad\qquad \right\},$$

$$Y_0 = \kappa \left\{ T_0 \psi'(t_0) + a_1 \frac{\partial \psi(t_0)}{\partial A_1} + a_2 \frac{\partial \psi(t_0)}{\partial A_2} \qquad\qquad \right\},$$

$$X_1 = \kappa \left\{ \qquad\qquad a_1 \frac{\partial \phi(t_1)}{\partial A_1} + a_2 \frac{\partial \phi(t_1)}{\partial A_2} + T_1 \phi'(t_1) \right\},$$

$$Y_1 = \kappa \left\{ \qquad\qquad a_1 \frac{\partial \psi(t_1)}{\partial A_1} + a_2 \frac{\partial \psi(t_1)}{\partial A_2} + T_1 \psi'(t_1) \right\},$$

being four equations for the potential determination of $T_0$, $T_1$, $a_1$, $a_2$, all of them finite quantities, while $X_0$, $Y_0$, $X_1$, $Y_1$, $\kappa$ are small quantities.

The determinant of the coefficients of $\kappa T_0$, $\kappa a_1$, $\kappa a_2$, $\kappa T_1$ on the right-hand side of these four equations is equal to

$$Z(t_1, t_0);$$

and therefore each of the quantities

$$\kappa T_0 Z(t_1, t_0), \quad \kappa a_1 Z(t_1, t_0), \quad \kappa a_2 Z(t_1, t_0), \quad \kappa T_1 Z(t_1, t_0)$$

is equal to a linear and homogeneous function of $X_0$, $Y_0$, $X_1$, $Y_1$, the coefficients being integral combinations of the third order in the derivatives of $\phi(t_0, A_1, A_2)$, $\phi(t_1, A_1, A_2)$, $\psi(t_0, A_1, A_2)$, and $\psi(t_1, A_1, A_2)$. If then $Z(t_1, t_0)$ is not zero, or is not itself a very small quantity of the same order of magnitude as any one of the four small quantities $X_0$, $Y_0$, $X_1$, $Y_1$, the resolved equations give values for $\kappa T_0$, $\kappa a_1$, $\kappa a_2$, $\kappa T_1$, of the same order of magnitude as $X_0$, $Y_0$, $X_1$, $Y_1$: that is, a consecutive characteristic can be drawn as required. Thus the possibility depends upon the magnitude of $Z(t_1, t_0)$.

We have seen that the conjugate of a point given by $t = t_0$ is the point given by the first root of $Z(\theta, t_0) = 0$ which is greater than $t_0$. If $t_1$ is appreciably less than $\theta$, that is, if the point on the characteristic curve given by $t_1$ falls short of the conjugate of the initial point by an arc-distance of a magnitude greater than the order of the small quantities $X_0$, $Y_0$, $X_1$, $Y_1$, then the consecutive curve can be drawn: and it is a unique curve. If however $t_1 = \theta$, the equations certainly do not give small values for $\kappa T_0$, $\kappa a_1$, $\kappa a_2$, $\kappa T_1$; and if $t_1$, while less than $\theta$, differs from it by a small quantity of the same order as $X_0$, $Y_0$, $X_1$, $Y_1$, no safe inference as to the magnitude of $\kappa T_0$, $\kappa a_1$, $\kappa a_2$, $\kappa T_1$ can be made, without investigating the values of the coefficients of $X_0$, $Y_0$, $X_1$, $Y_1$ in the expressions for those quantities.

It is unnecessary to consider values of $t_1$ greater than $\theta$, because the range of the characteristic curve is restricted (for reasons already adduced) not to extend beyond the conjugate of its initial point.

**69.** Lastly, part of the preceding analysis (so far as it applies solely to the curves at their extremities corresponding to the value $t_0$ of $t$) can be used to establish an assumption slightly wider than one which has already been made : *a characteristic curve, consecutive to a given characteristic, can be drawn through a point very near the latter so that the inclination of its tangent to the tangent of the characteristic is very small.*

We know that a characteristic curve can be drawn through any point with any assigned initial direction. For the present purpose we have to prove that a new characteristic, drawn as required, is consecutive to the given curve.

Taking the given curve in the usual form, let $A_1 + \kappa a_1$ and $A_2 + \kappa a_2$ be the constants for the new curve, where it must be proved that $\kappa a_1$ and $\kappa a_2$ occur as small quantities under the given data. Let the initial value $t$ for the given curve be $t_0$, as usual, and for the assumed curve be $t_0 + \kappa T_0$, so that the value of $t$ anywhere is $t + \kappa T$, where $T = T_0$ when $t = t_0$ (the value of $T$ will disappear from the analysis); we must prove that $\kappa T_0$ is small. Let the initial points of the two curves be $(x_0, y_0)$ and $(x_0 + X_0, y_0 + Y_0)$, where $X_0$ and $Y_0$ are small ; thus we have

$$X_0 = \kappa T_0 \phi'(t_0) + \kappa a_1 \frac{\partial \phi(t_0)}{\partial A_1} + \kappa a_2 \frac{\partial \phi(t_0)}{\partial A_2} + \ldots,$$

$$Y_0 = \kappa T_0 \psi'(t_0) + \kappa a_1 \frac{\partial \psi(t_0)}{\partial A_1} + \kappa a_2 \frac{\partial \psi(t_0)}{\partial A_2} + \ldots,$$

where the unexpressed terms are of the second and higher orders in the small quantity $\kappa$, and are negligible in comparison with the retained terms of the first order. Let $(x, y)$ be any point in the vicinity of $(x_0, y_0)$ along the range ; and let $(x + X, y + Y)$ be the displaced point on the new characteristic. We denote by $t$ the current variable at $(x, y)$, and by $t + \kappa T$ the current variable at $(x + X, y + Y)$ along the new curve; so that $T$ may be a function of $t$, and is equal to $T_0$ when $t = t_0$. Then

$$X = \kappa T \phi'(t) + \kappa a_1 \frac{\partial \phi(t)}{\partial A_1} + \kappa a_2 \frac{\partial \phi(t)}{\partial A_2} + \ldots,$$

$$Y = \kappa T \psi'(t) + \kappa a_1 \frac{\partial \psi(t)}{\partial A_1} + \kappa a_2 \frac{\partial \psi(t)}{\partial A_2} + \ldots,$$

and therefore

$$X' = \kappa T' \phi'(t) + \kappa T \phi''(t) + \kappa a_1 \frac{\partial \phi'(t)}{\partial A_1} + \kappa a_2 \frac{\partial \phi'(t)}{\partial A_2} + \ldots,$$

$$Y' = \kappa T' \psi'(t) + \kappa T \psi''(t) + \kappa a_1 \frac{\partial \psi'(t)}{\partial A_1} + \kappa a_2 \frac{\partial \psi'(t)}{\partial A_2} + \ldots,$$

where the unexpressed terms are negligible compared with the retained terms of the first order in $\kappa$. Hence

$$x'Y' - y'X' = Y'\phi'(t) - X'\psi'(t)$$
$$= \kappa T\{\phi'(t)\psi''(t) - \psi'(t)\phi''(t)\}$$
$$+ \kappa a_1\left\{\phi'(t)\frac{\partial\psi'(t)}{\partial A_1} - \psi'(t)\frac{\partial\phi'(t)}{\partial A_1}\right\} + \kappa a_2\left\{\phi'(t)\frac{\partial\psi'(t)}{\partial A_2} - \psi'(t)\frac{\partial\phi'(t)}{\partial A_2}\right\} + \ldots$$

The inclination of the tangents to the curves, at $(x+X, y+Y)$ and $(x, y)$ respectively, is

$$\tan^{-1}\frac{y'+Y'}{x'+X'} - \tan^{-1}\frac{y'}{x'}$$
$$= \tan^{-1}\frac{x'Y' - y'X'}{x'^2 + y'^2 + x'X' + y'Y'}.$$

At the initial places this inclination is small, and consequently $x'Y' - y'X'$ is small there, that is, when $t = t_0$ and therefore $T = T_0$; denoting its value at $t = t_0$ by a small quantity $\epsilon$, we have

$$\epsilon = \kappa T_0\{\phi'(t_0)\psi''(t_0) - \psi'(t_0)\phi''(t_0)\}$$
$$+ \kappa a_1\left\{\phi'(t_0)\frac{\partial\psi'(t_0)}{\partial A_1} - \psi'(t_0)\frac{\partial\phi'(t_0)}{\partial A_1}\right\} + \kappa a_2\left\{\phi'(t_0)\frac{\partial\psi'(t_0)}{\partial A_2} - \psi'(t_0)\frac{\partial\phi'(t_0)}{\partial A_2}\right\} + \ldots$$

Thus there are three equations connecting $\kappa T_0$, $\kappa a_1$, $\kappa a_2$ with $X_0$, $Y_0$, $\epsilon$, three given small quantities; in each of the expressions, there are linear terms in $\kappa T_0$, $\kappa a_1$, $\kappa a_2$. The determinant of the coefficients in the three sets of linear terms is

$$\begin{vmatrix} \phi'(t_0)\psi''(t_0) - \psi'(t_0)\phi''(t_0), & \phi'(t_0)\dfrac{\partial\psi'(t_0)}{\partial A_1} - \psi'(t_0)\dfrac{\partial\phi'(t_0)}{\partial A_1}, & \phi'(t_0)\dfrac{\partial\psi'(t_0)}{\partial A_2} - \psi'(t_0)\dfrac{\partial\phi'(t_0)}{\partial A_2} \\[2mm] \phi'(t_0) & \dfrac{\partial\phi(t_0)}{\partial A_1} & \dfrac{\partial\phi(t_0)}{\partial A_2} \\[2mm] \psi'(t_0) & \dfrac{\partial\psi(t_0)}{\partial A_1} & \dfrac{\partial\psi(t_0)}{\partial A_2} \end{vmatrix}$$

$$= \{\phi'(t_0)\psi''(t_0) - \psi'(t_0)\phi''(t_0)\}\left\{\frac{\partial\phi(t_0)}{\partial A_1}\frac{\partial\psi(t_0)}{\partial A_2} - \frac{\partial\phi(t_0)}{\partial A_2}\frac{\partial\psi(t_0)}{\partial A_1}\right\}$$
$$- \left\{\phi'(t_0)\frac{\partial\psi'(t_0)}{\partial A_1} - \psi'(t_0)\frac{\partial\phi'(t_0)}{\partial A_1}\right\}\omega(t_0)$$
$$+ \left\{\phi'(t_0)\frac{\partial\psi'(t_0)}{\partial A_2} - \psi'(t_0)\frac{\partial\phi'(t_0)}{\partial A_2}\right\}\chi(t_0).$$

The first line of the last right-hand side

$$= \left\{\phi'(t_0)\frac{\partial\psi(t_0)}{\partial A_1} - \psi'(t_0)\frac{\partial\phi(t_0)}{\partial A_1}\right\}\left\{\phi''(t_0)\frac{\partial\psi(t_0)}{\partial A_2} - \psi''(t_0)\frac{\partial\phi(t_0)}{\partial A_2}\right\}$$
$$- \left\{\phi'(t_0)\frac{\partial\psi(t_0)}{\partial A_2} - \psi'(t_0)\frac{\partial\phi(t_0)}{\partial A_2}\right\}\left\{\phi''(t_0)\frac{\partial\psi(t_0)}{\partial A_1} - \psi''(t_0)\frac{\partial\phi(t_0)}{\partial A_1}\right\}$$
$$= \left\{\phi''(t_0)\frac{\partial\psi(t_0)}{\partial A_2} - \psi''(t_0)\frac{\partial\phi(t_0)}{\partial A_2}\right\}\chi(t_0) - \left\{\phi''(t_0)\frac{\partial\psi(t_0)}{\partial A_1} - \psi''(t_0)\frac{\partial\phi(t_0)}{\partial A_1}\right\}\omega(t_0);$$

and therefore the whole of the right-hand side

$$= \chi(t_0) \frac{d}{dt_0} \left\{ \phi'(t_0) \frac{\partial \psi(t_0)}{\partial A_2} - \psi'(t_0) \frac{\partial \phi(t_0)}{\partial A_2} \right\}$$

$$- \omega(t_0) \frac{d}{dt_0} \left\{ \phi'(t_0) \frac{\partial \psi(t_0)}{\partial A_1} - \psi'(t_0) \frac{\partial \phi(t_0)}{\partial A_1} \right\}$$

$$= \chi(t_0) \frac{d\omega(t_0)}{dt_0} - \omega(t_0) \frac{d\chi(t_0)}{dt_0}$$

$$= \frac{C}{P_0},$$

by § 62. Now $C$ is a finite non-zero constant; and $P_0$, the value of $P$ at the beginning of the range, is finite and is of the same sign whatever value $t_0$ is initially chosen. The foregoing equations therefore give

$$\kappa T_0 \frac{C}{P_0}, \qquad \kappa a_1 \frac{C}{P_0}, \qquad \kappa a_2 \frac{C}{P_0},$$

as expressions that are integral functions of $X_0$, $Y_0$, $\epsilon$, beginning with linear terms in these three small quantities; and $\frac{C}{P_0}$ is finite, being different from zero.

Thus $\kappa a_1$, $\kappa a_2$, $\kappa T_0$ are small, and therefore the new characteristic is consecutive. We infer that a characteristic consecutive curve can be drawn as required.

### Characteristic curves through any two assigned points.

**70.** The limitation of the range of integration, so that it shall not extend as far as the conjugate of its initial point, has an important bearing on another issue, already established.

It has been seen that a characteristic curve is uniquely determined by an initial point and an initial direction, as data which yield the values of the arbitrary constants in the primitive of the characteristic equation; and no question arose concerning the range of the curve so determined. There is an implicit limitation in the theorem quoted as leading to the unique establishment of the curve; the theorem is an existence-theorem, and the validity depends upon the convergence of the power-series employed—that is, partly upon the so-called radius of convergence of that series.

Consider a range along the characteristic curve, extending from a point to its conjugate. By the definition of the conjugate, a consecutive characteristic curve can be drawn through the two points; if therefore we substitute (as often is done) a requirement, that the curve shall pass through two given points, for the requirement, that it have an initial position and direction, the characteristic curve would not be unique if the two points could be conjugate on any curve through them.

The insistence on the Jacobi test, that limits the range, excludes the admission of the possibility indicated; when the test is satisfied, the characteristic curve determined by the initial data remains unique for the substituted data.

To the discussion of this equivalent requirement we now proceed.

**71.** The property of conjugates upon a characteristic curve—which, in its analytical form, secures that the function $Z(t, t_0)$ of § 63 does not vanish within a range bounded by $t_0$ and its conjugate on the curve—enables us to compare the two modes of settling the curve. In one of these modes, which is the expression of the conditions used to secure a precise and unique primitive of the characteristic equation, the essential arbitrary constants in the general primitive are made specific, by the assignment of any initial arbitrary point on the curve and of any initial arbitrary direction through that point as determining the tangent there. In the other of the modes, the essential arbitrary constants are made specific, by requiring the curve to pass through two assigned arbitrary points. In the former mode, the constants are actually determined by the data. The latter mode is more descriptive in character; in order to complete the precise expression of the primitive, equations (not always even algebraical in form, because the general primitive is not always or often algebraical) have to be resolved. As is seen in the example of the catenary (§ 30, Ex. 1: § 75, Ex. 1), the resolution of the equations may provide results, which may be satisfactory, but among which discrimination is not made without further examination if there is more than one result; or the resolution of the equations may prove initially impossible in terms of real quantities. The case of failure, as to the catenary, is treated by reviewing an initial assumption that the curve necessarily is a catenary and that, in the assumption, a zero value for the arbitrary constant is necessarily excluded. The real significance of the difference is that the two modes of determination are not always equivalent in all circumstances; and the property of the function $Z(t, t_0)$ shews why the second mode of determination does not necessarily lead to the same result as the first.

**72.** To compare the modes, let $A$ be the initial point common to both; and let $B$ be the final point in the second mode. For the initial arbitrary direction through $A$, let the straight line $AB$ be taken; then, in its earlier course at least, the characteristic curve $K$ lies on one side of the line $AB$. With $B$ as centre and gradually increasing radius, let a circle be drawn until it touches $K$, say at a point $C$; and along $CB$ take a point $C'$ near $C$, so that $CC'$ is finite though small. Along $K$, let the value of $t$ at $A$ be $t_0$ and its value at $C$ be $T$. Then (§ 68) we can draw a consecutive characteristic through $A$ and $C'$, provided the function $Z(T, t_0)$ is not zero nor a small quantity of the same order of magnitude as the small length $C'B$. Thus, if $C$ is not near $A'$, the conjugate of $A$, and if it lies within $AA'$, a consecutive characteristic curve $AC'$ can be drawn; and the distance of $B$ from $AC'$ is less than the distance of $B$ from $AC$.

Now take $AC'$ as the first characteristic to be drawn. As before, with $B$ as centre and gradually increasing radius, let a circle be drawn until it touches the new curve $K'$, say at a point $D$. In $DB$, take a point $D'$ near $D$, so that $DD'$ is finite though small: and let the circle centre $B$ and radius $BD$ cut $CB$ in $C''$. We now can draw a characteristic curve through $A$ and $C''$, which is consecutive to $AC'$; and it is always possible to draw the new consecutive curve, provided $C'$ is not near $A''$, the conjugate of $A$ along the curve $AC'$, and provided $C'$ lies within the range $AA''$, bounded by the initial point $A$ and its conjugate. And the new curve, still a characteristic curve, passes nearer $B$ than does $AC'$, the distance from the preceding curve being finite though small.

We thus may pass from curve to curve in succession, each curve being drawn nearer to $B$ than its predecessor. The one persistent requirement, exacted as the condition of drawing a consecutive curve at each stage, is that the function $Z(t, t_0)$ must never be zero nor (at any stage) be of the same order as the small (though finite) distance taken nearer to $B$ as a point for that consecutive characteristic. When therefore the field is such that, at no stage in drawing the succession of characteristics, a conjugate of $A$ is approached within a small distance, we can pass from the initial arbitrarily drawn characteristic to a characteristic, still having $A$ as its initial point, and passing through the second point $B$. It follows that, when $B$ lies within a region of the field bounded by the locus of the conjugate of $A$ for different characteristic curves, a characteristic can certainly be drawn to pass through the two points $A$ and $B$ which are given initially.

It thus appears that, within certain limits as to the range of $B$ depending on the nature of the characteristic equation, the two modes of rendering the general primitive a precise curve are equivalent to one another, while, outside that range, their equivalence cannot be asserted*.

### Composite small variations.

**73.** Finally, one inference from the general result is to be noted. It adds nothing to the conditions, and it does not limit them. In character, its interest is not immediate but prospective, as regards the subsequent analysis in the discussion of relative maxima and minima (Chapter VIII).

In all the variations which have been imposed, we have dealt with only a single general arbitrary small variation of $x$ and of $y$ represented by $\kappa u$ and $\kappa v$. All our results have been deduced from the consideration of that single variation, quite general and quite arbitrary within the limits of continuity.

* An illustration of the possible range of the conjugate of an initial point, for varying initial directions, is given in § 75, Ex. 3, for the catenary.

But a small variation may be compounded of a number of small variations, equally general, and equally arbitrary; thus we could have

$$\kappa_1 u_1 + \kappa_2 u_2 + \ldots + \kappa_n u_n, \quad \kappa_1 v_1 + \kappa_2 v_2 + \ldots + \kappa_n v_n,$$

as a combination of arbitrary small variations, $\kappa_1, \ldots, \kappa_n$ being small arbitrary constants, and* $u_1$ and $v_1$, $u_2$ and $v_2$, $\ldots$, $u_n$ and $v_n$, being regular functions of $t$. It is not inconceivable that a combination of two such variations, distinct from one another, might have some combined discriminating effect upon the second variation, which is quadratic in the quantities $u$ and $v$ and their derivatives, though not upon first variation, which is linear in those elements.

To estimate the significance of the possibility, we note that the quantities $x$ and $y$, constituting the primitive of the characteristic equations, do not depend upon $\kappa$, $u$, $v$ and are not affected by them. The same remark applies to $\xi$ and $\eta$, constituting the primitive of the subsidiary characteristic equations; and it therefore applies also to $\zeta$ which is defined as $x_1 \eta - y_1 \xi$. If then we write

$$w_r = x_1 v_r - y_1 u_r,$$

(for $r = 1, \ldots, n$), denote the foregoing composite variations by $\kappa u$ and $\kappa v$ (where, if we please, we may take $\kappa$ equal to any multiple of $\kappa_1 + \ldots + \kappa_n$) and write

$$w = x_1 v - y_1 u,$$

we have

$$\kappa w = \kappa_1 w_1 + \ldots + \kappa_n w_n.$$

Now $\kappa w$, so defined, is an arbitrary small continuous variation; and all the results, that have been obtained, have been established for $\kappa w$. They hold for each variation taken by itself, such as $\kappa u_r$ and $\kappa v_r$, because it is merely a special instance of $\kappa u$ and $\kappa v$. That they hold for two or more such variations taken simultaneously, and that the requirements of such simultaneous variations lead to no additional conditions, can be seen at once from the full expression for the change of the integral. This change, due to the small variation $\kappa u$ and $\kappa v$, has been expressed in the form

$$\kappa \int \mathfrak{E} w \, dt + \tfrac{1}{2} \kappa^2 \int P \left( w_1 - \frac{z_1}{\zeta} w \right)^2 dt + R_3,$$

where $R_3$ is the aggregate of terms of the third and higher degrees in $\kappa$. Now $\kappa w = \sum_{r=1}^{n} \kappa_r w_r$, and $\zeta$ (with its derivative) is independent of the arbitrary constants $\kappa$; hence the change of the integral is equal to

$$\Sigma \left( \kappa_r \int \mathfrak{E} w_r \, dt \right) + \tfrac{1}{2} \int P \left\{ \frac{d}{dt} \Sigma \left( \kappa_r w_r' \right) - \frac{z_1}{\zeta} \Sigma \left( \kappa_r w_r \right) \right\}^2 dt + K_3.$$

Because $\mathfrak{E} = 0$, the first variation vanishes. The expression of the second variation takes account of the simultaneous variations $\kappa u_r$ and $\kappa v_r$. It remains

---

* For the immediate purpose, the subscript indices merely denote distinct variations and not derivatives of a single variation.

steadily of one sign when the Legendre test and the Jacobi test are satisfied. No new test, to meet the composite variation, is required in order that the second variation should remain of one sign.

Hence the analytical results, holding for one arbitrary variation, remain unamplified if any number of simultaneous variations are imposed.

## Summary of results.

**74.** We may now summarise the conditions which have been obtained as necessary for the existence of a maximum or a minimum in an integral

$$\int F(x, y, x_1, y_1)\, dt,$$

when the variables $x$ and $y$ are subject to weak variations of the general type. These necessary conditions correspond exactly to the conditions obtained for the special weak variations. Both of the variables are now subject to the changes, so that the form of the conditions differs from the form of conditions for the simpler type; and other properties have arisen incidentally which lie outside the considerations governing the former case.

The conditions still consist of three tests:

I. The *Euler* test: the characteristic equation $\mathfrak{E} = 0$, or the two equations $X = 0$ and $Y = 0$, must be satisfied by $x$ and $y$ as functions of $t$; and their values determine a characteristic curve (§§ 42, 44, 49).

II. The *Legendre* test: a quantity $P$, being the common value of

$$\frac{1}{x_1^2}\frac{\partial^2 F}{\partial y_1^2}, \qquad -\frac{1}{x_1 y_1}\frac{\partial^2 F}{\partial x_1 \partial y_1}, \qquad \frac{1}{y_1^2}\frac{\partial^2 F}{\partial x_1^2},$$

must keep an invariable sign throughout the range of the integral and may not vanish, the sign being positive for a minimum and negative for a maximum (§§ 44, 56).

III. The *Jacobi* test: the range of the integral must not extend over the whole stretch of variation represented by the arc of the characteristic curve between a point and its conjugate (§§ 64, 65).

Incidentally, it was shewn that the function $F$ satisfies the identity

$$x_1\frac{\partial F}{\partial x_1} + y_1\frac{\partial F}{\partial y_1} = F,$$

and that the second derivatives of $F$ are expressible in terms of a diminished number of magnitudes (§§ 37, 38, 51): that, if the limits of the integral are not given as fixed, they are determined by boundary conditions (§ 43): that, if the characteristic curve (supposed to be otherwise everywhere definite in direction) suffers a discontinuity of direction at a single free place, the quantities $\dfrac{\partial F}{\partial x_1}$ and $\dfrac{\partial F}{\partial y_1}$ are continuous at the place (§ 47); and properties of consecutive characteristic curves, useful in connection with the Jacobi test, have been established (§§ 68, 69).

Moreover, the three necessary tests, if satisfied, are sufficient to secure a maximum or a minimum under weak variations; but strong variations remain for consideration. The first (Euler) test, together with the boundary conditions, secures that the first variation of the integral vanishes. The second (Legendre) test and the third (Jacobi) test, taken together, secure that the second variation cannot vanish; according as the sign of this second variation is positive or is negative, a minimum or a maximum respectively is possessed by the integral.

*Note.* If it should happen that the Jacobi test is just not satisfied because the range extends to the conjugate, the second variation may vanish specially. The discrimination then can be made (as in the example on p. 33) by the examination of the third variation of the integral for that special variation of the characteristic curve.

**75.** Some examples will now be considered.

*Ex.* 1. To illustrate the preceding analysis, we consider again the problem of the catenoid already (§ 30, Ex. 1) discussed, to determine *the plane curve which, by revolution round the axis of x in the plane, provides the surface of minimum area.*

(A, i). As the integral is

$$2\pi \int y \, (1+y'^2)^{\frac{1}{2}} \, dx, \quad =2\pi \int y \, (x_1{}^2+y_1{}^2)^{\frac{1}{2}} \, dt,$$

we have

$$F=F(x, y, x_1, y_1)=y \, (x_1{}^2+y_1{}^2)^{\frac{1}{2}},$$

which is at once seen to verify the permanent identity of § 37. As $F$ does not involve $x$ explicitly, we choose the equation $X=0$ to obtain the characteristic curve, as being the easier one to integrate. It is

$$\frac{d}{dt} \left( \frac{\partial F}{\partial x_1} \right) = 0,$$

and therefore

$$\frac{\partial F}{\partial x_1} = c,$$

that is,

$$x_1 y \, (x_1{}^2+y_1{}^2)^{-\frac{1}{2}} = c :$$

or, if we take $\frac{y_1}{x_1} = \tan \psi$ as before, we have

$$y = c \sec \psi,$$

the intrinsic equation of the characteristic curve, a catenary. The variable $t$, even in implicit occurrence, has disappeared.

(A, ii). If we proceed from the equation $\mathfrak{E} = 0$, we here have

$$-\frac{x_1}{(x_1{}^2+y_1{}^2)^{\frac{1}{2}}} + (x_1 y_2 - y_1 x_2) \, \frac{y}{(x_1{}^2+y_1{}^2)^{\frac{3}{2}}} = 0,$$

that is,

$$y \, (x_1 y_2 - y_1 x_2) - x_1 \, (x_1{}^2+y_1{}^2) = 0.$$

But, always,

$$\frac{d}{dt} \left( \frac{y_1}{x_1} \right) = \frac{x_1 y_2 - y_1 x_2}{x_1{}^2},$$

so that

$$y \frac{d}{dt} \left( \frac{y_1}{x_1} \right) = \frac{1}{x_1} (x_1{}^2+y_1{}^2)$$

$$= y_1 \frac{x_1}{y_1} \left\{ 1 + \left( \frac{y_1}{x_1} \right)^2 \right\}$$

for the equation: or if $y_1 = x_1 \tan \psi$, this is

$$\frac{y_1}{y} = \psi_1 \tan \psi,$$

and therefore

$$y \cos \psi = c.$$

If we proceed from the equation $Y = 0$, which here is

$$(x_1{}^2 + y_1{}^2)^{\frac{1}{2}} - \frac{d}{dt}\left\{\frac{-yy_1}{(x_1{}^2 + y_1{}^2)^{\frac{1}{2}}}\right\} = 0,$$

this becomes

$$-\frac{d}{dt}(y \sin \psi) + \frac{y_1}{\sin \psi} = 0,$$

that is,

$$-y \cos \psi \cdot \psi_1 - y_1 \sin \psi + \frac{y_1}{\sin \psi} = 0;$$

and therefore

$$y_1 = y \psi_1 \tan \psi,$$

with the same result.

(A, iii). The only limitation, imposed upon the independent variable $t$, is that it shall be capable of continuous increase through the range. When account is taken of the form of the curve, and the positive direction of the tangent lies in the direction of increasing $x$, the variable $\psi$ is always increasing—it may begin by being negative. The variable $t$ can be changed arbitrarily, subject to its one limitation; so, as $\sec \psi$ is always positive for the range and is never numerically less than unity, we take

$$\sec \psi = \cosh T.$$

Then $\tan \psi = \sinh T$, and $y_1 = cT' \sinh T$; hence $x_1 = cT'$, and so we have

$$x = a + cT,$$
$$y = \quad c \cosh T,$$

where the function $T$ is to increase continually with $t$, but otherwise is arbitrary. Manifestly the simplest form is given by

$$\left.\begin{array}{l} x = a + ct \\ y = c \cosh t \end{array}\right\},$$

where an arbitrary constant implicitly additive to $t$ (as in § 49) is absorbed in $t$, and where the two essential arbitrary constants are $a$ and $c$.

(B). As regards the Legendre test, we have

$$\frac{\partial^2 F}{\partial y_1{}^2} = \frac{yx_1{}^2}{(x_1{}^2 + y_1{}^2)^{\frac{3}{2}}},$$

that is,

$$P = y (x_1{}^2 + y_1{}^2)^{-\frac{3}{2}},$$

the positive sign being taken for the radical $(x_1{}^2 + y_1{}^2)^{\frac{1}{2}}$ throughout the discussion. Thus $P$ is always positive. The Legendre test is satisfied; if other tests are satisfied, it allows a minimum.

(C). Next, for the Jacobi test. With the notation of the text, we have

$$a + ct = x = \phi(t, a, c), \qquad c \cosh t = y = \psi(t, a, c),$$

where $a$ does not occur in $\psi$. Hence

$$\chi(t) = \phi'(t)\frac{\partial \psi}{\partial a} - \psi'(t)\frac{\partial \phi}{\partial a} = -c \sinh t,$$

$$\omega(t) = \phi'(t)\frac{\partial \psi}{\partial c} - \psi'(t)\frac{\partial \phi}{\partial c} = c(\cosh t - t \sinh t);$$

and therefore

$$Z(t, t_0) = \omega(t_0)\chi(t) - \chi(t_0)\omega(t)$$
$$= c^2\{\sinh t_0(\cosh t - t\sinh t) - \sinh t(\cosh t_0 - t_0\sinh t_0)\},$$

which manifestly vanishes when $t = t_0$. For the conjugate of $t_0$, we require the first value of $t$ which, being greater than $t_0$, makes $Z(t, t_0)$ vanish. Now

$$\frac{dZ}{dt} = -c^2\cosh t\,(t\sinh t_0 + \cosh t_0 - t_0\sinh t_0),$$

the initial value of which is the negative quantity $-c^2\cosh^2 t_0$. Hence $Z(t, t_0)$, zero at $t_0$, becomes negative at once as $t$ increases from $t_0$; as it remains finite for finite values of $t$, it continues negative until the value $t_1$, the conjugate of $t$.

If $t_0$ is positive, then $\sinh t_0\sinh t$ (with the range $t > t_0$) cannot vanish; therefore, as

$$Z(t, t_0) = c^2\sinh t_0\sinh t\{\coth t - t - (\coth t_0 - t_0)\},$$

$Z$ can only vanish for a value of $t$, greater than $t_0$, such that

$$\coth t - t = \coth t_0 - t_0.$$

Within this range, $\coth t - t$ initially is equal to $\coth t_0 - t_0$, and is always decreasing because its derivative is negative, and it remains finite for all finite values of $t$; hence it cannot again resume its initial value. Consequently if $t_0$ be positive, there is no conjugate with a value of $t > t_0$: that is, no forward conjugate.

When $t_0$ is negative, $Z(t, t_0)$ becomes negative at once as $t$ increases from $t_0$; when $t$ is equal to 0, then $Z(t, t_0)$ is equal to $c^2\sinh t_0$, that is, it still is negative. Denoting the negative value of $t_0$ by $-t'$, the place $t_1$, where $Z$ first vanishes after the beginning of the range, is given by

$$\coth t_1 - t_1 = t' - \coth t'.$$

Whether $t' - \cosh t'$ is negative or positive, this equation gives one (and only one) positive root $t_1$: and it gives no negative root. (A negative value of $t' - \coth t'$ gives $B$ in the figure: a positive value gives $A$ in the figure.) The tangent to the curve at $t'$ is

$$Y - c\cosh t' = (X - a - ct')\sinh t',$$

so that the abscissa of the intersection of the tangent with the axis of $x$ is $a + c(t' - \coth t')$, the same as the abscissa of the intersection of the tangent at $t_1$ with that axis; the conjugate of the initial point $A$ (or $D$) is obtained by drawing the other tangent through $U$ (or $V$) to the curve. The construction of the conjugate is thus the same as in the former investigation (§ 30, Ex. 1).

When the initial point and its conjugate are symmetrically situated with regard to the axis, we have $t_1 = t'$; and the equation is

$$\coth t' = t',$$

the former limiting case.

(D). For an effective solution by the preceding analysis (with $c$ real, and different from zero), there must be a real catenary. Hence the initial data must be examined, to indicate whether they allow a real catenary; manifestly they admit no such curve if, for example, it is to join two points on opposite sides of the axis. The only alternative left is that $c$ should be zero, when the foregoing analysis does not apply; and the alternative must be considered independently.

We therefore take the case when the curve is to join two points on the same side of the axis; and we shall assume now (as distinct from the assumption in § 30, p. 31) that the points are not equidistant from the directrix. Taking the axis of $y$ midway between the points, we denote these by $(k, a)$ and $(h, -a)$; so that, as

$$x = a + ct, \qquad y = c \cosh t,$$

we have

$$k = \tfrac{1}{2}c \left( e^{\frac{a-a}{c}} + e^{-\frac{a-a}{c}} \right),$$

$$h = \tfrac{1}{2}c \left( e^{-\frac{a+a}{c}} + e^{\frac{a+a}{c}} \right),$$

being two equations for the determination of $c$ and $a$. We have

$$k e^{\frac{a}{c}} = \tfrac{1}{2}c \left( e^{2\frac{a}{c}} e^{-\frac{a}{c}} + e^{\frac{a}{c}} \right),$$

$$h e^{-\frac{a}{c}} = \tfrac{1}{2}c \left( e^{-\frac{2a}{c}} e^{-\frac{a}{c}} + e^{\frac{a}{c}} \right),$$

so that

$$\tfrac{1}{2}c \left( e^{2\frac{a}{c}} - e^{-2\frac{a}{c}} \right) e^{-\frac{a}{c}} = k e^{\frac{a}{c}} - h e^{-\frac{a}{c}},$$

$$\tfrac{1}{2}c \left( e^{2\frac{a}{c}} - e^{-2\frac{a}{c}} \right) e^{\frac{a}{c}} = -k e^{-\frac{a}{c}} + h e^{\frac{a}{c}};$$

and therefore, on eliminating $a$,

$$E(c) = \tfrac{1}{4}c^2 \left( e^{2\frac{a}{c}} - e^{-2\frac{a}{c}} \right)^2 + k^2 + h^2 - kh \left( e^{2\frac{a}{c}} + e^{-2\frac{a}{c}} \right) = 0,$$

an equation for the determination of $c$. For the existence of a real catenary or real catenaries, this equation must have a real root or real roots.

(E). Consider the march of $E(c)$, as the positive quantity $c$ ranges between zero and infinity. Let

$$cu = 2a,$$

so that the positive quantity $u$ ranges between infinity and 0; then

$$E(c) = \tfrac{1}{2}c^2 (\cosh 2u - 1) + (k - h)^2 - 2kh (\cosh u - 1).$$

Hence, when $c$ is very large and therefore $u$ is very small, $E(c)$ differs only slightly from

$$4a^2 + (k - h)^2,$$

and thus is positive. For any finite non-zero value of $c$ and therefore of $u$, $E(c)$ is finite. As $c$ becomes small and therefore $u$ becomes large, $E(c)$ grows large and is positive; and a zero value of $c$ makes $E(c)$ infinite. Hence as $E(0)$ is positive and $E(\infty)$ is positive, the number of real roots of $E(c)$, for $0 < c < \infty$, must be even and may be zero.

To find whether there are any real roots, we investigate the minimum values (if any) of $E(c)$; and so we need the roots of $E'(c)$. From the first expression for $E(c)$, we have

$$E'(c) = \left( e^{\frac{2a}{c}} - e^{-\frac{2a}{c}} \right) \left\{ \tfrac{1}{2}c \left( e^{\frac{2a}{c}} - e^{-\frac{2a}{c}} \right) - a \left( e^{\frac{2a}{c}} + e^{-\frac{2a}{c}} \right) + \frac{2ahk}{c^2} \right\}.$$

The factor $2 \sinh \dfrac{2a}{c}$ is finite when $0 < c < \infty$; so possible zeros of $E'(c)$ are the possible zeros of

$$G(c) = \tfrac{1}{2}c \left( e^{\frac{2a}{c}} - e^{-\frac{2a}{c}} \right) - a \left( e^{\frac{2a}{c}} + e^{-\frac{2a}{c}} \right) + \frac{2ahk}{c^2}$$

$$= 2au^2 \left\{ \frac{hk}{4a^2} - \tfrac{1}{3} + u^2 \left( \frac{1}{5!} - \frac{1}{4!} \right) + u^4 \left( \frac{1}{7!} - \frac{1}{6!} \right) + u^6 \left( \frac{1}{9!} - \frac{1}{8!} \right) + \cdots \right\}.$$

If then $\dfrac{hk}{4a^2} - \tfrac{1}{3}$ is negative, the quantity $G(c)$ is steadily negative for all the admissible values of $u$; and therefore $E'(c)$ also is steadily negative. In that case, $E(c)$ decreases continually from $\infty$ to $4a^2 + (k-h)^2$, and cannot vanish: $E(c)$ then has no real root. There is no real catenary.

If $\dfrac{hk}{4a^2} - \tfrac{1}{3}$ is positive, $G(c)$ is positive for a small value of $u$. As $u$ increases, the value of $G(c)$ decreases. Before $u^2$ attains the value $30\left(\dfrac{hk}{4a^2} - \tfrac{1}{3}\right)$, the quantity $G(c)$ has become negative; and thereafter it remains negative, as $u$ continues to increase. Thus $G(c)$ has one zero in the range in this event, and therefore $E'(c)$ has one zero. Thus $E(c)$ has a stationary value; a maximum is excluded by the range of values of $E(c)$; and therefore there is one value of $c$, which gives $E(c)$ a minimum value.

When this minimum value of $E(c)$ is positive, $E(c)$ never vanishes. There then is no real catenary.

When this minimum value of $E(c)$ is negative, let it arise for a value of $c$ denoted by $c_1$. Then $E(c)$ changes sign once for

$$0 < c < c_1,$$

from positive to negative; and it changes sign once for

$$c_1 < c < \infty,$$

from negative to positive. In this event, $E(c)$ has two roots. There are two real catenaries.

When the minimum value of $E(c)$ is zero, we have a single root of $E(c)$ at that place. There is one real catenary.

We must certainly have $hk > \tfrac{1}{3}a^2$ for the possibility of any real catenary. Though $G(c)$, and therefore $E'(c)$, then has one real root, it may not lead to a real root of $E(c)$. The existence of a real root, or of real roots, will depend upon the magnitude of the quantity $hk - \tfrac{1}{3}a^2$, necessarily positive.

(F). As an illustration, consider a case

$$hk = 4a^2.$$

The instance, when $k = h$, has already (§ 30, Ex. 1, p. 31) been considered; so we take $k$ and $h$ unequal, e.g.

$$k = \tfrac{9}{4}a, \qquad h = \tfrac{16}{9}a.$$

We now have

$$G(c) = 2a\left(u^2 + \frac{\sinh u}{u} - \cosh u\right);$$

and it is easy to verify that

    (i)   when $u$ is small, $G(c)$ is positive,

    (ii)  when $u^2 = 12\tfrac{1}{2}$, $G(c) = \cdot 014\ldots$,

    (iii)  when $u^2 = 13$, $G(c) = -\cdot 015\ldots$,

    (iv)  as $u^2$ increases further, $G(c)$ increases numerically and remains negative.

Hence $G(c)$ has one and only one real root, lying between $(\tfrac{100}{325})^{\frac{1}{2}}a$ and $(\tfrac{104}{325})^{\frac{1}{2}}a$; thus $E(c)$ has one minimum, and the minimum occurs for that root. It is easy to verify that, when $u^2 = 12\tfrac{1}{2}$, $E(c)$ is negative; hence $E(c)$ has one, and only one, root, such that

$$0 < c < (\tfrac{104}{325})^{\frac{1}{2}}a,$$

and it has one, and only one, root, such that

$$\left(\tfrac{104}{325}\right)^{\frac{1}{2}} a < c < \infty .$$

By actual calculation, we find the following:

(i)   $u^2 = 2^{\frac{1}{2}},$    $E(c)$ is positive;

(ii)   $u^2 = 2$ ,    $E(c)$ is negative;

(iii)   $u^2 = 18,$    $E(c)$ is negative;

(iv)   $u^2 = 20,$    $E(c)$ is positive.

We thus have two real catenaries.  As $a$ is not zero, their vertices do not lie on the same ordinate.

A reference to the Jacobi test shews that the catenary range $HV_2K$ provides a minimum, and that the catenary range $HV_1K$ does not provide a minimum.

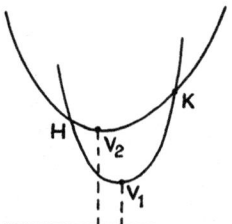

(G). We still have to consider the possibility that no real catenary can be drawn; (i) when the two points are on the same side of the directrix, but the equation $E(c)=0$ provides no real root : (ii) when the two points lie on opposite sides of the directrix.

The characteristic equation still has

$$\frac{\partial F}{\partial x_1} = \frac{x_1 y}{(x_1{}^2 + y_1{}^2)^{\frac{1}{2}}} = \text{constant}$$

as a first integral.  In the preceding analysis, the constant has been assumed to be different from zero.

We now take the alternative hypothesis that the constant is zero; and so we now have

$$\frac{\partial F}{\partial x_1} = \frac{x_1 y}{(x_1{}^2 + y_1{}^2)^{\frac{1}{2}}} = 0.$$

For a point on this curve off the axis of $x$, so that $y$ is not zero, we have $x_1 = 0$; that is, we must have parts $HM$ and $KN$ when the limiting points $H$ and $K$ are on the same side of $Ox$, and parts $HM$ and $LP$ when the limiting points $H$ and $L$ are on opposite sides.  From $M$ to $N$ in the one case, and from $M$ to $P$ in the other, the equation is satisfied by $y = 0$, that is, that part

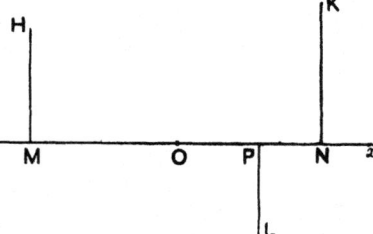

of the axis of $x$ is a portion of the curve.  Thus the broken line $HMNK$, or the broken line $HMPL$, provides the solution of the characteristic equation.

The points $M$ and $N$ (or $M$ and $P$) are 'free' points, in the sense that they are not fixed by assigned conditions; hence (§ 47) $\dfrac{\partial F}{\partial x_1}$ and $\dfrac{\partial F}{\partial y_1}$ remain unchanged at $M$ and $N$, (or at $M$ and $P$), in passing suddenly from one direction to the other.  But

$$\frac{\partial F}{\partial x_1} = \frac{x_1 y}{(x_1{}^2 + y_1{}^2)^{\frac{1}{2}}}, \qquad \frac{\partial F}{\partial y_1} = \frac{y y_1}{(x_1{}^2 + y_1{}^2)^{\frac{1}{2}}} = \frac{y}{\left(\dfrac{x_1{}^2}{y_1{}^2} + 1\right)^{\frac{1}{2}}} .$$

At $M$ for $HM$, we have $x_1 = 0$ and $y = 0$; so both these quantities vanish.  At $M$ for $MN$, we have $y = 0$ and $y_1 = 0$; so both these quantities vanish.  Hence the requirement is satisfied at $M$.  Similarly it is satisfied at $N$ (or at $P$).

For the Legendre test, we have

$$P = \frac{1}{y_1^2} \frac{\partial^2 F}{\partial x_1^2} = \frac{y}{(x_1^2 + y_1^2)^{\frac{3}{2}}},$$

with the positive sign for the radical, which is positive along $HM$ and $KN$: while, along $MN$, the contribution of a necessarily non-negative quantity to the original integral (composed of otherwise positive quantities) is zero and therefore cannot be reduced. Thus the test for a minimum is satisfied.

For the Jacobi test, there is no conjugate of an initial point on a straight line because it cannot be met again by a consecutive line through that point.

*Ex.* 2. Tangents, drawn to a catenary from a point $T$ on its directrix, meet it in the points $P$ and $Q$. Prove that any number of catenaries can be drawn having the same directrix and touching the lines $TP$ and $TQ$, and that all their vertices lie on a straight line through $T$.

Denoting the points of contact of one of these catenaries with $TP$ and $TQ$ by $P'$ and $Q'$ respectively, shew that $Q'$ lies within $TQ$ when $P'$ lies within $TP$.

Prove also that the surface engendered by the revolution (round the directrix) of the curve made up of the straight line $PP'$, the catenary arc $P'Q'$, and the straight line $Q'Q$, is equal to the surface engendered by the revolution (round the directrix) of the catenary arc $PQ$, and also is equal to the surface engendered by the revolution (round the directrix) of the curve made up of the straight lines $PT$ and $TQ$.

*Ex.* 3. Through a point $A$ on the axis $Oy$, catenaries are drawn having $Ox$ for directrix; and $OA = y_0$. Prove that the locus of the conjugate of $A$ for the different catenaries is given by the equations

$$\left. \begin{array}{l} \dfrac{\sinh^2 \theta}{\theta} \left( 1 - \dfrac{\coth \theta}{\theta} \right)^{\frac{1}{2}} = \dfrac{y - y_0}{x} \\[2ex] \dfrac{\cosh^2 \theta}{\theta} \left( 1 - \dfrac{\tanh \theta}{\theta} \right)^{\frac{1}{2}} = \dfrac{y + y_0}{x} \end{array} \right\},$$

where $\theta$ is a parameter increasing from sec $33° 32'$.

*Ex.* 4. Consider the *geodesics on a sphere* of radius $a$.

The coordinates of any point on the sphere are

$$x = a \sin \theta \cos \phi, \quad y = a \sin \theta \sin \phi, \quad z = a \cos \theta,$$

so that $\phi$ is the longitude and $\theta$ is the angular distance from the North Pole. As

$$ds^2 = a^2 (d\theta^2 + \sin^2 \theta \, d\phi^2),$$

the integral

$$\int (\theta'^2 + \phi'^2 \sin^2 \theta)^{\frac{1}{2}} \, dt$$

must be a minimum, so that

$$F = (\theta'^2 + \phi'^2 \sin^2 \theta)^{\frac{1}{2}}.$$

As $F$ does not involve $\phi$, a first integral of the characteristic equation is given by

$$\frac{\partial F}{\partial \phi'} = \text{constant},$$

that is,

$$\frac{\phi' \sin^2 \theta}{(\theta'^2 + \phi'^2 \sin^2 \theta)^{\frac{1}{2}}} = \sin a,$$

where $a$ is an arbitrary constant. Hence, for the characteristic curve,

$$d\phi = \frac{\sin a}{\sin \theta (\sin^2 \theta - \sin^2 a)^{\frac{1}{2}}} \, d\theta.$$

For reality, $\sin \theta$ cannot be less than $\sin a$; and therefore (regard being paid to the meaning of $\theta$), we must have $a \leqslant \theta \leqslant \pi - a$. Let

$$\cos \theta = \cos a \cos \psi \; ;$$

then

$$d\phi = \frac{\sin a}{1 - \cos^2 a \cos^2 \psi}\, d\psi = \frac{\sin a}{\sin^2 a + \tan^2 \psi}\, \sec^2 \psi \, d\psi,$$

and therefore

$$\tan \psi = \sin a \tan (\phi - \beta),$$

so that

$$\cos (\phi - \beta) = \frac{\tan a}{\tan \theta},$$

an integral equation of the characteristic curve. When this last equation is modified by the introduction of point-coordinates, it becomes

$$x \cos \beta + y \sin \beta = z \tan a,$$

shewing that the curve lies in a plane through the centre of the sphere: that is, it is a great circle.

The least and the greatest polar distances are given by $\theta' = 0$, that is, $\theta = a$ and $\theta = \pi - a$. The meridian through the pole of this geodesic great circle has $\beta$ for its longitude.

The *Legendre* test is satisfied; for

$$\frac{1}{\phi'^2}\frac{\partial^2 F}{\partial \theta'^2} = -\frac{1}{\theta' \phi'}\frac{\partial^2 F}{\partial \theta' \partial \phi'} = \frac{1}{\theta'^2}\frac{\partial^2 F}{\partial \phi'^2}, \quad = \frac{\sin^2 \theta}{(\theta'^2 + \phi'^2 \sin^2 \theta)^{\frac{3}{2}}},$$

is always positive; and therefore a great circle, along its range, admits a minimum.

The *Jacobi* test requires that the extent of the range, to admit a minimum, shall end before the limiting position of the intersection of a consecutive characteristic (here, a consecutive great circle) through the initial point. This intersection is diametrically opposite to this initial point; hence, for a minimum, the range must be less than half the great circle.

In the figure, let $N$ be the North Pole; $AP$ the geodesic great circle having $A$ for its point of highest latitude; and $NA$ the meridian through the pole of $NP$, so that $NAP$ is a right angle. Then

$$NA = a, \quad NP = \theta, \quad PNA = \phi - \beta, \quad AP = \psi;$$

the formulæ

$$\cos \theta = \cos a \cos \psi, \quad \tan \psi = \tan a \tan (\phi - \beta), \quad \cos (\phi - \beta) \tan \theta = \tan a,$$

are the usual formulæ of a right-angled triangle. Also, if $s$ be the arc measured from $A$,

$$\psi = \frac{s}{a}.$$

*Ex.* 5. *To find the geodesics on a surface of revolution.*

We take the axis of revolution to be the axis of $z$; and the surface itself to be given by the equation

$$x^2 + y^2 = 2f(z),$$

where $f$ is a given function of $z$. Any point on the surface can be represented by $r \cos \theta$, $r \sin \theta$, $z$, where $r^2 = 2f$; hence

$$\begin{aligned} ds^2 &= dx^2 + dy^2 + dz^2 \\ &= dr^2 + r^2 d\theta^2 + dz^2 \\ &= \left\{ \left(1 + \frac{f'^2}{2f}\right) z_1^2 + 2f\theta_1^2 \right\} dt^2. \end{aligned}$$

Thus

$$\int \left\{ \left(1 + \frac{f'^2}{2f}\right) z_1{}^2 + 2f\theta_1{}^2 \right\}^{\frac{1}{2}} dt$$

is to be a minimum. (Obviously a maximum is excluded from consideration.) We have

$$F(\theta, z, \theta_1, z_1) = \left\{ 2f\theta_1{}^2 + \left(1 + \frac{f'^2}{2f}\right) z_1{}^2 \right\}^{\frac{1}{2}}.$$

Of the two characteristic equations, we select

$$\frac{\partial F}{\partial \theta} - \frac{d}{dt}\left(\frac{\partial F}{\partial \theta_1}\right) = 0.$$

As $\theta$ does not occur explicitly in $F$, we have

$$\frac{\partial F}{\partial \theta_1} = \text{constant},$$

that is,

$$\frac{2f\theta_1}{\left\{ 2f\theta_1{}^2 + \left(1 + \frac{f'^2}{2f}\right) z_1{}^2 \right\}^{\frac{1}{2}}} = \text{constant};$$

and therefore

$$\frac{d\theta}{dz} = \frac{a}{2f}\left(\frac{2f + f'^2}{2f - a^2}\right)^{\frac{1}{2}},$$

where $a$ is an arbitrary constant. Thus the primitive is

$$\theta - a = \tfrac{1}{2} a \int \left(\frac{2f + f'^2}{2f - a^2}\right)^{\frac{1}{2}} \frac{dz}{f},$$

which must be associated with the equation $r^2 = 2f$, in order to provide the analytical equations of the characteristic curve.

For the Legendre test, we have

$$P = \frac{1}{z_1{}^2} \frac{\partial^2 F}{\partial \theta_1{}^2} = \frac{-1}{z_1 \theta_1} \frac{\partial^2 F}{\partial \theta_1 \partial z_1} = \frac{1}{\theta_1{}^2} \frac{\partial^2 F}{\partial z_1{}^2}$$

$$= \frac{2f + f'^2}{\left\{ \left(1 + \frac{f'^2}{2f}\right) z_1{}^2 + 2f\theta_1{}^2 \right\}^{\frac{3}{2}}}.$$

As $2f$ is necessarily positive (being the value of $r^2$), and as the positive sign is given to the radical, this quantity $P$ is always positive. Hence, if other necessary conditions are satisfied, the Legendre test admits a minimum.

For the Jacobi test, we have

$$\frac{\partial \theta}{\partial a} = 1,$$

$$\frac{\partial \theta}{\partial a} = \int \frac{(2f + f'^2)^{\frac{1}{2}}}{(2f - a^2)^{\frac{3}{2}}} dz.$$

If, then, we take the initial value of $z$ to be $z_0$, the range of the curve is limited by the conjugate (if any) given by a value $z_1$, such that

$$\int_{z_0}^{z_1} \frac{(2f + f'^2)^{\frac{1}{2}}}{(2f - a^2)^{\frac{3}{2}}} dz = 0,$$

or by the first of such values of $z_1$ greater than $z_0$, if there be more than one.

(A). Consider a *cylinder* given by $r=c$, a constant; thus $2f(z)=c^2$, and $f'=0$. The characteristic curve is given by

$$\theta - a = \frac{az}{c(c^2-a^2)^{\frac{1}{2}}},$$

together with $r=c$. The curve is a helix. For a real curve, $a$ must be less than $c$, say $a=c\cos\beta$; and then

$$\theta = a + \frac{z}{c}\cot\beta.$$

An analytical (but not geometrical) difficulty arises from this form of the equation. When $a$ and $\beta$ are to be determined by the initial and final positions, the initial position given by $\theta_1$ (though determinate geometrically) is indeterminate, because $\theta_1$ can be made subject to any additive multiple of $2\pi$; and similarly for the final position given by $\theta_2$. Hence there may be any number of helices on the cylinder through two assigned points: if $\vartheta_1, > 0 < 2\pi$, be the smallest value of $\theta_1$ for the initial position, and if $\vartheta_2, > 0 < 2\pi$, be the smallest value of $\theta_2$ for the final position, then

(i) if $\vartheta_2 > \vartheta_1$, while $z_1$ and $z_2$ are the initial and the final values of $z$, we have an unlimited number of values of $\cot\beta$, given by

$$\cot\beta = \frac{c}{z_2-z_1}(\vartheta_2-\vartheta_1+2p\pi),$$

where $p$ is any positive or negative whole number; and

(ii) if $\vartheta_2 < \vartheta_1$, the values of $\cot\beta$ are given by

$$\cot\beta = \frac{c}{z_2-z_1}(2\pi+\vartheta_2-\vartheta_1+2p\pi),$$

with the like significance for $z_1$, $z_2$, $p$.

Each value of $p$ gives a value to $\beta$, the inclination of the curve to the cross (circular) section of the cylinder.

The Legendre test for a minimum is satisfied for all the forms of $f$, and so is satisfied for the cylinder.

For the Jacobi test, we have

$$\frac{\partial\theta}{\partial a}=1, \quad \frac{\partial\theta}{\partial\beta}=-\frac{z}{c\sin^2\beta};$$

the critical quantity $\dfrac{\partial\theta}{\partial\beta} \div \dfrac{\partial\theta}{\partial a}$, beginning at a place given by $z_0$, never resumes its initial value along any helix: that is, along any selected helix, there is no limit. Each such helix provides a minimum; small variations of each helix provide an increase in the length of the curve. There are minima, each corresponding to each value of $p$; these minima are the numerical values of

$$\frac{z_2-z_1}{\sin\beta},$$

and the smallest of all the minima is provided by the largest numerical value of $\sin\beta$, among all the values of $\beta$, that is, by

$$\cot\beta' = c\frac{\vartheta_2 \sim \vartheta_1}{z_2 \sim z_1},$$

with $0 \leqslant \beta' \leqslant \frac{1}{2}\pi$.

(B). Consider a *sphere* given by

$$x^2+y^2+z^2=c^2,$$

so that $2f=c^2-z^2$. The first integral of the characteristic equation

$$\frac{d\theta}{dz} = \frac{ac}{(c^2-z^2)(c^2-a^2-z^2)^{\frac{1}{2}}},$$

the integral of which can be expressed in the form

$$(c^2 - a^2)^{\frac{1}{2}} \sin (\theta - a) = \frac{az}{(c^2 - z^2)^{\frac{1}{2}}}.$$

But $c^2 - z^2 = r^2$; and so this equation can be expressed in the form

$$(c^2 - a^2)^{\frac{1}{2}} r \sin (\theta - a) = az,$$

or, if $a = c \cos \beta$, this is

$$z = (y \cos a - x \sin a) \tan \beta.$$

The characteristic curve lies in a plane through the centre of the sphere: that is, it lies along a great circle.

The Legendre test is always satisfied for any form of the function $J$, and so is satisfied for the sphere: that is, other conditions being satisfied, the great circle admits a geodesic.

There remains the Jacobi test. In the foregoing equation, $a$ and $\beta$ are arbitrary constants, or $a$ and $a$ are the arbitrary constants. The primitive of the characteristic equation is

$$(c^2 - a^2)^{\frac{1}{2}} \sin (\theta - a) = \frac{az}{(c^2 - z^2)^{\frac{1}{2}}},$$

so that

$$\frac{\partial \theta}{\partial a} = 1, \quad (c^2 - a^2)^{\frac{1}{2}} \cos (\theta - a) \frac{\partial \theta}{\partial a} = \frac{z}{(c^2 - z^2)^{\frac{1}{2}}}.$$

Thus

$$\frac{\dfrac{\partial \theta}{\partial a}}{\dfrac{\partial \theta}{\partial a}} = \frac{c^2}{a (c^2 - a^2)} \tan (\theta - a);$$

and therefore the critical quantity, $\dfrac{\partial \theta}{\partial a} \div \dfrac{\partial \theta}{\partial a}$, resumes the value which it has at any initial place $\theta_1$, first when $\theta$ acquires the value $\theta_1 + \pi$, that is, at the other extremity of the diameter through the initial point. Thus, for a minimum, the range along a great circle is limited to half the circumference.

The result can be obtained by drawing (§ 65) a consecutive characteristic, that is, a consecutive great circle, through the initial point: the conjugate, being the next intersection of the two great circles, is the diametrically opposite point.

*Ex.* 6. Use the foregoing analysis (p. 106) relating to geodesics on a surface of revolution, to obtain the result (§ 34, Ex. 8) for a *paraboloid* $x^2 + y^2 = 2z$.

*Ex.* 7. Discuss the geodesics on a one-sheeted *hyperboloid* of revolution.

*Ex.* 8. Shew that geodesics (other than meridians) on an *oblate spheroid* undulate between two parallels of latitude equidistant from the equator.

Shew that, if any point on the surface be denoted by $z = c \cos \theta$, $(x^2 + y^2)^{\frac{1}{2}} = a \sin \theta$, the conjugate $\theta_1$ of a place $\theta_0$ is given by the equation

$$E (u_1 - u_0) = \frac{c^2 (u_1 - u_0)}{a^2 \cos^2 a + c^2 \sin^2 a} + \frac{\operatorname{sn} (u_1 - u_0)}{\operatorname{sn} u_1 \operatorname{sn} u_0},$$

where $a < \theta < \pi - a$, am $u_1 = \theta_1$, am $u_0 = \theta_0$, the modulus of the elliptic functions is $(a^2 - c^2)^{\frac{1}{2}} (a^2 \cos^2 a + c^2 \sin^2 a)^{-\frac{1}{2}} \cos a$, and $E$ denotes the second elliptic integral.

*Ex.* 9. Prove that a geodesic on the spheroid

$$\frac{x^2 + y^2}{a^2} + \frac{z^2}{b^2} = 1$$

can (in the notation of the Weierstrass elliptic functions) be represented by the equations

$$x + iy = \kappa \, \frac{\sigma (a+u)}{\sigma (u) \, \sigma (a)} \, e^{u \{\eta - \zeta (\omega + a)\}},$$

$$x - iy = \kappa \, \frac{\sigma (a-u)}{\sigma (u) \, \sigma (a)} \, e^{-u \{\eta - \zeta (\omega + a)\}},$$

$$z^2 = \lambda^2 \, \frac{\sigma (\omega'' + u) \, \sigma (\omega'' - u)}{\sigma^2 (u) \, \sigma^2 (a)},$$

where $\kappa$ and $\lambda$ are constants: and obtain equations sufficient to specify the constants of the elliptic functions*.

Ex. 10.  Shew that a geodesic upon a circular cone of semi-vertical angle $a$ projects orthogonally on a plane perpendicular to the axis into a curve

$$r \sin \{(\theta - \beta) \sin a\} = c,$$

where $c$ and $\beta$ are arbitrary constants.  Discuss the position of the conjugate of any point on the geodesic.

Ex. 11.  A fundamental property of *geodesics on any surface*—viz. the radius of circular curvature at any point coincides, in direction, with the normal to the surface at any point —can be deduced from the characteristic equations which express the minimum property. (The property can be established by purely geometrical considerations, applied to the position of equilibrium of a tight string on a smooth surface or to the path of a particle moving on such a surface under no forces.)

Let the surface have the equation $z = f(x, y)$; and denote, as usual, the partial first derivatives of $f(x, y)$ by $p$ and $q$, and its partial second derivatives by $r$, $s$, and $t$. Denoting the length of the elementary arc of the geodesic by $d\sigma$, we have

$$d\sigma = \{(1+p^2)\, x_1{}^2 + 2pq\, x_1 y_1 + (1+q^2)\, y_1{}^2\}^{\frac{1}{2}} dt = F dt;$$

and we have to make $\int F dt$ a minimum.  The characteristic equations $X = 0$ and $Y = 0$ determine the form of the curve.

The equation $X = 0$ is

$$\frac{\partial F}{\partial x} - \frac{d}{dt}\left(\frac{\partial F}{\partial x_1}\right) = 0.$$

We write

$$x' = \frac{dx}{d\sigma}, \quad x'' = \frac{d^2 x}{d\sigma^2}, \quad y' = \frac{dy}{d\sigma}, \quad y'' = \frac{d^2 y}{d\sigma^2},$$

so that

$$x' = \frac{x_1}{\sigma_1}, \quad \sigma_1 x'' = \frac{dx'}{dt}, \quad y' = \frac{y_1}{\sigma_1}, \quad \sigma_1 y'' = \frac{dy'}{dt}.$$

Now

$$\frac{\partial F}{\partial x} = \frac{1}{\sigma_1} \{x_1{}^2 pr + x_1 y_1 (ps + qr) + y_1{}^2 qs\} = \sigma_1 \{pr\, x'^2 + (ps + qr)\, x' y' + qs\, y'^2\};$$

and

$$\frac{\partial F}{\partial x_1} = \frac{1}{\sigma_1} \{(1+p^2)\, x_1 + pq\, y_1\} = (1+p^2)\, x' + pq\, y',$$

so that

$$\frac{d}{dt}\left(\frac{\partial F}{\partial x_1}\right) = \sigma_1 \{(1+p^2)\, x'' + pq\, y''\} + x'\, 2p\, (rx_1 + sy_1) + y' \{q\, (rx_1 + sy_1) + p\, (sx_1 + ty_1)\}$$

$$= \sigma_1 \{(1+p^2)\, x'' + pq\, y''\}$$

$$+ \frac{1}{\sigma_1} \{2pr\, x_1{}^2 + (3ps + qr)\, x_1 y_1 + (qs + pt)\, y_1{}^2\}.$$

---

* For a full discussion of these forms, see Halphen, *Fonctions elliptiques*, t. ii, pp. 238—243.

Hence the characteristic equation $X = 0$ becomes

$$(1 + p^2)\, x'' + pq\, y'' + pr\, x'^2 + 2ps\, x'y' + pt\, y'^2 = 0.$$

But from $z = f(x,\, y)$, we have

$$z'' = px'' + qy'' + rx'^2 + 2s\, x'y' + ty'^2\,;$$

and so the foregoing equation is

$$x'' + pz'' = 0.$$

Similarly, the equation $Y = 0$ becomes

$$y'' + qz'' = 0.$$

Consequently, we have

$$\frac{x''}{p} = \frac{y''}{q} = \frac{z''}{-1}\,,$$

along the geodesic—two equations, which are the analytical expression of the enunciated property.

Two immediate inferences can be made:

(I) The orthogonal projection of the geodesic made upon the plane of $z = 0$ is given by the equation

$$\frac{d^2y}{dx^2} = \left( q - p\, \frac{dy}{dx} \right) \left\{ r - 2s\, \frac{dy}{dx} + t \left( \frac{dy}{dx} \right)^2 \right\}\,;$$

(II) A first integral of the characteristic equation of geodesics upon a surface of revolution can always be obtained. For if the surface is $x^2 + y^2 = 2G\,(z)$, where $G$ is any appropriate function, we have

$$x + G'\,(z)\, p = 0, \quad y + G'\,(z)\, q = 0.$$

Hence one geodesic equation becomes

$$\frac{x''}{x} = \frac{y''}{y}\,;$$

and therefore we have a first integral given by

$$xy' - yx' = A.$$

*Note.* The development of the properties of geodesics soon passes from the range of application of the present calculus and becomes merged into the theory of differential geometry. A full discussion will be found in Darboux's *Théorie générale des surfaces*, vol. ii, pp. 402—437, vol. iii, pp. 113—192.

As regards geodesics on surfaces of revolution, in particular on quadrics of revolution, Halphen's *Fonctions elliptiques*, t. ii, ch. vi, may be consulted; whence it appears that the integral expressions of geodesics on the central quadrics of revolution (other than a cylinder or cone) involve elliptic functions. The integral expressions of geodesics on oblate (terrene) spheroids have some analogy with the trigonometry of a right-angled spherical triangle, with the substitution of elliptic functions for circular functions; and geodesics on an ellipsoid are analytically expressible in terms of hyper-elliptic functions[*].

*Weierstrass's theorem that a range must terminate at the conjugate of its origin.*

**76.** But one further proposition can be established, by way of imposing the definite limit upon the range of integration.

Suppose that all the tests, necessary to secure a maximum or a minimum for a given integral under weak variations, are satisfied: that is, the Euler test, the Legendre test, and the Jacobi test. The last of these is satisfied

---

[*] Fuller references are given in the author's *Lectures on Differential Geometry*, ch. v.

when the range of the integral does not extend as far as the conjugate of the initial place on the characteristic curve. It has been pointed out that, if the range of integration extends as far as this conjugate, one set of non-zero variations can be obtained which would allow the second variation to vanish, even though that second variation would vanish for no others; and that, then, it is necessary to examine the third variation because, if this third variation should not vanish for the particular set of variations, the integral would not possess a true maximum or a true minimum for weak variations*.

It was also indicated that, if the range were to extend beyond the conjugate, the same necessity would certainly arise, because weak variations could be framed, non-zero up to the conjugate, and zero onwards, which equally would make the second variation vanish. Further detailed consideration of the latter possibility is unnecessary because of a theorem, due to Weierstrass, that *no maximum or minimum can exist if the range of integration extends beyond the conjugate of the initial point*: in other words, if the range of integration includes within itself a complete range, bounded by two conjugate points, and is not limited to that range. The theorem can be established as follows.

**77.** It has been proved (§ 52) that the second variation can be expressed in the form $\frac{1}{2}\kappa^2 I_2$, where

$$I_2 = \int_{t_0}^{t} \left\{ P\left(\frac{dw}{dt}\right)^2 + Gw^2 \right\} dt,$$

where $w = x_1 v - y_1 u$; and the variation $\kappa w$ is to vanish at both limits of the integral, account having been taken of behaviour at the limits in making the first variation vanish. Now

$$\int P\left(\frac{dw}{dt}\right)^2 dt = \left[ wP\frac{dw}{dt} \right] - \int w\frac{d}{dt}\left( P\frac{dw}{dt} \right) dt,$$

and the terms at the limits vanish; thus† we can take $I_2$ in the form

$$I_2 = \int_{t_0}^{t} w\left\{ Gw - \frac{d}{dt}\left( P\frac{dw}{dt} \right) \right\} dt.$$

But, denoting any constant by $\epsilon$, the sign of $\epsilon$ being settled to be the same as that of $P$ and its (small) magnitude being settled later, we have

$$I_2 = \int_{t_0}^{t} \left\{ (P+\epsilon)\left(\frac{dw}{dt}\right)^2 + (G+\epsilon)w^2 \right\} dt - \epsilon \int_{t_0}^{t} \left\{ \left(\frac{dw}{dt}\right)^2 + w^2 \right\} dt.$$

As before

$$\int_{t_0}^{t} (P+\epsilon)\left(\frac{dw}{dt}\right)^2 dt = \left[ w(P+\epsilon)\frac{dw}{dt} \right]_{t_0}^{t} - \int w\frac{d}{dt}\left\{ (P+\epsilon)\frac{dw}{dt} \right\} dt.$$

---

* An example was given (§ 30, p. 33) arising out of the catenoid.
† See also § 58.

The terms at the limits vanish; and so

$$I_2 = \int_{t_0}^{t} w \left[ (G + \epsilon) w - \frac{d}{dt} \left\{ (P + \epsilon) \frac{dw}{dt} \right\} \right] dt - \epsilon \int_{t_0}^{t} \left\{ \left( \frac{dw}{dt} \right)^2 + w^2 \right\} dt.$$

Thus far, all that is required of $w$ is that it shall vanish at each limit of the integral. Now let a displacement $w$, say $\overline{w}$, be chosen which

(i) vanishes at each limit:

(ii) everywhere along the curve satisfies the equation

$$(G + \epsilon) w - \frac{d}{dt} \left\{ (P + \epsilon) \frac{dw}{dt} \right\} = 0:$$

and

(iii) is not zero everywhere.

Such a displacement would make $I_2$ consist solely of the second integral for that variation and would make its sign, hitherto settled by the sign of the unchanging quantity $P$, to be the same as that of $-\epsilon$, that is, by the opposite sign.

The quantity $\overline{w}$ can be determined. The equation to be satisfied is

$$(P + \epsilon) \frac{d^2 \overline{w}}{dt^2} + \frac{dP}{dt} \frac{d\overline{w}}{dt} - (G + \epsilon) \overline{w} = 0.$$

When $\epsilon$ is made zero, $\overline{w}$ becomes the integral of the old subsidiary characteristic equation vanishing when $t = t_0$, and thus is the old quantity $Z(t, t_0)$ of § 63, where

$$Z(t, t_0) = \omega(t_0) \chi(t) - \chi(t_0) \omega(t).$$

It will be remembered that $Z(t, t_0)$, continuous through the range, vanishes at $t_1$ for the first time after having vanished at $t_0$. Let

$$\overline{w} = Z(t, t_0) + \epsilon \Omega + [\epsilon^2],$$

$\epsilon$ being small, but as yet undetermined in magnitude: $\Omega$ is independent of $\epsilon$, and is to vanish at $t_0$: and $[\epsilon^2]$ represents an aggregate of second and higher powers of $\epsilon$, which will be negligible compared with $\epsilon \Omega$. The equation for $\Omega$ is

$$P \frac{d^2 \Omega}{dt^2} + \frac{dP}{dt} \frac{d\Omega}{dt} - G\Omega = Z - \frac{d^2 Z}{dt^2}.$$

**78.** The complementary function in the integral of this equation is

$$\Omega = A\chi(t) + B\omega(t).$$

To find the particular integral, we use the method of variation of parameters, and therefore make $A$ and $B$ functions of $t$. Then

$$\frac{d\Omega}{dt} = A \frac{d\chi}{dt} + B \frac{d\omega}{dt},$$

provided $A$ and $B$ are such that

$$\chi(t) \frac{dA}{dt} + \omega(t) \frac{dB}{dt} = 0.$$

When we substitute in the equation, we have

$$A \left( P \frac{d^2\chi}{dt^2} + \frac{dP}{dt} \frac{d\chi}{dt} - G\chi \right) + B \left( P \frac{d^2\omega}{dt^2} + \frac{dP}{dt} \frac{d\omega}{dt} - G\omega \right)$$

$$+ P \left( \frac{d\chi}{dt} \frac{dA}{dt} + \frac{d\omega}{dt} \frac{dB}{dt} \right) = Z - \frac{d^2Z}{dt^2}.$$

But $\chi(t)$ and $\omega(t)$ satisfy the subsidiary characteristic equation; hence

$$P \left( \frac{d\chi}{dt} \frac{dA}{dt} + \frac{d\omega}{dt} \frac{dB}{dt} \right) = Z - \frac{d^2Z}{dt^2}.$$

Consequently

$$\frac{1}{-\omega(t)} \frac{dA}{dt} = \frac{1}{\chi(t)} \frac{dB}{dt} = \frac{1}{P} \frac{Z - \dfrac{d^2Z}{dt^2}}{-\omega \dfrac{d\chi}{dt} + \chi \dfrac{d\omega}{dt}} = \frac{1}{C} \left( Z - \frac{d^2Z}{dt^2} \right),$$

where (§ 62) $C$ is a non-vanishing specific constant. Hence

$$A = \alpha - \frac{1}{C} \int_{t_0}^{t} \left\{ Z(\theta, t_0) - \frac{d^2}{d\theta^2} Z(\theta, t_0) \right\} \omega(\theta)\, d\theta,$$

$$B = \beta + \frac{1}{C} \int_{t_0}^{t} \left\{ Z(\theta, t_0) - \frac{d^2}{dt^2} Z(\theta, t_0) \right\} \chi(\theta)\, d\theta,$$

where $\alpha$ and $\beta$ are arbitrary constants, so far as this analysis is concerned. Now

$$A\chi(t) + B\omega(t)$$

is to vanish when $t = t_0$; that is,

$$\alpha\chi(t) + \beta\omega(t)$$

must vanish when $t = t_0$. Consequently $\alpha\chi(t) + \beta\omega(t)$ is a mere multiple of $Z(t, t_0)$, which can be absorbed into $Z(t, t_0)$ by taking a constant multiple of $\chi(t)$ or $\omega(t)$ in place of the initial $\chi(t)$ or $\omega(t)$. Hence

$$\Omega = \frac{1}{C} \omega(t) \int_{t_0}^{t} \left\{ Z(\theta, t_0) - \frac{d^2}{dt^2} Z(\theta, t_0) \right\} \chi(\theta)\, d\theta$$

$$- \frac{1}{C} \chi(t) \int_{t_0}^{t} \left\{ Z(\theta, t_0) - \frac{d^2}{dt^2} Z(\theta, t_0) \right\} \omega(\theta)\, d\theta$$

$$= - \frac{1}{C} \int_{t_0}^{t} Z(t, \theta) \left\{ Z(\theta, t_0) - \frac{d^2}{dt^2} Z(\theta, t_0) \right\} d\theta,$$

a function of $t$ which vanishes when $t = t_0$.

**79.** As regards the range of the original integral we know that, when the Euler test and the Legendre test are satisfied, it certainly possesses a maximum or a minimum under weak variations, provided also the Jacobi test of not extending to the conjugate of the initial point $t_0$ is satisfied. When the range extends as far as this conjugate $t_1$, we have seen that further investigation is wanted at once for the third variation; and such investigation can be effected in each particular instance. Now suppose that the range extends to a value of $t$ greater than $t_1$, say up to $T$, where $T > t_1$.

As $Z(t, t_0)$ vanishes at $t_1$, consider two places $t_1 - \tau$ and $t_1 + \tau$, where $\tau$ is positive and $t_1 + \tau \leqslant T$; and to arrange for even a small extension of the range beyond $t_1$, we take $\tau$ small. We know (§ 66) that the characteristic, which is consecutive to the central curve and passes through $t_0$ and $t_1$, crosses that central curve at each of these points. As $t_1$ is the first point of crossing after $t_0$, the quantities $Z(t_1 - \tau, t_0)$ and $Z(t_0 + \kappa T_0, t_0)$, where $\kappa$ and $T_0$ are positive, have the same sign; that is,

$$\frac{d}{dt} [Z(t, t_0)]_{t=t_1}, \quad \left[\frac{d}{dt} Z(t, t_0)\right]_{t=t_0}$$

have opposite signs. Let the former be denoted by $Q$; the latter (§ 67) is $-\dfrac{C}{P_0}$, where $P_0$ is the value of $P$ at $t_0$, the sign of $P$ being steady through the range. Thus $Q$ and $C/P$ have the same sign. Hence

$$\bar{w}(t_1 - \tau) = -\tau Q + \epsilon \Omega(t_1 - \tau) + [\epsilon^2, \tau^2],$$
$$\bar{w}(t_1 + \tau) = \tau Q + \epsilon \Omega(t_1 + \tau) + [\epsilon^2, \tau^2].$$

Now choose $\tau$ arbitrarily, keeping it small and making it positive. We have chosen $\epsilon$ (our arbitrary constant, also small) of the same sign as $P$ throughout the range. Finally, choose $\epsilon$ so small that, through a range of $t$ from $t_1 - \tau$ to $t_1 + \tau$,

$$|\epsilon \Omega(t)| < \tau |Q|.$$

If $\Omega(t_1)$ is positive, then

when $Q$ is positive, $\epsilon \Omega(t) - \tau Q$ is negative in that range,
$\epsilon \Omega(t) + \tau Q$ is positive in that range: and

when $Q$ is negative, $\epsilon \Omega(t) + \tau Q$ is negative in the range,
$\epsilon \Omega(t) - \tau Q$ is positive in the range.

If $\Omega(t_1)$ is negative, the same results hold. Hence, in every case,

$$-\tau Q + \epsilon \Omega(t_1 - \tau), \quad \tau Q + \epsilon \Omega(t_1 + \tau),$$

have opposite signs; that is, there is some place $t''$, which lies between $t_1 - \tau$ and $t_1 + \tau$, where $\bar{w}(t)$ vanishes. This place $t''$ may lie in the range between $t_1 - \tau$ and $t_1$, or it may lie outside that range and between $t_1$ and $t_1 + \tau$.

We therefore take the range of our integral from $t_0$ up to the place $t''$; and to make certain of including $t''$, we assume that the original range of the integral extends beyond $t_1$ to $T$.

A function $\bar{w}$ has now been chosen which

(i) vanishes at $t_0$, the lower limit of the integral,

(ii) ..............$t''$,......upper........................,

(iii) is not zero everywhere,

(iv) satisfies the equation $(G + \epsilon)\bar{w} - \dfrac{d}{dt}\left\{(P + \epsilon)\dfrac{d\bar{w}}{dt}\right\} = 0$ everywhere

along the curve.

We thus have an admissible variation, conforming to all the earlier conditions, and satisfying* additional conditions; and, for this variation,

$$I_2 = -\epsilon \int_{t_0}^{t''} \left\{ \left( \frac{d\overline{w}}{dt} \right)^2 + \overline{w}^2 \right\} dt.$$

Hence this value of $I_2$ is of a sign, opposite to that of $\epsilon$; and the sign of $\epsilon$ was chosen the same as that of $P$; so that this value of $I_2$ is of a sign opposite to that of $P$. But for all the other small variations that have been considered, the sign of the corresponding second variation was of a sign the same as that of $P$. Thus, in the circumstance that the range of integration of the original integral extends beyond the conjugate of the lower limit, there are variations, some of which give one sign to $I_2$ and others of which give the reverse sign to $I_2$, while for all of them the first variation vanishes. Hence the integral does not possess a true maximum or minimum : that is, if the range of the integral extends beyond a complete range, bounded by two conjugates on the characteristic curve.

Consequently *the Jacobian range is a strict limit for the integral.*

80. The theorem is due to Weierstrass†. It proves that, if the range of the integral certainly extends beyond the conjugate of the lower limit, the integral does not possess a maximum or a minimum; but it provides no decision on the issue, if the range merely extends as far as the conjugate. A discussion of the question, when the integral ranges exactly between conjugates, will be found in a memoir by Osgood‡.

---

* The non-zero value of $\overline{w}$ along the curve is due to the fact that the quantity

$$Z(t, t_0) - \frac{d^2}{dt^2} Z(t, t_0)$$

is manifestly not zero everywhere along the curve.

† It is published in accounts of the subject, based upon notes taken at his Berlin lectures; *e.g.*, see Harris Hancock's *Calculus of Variations (the Weierstrassian Theory)*, 1904, Chapter X.

‡ *Trans. Amer. Math. Soc.*, vol. ii (1901), pp. 166—182.

# CHAPTER III.

Integrals involving derivatives of the second order: special weak variations, by the method of Jacobi; general weak variations, by the method of Weierstrass.

The chapter[*] is concerned with integrals which, in their simplest initial form, involve one dependent variable together with its first and second derivatives. Some propositions are stated for integrals, when derivatives up to the general order $n$ occur. Many extensions almost suggest themselves, so far as regards mere form; and frequently the establishment of these self-suggested extensions can be made by formal enlargement of the analysis actually given. Thus the analysis, similar to that in §§ 88—96 and applicable to integrals containing derivatives up to order $n$, can be constructed without difficulty.

The first portion, comprising §§ 81—96, contains the discussion for the special weak variations of the dependent variable alone. The method, used in Chapter I, is again used.

The second portion, comprising §§ 97—128, contains the discussion for the general weak variations of the dependent variable and the independent variable simultaneously. The method used in Chapter II originating with Weierstrass has been used, in a necessarily amplified form.

### Integrals involving second derivatives.

**81.** We now proceed to the determination of criteria for the possession of maxima and minima of integrals, which involve the second derivative as well as the first derivative of the single dependent variable. At this stage, only those variations will be considered which are of the 'weak' type. Also, initially, the weak variations admitted will be those styled 'special,' being variations that affect the dependent variable only, the independent variable $x$ being free from the imposed arbitrary variations. We thus, in the first place, deal with an integral of the *second order*, defined as

$$I = \int F(x, y, y', y'')\, dx,$$

where $y'$ and $y''$ respectively denote the first derivative and the second derivative of the dependent variable $y$.

The subject of integration $F$ is assumed to remain continuous, so long as its arguments $x$, $y$, $y'$, $y''$ are continuous. It need not be everywhere a uniform function of its arguments, as it may (*e.g.*) be a radical. But, if it is not uniform, we shall take only such a range as will admit no branch-points: that is, the selected radical will be uniform within the range, so that it may be considered as expansible in regular power-series of appropriate variables.

---

[*] The substance of the chapter has already appeared in a memoir by the author, *Proc. Roy. Soc. Edin.*, vol. xlvi, part ii (1926), pp. 149—193. In connection with the first portion of the chapter, the memoirs by Hesse and by Clebsch, already quoted (p. 5), may be consulted: these memoirs dealing with integrals containing the derivative of general order.

*Variation of the integral.*

**82.** As before, the special weak variation adopted consists in the substitution of $y + \kappa v$ for $y$—with the corresponding substitutions for the derivatives—where $\kappa$, as usual, is so small that any power, which occurs, is negligible compared with lower powers; and where $v$ is a function of $x$, which can be arbitrarily assigned subject to the requirement of being regular (§ 9) within a given range of the variable $x$, supposed to increase continuously through that range. The limits of the integral may be fixed definitely by assigned data, or they may be required merely to conform to assigned conditions: in the latter event, their determination is part of the problem. To take the wider possibility into account from the beginning, we shall assume that, when the whole variation is imposed, the upper limit of the integral becomes $x_2 + \kappa \xi_2$ instead of $x_2$, and its lower limit becomes $x_1 + \kappa \xi_1$ instead of $x_1$. Thus the new value of the integral is

$$\int_{x_1 + \kappa \xi_1}^{x_2 + \kappa \xi_2} F(x, y + \kappa v, y' + \kappa v', y'' + \kappa v'') \, dx$$

$$= \left\{ \int_{x_2}^{x_2 + \kappa \xi_2} - \int_{x_1}^{x_1 + \kappa \xi_1} + \int_{x_1}^{x_2} \right\} F(x, y + \kappa v, y' + \kappa v', y'' + \kappa v'') \, dx.$$

Now (as in § 32) the first integral $\int_{x_2}^{x_2 + \kappa \xi_2} F \, dx$ differs from $\kappa \xi_2 F_2$, where $F_2$ denotes the value of $F(x, y, y', y'')$ at $x_2$, by quantities of the second degree and higher degrees in $\kappa$; we therefore take it as

$$\kappa \xi_2 F_2 + K_2',$$

with the customary significance of $K_2'$ as the aggregate of terms of the second and higher degrees in $\kappa$. Similarly, the second integral $\int_{x_1}^{x_1 + \kappa \xi_1} F \, dx$ differs from $\kappa \xi_1 F_1$, where $F_1$ denotes the value of $F(x, y, y', y'')$ at $x_1$, by quantities of the second degree and higher degrees in $\kappa$; we therefore take it as

$$\kappa \xi_1 F_1 + K_2'',$$

with the customary significance of $K_2''$.

Hence the increment of the original integral is

$$\kappa (\xi_2 F_2 - \xi_1 F_1) + K_2' - K_2''$$

$$+ \int_{x_1}^{x_2} \{ F(x, y + \kappa v, y' + \kappa v', y'' + \kappa v'') - F(x, y, y', y'') \} \, dx.$$

But the last integral

$$= \kappa \int_{x_1}^{x_2} \left( v \frac{\partial F}{\partial y} + v' \frac{\partial F}{\partial y'} + v'' \frac{\partial F}{\partial y''} \right) dx + K_2''',$$

where $K_2'''$ denotes an aggregate of terms involving the second and higher powers of $\kappa$. Writing $K_2 = K_2' - K_2'' + K_2'''$, so that $K_2$ is the complete aggregate of terms of the second and higher powers of $\kappa$ that occur in the whole increment of the integral, we have that increment equal to

$$\kappa \left\{ \xi_2 F_2 - \xi_1 F_1 + \int_{x_1}^{x_2} \left( v \frac{\partial F}{\partial y} + v' \frac{\partial F}{\partial y'} + v'' \frac{\partial F}{\partial y''} \right) dx \right\} + K_2,$$

$$= \kappa \left( \xi_2 F_2 - \xi_1 F_1 + I_1 \right) + K_2,$$

where

$$I_1 = \int_{x_1}^{x_2} \left( v \frac{\partial F}{\partial y} + v' \frac{\partial F}{\partial y'} + v'' \frac{\partial F}{\partial y''} \right) dx.$$

The whole term in the increment, which involves the first power of $\kappa$, is called the *first variation* of the integral.

### Vanishing of the first variation: characteristic equation.

**83.** If the first variation does not vanish, the quantity $K_2$ is relatively unimportant in magnitude. In that event, we can make the increment of the integral positive or negative at will, solely by changing the sign of $\kappa$. If the integral is to have a maximum or a minimum value, that increment must have a persistent sign (negative for a maximum, positive for a minimum); hence the hypothesis admitting the former possibility must be excluded, that is, the first variation of the integral must vanish. We therefore proceed to make the first variation vanish, for all variations which are possible: that is, for arbitrary small variations $\kappa v$ of $y$ throughout the range, and for terminal small variations as arbitrary as the conditions permit.

Now, with the assumptions as to the function $F$, we have

$$\frac{d}{dx}\left( v \frac{\partial F}{\partial y'} \right) - v \frac{d}{dx}\left( \frac{\partial F}{\partial y'} \right) = v' \frac{\partial F}{\partial y'}; \quad \frac{d}{dx}\left\{ v' \frac{\partial F}{\partial y''} - v \frac{d}{dx}\left( \frac{\partial F}{\partial y''} \right) \right\} = v'' \frac{\partial F}{\partial y''} - v \frac{d^2}{dx^2}\left( \frac{\partial F}{\partial y''} \right);$$

and therefore

$$I_1 = \int_{x_1}^{x_2} v \left\{ \frac{\partial F}{\partial y} - \frac{d}{dx}\left( \frac{\partial F}{\partial y'} \right) + \frac{d^2}{dx^2}\left( \frac{\partial F}{\partial y''} \right) \right\} dx + \left[ v \left\{ \frac{\partial F}{\partial y'} - \frac{d}{dx}\left( \frac{\partial F}{\partial y''} \right) \right\} + v' \frac{\partial F}{\partial y''} \right],$$

where the integral extends over the range, and the terms outside the sign of integration are to be taken at the limits. Hence the first variation of the original integral is

$$\kappa \int_{x_1}^{x_2} v \left\{ \frac{\partial F}{\partial y} - \frac{d}{dx}\left( \frac{\partial F}{\partial y'} \right) + \frac{d^2}{dx^2}\left( \frac{\partial F}{\partial y''} \right) \right\} dx$$

$$+ \kappa \left[ \xi F + v \left\{ \frac{\partial F}{\partial y'} - \frac{d}{dx}\left( \frac{\partial F}{\partial y''} \right) \right\} + v' \frac{\partial F}{\partial y''} \right];$$

and it is to vanish for all small variations, admissible within the range and at the limits.

First, consider variations which vanish at the limits but otherwise are arbitrary through the range. The terms at the limits then vanish, because $\xi$, $v$, $v'$ vanish at each limit; and the integral must then vanish also for all such variations which remain arbitrary within the range. This result can be attained only if the equation

$$\mathfrak{E} = \frac{\partial F}{\partial y} - \frac{d}{dx}\left(\frac{\partial F}{\partial y'}\right) + \frac{d^2}{dx^2}\left(\frac{\partial F}{\partial y''}\right) = 0$$

is satisfied everywhere along the range. Otherwise, if $\mathfrak{E}$ were sometimes positive and sometimes negative, a choice of $v$ to have the same sign as $\mathfrak{E}$ would make the integral positive, and a different choice of $v$ so as to have its sign different from that of $\mathfrak{E}$ would make the integral negative: that is, the integral could be made positive or negative at will, a possibility that is precluded.

Next, consider variations which do not vanish at the limits. Now that $\mathfrak{E}$ vanishes all along the range so that the integral in the first variation vanishes, and because the first variation as a whole vanishes, the terms at the limits must vanish; hence we must have

$$\left[\xi F + v\left\{\frac{\partial F}{\partial y'} - \frac{d}{dx}\left(\frac{\partial F}{\partial y''}\right)\right\} + v'\frac{\partial F}{\partial y''}\right] = 0.$$

This requirement may be satisfied automatically by assigned data. If not thus satisfied, it must be satisfied by inferences conforming to assigned conditions, affecting the limits only and determining the limits.

**84.** This critical equation (called the *characteristic equation*) $\mathfrak{E} = 0$ usually is of the fourth order, and not linear in $y$ and the derivatives of $y$. The only term involving the highest derivative of $y$ is

$$y''''\frac{\partial^2 F}{\partial y''^2},$$

a fact of importance in connection with the character of the primitive because of the property to be imposed later (§ 95) that $\dfrac{\partial^2 F}{\partial y''^2}$ is never to vanish within the range of the integral. The primitive is therefore of the form

$$y = y(x, A_1, A_2, A_3, A_4),$$

where $A_1, A_2, A_3, A_4$ are four arbitrary independent constants. When the form of the second variation (§ 88), as a function of $v, v', v''$, is considered in detail, this value of $y$ (with the consequent values of $y'$ and $y''$) must be substituted in the six coefficients $a_{00}, a_{01}, a_{02}, a_{11}, a_{12}, a_{22}$, which then can be regarded as functions of $x$ alone.

*Ex.* 1. Prove that, if
$$F(x, y, y', y'') = y'' f(x, y, y') + g(x, y, y'),$$
the equation $\mathfrak{E} = 0$ is only of the second order in $y$.

*Ex.* 2. Prove that, if the equation $\mathfrak{E} = 0$ is an actual identity and not a differential equation, then $F(x, y, y', y'')$ must be equal exactly to a quantity
$$y''\frac{\partial G}{\partial y'} + y'\frac{\partial G}{\partial y} + \frac{\partial G}{\partial x},$$
where $G$ can be any function of $x, y, y'$.

*Subsidiary characteristic equation: its primitive.*

**85.** The characteristic equation is satisfied by the foregoing value of $y$ for all values of the arbitrary constants $A_1$, $A_2$, $A_3$, $A_4$. It therefore still will be satisfied, when we effect any small variation upon these constants by substituting

$$A_1 + \kappa a_1, \quad A_2 + \kappa a_2, \quad A_3 + \kappa a_3, \quad A_4 + \kappa a_4$$

for $A_1$, $A_2$, $A_3$, $A_4$ respectively, the new constants $a_1$, $a_2$, $a_3$, $a_4$ themselves being arbitrary and independent. The constant $\kappa$ is so small that second and higher powers are negligible in value compared with the first power; hence, if

$$\eta = a_1 \frac{\partial y}{\partial A_1} + a_2 \frac{\partial y}{\partial A_2} + a_3 \frac{\partial y}{\partial A_3} + a_4 \frac{\partial y}{\partial A_4},$$

the consequent value of $y$ is $y + \kappa \eta + \ldots$, where the unexpressed part involves second and higher powers of $\kappa$. Thus $\kappa \eta$ can be regarded as a small variation of $y$, the value of $\eta$ being immediately derivable when the general value of $y$ is known.

As the fundamental change has originated in the arbitrary constants, for any set of which the characteristic equation is satisfied, that equation must remain invariable and characteristic, though individual parts are altered. The new values $\dfrac{\partial F}{\partial y}$, $\dfrac{\partial F}{\partial y'}$, $\dfrac{\partial F}{\partial y''}$ are

$$\frac{\partial F}{\partial y} + \kappa \frac{\partial \Omega}{\partial \eta} + \ldots, \quad \frac{\partial F}{\partial y'} + \kappa \frac{\partial \Omega}{\partial \eta'} + \ldots, \quad \frac{\partial F}{\partial y''} + \kappa \frac{\partial \Omega}{\partial \eta''} + \ldots,$$

where the unexpressed parts involve the second and higher powers of $\kappa$, and where

$$2\Omega = a_{00} \eta^2 + 2a_{01} \eta \eta' + 2a_{02} \eta \eta'' + a_{11} \eta'^2 + 2a_{12} \eta' \eta'' + a_{22} \eta''^2.$$

Consequently the new value of $\mathfrak{E}$ is

$$\mathfrak{E} + \kappa \left\{ \frac{\partial \Omega}{\partial \eta} - \frac{d}{dx} \left( \frac{\partial \Omega}{\partial \eta'} \right) + \frac{d^2}{dx^2} \left( \frac{\partial \Omega}{\partial \eta''} \right) \right\} + \ldots,$$

with the same significance as to unexpressed terms. The characteristic equation still is satisfied; and therefore

$$\kappa \left\{ \frac{\partial \Omega}{\partial \eta} - \frac{d}{dx} \left( \frac{\partial \Omega}{\partial \eta'} \right) + \frac{d^2}{dx^2} \left( \frac{\partial \Omega}{\partial \eta''} \right) \right\} + \ldots = 0.$$

Hence, as $\kappa$ becomes indefinitely small, this last equation is

$$\frac{\partial \Omega}{\partial \eta} - \frac{d}{dx} \left( \frac{\partial \Omega}{\partial \eta'} \right) + \frac{d^2}{dx^2} \left( \frac{\partial \Omega}{\partial \eta''} \right) = 0$$

in the limit.

This new equation in $\eta$, the coefficients in which are $a_{00}$, $a_{01}$, $a_{02}$, $a_{11}$, $a_{12}$, $a_{22}$ (all of them are functions of $x$), is linear and of the fourth order. Its primitive therefore involves four constants, arbitrary so far as the equation is concerned. But we have seen that

$$\eta = a_1 \frac{\partial y}{\partial A_1} + a_2 \frac{\partial y}{\partial A_2} + a_3 \frac{\partial y}{\partial A_3} + a_4 \frac{\partial y}{\partial A_4};$$

this expression for $\eta$ involves four arbitrary independent constants, $a_1, a_2, a_3, a_4$, which do not occur in the equation. If therefore the four quantities $\frac{\partial y}{\partial A_1}, \frac{\partial y}{\partial A_2}, \frac{\partial y}{\partial A_3}, \frac{\partial y}{\partial A_4}$ are such, that no linear relation subsists between them involving only constant coefficients, the foregoing expression for $\eta$ provides the primitive of the differential equation for $\eta$. This property, of linear independence among the four derivatives of $y$, will be established later (§ 112) and will be assumed in the meanwhile.

Thus we have a new equation derived from the characteristic equation; and its primitive can be derived from the primitive of the characteristic equation. Accordingly, the equation in $\eta$ is called the *subsidiary characteristic* equation, sometimes (more briefly) the *subsidiary* equation.

When the primitive of the characteristic equation is known, the primitive of the subsidiary equation can be derived without further integration. If however the primitive of the characteristic equation is not known, but only some integral (perhaps a very special integral) is known which may lead to a maximum or minimum, this special integral is to be substituted in the six coefficients $a_{00}, a_{01}, a_{02}, a_{11}, a_{12}, a_{22}$; and the subsidiary characteristic equation remains. Its primitive, which cannot now be obtained by the former process, must be obtained by the customary processes of integration.

In fact, to complete the analysis up to this stage, it is always necessary to integrate one of the two equations of the fourth order—the non-linear characteristic equation or the linear subsidiary characteristic equation; for, if the former has not been integrated in general and only a special integral is known, the latter cannot be used in order to construct the primitive of the subsidiary equation

### *Terminal conditions : four cases.*

**86.** Next, it is necessary to consider the inferences to be derived from the condition at the limits, as represented by the relation

$$\left[ \xi F + v \left\{ \frac{\partial F}{\partial y'} - \frac{d}{dx} \left( \frac{\partial F}{\partial y''} \right) \right\} + v' \frac{\partial F}{\partial y''} \right] = 0,$$

to be satisfied in connection with all variations.

We can select one variation, which leaves one extremity unchanged: that is, the position of the extremity and the direction of the extremity (given by the values of $y$ and $y'$) are fixed; so $\xi$, $v$, $v'$ are zero there. Hence we must have

$$\xi F + v \left\{ \frac{\partial F}{\partial y'} - \frac{d}{dx} \left( \frac{\partial F}{\partial y''} \right) \right\} + v' \frac{\partial F}{\partial y''} = 0$$

at the other extremity.

Similarly, we can select another variation not rigidly settled at the first extremity. For the second extremity, the condition just deduced is already required. Hence the same condition, now still remaining at the first extremity, is to be satisfied there.

Hence, as an initial (but not completely detailed) form of requirement, *the relation*

$$\xi F + v \left\{ \frac{\partial F}{\partial y'} - \frac{d}{dx} \left( \frac{\partial F}{\partial y''} \right) \right\} + v' \frac{\partial F}{\partial y''} = 0$$

*must be satisfied at each extremity.*

Various detailed forms emerge, according to the variety of character in the assigned terminal data. As the relation is to be satisfied at each extremity by itself, without reference to the other extremity, the data (and the consequent condition or conditions) are not necessarily the same for the two extremities. The result, appropriate to the data at one extremity, must be associated with the result appropriate to the other.

(i) Let fixed values be assigned to one extremity (the case really has already been considered).

We then have $\xi = 0$, $v = 0$, $v' = 0$ for that extremity; the relation is satisfied, without leaving any residual analytical condition.

(ii) Let the terminal values be required to satisfy a given relation

$$\chi(x, y, y') = 0,$$

this relation being the only restriction at the extremity. Now, at that extremity, we have supposed $x$ varied into $x + \kappa \xi$; let the corresponding values of $y$ and $y'$, connected with this relation, be changed into $y + \kappa \eta$ and $y' + \kappa \eta'$. Then the sole condition imposed upon $\xi, \eta, \eta'$, because of the assigned requirement, is

$$\frac{\partial \chi}{\partial x} \xi + \frac{\partial \chi}{\partial y} \eta + \frac{\partial \chi}{\partial y'} \eta' = 0.$$

We have to obtain the relation between $v$ and $\eta$, and the relation between $v'$ and $\eta'$, each such relation possibly involving $\xi$.

Now the relation between $v$ and $\eta$ can be obtained exactly as for the simpler case discussed in § 33. We take the same figure as before. The curve $EPN$ represents a current relation between $x$ and $y$; the dotted curve $...PQ...$ represents a terminal relation between them. In that figure,

$$QN = \kappa \eta, \quad SP = QR = \kappa v, \quad PN = \kappa \xi;$$

and we have

$$\kappa\eta = \kappa v + y'.\,\kappa\xi,$$

that is,

$$v = \eta - y'\xi.$$

The same diagram, with a different significance for ordinates, leads to the relation between $v'$ and $\eta'$. We now take $Y = y'$, and regard the curve $EPN$ as a current relation between $x$ and $Y$; the dotted curve $...PQ...$ now is a terminal relation between $x$ and $Y$, that is, between $x$ and $y'$. For this representation,

$QN = $ variation of $Y$ along $PQ = \kappa\eta'$,

$SP = QR = $ variation of the curve $EPN$ along the ordinate $= \kappa v'$,

$PN = \kappa\xi$,

$RN = PN \tan RPN = \kappa\xi \,.\, Y' = \kappa\xi y''$;

and thus

$$\kappa\eta' = QN = QR + RN = \kappa v' + \kappa\xi y'',$$

that is,

$$v' = \eta' - y''\xi.$$

With these values for $v$ and $v'$ in terms of $\eta$, $\eta'$, $\xi$, the terminal condition

$$\xi F + v\left\{\frac{\partial F}{\partial y'} - \frac{d}{dx}\left(\frac{\partial F}{\partial y''}\right)\right\} + v'\frac{\partial F}{\partial y''} = 0,$$

for the integral, becomes

$$\xi\left[F - y'\left\{\frac{\partial F}{\partial y'} - \frac{d}{dx}\left(\frac{\partial F}{\partial y''}\right)\right\} - y''\frac{\partial F}{\partial y''}\right] + \eta\left\{\frac{\partial F}{\partial y'} - \frac{d}{dx}\left(\frac{\partial F}{\partial y''}\right)\right\} + \eta'\frac{\partial F}{\partial y''} = 0.$$

But the only condition, representing assigned requirements, is

$$\frac{\partial\chi}{\partial x}\xi + \frac{\partial\chi}{\partial y}\eta + \frac{\partial\chi}{\partial y'}\eta' = 0.$$

Hence we have, as inferences from the requirement at the extremity,

$$\left.\begin{array}{r}F - y'\left\{\dfrac{\partial F}{\partial y'} - \dfrac{d}{dx}\left(\dfrac{\partial F}{\partial y''}\right)\right\} - y''\dfrac{\partial F}{\partial y''} = \lambda\dfrac{\partial\chi}{\partial x}\\[2ex]\dfrac{\partial F}{\partial y'} - \dfrac{d}{dx}\left(\dfrac{\partial F}{\partial y''}\right) = \lambda\dfrac{\partial\chi}{\partial y}\\[2ex]\dfrac{\partial F}{\partial y''} = \lambda\dfrac{\partial\chi}{\partial y'}\end{array}\right\},$$

together with the terminal condition

$$\chi(x, y, y') = 0,$$

that is, after the elimination of $\lambda$, there are three equations associated with the extremity of the range in question.

(iii) Let an extremity of the range be required to lie upon a given curve

$$h(x, y) = 0,$$

and let the direction at the extremity be fixed.

Because of the latter requirement, we have

$$y' = \text{given constant},$$

so that $v' = 0$ at the extremity: and the terminal condition becomes

$$\xi F + v \left\{\frac{\partial F}{\partial y'} - \frac{d}{dx}\left(\frac{\partial F}{\partial y''}\right)\right\} = 0,$$

that is,

$$\xi\left[F - y'\left\{\frac{\partial F}{\partial y'} - \frac{d}{dx}\left(\frac{\partial F}{\partial y''}\right)\right\}\right] + \eta\left\{\frac{\partial F}{\partial y'} - \frac{d}{dx}\left(\frac{\partial F}{\partial y''}\right)\right\} = 0.$$

The condition, representing assigned requirements, is now

$$\xi\frac{\partial h}{\partial x} + \eta\frac{\partial h}{\partial y} = 0;$$

and therefore, as requirements at the extremity, we have

$$\left.\begin{array}{r}F - y'\left\{\dfrac{\partial F}{\partial y'} - \dfrac{d}{dx}\left(\dfrac{\partial F}{\partial y''}\right)\right\} = \mu\dfrac{\partial h}{\partial x} \\[2mm] \dfrac{\partial F}{\partial y'} - \dfrac{d}{dx}\left(\dfrac{\partial F}{\partial y''}\right) = \mu\dfrac{\partial h}{\partial y}\end{array}\right\},$$

where $\mu$ is merely a multiplier; together with the terminal conditions

$$h(x, y) = 0, \quad y' = \text{given constant}.$$

(iv)  Let an extremity of the range be required to lie upon a given curve

$$k(x, y) = 0,$$

and let the direction at the extremity be unconditioned.

Then $v'$ is arbitrary and variable. The conditions at the extremity are

$$\left.\begin{array}{r}F - y'\left\{\dfrac{\partial F}{\partial y'} - \dfrac{d}{dx}\left(\dfrac{\partial F}{\partial y''}\right)\right\} = \rho\dfrac{\partial k}{\partial x} \\[2mm] \dfrac{\partial F}{\partial y'} - \dfrac{d}{dx}\left(\dfrac{\partial F}{\partial y''}\right) = \rho\dfrac{\partial k}{\partial y} \\[2mm] \dfrac{\partial F}{\partial y''} = 0\end{array}\right\},$$

together with the single terminal condition

$$k(x, y) = 0.$$

**87.** In every special kind of terminal relation, but with forms that vary from one kind to another, we have three conditions surviving at an extremity, and therefore six surviving conditions in all. As will be seen immediately, the primitive of the equation $\mathfrak{E} = 0$ involves four arbitrary constants. When any freedom is left to the position of an extremity of the range by the character of the imposed conditions, both the initial limit $x_1$ and the final limit $x_2$ have to be determined, as being quantities initially unknown. Thus, when the greatest freedom is allowed by the assigned conditions, there are six constants to be determined. But we have six relations from the terminal conditions; and thus all the quantities are potentially determinate.

When the assigned conditions are more precise and the limits $x_1$ and $x_2$ are actually known from the beginning, the four constants in the primitive are otherwise obtained (§ 111). And so for the various forms of condition that occur: either they impose limits, which initially are definitely fixed; or they lead to relations, arising out of imposed demands and definitely determining the limits as known.

Thus, in order to secure that the first variation of the original integral shall vanish for all admissible variations, we obtain a critical equation $\mathfrak{E} = 0$ and we have conditions which determine limits. These limits, henceforth to be regarded as known and fixed, admit no variation. The way is thus cleared for simplification in considering the second variation of the original integral.

The corresponding results for an integral, which involves a derivative of order higher than the second, are given in the following example.

*Ex.* 1. Prove that, if an integral $\int V\,dx$ of order $n$, where

$$V = f(x, y, y^{(1)}, \dots, y^{(n)})$$

and $y^{(m)}$ denotes $\dfrac{d^m y}{dx^m}$, has a maximum or minimum, then $y$ must satisfy the equation

$$E = \frac{\partial V}{\partial y} - \frac{d}{dx}\left\{\frac{\partial V}{\partial y^{(1)}}\right\} + \frac{d^2}{dx^2}\left\{\frac{\partial V}{\partial y^{(2)}}\right\} - \dots + (-1)^n \frac{d^n}{dx^n}\left\{\frac{\partial V}{\partial y^{(n)}}\right\} = 0 \ ;$$

and that this equation, usually of order $2n$, is of order $2n-2$ when

$$V = y^{(n)}\,\phi\,(x, y, y^{(1)}, \dots, y^{(n-1)}) + \psi\,(x, y, y^{(1)}, \dots, y^{(n-1)}).$$

Writing

$$P_r = \frac{\partial V}{\partial y^{(r)}} - \frac{d}{dx}\left\{\frac{\partial V}{\partial y^{(r+1)}}\right\} + \dots + (-1)^{n-r}\frac{d^{n-r}}{dx^{n-r}}\left\{\frac{\partial V}{\partial y^{(n)}}\right\}$$

for $r = 1, \dots, n$, prove that, if an extremity of the integral lies on a curve $F(x, y) = 0$ and be displaced to a position $x + \kappa X$, $y + \kappa Y$ upon that curve, then the condition

$$\{P_0 - y^{(1)}\,P_1 - y^{(2)}\,P_2 - \dots - y^{(n)}\,P_n\}\,X$$
$$+ P_1 Y + P_2 Y^{(1)} + \dots + P_n Y^{(n-1)} = 0$$

must be satisfied at that extremity, where

$$X\frac{\partial F}{\partial x} + Y\frac{\partial F}{\partial y} = 0,$$

$$\left\{\frac{\partial^2 F}{\partial x^2} + y^{(1)}\frac{\partial^2 F}{\partial x\,\partial y}\right\} X + \left\{\frac{\partial^2 F}{\partial x\,\partial y} + y^{(1)}\frac{\partial^2 F}{\partial y^2}\right\} Y + Y^{(1)}\frac{\partial F}{\partial y} = 0,$$

and so on.

Prove also that, if $V$ is the exact differential coefficient of a function of $x, y, y^{(1)}, \dots, y^{(n-1)}$, then $E = 0$ is the condition of integrability: and that, if $V$ is the exact second differential coefficient of a function of $x, y, y^{(1)}, \dots, y^{(n-2)}$, the additional condition

$$\frac{\partial V}{\partial y^{(1)}} - 2\frac{d}{dx}\left\{\frac{\partial V}{\partial y^{(2)}}\right\} + 3\frac{d^2}{dx^2}\left\{\frac{\partial V}{\partial y^{(3)}}\right\} - \dots + (-1)^{n-1}\frac{d^{n-1}}{dx^{n-1}}\left\{\frac{\partial V}{\partial y^{(n)}}\right\} = 0$$

must be satisfied.

*Ex.* 2. Prove (*a*) that, if the function $V$ in the preceding example does not involve $y$ explicitly, a first integral of the characteristic equation is

$$P_1 = A,$$

where $A$ is an arbitrary constant; (*b*) that, if it does not involve $x$ explicitly, a first integral is

$$V = B + y^{(1)} P_1 + y^{(2)} P_2 + \dots + y^{(n)} P_n,$$

where $B$ is another arbitrary constant: and (*c*) that these two integrals are independent of one another, so that they coexist, when $V$ does not involve $x$ or $y$ explicitly and does contain derivatives of $y$ higher than the first.

Is the latter statement correct, when $V$ contains only the derivative of the first order ?

### *The second variation, under special weak variations.*

**88.** We proceed to the consideration of the second variation of the original integral; but now it is necessary to consider it only in the circumstances established by the vanishing of the first variation. In particular, the limits of the integral are fixed (§ 87). Thus the variation of the integral is

$$\int_{x_1}^{x_2} \{ F(x, y + \kappa v, y' + \kappa v', y'' + \kappa v'') - F(x, y, y', y'') \} \, dx$$
$$= \tfrac{1}{2} \kappa^2 \int \mathbf{f}\mathbf{f} \, dx + K_3,$$

where

$$\mathbf{f}\mathbf{f} = a_{00} v^2 + 2a_{01} vv' + 2a_{02} vv'' + a_{11} v'^2 + 2a_{12} v'v'' + a_{22} v''^2,$$

in which the six coefficients $a_{00}$, $a_{01}$, $a_{02}$, $a_{11}$, $a_{12}$, $a_{22}$ respectively denote

$$\frac{\partial^2 F}{\partial y^2}, \quad \frac{\partial^2 F}{\partial y \, \partial y'}, \quad \frac{\partial^2 F}{\partial y \, \partial y''}, \quad \frac{\partial^2 F}{\partial y'^2}, \quad \frac{\partial^2 F}{\partial y' \, \partial y''}, \quad \frac{\partial^2 F}{\partial y''^2};$$

and where $K_3$ denotes the aggregate of terms arising out of the third power and higher powers of $\kappa$ in the expansion. The term involving the first power of $\kappa$ does not appear; it is the first variation of the integral and has been made to vanish.

The governing magnitude in this variation is the 'second' variation $\tfrac{1}{2} \kappa^2 \int \mathbf{f}\mathbf{f} \, dx$, unless it vanishes. We must, therefore, consider, in detail, the integral

$$\int \mathbf{f}\mathbf{f} \, dx :$$

and it will be noted that $\mathbf{f}\mathbf{f}$ is the same function of $v$, $v'$, $v''$ as $2\Omega$, connected (§ 85) with the subsidiary equation, is of $\eta$, $\eta'$, $\eta''$.

**89.** For this purpose, we take a quantity $U$, homogeneous and of the second degree in $v$ and $v'$, choosing it (if possible) so that

$$a_{22} \left( \mathbf{f}\mathbf{f} + \frac{dU}{dx} \right) = (a_{22} v'' + lv' + mv)^2,$$

where $l$ and $m$ are magnitudes also to be determined.

In the earliest discussion (§§ 19, 21) of the second variation of integrals originally involving derivatives of only the first order, the squared quantity in the modified subject of integration in the critical integral was such that, when $v$ was made a constant multiple of the associated integral of the subsidiary equation, the quantity vanished. By analogy, we are led to consider the possibility that the two quantities $l$ and $m$ should satisfy initially the sole condition

$$a_{22}\eta'' + l\eta' + m\eta = 0,$$

where $\eta$ is any integral of the subsidiary equation

$$\frac{\partial\Omega}{\partial\eta} - \frac{d}{dx}\left(\frac{\partial\Omega}{\partial\eta'}\right) + \frac{d^2}{dx^2}\left(\frac{\partial\Omega}{\partial\eta''}\right) = 0.$$

Other conditions, to render $l$ and $m$ precise, will be imposed; they need not be stated at this stage of the investigation.

On this possibility, we have

$$\frac{\partial\Omega}{\partial\eta''} = a_{02}\eta + a_{12}\eta' + a_{22}\eta''$$

$$= -\theta\eta' - \phi\eta,$$

where $\theta$ and $\phi$ are two quantities dependent upon $l$ and $m$ by the relations

$$l = a_{12} + \theta, \qquad m = a_{02} + \phi,$$

that is, $\theta$ and $\phi$ are not new unknown quantities, additional to $l$ and $m$. Next,

$$\frac{\partial\Omega}{\partial\eta'} - \frac{d}{dx}\left(\frac{\partial\Omega}{\partial\eta''}\right) = a_{01}\eta + a_{11}\eta' + a_{12}\eta'' + \frac{d}{dx}(\theta\eta' + \phi\eta)$$

$$= (a_{12} + \theta)\,\eta'' + (a_{11} + \theta' + \phi)\,\eta' + (a_{01} + \phi')\,\eta$$

$$= \frac{l}{a_{22}}(a_{22}\eta'' + l\eta' + m\eta) + \left(a_{11} + \theta' + \phi - \frac{l^2}{a_{22}}\right)\eta' + \left(a_{01} + \phi' - \frac{lm}{a_{22}}\right)\eta$$

$$= -\phi\eta' - \psi\eta,$$

if

$$l^2 = a_{22}(a_{11} + \theta' + 2\phi), \qquad lm = a_{22}(a_{01} + \phi' + \psi):$$

the former of these equations is a second relation (apparently) affecting $l$ and $m$, while the latter defines a new quantity $\psi$. And now

$$\frac{\partial\Omega}{\partial\eta} - \frac{d}{dx}\left(\frac{\partial\Omega}{\partial\eta'}\right) + \frac{d^2}{dx^2}\left(\frac{\partial\Omega}{\partial\eta''}\right)$$

$$= a_{00}\eta + a_{01}\eta' + a_{02}\eta'' + \frac{d}{dx}(\phi\eta' + \psi\eta)$$

$$= (a_{02} + \phi)\,\eta'' + (a_{01} + \phi' + \psi)\,\eta' + (a_{00} + \psi')\,\eta$$

$$= m\eta'' + \frac{lm}{a_{22}}\,\eta' + (a_{00} + \psi')\,\eta$$

$$= \frac{m}{a_{22}}(a_{22}\eta'' + l\eta' + m\eta) + \left(a_{00} + \psi' - \frac{m^2}{a_{22}}\right)\eta.$$

As the left-hand side vanishes and as $\eta$ is not zero, we have

$$m^2 = a_{22}(a_{00} + \psi').$$

It therefore follows that, if $\eta$ denote any integral of the subsidiary characteristic equation, and if $l$, $m$, $\theta$, $\phi$, $\psi$ are such that

$$a_{22}\eta'' + l\eta' + m\eta = 0,$$

$$l = a_{12} + \theta, \qquad m = a_{02} + \phi,$$

$$l^2 = a_{22}(a_{11} + \theta' + 2\phi), \quad lm = a_{22}(a_{01} + \phi' + \psi), \quad m^2 = a_{22}(a_{00} + \psi'),$$

then

$$(a_{22}v'' + lv' + mv)^2$$
$$= a_{22}^2 v''^2 + 2a_{22}(a_{12} + \theta)v''v' + 2a_{22}(a_{02} + \phi)v''v$$
$$\qquad + a_{22}(a_{11} + \theta' + \phi)v'^2 + 2a_{22}(a_{01} + \phi' + \psi)v'v + a_{22}(a_{00} + \psi')v^2$$
$$= a_{22}\left(\mathbf{F} + \frac{dU}{dx}\right),$$

where

$$U = \theta v'^2 + 2\phi vv' + \psi v^2.$$

*Determination of two quantities, for the expression of the second variation.*

**90.** Thus far, we have considered only a single integral $\eta$ of the subsidiary characteristic equation in connection with the two quantities $l$ and $m$; and no limitation has been imposed on the integral. We proceed to prove that all the determining relations affecting $l$ and $m$ (and the quantities $\theta$, $\phi$, $\psi$) are satisfied, when two distinct integrals of that equation are taken, subject to a single condition connected with the subsidiary equation itself.

Let two such integrals be denoted by $\alpha$ and $\beta$, so that we have

$$a_{22}\alpha'' + l\alpha' + m\alpha = 0, \quad a_{22}\beta'' + l\beta' + m\beta = 0.$$

In order that $l$ and $m$ may be determinate, $\alpha\beta' - \alpha'\beta$ must not be zero, a requirement satisfied because $\alpha$ and $\beta$ are two distinct integrals such that $\beta$ is not a mere constant multiple of $\alpha$. In order that $l$ may not be zero always, we must have $\alpha\beta'' - \alpha''\beta$ not always zero, so that $\alpha\beta' - \alpha'\beta$ must not be a constant. And in order that $m$ may not be zero always, we must have $\alpha'\beta'' - \alpha''\beta'$ not always zero, so that we do not have $\beta = k\alpha + h$, where $h$ and $k$ are constants: in other words, the subsidiary equation must not admit any single integral that is a mere constant. In connection with the equations for $l$ and $m$, it is convenient to write

$$a_{11} - 2a_{02} - a_{12}' = b, \quad a_{00} - a_{01}' + a_{02}'' = c;$$

and then the subsidiary characteristic equation is

$$a_{22}\eta'''' + 2a_{22}'\eta''' + (a_{22}'' - b)\eta'' - b'\eta' + c\eta = 0,$$

so that the exclusion of the possibility $\beta = k\alpha + h$, where $\alpha$ and $\beta$ are two integrals of this equation, merely requires that $c$ must not vanish, clearly not a limitation of a general kind.

**91.** First, as regards the indicated condition connected with the equation itself. The equation, being linear in the dependent variable $\eta$ and of the fourth order, possesses four linearly independent integrals: and the primitive is a linear combination of these four integrals with arbitrary constant coefficients. Let $\alpha$ and $\beta$ be any two of the four; and introduce new quantities $z$ and $\mu$ by the definitions

$$z = \alpha\beta' - \alpha'\beta, \quad \mu = \alpha'\beta'' - \alpha''\beta'.$$

Then

$$z'' = \alpha\beta''' - \alpha'''\beta + \mu,$$
$$z''' = \alpha\beta'''' - \alpha''''\beta + 2\mu',$$
$$\mu' = \alpha'\beta''' - \alpha'''\beta',$$
$$\mu'' = \alpha'\beta'''' - \alpha''''\beta' + \alpha''\beta''' - \alpha'''\beta''.$$

One property, to be used immediately (p. 131), may be noted in passing. If $\sigma$ denote $\alpha''\beta''' - \alpha'''\beta''$, we have

$$\sigma = \alpha''\beta''' - \alpha'''\beta'',$$
$$\mu' = \alpha'\beta''' - \alpha'''\beta',$$
$$z'' - \mu = \alpha\beta''' - \alpha'''\beta,$$

and therefore

$$\begin{vmatrix} \sigma & , & \alpha'' & , & \beta'' \\ \mu' & , & \alpha' & , & \beta' \\ z'' - \mu & , & \alpha & , & \beta \end{vmatrix} = 0,$$

that is,

$$-\sigma z + \mu' z' - (z'' - \mu)\mu = 0,$$

so that

$$z(\alpha''\beta''' - \alpha'''\beta'') = \mu' z' + \mu^2 - \mu z''.$$

As $\alpha$ and $\beta$ are integrals of the subsidiary characteristic equation,

$$a_{22}\beta'''' + 2a_{22}'\beta''' + (a_{22}'' - b)\beta'' - b'\beta' + c\beta = 0,$$
$$a_{22}\alpha'''' + 2a_{22}'\alpha''' + (a_{22}'' - b)\alpha'' - b'\alpha' + c\alpha = 0.$$

Multiplying the first by $\alpha$, the second by $\beta$, and subtracting, we have

$$a_{22}(z''' - 2\mu') + 2a_{22}'(z'' - \mu) + (a_{22}'' - b)z' - b'z = 0.$$

Thus

$$\frac{d}{dt}\{a_{22}(z'' - 2\mu) + a_{22}'z' - bz\} = 0,$$

and therefore

$$a_{22}(z'' - 2\mu) + a_{22}'z' - bz = C,$$

where $C$ is a constant, which manifestly is specific and not arbitrary, because all the quantities on the left-hand side are specific.

**92.** We have taken $\alpha$ and $\beta$ to be any two linearly independent integrals, out of a complete set. We now proceed to prove that, if $C$ be not zero for any two initially selected integrals, two can be chosen so as to make the specific constant zero. Let $\alpha$ and $\gamma$ be a pair with $z_1$, $\mu_1$, $C_1$ as the values of $z$, $\mu$, $C$; and let $\alpha$ and $\delta$ be another pair (where $\delta$ is distinct from $\gamma$ and is not a linear combination of $\alpha$ and $\gamma$ with constant coefficients), this second pair having $z_2$, $\mu_2$, $C_2$ as the values of $z$, $\mu$, $C$. (It is assumed that neither $C_1$ nor $C_2$ is zero; for, if either were zero, the desired combination would be to hand.) Then

$$a_{22}(z_1'' - 2\mu_1) + a_{22}'z_1' - bz_1 = C_1,$$
$$a_{22}(z_2'' - 2\mu_2) + a_{22}'z_2' - bz_2 = C_2.$$

Let

$$\beta = C_2\gamma - C_1\delta\,;$$

then

$$z \ = C_2 z_1 \ - C_1 z_2\,,$$
$$z' = C_2 z_1' - C_1 z_2'\,,$$
$$z'' = C_2 z_1'' - C_1 z_2''\,,$$
$$\mu = C_2 \mu_1 - C_1 \mu_2\,,$$

and therefore

$$a_{22}(z'' - 2\mu) + a_{22}'z' - bz = 0.$$

But $\beta$, thus chosen as a linear combination of the two integrals $\gamma$ and $\delta$ with constant coefficients, is itself an integral of the characteristic equation; and it is distinct from $\alpha$, because $\delta$ is not a linear combination of $\alpha$ and $\gamma$.

Thus two integrals $\alpha$ and $\beta$ can always be chosen, so that the specific constant $C$ is zero—which is the property in question. Moreover, the choice manifestly is not unique.

**93.** We now proceed to shew that all the equations for $l$, $m$, $\theta$, $\phi$, $\psi$ are satisfied by the equations

$$a_{22}\alpha'' + l\alpha' + m\alpha = 0, \quad a_{22}\beta'' + l\beta' + m\beta = 0,$$

defining $l$ and $m$ in terms of the two integrals $\alpha$ and $\beta$ with the property just established. From these two equations, it follows that

$$l = -a_{22}\frac{z'}{z}, \quad m = a_{22}\frac{\mu}{z}.$$

The other three equations are

$$l^2 = a_{22}(a_{11} + \theta' + 2\phi) = a_{22}(b + l' + 2m),$$
$$lm = a_{22}(a_{01} + \phi' + \psi) = a_{22}(a_{01} - a_{02}' + m' + \psi),$$
$$m^2 = a_{22}(a_{00} + \psi').$$

When the values of $l$ and $m$ are substituted in the first of these equations, it becomes

$$a_{22}^2\frac{z'^2}{z^2} = a_{22}\left(b + a_{22}\frac{z'^2}{z^2} - a_{22}\frac{z''}{z} - a_{22}'\frac{z'}{z} + 2a_{22}\frac{\mu}{z}\right),$$

which is satisfied because

$$a_{22}(z'' - 2\mu) + a_{22}'z' - bz = 0.$$

When the values of $l$ and $m$ are substituted in the second of these equations, it determines a quantity $\psi$ in the form

$$-\psi = a_{01} - a_{02}' + \frac{1}{z}(a_{22}'\mu + a_{22}\mu');$$

that is, with this value for $\psi$, the second equation is satisfied.

When the value of $m$ and this value of $\psi$ are substituted in the third equation, it becomes

$$a_{00} - a_{22}\frac{\mu^2}{z^2} = -\psi'$$
$$= a_{01}' - a_{02}'' + \frac{1}{z}(a_{22}''\mu + 2a_{22}'\mu' + a_{22}\mu'') - \frac{z'}{z^2}(a_{22}'\mu + a_{22}\mu'),$$

that is,

$$c = a_{22}\frac{\mu^2}{z^2} + \frac{1}{z}(a_{22}''\mu + 2a_{22}'\mu' + a_{22}\mu'') - \frac{z'}{z^2}(a_{22}'\mu + a_{22}\mu').$$

Now, taking the two equations satisfied singly by $\beta$ and $\alpha$ because they are integrals of the subsidiary characteristic equation, multiplying the first by $\alpha'$, the second by $\beta'$, and subtracting, we have

$$a_{22}(\alpha'\beta'''' - \alpha''''\beta') + 2a_{22}'(\alpha'\beta''' - \alpha'''\beta') + (a_{22}'' - b)(\alpha'\beta'' - \alpha''\beta') + c(\alpha'\beta - \alpha\beta') = 0,$$

and therefore

$$a_{22}\{\mu'' - (\alpha''\beta''' - \alpha'''\beta'')\} + 2a_{22}'\mu' + (a_{22}'' - b)\mu - cz = 0.$$

Now

$$a_{22}z(\alpha''\beta''' - \alpha'''\beta'') = a_{22}(\mu'z + \mu^2 - \mu z''),$$

always, whatever integrals $\alpha$ and $\beta$ be chosen; and for the present choice,

$$a_{22}z'' = 2a_{22}\mu - a_{22}'z' + bz;$$

thus

$$a_{22}z(\alpha''\beta''' - \alpha'''\beta'') = a_{22}\mu'z + a_{22}\mu^2 - \mu(2a_{22}\mu - a_{22}'z' + bz)$$
$$= a_{22}(\mu'z - \mu^2) + a_{22}'\mu z' - b\mu z.$$

Therefore

$$a_{22}\mu'' - \frac{1}{z}\{a_{22}(\mu'z - \mu^2) + a_{22}'\mu z' - b\mu z\} + 2a_{22}'\mu' + (a_{22}'' - b)\mu - cz = 0,$$

that is, on dividing by $z$,

$$a_{22}\frac{\mu^2}{z^2} + \frac{1}{z}(a_{22}\mu'' + 2a_{22}'\mu' + a_{22}''\mu) - \frac{z'}{z^2}(a_{22}\mu' + a_{22}'\mu) - c = 0.$$

Thus the third of the three equations is satisfied; and the five quantities $l$, $m$, $\theta$, $\phi$, $\psi$ are expressed in terms of $\alpha$ and $\beta$ by the equations

$$l = -a_{22}\frac{z'}{z},$$

$$m = a_{22}\frac{\mu}{z},$$

$$\theta = -a_{12} - a_{22}\frac{z'}{z},$$

$$\phi = -a_{02} + a_{22}\frac{\mu}{z},$$

$$\psi = -a_{01} - a_{02}' - \frac{1}{z}(a_{22}'\mu + a_{22}\mu'),$$

where

$$z = \alpha\beta' - \alpha'\beta, \quad \mu = \alpha'\beta'' - \alpha''\beta',$$

and where the condition

$$a_{22}(z'' - 2\mu) + a_{22}'z' - bz = 0$$

is satisfied.

**94.** The preceding analysis can be modified and restated as follows. We have

$$\mathfrak{F} = a_{22}v''^2 + bv'^2 + cv^2 + \frac{d}{dt}\{a_{12}v'^2 + 2a_{02}vv' + (a_{01} - a_{02}')v^2\},$$

where, as before,

$$b = a_{11} - a_{12}' - 2a_{02}, \quad c = a_{00} - a_{01}' + a_{02}''.$$

Also

$$\frac{d}{dt}(lv'^2 + 2mvv' + nv^2)$$
$$= 2lv'v'' + 2mvv'' + (l' + 2m)v'^2 + (2m' + 2n)vv' + n'v^2;$$

and therefore

$$a_{22}\left[\mathfrak{F} - \frac{d}{dt}\{(a_{12} - l)v'^2 + 2(a_{02} - m)vv' + (a_{01} - a_{02}' - n)v^2\}\right]$$
$$= a_{22}^2 v''^2 + 2a_{22}lv'v'' + 2a_{22}mvv''$$
$$\quad + a_{22}(b + l' + 2m)v'^2 + 2a_{22}(m' + n)vv' + a_{22}(c + n')v^2$$
$$= (a_{22}v'' + lv' + mv)^2,$$

provided

$$l^2 = a_{22}(b + l' + 2m),$$
$$lm = a_{22}(m' + n),$$
$$m^2 = a_{22}(c + n').$$

This is the earlier form (§ 89) of the result, when we take

$$a_{12} - l = \theta, \quad a_{02} - m = \phi, \quad a_{01} - a_{02}' - n = \psi.$$

The remainder of the analysis, for the establishment of the equations

$$a_{22}\alpha'' + l\alpha' + m\alpha = 0, \quad a_{22}\beta'' + l\beta' + m\beta = 0,$$

with the relation between $\alpha$ and $\beta$, is unaltered.

*Normal form of second variation : the Legendre test.*

**95.** It follows that

$$a_{22}\left(\mathfrak{F} + \frac{dU}{dx}\right) = (a_{22}v'' + lv' + mv)^2;$$

and therefore, on inserting the values of $l$ and $m$,

$$\mathfrak{F} + \frac{dU}{dx} = \frac{a_{22}}{(\alpha\beta' - \alpha'\beta)^2}\begin{vmatrix} v'' & v' & v \\ \alpha'' & \alpha' & \alpha \\ \beta'' & \beta' & \beta \end{vmatrix}^2$$
$$= a_{22}\,\square,$$

where

$$(\alpha\beta' - \alpha'\beta)^2 \square = \left| \begin{array}{ccc} v'', & v', & v \\ \alpha'', & \alpha', & \alpha \\ \beta'', & \beta', & \beta \end{array} \right|^2.$$

The quantity determining the second variation of the original integral $\int F(x, y, y', y'')\,dx$ is

$$\int \mathscr{F}\,dx.$$

But

$$\int \frac{dU}{dx}\,dx = [U] = [\theta v'^2 + 2\phi v v' + \psi v^2],$$

taken at the limits of the integral. As these limits have become fixed by the inferences from the vanishing of the first variation, we have $v = 0$ and $v' = 0$ at each limit; and thus $[U] = 0$. Hence the second variation

$$= \tfrac{1}{2}\kappa^2 \int \mathscr{F}\,dx$$

$$= \tfrac{1}{2}\kappa^2 \int a_{22} \square\,dx;$$

and the whole variation of the integral is

$$\tfrac{1}{2}\kappa^2 \int a_{22} \square\,dx + K,$$

where $K$ is the aggregate of terms involving the third and higher powers of $\kappa$.

Thus the integral $\int a_{22}\square\,dx$ is critical for the second variation. If it is positive for all admissible variations $v$, all such variations make the increment of the original integral an increase: subject to any other conditions that may occur, the original integral is a minimum. If it is negative for all admissible variations, all such variations make the increment of the original integral a decrease: subject to any other conditions that may occur, the original integral is a minimum. Hence $\int a_{22}\square\,dx$ is to be regularly positive or regularly negative for all arbitrary admissible variations.

Then $a_{22}$ must have a persistent sign throughout the range. For, if $a_{22}$ were sometimes positive, sometimes negative, then by choosing $v = 0$ when $a_{22}$ is negative, we make $\int a_{22}\square\,dx$ positive; and by choosing $v = 0$ when $a_{22}$ is positive, we make $\int a_{22}\square\,dx$ negative; that is, the integral in question can be made positive or negative at will, if $a_{22}$ can change its sign through the range. Hence we infer, as a necessary requirement, that the quantity $a_{22}$ must be monotonic: that is, $\dfrac{\partial^2 F}{\partial y''^2}$ *must have a constant sign throughout the range of the integral along the characteristic curve.* This test, an extension of the Legendre test, for the former case, may be called the Legendre test.

### *The Jacobi test.*

**96.** But the quantity $\square$ must still be considered. Apparently, it is a positive quantity. The magnitudes $\alpha$ and $\beta$, which occur in its expression, are linearly independent, so that no difficulty arises from the quantity $\alpha\beta' - \alpha'\beta$, or from the (excluded) possibility that the second and third lines of the determinant might effectively be the same. It vanishes of course when $v = 0$: in that event, there is no variation. It may however vanish if some variation $v$ is possible such that through the range

$$v = A\alpha + B\beta,$$

where $A$ and $B$ are constants. The conditions now require that $v$ shall vanish at the limits of the range. If, therefore, the range is such as to allow constants $A$ and $B$ to be chosen, so that

$$A\alpha + B\beta$$

shall in general be distinct from zero, shall vanish at the lower limit, and shall vanish also at the upper limit (or earlier), then the non-zero variation $v = A\alpha + B\beta$ makes $\square$ vanish. For that variation, the second variation of the original integral $\int F(x, y, y', y'')\, dx$ vanishes: we cannot assert the existence of a maximum or a minimum, without examining the quantity $K$.

If, however, the range of the integral along the characteristic curve, beginning at a lower limit where a quantity $A\alpha + B\beta$ vanishes, does not extend as far as the nearest place where that quantity again vanishes, then the indicated variation cannot occur. There then is no non-zero variation $v$ which annihilates $\square$; and the former argument now remains valid.

The further significance of this result will be indicated later (§§ 124–126), when we come to consider the conjugate of the lower limit of the integral as a point on the characteristic curve. Meanwhile, we have the initial form of the *Jacobi test limiting the range of the integral.*

### *Summary of tests.*

Summarising, we have three tests, thus far:

(i)  the characteristic equation (the *Euler* test), leading to the characteristic curve;

(ii)  the non-evanescence of $\dfrac{\partial^2 F}{\partial y''^2}$ along the range (the *Legendre* test);

(iii)  the limitation of the range (the *Jacobi* test).

*General weak variations: the Weierstrass method.*

**97.** We now pass to the consideration of the more extensive type of variation, still required to be 'weak,' wherein both the dependent variable and the independent variable are subject, separately or simultaneously, to small unrelated changes.

To facilitate the representation and expression of such variations, it is convenient, as before (§ 36), to make $x$ and $y$ functions of another variable $t$ which is parametric along the characteristic curve. Initially at least, this new variable will not be restricted to represent any element of the curve so essentially intrinsic as the length of the arc from a fixed point. The introduction of the variable $t$ changes the expression of the integral itself and substantially modifies the analysis; a reversion to the previous form can always be made, so far as some results are concerned, by adopting $x$ as the independent variable $t$.

We write

$$x_m = \frac{d^m x}{dt^m}, \quad y_n = \frac{d^n y}{dt^n},$$

for integer values of $m$ and of $n$, while $x_0$ is $x$ and $y_0$ is $y$. To express partial differentiations of any function $G$ involving $x, x_1, x_2, ..., y, y_1, y_2, ...$, we write

$$\frac{\partial^2 G}{\partial y_m \partial y_n} = c_{mn}, \quad \frac{\partial^2 G}{\partial x_m \partial x_n} = \gamma_{mn},$$

for $m, n = 0, 1, 2$; so that there are six coefficients $c$ (because $c_{mn} = c_{nm}$), and six coefficients $\gamma$ (because $\gamma_{mn} = \gamma_{nm}$). Also we write

$$\frac{\partial^2 G}{\partial x_m \partial y_n} = k_{mn},$$

for $m = 0, 1, 2$, and $n = 0, 1, 2$; so that (as $k_{mn}$ and $k_{nm}$ are not the same, when $m$ and $n$ are different) there are nine coefficients $k$.

To transform the integral $\int F(x, y, y', y'') \, dx$, we have

$$y' = \frac{y_1}{x_1}, \quad y'' = \frac{1}{x_1^3}(x_1 y_2 - y_1 x_2);$$

and we write

$$G(x, x_1, x_2, y, y_1, y_2) = x_1 F(x, y, y', y'')$$

$$= x_1 F\left(x, y, \frac{y_1}{x_1}, \frac{x_1 y_2 - y_1 x_2}{x_1^3}\right),$$

so that $G$ is a function of the same character in its arguments as the original function $F$. The integral becomes

$$\int G(x, x_1, x_2, y, y_1, y_2) \, dt.$$

We assume that $t$ increases steadily throughout the range from the lower limit to the upper limit of the integral; thus $x_1$ does not become infinite.

The function $F$ is a function of only four arguments, while $G$ (as expressed) is a function of six arguments. Thus $G$ cannot be a quite general function of those six, and therefore the limitations on generality must be indicated explicitly. We have

$$\frac{\partial G}{\partial x_1} = F - \frac{y_1}{x_1}\frac{\partial F}{\partial y'} + \left(\frac{2y_2}{x_1^2} - \frac{3y_1 x_2}{x_1^3}\right)\frac{\partial F}{\partial y''},$$

$$\frac{\partial G}{\partial y_1} = \frac{\partial F}{\partial y'} - \frac{x_2}{x_1^2}\frac{\partial F}{\partial y''},$$

$$\frac{\partial G}{\partial x_2} = -\frac{y_1}{x_1^2}\frac{\partial F}{\partial y''},$$

$$\frac{\partial G}{\partial y_2} = \frac{1}{x_1}\frac{\partial F}{\partial y''}.$$

By eliminating $\dfrac{\partial F}{\partial y'}$, $\dfrac{\partial F}{\partial y''}$, and $F$, using the relation $G = x_1 F$, we have two (and only two) identical relations satisfied by $G$; and these, which will be called the fundamental identities, can be taken in the form*

$$\left.\begin{aligned} G &= x_1\frac{\partial G}{\partial x_1} + y_1\frac{\partial G}{\partial y_1} + 2x_2\frac{\partial G}{\partial x_2} + 2y_2\frac{\partial G}{\partial y_2}\\ 0 &= x_1\frac{\partial G}{\partial x_2} + y_1\frac{\partial G}{\partial y_2} \end{aligned}\right\}.$$

These two identities are the expression of the limitations upon $G$ as a function of its six arguments.

### Variation of the integral : the ' first' variation.

**98.** To obtain the conditions for a maximum or a minimum of the integral $\int G\,dt$, the variables $x$ and $y$ are subjected simultaneously to small variations so that they become $x + \kappa u$ and $y + \kappa v$, where $\kappa$ is the usual small arbitrary constant, while $u$ and $v$ are arbitrary functions of $t$, regular within the range

---

* For another mode of derivation, based upon the fact that the integral $\int G\,dt$, which (as in § 38) is an invariant for changes of $t$, is to be unaltered by infinitesimal changes in $t$, see § 105 *post*.

It is to be noted that if, instead of regarding the two identities as relations satisfied by every given function $G$ arising as in the text, the two relations are regarded as simultaneous partial differential equations satisfied by an unknown variable $G$, the most general primitive of those two equations is

$$G = x_1 F\left(x,\ y,\ \frac{y_1}{x_1},\ \frac{x_1 y_2 - y_1 x_2}{x_1^3}\right),$$

where, so far as concerns the two differential equations, $F$ denotes the most general arbitrary function of its four arguments.

admitted, and independent of one another along the range. The values of $t$, which are the limits of the integral, are taken to be fixed, as in § 42.

The new value of the integral is

$$\int G\left(x + \kappa u, \; x_1 + \kappa u_1, \; x_2 + \kappa u_2, \; y + \kappa v, \; y_1 + \kappa v_1, \; y_2 + \kappa v_2\right) dt.$$

Hence the increment of the integral is

$$\kappa I_1 + \tfrac{1}{2}\kappa^2 \int \mathfrak{G}\, dt + K_3,$$

where

$$I_1 = \int \left( u\frac{\partial G}{\partial x} + u_1\frac{\partial G}{\partial x_1} + u_2\frac{\partial G}{\partial x_2} + v\frac{\partial G}{\partial y} + v_1\frac{\partial G}{\partial y_1} + v_2\frac{\partial G}{\partial y_2} \right) dt\,;$$

where $K_3$ denotes the aggregate of terms involving the third power and higher powers of $\kappa$; and where $\mathfrak{G}$ is the combined aggregate of all the terms of the second degree in $u$, $u_1$, $u_2$, $v$, $v_1$, $v_2$, viz.

$$\begin{aligned}
\mathfrak{G} = \;\; & c_{00}v^2 + 2c_{01}vv_1 + 2c_{02}vv_2 + c_{11}v_1{}^2 + 2c_{12}v_1v_2 + c_{22}v_2{}^2 \\
&+ \gamma_{00}u^2 + 2\gamma_{01}uu_1 + 2\gamma_{02}uu_2 + \gamma_{11}u_1{}^2 + 2\gamma_{12}u_1u_2 + \gamma_{22}u_2{}^2 \\
&+ 2k_{00}uv + 2k_{01}uv_1 + 2k_{02}uv_2 \\
&+ 2k_{10}u_1v + 2k_{11}u_1v_1 + 2k_{12}u_1v_2 \\
&+ 2k_{20}u_2v + 2k_{21}u_2v_1 + 2k_{22}u_2v_2.
\end{aligned}$$

Proceeding as before in § 83, we have

$$\int \left( u\frac{\partial G}{\partial x} + u_1\frac{\partial G}{\partial x_1} + u_2\frac{\partial G}{\partial x_2} \right) dt = \int uX\, dt + \left[ u\left\{ \frac{\partial G}{\partial x_1} - \frac{d}{dt}\left( \frac{\partial G}{\partial x_2} \right) \right\} + u_1\frac{\partial G}{\partial x_2} \right],$$

$$\int \left( v\frac{\partial G}{\partial y} + v_1\frac{\partial G}{\partial y_1} + v_2\frac{\partial G}{\partial y_2} \right) dt = \int vY\, dt + \left[ v\left\{ \frac{\partial G}{\partial y_1} - \frac{d}{dt}\left( \frac{\partial G}{\partial y_2} \right) \right\} + v_1\frac{\partial G}{\partial y_2} \right],$$

where

$$\left.\begin{aligned}
X &= \frac{\partial G}{\partial x} - \frac{d}{dt}\left( \frac{\partial G}{\partial x_1} \right) + \frac{d^2}{dt^2}\left( \frac{\partial G}{\partial x_2} \right) \\
Y &= \frac{\partial G}{\partial y} - \frac{d}{dt}\left( \frac{\partial G}{\partial y_1} \right) + \frac{d^2}{dt^2}\left( \frac{\partial G}{\partial y_2} \right)
\end{aligned}\right\},$$

and the terms outside the sign of integration are taken at the limits of the range of the variable $t$.

The customary argument concerning the maximum or minimum of the integral requires the total increment of the integral to have one uniform sign, persistent through all the possible small variations which can be imposed upon the variables $x$ and $y$. Consequently the first increment $I_1$, containing only the first power of $\kappa$ which can be changed at will, must vanish for all such variations.

*The two characteristic equations.*

**99.** Now $I_1$ consists of two parts: an integral, and a set of terms at the limits.

Consider, first, the possible variations which, arbitrary along the range, are chosen to be such that $u, u_1, v, v_1$ are zero at each limit; thus, at each limit, the position and the directions of $x$ and $y$ are kept fixed. For these variations, the limit-terms vanish; and $I_1$ reduces to the integral

$$\int (uX + vY)\, dt.$$

Consequently, to achieve the purpose, this integral must vanish; and, in it, the quantities $u$ and $v$ are arbitrary regular functions of $x$, independent of one another. The only way, in which this requirement can be met, is by having the two equations

$$X = 0, \quad Y = 0,$$

satisfied everywhere along the range. Otherwise, by choosing $u$ positive or negative as $X$ is positive or negative, and choosing $v$ similarly positive or negative as $Y$ is positive or negative, we could make the integral positive: and by choosing $u$ negative or positive as $X$ is positive or negative, and choosing $v$ similarly negative or positive as $Y$ is positive or negative, we could make the integral negative. These possibilities are to be excluded; hence the two equations

$$X = 0, \quad Y = 0,$$

must be satisfied. They are the *characteristic equations*. It will be noted that, in their expression, the variable $t$ does not occur explicitly.

*Conditions at the limits.*

**100.** Assuming now that these necessary equations are satisfied, the integral $I_1$ for all other variations reduces to the aggregates of terms at the limits. But variations of $x$ and $y$ are possible, vanishing at either limit though not at both. When these are taken for the limits in turn, we infer that the magnitude

$$u \left\{ \frac{\partial G}{\partial x_1} - \frac{d}{dt}\left(\frac{\partial G}{\partial x_2}\right) \right\} + u_1 \frac{\partial G}{\partial x_2} + v \left\{ \frac{\partial G}{\partial y_1} - \frac{d}{dt}\left(\frac{\partial G}{\partial y_2}\right) \right\} + v_1 \frac{\partial G}{\partial y_2}$$

must vanish at each of the limits of the integral.

It is convenient to conclude, at once, the consideration of these conditions. They require that the equation

$$u \left\{ \frac{\partial G}{\partial x_1} - \frac{d}{dt}\left(\frac{\partial G}{\partial x_2}\right) \right\} + u_1 \frac{\partial G}{\partial x_2} + v \left\{ \frac{\partial G}{\partial y_1} - \frac{d}{dt}\left(\frac{\partial G}{\partial y_2}\right) \right\} + v_1 \frac{\partial G}{\partial y_2} = 0$$

shall be satisfied at each limit. As before (§ 86), various analytical conditions emerge according to the character of the assigned data.

(i) Let the extremity be rigidly settled, so that $x$, $y$, $y'$ are not subject to variation. Then $u = 0$, $v = 0$; the variation in $y'$ is

$$\frac{y_1 + \kappa v_1}{x_1 + \kappa u_1} - \frac{y_1}{x_1}, \quad = \kappa \frac{x_1 v_1 - y_1 u_1}{x_1 (x_1 + \kappa u_1)},$$

so that, as this variation is to be zero at the limit, we there have

$$x_1 v_1 - y_1 u_1 = 0.$$

In the foregoing equation, we take $u = 0$, $v = 0$; the other terms are

$$u_1 \frac{\partial G}{\partial x_2} + v_1 \frac{\partial G}{\partial y_2}$$

$$= \frac{1}{x_1} (x_1 v_1 - y_1 u_1) \frac{\partial G}{\partial y_2},$$

on account of the identity $x_1 \dfrac{\partial G}{\partial x_2} + y_1 \dfrac{\partial G}{\partial y_2} = 0$. The terminal equation is satisfied without leaving any residual condition.

(ii) Let the quantities $x$, $y$, $y'$ be subject to a condition

$$H(x, y, y') = 0$$

at a limit. Then the elements of the small variation at that limit are subject to the single equation

$$u \frac{\partial H}{\partial x} + v \frac{\partial H}{\partial y} + \frac{1}{x_1^2} (x_1 v_1 - y_1 u_1) \frac{\partial H}{\partial y'} = 0,$$

keeping only terms of the first degree in $\kappa$, as we are dealing with the first variation. But the terminal condition for the integral is

$$u \left\{ \frac{\partial G}{\partial x_1} - \frac{d}{dt} \left( \frac{\partial G}{\partial x_2} \right) \right\} + v \left\{ \frac{\partial G}{\partial y_1} - \frac{d}{dt} \left( \frac{\partial G}{\partial y_2} \right) \right\} + \frac{1}{x_1} (x_1 v_1 - y_1 u_1) \frac{\partial G}{\partial y_2} = 0;$$

and there is only the foregoing linear relation connecting the otherwise arbitrary quantities $u$, $v$, $x_1 v_1 - y_1 u_1$. Hence there is some quantity $\lambda$, such that

$$\left. \begin{aligned} \frac{\partial G}{\partial x_1} - \frac{d}{dt} \left( \frac{\partial G}{\partial x_2} \right) &= \lambda \frac{\partial H}{\partial x} \\ \frac{\partial G}{\partial y_1} - \frac{d}{dt} \left( \frac{\partial G}{\partial y_2} \right) &= \lambda \frac{\partial H}{\partial y} \\ x_1 \frac{\partial G}{\partial y_2} &= \lambda \frac{\partial H}{\partial y'} \end{aligned} \right\},$$

together with the equation

$$H\left( x, y, \frac{y_1}{x_1} \right) = 0,$$

in effect, three equations when $\lambda$ is eliminated.

(iii)  Let the extremity be required to lie upon a curve

$$h(x, y) = 0$$

and be rigidly fixed in direction.  Then, at the limit, we have

$$u\frac{\partial h}{\partial x} + v\frac{\partial h}{\partial y} = 0, \quad x_1 v_1 - y_1 u_1 = 0;$$

and so there exists a quantity $\mu$ such that

$$\left.\begin{aligned}
\frac{\partial G}{\partial x_1} - \frac{d}{dt}\left(\frac{\partial G}{\partial x_2}\right) &= \mu\frac{\partial h}{\partial x} \\
\frac{\partial G}{\partial y_1} - \frac{d}{dt}\left(\frac{\partial G}{\partial y_2}\right) &= \mu\frac{\partial h}{\partial y}
\end{aligned}\right\},$$

together with the equations

$$h(x, y) = 0, \quad y' = \text{given constant.}$$

(iv)  Let the extremity be required to lie upon a curve

$$k(x, y) = 0$$

and be perfectly free in direction.  Then the arbitrary quantities $u$ and $v$ are subject to the single condition

$$u\frac{\partial k}{\partial x} + v\frac{\partial k}{\partial y} = 0,$$

while the arbitrary quantity $x_1 v_1 - y_1 u_1$ remains unconditioned.  Then

$$\left.\begin{aligned}
\frac{\partial G}{\partial x_1} - \frac{d}{dt}\left(\frac{\partial G}{\partial x_2}\right) &= \rho\frac{\partial k}{\partial x} \\
\frac{\partial G}{\partial y_1} - \frac{d}{dt}\left(\frac{\partial G}{\partial y_2}\right) &= \rho\frac{\partial k}{\partial y} \\
\frac{\partial G}{\partial y_2} &= 0
\end{aligned}\right\},$$

together with the equation

$$k(x, y) = 0.$$

In every type of boundary condition, except the first (where all the variations are zero at the limits), we have three conditions at an extremity, amounting to six in all.  Their significance is the same as before: they serve to determine the constants in the primitive and (by means of the values of $x$) the actual positions of the extremities.  When these are thus settled, no further variations at either extremity are possible; we then may regard them as definitely settled by the first variation.

This settlement completed, we can consider the second variation of the integral for all variations which are subject to fixed extremities, that is, which are such that $u$, $v$, $x_1 v_1 - y_1 u_1$ are zero.  It will be convenient to consider a quantity

$$w = x_1 v - y_1 u,$$

alike for the current variation and for the terminal variation; as

$$\frac{dw}{dt} = x_1 v_1 - y_1 u_1 + x_2 v - y_2 u,$$

it follows that, at any fixed extremity (and we now consider each extremity as fixed), the quantities $w$ and $\dfrac{dw}{dt}$ vanish.

**101.** These results, and the results obtained (§ 86) for the special variation, can be harmonised as follows. We have

$$G = x_1 F(x, y, y', y''),$$

$$y' = \frac{y_1}{x_1}, \quad y'' = \frac{1}{x_1{}^3}(x_1 y_2 - y_1 x_2),$$

so that

$$\frac{\partial G}{\partial x_2} = -\frac{y_1}{x_1{}^2}\frac{\partial F}{\partial y''}, \quad \frac{\partial G}{\partial y_2} = \frac{1}{x_1}\frac{\partial F}{\partial y''};$$

and therefore

$$x_1 \frac{\partial G}{\partial y_2} = \frac{\partial F}{\partial y''}.$$

Again,

$$\frac{\partial G}{\partial y_1} = \frac{\partial F}{\partial y'} - \frac{x_2}{x_1{}^2}\frac{\partial F}{\partial y''},$$

$$\frac{d}{dt}\left(\frac{\partial G}{\partial y_2}\right) = \frac{d}{dt}\left(\frac{1}{x_1}\frac{\partial F}{\partial y''}\right)$$

$$= \frac{1}{x_1}\frac{d}{dt}\left(\frac{\partial F}{\partial y''}\right) - \frac{x_2}{x_1{}^2}\frac{\partial F}{\partial y''}$$

$$= \frac{d}{dx}\left(\frac{\partial F}{\partial y''}\right) - \frac{x_2}{x_1{}^2}\frac{\partial F}{\partial y''};$$

and therefore

$$\frac{\partial G}{\partial y_1} - \frac{d}{dt}\left(\frac{\partial G}{\partial y_2}\right) = \frac{\partial F}{\partial y'} - \frac{d}{dx}\left(\frac{\partial F}{\partial y'}\right).$$

Next,

$$G = x_1 \frac{\partial G}{\partial x_1} + y_1 \frac{\partial G}{\partial y_1} + 2x_2 \frac{\partial G}{\partial x_2} + 2y_2 \frac{\partial G}{\partial y_2}$$

$$= x_1 \frac{\partial G}{\partial x_1} + y_1 \frac{\partial G}{\partial y_1} + 2x_1 y'' \frac{\partial F}{\partial y''}.$$

Also, $G = x_1 F$; hence

$$\frac{\partial G}{\partial x_1} = F - y'\frac{\partial G}{\partial y_1} - 2y''\frac{\partial F}{\partial y''}.$$

Again,

$$\frac{\partial G}{\partial x_2} = -\frac{y_1}{x_1}\frac{\partial G}{\partial y_2} = -y'\frac{\partial G}{\partial y_2};$$

and therefore

$$\frac{d}{dt}\left(\frac{\partial G}{\partial x_2}\right) = -y'\frac{d}{dt}\left(\frac{\partial G}{\partial y_2}\right) - \frac{\partial G}{\partial y_2}x_1 y''$$

$$= -y'\frac{d}{dt}\left(\frac{\partial G}{\partial y_2}\right) - y''\frac{\partial F}{\partial y''};$$

hence

$$\frac{\partial G}{\partial x_1} - \frac{d}{dt}\left(\frac{\partial G}{\partial x_2}\right) = F - y'\left\{\frac{\partial G}{\partial y_1} - \frac{d}{dt}\left(\frac{\partial G}{\partial y_2}\right)\right\} - y''\frac{\partial F}{\partial y''}$$

$$= F - y'\left\{\frac{\partial F}{\partial y'} - \frac{d}{dx}\left(\frac{\partial F}{\partial y''}\right)\right\} - y''\frac{\partial F}{\partial y''}.$$

When these equivalent values for combinations of derivatives of $G$ and combinations of derivatives of $F$ are used, the two sets of terminal relations are seen to agree for the corresponding types of terminal condition.

*Ex.* 1.  The integral $\int f(x, y, y', y'', y''')\, dx$ is transformed to an integral

$$\int F(x, x_1, x_2, x_3, y, y_1, y_2, y_3)\, dt$$

by changing the current variable to $t$, so that both $x$ and $y$ are functions of $t$.  Prove that the function $F$ satisfies the three identities

$$\left.\begin{aligned}
F &= x_1\frac{\partial F}{\partial x_1} + y_1\frac{\partial F}{\partial y_1} + 2x_2\frac{\partial F}{\partial x_2} + 2y_2\frac{\partial F}{\partial y_2} + 3x_3\frac{\partial F}{\partial x_3} + 3y_3\frac{\partial F}{\partial y_3} \\
0 &= x_1\frac{\partial F}{\partial x_2} + y_1\frac{\partial F}{\partial y_2} + 3x_2\frac{\partial F}{\partial x_3} + 3y_2\frac{\partial F}{\partial y_3} \\
0 &= x_1\frac{\partial F}{\partial x_3} + y_1\frac{\partial F}{\partial y_3}
\end{aligned}\right\}.$$

*Ex.* 2.  Prove that, if the preceding integral $\int F(x, x_1, x_2, x_3, y, y_1, y_2, y_3)\, dt$ is to be a maximum or minimum, $x$ and $y$ as functions of $t$ must satisfy the two characteristic equations

$$\frac{\partial F}{\partial x} - \frac{d}{dt}\left(\frac{\partial F}{\partial x_1}\right) + \frac{d^2}{dt^2}\left(\frac{\partial F}{\partial x_2}\right) - \frac{d^3}{dt^3}\left(\frac{\partial F}{\partial x_3}\right) = 0,$$

$$\frac{\partial F}{\partial y} - \frac{d}{dt}\left(\frac{\partial F}{\partial y_1}\right) + \frac{d^2}{dt^2}\left(\frac{\partial F}{\partial y_2}\right) - \frac{d^3}{dt^3}\left(\frac{\partial F}{\partial y_3}\right) = 0;$$

and verify, by means of the preceding identities, that these two equations are satisfied in virtue of a single equation which is the equivalent of

$$\frac{\partial f}{\partial y} - \frac{d}{dx}\left(\frac{\partial f}{\partial y'}\right) + \frac{d^2}{dx^2}\left(\frac{\partial f}{\partial y''}\right) - \frac{d^3}{dx^3}\left(\frac{\partial f}{\partial y'''}\right) = 0.$$

Find the terminal condition (or conditions) to be satisfied at a free limit ; and identify it (or them) with the result stated in Ex. 1, § 87.

### *Continuity of four magnitudes through a free discontinuity on the characteristic curve.*

**102.**  One further inference (similar to the proposition in § 47) may be derived from the consideration of the terms at the limits of the original integral, as follows :

*At a free place on the characteristic curve where, while the arc of the curve is continuous, there is discontinuity of curvature or of direction or of both curvature and direction, the quantities*

$$\frac{\partial G}{\partial x_1} - \frac{d}{dt}\left(\frac{\partial G}{\partial x_2}\right), \quad \frac{\partial G}{\partial x_2}, \quad \frac{\partial G}{\partial y_1} - \frac{d}{dt}\left(\frac{\partial G}{\partial y_2}\right), \quad \frac{\partial G}{\partial y_2},$$

*are continuous in value.*

Let $P$ be such a place\*, given by a value $T$ of $t$. Let $Q''$ (for the value $t''$) be a place immediately beyond $P$, and $Q'$ (for the value $t'$) be a place immediately behind $P$.

Consider a variation of the characteristic curve $AQ'PQ''B$ such that $u$ and $u_1$ (as also $v$ and $v_1$) are zero along $AQ'$; are zero along $Q''B$; and along $Q'PQ''$,

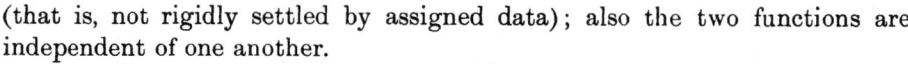

$$ u = (t - t')^2 (t'' - t)^2\, U(t), \quad v = (t - t')^2 (t'' - t)^2\, V(t), $$

where $U(t)$ and $V(t)$ are arbitrary regular functions of $t$ in the vicinity of $T$. These functions $U(t)$ and $V(t)$ are not zero, because the place $P$ is supposed to be free (that is, not rigidly settled by assigned data); also the two functions are independent of one another.

For variations of this type, continuous along the whole range, the first variation of the original integral along $AQ'$ is zero, and is also zero along $Q''B$; because both $u$ and $v$ are steadily zero. The first variation along the whole path is to be zero; and therefore the contribution by the portion $Q'PQ''$ must be zero. Thus we must have the relation

$$ \int_{t'}^{T} (uX + vY)\, dt $$

$$ + \left[ u \left\{ \frac{\partial G}{\partial x_1} - \frac{d}{dt} \left( \frac{\partial G}{\partial x_2} \right) \right\} + u_1 \frac{\partial G}{\partial x_2} + v \left\{ \frac{\partial G}{\partial y_1} - \frac{d}{dt} \left( \frac{\partial G}{\partial y_2} \right) \right\} + v_1 \frac{\partial G}{\partial y_2} \right]_{t'}^{T} $$

$$ + \int_{T}^{t''} (uX + vY)\, dt $$

$$ + \left[ u \left\{ \frac{\partial G}{\partial x_1} - \frac{d}{dt} \left( \frac{\partial G}{\partial x_2} \right) \right\} + u_1 \frac{\partial G}{\partial x_2} + v \left\{ \frac{\partial G}{\partial y_1} - \frac{d}{dt} \left( \frac{\partial G}{\partial y_2} \right) \right\} + v_1 \frac{\partial G}{\partial y_2} \right]_{T}^{t''} = 0. $$

Along $Q'P$, we have $X = 0$ and $Y = 0$, and likewise along $PQ''$. Hence the integral in the first line vanishes, and likewise the integral in the third line.

For the particular variations selected, $u = 0$, $u_1 = 0$, $v = 0$, $v_1 = 0$ when $t = t'$, and also when $t = t''$. Hence the terms at the lower limit in the second line vanish, and the terms at the upper limit in the fourth line vanish. We denote by $\theta_-$ the value of a quantity $\theta$ in the range $Q'P$ in the limit when $t$ approaches $T$ from $t'$ along $Q'P$; and we denote by $\theta_+$ the value of the quantity $\theta$ in the range $PQ''$ as $t$ approaches $T$ from $t''$ along $Q''P$. The foregoing requirement is

$$ u \left\{ \frac{\partial G}{\partial x_1} - \frac{d}{dt} \left( \frac{\partial G}{\partial x_2} \right) \right\}_- + u_1 \left( \frac{\partial G}{\partial x_2} \right)_- + v \left\{ \frac{\partial G}{\partial y_1} - \frac{d}{dt} \left( \frac{\partial G}{\partial y_2} \right) \right\}_- + v_1 \left( \frac{\partial G}{\partial y_2} \right)_- $$

$$ - \left[ u \left\{ \frac{\partial G}{\partial x_1} - \frac{d}{dt} \left( \frac{\partial G}{\partial x_2} \right) \right\}_+ + u_1 \left( \frac{\partial G}{\partial x_2} \right)_+ + v \left\{ \frac{\partial G}{\partial y_1} - \frac{d}{dt} \left( \frac{\partial G}{\partial y_2} \right) \right\}_+ + v_1 \left( \frac{\partial G}{\partial y_2} \right)_+ \right] = 0 ; $$

---

\* As in § 46, the characteristic curve generally is continuous in direction, curvature, and other intrinsic properties, because the functions $x(t)$ and $y(t)$ satisfying the characteristic equation are assumed to be analytic.

and this requirement must be met for all forms of the functions $u$ and $v$ specified, the quantities $u$, $u_1$, $v$, $v_1$ acquiring their values at the place $T$.

First, let $v$ be a persistent zero through the range $Q'PQ''$, so that $V(t)$ is zero for all values of $t$ in the range; thus $v = 0$ and $v_1 = 0$ for this selection. At the same time, let the arbitrary function $U(t)$ be chosen, so that

$$U(t) = A + B(t - T) + (t - T)^2 \phi(t),$$

where $\phi(t)$ is a regular function of $t$, and $A$, $B$ are arbitrary constants at our disposal; thus, for this selection,

$$U(T) = A, \quad \{U'(t)\}_{t=T} = B.$$

Then, at $T$,

$$u = (T - t')^2 (t'' - T)^2 A,$$

$$u_1 = (T - t')^2 (t'' - T)^2 \left\{ A \frac{2(t'' + t' - 2T)}{(T - t')(t'' - T)} + B \right\};$$

and the condition gives

$$u \left[ \left\{ \frac{\partial G}{\partial x_1} - \frac{d}{dt} \left( \frac{\partial G}{\partial x_2} \right) \right\}_- - \left\{ \frac{\partial G}{\partial x_1} - \frac{d}{dt} \left( \frac{\partial G}{\partial x_2} \right) \right\}_+ \right] + u_1 \left[ \left( \frac{\partial G}{\partial x_2} \right)_- - \left( \frac{\partial G}{\partial x_2} \right)_+ \right] = 0,$$

while $A$ and $B$ are arbitrary constants. When, as one choice, we take

$$B + A \frac{2(t'' + t' - 2T)}{(T - t')(t'' - T)} = 0,$$

the condition can only be satisfied if

$$\left\{ \frac{\partial G}{\partial x_1} - \frac{d}{dt} \left( \frac{\partial G}{\partial x_2} \right) \right\}_- = \left\{ \frac{\partial G}{\partial x_1} - \frac{d}{dt} \left( \frac{\partial G}{\partial x_2} \right) \right\}_+.$$

Thereafter, for any other choice of $B$, the condition can only be satisfied if

$$\left( \frac{\partial G}{\partial x_2} \right)_- = \left( \frac{\partial G}{\partial x_2} \right)_+.$$

Secondly, as the terms involving $u$ and $u_1$ now disappear from the condition whatever be the choice of $u$, the condition now involves $v$ and $v_1$ alone. By a corresponding choice for the function $V(t)$ in the assumed form of $v$, we have

$$\left\{ \frac{\partial G}{\partial y_1} - \frac{d}{dt} \left( \frac{\partial G}{\partial y_2} \right) \right\}_- = \left\{ \frac{\partial G}{\partial y_1} - \frac{d}{dt} \left( \frac{\partial G}{\partial y_2} \right) \right\}_+,$$

$$\left( \frac{\partial G}{\partial y_2} \right)_- = \left( \frac{\partial G}{\partial y_2} \right)_+.$$

*Note.* (α) If however the place $T$, where the hypothetical discontinuity may occur, is a definitely and rigidly fixed place, that is, fixed in position and with assigned direction at the position, the inferences cannot be drawn. For then we have

$$u = 0, \quad v = 0, \quad x_1 v_1 - y_1 u_1 = 0,$$

at $T$; and the requirement is satisfied without any residual inference.

($\beta$) Again, if the place $T$, where the hypothetical discontinuity may occur, is definitely fixed in position without any condition imposed on the direction, at such a place

$$u = 0, \quad v = 0,$$

while $x_1 v_1 - y_1 u_1$ is arbitrary. The requirement is

$$u_1 \left( \frac{\partial G}{\partial x_2} \right)_- + v_1 \left( \frac{\partial G}{\partial y_2} \right)_- = u_1 \left( \frac{\partial G}{\partial x_2} \right)_+ + v_1 \left( \frac{\partial G}{\partial y_2} \right)_+ .$$

But we have

$$x_1 \frac{\partial G}{\partial x_2} + y_1 \frac{\partial G}{\partial y_2} = 0$$

along $Q'P$ up to $P$, and so

$$x_1 \left( \frac{\partial G}{\partial x_2} \right)_- + y_1 \left( \frac{\partial G}{\partial y_2} \right)_- = 0 ;$$

and similarly

$$x_1 \left( \frac{\partial G}{\partial x_2} \right)_+ + y_1 \left( \frac{\partial G}{\partial y_2} \right)_+ = 0.$$

Thus the requirement becomes

$$\frac{1}{x_1} (x_1 v_1 - y_1 u_1) \left\{ \left( \frac{\partial G}{\partial y_2} \right)_- - \left( \frac{\partial G}{\partial y_2} \right)_+ \right\} = 0,$$

while $x_1 v_1 - y_1 u_1$ at $T$ is arbitrary. Hence

$$\left( \frac{\partial G}{\partial y_2} \right)_- = \left( \frac{\partial G}{\partial y_2} \right)_+ ;$$

and thence

$$\left( \frac{\partial G}{\partial x_2} \right)_- = \left( \frac{\partial G}{\partial x_2} \right)_+ .$$

($\gamma$) But when the place $T$ is completely free, all the four quantities specified in the proposition are continuous.

*The two characteristic equations of § 99 are equivalent to a single equation.*

**103.** Before the consideration of the second variation of the original integral, we return to the two characteristic equations

$$X = 0, \quad Y = 0.$$

It is assumed that they determine $x$ and $y$ as analytic functions of the variable $t$. When $t$ is eliminated between the equations giving the values, we have a relation between $x$ and $y$ which is the equation of the characteristic curve. Now $t$ is a parametric variable, not essential so far as the curve is concerned; and therefore it may be expected that the elimination of $t$ would leave a single characteristic equation. It is possible to prove that the two equations are, in fact, equivalent to a single equation, though, for subsequent analysis it is convenient to retain both of them.

Because $G$ is a function of $x$, $x_1$, $x_2$, $y$, $y_1$, $y_2$, we have

$$\frac{dG}{dt} = x_1 \frac{\partial G}{\partial x} + x_2 \frac{\partial G}{\partial x_1} + x_3 \frac{\partial G}{\partial x_2} + y_1 \frac{\partial G}{\partial y} + y_2 \frac{\partial G}{\partial y_1} + y_3 \frac{\partial G}{\partial y_2}$$

$$= x_1 \left\{ X + \frac{d}{dt}\left(\frac{\partial G}{\partial x_1}\right) - \frac{d^2}{dt^2}\left(\frac{\partial G}{\partial x_2}\right) \right\} + x_2 \frac{\partial G}{\partial x_1} + x_3 \frac{\partial G}{\partial x_2}$$

$$+ y_1 \left\{ Y + \frac{d}{dt}\left(\frac{\partial G}{\partial y_1}\right) - \frac{d^2}{dt^2}\left(\frac{\partial G}{\partial y_2}\right) \right\} + y_2 \frac{\partial G}{\partial y_1} + y_3 \frac{\partial G}{\partial y_2}$$

$$= x_1 X + y_1 Y - \frac{d^2}{dt^2}\left( x_1 \frac{\partial G}{\partial x_2} + y_1 \frac{\partial G}{\partial y_2} \right)$$

$$+ \frac{d}{dt}\left( x_1 \frac{\partial G}{\partial x_1} + y_1 \frac{\partial G}{\partial y_1} + 2x_2 \frac{\partial G}{\partial x_2} + 2y_2 \frac{\partial G}{\partial y_2} \right);$$

and therefore, owing to the two identities

$$\left. \begin{aligned} G &= x_1 \frac{\partial G}{\partial x_1} + y_1 \frac{\partial G}{\partial y_1} + 2x_2 \frac{\partial G}{\partial x_2} + 2y_2 \frac{\partial G}{\partial y_2} \\ 0 &= x_1 \frac{\partial G}{\partial x_2} + y_1 \frac{\partial G}{\partial y_2} \end{aligned} \right\},$$

it follows that

Thus we can have

$$x_1 X + y_1 Y = 0.$$

$$Y = x_1 E, \quad X = -y_1 E;$$

and there is a single characteristic equation $E = 0$, the explicit form of which is obtained later (§ 105, p. 151). But, as already stated, the simultaneous equations $X = 0$, $Y = 0$, may prove more useful than the single equation $E = 0$.

### *The identities* (§ 97) *satisfied by* $G$ : *invariantive forms.*

**104.** At the beginning of the investigation, a non-intrinsic new independent variable $t$ was chosen, in terms of which $x$ and $y$ and their derivatives were to be expressible; and there is a convenience in having some non-intrinsic variable, so long as the characteristic curve is unknown. But when once the curve can be regarded as known, and its properties are to be derived from its known character, advantage may accrue from the choice of the arc-length (that is, the length of the arc measured along the curve from some fixed point) as the independent variable. The forms, that result from the choice of $s$, may be regarded as canonical.

Meanwhile, let another variable $\theta$ be taken; we write

$$\bar{x}_1 = \frac{dx}{d\theta} = x_1 \frac{dt}{d\theta} = mx_1, \qquad \bar{y}_1 = \frac{dy}{d\theta} = my_1,$$

$$\bar{x}_2 = \frac{d^2x}{d\theta^2} = m^2 x_2 + m\frac{dm}{dt} x_1, \qquad \bar{y}_2 = \frac{d^2y}{d\theta^2} = m^2 y_2 + m\frac{dm}{dt} y_1,$$

where

$$m = \frac{dt}{d\theta}.$$

The forms of $F\left(x,\, y,\, \frac{dy}{dx},\, \frac{d^2y}{dx^2}\right) dx$, after the respective changes, are

$$G\left(x,\, y,\, \bar{x}_1,\, \bar{y}_1,\, \bar{x}_2,\, \bar{y}_2\right) d\theta, \quad G\left(x,\, y,\, x_1,\, y_1,\, x_2,\, y_2\right) dt,$$

which are necessarily equal to one another; hence

$$G\left(x,\, y,\, \bar{x}_1,\, \bar{y}_1,\, \bar{x}_2,\, \bar{y}_2\right) = mG\left(x,\, y,\, x_1,\, y_1,\, x_2,\, y_2\right);$$

and therefore

$$G\left(x,\, y,\, mx_1,\, my_1,\, m^2x_2 + m\frac{dm}{dt}x_1,\, m^2y_2 + m\frac{dm}{dt}y_1\right) = mG\left(x,\, y,\, x_1,\, y_1,\, x_2,\, y_2\right).$$

Here $t$ and $\theta$ are arbitrary variables; by appropriate selections some useful inferences can be derived.

(a)  Let $m = -1$; then

$$G\left(x,\, y,\, -x_1,\, -y_1,\, x_2,\, y_2\right) = -G\left(x,\, y,\, x_1,\, y_1,\, x_2,\, y_2\right),$$

a result that will be useful when strong variations are considered.

(b)  Let $m = 1 + \epsilon$, where $\epsilon$ is a small arbitrary variable quantity such that $\frac{d\epsilon}{dt}$ is a small quantity of the same order of magnitude as $\epsilon$. Then

$$G\left(x,\, y,\, x_1 + \epsilon x_1,\, y_1 + \epsilon y_1,\, x_2 + 2\epsilon x_2 + x_1\frac{d\epsilon}{dt} + \dots,\, y_2 + 2\epsilon y_2 + y_1\frac{d\epsilon}{dt} + \dots\right)$$
$$= (1 + \epsilon)\, G\left(x,\, y,\, x_1,\, y_1,\, x_2,\, y_2\right),$$

where the unexpressed terms on the left-hand side are of the second order of small quantities.  Expanding in powers of the arbitrary small quantity $\epsilon$ and of $\frac{d\epsilon}{dt}$, and equating the coefficients of $\epsilon$ and of $\frac{d\epsilon}{dt}$ on the two sides, we have

$$\left.\begin{aligned} G &= x_1\frac{\partial G}{\partial x_1} + y_1\frac{\partial G}{\partial y_1} + 2x_2\frac{\partial G}{\partial x_2} + 2y_2\frac{\partial G}{\partial y_2} \\ 0 &= x_1\frac{\partial G}{\partial x_2} + y_1\frac{\partial G}{\partial y_2} \end{aligned}\right\},$$

the fundamental identities satisfied by $G$. Thus the two fundamental identities are a partial expression of the invariance of the differential element

$$G\left(x,\, y,\, x_1,\, y_1,\, x_2,\, y_2\right) dt$$

for all changes of the independent variable $t$.

(c)  When the characteristic curve is such that the independent variable may be taken to be the length $s$ of the arc measured from a fixed point, and we write

$$\frac{dx}{ds} = x', \quad \frac{dy}{ds} = y', \quad \frac{d^2x}{ds^2} = x'', \quad \frac{d^2y}{ds^2} = y'',$$

we have

$$G\left(x,\, y,\, x',\, y',\, x'',\, y''\right) = t'G\left(x,\, y,\, x_1,\, y_1,\, x_2,\, y_2\right);$$

and the fundamental identities are

$$G = x' \frac{\partial G}{\partial x'} + y' \frac{\partial G}{\partial y'} + 2x'' \frac{\partial G}{\partial x''} + 2y'' \frac{\partial G}{\partial y''} \Bigg\}\,,$$

$$0 = x' \frac{\partial G}{\partial x''} + y' \frac{\partial G}{\partial y''}$$

together with

$$x'^2 + y'^2 = 1, \quad x'x'' + y'y'' = 0.$$

*Ex.* Let

$$\Theta = G\,(x, y, \bar{x}_1, \bar{y}_1, \bar{x}_2, \bar{y}_2), \quad T = G\,(x, y, x_1, y_1, x_2, y_2),$$

so that

$$\Theta = mT\,;$$

prove that

$$\frac{\partial \Theta}{\partial x} - \frac{d}{d\theta}\left(\frac{\partial \Theta}{\partial \bar{x}_1}\right) + \frac{d^2}{d\theta^2}\left(\frac{\partial \Theta}{\partial \bar{x}_2}\right) = m \left\{ \frac{\partial T}{\partial x} - \frac{d}{dt}\left(\frac{\partial T}{\partial x_1}\right) + \frac{d^2}{dt^2}\left(\frac{\partial T}{\partial x_2}\right)\right\},$$

$$\frac{\partial \Theta}{\partial y} - \frac{d}{d\theta}\left(\frac{\partial \Theta}{\partial \bar{y}_1}\right) + \frac{d^2}{d\theta^2}\left(\frac{\partial \Theta}{\partial \bar{y}_2}\right) = m \left\{ \frac{\partial T}{\partial y} - \frac{d}{dt}\left(\frac{\partial T}{\partial y_1}\right) + \frac{d^2}{dt^2}\left(\frac{\partial T}{\partial y_2}\right)\right\}.$$

Thus $X$ and $Y$ are invariantive under change of the independent variable.

### Relations among the second derivatives of $G$.

**105.** We now proceed to consider the integral $\int \mathfrak{G}\, dt$ in the second variation of the original integral. The variational quantities $u$ and $v$, which occur
in $\mathfrak{G}$, are arbitrary regular functions of $t$. As the limits of the integral are
completely specified by the enforced vanishing of the first variation, $u$ and $v$
(as well as their derivatives) vanish at the limits. Let $\square$ denote any homogeneous quadratic function of $u$, $v$, $u_1$, $v_1$, so that $\square$ must vanish at each
limit. The integral in the second variation, which can be expressed in the
form

$$\int\left(\mathfrak{G} - \frac{d\square}{dt}\right) dt + [\square],$$

is equal to the integral $\int\left(\mathfrak{G} - \dfrac{d\square}{dt}\right) dt$. The quantity $\square$ will be chosen so as
to simplify the expression under the integral sign, though it does not affect
the complete value of the second variation.

In securing this modification to a normal simplified form, special account
must be taken of the coefficients of the combinations of $u$, $v$, $u'$, $v'$, $u''$, $v''$
which occur in $\mathfrak{G}$. These coefficients (as defined in § 97) are not independent
of one another; certain relations subsist among them, owing to the two
identities satisfied by the function $\mathfrak{G}$. As in § 51, a knowledge of these
relations is a convenient preliminary to the reduction of the second variation.

(i) From the identity

$$x_1 \frac{\partial G}{\partial x_2} + y_1 \frac{\partial G}{\partial y_2} = 0,$$

partially differentiated with respect to $x_2$ and $y_2$ successively, we have

$$x_1\gamma_{22} + y_1 k_{22} = 0, \quad x_1 k_{22} + y_1 c_{22} = 0.$$

We therefore may take a quantity $Q$, such that

$$c_{22} = x_1{}^2 Q, \quad k_{22} = -x_1 y_1 Q, \quad \gamma_{22} = y_1{}^2 Q.$$

It is easy to verify that

$$x_1{}^5 Q = \frac{\partial^2 F}{\partial y''_2}.$$

Thus, as $x_1$ does not vanish or become infinite in the range, because $x$ in the original form and $t$ in the present form increase continuously, $Q$ reverts to the quantity $a_{22}$ (of § 88) when $x$ is made the independent variable.

(ii) When the same identity is differentiated partially with respect to $x_1$ and $y_1$ successively, we have

$$\frac{\partial G}{\partial x_2} + x_1\gamma_{12} + y_1 k_{12} = 0, \quad \frac{\partial G}{\partial y_2} + x_1 k_{21} + y_1 c_{12} = 0.$$

When the other identity

$$G = x_1\frac{\partial G}{\partial x_1} + y_1\frac{\partial G}{\partial y_1} + 2x_2\frac{\partial G}{\partial x_2} + 2y_2\frac{\partial G}{\partial y_2}$$

is differentiated partially with respect to $x_2$ and $y_2$ successively, we have

$$0 = \frac{\partial G}{\partial x_2} + x_1\gamma_{12} + y_1 k_{21} + 2x_2\gamma_{22} + 2y_2 k_{22},$$

$$0 = \frac{\partial G}{\partial y_2} + x_1 k_{12} + y_1 c_{12} + 2x_2 k_{22} + 2y_2 c_{22}.$$

Combining the two first equations of these two sets, we have

$$y_1(k_{21} - k_{12}) + 2y_1(x_2 y_1 Q - x_1 y_2 Q) = 0,$$

that is,

$$k_{12} + 2x_1 y_2 Q = k_{21} + 2x_2 y_1 Q.$$

The same result follows from combining the two second equations of the two sets. Accordingly, new quantities $L$, $M$, $N$ are taken, such that

$$\left.\begin{aligned}
\gamma_{12} &= L + 2y_1 y_2 Q \\
k_{12} &= M - 2x_1 y_2 Q \\
k_{21} &= M - 2y_1 x_2 Q \\
c_{12} &= N + 2x_1 x_2 Q
\end{aligned}\right\}.$$

With these definitions of $L$, $M$, $N$, we retain the two relations

$$Lx_1 + My_1 = -\frac{\partial G}{\partial x_2}, \quad Mx_1 + Ny_1 = -\frac{\partial G}{\partial y_2};$$

and then full account is taken of the four derived equations.

(iii) When the identity

$$G = x_1\frac{\partial G}{\partial x_1} + y_1\frac{\partial G}{\partial y_1} + 2x_2\frac{\partial G}{\partial x_2} + 2y_2\frac{\partial G}{\partial y_2}$$

is differentiated partially with respect to $x_1$ and $y_1$ successively, we have

$$0 = x_1\gamma_{11} + y_1 k_{11} + 2x_2\gamma_{12} + 2y_2 k_{12}$$
$$= x_1(\gamma_{11} - 4y_2^2 Q) + y_1(k_{11} + 4x_2 y_2 Q) + 2Lx_2 + 2My_2,$$
$$0 = x_1(k_{11} + 4x_2 y_2 Q) + y_1(c_{11} - 4x_2^2 Q) + 2Mx_2 + 2Ny_2.$$

We introduce quantities $R$, $S$, $T$, defined by the relations

$$\left.\begin{aligned} \gamma_{11} &= R + 4y_2^2 Q \\ k_{11} &= S - 4x_2 y_2 Q \\ c_{11} &= T + 4x_2^2 Q \end{aligned}\right\};$$

these new derived equations become

$$Rx_1 + Sy_1 + 2Lx_2 + 2My_2 = 0, \quad Sx_1 + Ty_1 + 2Mx_2 + 2Ny_2 = 0.$$

(iv)  When the identity

$$x_1\frac{\partial G}{\partial x_2} + y_1\frac{\partial G}{\partial y_2} = 0$$

is differentiated partially with respect to $x$ and $y$ successively, we have

$$x_1\gamma_{02} + y_1 k_{02} = 0, \quad x_1 k_{20} + y_1 c_{02} = 0.$$

We introduce quantities $C$ and $D$, defined by the equations

$$\left.\begin{aligned} \gamma_{02} &= y_1 C \\ k_{02} &= -x_1 C \end{aligned}\right\}, \quad \left.\begin{aligned} k_{20} &= -y_1 D \\ c_{02} &= x_1 D \end{aligned}\right\};$$

but sometimes it will prove preferable to revert to the actual coefficients $\gamma_{02}$, $k_{02}$, $k_{20}$, $c_{02}$.

(v)  Differentiating, completely with regard to $t$, the relation

$$Lx_1 + My_1 = -\frac{\partial G}{\partial x_2}$$

obtained in (ii) above, we have

$$-(x_1 L' + y_1 M' + x_2 L + y_2 M) = \frac{d}{dt}\left(\frac{\partial G}{\partial x_2}\right)$$
$$= x_1\gamma_{02} + y_1 k_{20} + x_2\gamma_{12} + y_2 k_{21} + x_3\gamma_{22} + y_3 k_{22}$$
$$= x_1 y_1 C - y_1^2 D + x_2 L + y_2 M + y_1(y_1 x_3 - x_1 y_3) Q;$$

that is, on the substitution of $-(Rx_1 + Sy_1)$ for $2x_2 L + 2y_2 M$ from (iii),

$$x_1\{R - L' - y_1(C - y_3 Q)\} + y_1\{S - M' + y_1(D - x_3 Q)\} = 0.$$

Proceeding similarly from the relation

$$Mx_1 + Ny_1 = -\frac{\partial G}{\partial y_2},$$

also obtained in (ii) above, we find

$$x_1\{S - M' + x_1(C - y_3 Q)\} + y_1\{T - N' - x_1(D - x_3 Q)\} = 0.$$

The two relations, thus deduced, shew that a quantity $J$ can be taken such that

$$\left.\begin{array}{l} R-L'=- \quad y_1{}^2 J + 2y_1(C-y_3 Q) \\ S-M'= \quad x_1 y_1 J - \quad x_1(C-y_3 Q) - \quad y_1(D-x_3 Q) \\ T-N'=- \quad x_1{}^2 J \qquad\qquad\qquad + 2x_1(D-x_3 Q) \end{array}\right\}.$$

(vi) Differentiating, partially with regard to $y$, the identity

$$G = x_1\frac{\partial G}{\partial x_1} + y_1\frac{\partial G}{\partial y_1} + 2x_2\frac{\partial G}{\partial x_2} + 2y_2\frac{\partial G}{\partial y_2},$$

we have

$$\frac{\partial G}{\partial y} = x_1 k_{10} + y_1 c_{01} + 2x_2 k_{20} + 2y_2 c_{02}.$$

Also

$$\frac{d}{dt}\left(\frac{\partial G}{\partial y_1}\right) = x_1 k_{01} + y_1 c_{01} + x_2 k_{11} + y_2 c_{11} + x_3 k_{21} + y_3 c_{12},$$

$$\frac{d^2}{dt^2}\left(\frac{\partial G}{\partial y_2}\right) = \frac{d}{dt}(x_1 k_{02} + y_1 c_{02} + x_2 k_{12} + y_2 c_{12} + x_3 k_{22} + y_3 c_{22}).$$

Hence, writing $\dfrac{dk_{02}}{dt} = k_{02}'$ and similarly for the other quantities, we have

$$Y = \frac{\partial G}{\partial y} - \frac{d}{dt}\left(\frac{\partial G}{\partial y_1}\right) + \frac{d^2}{dt^2}\left(\frac{\partial G}{\partial y_2}\right)$$

$$= x_1(k_{10} - k_{01} + k_{02}') + x_2(2k_{20} - k_{11} + k_{02} + k_{12}') + x_3(-k_{21} + k_{12} + k_{22}') + x_4 k_{22}$$
$$+ y_1 c_{02}' + y_2(2c_{02} - c_{11} + c_{02} + c_{12}') + y_3 c_{22}' + y_4 c_{22}.$$

When substitution is made for the various coefficients $c_{lm}$ and $k_{lm}$ in terms of the quantities that have been introduced, we find

$$Y = x_1 E,$$

where

$$E = k_{10} - k_{01} + k_{02}' - k_{20}'$$
$$+ (x_1 y_4 - x_4 y_1 - x_2 y_3 + x_3 y_2)Q + (x_1 y_3 - x_3 y_1)Q' + (x_1 y_2 - x_2 y_1)J.$$

Proceeding in the same way to a similar expression for $X$, denoting

$$\frac{\partial G}{\partial x} - \frac{d}{dt}\left(\frac{\partial G}{\partial x_1}\right) + \frac{d^2}{dt^2}\left(\frac{\partial G}{\partial x_2}\right),$$

we find

$$X = -y_1 E.$$

We thus verify the result, already (§ 103) established, that the two equations $X = 0$ and $Y = 0$ are satisfied in virtue of a single equation, in virtue of the identities satisfied by $G$. This single equation is

$$E = 0;$$

it is the *characteristic equation* (constituting the *Euler* test). Its primitive gives, in effect, the *characteristic curve*.

The characteristic equation can be expressed in the form

$$k_{01} - k_{02}' - Q(x_1 y_4 - x_2 y_3) - Q' x_1 y_3 - J x_1 y_2$$
$$= k_{10} - k_{20}' - Q(y_1 x_4 - y_2 x_3) - Q' y_1 x_3 - J y_1 x_2.$$

We therefore may take a quantity $\Gamma$, such that

$$\left.\begin{array}{l} k_{01} - k_{02}' = \Gamma + Q\,(x_1 y_4 - x_2 y_3) + Q' x_1 y_3 + J x_1 y_2 \\ k_{10} - k_{20}' = \Gamma + Q\,(y_1 x_4 - y_2 x_3) + Q' y_1 x_3 + J y_1 x_2 \end{array}\right\},$$

the equivalence of the two values of $\Gamma$ representing the characteristic equation.

(vii) Lastly, proceeding from the relation

$$\frac{\partial G}{\partial y} = x_1 k_{10} + y_1 c_{01} + 2 x_2 k_{20} + 2 y_2 c_{02}$$

derived at the beginning of (vi), and differentiating both sides completely with regard to $t$, we have

$$x_1 k_{00} + y_1 c_{00} + x_2 k_{10} + y_2 c_{01} + x_3 k_{20} + y_3 c_{02}$$
$$= x_1 k_{10}' + x_2 k_{10} + y_1 c_{01}' + y_2 c_{01} + 2 x_2 k_{20}' + 2 x_3 k_{20} + 2 y_2 c_{02}' + 2 y_3 c_{02}.$$

Hence

$$x_1\,(k_{00} - k_{10}' + k_{20}'') + y_1\,(c_{00} - c_{01}' + c_{02}'')$$
$$= x_1 k_{20}'' + 2 x_2 k_{20}' + x_3 k_{20} + y_1 c_{02}'' + 2 y_2 c_{02}' + y_3 c_{02}$$
$$= \frac{d^2}{dt^2}\,(x_1 k_{20} + y_1 c_{02}) = 0.$$

Proceeding similarly from the relation

$$\frac{\partial G}{\partial x} = x_1 \gamma_{01} + y_1 k_{01} + 2 x_2 \gamma_{02} + 2 y_2 k_{02},$$

obtained by differentiating the identity

$$G = x_1 \frac{\partial G}{\partial x_1} + y_1 \frac{\partial G}{\partial y_1} + 2 x_2 \frac{\partial G}{\partial x_2} + 2 y_2 \frac{\partial G}{\partial y_2}$$

partially with respect to $x$, we find

$$x_1\,(\gamma_{00} - \gamma_{01}' + \gamma_{02}'') + y_1\,(k_{00} - k_{01}' + k_{02}'') = 0.$$

We thus can take two quantities $U$ and $V$, such that

$$\gamma_{00} - \gamma_{01}' + \gamma_{02}'' = y_1 U, \qquad k_{00} - k_{01}' + k_{02}'' = -x_1 U,$$
$$k_{00} - k_{10}' + k_{20}'' = -y_1 V, \qquad c_{00} - c_{01}' + c_{02}'' = x_1 V.$$

Equating the two values of $k_{00}$, we have

$$y_1 V - x_1 U = k_{10}' - k_{01}' - k_{20}'' + k_{02}''$$
$$= -(x_1 y_2 - x_2 y_1)\,J' - (x_1 y_3 - x_3 y_1)\,J$$
$$\quad - Q\,(x_1 y_5 - x_5 y_1) - 2 Q'\,(x_1 y_4 - x_4 y_1) - Q''\,(x_1 y_3 - x_3 y_1),$$

by means of the characteristic equation $F = 0$; thus

$$y_1\left\{ V - \frac{d}{dt}\,(x_2 J) - \frac{d^2}{dt^2}\,(x_3 Q) \right\} = x_1 \left\{ U - \frac{d}{dt}\,(y_2 J) - \frac{d^2}{dt^2}\,(y_3 Q) \right\}.$$

Hence a quantity $\Lambda$ exists, such that

$$V = x_1 \Lambda + \frac{d}{dt}\,(x_2 J) + \frac{d^2}{dt^2}\,(x_3 Q), \quad U = y_1 \Lambda + \frac{d}{dt}\,(y_2 J) + \frac{d^2}{dt^2}\,(y_3 Q);$$

and therefore, when these values of $U$ and $V$ are used, we have

$$
\left.
\begin{aligned}
c_{00} - c_{01}' + c_{02}'' &= \phantom{-} x_1\left\{ x_1\Lambda + \frac{d}{dt}(x_2 J) + \frac{d^2}{dt^2}(x_3 Q) \right\} \\
k_{00} - k_{10}' + k_{20}'' &= -y_1\left\{ x_1\Lambda + \frac{d}{dt}(x_2 J) + \frac{d^2}{dt^2}(x_3 Q) \right\} \\
k_{00} - k_{01}' + k_{02}'' &= -x_1\left\{ y_1\Lambda + \frac{d}{dt}(y_2 J) + \frac{d^2}{dt^2}(y_3 Q) \right\} \\
\gamma_{00} - \gamma_{01}' + \gamma_{02}'' &= \phantom{-} y_1\left\{ y_1\Lambda + \frac{d}{dt}(y_2 J) + \frac{d^2}{dt^2}(y_3 Q) \right\}
\end{aligned}
\right\}.
$$

These various equations constitute the aggregate of relations among the coefficients $\gamma_{mn}$, $k_{mn}$, $c_{mn}$, to be used in the transformation of the second variation.

### First normal form of the ' second ' variation.

**106.** In order to modify the expression for $\mathfrak{G}$, which occurs in that second variation, we introduce the former linear combination $w$ of $u$ and $v$, defined by the relation

$$ w = x_1 v - y_1 u\,; $$

so that $w$ is a measure of the deviation* along the normal to the curve caused by the small variation $\kappa u$ and $\kappa v$. (On this account, the quantity $w$ will occasionally be called the *deviation*.) We have

$$
\begin{aligned}
w' &= x_1 v_1 - y_1 u_1 + x_2 v - y_2 u, \\
w'' &= x_1 v_2 - y_1 u_2 + 2(x_2 v_1 - y_2 u_1) + x_3 v - y_3 u,
\end{aligned}
$$

and similarly for higher derivatives of $w$. As the limits can now be regarded as fixed, either from the express original data or by inferences deduced from original conditions satisfied in the process of making the first variation vanish, the quantities $u$ and $v$ (as well as their derivatives) vanish at each of the limits.

The expression for $\mathfrak{G}$, as already indicated in § 105, is to be modified by using the relation

$$ \int \mathfrak{G}\,dt = \int\left( \mathfrak{G} - \frac{d\square}{dt} \right)dt + [\square], $$

where $\square$ is an aggregate of terms homogeneous and of the second degree in $u$, $v$, $u_1$, $v_1$ (but not $u_2$ nor $v_2$), chosen so as to reduce $\mathfrak{G} - \dfrac{d\square}{dt}$ to what seems the simplest form for our purpose. This form is linear in $w''^2$, $w'^2$, $w^2$; that is, it is devoid of terms in $w''w'$, $w''w$, $w'w$; it involves no combinations of $u$

---

* If the independent variable $t$ were the arc $s$ of the curve measured along the curve from a fixed point, $\kappa w$ would be the actual deviation along the normal; but, as already stated, it is convenient to use a free independent variable such as $t$, distinct from $s$.

and $v$, and their derivatives, other than those which occur in $w''$, $w'$, $w$. But this characteristic simplicity of form does not apply to $\square$; it is a combination of $u$, $v$, $u_1$, $v_1$ not expressible solely in terms of $w$ and $w'$. As, however, after the transformation of the expression $\mathfrak{G} - \dfrac{d\square}{dt}$, the quantity $\square$ arises in our analysis solely at the limits, at each of which it vanishes, this lack of expressibility in terms of $w$ and $w'$ is of no significance*.

**107.** The quantity $\square$ is taken to be

$$\square = \alpha v_1{}^2 + 2\beta v_1 u_1 + \gamma u_1{}^2 + \rho v^2 + 2\sigma uv + \tau u^2$$
$$+ 2\theta vv_1 + 2\phi vu_1 + 2\psi uv_1 + 2\chi uu_1,$$

where the coefficients $\alpha$, $\beta$, ..., $\psi$, $\chi$ do not involve $u$, $v$, $u_1$, $v_1$, and are functions of $t$ at our disposal. In $\mathfrak{G}$, the values of $x$ and $y$ (and their derivatives), given as functions of $t$ by the primitive of the characteristic equations, are supposed substituted in all the coefficients $\gamma_{mn}$, $k_{mn}$, $c_{mn}$. We write **G** as the equivalent of $\mathfrak{G} - \dfrac{d\square}{dt}$, which (for brevity) will be called the earlier form. We proceed to shew that, by choice of the coefficients $\alpha$, $\beta$, ..., $\psi$, $\chi$, it is possible to have the expression

$$\mathbf{G} = Qw''^2 + Iw'^2 + Kw^2,$$

which (for brevity) will be called the later form; here, $I$ and $K$ are functions that do not involve $u$, $v$, $u_1$, $v_1$.

To effect the equivalence of the two forms, we make the terms, in selected sets, the same in the forms, as follows:

(a) The terms in $u_2{}^2$, $u_2 v_2$, $v_2{}^2$ in the earlier form are

$$= \gamma_{22} u_2{}^2 + 2k_{22} u_2 v_2 + c_{22} v_2{}^2$$
$$= Q\,(x_1 v_2 - y_1 u_2)^2:$$

that is, they are the same as the corresponding set of terms in the later form.

(b) The terms in $u_1 u_2$, $u_1 v_2$, $v_1 u_2$, $v_1 v_2$ in the earlier form are

$$= 2\gamma_{12} u_1 u_2 + 2k_{12} u_1 v_2 + 2k_{21} u_2 v_1 + 2c_{12} v_1 v_2$$
$$- \{2\gamma u_1 u_2 + 2\beta\,(u_1 v_2 + u_2 v_1) + 2\alpha v_1 v_2\}$$
$$= 2\,\{(L - \gamma)\,u_1 u_2 + (M - \beta)\,(u_1 v_2 + u_2 v_1) + (N - \alpha)\,v_1 v_2\}$$
$$+ 4Q\,(x_1 v_2 - y_1 u_2)\,(x_2 v_1 - y_2 u_1).$$

These are the same as the corresponding set of terms in the later form, provided

$$\gamma = L, \quad \beta = M, \quad \alpha = N:$$

which, accordingly, will be taken as the values of $\alpha$, $\beta$, $\gamma$.

---

* Had it been of significance, we should have introduced the quantity $T$, $= x_1 u + y_1 v$ (of § 52, Note). As we shall establish the reduced form of $\mathfrak{G} - \dfrac{d\square}{dt}$ in a shape which is independent of $T$, and as $\square$ vanishes at each limit, this quantity $T$ remains latent and devoid of explicit influence.

(c) The terms in $uu_2$, $uv_2$, $vu_2$, $vv_2$ in the earlier form are

$$= 2\gamma_{02}uu_2 + 2k_{02}uv_2 + 2k_{20}u_2v + 2c_{02}vv_2$$
$$- (2\theta vv_2 + 2\phi vu_2 + 2\psi uv_2 + 2\chi uu_2).$$

The corresponding terms in the later form are

$$2Qx_1x_3vv_2 - 2Qy_1x_3vu_2 - 2Qx_1y_3uv_2 + 2Qy_1y_3uu_2.$$

These two sets of terms are the same if

$$Qx_1x_3 = c_{02} - \theta = \quad x_1D - \theta,$$
$$- Qy_1x_3 = k_{20} - \phi = -y_1D - \phi,$$
$$- Qx_1y_3 = k_{02} - \psi = -x_1C - \psi,$$
$$Qy_1y_3 = \gamma_{02} - \chi = \quad y_1C - \chi,$$

that is, if

$$\left.\begin{array}{l}\theta = \quad x_1(D - x_3Q) = c_{02} - x_1x_3Q \\ \phi = -y_1(D - x_3Q) = k_{20} + y_1x_3Q \\ \psi = -x_1(C - y_3Q) = k_{02} + x_1y_3Q \\ \chi = \quad y_1(C - y_3Q) = \gamma_{02} - y_1y_3Q\end{array}\right\} :$$

which, accordingly, will be taken as the values of $\theta$, $\phi$, $\psi$, $\chi$.

(d) The terms in $u_1^2$, $u_1v_1$, $v_1^2$ in the earlier form are

$$= \gamma_{11}u_1^2 + 2k_{11}u_1v_1 + c_{11}v_1^2$$
$$- \{(\alpha' + 2\theta)v_1^2 + (2\beta' + 2\phi + 2\psi)u_1v_1 + (\gamma' + 2\chi)u_1^2\}.$$

The corresponding terms in the later form are

$$4Q(x_2v_1 - y_2u_1)^2 + I(x_1v_1 - y_1u_1)^2.$$

Equating the coefficients of $v_1^2$, we have

$$c_{11} - \alpha' - 2\theta = -x_1^2J + 4x_2^2Q,$$

on substituting the expressions obtained for $c_{11}$, $\alpha$, $\theta$; and similarly for the coefficients of $u_1v_1$, $u_1^2$. The two sets of terms are the same if

$$I = -J,$$

which, accordingly, will be taken as the value of $I$.

(e) The terms in $uu_1$, $uv_1$, $vu_1$, $vv_1$ in the earlier form are

$$= 2\gamma_{01}uu_1 + 2k_{01}uv_1 + 2k_{10}u_1v + 2c_{01}vv_1$$
$$- \{(2\theta' + 2\rho)vv_1 + (2\psi' + 2\sigma)uv_1 + (2\phi' + 2\sigma)u_1v + (2\chi' + 2\tau)uu_1\}.$$

The corresponding terms in the later form are

$$4Q(x_2v_1 - y_2u_1)(x_3v - y_3u) + 2J(x_2v - y_2u)(x_1v_1 - y_1u_1).$$

These two sets of terms are the same if

$$c_{01} - \theta' - \rho = \quad 2Qx_2x_3 - Jx_1x_2,$$
$$k_{10} - \phi' - \sigma = -2Qx_3y_2 + Jx_2y_1,$$
$$k_{01} - \psi' - \sigma = -2Qx_2y_3 + Jx_1y_2,$$
$$\gamma_{01} - \chi' - \tau \doteq \quad 2Qy_2y_3 - Jy_1y_2,$$

that is, if

$$\left.\begin{aligned}
\rho &= c_{01} - c_{02}' + Q(x_1 x_4 - x_2 x_3) + Q' x_1 x_3 + J x_1 x_2 \\
\sigma &= k_{10} - k_{20}' - Q(x_4 y_1 - x_3 y_2) - Q' x_3 y_1 - J x_2 y_1 \\
\sigma &= k_{01} - k_{02}' - Q(x_1 y_4 - x_2 y_3) - Q' x_1 y_3 - J x_1 y_2 \\
\tau &= \gamma_{01} - \gamma_{02}' + Q(y_1 y_4 - y_2 y_3) + Q' y_1 y_3 + J y_1 y_2
\end{aligned}\right\}.$$

The two expressions for $\sigma$ are equivalent to one another, on account of the single characteristic equation $E = 0$; and each is equal to $\Gamma$. Accordingly, these relations will be taken as defining the values of $\rho$, $\sigma$, $\tau$.

($f$) The terms in $u^2$, $uv$, $v^2$ in the earlier form are

$$(c_{00} - \rho') v^2 + 2(k_{00} - \sigma') uv + (\gamma_{00} - \tau') u^2.$$

The corresponding terms in the later form are

$$Q(x_3 v - y_3 u)^2 - J(x_2 v - y_2 u)^2 + K(x_1 v - y_1 u)^2.$$

In order that the two sets of terms may be the same, we must have

$$\begin{aligned}
c_{00} - \rho' &= \quad Q x_3^2 \quad - J x_2^2 \quad + K x_1^2, \\
k_{00} - \sigma' &= - Q x_3 y_3 + J x_2 y_2 - K x_1 y_1, \\
\gamma_{00} - \tau' &= \quad Q y_3^2 \quad - J y_2^2 \quad + K y_1^2.
\end{aligned}$$

When the foregoing value of $\rho$ is substituted in the first of these relations, as well as the value of $c_{00} - c_{01}' + c_{02}''$, it is satisfied if

$$K = \Lambda.$$

The same condition allows the second relation to be satisfied when the prior value of $\sigma$ is substituted as well as the value of $k_{00} - k_{10}' + k_{20}''$: and also to be satisfied when the later value of $\sigma$ is substituted as well as the value of $k_{00} - k_{01}' + k_{02}''$. And the same condition allows the third relation to be satisfied when the value of $\tau$ is substituted as well as the value of $\gamma_{00} - \gamma_{01}' + \gamma_{02}''$. Accordingly, we take

$$K = \Lambda,$$

as giving the value of $K$ in the expression for $\mathbf{G}$.

Summarising these results, we have

$$\square = L u_1^2 + 2 M u_1 v_1 + N v_1^2 + \rho v^2 + 2 \sigma u v + \tau u^2$$
$$+ 2 \{ (\gamma_{02} - y_1 y_3 Q) u u_1 + (k_{02} + x_1 y_3 Q) u v_1 + (k_{20} + x_3 y_1 Q) v u_1 + (c_{02} - x_1 x_3 Q) v v_1 \},$$

$$\mathfrak{G} - \frac{d\square}{dt} = \mathbf{G} = Q w''^2 - J w'^2 + \Lambda w^2.$$

Other forms can be given to $\square$; as $\square$ vanishes at the limits, these alternatives cease to be significant.

The first modified form of the integral in the second variation is

$$\int \mathbf{G}\, dt.$$

*Subsidiary characteristic equation.*

**108.** The critical expression, for the second variation of an integral involving derivatives of the first order in Chapter II, was constructed by reference to the subsidiary characteristic equation—that is, the equation obtained for a small variation of a characteristic curve which leaves it a characteristic curve. The critical expression for the second variation of the integral now under consideration can be constructed by reference to the analogous equation for the present case.

The equation $E = 0$ is a differential equation for the determination of the characteristic curve, as representing the relation between $x$ and $y$. The two equations $X = 0$ and $Y = 0$ are differential equations for the determination of $x$ and $y$ as analytic functions of an independent variable $t$; and the elimination of $t$ between the equations, expressing these functional forms, leads to the characteristic curve. In either form, whether $E = 0$ be adopted, or $X = 0$ and $Y = 0$ be adopted, the primitive contains four essential arbitrary independent constants; and the differential equation, or the differential equations, must be satisfied whatever be the values of these constants.

Now it is possible to pass from a characteristic curve to another (and adjacent) characteristic curve by means of a small variation. Let this be such that the point $(x, y)$ on the former is displaced to the position $(x + \kappa\xi, y + \kappa\eta)$ on the latter, where $\xi$ and $\eta$ are such that the characteristic equations are satisfied by $x + \kappa\xi, y + \kappa\eta$. To express this requirement concerning the small variation represented by $\kappa\xi$ and $\kappa\eta$, let $2\Omega$ denote the same function of $\xi$ and $\eta$ as $\mathfrak{G}$ is of $u$ and $v$, so that

$$
\begin{aligned}
2\Omega = \quad & c_{00}\eta^2 + 2c_{01}\eta\eta_1 + 2c_{02}\eta\eta_2 \;+ c_{11}\eta_1{}^2 + 2c_{12}\eta_1\eta_2 + c_{22}\eta_2{}^2 \\
& + \gamma_{00}\xi^2 + 2\gamma_{01}\xi\xi_1 + 2\gamma_{02}\xi\xi_2 \;+ \gamma_{11}\xi_1{}^2 + 2\gamma_{12}\xi_1\xi_2 + \gamma_{22}\xi_2{}^2 \\
& + 2k_{00}\xi\eta + 2k_{01}\xi\eta_1 + 2k_{02}\xi\eta_2 \\
& + 2k_{10}\xi_1\eta + 2k_{11}\xi_1\eta_1 + 2k_{12}\xi_1\eta_2 \\
& + 2k_{20}\xi_2\eta + 2k_{21}\xi_2\eta_1 + 2k_{22}\xi_2\eta_2.
\end{aligned}
$$

As $x$ and $y$ are subjected to small variations, the quantities $X, Y, E$ will undergo consequent variations; let them become $X + \kappa\mathbf{X}, Y + \kappa\mathbf{Y}, E + \kappa\mathbf{E}$ respectively.

In consequence of these variations, $\dfrac{\partial G}{\partial x}$ changes to

$$
\frac{\partial G}{\partial x} + \frac{\partial^2 G}{\partial x^2}\kappa\xi + \frac{\partial^2 G}{\partial x\partial x_1}\kappa\xi_1 + \frac{\partial^2 G}{\partial x\partial x_2}\kappa\xi_2 + \frac{\partial^2 G}{\partial x\partial y}\kappa\eta + \frac{\partial^2 G}{\partial x\partial y_1}\kappa\eta_1 + \frac{\partial^2 G}{\partial x\partial y_2}\kappa\eta_2;
$$

thus

$$
\frac{\partial G}{\partial x} \text{ becomes } \frac{\partial G}{\partial x} + \kappa\frac{\partial \Omega}{\partial \xi}.
$$

Similarly

$$\frac{\partial G}{\partial x_1} \text{ becomes } \frac{\partial G}{\partial x_1} + \kappa \frac{\partial \Omega}{\partial \xi_1},$$

$$\frac{\partial G}{\partial x_2} \cdots\cdots\cdots \frac{\partial G}{\partial x_2} + \kappa \frac{\partial \Omega}{\partial \xi_2}.$$

Hence $X$, which is

$$\frac{\partial G}{\partial x} - \frac{d}{dt}\left(\frac{\partial G}{\partial x_1}\right) + \frac{d^2}{dt^2}\left(\frac{\partial G}{\partial x_2}\right),$$

becomes

$$X + \kappa \left\{ \frac{\partial \Omega}{\partial \xi} - \frac{d}{dt}\left(\frac{\partial \Omega}{\partial \xi_1}\right) + \frac{d^2}{dt^2}\left(\frac{\partial \Omega}{\partial \xi_2}\right) \right\}.$$

But $X$ becomes $X + \kappa \mathbf{X}$; so we have

$$\mathbf{X} = \frac{\partial \Omega}{\partial \xi} - \frac{d}{dt}\left(\frac{\partial \Omega}{\partial \xi_1}\right) + \frac{d^2}{dt^2}\left(\frac{\partial \Omega}{\partial \xi_2}\right).$$

In the same way, we find

$$\mathbf{Y} = \frac{\partial \Omega}{\partial \eta} - \frac{d}{dt}\left(\frac{\partial \Omega}{\partial \eta_1}\right) + \frac{d^2}{dt^2}\left(\frac{\partial \Omega}{\partial \eta_2}\right).$$

Now the characteristic equations, which originally are $X = 0$ and $Y = 0$, have still to be satisfied. We therefore must have

$$X + \kappa \mathbf{X} = 0, \quad Y + \kappa \mathbf{Y} = 0;$$

hence

$$\mathbf{X} = 0, \quad \mathbf{Y} = 0,$$

that is, the variations $\kappa \xi$ and $\kappa \eta$, which change a given characteristic curve into an adjacent characteristic curve, satisfy (and are determined by) the equations

$$\left.\begin{aligned}
\mathbf{X} &= \frac{\partial \Omega}{\partial \xi} - \frac{d}{dt}\left(\frac{\partial \Omega}{\partial \xi_1}\right) + \frac{d^2}{dt^2}\left(\frac{\partial \Omega}{\partial \xi_2}\right) = 0 \\
\mathbf{Y} &= \frac{\partial \Omega}{\partial \eta} - \frac{d}{dt}\left(\frac{\partial \Omega}{\partial \eta_1}\right) + \frac{d^2}{dt^2}\left(\frac{\partial \Omega}{\partial \eta_2}\right) = 0
\end{aligned}\right\}.$$

As these equations are derived from the characteristic equations by means solely of small variations that leave them still satisfied, the new equations $\mathbf{X} = 0$ and $\mathbf{Y} = 0$ are called the *subsidiary characteristic equations*.

*Single subsidiary equation, having the 'deviation' as variable.*

**109.** It was proved that

$$X = -y_1 E, \quad Y = x_1 E.$$

Hence

$$X + \kappa \mathbf{X} = -(y_1 + \kappa \eta_1)(E + \kappa \mathbf{E}),$$
$$Y + \kappa \mathbf{Y} = (x_1 + \kappa \xi_1)(E + \kappa \mathbf{E});$$

or as, in taking $\kappa\mathbf{X}$ and $\kappa\mathbf{Y}$ as the variations of $X$ and $Y$ consequent upon small variations $\kappa\xi$ and $\kappa\eta$ of $x$ and $y$, the second and higher powers of $\kappa$ have to be neglected, we have

$$\mathbf{X} = -\,\eta_1 E - y_1\mathbf{E}, \quad \mathbf{Y} = \xi_1 E + x_1\mathbf{E}$$
$$= -\,y_1\mathbf{E}, \qquad\qquad = x_1\mathbf{E},$$

when we regard the equation $E = 0$. Thus if we deal with the single characteristic equation $E = 0$, the subsidiary equation $\mathbf{E} = 0$ should be deducible from $\mathbf{X} = 0$ and from $\mathbf{Y} = 0$: the deduction is as follows.

From the expression for $\Omega$, we have

$$\frac{\partial\Omega}{\partial\eta_1} - \frac{d}{dt}\left(\frac{\partial\Omega}{\partial\eta_2}\right) = -\,k_{22}'\xi_3 - (k_{22}' + k_{12} - k_{21})\,\xi_2 + (k_{11} - k_{02} - k_{12}')\,\xi_1 + (k_{01} - k_{02}')\,\xi$$
$$-\,c_{22}\eta_3 + c_{22}'\eta_2 + (c_{11} - c_{02} - c_{12}')\,\eta_1 + (c_{01} - c_{02}')\,\eta\,;$$

and therefore

$$\frac{\partial\Omega}{\partial\eta} - \frac{d}{dt}\left(\frac{\partial\Omega}{\partial\eta_1}\right) + \frac{d^2}{dt^2}\left(\frac{\partial\Omega}{\partial\eta_2}\right)$$
$$= \quad k_{22}\xi_4 + (2k_{22}' + k_{12} - k_{21})\,\xi_3 + (k_{22}'' + 2k_{12}' - k_{21}' + k_{20} + k_{02} - k_{11})\,\xi_2$$
$$+ (k_{10} - k_{01} + 2k_{02}' - k_{11}' + k_{12}'')\,\xi_1 + (k_{00} - k_{01}' + k_{02}'')\,\xi$$
$$+\,c_{22}\eta_4 + 2c_{22}'\eta_3 + (c_{22}' + 2c_{02} + c_{12}' - c_{11})\,\eta_2$$
$$+ (2c_{02}' - c_{11}' + c_{12}'')\,\eta_1 + (c_{00} - c_{01}' + c_{02}'')\,\eta.$$

Let the expressions, which have been obtained for the coefficients $c_{mn}$ and $k_{mn}$, be substituted in the right-hand side: then

(i)  the terms in $\xi_4$ and $\eta_4$ are

$$x_1 Q\,(x_1\eta_4 - y_1\xi_4)\,;$$

(ii)  the terms in $\xi_3$ and $\eta_3$ are

$$4x_1 Q\,(x_2\eta_3 - y_2\xi_3) + 2x_1 Q'\,(x_1\eta_3 - y_1\xi_3)\,;$$

(iii)  the terms in $\xi_2$ and $\eta_2$ are

$$6x_1 Q\,(x_3\eta_2 - y_3\xi_2) + 6x_1 Q'\,(x_2\eta_2 - y_2\xi_2) + x_1 Q''\,(x_1\eta_2 - y_1\xi_2)$$
$$+ x_1 J\,(x_1\eta_2 - y_1\xi_2)\,;$$

(iv)  the terms in $\xi_1$ and $\eta_1$ are

$$4x_1 Q\,(x_4\eta_1 - y_4\xi_1) + 6x_1 Q'\,(x_3\eta_1 + y_3\xi_1) + 2x_1 Q''\,(x_2\eta_1 - y_2\xi_1)$$
$$+ 2x_1 J\,(x_2\eta_1 - y_2\xi_1) + x_1 J'\,(x_1\eta_1 - y_1\xi_1)\,;$$

(v)  the terms in $\xi$ and $\eta$ are

$$x_1 Q\,(x_5\eta - y_5\xi) + 2x_1 Q'\,(x_4\eta - y_4\xi) + x_1 Q''\,(x_3\eta - y_3\xi)$$
$$+ x_1 J\,(x_3\eta - y_3\xi) + x_1 J'\,(x_2\eta - y_2\xi) + x_1 \Lambda\,(x_1\eta - y_1\xi).$$

Just as, in connection with the general variation $\kappa u$ and $\kappa v$, it proved convenient (§ 106) to introduce a quantity $w$ such that

$$w = x_1 v - y_1 u,$$

so, in connection with the characteristic variation $\kappa\xi$ and $\kappa\eta$, it proves convenient to introduce a quantity $\zeta$ such that

$$\zeta = x_1\eta - y_1\xi.$$

With this significance for $\zeta$, we have

$$\mathbf{Y} = \frac{\partial\Omega}{\partial\eta} - \frac{d}{dt}\left(\frac{\partial\Omega}{\partial\eta_1}\right) + \frac{d^2}{dt^2}\left(\frac{\partial\Omega}{\partial\eta_2}\right)$$
$$= x_1\{Q\zeta'''' + 2Q'\zeta''' + (Q'' + J)\,\zeta'' + J'\zeta' + \Lambda\zeta\}.$$

Proceeding in the same way for $\mathbf{X}$, we find

$$\mathbf{X} = \frac{\partial\Omega}{\partial\xi} - \frac{d}{dt}\left(\frac{\partial\Omega}{\partial\xi_1}\right) + \frac{d^2}{dt^2}\left(\frac{\partial\Omega}{\partial\xi_2}\right)$$
$$= -y_1\{Q\zeta'''' + 2Q'\zeta''' + (Q'' + J)\,\zeta'' + J'\zeta' + \Lambda\zeta\}.$$

But we have seen that

$$\mathbf{Y} = x_1\mathbf{E}, \quad \mathbf{X} = -y_1\mathbf{E},$$

and that $\mathbf{X} = 0$, $\mathbf{Y} = 0$; hence

$$\mathbf{E} = Q\zeta'''' + 2Q'\zeta''' + (Q'' + J)\,\zeta'' + J'\zeta' + \Lambda\zeta = 0.$$

We call $\mathbf{E} = 0$ the subsidiary characteristic equation.

### Final normal form of the 'second' variation.

**110.** We now can proceed at once to the final reduced form of the second variation.

The earlier analysis (in §§ 88, 89, 93) now applies with a mere change of symbols, and leads to the formal result. We have

$$Q\left\{\mathbf{G} + \frac{d}{dt}\left(lw'^2 + 2mww' + nw^2\right)\right\}$$
$$= Q^2w''^2 + 2Q\,lw'w'' + 2Qmww'' + Q(-J + l' + 2m)\,w'^2$$
$$\qquad\qquad + 2Q(n + m')\,ww' + Q(\Lambda + n')\,w^2$$
$$= (Qw'' + lw' + mw)^2,$$

provided $l$, $m$, $n$ are such that

$$\left.\begin{aligned} l^2 &= Q(-J + l' + 2m) \\ lm &= Q(n + m') \\ m^2 &= Q(\Lambda + n') \end{aligned}\right\}.$$

Let $\alpha$ and $\beta$ be two linearly independent integrals of the equation $\mathbf{E} = 0$ such that, if $z$ denote $\alpha\beta' - \alpha'\beta$, then $z$ is not a constant quantity and the quantity

$$Qz'' + Q'z' + Jz - 2Q(\alpha'\beta'' - \alpha''\beta'),$$

which is always constant for any two integrals, is made zero by choice of $\alpha$ and $\beta$, a choice that always is possible. Then

$$\left.\begin{aligned} Q\alpha'' + l\alpha' + m\alpha &= 0 \\ Q\beta'' + l\beta' + m\beta &= 0 \end{aligned}\right\}$$

lefine the values of $l$ and $m$, so that

$$l = -Q\frac{z'}{z}$$

$$2m = \frac{1}{z}(Qz'' + Q'z' + Jz) = \frac{1}{z}\left\{\frac{d}{dt}(Qz') + Jz\right\}$$

$$2n = -\frac{1}{z}\left\{\frac{d^2}{dt^2}(Qz') + \frac{d}{dt}(Jz)\right\} = \frac{1}{z}\frac{d}{dt}\{Q(\alpha'\beta'' - \alpha''\beta')\}$$

With these values

$$\mathbf{G} + \frac{d}{dt}(lw'^2 + 2mww' + nw^2) = \frac{1}{Q}(Qw'' + lw' + mw)^2.$$

Now

$$[lw'^2 + 2mww' + nw^2] = 0$$

when taken at the limits of the integral, because, as the limits are now rigidly ixed, $w$ and $w'$ vanish at each limit. Hence, finally,

$$\tfrac{1}{2}\kappa^2\int \mathbf{G}\,dt = \tfrac{1}{2}\kappa^2\int \frac{1}{Q}(Qw'' + lw' + mw)^2\,dt$$

$$= \tfrac{1}{2}\kappa^2\int \frac{Q}{(\alpha\beta'-\alpha'\beta)^2}\begin{vmatrix} w'', & w', & w \\ \alpha'', & \alpha', & \alpha \\ \beta'', & \beta' & \beta \end{vmatrix}^2 dt,$$

where $\alpha$ and $\beta$ are two integrals of the subsidiary characteristic equation $\mathbf{E} = 0$ possessing the specified properties.

The right-hand side is the critical normal form of the second variation of the original integral.

### The central characteristic equations: their primitive.

**111.** The original characteristic equation (§ 84) is of the fourth order in $y$, provided the original function $F(x, y, y', y'')$ is not linear in $y''$. When $F$ is linear in $y''$, the characteristic equation actually is only of the second order in $y$, because the quantities $y''''$ and $y'''$ do not occur; but in that event, the quantity $\dfrac{\partial^2 F}{\partial y''^2}$, and therefore $Q$, would be a permanent zero, a possibility excluded for an entirely different reason (§ 95). We therefore omit the contingency from further consideration.

In the original characteristic equation $E = 0$, the only term involving the highest derivative of $y$ is

$$y''''\frac{\partial^2 F}{\partial y''^2}.$$

In the characteristic equations $X = 0$ and $Y = 0$, the terms involving the highest derivatives of $x$ and $y$ are

$$-y_1Q(x_1y_4 - y_1x_4) \text{ and } x_1Q(x_1y_4 - y_1x_4)$$

respectively. It is clear from the expression just obtained for the second variation of the integral—we shall return to its fuller consideration hereafter—that, if a maximum or a minimum is to be possessed, the quantity $Q$ here $\left(\text{like the quantity } \dfrac{\partial^2 F}{\partial y''^2} \text{ in the earlier discussion}\right)$ must not vanish within the range. The character of the primitive* is then simple. That primitive contains four arbitrary independent constants. These are made determinate, by the assignment of four arbitrary independent magnitudes as the values of $y, y', y'', y'''$ for any assigned initial value $a$ of $x$ within the range.

The same result ensues, when we take $x$ and $y$ as functions of an independent variable $t$. The primitive of the equations $X = 0$ and $Y = 0$ is of the form

$$x = \phi(t, A_1, A_2, A_3, A_4), \quad y = \psi(t, A_1, A_2, A_3, A_4),$$

where $A_1, A_2, A_3, A_4$ are the essential independent arbitrary constants to be made determinate, by the assignment of initial values for $x, y$ and for their derivatives at some initial value $t_0$ of $t$. Thus we should have equations of the form

$$a = \phi(t_0, A_1, A_2, A_3, A_4), \quad b = \psi(t_0, A_1, A_2, A_3, A_4),$$

$$m = \frac{\psi'(t_0)}{\phi'(t_0)},$$

$$k = \frac{1}{\phi'^3(t_0)}\{\phi'(t_0)\psi''(t_0) - \psi'(t_0)\phi''(t_0)\},$$

$$l = \frac{1}{\phi'^5(t_0)}[\phi'(t_0)\{\phi'(t_0)\psi'''(t_0) - \psi'(t_0)\phi'''(t_0)\} \\ - 3\phi''(t_0)\{\phi'(t_0)\psi''(t_0) - \psi'(t_0)\phi''(t_0)\}].$$

When $t_0$ is eliminated, we have relations of the form

$$\begin{aligned} b &= b(A_1, A_2, A_3, A_4) \\ m &= m(A_1, A_2, A_3, A_4) \\ k &= k(A_1, A_2, A_3, A_4) \\ l &= l(A_1, A_2, A_3, A_4) \end{aligned},$$

expressing the one set of independent arbitrary constants $b, m, k, l$ in terms of the other set of independent arbitrary constants $A_1, A_2, A_3, A_4$; and conversely. Thus neither of the Jacobians

$$J\left(\frac{b, m, k, l}{A_1, A_2, A_3, A_4}\right), \quad J\left(\frac{A_1, A_2, A_3, A_4}{b, m, k, l}\right),$$

can vanish.

---

* So far as concerns the differential equation alone, without regard to other requirements, $\dfrac{\partial^2 F}{\partial y''^2}$ might have isolated zeros of limited degree and the primitive still be relatively simple: see my *Theory of Differential Equations*, vol. iv, § 31.

**112.** Moreover, when $A_1, A_2, A_3, A_4$, and $t_0$, as well as $t$, are eliminated among the preceding equations combined with

$$x = \phi(t, A_1, A_2, A_3, A_4), \quad y = \psi(t, A_1, A_2, A_3, A_4),$$

we have a relation

$$y = y(x, b, m, k, l),$$

which is the equation of the characteristic curve when the initially assigned quantities $b, m, k, l$ are forced into evidence. Conversely, when the value $x = \phi(t, A_1, A_2, A_3, A_4)$, as well as the values of $b, m, k, l$ are inserted in this equation of the curve, the result is to give the equation

$$y = \psi(t, A_1, A_2, A_3, A_4).$$

Owing to the significance of $b, m, k, l$ as the values of $y, y', y'', y'''$ for an initial value $a$ of $x$ within the range, we have the expression of $y$ in the vicinity of $a$ given by

$$y = b + m(x-a) + \frac{1}{2!}k(x-a)^2 + \frac{1}{3!}l(x-a)^3 + (x-a)^4 R(x-a, b, m, k, l),$$

where $R$ is a regular function of its arguments, this relation being the expression of the theorem as to the existence of the unique uniform primitive satisfying the initial conditions. Thus

$$\frac{\partial y}{\partial b} = 1 + (x-a)^4 \frac{\partial R}{\partial b},$$

$$\frac{\partial y}{\partial m} = x - a + (x-a)^4 \frac{\partial R}{\partial m},$$

$$\frac{\partial y}{\partial k} = \frac{1}{2!}(x-a)^2 + (x-a)^4 \frac{\partial R}{\partial k},$$

$$\frac{\partial y}{\partial l} = \frac{1}{3!}(x-a)^3 + (x-a)^4 \frac{\partial R}{\partial l}.$$

Manifestly, no identical linear relation

$$\gamma_1 \frac{\partial y}{\partial b} + \gamma_2 \frac{\partial y}{\partial m} + \gamma_3 \frac{\partial y}{\partial k} + \gamma_4 \frac{\partial y}{\partial l} = 0,$$

where $\gamma_1, \gamma_2, \gamma_3, \gamma_4$ are non-zero constants, can exist among the four derivatives

$$\frac{\partial y}{\partial b}, \quad \frac{\partial y}{\partial m}, \quad \frac{\partial y}{\partial k}, \quad \frac{\partial y}{\partial l}.$$

### Primitive of the subsidiary equation.

**113.** The characteristic equation $E = 0$ is satisfied whatever be the arbitrary constants $A_1, A_2, A_3, A_4$; it therefore remains satisfied when these are changed into $A_r + \kappa a_r (r = 1, 2, 3, 4)$ respectively, the constants $a_1, a_2, a_3, a_4$ themselves being arbitrary. The effect of this modification is to change $x$ and $y$ into $x + \kappa\xi$ and $y + \kappa\eta$ respectively where, on the hypothesis that $\kappa$ is very small,

$$\xi = a_1\phi_1(t) + a_2\phi_2(t) + a_3\phi_3(t) + a_4\phi_4(t) \Big\}$$
$$\eta = a_1\psi_1(t) + a_2\psi_2(t) + a_3\psi_3(t) + a_4\psi_4(t) \Big\}$$ ,

in which, for $r = 1, 2, 3, 4$,

$$\phi_r(t) = \frac{\partial\phi(t, A_1, A_2, A_3, A_4)}{\partial A_r}, \quad \psi_r(t) = \frac{\partial\psi(t, A_1, A_2, A_3, A_4)}{\partial A_r}.$$

Hence these quantities $\xi$ and $\eta$ satisfy the equations

$$\mathbf{X} = 0, \quad \mathbf{Y} = 0.$$

We had to deal with a quantity $\zeta$, defined by the relation

$$\zeta = x_1\eta - y_1\xi;$$

hence, writing

$$\theta_r(t) = \frac{\partial\phi}{\partial t}\psi_r(t) - \frac{\partial\psi}{\partial t}\phi_r(t) = \phi'(t)\psi_r(t) - \psi'(t)\phi_r(t),$$

for $r = 1, 2, 3, 4$, the value of $\zeta$ is given by

$$\zeta = a_1\theta_1(t) + a_2\theta_2(t) + a_3\theta_3(t) + a_4\theta_4(t).$$

Thus $\zeta$ satisfies the equation

$$\mathbf{E} = Q\zeta'''' + 2Q'\zeta''' + (Q'' + J)\zeta'' + J'\zeta' + \Lambda\zeta = 0,$$

and its expression as a function of $t$ contains four arbitrary constants $a_1, a_2, a_3, a_4$ linearly. If therefore the four functions $\theta_1(t), \theta_2(t), \theta_3(t), \theta_4(t)$ are linearly independent of one another, the foregoing value of $\zeta$ provides the primitive of $\mathbf{E} = 0$. This linear independence can be established as follows.

When we take the form of the characteristic primitive to be

$$y = y(x, b, m, k, l),$$

and the equivalent forms to be

$$x = \phi(t, A_1, A_2, A_3, A_4), \quad y = \psi(t, A_1, A_2, A_3, A_4),$$

together with the equations connecting the set of arbitrary constants $b, m, k, l$ and the set of arbitrary constants $A_1, A_2, A_3, A_4$, we have

$$\frac{\partial y}{\partial A_r} = \frac{dy}{dx}\frac{\partial x}{\partial A_r} + \frac{\partial y}{\partial b}\frac{\partial b}{\partial A_r} + \frac{\partial y}{\partial m}\frac{\partial m}{\partial A_r} + \frac{\partial y}{\partial k}\frac{\partial k}{\partial A_r} + \frac{\partial y}{\partial l}\frac{\partial l}{\partial A_r},$$

that is,

$$\psi_r(t) = \frac{\psi'(t)}{\phi'(t)}\phi_r(t) + \frac{\partial y}{\partial b}\frac{\partial b}{\partial A_r} + \frac{\partial y}{\partial m}\frac{\partial m}{\partial A_r} + \frac{\partial y}{\partial k}\frac{\partial k}{\partial A_r} + \frac{\partial y}{\partial l}\frac{\partial l}{\partial A_r},$$

and therefore

$$\theta_r(t) = \phi'(t)\psi_r(t) - \psi'(t)\phi_r(t)$$
$$= \phi'(t)\left\{\frac{\partial y}{\partial b}\frac{\partial b}{\partial A_r} + \frac{\partial y}{\partial m}\frac{\partial m}{\partial A_r} + \frac{\partial y}{\partial k}\frac{\partial k}{\partial A_r} + \frac{\partial y}{\partial l}\frac{\partial l}{\partial A_r}\right\},$$

for $r = 1, 2, 3, 4$. If, then, a permanent linear relation could exist among the four functions $\theta_1(t), \theta_2(t), \theta_3(t), \theta_4(t)$ with constant coefficients in the form

$$\mu_1 \theta_1(t) + \mu_2 \theta_2(t) + \mu_3 \theta_3(t) + \mu_4 \theta_4(t) = 0,$$

where $\mu_1, \mu_2, \mu_3, \mu_4$ are constant, we should have

$$\frac{\partial y}{\partial b} \sum_{r=1}^{4} \mu_r \frac{\partial b}{\partial A_r} + \frac{\partial y}{\partial m} \sum_{r=1}^{4} \mu_r \frac{\partial m}{\partial A_r} + \frac{\partial y}{\partial k} \sum_{r=1}^{4} \mu_r \frac{\partial k}{\partial A_r} + \frac{\partial y}{\partial l} \sum_{r=1}^{4} \mu_r \frac{\partial l}{\partial A_r} = 0,$$

because $\phi'(t)$ is not zero. It has been proved (§ 112) that no such linear relation with constant coefficients among the four quantities $\dfrac{\partial y}{\partial b}, \dfrac{\partial y}{\partial m}, \dfrac{\partial y}{\partial k}, \dfrac{\partial y}{\partial l}$ can subsist. Hence the four coefficients in the preceding relation must vanish; and therefore

$$\sum_{r=1}^{4} \mu_r \frac{\partial b}{\partial A_r} = 0, \quad \sum_{r=1}^{4} \mu_r \frac{\partial m}{\partial A_r} = 0, \quad \sum_{r=1}^{4} \mu_r \frac{\partial k}{\partial A_r} = 0, \quad \sum_{r=1}^{4} \mu_r \frac{\partial l}{\partial A_r} = 0.$$

The determinant of the coefficients of $\mu_1, \mu_2, \mu_3, \mu_4$ in these four equations is the Jacobian

$$J\left(\frac{b, m, k, l}{A_1, A_2, A_3, A_4}\right),$$

which does not vanish; consequently, the equations are only satisfied by

$$\mu_1 = 0, \quad \mu_2 = 0, \quad \mu_3 = 0, \quad \mu_4 = 0.$$

Therefore the four functions $\theta_1(t), \theta_2(t), \theta_3(t), \theta_4(t)$ are linearly independent. The constants $a_1, a_2, a_3, a_4$ are arbitrary independent constants. Hence the primitive of the subsidiary characteristic equation $\mathbf{E} = 0$ is given by

$$\zeta = a_1 \theta_1(t) + a_2 \theta_2(t) + a_3 \theta_3(t) + a_4 \theta_4(t).$$

### Properties of the integrals of the subsidiary equation.

**114.** Next, these functions $\theta_1(t), \theta_2(t), \theta_3(t), \theta_4(t)$ are separate and distinct integrals of the subsidiary equation $\mathbf{E} = 0$, so that

$$Q\theta_r'''' + 2Q'\theta_r''' + (Q'' + J)\theta_r'' + J'\theta_r' + \Lambda\theta_r = 0,$$

for $r = 1, 2, 3, 4$. Therefore

$$Q \begin{vmatrix} \theta_1 \,, & \theta_2 \,, & \theta_3 \,, & \theta_4 \\ \theta_1' \,, & \theta_2' \,, & \theta_3' \,, & \theta_4' \\ \theta_1'' \,, & \theta_2'' \,, & \theta_3'' \,, & \theta_4'' \\ \theta_1'''' , & \theta_2'''' , & \theta_3'''' , & \theta_4'''' \end{vmatrix} + 2Q' \begin{vmatrix} \theta_1 \,, & \theta_2 \,, & \theta_3 \,, & \theta_4 \\ \theta_1' \,, & \theta_2' \,, & \theta_3' \,, & \theta_4' \\ \theta_1'' \,, & \theta_2'' \,, & \theta_3'' \,, & \theta_4'' \\ \theta_1''' , & \theta_2''' , & \theta_3''' , & \theta_4''' \end{vmatrix} = 0,$$

so that

$$Q^2 \begin{vmatrix} \theta_1 \,, & \theta_2 \,, & \theta_3 \,, & \theta_4 \\ \theta_1' \,, & \theta_2' \,, & \theta_3' \,, & \theta_4' \\ \theta_1'' \,, & \theta_2'' \,, & \theta_3'' \,, & \theta_4'' \\ \theta_1''' , & \theta_2''' , & \theta_3''' , & \theta_4''' \end{vmatrix} = B,$$

where $B$ is a constant. On the one hand, $B$ may not vanish: because the determinant does not vanish, owing to the property that no linear relation with constant coefficients subsists among the functions $\theta_1$, $\theta_2$, $\theta_3$, $\theta_4$: and because, as appears from other considerations, $Q$ does not vanish within the range. On the other hand, $B$ is not an arbitrary constant; for $Q$, $\theta_1$, $\theta_2$, $\theta_3$, $\theta_4$ are specific functions. Thus $B$ is a specific non-zero constant.

Owing to the fact that $B$ is not zero, it follows that

(a)  The functions $\theta_1$, $\theta_2$, $\theta_3$, $\theta_4$ cannot vanish simultaneously for any value of $t$ within the range:

(b)  The set of their first derivatives cannot vanish simultaneously, nor can any set of their derivatives of the same order vanish simultaneously:

(c)  Owing to the invariantive character of the determinant, any set of four linearly independent integrals as a fundamental system can take the place of the set $\theta_1$, $\theta_2$, $\theta_3$, $\theta_4$; the only effect of such a change is to change the specific constant $B$.

Denoting the determinant by $W(t)$, we have
$$Q^2(t) \, W(t) = B;$$
and therefore
$$\frac{W(t)}{W(t_0)} = \frac{Q^2(t_0)}{Q^2(t)},$$
where $t_0$ and $t$ are any two values in the range.

**115.**  Again, as we have
$$Q\theta_1'''' + 2Q'\theta_1''' + (Q'' + J)\,\theta_1'' + J'\theta_1' + \Lambda\theta_1 = 0,$$
$$Q\theta_2'''' + 2Q'\theta_2''' + (Q'' + J)\,\theta_2'' + J'\theta_2' + \Lambda\theta_2 = 0,$$
it follows that
$$Q\,(\theta_1\theta_2'''' - \theta_2\theta_1'''') + 2Q'\,(\theta_1\theta_2''' - \theta_2\theta_1''') + Q''\,(\theta_1\theta_2'' - \theta_2\theta_1'')$$
$$+ J\,(\theta_1\theta_2'' - \theta_2\theta_1'') + J'\,(\theta_1\theta_2' - \theta_2\theta_1') = 0,$$
that is,
$$\frac{d}{dt}\{Q\,(\theta_1\theta_2''' - \theta_2\theta_1''' - \theta_1'\theta_2'' + \theta_2'\theta_1'') + Q'\,(\theta_1\theta_2'' - \theta_2\theta_1'') + J\,(\theta_1\theta_2' - \theta_2\theta_1')\} = 0,$$
and therefore
$$Q\,(\theta_1\theta_2''' - \theta_2\theta_1''' - \theta_1'\theta_2'' + \theta_2'\theta_1'') + Q'\,(\theta_1\theta_2'' - \theta_2\theta_1'') + J\,(\theta_1\theta_2' - \theta_2\theta_1') = A,$$
where $A$ is a constant, not arbitrary because $J$, $Q$, $\theta_1$, $\theta_2$ are specific quantities; and $A$ may be zero.

When $A$ is not zero, we combine $\theta_1$ and $\theta_3$ in the same way as $\theta_1$ and $\theta_2$ have just been combined; and we find
$$Q\,(\theta_1\theta_3''' - \theta_3\theta_1''' - \theta_1'\theta_3'' + \theta_3'\theta_1'') + Q'\,(\theta_1\theta_3'' - \theta_3\theta_1'') + J\,(\theta_1\theta_3' - \theta_3\theta_1') = C,$$
where $C$ is a constant, again not arbitrary because $J$, $Q$, $\theta_1$, $\theta_3$ are specific quantities; and $C$ may be zero.

If neither $A$ nor $C$ is zero, we take
$$\vartheta(t) = C\theta_2(t) - A\theta_3(t),$$
so that, as $\theta_1(t)$, $\theta_2(t)$, $\theta_3(t)$ are linearly independent, $\vartheta(t)$ is distinct from $\theta_1(t)$; but, as it is a linear combination of $\theta_2(t)$ and $\theta_3(t)$, it is an integral of the equation $E = 0$. We now have
$$Q(\theta_1\vartheta''' - \vartheta\theta_1''' - \theta_1'\vartheta'' + \vartheta'\theta_1'') + Q'(\theta_1\vartheta'' - \vartheta\theta_1'') + J(\theta_1\vartheta' - \vartheta\theta_1') = 0.$$

If $A$ is zero, we have a combination $\theta_1$ and $\theta_2$ which makes the specific constant zero.

If $C$ is zero, we have a combination $\theta_1$ and $\theta_3$ which makes the specific constant zero.

If $A$ is not zero, and if $C$ also is not zero, we have a combination $\theta_1$ and $\vartheta$ which makes the specific constant zero.

Thus we can always obtain two integrals of the equation $\mathbf{E} = 0$ which will make the specific constant zero. Denoting these by $\alpha$ and $\beta$, we have
$$Q(\alpha\beta''' - \alpha'''\beta - \alpha'\beta'' - \alpha''\beta') + Q'(\alpha\beta'' - \alpha''\beta) + J(\alpha\beta' - \alpha'\beta) = 0.$$
Manifestly the choice is not unique.

Moreover, we can at once tell whether the specific constant is zero or not zero, for any selected combination whatever. We need only substitute the values of the selected integrals at any point in the range (frequently, it is convenient to choose the initial point); the specific constant is the value of the quantity at that point.

### Consecutive characteristic curve.

**116.** The original characteristic curve is uniquely determined by the assignment of initial arbitrary values of $y$, $y'$, $y''$, $y'''$; and another characteristic curve in the immediate vicinity is also uniquely determined by the values of $\xi$ and $\eta$ (and therefore of $\zeta$) at the initial place. This neighbouring characteristic curve will be called *consecutive* when, passing through an initial point on the original (or central) characteristic, it has the same tangent, the same curvature, but not the same arc-rate of curvature, at that initial point as the original curve. As the arc-rates of curvature are distinct, the two characteristics are distinct.

Let the deviation of the consecutive characteristic from the central curve be $\kappa\zeta$, so that
$$\zeta = c_1\theta_1(t) + c_2\theta_2(t) + c_3\theta_3(t) + c_4\theta_4(t),$$
where the coefficients $c$ (save as to an unessential common factor which merely affects the scale of the deviation) are determined by the initial conditions. Let the initial value of $t$ for the central curve be $t_0$, and for the consecutive curve be $t_0 + \kappa\tau_0$: then
$$\xi_0 = \tau_0\phi'(t_0) + c_1\phi_1(t_0) + c_2\phi_2(t_0) + c_3\phi_3(t_0) + c_4\phi_4(t_0),$$
$$\eta_0 = \tau_0\psi'(t_0) + c_1\psi_1(t_0) + c_2\psi_2(t_0) + c_3\psi_3(t_0) + c_4\psi_4(t_0),$$

and therefore

$$\zeta_0 = \phi'(t_0)\eta_0 - \psi'(t_0)\xi_0$$
$$= c_1\theta_1(t_0) + c_2\theta_2(t_0) + c_3\theta_3(t_0) + c_4\theta_4(t_0).$$

The consecutive curve is to pass through the initial point; hence $\xi_0 = 0$, $\eta_0 = 0$. Consequently $\zeta_0 = 0$; when we retain $\zeta_0 = 0$ as an initial equation, either of the equations $\xi_0 = 0$, $\eta_0 = 0$, determines $\tau_0$ for which, however, there is no further explicit use. Accordingly, the coincidence of the initial points of the two curves requires the condition

$$c_1\theta_1(t_0) + c_2\theta_2(t_0) + c_3\theta_3(t_0) + c_4\theta_4(t_0) = 0.$$

The tangents to the two curves are to coincide at the common initial point. Hence, at that point,

$$\frac{y_1 + \kappa\eta_1}{x_1 + \kappa\xi_1} = \frac{y_1}{x_1};$$

and therefore

$$x_1\eta_1 - y_1\xi_1 = 0$$

at the point. But, there, $\xi = 0$ and $\eta = 0$; and so

$$\zeta'(t_0) = [x_1\eta_1 - y_1\xi_1 + x_2\eta - y_2\xi]_{t=t_0} = 0,$$

that is, the coincidence of the tangents to the two curves at the common initial point requires the condition

$$c_1\theta_1'(t_0) + c_2\theta_2'(t_0) + c_3\theta_3'(t_0) + c_4\theta_4'(t_0) = 0.$$

(In passing, it should be noted that

$$\frac{\eta_1}{y_1} = \frac{\xi_1}{x_1}, = p,$$

at the initial point.)

The curvature of the two curves is to be the same at the common initial point. Consequently, the magnitude

$$(x_1^2 + y_1^2)^{-\frac{3}{2}}(x_1y_2 - x_2y_1)$$

must have a zero increment, when the small variations $\kappa\xi$ and $\kappa\eta$ are effected on $x$ and $y$. Under this variation, the new value of $(x_1^2 + y_1^2)^{-\frac{3}{2}}$ is

$$(x_1^2 + y_1^2)^{-\frac{3}{2}}\left(1 - 3\kappa\frac{x_1\xi_1 + y_1\eta_1}{x_1^2 + y_1^2}\right)$$

up to the first power of $\kappa$ inclusive, that is, it is

$$(x_1^2 + y_1^2)^{-\frac{3}{2}}(1 - 3\kappa p).$$

The new value of $x_1y_2 - x_2y_1$ is

$$x_1y_2 - x_2y_1 + \kappa(x_1\eta_2 - y_1\xi_2 + y_2\xi_1 - x_2\eta_1).$$

But

$$-p(x_1y_2 - x_2y_1) = x_2\eta_1 - y_2\xi_1;$$

and so the new value of the expression for the curvature is

$$(x_1^2 + y_1^2)^{-\frac{3}{2}}[x_1y_2 - x_2y_1 + \kappa\{x_1\eta_2 - y_1\xi_2 + 2(x_2\eta_1 - y_2\xi_1)\}].$$

As the increment is to be zero at the initial point, we have

$$x_1 \eta_2 - y_1 \xi_2 + 2 (x_2 \eta_1 - y_2 \xi_1) = 0,$$

at the initial point; and therefore

$$\zeta''(t_0) = [x_1 \eta_2 - y_1 \xi_2 + 2 (x_2 \eta_1 - y_2 \xi_1) + x_3 \eta - y_3 \xi]_{t=t_0} = 0,$$

that is, the equality of the curvatures of the two curves at the common initial point requires the condition

$$c_1 \theta_1''(t_0) + c_2 \theta_2''(t_0) + c_3 \theta_3''(t_0) + c_4 \theta_4''(t_0) = 0.$$

The three specified conditions determine the ratios of the coefficients $c_1$, $c_2$, $c_3$, $c_4$, which occur in the expression for $\zeta(t)$, where $\kappa \zeta(t)$ is the deviation of the consecutive characteristic from the original curve. Manifestly, we have

$$\zeta(t) = \delta \begin{vmatrix} \theta_1(t) , & \theta_2(t) , & \theta_3(t) , & \theta_4(t) \\ \theta_1(t_0) , & \theta_2(t_0) , & \theta_3(t_0) , & \theta_4(t_0) \\ \theta_1'(t_0) , & \theta_2'(t_0), & \theta_3'(t_0), & \theta_4'(t_0) \\ \theta_1''(t_0), & \theta_2''(t_0), & \theta_3''(t_0), & \theta_4''(t_0) \end{vmatrix},$$

where $\delta$ is a non-vanishing quantity independent of $t$.

But the arc-rate of curvature must not be the same for the consecutive characteristic as for the original characteristic at the common initial point. A measure of the difference is $\kappa \zeta'''(t_0)$, that is, it is

$$\kappa \delta \begin{vmatrix} \theta_1'''(t_0), & \theta_2'''(t_0), & \theta_3'''(t_0), & \theta_4'''(t_0) \\ \theta_1(t_0) , & \theta_2(t_0) , & \theta_3(t_0) , & \theta_4(t_0) \\ \theta_1'(t_0) , & \theta_2'(t_0) , & \theta_3'(t_0) , & \theta_4'(t_0) \\ \theta_1''(t_0), & \theta_2''(t_0), & \theta_3''(t_0), & \theta_4''(t_0) \end{vmatrix}.$$

This quantity is

$$- \kappa \delta B Q^{-2}(t_0),$$

which does not vanish; thus the distinctness of the consecutive characteristic from the original characteristic is secured.

### Conjugate of initial point.

**117.** The consecutive characteristic may or may not cut the original characteristic in a point distinct from the initial point. Possibly, they may intersect in a number of such points. Let the first of such points (for increasing values of $t$), or the actual point if there be only one, be given by the value $t_1$, so that $t_1 > t_0$. Then $t_1$ is the smallest value of $t$ greater than $t_0$ such that $\zeta(t)$ vanishes there: or, if we write $\zeta(t_0, t)$ in place of $\zeta(t)$, it is the smallest value of $t_1$, $> t_0$, such that

$$\zeta(t_0, t_1) = \delta \begin{vmatrix} \theta_1(t_1) , & \theta_2(t_1) , & \theta_3(t_1) , & \theta_4(t_1) \\ \theta_1(t_0) , & \theta_2(t_0) , & \theta_3(t_0) , & \theta_4(t_0) \\ \theta_1'(t_0), & \theta_2'(t_0), & \theta_3'(t_0), & \theta_4'(t_0) \\ \theta_1''(t_0), & \theta_2''(t_0), & \theta_3''(t_0), & \theta_4''(t_0) \end{vmatrix} = 0.$$

Here $\delta$ is a mere non-zero unessential constant, independent of $t_0$ and $t_1$. The point on the characteristic curve, determined by this value of $t_1$, is called the *conjugate* of the point determined by the value of $t_0$.

As the determinant in the expression for $\zeta(t_0, t_1)$ is invariantive under the linear transformations to which a fundamental set of integrals of the equation $\mathbf{E} = 0$ is subject, manifestly any four linearly independent integrals of $\mathbf{E}$ suffice for the construction of the equation $\zeta(t_0, t_1) = 0$. Thus the settlement of the conjugate of the initial point $t_0$ is not affected by the particular choice of fundamental integrals $\theta_1, \theta_2, \theta_3, \theta_4$ of the subsidiary characteristic. Also, the position of $t_1$ is independent of $\delta$, which affects only the scale of the deviation of the consecutive from the original characteristic.

Finally, $\zeta(t_0, t)$ is a regular function of $t$. After vanishing when $t = t_0$, it does not again vanish for increasing values of $t$, until $t = t_1$; that is, for values of $t$ such that $t_0 < t < t_1$, the function $\zeta(t_0, t)$ keeps its sign unaltered.

### *Property of the characteristic range between conjugates.*

**118.** Before passing to consider the relation between the second variation of the integral and the range of integration from an initial point up to the conjugate, it is convenient to establish the proposition :

*The range along a characteristic from a point to its conjugate cannot include a similar range.*

Let $C'$ be the conjugate of a point $C$ on a characteristic curve $...CDPC'...$; we are to prove that the conjugate of a point $D$ lying between $C$ and $C'$ (but not at $C$ nor at $C'$) lies without the range.

The consecutive characteristic through $C$ is given by its deviation $\kappa\zeta(t_0, t)$, where

$$\zeta(t_0, t) = \delta \begin{vmatrix} \theta_1(t) & , & \theta_2(t) & , & \theta_3(t) & , & \theta_4(t) \\ \theta_1(t_0) & , & \theta_2(t_0) & , & \theta_3(t_0) & , & \theta_4(t_0) \\ \theta_1'(t_0) & , & \theta_2'(t_0) & , & \theta_3'(t_0) & , & \theta_4'(t_0) \\ \theta_1''(t_0) & , & \theta_2''(t_0) & , & \theta_3''(t_0) & , & \theta_4''(t_0) \end{vmatrix}$$

$$= c_1\theta_1(t) + c_2\theta_2(t) + c_3\theta_3(t) + c_4\theta_4(t).$$

First, consider a point $D$ within the range $CC'$, lying very near $C$ and given by a value $t_0 + \epsilon T_0$ of $t$, where $\epsilon$ and $T_0$ are positive and $\epsilon$ is small. Then

$$\zeta(t_0, t_0 + \epsilon T_0) = \tfrac{1}{6}\epsilon\delta \begin{vmatrix} \theta_1'''(t_0), & \theta_2'''(t_0), & \theta_3'''(t_0), & \theta_4'''(t_0) \\ \theta_1(t_0) & , & \theta_2(t_0) & , & \theta_3(t_0) & , & \theta_4(t_0) \\ \theta_1'(t_0) & , & \theta_2'(t_0) & , & \theta_3'(t_0) & , & \theta_4'(t_0) \\ \theta_1''(t_0), & \theta_2''(t_0), & \theta_3''(t_0), & \theta_4''(t_0) \end{vmatrix} + \dots,$$

where the unexpressed terms involve the fourth and higher powers of the small quantity $\epsilon$: hence, effectively,

$$\zeta(t_0, t_0 + \epsilon T_0) = -\tfrac{1}{6}\epsilon^3 \delta T_0{}^3 W(t_0).$$

Also $W(t_0)$, never vanishing within any range of the characteristic curve we are considering, has the same sign whatever initial value $t_0$ be chosen for $t$.

Next, consider a point $P$ within the range $CC'$, lying very near $C'$ and given by a value $t_1 - \eta T_1$, where $\eta$ and $T_1$ are positive and $\eta$ is small. Then

$$\zeta(t_0, t_1 - \eta T_1) = -\eta \delta T_1 \begin{vmatrix} \theta_1{}'(t_1), & \theta_2{}'(t_1), & \theta_3{}'(t_1), & \theta_4{}'(t_1) \\ \theta_1(t_0), & \theta_2(t_0), & \theta_3(t_0), & \theta_4(t_0) \\ \theta_1{}'(t_0), & \theta_2{}'(t_0), & \theta_3{}'(t_0), & \theta_4{}'(t_0) \\ \theta_1{}''(t_0), & \theta_2{}''(t_0), & \theta_3{}''(t_0), & \theta_4{}''(t_0) \end{vmatrix} + \dots$$

$$= -\eta \delta T_1 M + \dots,$$

where $M$ denotes the determinant, and the unexpressed terms involve the second and higher powers of the small quantity $\eta$. Hence, effectively,

$$\zeta(t_0, t_1 - \eta T_1) = -\eta \delta T_1 M.$$

But $\zeta(t_0, t)$ has the same sign at $P$ as it has at $D$, because it does not vanish between $C$ and $C'$. Hence, as $M$ is not zero, it has the same sign as $W(t_0)$. But $W(t_0)$ has a persistent sign whatever initial point $t_0$ be chosen; hence $M$ also has a persistent sign, for any choice of initial point $t_0$ and the consequent conjugate $t_1$.

Now take the point $D$ as the initial point for another range of the characteristic curve, the value of $t$ at $D$ being $t_0 + \epsilon T_0$; and let $D'$ be the conjugate of $D$. If $\epsilon$ were zero, so that $D$ would coincide with $C$, then $D'$ would coincide with $C'$; hence, when $\epsilon$ is not zero but still small, so that $D$ is merely very near $C$, then $D'$ must be near $C'$. But $D'$ cannot coincide with $C'$: if it could, we should have

$$\zeta(t_0 + \epsilon T_0, t_1) = 0,$$

that is,

$$\epsilon \delta T_0 \begin{vmatrix} \theta_1(t_1), & \theta_2(t_1), & \theta_3(t_1), & \theta_4(t_1) \\ \theta_1(t_0), & \theta_2(t_0), & \theta_3(t_0), & \theta_4(t_0) \\ \theta_1{}'(t_0), & \theta_2{}'(t_0), & \theta_3{}'(t_0), & \theta_4{}'(t_0) \\ \theta_1{}'''(t_0), & \theta_2{}'''(t_0), & \theta_3{}'''(t_0), & \theta_4{}'''(t_0) \end{vmatrix} + \dots = 0,$$

where the unexpressed terms involve the second and higher powers of $\epsilon$. From the equations

$$c_1 \theta_1(t_1) + c_2 \theta_2(t_1) + c_3 \theta_3(t_1) + c_4 \theta_4(t_1) = 0,$$

$$c_1 \theta_1(t_0) + c_2 \theta_2(t_0) + c_3 \theta_3(t_0) + c_4 \theta_4(t_0) = 0,$$

$$c_1 \theta_1{}'(t_0) + c_2 \theta_2{}'(t_0) + c_3 \theta_3{}'(t_0) + c_4 \theta_4{}'(t_0) = 0,$$

we have

$$\underset{\begin{vmatrix} \theta_2(t_1), & \theta_3(t_1), & \theta_4(t_1) \\ \theta_2(t_0), & \theta_3(t_0), & \theta_4(t_0) \\ \theta_2'(t_0), & \theta_3'(t_0), & \theta_4'(t_0) \end{vmatrix}}{c_1} = \underset{\begin{vmatrix} \theta_3(t_1), & \theta_4(t_1), & \theta_1(t_1) \\ \theta_3(t_0), & \theta_4(t_0), & \theta_1(t_0) \\ \theta_3'(t_0), & \theta_4'(t_0), & \theta_1'(t_0) \end{vmatrix}}{-c_2}$$

$$= \underset{\begin{vmatrix} \theta_4(t_1), & \theta_1(t_1), & \theta_2(t_1) \\ \theta_4(t_0), & \theta_1(t_0), & \theta_2(t_0) \\ \theta_4'(t_0), & \theta_1'(t_0), & \theta_2'(t_0) \end{vmatrix}}{c_3} = \underset{\begin{vmatrix} \theta_1(t_1), & \theta_2(t_1), & \theta_3(t_1) \\ \theta_1(t_0), & \theta_2(t_0), & \theta_3(t_0) \\ \theta_1'(t_0), & \theta_2'(t_0), & \theta_3'(t_0) \end{vmatrix}}{-c_4} = P,$$

where $P$ is a finite non-vanishing quantity. If the foregoing equation were satisfied, it would require the relation

$$\frac{\epsilon T_0}{P} \{ c_1 \theta_1'''(t_0) + c_2 \theta_2'''(t_0) + c_3 \theta_3'''(t_0) + c_4 \theta_4'''(t_0) \} + \ldots = 0.$$

We know (§ 114) that the quantity

$$c_1 \theta_1'''(t_0) + c_2 \theta_2'''(t_0) + c_3 \theta_3'''(t_0) + c_4 \theta_4'''(t_0)$$

$$= P \begin{vmatrix} \theta_1'''(t_0), & \theta_2'''(t_0), & \theta_3'''(t_0), & \theta_4'''(t_0) \\ \theta_1(t_0), & \theta_2(t_0), & \theta_3(t_0), & \theta_4(t_0) \\ \theta_1'(t_0), & \theta_2'(t_0), & \theta_3'(t_0), & \theta_4'(t_0) \\ \theta_1''(t_0), & \theta_2''(t_0), & \theta_3''(t_0), & \theta_4''(t_0) \end{vmatrix}$$

$$= - P W (t_0),$$

and therefore cannot vanish. The foregoing equation cannot be satisfied; and therefore the quantity

$$\zeta(t_0 + \epsilon T_0, t_1)$$

cannot vanish. Hence $D'$ cannot coincide with $C'$.

**119.** Thus, when $\epsilon$ is very small, $D'$ is very slightly distant from $C'$ and distinct from it. Let the value of $t$ at $D'$ be $t_1 + \epsilon \Theta$, where $D'$ is beyond or behind $C'$ according as $\Theta$ is positive or negative. As $D'$ is the conjugate of $D$, we have

$$\zeta(t_0 + \epsilon T_0, t_1 + \epsilon \Theta) = 0,$$

that is,

$$\zeta(t_0, t_1) + \epsilon T_0 \frac{\partial \zeta(t_0, t_1)}{\partial t_0} + \epsilon \Theta \frac{\partial \zeta(t_0, t_1)}{\partial t_1} + \ldots = 0,$$

where the unexpressed terms are of the second and higher degrees in $\epsilon$. Also $\zeta(t_0, t_1) = 0$, because $C$ and $C'$ are conjugate; hence, effectively

$$T_0 \frac{\partial \zeta(t_0, t_1)}{\partial t_0} + \Theta \frac{\partial \zeta(t_0, t_1)}{\partial t_1} = 0.$$

By the preceding analysis,

$$\frac{\partial \zeta(t_0, t_1)}{\partial t_0} = \delta \begin{vmatrix} \theta_1(t_1) , & \theta_2(t_1) , & \theta_3(t_1) , & \theta_4(t_1) \\ \theta_1(t_0) , & \theta_2(t_0) , & \theta_3(t_0) , & \theta_4(t_0) \\ \theta_1'(t_0) , & \theta_2'(t_0) , & \theta_3'(t_0) , & \theta_4'(t_0) \\ \theta_1'''(t_0), & \theta_2'''(t_0), & \theta_3'''(t_0), & \theta_4'''(t_0) \end{vmatrix}$$

$$= -\frac{\delta}{P} \{c_1 \theta_1'''(t_0) + c_2 \theta_2'''(t_0) + c_3 \theta_3'''(t_0) + c_4 \theta_4'''(t_0)\}$$

$$= \frac{\delta}{P} W(t_0);$$

and

$$\frac{\partial \zeta(t_0, t_1)}{\partial t_1} = \delta M;$$

consequently

$$T_0 \frac{W(t_0)}{P} + \Theta M = 0,$$

and therefore

$$\frac{\Theta}{T_0} = -\frac{W(t_0)}{M} \frac{1}{P}.$$

Now $W(t_0)$ and $M$ have the same sign; and' $P$ is a finite non-vanishing quantity. Thus $\Theta$ does not vanish.

If it were possible that $\Theta$ could be negative, then $D'$ would lie within $CC'$, the arc $C'D'$ being of the same order of magnitude as $CD$: and $DD'$ would be a complete range.

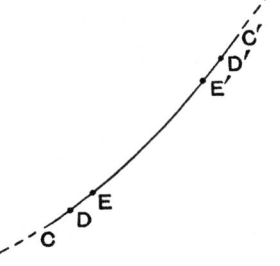

But at $D$, the quantity $W(t)$ is of the same sign as it is at $C$ and it does not vanish; the quantity $M$ at $D'$ is of the same sign as $W(t)$ at $D$. Hence taking a point $E$ within the range $DD'$ near $C$, we should have a complete range $EE'$ within $DD'$, where the arc $D'E'$ is of the same order of magnitude as $DE$. And so on, from complete range to complete range: each would be included within the preceding range, and the diminutions at the two extremities of any range would be of the same order of magnitude.

We should then be able to diminish the finite range to any extent, and attain a complete range between a place $t'$ and a place $t''$, $= t' + \epsilon T'$, where $T'$ is finite. But we have seen that

$$\zeta(t', t' + \epsilon T') = -\tfrac{1}{6} \epsilon^3 \delta T'^3 W(t') + \dots,$$

where $W(t')$ is of the same sign as $W(t_0)$ and cannot vanish; the fundamental property, defining the complete range—viz. that $\zeta(t', t'') = 0$, where $t'$ and $t''$ mark the boundaries, of the complete range—would be violated for such a range.

Hence $\Theta$ cannot be negative. It is not zero. Therefore it must be positive. That is, the point $D''$, the conjugate of $D$, lies beyond $C'$ when $D$ lies within $CC'$ and near $C$; and the arc $C'D''$ is of the same order of magnitude as $CD$.

Again, take a point $E$, lying within $DD''$ and near $D$; its conjugate $E''$ on the characteristic lies beyond $D''$, the arc $D''E''$ being of the same order of magnitude as $DE$.

And so on in succession. We infer the result that no complete range can be included within any other complete range: the conjugate of any point within a complete range lies outside that range.

*Note* 1. The quantity $P$, defined in connection with the coefficients $c$, is such that

$$c_1 = P \begin{vmatrix} \theta_2(t_1), & \theta_3(t_1), & \theta_4(t_1) \\ \theta_2(t_0), & \theta_3(t_0), & \theta_4(t_0) \\ \theta_2'(t_0), & \theta_3'(t_0), & \theta_4'(t_0) \end{vmatrix} ;$$

and so for the other coefficients $c_2, c_3, c_4$. But, in the expression for $\zeta(t_0, t)$, we have

$$c_1 = \delta \begin{vmatrix} \theta_2(t_0), & \theta_3(t_0), & \theta_4(t_0) \\ \theta_2'(t_0), & \theta_3'(t_0), & \theta_4'(t_0) \\ \theta_2''(t_0), & \theta_3''(t_0), & \theta_4''(t_0) \end{vmatrix} ;$$

and therefore

$$P \begin{vmatrix} \theta_2(t_1), & \theta_3(t_1), & \theta_4(t_1) \\ \theta_2(t_0), & \theta_3(t_0), & \theta_4(t_0) \\ \theta_2'(t_0), & \theta_3'(t_0), & \theta_4'(t_0) \end{vmatrix} = \delta \begin{vmatrix} \theta_2(t_0), & \theta_3(t_0), & \theta_4(t_0) \\ \theta_2'(t_0), & \theta_3'(t_0), & \theta_4'(t_0) \\ \theta_2''(t_0), & \theta_3''(t_0), & \theta_4''(t_0) \end{vmatrix} ,$$

together with three similar expressions arising out of the different expression for $c_2$, for $c_3$, and for $c_4$.

It is easy to verify that the four values of $P$, given by these relations, are equal to one another because of the equation

$$\zeta(t_0, t_1) = 0,$$

defining the conjugate of the initial point on the characteristic range.

*Note* 2. Further, the preceding argument shews that $P$, there defined, is a negative quantity; for $\Theta$ and $T_0$ these must have the same sign, and in the equation

$$\frac{\Theta}{T_0} = -\frac{W(t_0)}{M} \frac{1}{P},$$

$M$ and $W(t_0)$ have the same sign.

*Adjacent characteristic curves : sub-consecutive curve.*

**120.**  We say that a characteristic curve is *adjacent* (instead of *consecutive*) to an original (or central) characteristic curve, when it touches the latter at an initial common point but has not the same curvature at that point. Let $g_1, g_2, g_3, g_4$ be the constants for an adjacent curve through an initial point $t_0$, determined by the assignment of $\kappa\rho_0$ as a measure of the change from the curvature of the original character at $t_0$; then, if this adjacent curve is given by

$$\zeta = g_1\theta_1(t) + g_2\theta_2(t) + g_3\theta_3(t) + g_4\theta_4(t),$$

we have

$$0 = g_1\theta_1(t_0) + g_2\theta_2(t_0) + g_3\theta_3(t_0) + g_4\theta_4(t_0),$$
$$0 = g_1\theta_1{}'(t_0) + g_2\theta_2{}'(t_0) + g_3\theta_3{}'(t_0) + g_4\theta_4{}'(t_0),$$
$$\rho_0 = g_1\theta_1{}''(t_0) + g_2\theta_2{}''(t_0) + g_3\theta_3{}''(t_0) + g_4\theta_4{}''(t_0).$$

As $\rho_0$ is not zero, the adjacent curve is not a consecutive curve.

This adjacent curve may cut the original characteristic in one point, or in more than one point, after the initial point—that is, for values of $t$ increasing from $t_0$. Let the first of these be given by the value $t_2$ of $t$, where $t_2 > t_0$, so that we have

$$0 = g_1\theta_1(t_2) + g_2\theta_2(t_2) + g_3\theta_3(t_2) + g_4\theta_4(t_2);$$

the adjacent characteristic is represented by the equation

$$\begin{vmatrix} \zeta(t) & \theta_1{}''(t_0), & \theta_2{}''(t_0), & \theta_3{}''(t_0), & \theta_4{}''(t_0) \\ \theta_1(t_0), & \theta_2(t_0), & \theta_3(t_0), & \theta_4(t_0) \\ \theta_1{}'(t_0), & \theta_2{}'(t_0), & \theta_3{}'(t_0), & \theta_4{}'(t_0) \\ \theta_1(t_2), & \theta_2(t_2), & \theta_3(t_2), & \theta_4(t_2) \end{vmatrix} = \rho_0 \begin{vmatrix} \theta_1(t), & \theta_2(t), & \theta_3(t), & \theta_4(t) \\ \theta_1(t_0), & \theta_2(t_0), & \theta_3(t_0), & \theta_4(t_0) \\ \theta_1{}'(t_0), & \theta_2{}'(t_0), & \theta_3{}'(t_0), & \theta_4{}'(t_0) \\ \theta_1(t_2), & \theta_2(t_2), & \theta_3(t_2), & \theta_4(t_2) \end{vmatrix}.$$

So far, the place $t_2$ can be chosen at will; because one arbitrary element in $\zeta(t)$ has been left, a survival after the assignment of the three initial conditions, as regards position and tangency (both unchanged) and curvature (changed). Among these adjacent curves, let that one be selected which touches the original curve at the next point of meeting after $t_0$. Denoting this value of $t$ for this point still by $t_2$, we have $\dfrac{d\zeta}{dt} = 0$ when $t = t_2$; that is, the equation

$$\Delta(t_0, t_2) = \begin{vmatrix} \theta_1(t_0), & \theta_2(t_0), & \theta_3(t_0), & \theta_4(t_0) \\ \theta_1{}'(t_0), & \theta_2{}'(t_0), & \theta_3{}'(t_0), & \theta_4{}'(t_0) \\ \theta_1(t_2), & \theta_2(t_2), & \theta_3(t_2), & \theta_4(t_2) \\ \theta_1{}'(t_2), & \theta_2{}'(t_2), & \theta_3{}'(t_2), & \theta_4{}'(t_2) \end{vmatrix} = 0$$

is satisfied. It may be that $\Delta(t_0, t_2) = 0$, as an equation to determine $t_2$, has more than a single root $> t_0$; if so, we shall assume $t_2$ to be the smallest of such roots.

A characteristic curve, touching an original characteristic curve in two distinct points $t_0$ and $t_2$, instead of osculating it at $t_0$ and merely meeting it again in a distinct point $t_1$, will be called a *sub-consecutive* characteristic.

### Equation of sub-consecutive characteristic.

**121.** The analytical expression of a sub-consecutive characteristic, when explicit account is taken initially of the fact that it touches the characteristic at $t_0$, is

$$\zeta(t) = G \begin{vmatrix} \theta_1(t), & \theta_2(t), & \theta_3(t), & \theta_4(t) \\ \theta_1(t_0), & \theta_2(t_0), & \theta_3(t_0), & \theta_4(t_0) \\ \theta_1'(t_0), & \theta_2'(t_0), & \theta_3'(t_0), & \theta_4'(t_0) \\ \theta_1(t_2), & \theta_2(t_2), & \theta_3(t_2), & \theta_4(t_2) \end{vmatrix};$$

and the relation, securing tangency to the characteristic at $t_2$, is

$$\Delta(t_0, t_2) = 0.$$

When explicit account is taken initially of the fact, that the sub-consecutive characteristic touches the characteristic at $t_2$, the analytical expression of the sub-consecutive curve is

$$\zeta(t) = H \begin{vmatrix} \theta_1(t), & \theta_2(t), & \theta_3(t), & \theta_4(t) \\ \theta_1(t_0), & \theta_2(t_0), & \theta_3(t_0), & \theta_4(t_0) \\ \theta_1(t_2), & \theta_2(t_2), & \theta_3(t_2), & \theta_4(t_2) \\ \theta_1'(t_2), & \theta_2'(t_2), & \theta_3'(t_2), & \theta_4'(t_2) \end{vmatrix},$$

together with the same relation

$$\Delta(t_0, t_2) = 0,$$

which now secures the tangency to the characteristic $t_0$.

As the two expressions for $\zeta(t)$ represent the sub-consecutive curve, each taken in conjunction with the equation $\Delta(t_0, t_2) = 0$, it is to be expected that the two expressions will agree. To secure full agreement of the two expressions along the whole curve, it is necessary that the coefficients of $\theta_1(t)$ in the two expressions must be the same : and likewise for the two coefficients of $\theta_2(t)$, of $\theta_3(t)$, and of $\theta_4(t)$, respectively. Consequently, there must be four relations of the form

$$G \begin{vmatrix} \theta_2(t_0), & \theta_3(t_0), & \theta_4(t_0) \\ \theta_2'(t_0), & \theta_3'(t_0), & \theta_4'(t_0) \\ \theta_2(t_2), & \theta_3(t_2), & \theta_4(t_2) \end{vmatrix} = H \begin{vmatrix} \theta_2(t_0), & \theta_3(t_0), & \theta_4(t_0) \\ \theta_2(t_2), & \theta_3(t_2), & \theta_4(t_2) \\ \theta_2'(t_2), & \theta_3'(t_2), & \theta_4'(t_2) \end{vmatrix};$$

and they must be consistent with one another, in virtue of the relation $\Delta(t_0, t_2) = 0$ connecting the variables $t_0$ and $t_2$.

Let

$$\begin{Vmatrix} \theta_1(t_0), & \theta_2(t_0), & \theta_3(t_0), & \theta_4(t_0) \\ \theta_1(t_2), & \theta_2(t_2), & \theta_3(t_2), & \theta_4(t_2) \end{Vmatrix} = c_{12}, \ c_{13}, \ c_{14}, \ c_{23}, \ c_{24}, \ c_{34};$$

$$\begin{Vmatrix} \theta_1'(t_0), & \theta_2'(t_0), & \theta_3'(t_0), & \theta_4'(t_0) \\ \theta_1'(t_2), & \theta_2'(t_2), & \theta_3'(t_2), & \theta_4'(t_2) \end{Vmatrix} = \gamma_{12}, \ \gamma_{13}, \ \gamma_{14}, \ \gamma_{23}, \ \gamma_{24}, \ \gamma_{34}.$$

Then, if

$$G\,\theta_r'(t_0) + H\,\theta_r'(t_2) = K_r,$$

for $r = 1, 2, 3, 4$, the foregoing relation becomes

$$K_2 c_{34} + K_3 c_{42} + K_4 c_{23} = 0.$$

There are three other such relations. One of them is

$$K_1 c_{34} + K_3 c_{41} + K_4 c_{13} = 0.$$

Eliminating $K_4$ between these two modified expressions of relation, and using the identity

$$c_{41} c_{23} + c_{42} c_{31} + c_{43} c_{12} = 0,$$

we find

$$K_1 c_{23} + K_2 c_{31} + K_3 c_{12} = 0.$$

In a similar way, we deduce from those two relations

$$K_1 c_{24} + K_2 c_{41} + K_4 c_{12} = 0.$$

Again, eliminating the ratio $G : H$ between the two relations

$$K_2 c_{34} + K_3 c_{42} + K_4 c_{23} = 0, \quad K_1 c_{34} + K_3 c_{41} + K_4 c_{13} = 0,$$

we have

$$\{\theta_2'(t_0)\, c_{34} + \theta_3'(t_0)\, c_{42} + \theta_4'(t_0)\, c_{23}\} \{\theta_1'(t_2)\, c_{34} + \theta_3'(t_2)\, c_{41} + \theta_4'(t_2)\, c_{13}\}$$
$$= \{\theta_1'(t_0)\, c_{34} + \theta_3'(t_0)\, c_{41} + \theta_4'(t_0)\, c_{13}\} \{\theta_2'(t_2)\, c_{34} + \theta_3'(t_2)\, c_{42} + \theta_4'(t_2)\, c_{23}\},$$

that is,

$$c_{34}\, (\gamma_{12} c_{34} + \gamma_{23} c_{14} + \gamma_{31} c_{24} + \gamma_{34} c_{12} + \gamma_{14} c_{23} + \gamma_{24} c_{31}) = 0,$$

or

$$-c_{34}\Delta\,(t_0, t_2) = 0,$$

which is satisfied. Thus the four relations, which express the equivalence of the two forms for $\zeta(t)$, are consistent with one another. All of them, when account is taken of the equation $\Delta(t_0, t_2) = 0$, are satisfied, if $G$ and $H$ are connected by a single one of them, such as

$$K_2 c_{34} + K_3 c_{42} + K_4 c_{23} = 0.$$

**122.** Take a point on the sub-consecutive curve between $t_0$ and $t_2$, and very near to $t_0$. Then from the $G$-expression for $\zeta$ we have, effectively,

$$\zeta(t) = \tfrac{1}{2}(t - t_0)^2\, G \begin{vmatrix} \theta_1''(t_0), & \theta_2''(t_0), & \theta_3''(t_0), & \theta_4''(t_0) \\ \theta_1(t_0), & \theta_2(t_0), & \theta_3(t_0), & \theta_4(t_0) \\ \theta_1'(t_0), & \theta_2'(t_0), & \theta_3'(t_0), & \theta_4'(t_0) \\ \theta_1(t_2), & \theta_2(t_2), & \theta_3(t_2), & \theta_4(t_2) \end{vmatrix},$$

and from the $H$-expression for $\zeta$ we have, effectively,

$$\zeta(t) = \tfrac{1}{2}(t - t_0)^2 H \begin{vmatrix} \theta_1''(t_0), & \theta_2''(t_0), & \theta_3''(t_0), & \theta_4''(t_0) \\ \theta_1(t_0), & \theta_2(t_0), & \theta_3(t_0), & \theta_4(t_0) \\ \theta_1(t_2), & \theta_2(t_2), & \theta_3(t_2), & \theta_4(t_2) \\ \theta_1'(t_2), & \theta_2'(t_2), & \theta_3'(t_2), & \theta_4'(t_2) \end{vmatrix},$$

the coefficient of $t - t_0$ in the $H$-expression for $\zeta$ being the vanishing quantity $\Delta(t_0, t_2)$. The two expressions for $\zeta(t)$ in the near vicinity of $t_0$ agree.

Similarly, for a place in the range in the near vicinity of $t_2$, we have

$$\zeta(t) = \tfrac{1}{2}(t_2 - t)^2 G \begin{vmatrix} \theta_1''(t_2), & \theta_2''(t_2), & \theta_3''(t_2), & \theta_4''(t_2) \\ \theta_1(t_0), & \theta_2(t_0), & \theta_3(t_0), & \theta_4(t_0) \\ \theta_1'(t_0), & \theta_2'(t_0), & \theta_3'(t_0), & \theta_4'(t_0) \\ \theta_1(t_2), & \theta_2(t_2), & \theta_3(t_2), & \theta_4(t_2) \end{vmatrix},$$

and

$$= \tfrac{1}{2}(t_2 - t)^2 H \begin{vmatrix} \theta_1''(t_2), & \theta_2''(t_2), & \theta_3''(t_2), & \theta_4''(t_2) \\ \theta_1(t_0), & \theta_2(t_0), & \theta_3(t_0), & \theta_4(t_0) \\ \theta_1(t_2), & \theta_2(t_2), & \theta_3(t_2), & \theta_4(t_2) \\ \theta_1'(t_2), & \theta_2'(t_2), & \theta_3'(t_2), & \theta_4'(t_2) \end{vmatrix},$$

from the two expressions respectively, the coefficient of $t_2 - t$ in the $G$-expression for $\zeta$ being the vanishing quantity $-\Delta(t_0, t_2)$. These two expressions for $\zeta(t)$ in the near vicinity of $t_2$ agree.

Manifestly a sub-consecutive characteristic does not cross the characteristic curve with which it is associated.

### *Property concerning a consecutive characteristic.*

**123.** One other proposition, connected with characteristic curves, will be established here; it is required, later, in the discussion of small variations that are not weak.

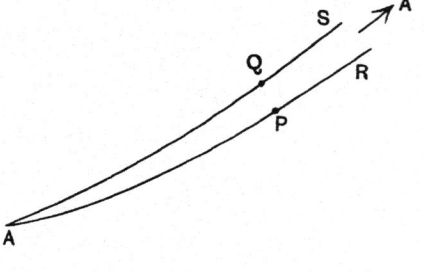

Let $APR\ldots$ be a characteristic curve on which $A'$ is the conjugate of $A$; and let $P$ be any point on the curve within the range $AA'$. Take any point $Q$, off the curve, and contiguous to $P$; then one, and only one, consecutive characteristic curve can be drawn from the same initial point $A$ so as to pass through $Q$.

If possible, let the consecutive characteristic curve be represented by

$$\zeta = k_1\theta_1(t) + k_2\theta_2(t) + k_3\theta_3(t) + k_4\theta_4(t);$$

because it is consecutive to $APR...$, we have

$$0 = k_1\theta_1\ (t_0) + k_2\theta_2\ (t_0) + k_3\theta_3\ (t_0) + k_4\theta_4\ (t_0),$$
$$0 = k_1\theta_1'\ (t_0) + k_2\theta_2'\ (t_0) + k_3\theta_3'\ (t_0) + k_4\theta_4'\ (t_0),$$
$$0 = k_1\theta_1''(t_0) + k_2\theta_2''(t_0) + k_3\theta_3''(t_0) + k_4\theta_4''(t_0),$$

where $t_0$ is the value of $t$ at $A$. At $P$, let $t'$ be the value of $t$ on the original curve, and $t' + \kappa T'$ its value at $Q$ on the hypothetical consecutive characteristic. If $x$ and $y$ be the coordinates of $P$, those of $Q$ can be represented by $x + \kappa X'$ and $y + \kappa Y'$; thus

$$X' = \phi'\ (t')\ T' + k_1\phi_1\ (t') + k_2\phi_2\ (t') + k_3\phi_3(t') + k_4\phi_4\ (t'),$$
$$Y' = \psi'\ (t')\ T' + k_1\psi_1\ (t') + k_2\psi_2\ (t') + k_3\psi_3(t') + k_4\psi_4\ (t'),$$

and therefore

$$\phi'\ (t')\ Y' - \psi'\ (t')\ X' = k_1\theta_1\ (t') + k_2\theta_2\ (t') + k_3\theta_3\ (t') + k_4\theta_4\ (t').$$

Resolving the four equations for $k_1$, $k_2$, $k_3$, $k_4$, we have

$$\zeta(t_0,\ t')\ k_r = \{\phi'\ (t')\ Y' - \psi'\ (t')\ X'\}\ c_r,$$

for $r = 1,\ 2,\ 3,\ 4$, where $c_1, c_2, c_3, c_4$ are the coefficients in the expression of the consecutive characteristic curve in § 118, and $\zeta(t_0,\ t)$ is the former expression for the deviation in that consecutive curve.

As $A'$, given by $t = t_1$, is the conjugate of $A$, so that $t_1$ is the first value of $t_1 (> t_0)$ satisfying the equation $\zeta(t_0,\ t_1) = 0$, and as $t'$ is intermediate in value between $t_0$ and $t_1$, the quantity $\zeta(t_0,\ t')$ does not vanish; hence $k_1, k_2, k_3, k_4$ are finite and definite. The consecutive characteristic can be drawn as required.

*Ex.* Let $B$ be a point on the characteristic in the range $AP$, and let $C$ be a point contiguous to $B$. Prove that it is possible to draw a characteristic curve through $C$ and $Q$, when the difference between the directions of the tangents at $B$ and $C$, and the difference between the curvatures at $B$ and $C$, are of the same order of magnitude as the small variation displacing $B$ to $C$ and $P$ to $Q$.

*Discussion of the final normal form of the 'second' variation.*

**124.** We now come to the normal reduced form for the second variation of the original integral, which has been proved (§ 110) to be

$$\tfrac{1}{2}\kappa^2 \int \frac{Q}{(\alpha\beta' - \alpha'\beta)^2} \begin{vmatrix} w'', & w', & w \\ \alpha'', & \alpha', & \alpha \\ \beta'', & \beta', & \beta \end{vmatrix}^2 dt.$$

Here, $\alpha$ and $\beta$ are two linearly independent integrals of the subsidiary characteristic equation $\mathbf{E} = 0$ such that the quantity

$$Q\ (\alpha\beta''' - \alpha'''\beta - \alpha'\beta'' + \alpha''\beta') + Q'\ (\alpha\beta'' - \alpha''\beta) + J\ (\alpha\beta' - \alpha'\beta),$$

known to be a constant for any two integrals, is made equal to zero by the choice of $\alpha$ and $\beta$. For the discussion, we shall need the properties of adjacent (but not sub-consecutive) characteristics and, ultimately, the properties of the consecutive characteristic.

The first variation of the original integral has been made to vanish. In vanishing, it secured fixed limits for the integral from assigned conditions, if these were not already fixed by the preliminary assigned data. Thus all the variations, which now may be considered, must be such as to vanish, at the lower limit of the integral and also at the upper limit of the integral.

Consider, first, the lower limit of the integral, being the value $t_0$ of $t$ for the initial point of the range on the characteristic. We take a couple of independent adjacent curves, represented by small variations $\kappa\alpha$ and $\kappa\beta$. If the coefficients for $\alpha$ be $m_1$, $m_2$, $m_3$, $m_4$, and those for $\beta$ be $n_1$, $n_2$, $n_3$, $n_4$, then

$$m_1\theta_1(t_0) + m_2\theta_2(t_0) + m_3\theta_3(t_0) + m_4\theta_4(t_0) = 0,$$
$$m_1\theta_1'(t_0) + m_2\theta_2'(t_0) + m_3\theta_3'(t_0) + m_4\theta_4'(t_0) = 0,$$

because the characteristic represented by the deviation $\kappa\alpha$ is adjacent to the central curve; and

$$n_1\theta_1(t_0) + n_2\theta_2(t_0) + n_3\theta_3(t_0) + n_4\theta_4(t_0) = 0,$$
$$n_1\theta_1'(t_0) + n_2\theta_2'(t_0) + n_3\theta_3'(t_0) + n_4\theta_4'(t_0) = 0,$$

because the characteristic represented by the deviation $\kappa\beta$ is adjacent to the central curve. The two former relations give $\alpha = 0$ and $\alpha' = 0$ at $t_0$, and the two latter relations give $\beta = 0$ and $\beta' = 0$ at $t_0$; hence the quantity

$$Q(\alpha\beta''' - \alpha'''\beta - \alpha'\beta'' + \alpha''\beta') + Q'(\alpha\beta'' - \alpha''\beta) + J(\alpha\beta' - \alpha'\beta)$$

is zero at $t = t_0$, and so (because it is a constant along the range) it is equal to zero for all values of $t$. Thus the two selected integrals $\alpha$ and $\beta$ of the equation $\mathbf{E} = 0$ satisfy the prescribed demand; and the second variation is expressible in the foregoing form.

**125.** For the adjacent curve represented by $\alpha$, let $\rho_1$ and $\sigma_1$ be the values of $\alpha''$ and $\alpha'''$ at the initial point; and for the adjacent curve represented by $\beta$, let $\rho_2$ and $\sigma_2$ be the values of $\beta''$ and $\beta'''$ at the initial point: then

$$\rho_1 = m_1\theta_1''(t_0) + m_2\theta_2''(t_0) + m_3\theta_3''(t_0) + m_4\theta_4''(t_0) \Big\}$$
$$\sigma_1 = m_1\theta_1'''(t_0) + m_2\theta_2'''(t_0) + m_3\theta_3'''(t_0) + m_4\theta_4'''(t_0) \Big\}$$
$$\rho_2 = n_1\theta_1''(t_0) + n_2\theta_2''(t_0) + n_3\theta_3''(t_0) + n_4\theta_4''(t_0) \Big\}$$
$$\sigma_2 = n_1\theta_1'''(t_0) + n_2\theta_2'''(t_0) + n_3\theta_3'''(t_0) + n_4\theta_4'''(t_0) \Big\}$$

The determinant of the coefficients in the four linear equations, in which $m_1$, $m_2$, $m_3$, $m_4$ occur, is $-W(t_0)$, a non-vanishing quantity; and likewise for the four linear equations in which $n_1$, $n_2$, $n_3$, $n_4$ occur. The coefficients $m$, and the coefficients $n$, are thus uniquely determinate; they determine two adjacent curves given by

$$\alpha = m_1\theta_1(t) + m_2\theta_2(t) + m_3\theta_3(t) + m_4\theta_4(t),$$
$$\beta = n_1\theta_1(t) + n_2\theta_2(t) + n_3\theta_3(t) + n_4\theta_4(t).$$

Now consider a characteristic curve, represented by the combination

$$\rho_2 \alpha - \rho_1 \beta.$$

The coefficients for this characteristic are $\rho_2 m_r - \rho_1 n_r$ (for $r = 1, 2, 3, 4$). These four coefficients satisfy the relations

$$0 = \sum_{r=1}^{4} (\rho_2 m_r - \rho_1 n_r) \, \theta_r(t_0),$$

$$0 = \sum_{r=1}^{4} (\rho_2 m_r - \rho_1 n_r) \, \theta_r'(t_0),$$

$$0 = \sum_{r=1}^{4} (\rho_2 m_r - \rho_1 n_r) \, \theta_r''(t_0).$$

We must not have

$$\sum_{r=1}^{4} (\rho_2 m_r - \rho_1 n_r) \, \theta_r'''(t_0)$$

a vanishing quantity. Otherwise, as the determinant of the coefficients of the four quantities $\rho_2 m_r - \rho_1 n_r$ is $W(t_0)$ which is not a vanishing quantity, we should have

$$\rho_2 m_r - \rho_1 n_r = 0,$$

for $r = 1, 2, 3, 4$; and then $\rho_2 \alpha - \rho_1 \beta$ would be zero for all values of $t$, so that $\alpha$ and $\beta$ would not be linearly independent. Hence

$$\rho_2 \sigma_1 - \rho_1 \sigma_2$$

must not be zero. And now, the combination $\rho_2 \alpha - \rho_1 \beta$, instead of being a zero combination, gives a consecutive characteristic curve.

### The Jacobi test.

**126.** Now the first value of $t$, greater than $t_0$, for which a consecutive characteristic meets the original characteristic, is the value denoted by $t_1$. If therefore the range of the integral, beginning at $t_0$, should extend as far as $t_1$, we could have a non-zero variation $\rho_2 \alpha - \rho_1 \beta$ of the characteristic. That variation vanishes at $t_0$, because $\alpha$ and $\beta$ represent adjacent curves; it vanishes at $t_1$, because $\rho_2 \alpha - \rho_1 \beta$ represents a consecutive characteristic; and it is not zero in the course of the range between $t_0$ and $t_1$. In these circumstances, we could take

$$w = \rho_2 \alpha - \rho_1 \beta;$$

such an assumption would meet the requirements of vanishing at each extremity of the range, and of being distinct from zero throughout the whole of the range save at the extremities. But $\rho_1$ and $\rho_2$ are constants; and therefore, for such a value of $w$, the magnitude

$$\begin{vmatrix} w'', & w', & w \\ \alpha'', & \alpha', & \alpha \\ \beta'', & \beta', & \beta \end{vmatrix}$$

vanishes, while $\alpha\beta' - \alpha'\beta$ is not zero because $\alpha$ and $\beta$ are linearly independent. Should this occur—that is, by selecting a particular non-zero variation of the characteristic curve—the second variation vanishes.

But if a maximum or a minimum is to be possessed by an integral, it must arise for a stationary conformation settled by the characteristic curve. The maximum or minimum is certainly possessed, if no admissible small variation can occur which makes the second variation vanish. (It might be that admissible small variations could occur, of the type just indicated, for which the second variation vanishes. In that eventuality, the third variation would have to vanish for those variations if the maximum or minimum is to be possessed; and the fourth variation must not vanish for them. The case is not sufficiently general to justify discussion here; when it arises in any particular instance—as in Ex. 1, § 30, p. 33—it can be considered by a special discussion. It will therefore be omitted from further consideration in the present general argument.) We therefore exclude the possibility which renders small variations of the foregoing type admissible: that is, $\rho_2\alpha - \rho_1\beta$ which vanishes at the beginning of the range must not again vanish within the range and, as has been proved, it cannot vanish earlier than at the conjugate of the initial point. Consequently

*the range of the integral must not extend as far as the conjugate of the initial point on the characteristic curve.*

We thus obtain the range-test (the *Jacobi* test), limiting the range.

### The Legendre test.

**127.** There still remains the quantity $Q$ for consideration, as it occurs in the subject of integration in the integral expressing the second variation.

If $Q$ could be sometimes positive and sometimes negative, then, by taking $w$ zero over a range where $Q$ is negative and distinct from zero over a range where $Q$ is positive, the second variation could be made positive; and, by taking $w$ zero over a range where $Q$ is positive and distinct from zero over a range where $Q$ is negative, the second variation could be made negative. But the second variation must have a persistent sign, if the integral is to possess a maximum or a minimum. The possibility, just indicated, invalidates the existence of either, and must therefore be precluded from occurrence in connection with a favourable issue for the problem. Thus $Q$ must be either always positive through the range (and then a minimum can exist, subject to the other conditions) or it must be always negative through the range (and then a maximum can exist, subject to the other conditions).

We thus obtain the test (the *Legendre* test):

*The quantity $Q(t)$ must have a uniform sign throughout the whole range of the integral.*

*Summary of conditions for weak variations.*

**128.** We thus have established the following three tests for the existence of a maximum or of a minimum of an integral when its variable quantities are subject to weak variations:

(A) The curve, along which the integration takes place, must satisfy the characteristic equations.

> (This test may be called the Euler test, as Euler was the first to obtain the critical differential equation in virtue of which the characteristic equations are satisfied. Sometimes it is called the Lagrange test, because of his more extended investigations at a date later than Euler's first investigation.)

> Moreover, either the terminal data explicitly, or inferences from terminal descriptive conditions onstructively, determine the limits of the integral as known quantities which, once known, are immune from variation.

(B) The range of the integral along a characteristic curve must not extend as far as the conjugate of the initial point of the range.

> (This is the Jacobi test. In any exceptional instance when the range is extended up to the conjugate, the second variation of the integral can be made to vanish for a particular type of small variation of the variables; the higher variations of the integral, for that type of variation of the variables, must then be investigated before a declaration can be made in favour of the existence of a maximum or of a minimum.)

(C) The quantity, which has been denoted by $Q$ or $Q(t)$, must not vanish nor become infinite throughout the range of integration: it must keep a uniform sign throughout.

> (This is the Legendre test. It is conceivable that, at isolated points, $Q$ may vanish; but the zero then must be of even order, and that order must not exceed four, in the present case: also, in that event, there would be limitations on the quantities $J$ and $\Lambda$. It is simpler to set these special possibilities aside, examining them only when they arise in particular instances.)

The three conditions are necessary conditions, arising out of the discussion of the integral under weak variations.

When satisfied, the conditions are sufficient to secure a maximum or a minimum for weak variations. For the first condition (the Euler test, when satisfied), together with the accompanying inferences as regards the terminals, secures that the first variation of the integral vanishes for weak variations. The third condition (the Legendre test) and the second condition (the Jacobi test), when they are taken together and are satisfied, secure that the second variation of the integral is either definitely positive in all circumstances, or definitely negative in all circumstances, for weak variations. When the definite sign is positive, the integral possesses a minimum: when the definite sign is negative, the integral possesses a maximum.

**129.** An example follows, in which the tests for weak variations are satisfied, yet for which a real minimum does not exist.

*Ex. Find the curve (if any) such that the area, comprised between* (i) *any arc,* (ii) *the radii of curvature at the extremities of the arc, and* (iii) *its evolute, is a minimum.*

The area in question is

$$\tfrac{1}{2}\int \rho \, ds$$

$$=\tfrac{1}{2}\int \frac{(1+y'^2)^2}{y''}\,dx \; ;$$

so that we may take

$$f(x, y, y', y'')=\frac{(1+y'^2)^2}{y''} \, .$$

(I) As $f$ does not involve $y$ explicitly, a first integral of the characteristic equation is

$$\frac{\partial f}{\partial y'}-\frac{d}{dx}\left(\frac{\partial f}{\partial y''}\right)=\text{constant},$$

that is,

$$4\,\frac{y'\,(1+y'^2)}{y''}+\frac{d}{dx}\left(\frac{1+y'^2}{y''}\right)^2=B.$$

But

$$\frac{d}{dx}\left\{\frac{(1+y'^2)^2}{y''}\right\}=4y'\,(1+y'^2)-\frac{(1+y'^2)^2}{y''^2}\,y'''$$

$$=By''-y''\frac{d}{dx}\left(\frac{1+y'^2}{y''}\right)^2-\frac{(1+y'^2)^2}{y''^2}\,y'''$$

$$=By''-\frac{d}{dx}\left\{\frac{(1+y'^2)^2}{y''}\right\} \; ;$$

and therefore

$$2\,\frac{(1+y'^2)^2}{y''}=A+By' :$$

that is, changing the arbitrary constants $A$ and $B$ to $2a$ and $2b$,

$$\frac{(1+y'^2)^{\frac{3}{2}}}{y''}=\frac{a}{(1+y'^2)^{\frac{1}{2}}}+\frac{by'}{(1+y'^2)^{\frac{1}{2}}},$$

or

$$\frac{ds}{d\psi}=a\cos\psi+b\sin\psi,$$

where, as usual, we have

$$y'=\tan\psi.$$

Thus

$$\frac{dx}{d\psi}=(a\cos\psi+b\sin\psi)\cos\psi, \qquad \frac{dy}{d\psi}=(a\cos\psi+b\sin\psi)\sin\psi \; ;$$

and therefore

$$\left.\begin{array}{l}4\,(x-h)=2a\psi+a\sin 2\psi-b\cos 2\psi\\[4pt]4\,(y-k)=2b\psi-a\cos 2\psi-b\sin 2\psi\end{array}\right\},$$

where $h$, $k$ are arbitrary constants. We thus have a primitive, in which the four essential arbitrary constants are $a$, $b$, $h$, $k$.

(II) If $\kappa X$ and $\kappa Y$ denote the displacements at the limit, the equation

$$\left[ f - y' \left\{ \frac{\partial f}{\partial y'} - \frac{d}{dx} \left( \frac{\partial f}{\partial y''} \right) \right\} - y'' \frac{\partial f}{\partial y''} \right] X + Y \left\{ \frac{\partial f}{\partial y'} - \frac{d}{dx} \left( \frac{\partial f}{\partial y''} \right) \right\} + Y' \frac{\partial f}{\partial y''} = 0$$

must be satisfied there. We shall assume that the required curve is to join two fixed points, so that each extremity is fixed in position ; we thus have

$$X = 0, \quad Y = 0,$$

at the extremity, and therefore the condition becomes

$$Y' \frac{\partial f}{\partial y''} = 0$$

at that extremity.

(i) If the direction at the extremity be fixed, as well as the position, $Y' = 0$ : no further condition emerges from the limiting term, and the value of $\psi$ is then known as a given datum.

(ii) If the direction at the extremity is free, so that there is a condition to be satisfied for all directions there, then

$$\frac{\partial f}{\partial y''} = 0,$$

that is, $(1 + y'^2)^2 \, y''^{-2}$ is zero there, and the radius of curvature is zero, that is, the extremity is a cusp, or, reverting to the current equation

$$\rho = a \cos \psi + b \sin \psi,$$

we should have

$$a \cos \psi + b \sin \psi = 0$$

at an extremity, when it is fixed in position and free in direction.

(iii) Let one extremity be the point $(x_0, y_0)$, and let $\psi_0$ be the value of $\psi$ there: the other, the point $(x_1, y_1)$, and $\psi_1$ the value of $\psi$ there : where $\psi_0$ and $\psi_1$ may be regarded as known, if initial directions are given : or, otherwise, are subject to the foregoing external condition.

In the former case, the four constants $a$, $b$, $h$, $k$ are determined by the conditions

$$\left. \begin{aligned} 4 \, (x_0 - h) &= 2a\psi_0 + a \sin 2\psi_0 - b \cos 2\psi_0 \\ 4 \, (y_0 - k) &= 2b\psi_0 - a \cos 2\psi_0 - b \sin 2\psi_0 \\ 4 \, (x_1 - h) &= 2a\psi_1 + a \sin 2\psi_1 - b \cos 2\psi_1 \\ 4 \, (y_1 - k) &= 2b\psi_1 - a \cos 2\psi_1 - b \sin 2\psi_1 \end{aligned} \right\} :$$

and, in these four equations, $\psi_0$ and $\psi_1$ are known.

In the latter case, we have the same four formal equations of condition, where now $\psi_0$ and $\psi_1$ are not initially given. But now we also have the two terminal conditions

$$a \cos \psi_0 + b \sin \psi_0 = 0, \quad a \cos \psi_1 + b \sin \psi_1 = 0,$$

arising out of the evanescence of the first variation of the integral. The results $a = 0$, $b = 0$, deducible from these last two conditions, are out of the question ; and so we have

$$\sin (\psi_1 - \psi_0) = 0,$$

that is, we can take (as the simplest case) $\psi_1 - \psi_0 = \pi$, or (as a more general case) $\psi_1 - \psi_0 = m\pi$, where $m$ is an integer.

We at once have

$$\frac{y_1 - y_0}{x_1 - x_0} \tan \psi_0 + 1 = 0$$

which, with the condition $\psi_1 - \psi_0 = \pi$, shews that the tangents to the curve at the fixed points are perpendicular to the line joining the fixed extremities. When we write

$$a = c \sin \psi_0, \quad b = -c \cos \psi_0,$$

the equations of the curve are

$$4 (x - h) = c \cos (2\psi - \psi_0) + 2c\psi \sin \psi_0 \bigg\} \ ;$$
$$4 (y - k) = c \sin (2\psi - \psi_0) - 2c\psi \cos \psi_0 \bigg\}$$

and the terminal conditions give

$$4 (x_0 - h) = c \cos \psi_0 + 2c\psi_0 \sin \psi_0$$
$$4 (y_0 - k) = c \sin \psi_0 - 2c\psi_0 \cos \psi_0 \Bigg\} \ .$$
$$4 (x_1 - h) = c \cos \psi_0 + 2c (\pi + \psi_0) \sin \psi_0 \Bigg\}$$
$$4 (y_1 - k) = c \sin \psi_0 - 2c (\pi + \psi_0) \cos \psi_0 \Bigg\}$$

The curve is a cycloid having the line joining the fixed points for base ; and the fixed points are cusps.

If $m$ is greater than unity, the curve is made up of $m$ equal cycloids, all along that base and having all the cusps along the base.

(III) As regards the Legendre test, we have

$$f = \tfrac{1}{2} \frac{(1 + y'^2)^2}{y''} \ ;$$

and, naturally, the area is to be taken positive, so we may take $y''$ positive *. Now

$$\frac{\partial^2 f}{\partial y''^2} = \frac{(1 + y'^2)^2}{y''^3} ,$$

a positive quantity ; hence, if other tests are satisfied, the Legendre test will allow the integral to possess a minimum.

(IV) To apply the Jacobi test, we must construct the equation which is to settle the conjugate (say $\psi_1$) of an initial value of $\psi$ (say $\psi_0$). Merely changing the arbitrary constants $a$ and $b$, we have the primitive of the characteristic equation in the form

$$x = h + a (2\psi + \sin 2\psi) - b \cos 2\psi,$$
$$y = k + b (2\psi - \sin 2\psi) - a \cos 2\psi.$$

We have

$$\theta_1 = \frac{\partial x}{\partial \psi} \frac{\partial y}{\partial h} - \frac{\partial y}{\partial \psi} \frac{\partial x}{\partial h}$$
$$= - 4 (a \cos \psi + b \sin \psi) \cos \psi = - T \cos \psi,$$

$$\theta_2 = \frac{\partial x}{\partial \psi} \frac{\partial y}{\partial k} - \frac{\partial y}{\partial \psi} \frac{\partial x}{\partial k}$$
$$= 4 (a \cos \psi + b \sin \psi) \sin \psi = T \sin \psi,$$

$$\theta_3 = \frac{\partial x}{\partial \psi} \frac{\partial y}{\partial a} - \frac{\partial y}{\partial \psi} \frac{\partial x}{\partial a} \qquad\quad = - T (\cos \psi + 2\psi \sin \psi),$$

$$\theta_4 = \frac{\partial x}{\partial \psi} \frac{\partial y}{\partial b} - \frac{\partial y}{\partial \psi} \frac{\partial x}{\partial b} \qquad\quad = - T (\sin \psi - 2\psi \cos \psi),$$

where

$$T = 4 (a \cos \psi + b \sin \psi).$$

The critical equation is

$$\begin{vmatrix} \theta_1 (\psi_0), & \theta_2 (\psi_0), & \theta_3 (\psi_0), & \theta_4 (\psi_0) \\ \theta_1' (\psi_0), & \theta_2' (\psi_0), & \theta_3' (\psi_0), & \theta_4' (\psi_0) \\ \theta_1'' (\psi_0), & \theta_2'' (\psi_0), & \theta_3'' (\psi_0), & \theta_4'' (\psi_0) \\ \theta_1 (\psi_1), & \theta_2 (\psi_1), & \theta_3 (\psi_1), & \theta_4 (\psi_1) \end{vmatrix} = 0.$$

* Were $y''$ negative, the radius of curvature would be taken equal to $- (1 + y'^2)^{\frac{3}{2}} \dfrac{1}{y''}$ for the purpose of making $\tfrac{1}{2}\rho \, ds$ to be positive: and the Legendre test would then be satisfied as in the test.

Substituting, and reducing, we have the left-hand side equal to

$$512 (a \cos \psi_0 + b \sin \psi_0)^3 (a \cos \psi_1 + b \sin \psi_1) \sin (\psi_1 - \psi_0) \{(\psi_1 - \psi_0) \cot (\psi_1 - \psi_0) - 1\}.$$

We have to consider the roots of this equation. The factors, which give rise to

$$a \cos \psi_0 + b \sin \psi_0 = 0, \quad a \cos \psi_1 + b \sin \psi_1 = 0,$$

as in the case where the end directions are free, and as connected with the equations

$$\frac{dx}{d\psi} = 4 (a \cos \psi + b \sin \psi) \cos \psi, \quad \frac{dy}{d\psi} = 4 (a \cos \psi + b \sin \psi) \sin \psi,$$

indicate cusps of the cycloid. Thus the equation precludes a range which actually includes a cusp.

The equation

$$(\psi - \psi_0) \cot (\psi - \psi_0) - 1 = 0,$$

for a range $\pi + \psi_0 \geqq \psi \geqq \psi_0$, has only one real root, viz. $\psi = \psi_0$, merely giving the initial point of one complete branch of the cycloid ; the factor

$$(\psi_1 - \psi_0) \cot (\psi_1 - \psi_0) - 1$$

provides no conjugate in the range.

The equation

$$\sin (\psi_1 - \psi_0) = 0$$

gives $\psi_1 = \pi + \psi_0$, as the first value of $\psi_1$ greater than $\psi_0$ and satisfying the equation ; the factor therefore, in the extreme case, merely indicates a cusp as the conjugate of a preceding cusp.

It thus appears, when account is taken of all the equations into which the critical equation can be resolved, that the range of the cycloid must not include a cusp. The inference therefore would be that, if a curve is to be drawn joining two fixed points with fixed directions at each point, we take a cycloid through those points so that no cusp of the curve shall occur within the range : while, if the directions at the extreme points are not fixed, we obtain a minimum area though the Jacobi condition is just not satisfied.

*Note.* In the latter case, it would follow that, for one special type of variation, the second variation of the integral vanishes ; as an immediate preliminary, it would be necessary to consider the third variation of the integral.

(V) Other considerations shew that, for even the most generally satisfactory case where the Euler test, the Legendre test, and the Jacobi test are satisfied, there is no real minimum.

The fact is that only weak variations have been taken into account. The characteristic curve, thence derived through the Euler test, is a cycloidal arc $ACB$: to simplify the discussion, we may regard the directions at $A$ and $B$ as fixed at the cusp.

Take any point $G$ near $A$ in the line $AB$ ; and draw a small cycloid on $AG$ as base. Also draw a cycloid on $GB$ as base. The two cycloids on $AG$, $GB$ lie within $ACB$ ; they constitute a small variation of $ACB$. The area, associated with the double cycloidal arc from $A$ to $G$ and from $G$ to $B$, is less than the area associated with the arc $ACB$.

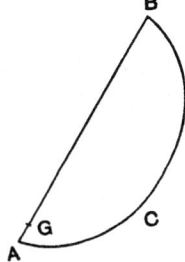

But the variation from $ACB$ to $AG$, while small in the magnitude of the displacement of corresponding points, is not a weak variation;

the changes in the direction of the tangent and in the curvature are violent. Thus a strong variation has led to a reduction in the minimum.

We can see otherwise that the area produced by the cycloidal arc can be reduced. Divide $AB$ into $n$ equal parts, one of which is $AG$; on each such part as diameter, let a semi-circle be drawn. The total area, under the requirements of the problem, and provided by the $n$ semi-circles, is

$$= n \cdot \tfrac{1}{2} \pi \left(\frac{AB}{2n}\right)^2, \quad i.e. \ = \frac{\pi}{8n} AB^2,$$

which, with increase of $n$, can be reduced indefinitely.

# CHAPTER IV.

*Integrals of the first order in two dependent variables.*

**130.** We now proceed to deal with integrals which involve two dependent variables and one independent variable, denoted by $y$, $z$, and $x$ respectively. We begin with the simplest problem; only the first derivatives of the dependent variables $\left(\text{viz. } y', = \dfrac{dy}{dx}; \text{ and } z', = \dfrac{dz}{dx}\right)$ will be supposed to occur. It will be convenient to associate a geometrical interpretation with the analysis for all such problems; manifestly, skew curves offer the natural medium. In the geometry of such curves there is less tendency, than in plane curves, to a preferential selection of one variable (such as $x$), over the other two, to rank as the independent variable; and so there will be added reason for ultimately considering $x$, $y$, $z$ as functions of a new independent variable. Such an ultimate selection will, as before, allow the small variations to be made in all the variables, instead of merely special variations to be made upon the initially selected dependent variable. Then (as in the preceding chapters) possible discontinuities, as of direction, may be discussed. Moreover, the selection simplifies the expression of properties of tangency and curvature.

As, however, certain results, fundamental in character, are derivable from the analytically simpler supposition, which makes $y$ and $z$ functions of $x$, and subjects $y$ and $z$ to small variations, without regard to small variations of $x$, it is advantageous to derive the corresponding results at once. They will indicate the type of test required for the general weak variations.

We accordingly consider an integral

$$I = \int f(x, y, y', z, z') \, dx,$$

the limits of which may be definitely fixed or are to be deducible from assigned data. The problem is to determine the conditions to be satisfied in order that $I$ may possess a maximum or a minimum.

*Variation of the integral under special variations.*

**131.** We begin with special weak variations. We therefore assume $y + \kappa v$ as a varied value of $y$, and $z + \kappa w$ as a varied value of $z$. Here $\kappa$ is an arbitrary constant quantity, of sign that can be taken at will, and of magnitude sufficiently small to render any power negligible in comparison with every lower power. The quantities $v$ and $w$ are arbitrary regular functions of $x$; and we assume that $\kappa v'$ and $\kappa w'$ are small quantities. If the limits are fixed, they remain unaffected for variations made upon $I$. If the limits are to be deduced from assigned data—thus, if the extremities of the curve, representing $y$ and $z$ as functions of $x$, are to lie upon a given curve or curves, or upon a given surface or surfaces—the varied upper limit may be taken with $x_2 + \kappa X_2$ as the varied value of $x_2$, and the varied lower limit may be taken with $x_0 + \kappa X_0$ as the varied value of $x_0$. In this event, $X_2$ and the values of $v$ and of $w$ at $x_2$ must conform to the data at the upper limit, while $X_0$ and the values of $v$ and of $w$ at $x_0$ must conform to the data at the lower limit. Further, the function $f(x, y + \kappa v, y' + \kappa v', z + \kappa w, z' + \kappa w')$ is assumed to be expansible in a series proceeding in powers of $\kappa$. Then, if $J$ denote the integral arising from the complete variation of $I$, we have

$$J - I = \int_{x_0 + \kappa X_0}^{x_2 + \kappa X_2} f(x, y + \kappa v, y' + \kappa v', z + \kappa w, z' + \kappa w') \, dx$$

$$- \int_{x_0}^{x_2} f(x, y, y', z, z') \, dx$$

$$= \int_{x_0}^{x_2} \{ f(x, y + \kappa v, y' + \kappa v', z + \kappa w, z' + \kappa w') - f(x, y, y', z, z') \} \, dx$$

$$+ \int_{x_2}^{x_2 + \kappa X_2} f(x, y + \kappa v, y' + \kappa v', z + \kappa w, z' + \kappa w') \, dx$$

$$- \int_{x_0}^{x_0 + \kappa X_0} f(x, y + \kappa v, y' + \kappa v', z + \kappa w, z' + \kappa w') \, dx.$$

The integral in the first line of the right-hand side is

$$= \kappa \int_{x_0}^{x_2} \left( v \frac{\partial f}{\partial y} + v' \frac{\partial f}{\partial y'} + w \frac{\partial f}{\partial z} + w' \frac{\partial f}{\partial z'} \right) dx + K_2 = \kappa I_1 + K_2,$$

where $K_2$ is the aggregate of terms involving second and higher powers of $\kappa$. Now

$$\int_{x_0}^{x_2} v' \frac{\partial f}{\partial y'} \, dx = \left[ v \frac{\partial f}{\partial y'} \right]_{x_0}^{x_2} - \int_{x_0}^{x_2} v \frac{d}{dx} \left( \frac{\partial f}{\partial y'} \right) dx,$$

$$\int_{x_0}^{x_2} w' \frac{\partial f}{\partial z'} \, dx = \left[ w \frac{\partial f}{\partial z'} \right]_{x_0}^{x_2} - \int_{x_0}^{x_2} w \frac{d}{dx} \left( \frac{\partial f}{\partial z'} \right) dx;$$

and therefore

$$I_1 = \int_{x_0}^{x_2} \left[ v \left\{ \frac{\partial f}{\partial y} - \frac{d}{dx} \left( \frac{\partial f}{\partial y'} \right) \right\} + w \left\{ \frac{\partial f}{\partial z} - \frac{d}{dx} \left( \frac{\partial f}{\partial z'} \right) \right\} \right] dx + \left[ v \frac{\partial f}{\partial y'} + w \frac{\partial f}{\partial z'} \right]_{x_0}^{x_2}.$$

Let $f_0$ and $f_2$ denote the respective values of $f(x, y, y', z, z')$ at $x_0$ and $x_2$; then (as in § 32)

$$\int_{x_2}^{x_2+\kappa X_2} f(x, y+\kappa v, y'+\kappa v', z+\kappa w, z'+\kappa w') \, dx = \kappa X_2 f_2 + K_2',$$

$$\int_{x_0}^{x_0+\kappa X_0} f(x, y+\kappa v, y'+\kappa v', z+\kappa w, z'+\kappa w') \, dx = \kappa X_0 f_0 + K_2'',$$

where $K_2'$ and $K_2''$ denote aggregates of quantities involving the second and higher powers of $\kappa$.

**132.** We have to bring $\kappa X$, $\kappa v$, $\kappa w$ into relations with one another in connection with the curve or the surface upon which a displaced limit stands: and the necessary relations must be estab-lished for each limit. In the figure, $P$ is the point of a curve at the limit which is subjected to a displacement taking it to $Q$. Through $P$, we draw $PA$, $PB$, $PC$, parallel to the axes of $x$, $y$, $z$. Through $Q$ we draw planes $QDA$ and $QDB$, parallel to the planes $Oyz$ and $Ozx$. The curve which is being varied is $...PR...$ cutting the plane $QDA$ in $R$; so that, if the arc $PR = \sigma$ and if $\Delta$ denote $(1 + y'^2 + z'^2)^{\frac{1}{2}}$, we have

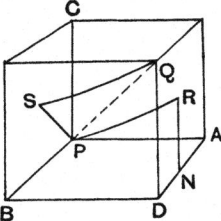

$$PA = \frac{1}{\Delta}\sigma, \quad AN = \frac{y'}{\Delta}\sigma, \quad NR = \frac{z'}{\Delta}\sigma.$$

The curve along which the limiting point $P$ is compelled to move is $PQ$, the dotted curve: or, if the limiting point $P$ is compelled to move on a given surface, then $...PQ...$ is some curve (and, indeed, may be any curve) on that surface. As $Q$ is the displaced position of $P$ on this hypothesis, let the coordinates of $Q$ relative to $P$ be $kX$, $kY$, $kZ$, so that relations (or a relation) will exist between $kX$, $kY$, $kZ$, determined by the equations of the curve $...PQ...$ (or the equation of the surface on which $...PQ...$ lies). The small variation $\kappa v$ and $\kappa w$, to which $P$ is subjected by variation of the curve that is to provide the maximum or minimum, takes $P$ to $S$, where $S$ is in the plane $BPC$. Up to small quantities of the first order, $QR$ is parallel and equal to $PS$; and therefore the coordinates of $Q$ relative to $R$ are equal to the coordinates of $S$ relative to $P$. Thus

$$\kappa v = AD - AN, \quad \kappa w = QD - RN.$$

But $PA = kX$, $PB = kY = AD$, $PC = kZ = QD$; also

$$AN = \frac{y'}{\Delta}\sigma = Ky'X, \quad NR = \frac{z'}{\Delta}\sigma = Kz'X;$$

and therefore

$$\kappa v = \kappa Y - \kappa y'X, \quad \kappa w = \kappa Z - \kappa z'X,$$

that is, at an extremity of the integral,

$$v = Y - y'X, \quad w = Z - z'X.$$

*Note.* These results can also be established as follows. When $P$ is taken temporarily as origin, the coordinates of $Q$ are $\kappa X, \kappa Y, \kappa Z$; and the equations of the tangent at $Q$ are

$$\xi - \kappa X = \frac{\eta - \kappa Y}{y'} = \frac{\zeta - \kappa Z}{z'},$$

where $\xi, \eta, \zeta$ are current coordinates along the tangent, the equations being accurate up to the first power of $\kappa$ inclusive. Now $S$ is a point on this tangent, to the same order of accuracy; and its coordinates are $0, \kappa v, \kappa w$. Hence

$$v = Y - y'X, \quad w = Z - z'X.$$

**133.** Gathering together the results, we now have

$$J - I = \kappa \int_{x_0}^{x_2} \left[ v \left\{ \frac{\partial f}{\partial y} - \frac{d}{dx} \left( \frac{\partial f}{\partial y'} \right) \right\} + w \left\{ \frac{\partial f}{\partial z} - \frac{d}{dx} \left( \frac{\partial f}{\partial z'} \right) \right\} \right] dx$$

$$+ \kappa \left[ X \left( f - y' \frac{\partial f}{\partial y'} - z' \frac{\partial f}{\partial z'} \right) + Y \frac{\partial f}{\partial y'} + Z \frac{\partial f}{\partial z'} \right]_{x_0}^{x_2} + R_2,$$

where $R_2 = K_2 + K_2' + K_2''$, being the aggregate of all the parts of $J - I$ which involve terms of the second and higher powers of $\kappa$ arising from the expansion. In this expression, $\kappa v$ and $\kappa w$ in the integral are the arbitrary small variations of the points on the curve, while $\kappa X, \kappa Y, \kappa Z$ represent the displacement of a limit upon the terminal curve or the terminal surface: and variations of every type are admissible and must be taken into account, arbitrary, independent of one another, subject to the conditions (i), that $v$ and $w$ are regular functions of $x$; (ii), that $X, Y, Z$ are consistent with the maintenance of a terminal position; and (iii), continuity between $v$ and $w$, and $X, Y, Z$ at the terminal positions.

*The two characteristic equations, and the terminal conditions.*

**134.** By the usual argument relating to the possession of a maximum or minimum, the 'first' variation in $J - I$ must vanish; and it must vanish for all small variations of the curve.

Consider, first, a small variation $\kappa v$ and $\kappa w$ for $y$ and $z$, with real variations at the extremities. The terms included in the expression $\left[ \quad \right]_{x_0}^{x_2}$ vanish for such variations, because $X, Y, Z$ are zero at each limit. Hence

$$\int_{x_0}^{x_2} \left[ v \left\{ \frac{\partial f}{\partial y} - \frac{d}{dx} \left( \frac{\partial f}{\partial y'} \right) \right\} + w \left\{ \frac{\partial f}{\partial z} - \frac{d}{dx} \left( \frac{\partial f}{\partial z'} \right) \right\} \right] dx = 0,$$

for all arbitrary regular functions $v$ and $w$ independent of one another. The conditions, necessary and sufficient to meet this requirement, are that the equations

$$\mathfrak{Y} = \frac{\partial f}{\partial y} - \frac{d}{dx}\left(\frac{\partial f}{\partial y'}\right) = 0 \Bigg\rangle$$
$$\mathfrak{Z} = \frac{\partial f}{\partial z} - \frac{d}{dx}\left(\frac{\partial f}{\partial z'}\right) = 0 \Bigg/$$

shall be satisfied everywhere along the curve. Were it not so, by choice of $v$ where $\mathfrak{Y}$ is not zero, and of $w$ where $\mathfrak{Z}$ is not zero, the integral could be made positive or negative at will, contrary to the necessity that it shall vanish. The two equations are the *characteristic equations*. Their primitive, being the complete simultaneous integral of the two equations, determines $y$ and $z$ as functions of $x$; it thus determines a curve in space, which is called the *characteristic curve*.

We shall now suppose that we are dealing with such a curve, so that the characteristic equations are satisfied. Consider, next, a small variation, which is zero at the lower limit of the integral and not zero at the upper limit. Because the curve is now restricted to be a characteristic curve, the portion of $J - I$ depending upon the integral vanishes. The requirement, that the first variation shall completely vanish, can now be met only if the relation

$$X\left(f - y'\frac{\partial f}{\partial y'} - z'\frac{\partial f}{\partial z'}\right) + Y\frac{\partial f}{\partial y'} + Z\frac{\partial f}{\partial z'} = 0$$

is satisfied at the upper limit.

Similarly, by considering a small variation not zero at the lower limit, we infer that the same relation must be satisfied at that limit.

Consequently, in order that the first variation in $J - I$ shall vanish, it is necessary and sufficient (i), that the characteristic equations shall be satisfied, thus determining the characteristic curve : and (ii), that the condition

$$X\left(f - y'\frac{\partial f}{\partial y'} - z'\frac{\partial f}{\partial z'}\right) + Y\frac{\partial f}{\partial y'} + Z\frac{\partial f}{\partial z'} = 0$$

shall be satisfied at each limit separately, in accordance with some terminal property or properties for the characteristic curve.

### Hamilton form of the characteristic equations.

**135.** The characteristic equations for the present problem, when $f$ does not explicitly involve the independent variable, as also for many such problems involving more than two dependent variables, can be expressed in the Hamilton form which is used in theoretical dynamics. We write

$$\frac{\partial f}{\partial y'} = \eta, \quad \frac{\partial f}{\partial z'} = \zeta,$$

taking $\eta$ and $\zeta$ as new variables, being of course dependent variables; then the characteristic equations are

$$\frac{\partial f}{\partial y} = \frac{d}{dx}\left(\frac{\partial f}{\partial y'}\right) = \eta', \quad \frac{\partial f}{\partial z} = \frac{d}{dx}\left(\frac{\partial f}{\partial z'}\right) = \zeta'.$$

Thus $y$, $y'$, $z$, $z'$ are expressible in terms of $\eta$, $\eta'$, $\zeta$, $\zeta'$. We now introduce a new function $H$, by the definition

$$H = y'\frac{\partial f}{\partial y'} + z'\frac{\partial f}{\partial z'} - f$$

$$= y'\eta + z'\zeta - f.$$

As $f$ does not now explicitly involve $x$, we have

$$dH = (y'\,d\eta + \eta\,dy') + (z'\,d\zeta + \zeta\,dz')$$

$$- \left(\frac{\partial f}{\partial y'}\,dy' + \frac{\partial f}{\partial y}\,dy + \frac{\partial f}{\partial z'}\,dz' + \frac{\partial f}{\partial z}\,dz\right)$$

$$= y'\,d\eta + z'\,d\zeta - \eta'\,dy - \zeta'\,dz.$$

Let the function $H$, initially defined with reference to $f$ and the arguments in $f$, be transformed by the equations

$$\frac{\partial f}{\partial y'} = \eta, \quad \frac{\partial f}{\partial z'} = \zeta,$$

so as to become a function of $y$, $z$, $\eta$, $\zeta$. In this form,

$$dH = \frac{\partial H}{\partial \eta}\,d\eta + \frac{\partial H}{\partial \zeta}\,d\zeta + \frac{\partial H}{\partial y}\,dy + \frac{\partial H}{\partial z}\,dz.$$

The two forms, obtained for $dH$, are equivalent for all variations. Hence

$$\frac{\partial H}{\partial \eta} = y', \quad \frac{\partial H}{\partial \zeta} = z', \quad \frac{\partial H}{\partial y} = -\eta', \quad \frac{\partial H}{\partial z} = -\zeta';$$

and therefore

$$\frac{dy}{\dfrac{\partial H}{\partial \eta}} = \frac{dz}{\dfrac{\partial H}{\partial \zeta}} = \frac{d\eta}{-\dfrac{\partial H}{\partial y}} = \frac{d\zeta}{-\dfrac{\partial H}{\partial z}} = dx,$$

which is the Hamilton form.

### *Nature of the primitive of the two characteristic equations.*

**136.** If we are to obtain the most general solution of the problem, we must obtain the primitive of the characteristic equations. (It might happen that a particular solution of the problem could be provided by a special integral of those equations.) Now the equations are

$$\frac{\partial^2 f}{\partial y'^2}y'' + \frac{\partial^2 f}{\partial y'\partial z'}z'' + g\,(x, y, y', z, z') = 0,$$

$$\frac{\partial^2 f}{\partial y'\partial z'}y'' + \frac{\partial^2 f}{\partial z'^2}z'' + h\,(x, y, y', z, z') = 0,$$

leading to equations

$$\theta y'' = G\,(x,\,y,\,z,\,y',\,z'), \quad \theta z'' = H\,(x,\,y,\,z,\,y',\,z'),$$

provided the quantity $\theta$, where

$$\theta = \frac{\partial^2 f}{\partial y'^2}\frac{\partial^2 f}{\partial z'^2} - \left(\frac{\partial^2 f}{\partial y' \partial z'}\right)^2,$$

does not vanish. Then the elimination of $z$ and its derivatives leads to an ordinary differential equation of the fourth order in $y$; and the new equation has a primitive of the form

$$y = \phi\,(x,\,A_1,\,A_2,\,A_3,\,A_4),$$

where $A_1$, $A_2$, $A_3$, $A_4$ are arbitrary constants. When this value of $y$, once known, is substituted in the two equations

$$y'' = G, \quad y''' = \frac{\partial G}{\partial x} + \frac{\partial G}{\partial y}y' + \frac{\partial G}{\partial z}z' + \frac{\partial G}{\partial y'}\frac{G}{\theta} + \frac{\partial G}{\partial z'}\frac{H}{\theta},$$

and $z'$ is then eliminated between these two equations, we have

$$z = \psi\,(x,\,A_1,\,A_2,\,A_3,\,A_4).$$

The primitive of the characteristic equations manifestly is

$$\left. \begin{aligned} y &= \phi\,(x,\,A_1,\,A_2,\,A_3,\,A_4) \\ z &= \psi\,(x,\,A_1,\,A_2,\,A_3,\,A_4) \end{aligned} \right\},$$

which, accordingly, are the equations of the characteristic curve.

*Note.* Later, it will be proved that, for the possession of a maximum or a minimum, the quantity

$$\frac{\partial^2 f}{\partial y'^2}\frac{\partial^2 f}{\partial z'^2} - \left(\frac{\partial^2 f}{\partial y' \partial z'}\right)^2$$

must usually be positive and must never be negative; but, in exceptional circumstances and subject to further conditions attaching to the function $f$ it may vanish. If it should happen that $\theta = 0$, the two initial characteristic equations are equivalent to

$$\frac{\partial^2 f}{\partial y'^2}y'' + \frac{\partial^2 f}{\partial y' \partial z'}z'' + g\,(x,\,y,\,y',\,z,\,z') = 0,$$

$$k\,(x,\,y,\,y',\,z,\,z') = 0.$$

Differentiating the second equation and combining the result with the first, we obtain equations

$$y'' + P\,(x,\,y,\,y',\,z,\,z') = 0, \quad z'' + Q\,(x,\,y,\,y',\,z,\,z') = 0,$$

simultaneously with $k\,(x,\,y,\,y',\,z,\,z') = 0$. A differentiation of $y'' + P = 0$, leads to a relation

$$y''' + R\,(x,\,y,\,y',\,z,\,z') = 0;$$

and we then have, on the elimination of $z$ and $z'$, an ordinary differential equation of the third order in $y$. The primitive of the characteristic equations becomes

$$y = \chi\,(x,\,C_1,\,C_2,\,C_3), \quad z = \omega\,(x,\,C_1,\,C_2,\,C_3),$$

which then constitute the equations of the characteristic curve.

As however the occurrence is exceptional when a maximum or a minimum could be possessed, the detailed examination may be left for any problem when it actually arises.

We shall therefore assume that $\theta$ is always positive.

*Ex.* Prove that each of the characteristic equations is only of the first order when $f(x, y, y', z, z')$ is of the form
$$y'F(x, y, z) + z'G(x, y, z) + H(x, y, z);$$
and conversely.

Find the form of $f(x, y, y', z, z')$, when the characteristic equations are satisfied identically.

**137.** As regards the determination of the four arbitrary constants that occur in the primitive, various cases may arise.

(i) Two fixed extremities may be given: that is, we have $y = y_0$ and $z = z_0$ when $x = x_0$, and $y = y_2$ and $z = z_2$ when $x = x_2$. Thus
$$\left.\begin{array}{l} y_0 = \phi(x_0, A_1, A_2, A_3, A_4) \\ z_0 = \psi(x_0, A_1, A_2, A_3, A_4) \\ y_2 = \phi(x_2, A_1, A_2, A_3, A_4) \\ z_2 = \psi(x_2, A_1, A_2, A_3, A_4) \end{array}\right\},$$
equations potentially sufficient to determine $A_1, A_2, A_3, A_4$.

(ii) An extremity may be fixed in position and direction: that is, the values of $y, z, y', z'$ are given when $x = x_0$. Again, there are four equations to determine the constants.

(iii) One extremity may be fixed in position, and the other be required to lie on a given curve
$$y = F(x), \quad z = G(x).$$
At that extremity, say $x = x_2$ where $x_2$ is not known beforehand, we have
$$Y = F'(x_2)X, \quad Z = G'(x_2)X,$$
so that, from the terminal condition at a mobile extremity, we have
$$f - y'\frac{\partial f}{\partial y'} - z'\frac{\partial f}{\partial z'} + F'(x)\frac{\partial f}{\partial y'} + G'(x)\frac{\partial f}{\partial z'} = 0$$
when $x = x_2$. With
$$F(x_2) = \phi(x_2, A_1, A_2, A_3, A_4),$$
$$G(x_2) = \psi(x_2, A_1, A_2, A_3, A_4),$$
we have three equations. Also
$$y_0 = \phi(x_0, A_1, A_2, A_3, A_4),$$
$$z_0 = \psi(x_0, A_1, A_2, A_3, A_4),$$
making five equations in all, potentially sufficient to determine
$$x_2, A_1, A_2, A_3, A_4.$$

(iv) One extremity may be fixed in position, and the other be required to lie on a given surface

$$S\,(x,\,y,\,z) = 0.$$

Let the extremity be $x_2,\, y_2,\, z_2$, so that

$$S\,\{x_2,\, \phi\,(x_2),\, \psi\,(x_2)\} = 0.$$

Also, at $x_2$,

$$\frac{\partial S}{\partial x}\,X + \frac{\partial S}{\partial y}\,Y + \frac{\partial S}{\partial z}\,Z = 0,$$

while the terminal condition at a mobile extremity is

$$X\left(f - y'\frac{\partial f}{\partial y'} - z'\frac{\partial f}{\partial z'}\right) + Y\frac{\partial f}{\partial y'} + Z\frac{\partial f}{\partial z'} = 0,$$

for all variations $X,\, Y,\, Z$ along the surface $S = 0$. Hence, at $x_2$,

$$\frac{f - y'\dfrac{\partial f}{\partial y'} - z'\dfrac{\partial f}{\partial z'}}{\dfrac{\partial S}{\partial x}} = \frac{\dfrac{\partial f}{\partial y'}}{\dfrac{\partial S}{\partial y}} = \frac{\dfrac{\partial f}{\partial z'}}{\dfrac{\partial S}{\partial z}},$$

which, with $S = 0$, are three equations. At the fixed extremity,

$$y_0 = \phi\,(x_0,\, A_1,\, A_2,\, A_3,\, A_4),$$
$$z_0 = \psi\,(x_0,\, A_1,\, A_2,\, A_3,\, A_4).$$

Thus there are five equations in all, potentially sufficient to determine

$$x_2,\, A_1,\, A_2,\, A_3,\, A_4.$$

(v) Similarly for other cases, when both extremities are mobile separately, upon a curve or upon a surface.

### Forms when a general first integral can be obtained.

**138.** There are some forms, not unimportant, for which a first stage toward the integration of the characteristic equations can be attained.

(a) Let $f$ not explicitly involve $y$, or not explicitly involve $z$, or explicitly involve neither of them.

For the first of these, we have

$$\frac{\partial f}{\partial y'} = B,$$

as a first integral; for the second, we have

$$\frac{\partial f}{\partial z'} = C,$$

as a first integral; for the third, we have

$$\frac{\partial f}{\partial y'} = B, \quad \frac{\partial f}{\partial z'} = C,$$

as two first integrals.

(*b*) Let $f$ not explicitly involve $x$. Then

$$\frac{df}{dx} = \frac{\partial f}{\partial y} y' + \frac{\partial f}{\partial y'} y'' + \frac{\partial f}{\partial z} z' + \frac{\partial f}{\partial z'} z''$$

$$= \left\{ y' \frac{d}{dx}\left(\frac{\partial f}{\partial y'}\right) + y'' \frac{\partial f}{\partial y'} \right\} + \left\{ z' \frac{d}{dx}\left(\frac{\partial f}{\partial z'}\right) + z'' \frac{\partial f}{\partial z'} \right\},$$

from the characteristic equations; and therefore we have

$$f - y' \frac{\partial f}{\partial y'} - z' \frac{\partial f}{\partial z'} = A,$$

as a first integral.

(*c*) But if $f$ does not explicitly involve $x$ or $y$ or $z$, we do not acquire three first integrals by combining the two cases of (*a*) with the case (*b*); the three constants $A$, $B$, $C$, each arbitrary, are not independent, the relation between them being obtained by the elimination of $y'$ and $z'$, the only quantities which now occur, between the three equations. Thus, if

$$f = (1 + y'^2 + z'^2)^{\frac{1}{2}},$$

we have $A^2 + B^2 + C^2 = 1$.

(*d*) The simplest (and perhaps the most important) case of all occurs, when $f$ is of the form

$$f = \mu(x, y, z)(1 + y'^2 + z'^2)^{\frac{1}{2}}$$

$$= \mu(1 + y'^2 + z'^2)^{\frac{1}{2}},$$

where $\mu$ is a function of $x$, $y$, $z$. Then the equations are

$$\frac{\partial \mu}{\partial y}(1 + y'^2 + z'^2)^{\frac{1}{2}} - \frac{d}{dx}\left\{\frac{\mu y'}{(1 + y'^2 + z'^2)^{\frac{1}{2}}}\right\} = 0,$$

$$\frac{\partial \mu}{\partial z}(1 + y'^2 + z'^2)^{\frac{1}{2}} - \frac{d}{dx}\left\{\frac{\mu z'}{(1 + y'^2 + z'^2)^{\frac{1}{2}}}\right\} = 0.$$

When the length of the arc is made the variable, so that

$$\frac{ds}{dx} = (1 + y'^2 + z'^2)^{\frac{1}{2}},$$

these equations are

$$\frac{\partial \mu}{\partial y} - \frac{d}{ds}\left(\mu \frac{dy}{ds}\right) = 0,$$

$$\frac{\partial \mu}{\partial z} - \frac{d}{ds}\left(\mu \frac{dz}{ds}\right) = 0.$$

Let

$$U = \frac{\partial \mu}{\partial x} - \frac{d}{ds}\left(\mu \frac{dx}{ds}\right).$$

Multiplying this equation by $\dfrac{dx}{ds}$, the two characteristic equations by $\dfrac{dy}{ds}$ and $\dfrac{dz}{ds}$ respectively, and adding, we have

$$U\frac{dx}{ds} = \frac{dx}{ds}\left\{\frac{\partial\mu}{\partial x} - \frac{d}{ds}\left(\mu\frac{dx}{ds}\right)\right\} + \frac{dy}{ds}\left\{\frac{\partial\mu}{\partial y} - \frac{d}{ds}\left(\mu\frac{dy}{ds}\right)\right\} + \frac{dz}{ds}\left\{\frac{\partial\mu}{\partial z} - \frac{d}{ds}\left(\mu\frac{dz}{ds}\right)\right\}$$

$$= \frac{\partial\mu}{\partial x}\frac{dx}{ds} + \frac{\partial\mu}{\partial y}\frac{dy}{ds} + \frac{\partial\mu}{\partial z}\frac{dz}{ds} - \frac{d\mu}{ds},$$

because

$$\left(\frac{dx}{ds}\right)^2 + \left(\frac{dy}{ds}\right)^2 + \left(\frac{dz}{ds}\right)^2 = 1, \quad \frac{dx}{ds}\frac{d^2x}{ds^2} + \frac{dy}{ds}\frac{d^2y}{ds^2} + \frac{dz}{ds}\frac{d^2z}{ds^2} = 0.$$

Thus

$$U\frac{dx}{ds} = 0,$$

and therefore

$$U = 0.$$

We thus have three equations, symmetric in $x$, $y$, $z$: they will be discussed later*.

*Ex.* 1. Shew that the shortest curve from one fixed curve (or surface) to another fixed curve (or surface) in space is a straight line which, at each extremity, is normal to the curve (or surface).

Prove that the shortest distance between the circle $x^2+y^2=a^2$, $z=0$, and the hyperbola $z^2-y^2=c^2$, $x=0$, is $(c^2+\frac{1}{2}a^2)^{\frac{1}{2}}$.

*Ex.* 2. Find the integral equations of the curve, joining two points not lying in a plane with the axis of $x$, such that its moment about the axis of $x$ is a minimum.

(I) We are to minimise the integral

$$\int (y^2+z^2)^{\frac{1}{2}} (1+y'^2+z'^2)^{\frac{1}{2}}\, dx.$$

The characteristic equations are

$$\frac{y(1+y'^2+z'^2)^{\frac{1}{2}}}{(y^2+z^2)^{\frac{1}{2}}} - \frac{d}{dx}\left\{\frac{y'(y^2+z^2)^{\frac{1}{2}}}{(1+y'^2+z'^2)^{\frac{1}{2}}}\right\} = 0,$$

$$\frac{z(1+y'^2+z'^2)^{\frac{1}{2}}}{(y^2+z^2)^{\frac{1}{2}}} - \frac{d}{dx}\left\{\frac{z'(y^2+z^2)^{\frac{1}{2}}}{(1+y'^2+z'^2)^{\frac{1}{2}}}\right\} = 0.$$

Because $f$, the subject of integration, does not involve $x$ explicitly, there is a first integral

$$f - y'\frac{\partial f}{\partial y'} - z'\frac{\partial f}{\partial z'} = c,$$

where $c$ is an arbitrary constant: that is,

$$(y^2+z^2)^{\frac{1}{2}} = c\,(1+y'^2+z'^2)^{\frac{1}{2}}.$$

When this integral is used, the characteristic equations become

$$\frac{y}{c} - c\frac{d^2y}{dx^2} = 0, \quad \frac{z}{c} - c\frac{d^2z}{dx^2} = 0.$$

---

* They will appear (§ **175**) as the characteristic equations when $x$, $y$, $z$ are simultaneously affected by small variations.

Hence

$$y = a \cosh \frac{x}{c} + b \sinh \frac{x}{c} \Bigg\}\ ,$$
$$z = a' \cosh \frac{x}{c} + b' \sinh \frac{x}{c} \Bigg\}$$

where $a$, $b$, $a'$, $b'$ are arbitrary constants. Substitution in the integral

$$(y^2 + z^2)^{\frac{1}{2}} = c\,(1 + y'^2 + z'^2)^{\frac{1}{2}}$$

leads to the relation

$$a^2 - b^2 + a'^2 - b'^2 = c^2,$$

so that there are four independent arbitrary constants, the five arbitrary constants being connected by this single relation.

When new constants $\alpha$, $\beta$, $\gamma$ are introduced*, such that

$$a = c \cos \alpha \cosh \frac{\beta}{c}, \qquad a' = c \sin \alpha \cosh \frac{\gamma}{c},$$

$$b = c \cos \alpha \sinh \frac{\beta}{c}, \qquad b' = c \sin \alpha \sinh \frac{\gamma}{c},$$

which satisfy the relation and now leave $c$, $\alpha$, $\beta$, $\gamma$ as the four independent arbitrary constants, the equations of the curve are

$$y = c \cos \alpha \cosh \frac{x+\beta}{c}, \qquad z = c \sin \alpha \cosh \frac{x+\gamma}{c}.$$

Manifestly the curve is not a plane curve unless $\beta = \gamma$: when $\beta = \gamma$, the curve lies in the particular plane $y \sin \alpha - z \cos \alpha = 0$ through the axis of $x$.

The projection on any plane through the axis of $x$ is, in that plane, an orthogonal projection of a catenary. In the foregoing case $\beta = \gamma$, the curve, in the plane $y \sin \alpha - z \cos \alpha = 0$, is the catenary $y \cos \alpha + z \sin \alpha = c \cosh \frac{x+\beta}{c}$. The projection on a plane, perpendicular to the axis of $x$, is the hyperbola

$$2yz \sin \alpha \cos \alpha \cosh \frac{\beta - \gamma}{c} - y^2 \sin^2 \alpha - z^2 \cos^2 \alpha = c^2 \sin^2 \alpha \cos^2 \alpha \sinh^2 \frac{\beta - \gamma}{c},$$

which, in the foregoing particular case, becomes the repeated line $y \sin \alpha - z \cos \alpha = 0$.

(II) We thus have the characteristic curve. To complete the discussion, so far as weak variations are concerned, we anticipate the establishment of the further tests.

For the Legendre test (§ 148), we have

$$\frac{\partial^2 g}{\partial y'^2} = (y^2 + z^2)^{\frac{1}{2}} \frac{1 + z'^2}{(1 + y'^2 + z'^2)^{\frac{3}{2}}},$$

$$\frac{\partial^2 g}{\partial z'^2} = (y^2 + z^2)^{\frac{1}{2}} \frac{1 + y'^2}{(1 + y'^2 + z'^2)^{\frac{3}{2}}},$$

$$\frac{\partial^2 g}{\partial y' \partial z'} = (y^2 + z^2)^{\frac{1}{2}} \frac{-y'z'}{(1 + y'^2 + z'^2)^{\frac{3}{2}}},$$

hence $\dfrac{\partial^2 g}{\partial y'^2} > 0$, $\dfrac{\partial^2 g}{\partial z'^2} > 0$, $\dfrac{\partial^2 g}{\partial y'^2} \dfrac{\partial^2 g}{\partial z'^2} > \left( \dfrac{\partial^2 g}{\partial y' \partial z'} \right)^2$, and therefore (so far as the Legendre test is concerned) the solution admits a minimum.

* On the assumption that no one of the constants $a$, $b$, $a'$, $b'$ is zero, such as may happen in a particular case.

To apply the Jacobi test (§ 149), the conjugate (say $x_1$) of the initial point (say $x_0$) is required. It can be obtained as the limiting position of the intersection of the characteristic curve with a consecutive characteristic. Using, first, the form

$$y = a \cosh \frac{x}{c} + b \sinh \frac{x}{c}, \qquad z = a' \cosh \frac{x}{c} + b' \sinh \frac{x}{c},$$

with the relation

$$a^2 - b^2 + a'^2 - b'^2 = c^2,$$

we apply a small variation to the constants by taking $a + \kappa A$, $b + \kappa B$, $a' + \kappa A'$, $b' + \kappa B'$, $c + \kappa C$; and we denote the consequently varied values of $y$ and $z$ by $y + \kappa Y$ and $z + \kappa Z$. Then

$$Y = A \cosh \frac{x}{c} + B \sinh \frac{x}{c} - \frac{x}{c^2} C \left( a \sinh \frac{x}{c} + b \cosh \frac{x}{c} \right)$$

$$= A \cosh \frac{x}{c} + B \sinh \frac{x}{c} - \frac{x}{c} C y',$$

$$Z = A' \cosh \frac{x}{c} + B' \sinh \frac{x}{c} - \frac{x}{c} C z'.$$

But for the initial value $x_0$ of $x$, we have $Y = 0$ and $Z = 0$; and for the conjugate value $x_1$, the intersection of the consecutive characteristic with the original curve, we again have $Y = 0$ and $Z = 0$. Thus, because $Y = 0$ at the two places,

$$A \cosh \frac{x_1}{c} + B \sinh \frac{x_1}{c} - \frac{x_1}{c} C y_1' = 0,$$

$$A \cosh \frac{x_0}{c} + B \sinh \frac{x_0}{c} - \frac{x_0}{c} C y_0' = 0,$$

and therefore

$$\frac{A}{x_0 y_0' \sinh \frac{x_1}{c} - x_1 y_1' \sinh \frac{x_0}{c}} = \frac{B}{x_1 y_1' \cosh \frac{x_0}{c} - x_0 y_0' \cosh \frac{x_1}{c}} = \frac{-\frac{1}{c} C}{\sinh \frac{x_0 - x_1}{c}}.$$

Similarly, because $Z = 0$ at the two places,

$$\frac{A'}{x_0 z_0' \sinh \frac{x_1}{c} - x_1 z_1' \sinh \frac{x_0}{c}} = \frac{B'}{x_1 z_1' \cosh \frac{x_0}{c} - x_0 z_0' \cosh \frac{x_1}{c}} = \frac{-\frac{1}{c} C}{\sinh \frac{x_0 - x_1}{c}}.$$

When these are substituted in the relation

$$aA - bB + a'A' - b'B' = cC$$

arising out of the relation among the five arbitrary constants by varying these constants, we have

$$x_0 y_0' c y_1' - x_1 y_1' c y_0' + x_0 z_0' c z_1' - x_1 z_1' c z_0' + c^2 \sinh \frac{x_0 - x_1}{c} = 0,$$

that is,

$$(x_0 - x_1)(y_0' y_1' + z_0' z_1') + c \sinh \frac{x_0 - x_1}{c} = 0,$$

an equation for determining $x_1$ in terms of $x_0$.

When the later expressions for the constants are used, we have

$$y' = \cos \alpha \sinh \frac{x + \beta}{c}, \qquad z' = \sin \alpha \sinh \frac{x + \gamma}{c};$$

and so the equation determining the conjugate is

$$\frac{x_0 - x_1}{c} \left\{ \cos^2 \alpha \sinh \frac{x_0 + \beta}{c} \sinh \frac{x_1 + \beta}{c} + \sin^2 \alpha \sinh \frac{x_0 + \gamma}{c} \sinh \frac{x_1 + \gamma}{c} \right\} + \sinh \frac{x_0 - x_1}{c} = 0.$$

This can also be taken in the form

$$\frac{x_0 - x_1}{c} \left\{ \frac{\cos^2 \alpha}{\coth \frac{x_1+\beta}{c} - \coth \frac{x_0+\beta}{c}} + \frac{\sin^2 \alpha}{\coth \frac{x_1+\gamma}{c} - \coth \frac{x_0+\gamma}{c}} \right\} + 1 = 0,$$

or

$$\cos^2 \alpha \left\{ 1 - \frac{\frac{x_1-x_0}{c}}{\coth \frac{x_1+\beta}{c} - \coth \frac{x_0+\beta}{c}} \right\} + \sin^2 \alpha \left\{ 1 - \frac{\frac{x_1-x_0}{c}}{\coth \frac{x_1+\gamma}{c} - \coth \frac{x_0+\gamma}{c}} \right\} = 0.$$

In the particular case when $\beta = \gamma$, the curve becomes the plane curve

$$\eta = y \cos \alpha + z \sin \alpha = c \cosh \frac{x+\beta}{c},$$

the plane of the curve passing through the axis of $x$; the necessary boundary condition is

$$\frac{y_0}{z_0} = \frac{y_1}{z_1} = \frac{y}{z};$$

and the relation giving the conjugate is

$$\frac{x_1 - x_0}{c} = \coth \frac{x_1+\beta}{c} - \coth \frac{x_0+\beta}{c},$$

in effect, the old relation (§ 30, Ex. 1, p. 31) which determines the complete range upon the characteristic catenary.

*Ex.* 3. From a given point, in one of two parallel planes, a curve of given length is drawn to some point in the other plane. From every point on the curve a perpendicular is drawn to the first plane, these perpendiculars forming a cylindrical surface. The area of the intercepted portion of this surface is to be a maximum.

Prove that the curve must be a circle; and find the relation between the circular arc and the planes, at each of the extremities.

*Ex.* 4. Find the equations of a curve joining two points, such that the moment of inertia of the curve about the axis of $x$ is a minimum, when the points do not lie in one plane with the axis of $x$.

The integral to be minimised is

$$\int (y^2 + z^2)(1 + y'^2 + z'^2)^{\frac{1}{2}} \, dx \, ;$$

the characteristic equations are

$$2y (1 + y'^2 + z'^2)^{\frac{1}{2}} - \frac{d}{dx} \left\{ \frac{y'(y^2 + z^2)}{(1 + y'^2 + z'^2)^{\frac{1}{2}}} \right\} = 0,$$

$$2z (1 + y'^2 + z'^2)^{\frac{1}{2}} - \frac{d}{dx} \left\{ \frac{z'(y^2 + z^2)}{(1 + y'^2 + z'^2)^{\frac{1}{2}}} \right\} = 0.$$

As $f$, the subject of integration, does not involve $x$ explicitly, there is a first integral

$$f - y' \frac{\partial f}{\partial y'} - z' \frac{\partial f}{\partial z'} = \text{constant} = a^2,$$

that is,

$$\frac{y^2 + z^2}{(1 + y'^2 + z'^2)^{\frac{1}{2}}} = a^2.$$

When this integral is used, the characteristic equations become

$$2y (y^2 + z^2) - a^4 \frac{d^2 y}{dx^2} = 0, \qquad 2z (y^2 + z^2) - a^4 \frac{d^2 z}{dx^2} = 0.$$

Of these equations, there is an integral

$$z \frac{dy}{dx} - y \frac{dz}{dx} = c,$$

where $c$ is an arbitrary constant.

For further integration, take a new variable $u$ given by

$$y^2 + z^2 = a^2 u,$$

so that the actual value of $u$ is $(1 + y'^2 + z'^2)^{\frac{1}{2}}$, that is, $\frac{ds}{dx}$; but, at this stage, there is no advantage in the specific use of the variable $s$. Then

$$yy' + zz' = \tfrac{1}{2} a^2 u'.$$

But

$$(y^2 + z^2)(y'^2 + z'^2) = (yy' + zz')^2 + (zy' - yz')^2 ;$$

and therefore

$$a^2 u (u^2 - 1) = \tfrac{1}{4} a^4 u'^2 + c^2.$$

Write

$$\frac{4c^2}{a^2} = g,$$

so that $g$ may be taken as a new constant instead of $c$: then

$$a^2 \left( \frac{du}{dx} \right)^2 = 4u^3 - 4u - g,$$

the differential equation of Weierstrass elliptic functions with 4 and $g$ for invariants. Thus

$$u = \wp \left( \frac{x - x_0}{a} \right),$$

or, if we write

$$x = x_0 + at,$$

we have

$$u = \wp(t).$$

Again, let

$$y = r \sin \theta, \qquad z = r \cos \theta,$$

where $r^2 = a^2 u$; then from the equation $z \dfrac{dy}{dx} - y \dfrac{dz}{dx} = c$, we have

$$r^2 \frac{d\theta}{dx} = c,$$

that is,

$$d\theta = \frac{c\, dx}{r^2} = \frac{c}{a} \frac{dt}{\wp(t)}.$$

Let $\tau$ be a value of $t$ such that $\wp(t) = 0$, so that

$$\wp'(\tau) = (-g)^{\frac{1}{2}} = 2i \frac{c}{a} ;$$

then

$$2i\, d\theta = \frac{\wp'(\tau)}{\wp(t) - \wp(\tau)} dt ;$$

and therefore

$$2i(\theta - \beta) = 2t\zeta(\tau) - \log \sigma(t + \tau) + \log \sigma(t - \tau),$$

that is,

$$e^{2i(\theta - \beta)} = e^{2t\zeta(\tau)} \frac{\sigma(t - \tau)}{\sigma(t + \tau)}.$$

Thus the primitive of the characteristic equations is

$$\left.\begin{aligned}
x &= x_0 + at \\
y &= r \sin \theta, \quad z = r \cos \theta \\
r^2 &= a^2 \wp (t) \\
&= a^2 \{\wp (t) - \wp (\tau)\} \\
&= a^2 \, \frac{\sigma (\tau - t) \, \sigma (\tau + t)}{\sigma^2 (\tau) \, \sigma^2 (t)} \\
e^{2i(\theta - \beta)} &= e^{2t} \zeta(\tau) \, \frac{\sigma (t - \tau)}{\sigma (t + \tau)}
\end{aligned}\right\}.$$

The four arbitrary constants are $x_0$, $a$, $\beta$, with either $c$ or $\tau$ (which are equivalent to one another).

As in the last example but one, the equations of the characteristic curve have been obtained. The tests, as to whether it provides an actual minimum, must be deferred; their formation constitutes an exercise in elliptic functions.

### Subsidiary characteristic equations, and their primitive.

**139.** The characteristic equations are

$$\frac{\partial f}{\partial y} - \frac{d}{dx}\left(\frac{\partial f}{\partial y'}\right) = 0, \quad \frac{\partial f}{\partial z} - \frac{d}{dx}\left(\frac{\partial f}{\partial z'}\right) = 0;$$

and their primitive is of the form

$$y = \phi (x, A_1, A_2, A_3, A_4), \quad z = \psi (x, A_1, A_2, A_3, A_4),$$

where $A_1$, $A_2$, $A_3$, $A_4$ are arbitrary constants.

When variations of integrals involving only one dependent variable with one independent variable were discussed, it was found convenient to consider a 'consecutive' characteristic curve: and advantage arose for two issues, one determining the limit (if any) to the range of the integral for the conservation of a maximum or minimum property, the other leading to the construction of a critical normal form for the second variation of the integral. The same advantages can accrue in this investigation.

Accordingly, we consider a small variation of the foregoing primitive, so that it shall still satisfy the characteristic equations. We submit the arbitrary constants $A_1$, $A_2$, $A_3$, $A_4$ to small arbitrary variations by changing them into $A_1 + \kappa a_1$, $A_2 + \kappa a_2$, $A_3 + \kappa a_3$, $A_4 + \kappa a_4$, where $a_1$, $a_2$, $a_3$, $a_4$ are themselves independent arbitrary constants, $\kappa$ being the usual small arbitrary constant. When the changed values of $y$ and $z$ are denoted by $y + \kappa \eta$ and $z + \kappa \zeta$, we have

$$\left.\begin{aligned}
\eta &= a_1 \frac{\partial \phi}{\partial A_1} + a_2 \frac{\partial \phi}{\partial A_2} + a_3 \frac{\partial \phi}{\partial A_3} + a_4 \frac{\partial \phi}{\partial A_4} \\
\zeta &= a_1 \frac{\partial \psi}{\partial A_1} + a_2 \frac{\partial \psi}{\partial A_2} + a_3 \frac{\partial \psi}{\partial A_3} + a_4 \frac{\partial \psi}{\partial A_4}
\end{aligned}\right\} :$$

while the new values $y + \kappa \eta$ and $z + \kappa \zeta$ still satisfy the original characteristic equations.

The function $f$ involves five arguments $x$, $y$, $y'$, $z$, $z'$. Derivatives with regard to these will be denoted by subscripts $0, 2, 3, 4, 5$ respectively, attached to $f$, the notation being used here only for second derivatives as given by the definitions

$$\frac{\partial^2 f}{\partial y'^2} = f_{33}, \quad \frac{\partial^2 f}{\partial y' \partial z'} = f_{35}, \quad \frac{\partial^2 f}{\partial z'^2} = f_{55}, \quad \frac{\partial^2 f}{\partial y^2} = f_{22},$$

and so on. We need the modified form of the equation, under the change of $y$ and $z$ into $y + \kappa \eta$ and $z + \kappa \zeta$, subject to the restriction that the equation remains the same. The modified forms of $\dfrac{\partial f}{\partial y}$, $\dfrac{\partial f}{\partial y'}$, $\dfrac{\partial f}{\partial z}$, $\dfrac{\partial f}{\partial z'}$, are

$$\frac{\partial f}{\partial y} + \kappa \left( f_{22} \eta + f_{23} \eta' + f_{24} \zeta + f_{25} \zeta' \right) + \dots,$$

$$\frac{\partial f}{\partial y'} + \kappa \left( f_{23} \eta + f_{33} \eta' + f_{34} \zeta + f_{35} \zeta' \right) + \dots,$$

$$\frac{\partial f}{\partial z} + \kappa \left( f_{24} \eta + f_{34} \eta' + f_{44} \zeta + f_{45} \zeta' \right) + \dots,$$

$$\frac{\partial f}{\partial z'} + \kappa \left( f_{25} \eta + f_{35} \eta' + f_{45} \zeta + f_{55} \zeta' \right) + \dots,$$

respectively: or, if

$$2\Omega = f_{22} \eta^2 + 2f_{24} \eta \zeta + f_{44} \zeta^2$$
$$+ 2 \left( f_{23} \eta \eta' + f_{25} \eta \zeta' + f_{34} \eta' \zeta + f_{45} \zeta \zeta' \right)$$
$$+ f_{33} \eta'^2 + 2f_{35} \eta' \zeta' + f_{55} \zeta'^2,$$

the four modified forms are

$$\frac{\partial f}{\partial y} + \kappa \frac{\partial \Omega}{\partial \eta} + \dots, \quad \frac{\partial f}{\partial y'} + \kappa \frac{\partial \Omega}{\partial \eta'} + \dots, \quad \frac{\partial f}{\partial z} + \kappa \frac{\partial \Omega}{\partial \zeta} + \dots, \quad \frac{\partial f}{\partial z'} + \kappa \frac{\partial \Omega}{\partial \zeta'} + \dots,$$

where, in every instance, the unexpressed terms constitute aggregates of terms in the second and higher powers of $\kappa$. But the characteristic equations are to remain unaltered. Hence

$$\frac{\partial f}{\partial y} + \kappa \frac{\partial \Omega}{\partial \eta} + \dots - \frac{d}{dx} \left( \frac{\partial f}{\partial y'} + \kappa \frac{\partial \Omega}{\partial \eta'} + \dots \right) = 0,$$

that is, in the limit when $\kappa$ tends to zero,

$$\frac{\partial \Omega}{\partial \eta} - \frac{d}{dx} \left( \frac{\partial \Omega}{\partial \eta'} \right) = 0 \,;$$

and similarly

$$\frac{\partial \Omega}{\partial \zeta} - \frac{d}{dx} \left( \frac{\partial \Omega}{\partial \zeta'} \right) = 0.$$

In attaining this result, we have supposed that the original characteristic equations are satisfied, and that therefore the quantities $y$ and $z$ (with their derivatives) are given by the primitive which expresses the characteristic curve; hence all the coefficients $f_{mn}$ are to be regarded as functions of $x$. Thus the two equations just obtained, viz.,

$$\frac{\partial \Omega}{\partial \eta} - \frac{d}{dx}\left(\frac{\partial \Omega}{\partial \eta'}\right) = 0, \quad \frac{\partial \Omega}{\partial \zeta} - \frac{d}{dx}\left(\frac{\partial \Omega}{\partial \zeta'}\right) = 0,$$

or, in full,

$$\left. \begin{aligned} f_{33}\eta'' + f_{35}\zeta'' &= (f_{22} - f_{23}')\,\eta & -f_{33}'\eta' + (f_{24} - f_{34}')\,\zeta + (f_{25} - f_{34} - f_{35}')\,\zeta' \\ f_{35}\eta'' + f_{55}\zeta'' &= (f_{24} - f_{25}')\,\eta + (f_{34} - f_{25} - f_{35}')\,\eta' + (f_{44} - f_{45}')\,\zeta - f_{55}'\zeta' \end{aligned} \right\},$$

are two linear equations in $\eta$ and $\zeta$ with variable coefficients. On the assumption (§ 136, *Note*) of a non-vanishing quantity $f_{33}f_{55} - f_{35}{}^2$, their form shews that the primitive involves four arbitrary constants. But we have obtained values

$$\left. \begin{aligned} \eta &= a_1 \frac{\partial \phi}{\partial A_1} + a_2 \frac{\partial \phi}{\partial A_2} + a_3 \frac{\partial \phi}{\partial A_3} + a_4 \frac{\partial \phi}{\partial A_4} \\ \zeta &= a_1 \frac{\partial \psi}{\partial A_1} + a_2 \frac{\partial \psi}{\partial A_2} + a_3 \frac{\partial \psi}{\partial A_3} + a_4 \frac{\partial \psi}{\partial A_4} \end{aligned} \right\},$$

where $a_1$, $a_2$, $a_3$, $a_4$ are independent arbitrary constants. Hence these two equations, when combined, can constitute the primitive of

$$\frac{\partial \Omega}{\partial \eta} - \frac{d}{dx}\left(\frac{\partial \Omega}{\partial \eta'}\right) = 0, \quad \frac{\partial \Omega}{\partial \zeta} - \frac{d}{dx}\left(\frac{\partial \Omega}{\partial \zeta'}\right) = 0.$$

The quantities $y + \kappa\eta$ and $z + \kappa\zeta$ determine a consecutive characteristic curve. Accordingly, as before, these new equations, satisfied by $\Omega$, are called the *subsidiary characteristic equations*.

*Note.* The primitive of the subsidiary equations has been derived from the primitive of the original equations. If however that original primitive is not known, but only a special integral is known and has to be tested, the special integral is substituted in the form of the subsidiary equations as they stand; and these equations must then be integrated, without assistance from the special integral except in so far as it may contain some arbitrary elements [*].

*Lemma necessary for the establishment of the primitive of the subsidiary equations.*

**140.** In declaring that the values of $\eta$ and $\zeta$ can constitute the primitive of the subsidiary equations, it is tacitly assumed that the four quantities $\dfrac{\partial \phi}{\partial A_r}$ $(r = 1, 2, 3, 4)$ are linearly independent, and that the four quantities

---

[*] In the most restricted instance, the special integral will contain no arbitrary element; no assistance towards the construction of the primitive of the subsidiary equations can thence be derived. In other instances, a special integral may contain some arbitrary elements, though not enough of them to make it the primitive; in that event, some assistance towards the construction of the primitive could then be derived, of a range limited by the amount of generality of the special integral.

$\frac{\partial \psi}{\partial A_r}$ $(r = 1, 2, 3, 4)$ also are linearly independent. The assumption must be justified.

We have seen that the elimination of $z$ between the two fundamental equations leads to an equation of the fourth order, deduced through

$$\theta y'' = G(x, y, y', z, z'), \quad \theta z'' = H(x, y, y', z, z'),$$

and arising from the elimination of $z$ and $z'$ between

$$\theta y'' = G(x, y, y', z, z'),$$
$$\theta^2 y''' = K(x, y, y', z, z'),$$
$$\theta^3 y'''' = L(x, y, y', z, z'),$$

where $K$ and $L$ are similar in form to $G$ and $H$, and where $\theta$ denotes $f_{33}f_{55} - f_{35}^2$. Thus the equation will be of the form

$$\theta^3 y'''' = M(x, y, y', y'', y'''),$$

where $\theta$ is always positive. Then the customary existence-theorem * provides a unique integral of this equation defined by the initial data that, for an assumed value $x = a$, the quantities $y, y', y'', y'''$ respectively assume arbitrarily assigned values $B_1, B_2, B_3, B_4$, independent of one another, provided that the function $M$ is finite for these values and is regular in their immediate vicinity. This unique integral is consequently of the form

$$y = B_1 + B_2(x-a) + \tfrac{1}{2}B_3(x-a)^2 + \tfrac{1}{6}B_4(x-a)^3 + (x-a)^4 R(x-a, B_1, B_2, B_3, B_4),$$

where $R$ is a regular function of $x - a$. Thus, as

$$y = \phi(x, B_1, B_2, B_3, B_4),$$

we have

$$\frac{\partial \phi}{\partial B_1} = \qquad\quad 1 + (x-a)^4 \frac{\partial R}{\partial B_1},$$

$$\frac{\partial \phi}{\partial B_2} = \qquad x - a + (x-a)^4 \frac{\partial R}{\partial B_2},$$

$$\frac{\partial \phi}{\partial B_3} = \tfrac{1}{2}(x-a)^2 + (x-a)^4 \frac{\partial R}{\partial B_3},$$

$$\frac{\partial \phi}{\partial B_4} = \tfrac{1}{6}(x-a)^3 + (x-a)^4 \frac{\partial R}{\partial B_4}.$$

Clearly no linear and homogeneous relation of the type

$$b_1 \frac{\partial \phi}{\partial B_1} + b_2 \frac{\partial \phi}{\partial B_2} + b_3 \frac{\partial \phi}{\partial B_3} + b_4 \frac{\partial \phi}{\partial B_4} = 0$$

exists among the four quantities, the coefficients $b$ being constants; and consequently the expression for $\eta$ is satisfactory as containing four (and not fewer than four) independent arbitrary constants.

* See my *Theory of Differential Equations*, vol. iii, ch. xi, ch. xiii.

Similarly, when we eliminate $y$ between the two fundamental equations, we obtain an equation of the fourth order in $z$. Of this equation there exists a unique integral, defined in the same manner as the foregoing integral $y$, by means of arbitrarily assigned initial values for $z$, $z'$, $z''$, $z'''$; the integral is regular in the vicinity of $x$; and it has the form

$$z = C_1 + C_2(x-a) + \tfrac{1}{2}C_3(x-a)^2 + \tfrac{1}{6}C_4(x-a)^3 + (x-a)^4 S(x-a, C_1, C_2, C_3, C_4),$$

where $S$ is a regular function of $x - a$. As before, there exists no linear and homogeneous relation of the type

$$c_1 \frac{\partial \psi}{\partial C_1} + c_2 \frac{\partial \psi}{\partial C_2} + c_3 \frac{\partial \psi}{\partial C_3} + c_4 \frac{\partial \psi}{\partial C_4} = 0,$$

where the coefficients $c$ are constants.

In order that the value of $y$ and the value of $z$ thus obtained may be simultaneous integrals of the equations

$$\theta y'' = G(x, y, y', z, z'), \quad \theta z'' = H(x, y, y', z, z'),$$

the four independent constants $B_1$, $B_2$, $B_3$, $B_4$ and the four independent constants $C_1$, $C_2$, $C_3$, $C_4$ must be brought into relations with one another, so as to leave only four independent constants in all. To indicate these relations, we can proceed as follows. The equations can be written in the forms

$$y' = p, \qquad\qquad z' = q,$$
$$\theta p' = G(x, y, p, z, q), \quad \theta q' = H(x, y, p, z, q);$$

a set of four equations of the first order. Here, $\theta$ never vanishes within the range allowed to the variable $x$; and the functions $G$ and $H$ are regular, for values of $x, y, p, z, q$, in the vicinity of $a, B_1, B_2, C_1, C_2$ respectively. By the existence-theorem applying to such equations, there exists a unique set of regular integrals

$$y = B_1 + B_2(x-a) + \tfrac{1}{2}B_3(x-a)^2 + \tfrac{1}{6}B_4(x-a)^3 + (x-a)^4 R(x-a, B_1, B_2, B_3, B_4),$$
$$z = C_1 + C_2(x-a) + \tfrac{1}{2}C_3(x-a)^2 + \tfrac{1}{6}C_4(x-a)^3 + (x-a)^4 S(x-a, C_1, C_2, C_3, C_4),$$
$$p = B_2 + B_3(x-a) + \cdots,$$
$$q = C_2 + C_3(x-a) + \cdots,$$

where the independent constants $B_3$ and $B_4$ are functions of $B_1, B_2, C_1, C_2$, and the independent constants $C_3$ and $C_4$ also are functions of $B_1, B_2, C_1, C_2$. To compare with the general theorem, we take

$$B_1 = A_1, \quad B_2 = A_2, \quad C_1 = A_3, \quad C_2 = A_4,$$

so that the primitive of the two characteristic equations will be regarded as determined by independent arbitrary initially assigned values of $y$, $y'$, $z$, $z'$ for an initial value of $x$. Thus

$$\frac{\partial y}{\partial A_1} = 1 + \tfrac{1}{2}\frac{\partial B_3}{\partial A_1}(x-a)^2 + \tfrac{1}{6}\frac{\partial B_4}{\partial A_1}(x-a)^3 + (x-a)^4\left(\frac{\partial R}{\partial A_1} + \frac{\partial R}{\partial B_3}\frac{\partial B_3}{\partial A_1} + \frac{\partial R}{\partial B_4}\frac{\partial B_4}{\partial A_1}\right),$$

$$\frac{\partial y}{\partial A_2} = x - a + \tfrac{1}{2}\frac{\partial B_3}{\partial A_2}(x-a)^2 + \tfrac{1}{6}\frac{\partial B_4}{\partial A_2}(x-a)^3 + (x-a)^4\left(\frac{\partial R}{\partial A_2} + \frac{\partial R}{\partial B_3}\frac{\partial B_3}{\partial A_2} + \frac{\partial R}{\partial B_4}\frac{\partial B_4}{\partial A_2}\right),$$

$$\frac{\partial y}{\partial C_1} = \tfrac{1}{2}\frac{\partial B_3}{\partial C_1}(x-a)^2 + \tfrac{1}{6}\frac{\partial B_4}{\partial C_1}(x-a)^3 + (x-a)^4\left(\frac{\partial R}{\partial C_1} + \frac{\partial R}{\partial B_3}\frac{\partial B_3}{\partial C_1} + \frac{\partial R}{\partial B_4}\frac{\partial B_4}{\partial C_1}\right),$$

$$\frac{\partial y}{\partial C_2} = \tfrac{1}{2}\frac{\partial B_3}{\partial C_2}(x-a)^2 + \tfrac{1}{6}\frac{\partial B_4}{\partial C_2}(x-a)^3 + (x-a)^4\left(\frac{\partial R}{\partial C_2} + \frac{\partial R}{\partial B_3}\frac{\partial B_3}{\partial C_2} + \frac{\partial R}{\partial B_4}\frac{\partial B_4}{\partial C_2}\right).$$

If a linear relation

$$\lambda_1\frac{\partial y}{\partial A_1} + \lambda_2\frac{\partial y}{\partial A_2} + \lambda_3\frac{\partial y}{\partial C_1} + \lambda_4\frac{\partial y}{\partial C_2} = 0$$

could exist with constant coefficients $\lambda$, we clearly must have

$$\lambda_1 = 0, \quad \lambda_2 = 0, \quad \lambda_3\frac{\partial B_3}{\partial C_1} + \lambda_4\frac{\partial B_3}{\partial C_2} = 0, \quad \lambda_3\frac{\partial B_4}{\partial C_1} + \lambda_4\frac{\partial B_4}{\partial C_2} = 0.$$

The relation

$$\frac{\partial B_3}{\partial C_1}\frac{\partial B_4}{\partial C_2} - \frac{\partial B_4}{\partial C_1}\frac{\partial B_3}{\partial C_2} = 0$$

cannot be satisfied; for it would imply that, between the equations

$$B_3 = \text{function of } A_1, A_2, C_1, C_2, \quad B_4 = \text{function of } A_1, A_2, C_1, C_2,$$

$C_1$ and $C_2$ can be eliminated simultaneously, and so secure a relation

$$g(A_1, A_2, B_3, B_4) = 0:$$

contrary to the hypothesis, that $B_1 (= A_1)$, $B_2 (= A_2)$, $B_3$, $B_4$ are independent of one another. Hence we must have $\lambda_3 = 0$, $\lambda_4 = 0$; and there is no linear relation between $\dfrac{\partial y}{\partial A_1}$, $\dfrac{\partial y}{\partial A_2}$, $\dfrac{\partial y}{\partial C_1}$, $\dfrac{\partial y}{\partial C_2}$.

Similarly, there is no linear relation between $\dfrac{\partial z}{\partial A_1}$, $\dfrac{\partial z}{\partial A_2}$, $\dfrac{\partial z}{\partial C_1}$, $\dfrac{\partial z}{\partial C_2}$; and therefore the expressions, obtained in § 139 for $\eta$ and $\zeta$, stand as the primitive of the subsidiary equation.

### *Property of a pair of integral-sets of the subsidiary equation.*

**141.** Discussion of the geometrical properties and possibilities of the primitive of the original characteristic equations, especially in connection with the primitive of the subsidiary characteristic equations, will be deferred until the general (Weierstrass) weak variations of $x$, $y$, $z$ simultaneously have been considered.

There is, however, one analytical property of the integrals of the subsidiary equations which should be established at once, because it will be useful in the reduction of the second variation of $I$ to a normal form. The property is as follows:

*If $\eta_1$ and $\zeta_1$, $\eta_2$ and $\zeta_2$, be a couple of independent* integral-sets of the subsidiary equations, then the quantity*

$$(\eta_2\eta_1' - \eta_1\eta_2')f_{33} + (\eta_2\zeta_1' - \zeta_1\eta_2' + \zeta_2\eta_1' - \eta_1\zeta_2')f_{35} + (\zeta_2\zeta_1' - \zeta_1\zeta_2')f_{55}$$
$$+ (\eta_1\zeta_2 - \eta_2\zeta_1)(f_{25} - f_{34})$$

*is always constant, whatever independent solutions be chosen; and it is always possible to select† two integral-sets such that the constant is zero.*

We denote the quantity by $Q$: the verification is direct. We form $\dfrac{dQ}{dx}$, remembering that the coefficients $f_{mn}$ are functions of $x$ alone. The terms involving $\eta_1''$, $\zeta_1''$, $\eta_2''$, $\zeta_2''$ are

$$= (\eta_2\eta_1'' - \eta_1\eta_2'')f_{33} + (\eta_2\zeta_1'' - \zeta_1\eta_2'' + \zeta_2\eta_1'' - \eta_1\zeta_2'')f_{35} + (\zeta_2\zeta_1'' - \zeta_1\zeta_2'')f_{55}$$
$$= \eta_2(f_{33}\eta_1'' + f_{35}\zeta_1'') - \eta_1(f_{33}\eta_2'' + f_{35}\zeta_2'') + \zeta_2(f_{35}\eta_1'' + f_{55}\zeta_1'') - \zeta_1(f_{35}\eta_2'' + f_{55}\zeta_2'').$$

We substitute for $f_{33}\eta_1'' + f_{35}\zeta_1''$, $f_{35}\eta_1'' + f_{55}\zeta_1''$, $f_{33}\eta_2'' + f_{35}\zeta_2''$, $f_{35}\eta_2'' + f_{55}\zeta_2''$ from the second (the full) form of the characteristic equations in § 139; we associate, with the result, the remaining terms in $\dfrac{dQ}{dx}$; and we find, after reduction,

$$\frac{dQ}{dx} = 0$$

identically. Hence $Q$ is a constant, which may be zero.

If the constant is not zero but equal to $m$, then $m$ is not a new arbitrary constant; it is specific, depending upon the selection of $\eta_1$ and $\zeta_1$, $\eta_2$ and $\zeta_2$. In this event, we take a new combination $\eta_1$ and $\zeta_1$, $\eta_3$ and $\zeta_3$, where the set $\eta_3$ and $\zeta_3$ is linearly independent both of the set $\eta_1$ and $\zeta_1$, and of the set $\eta_2$ and $\zeta_2$. For this new combination also, $Q$ is a constant which may be zero. If the new constant is zero, the desired combination is to hand. If the new constant is not zero but has a specific value $n$, then we consider a combination $\eta_1$ and $\zeta_1$ with $n\eta_2 - m\eta_3$ and $n\zeta_2 - m\zeta_3$. Since the quantity $Q$ is linear in $\eta_1$ and its derivatives, and also in $\zeta_1$ and its derivatives, as well as in $\eta_2$ and $\zeta_2$, the quantity $Q$ for the new combination is equal to

$$n \cdot m - m \cdot n,$$

---

* The independence means that $\kappa\eta_1$ and $\kappa\zeta_1$ constitute a small variation of $y$ and $z$, distinct from a small variation $\kappa\eta_2$ and $\kappa\zeta_2$, in this sense: the three points $(x, y, z)$, $(x, y + \kappa\eta_1, z + \kappa\zeta_1)$, $(x, y + \kappa\eta_2, z + \kappa\zeta_2)$, do not lie in a straight line. Otherwise $\kappa\eta_2$ and $\kappa\zeta_2$ would give a small variation, which is merely a magnification (and nothing but a magnification, without change of direction) of the variation given by $\kappa\eta_1$ and $\kappa\zeta_1$. Obviously the analytical condition is that $\eta_1\zeta_2 - \eta_2\zeta_1$ does not vanish.

† The selection, moreover, is not unique; but uniqueness is not of importance in the immediate application.

that is, it is zero. Thus the required combination can be obtained, from any two linearly independent combinations having a common integral-set but not satisfying the requirement.

Manifestly other suitable selections can be constructed out of linear combinations of the four sets $\eta_1$ and $\zeta_1$, $\eta_2$ and $\zeta_2$, $\eta_3$ and $\zeta_3$, $\eta_4$ and $\zeta_4$.

**142.** A more symmetrical form can be given to the result. Let $2\Omega_1$ denote the value of $2\Omega$ when $\eta = \eta_1$ and $\zeta = \zeta_1$, and $2\Omega_2$ its value when $\eta = \eta_2$ and $\zeta = \zeta_2$. Then

$$\eta_1\frac{\partial\Omega_2}{\partial\eta_2'} + \zeta_1\frac{\partial\Omega_2}{\partial\zeta_2'} = \eta_1(f_{23}\eta_2 + f_{34}\zeta_2 + f_{33}\eta_2' + f_{35}\zeta_2') + \zeta_1(f_{25}\eta_2 + f_{45}\zeta_2 + f_{35}\eta_2' + f_{55}\zeta_2'),$$

$$\eta_2\frac{\partial\Omega_1}{\partial\eta_1'} + \zeta_2\frac{\partial\Omega_1}{\partial\zeta_1'} = \eta_2(f_{23}\eta_1 + f_{34}\zeta_1 + f_{33}\eta_1' + f_{35}\zeta_1') + \zeta_2(f_{25}\eta_1 + f_{45}\zeta_1 + f_{35}\eta_1' + f_{55}\zeta_1').$$

The two expressions on the right-hand side are equal because $Q = 0$; and therefore the relation can be expressed in the form

$$\eta_1\frac{\partial\Omega_2}{\partial\eta_2'} + \zeta_1\frac{\partial\Omega_2}{\partial\zeta_2'} = \eta_2\frac{\partial\Omega_1}{\partial\eta_1'} + \zeta_2\frac{\partial\Omega_1}{\partial\zeta_1'}.$$

We can enunciate the result thus:

*If $\eta_1$ and $\zeta_1$, $\eta_2$ and $\zeta_2$, be a couple of independent integral-sets of the subsidiary characteristic equations, they can always be selected so as to satisfy the equation*

$$\eta_1\frac{\partial\Omega_2}{\partial\eta_2'} + \zeta_1\frac{\partial\Omega_2}{\partial\zeta_2'} = \eta_2\frac{\partial\Omega_1}{\partial\eta_1'} + \zeta_2\frac{\partial\Omega_1}{\partial\zeta_1'},$$

*where $\Omega_1$ and $\Omega_2$ denote the values of $\Omega$, when $\eta = \eta_1$ and $\zeta = \zeta_1$, and when $\eta = \eta_2$ and $\zeta = \zeta_2$, respectively.*

**143.** Owing to the importance of the result, another proof[*] is given in establishment of the latter form. As $\Omega_1$ is homogeneous of the second order in $\eta_1$, $\eta_1'$, $\zeta_1$, $\zeta_1'$, and $\Omega_2$ is homogeneous of the second order in $\eta_2$, $\eta_2'$, $\zeta_2$, $\zeta_2'$, we have

$$\eta_1\frac{\partial\Omega_2}{\partial\eta_2} + \zeta_1\frac{\partial\Omega_2}{\partial\zeta_2} + \eta_1'\frac{\partial\Omega_2}{\partial\eta_2'} + \zeta_1'\frac{\partial\Omega_2}{\partial\zeta_2'} = \eta_2\frac{\partial\Omega_1}{\partial\eta_1} + \zeta_2\frac{\partial\Omega_1}{\partial\zeta_1} + \eta_2'\frac{\partial\Omega_1}{\partial\eta_1'} + \zeta_2'\frac{\partial\Omega_1}{\partial\zeta_1'}$$

identically. From the characteristic equations in $\Omega_1$ and $\Omega_2$, we have

$$\frac{\partial\Omega_2}{\partial\eta_2} - \frac{d}{dx}\left(\frac{\partial\Omega_2}{\partial\eta_2'}\right) = 0, \qquad \frac{\partial\Omega_2}{\partial\zeta_2} - \frac{d}{dx}\left(\frac{\partial\Omega_2}{\partial\zeta_2'}\right) = 0,$$

$$\frac{\partial\Omega_1}{\partial\eta_1} - \frac{d}{dx}\left(\frac{\partial\Omega_1}{\partial\eta_1'}\right) = 0, \qquad \frac{\partial\Omega_1}{\partial\zeta_1} - \frac{d}{dx}\left(\frac{\partial\Omega_1}{\partial\zeta_1'}\right) = 0.$$

The foregoing identity, by the use of these characteristic equations, becomes the equation

$$\eta_1\frac{d}{dx}\left(\frac{\partial\Omega_2}{\partial\eta_2'}\right) + \eta_1'\frac{\partial\Omega_2}{\partial\eta_2'} + \zeta_1\frac{d}{dx}\left(\frac{\partial\Omega_2}{\partial\zeta_2'}\right) + \zeta_1'\frac{\partial\Omega_2}{\partial\zeta_2'}$$

$$= \eta_2\frac{d}{dx}\left(\frac{\partial\Omega_1}{\partial\eta_1'}\right) + \eta_2'\frac{\partial\Omega_1}{\partial\eta_1'} + \zeta_2\frac{d}{dx}\left(\frac{\partial\Omega_1}{\partial\zeta_1'}\right) + \zeta_2'\frac{\partial\Omega_1}{\partial\zeta_1'},$$

---

[*] It is due to Clebsch, *Crelle*, t. lv (1858), p. 260; he establishes it for $n$ dependent variables.

that is,

$$\frac{d}{dx}\left(\eta_1\frac{\partial\Omega_2}{\partial\eta_2{'}}+\zeta_1\frac{\partial\Omega_2}{\partial\zeta_2{'}}\right)=\frac{d}{dx}\left(\eta_2\frac{\partial\Omega_1}{\partial\eta_1{'}}+\zeta_2\frac{\partial\Omega_1}{\partial\zeta_1{'}}\right).$$

Therefore

$$\eta_1\frac{\partial\Omega_2}{\partial\eta_2{'}}+\zeta_1\frac{\partial\Omega_2}{\partial\zeta_2{'}}-\left(\eta_2\frac{\partial\Omega_1}{\partial\eta_1{'}}+\zeta_2\frac{\partial\Omega_1}{\partial\zeta_1{'}}\right)=\text{constant};$$

and the constant may be zero.

If the constant be not zero, we proceed as before. We choose another integral-set $\eta_3$ and $\zeta_3$, independent of $\eta_2$ and $\zeta_2$. For the new combination of $\eta_1$ and $\zeta_1$ with $\eta_3$ and $\zeta_3$, the constant may be zero; if it is not zero, we take a combination of $\eta_1$ and $\zeta_1$ and $m\eta_2+n\eta_3$ with $m\zeta_2+n\zeta_3$; and we can choose $m:n$ so that the new constant is zero. Thus there are integral-sets of the characteristic equations such that

$$\eta_1\frac{\partial\Omega_2}{\partial\eta_2{'}}+\zeta_1\frac{\partial\Omega_2}{\partial\zeta_2{'}}=\eta_2\frac{\partial\Omega_1}{\partial\eta_1{'}}+\zeta_2\frac{\partial\Omega_1}{\partial\zeta_1{'}}.$$

### *The 'second' variation.*

**144.** As the first variation of the original integral has been made to vanish, partly by means of the characteristic equations which determine the characteristic curve, and partly through the terminal conditions which have fixed the limits of the integral should these not be assigned by initial data, the expression for $J-I$ is governed by the terms in $K_2$ which are of the second degree in $\kappa$. We have

$$K_2=\tfrac{1}{2}\kappa^2\int\Theta\,dx+K_3,$$

where $K_3$ is the aggregate of terms involving the third and higher powers of $\kappa$, and where

$$\begin{aligned}\Theta=&f_{33}v'^2+2f_{35}v'w'+f_{55}w'^2\\&+2\left(f_{23}vv'+f_{34}wv'+f_{25}vw'+f_{45}ww'\right)\\&+f_{22}v^2+2f_{24}vw+f_{44}w^2,\end{aligned}$$

so that $\Theta$ is the same function of $v$, $w$ and their derivatives, as (§ 139) $2\Omega$ is of $\eta$, $\zeta$ and their derivatives.

The expression for $\Theta$, in which the coefficients $f_{mn}$ are known functions of $x$ because variations from the characteristic curve are under consideration, and $y$ and $z$ are known functions of $x$ along that curve, can be modified into a more useful form exhibiting the significance of $\tfrac{1}{2}\kappa^2\int\Theta\,dx$, often called the *second variation*. The character of this modified form is suggested by the character of the corresponding forms in the case of the former integrals, which have involved only one dependent variable. A different analytical method will be adopted, principally to shew that we are not bound or restricted to a single process.

*Equations for transformation of the variation.*

**145.** Accordingly, we take two variable quantities $\eta$ and $\zeta$—they will be identified later with the quantities $\eta$ and $\zeta$ already used, but the identification is not assumed initially. Concurrently, we define four quantities $\alpha$, $\lambda$, $\mu$, $\beta$, subject initially solely to the relations

$$\eta' + \alpha\eta + \lambda\zeta = 0, \quad \zeta' + \mu\eta + \beta\zeta = 0.$$

Then

$$f_{33}\eta' + f_{35}\zeta' + f_{23}\eta + f_{34}\zeta$$
$$= \eta(f_{23} - \alpha f_{33} - \mu f_{35}) + \zeta(f_{34} - \lambda f_{33} - \beta f_{35})$$
$$= -A\eta - B\zeta,$$

if quantities $A$ and $B$ are introduced by the definitions

$$\left. \begin{array}{l} f_{23} + A = \alpha f_{33} + \mu f_{35} \\ f_{34} + B = \lambda f_{33} + \beta f_{35} \end{array} \right\}.$$

Then

$$f_{23}\eta' + f_{25}\zeta' + f_{22}\eta + f_{24}\zeta - \frac{d}{dx}(f_{33}\eta' + f_{35}\zeta' + f_{23}\eta + f_{34}\zeta)$$
$$= \eta(f_{22} - \alpha f_{23} - \mu f_{25} + A' - A\alpha - B\mu)$$
$$\qquad + \zeta(f_{24} - \lambda f_{23} - \beta f_{25} + B' - A\lambda + B\beta)$$
$$= 0,$$

provided we have

$$\left. \begin{array}{l} f_{22} + A' = \alpha^2 f_{33} + 2\alpha\mu f_{35} + \mu^2 f_{55} \\ f_{24} + B' = \alpha\lambda f_{33} + (\alpha\beta + \lambda\mu)f_{35} + \mu\beta f_{55} \end{array} \right\}.$$

Similarly,

$$f_{35}\eta' + f_{55}\zeta' + f_{25}\eta + f_{45}\zeta = -B\eta - C\zeta,$$

if

$$\left. \begin{array}{l} f_{25} + B = \alpha f_{35} + \mu f_{55} \\ f_{45} + C = \lambda f_{35} + \beta f_{55} \end{array} \right\};$$

and then

$$f_{34}\eta' + f_{45}\zeta' + f_{24}\eta + f_{44}\zeta - \frac{d}{dx}(f_{35}\eta' + f_{55}\zeta' + f_{25}\eta + f_{45}\zeta) = 0,$$

provided we have

$$\left. \begin{array}{l} f_{24} + B' = \alpha\lambda f_{33} + (\alpha\beta + \lambda\mu)f_{35} + \mu\beta f_{55} \\ f_{44} + C' = \lambda^2 f_{33} + 2\lambda\beta f_{35} + \beta^2 f_{55} \end{array} \right\}.$$

But the equations which have been obtained are, in fact,

$$\left. \begin{array}{l} \dfrac{\partial\Omega}{\partial\eta} - \dfrac{d}{dx}\left(\dfrac{\partial\Omega}{\partial\eta'}\right) = 0 \\[2mm] \dfrac{\partial\Omega}{\partial\zeta} - \dfrac{d}{dx}\left(\dfrac{\partial\Omega}{\partial\zeta'}\right) = 0 \end{array} \right\}.$$

They hold, provided the equations

$$\left.\begin{aligned}
f_{23} + A &= \alpha f_{33} + \mu f_{35} \\
f_{34} + B &= \lambda f_{33} + \beta f_{35} \\
f_{25} + B &= \alpha f_{35} + \mu f_{55} \\
f_{45} + C &= \lambda f_{35} + \beta f_{55}
\end{aligned}\right\},$$

$$\left.\begin{aligned}
f_{22} + A' &= \alpha^2 f_{33} + 2\alpha\mu f_{35} + \mu^2 f_{55} \\
f_{24} + B' &= \alpha\lambda f_{33} + (\alpha\beta + \lambda\mu) f_{35} + \mu\beta f_{55} \\
f_{44} + C' &= \lambda^2 f_{33} + 2\lambda\beta f_{35} + \beta^2 f_{55}
\end{aligned}\right\},$$

are satisfied. Taking these equations as equations of definition, we at once infer that the quantities $\eta$ and $\zeta$, which have been introduced, are in fact the original quantities $\eta$ and $\zeta$ which arose in connection with the small variation of the characteristic curve. Moreover, we have

$$\frac{\partial\Omega}{\partial\eta'} = -A\eta - B\zeta, \qquad \frac{\partial\Omega}{\partial\zeta'} = -B\eta - C\zeta,$$

which are derived from the foregoing equations.

Thus there are, in all, nine defining equations, viz.

(i) the two initial equations $\eta' + \alpha\eta + \lambda\zeta = 0$, $\zeta' + \mu\eta + \beta\zeta = 0$; and

(ii) the seven later equations, involving $A$, $B$, $C$, $A'$, $B'$, $C'$;

and there are, in all, nine initially unknown quantities, viz. $\eta$, $\zeta$, $\alpha$, $\lambda$, $\mu$, $\beta$, $A$, $B$, $C$. Consequently the equations are potentially sufficient for the determination of the unknown quantities.

**146.** It has been proved that the quantities $\eta$ and $\zeta$ satisfy the subsidiary characteristic equations; and, so far as the analysis has been concerned, they can be any integral-set of those equations. But we can now prove that all the quantities $\alpha$, $\lambda$, $\mu$, $\beta$, $A$, $B$, $C$ can be expressed simply in terms of two integral-sets $\eta_1$ and $\zeta_1$, $\eta_2$ and $\zeta_2$, which satisfy the relation (§ 142)

$$\eta_1 \frac{\partial\Omega_2}{\partial\eta_2'} + \zeta_1 \frac{\partial\Omega_2}{\partial\zeta_2'} = \eta_2 \frac{\partial\Omega_1}{\partial\eta_1'} + \zeta_2 \frac{\partial\Omega_1}{\partial\zeta_1'}.$$

With two integral-sets $\eta_1$ and $\zeta_1$, $\eta_2$ and $\zeta_2$, if the equations can subsist for both sets, we have

$$\left.\begin{aligned}
\eta_1' + \alpha\eta_1 + \lambda\zeta_1 &= 0 \\
\eta_2' + \alpha\eta_2 + \lambda\zeta_2 &= 0
\end{aligned}\right\}, \qquad \left.\begin{aligned}
\zeta_1' + \mu\eta_1 + \beta\zeta_1 &= 0 \\
\zeta_2' + \mu\eta_2 + \beta\zeta_2 &= 0
\end{aligned}\right\};$$

and therefore, if

$$\rho = \eta_1\zeta_2 - \zeta_1\eta_2,$$

so that $\rho$ is neither zero nor a non-zero constant, we have

$$\left.\begin{aligned}
\rho\alpha &= \zeta_1\eta_2' - \zeta_2\eta_1' \\
\rho\lambda &= \eta_2\eta_1' - \eta_1\eta_2' \\
\rho\mu &= \zeta_1\zeta_2' - \zeta_2\zeta_1' \\
\rho\beta &= \eta_2\zeta_1' - \eta_1\zeta_2'
\end{aligned}\right\},$$

giving the values of $\alpha$, $\lambda$, $\mu$, $\beta$.

Again,

$$\frac{\partial \Omega_1}{\partial \eta_1'} = - A\eta_1 - B\zeta_1 \Bigg\} , \qquad \frac{\partial \Omega_1}{\partial \zeta_1'} = - B\eta_1 - C\zeta_1 \Bigg\} ;$$
$$\frac{\partial \Omega_2}{\partial \eta_2'} = - A\eta_2 - B\zeta_2 \Bigg\} \qquad \frac{\partial \Omega_2}{\partial \zeta_2'} = - B\eta_2 - C\zeta_2 \Bigg\}$$

and therefore

$$-\rho A = \zeta_2 \frac{\partial \Omega_1}{\partial \eta_1'} - \zeta_1 \frac{\partial \Omega_2}{\partial \eta_2'}\Bigg\}$$
$$-\rho B = \eta_1 \frac{\partial \Omega_2}{\partial \eta_2'} - \eta_2 \frac{\partial \Omega_1}{\partial \eta_1'}\Bigg\}$$
$$-\rho B = \zeta_2 \frac{\partial \Omega_1}{\partial \zeta_1'} - \zeta_1 \frac{\partial \Omega_2}{\partial \zeta_2'}\Bigg\} ,$$
$$-\rho C = \eta_1 \frac{\partial \Omega_2}{\partial \zeta_2'} - \eta_2 \frac{\partial \Omega_1}{\partial \zeta_1'}\Bigg\}$$

the two values of $B$ being equal, because of the relation which connects the two integral-sets $\eta_1$ and $\zeta_1$, $\eta_2$ and $\zeta_2$. These values have to satisfy the seven relations which involve $A$, $B$, $C$, $A'$, $B'$, $C'$.

We have

$$\rho f_{23} - \zeta_2 \frac{\partial \Omega_1}{\partial \eta_1'} + \zeta_1 \frac{\partial \Omega_2}{\partial \eta_2'} = \rho \left( f_{23} + A \right)$$
$$= \rho \alpha f_{33} + \rho \mu f_{35}$$
$$= f_{33} \left( \zeta_1 \eta_2' - \zeta_2 \eta_1' \right) + f_{35} \left( \zeta_1 \zeta_2' - \zeta_2 \zeta_1' \right),$$

which, on substituting the two integral-sets in

$$\frac{\partial \Omega}{\partial \eta'} = f_{33} \eta' + f_{35} \zeta' + f_{23} \eta + f_{34} \zeta,$$

is satisfied identically. Again, the relation

$$f_{25} + B = \alpha f_{35} + \mu f_{55},$$

similarly treated with the use of

$$\rho B = \zeta_1 \frac{\partial \Omega_2}{\partial \zeta_2'} - \zeta_2 \frac{\partial \Omega_1}{\partial \zeta_1'},$$

is satisfied identically; and so for the other two relations, involving $B$ and $C$.

Next, we have

$$\rho A = \zeta_1 \frac{\partial \Omega_2}{\partial \eta_2'} - \zeta_2 \frac{\partial \Omega_1}{\partial \eta_1'};$$

and therefore

$$\rho A' + \rho' A = \zeta_1 \frac{d}{dx}\left(\frac{\partial \Omega_2}{\partial \eta_2'}\right) - \zeta_2 \frac{d}{dx}\left(\frac{\partial \Omega_1}{\partial \eta_1'}\right) + \zeta_1' \frac{\partial \Omega_2}{\partial \eta_2'} - \zeta_2' \frac{\partial \Omega_1}{\partial \eta_1'}$$
$$= \zeta_1 \frac{\partial \Omega_2}{\partial \eta_2} - \zeta_2 \frac{\partial \Omega_1}{\partial \eta_1} + \zeta_1' \frac{\partial \Omega_2}{\partial \eta_2'} - \zeta_2' \frac{\partial \Omega_1}{\partial \eta_1'}.$$

Now

$$\frac{\partial \Omega_1}{\partial \eta_1} = f_{23}\eta_1{}' + f_{25}\zeta_1{}' + f_{22}\eta_1 + f_{24}\zeta_1$$
$$= (f_{22} - \alpha f_{23} - \mu f_{25})\,\eta_1 + (f_{24} - \lambda f_{23} - \beta f_{35})\,\zeta_1$$
$$= (A\alpha + B\mu - A')\,\eta_1 + (A\lambda + B\beta - B')\,\zeta_1 ;$$

and

$$\frac{\partial \Omega_2}{\partial \eta_2} = (A\alpha + B\mu - A')\,\eta_2 + (A\lambda + B\beta - B')\,\zeta_2 .$$

Therefore

$$\zeta_1 \frac{\partial \Omega_2}{\partial \eta_2} - \zeta_2 \frac{\partial \Omega_1}{\partial \eta_1} = - (A\alpha + B\mu - A')\,\rho .$$

Also

$$\zeta_1{}' \frac{\partial \Omega_2}{\partial \eta_2{}'} - \zeta_2{}' \frac{\partial \Omega_1}{\partial \eta_1{}'} = - \zeta_1{}' (A\eta_2 + B\zeta_2) + \zeta_2{}' (A\eta_1 + B\zeta_1)$$
$$= - A\,(\zeta_1{}'\eta_2 - \zeta_2{}'\eta_1) - B\,(\zeta_1{}'\zeta_2 - \zeta_1\zeta_2{}')$$
$$= - \rho\,(A\beta - B\mu) ;$$

and thus the right-hand side is equal to

$$\rho \{A' - A\,(\alpha + \beta)\}.$$

But

$$\rho' = \eta_1 \zeta_2{}' + \zeta_2 \eta_1{}' - \eta_2 \zeta_1{}' - \zeta_1 \eta_2{}'$$
$$= - \eta_1 (\mu\eta_2 + \beta\zeta_2) - \zeta_2 (\alpha\eta_1 + \lambda\zeta_1) + \eta_2 (\mu\eta_1 + \beta\zeta_1) + \zeta_1 (\alpha\eta_2 + \lambda\zeta_2)$$
$$= - (\alpha + \beta)\,\rho ;$$

and so the equation, which involves $A'$, is satisfied identically. Similarly for the other two equations, which involve $B'$ and $C'$ respectively.

Thus all the equations are satisfied by the values of $\alpha$, $\lambda$, $\mu$, $\beta$, $A$, $B$, $C$, when these respectively acquire the values expressed in terms of the two independent integral-sets $\eta_1$ and $\zeta_1$, $\eta_2$ and $\zeta_2$.

### Normal form of the second variation.

**147.** With these values, we have

$$\Theta + \frac{d}{dx}\,(Av^2 + 2Bvw + Cw^2) = \Delta,$$

where

$$\Delta = f_{33}v'^2 + 2f_{35}v'w' + f_{55}w'^2$$
$$\quad + 2\,(f_{23} + A)\,vv' + 2\,(f_{34} + B)\,wv' + 2\,(f_{25} + B)\,vw' + 2\,(f_{45} + C)\,ww'$$
$$\quad + (f_{22} + A')\,v^2 + 2\,(f_{24} + B')\,vw + (f_{44} + C')\,w^2$$
$$= f_{33}\,(v' + \alpha v + \lambda w)^2 + 2f_{35}\,(v' + \alpha v + \lambda w)\,(w' + \mu v + \beta w) + f_{55}\,(w' + \mu v + \beta w)^2 ;$$

and the values of $\alpha$, $\lambda$, $\mu$, $\beta$ are known. The limits of the original integral have been fixed by the terminal conditions connected with the vanishing of the first variation; and therefore $v$ and $w$ must be zero at each limit, that is,

$$Av^2 + 2Bvw + Cw^2$$

vanishes at each limit. Hence

$$\int \Delta \, dx = \int \left\{ \Theta + \frac{d}{dx} (Av^2 + 2Bvw + Cw^2) \right\} dx$$

$$= \int \Theta \, dx + [Av^2 + 2Bvw + Cw^2]$$

$$= \int \Theta \, dx.$$

Consequently, the second variation of the original integral is equal to

$$\tfrac{1}{2} \kappa^2 \int \Delta \, dx,$$

where $\Delta$ has the foregoing value. The coefficients $\alpha$, $\lambda$, $\mu$, $\beta$ in the arguments of $\Delta$ are such that

$$\left. \begin{array}{l} \eta_1' + \alpha \eta_1 + \lambda \zeta_1 = 0 \\ \eta_2' + \alpha \eta_2 + \lambda \zeta_2 = 0 \end{array} \right\}, \qquad \left. \begin{array}{l} \zeta_1' + \mu \eta_1 + \beta \zeta_1 = 0 \\ \zeta_2' + \mu \eta_2 + \beta \zeta_2 = 0 \end{array} \right\},$$

where $\eta_1$ and $\zeta_1$, $\eta_2$ and $\zeta_2$, are two independent integral-sets of the subsidiary characteristic equations such that

$$\left| \begin{array}{cc} \dfrac{\partial \Omega_1}{\partial \eta_1'}, & \dfrac{\partial \Omega_2}{\partial \eta_2'} \\[2mm] \eta_1, & \eta_2 \end{array} \right| + \left| \begin{array}{cc} \dfrac{\partial \Omega_1}{\partial \zeta_1'}, & \dfrac{\partial \Omega_2}{\partial \zeta_2'} \\[2mm] \zeta_1, & \zeta_2 \end{array} \right| = 0.$$

In order to assure a maximum or a minimum for the original integral, $\int \Delta \, dx$ must have a steady sign (always negative or always positive) for all possible variations, and must not vanish.

### Discussion of second variation: Legendre test.

**148.** If $\Delta$ could be sometimes of one sign and sometimes of another in the range for different variations, we could make $\int \Delta \, dx$ acquire the former sign by choosing zero variations for that part of the range where $\Delta$ could have the other sign; and we could make $\int \Delta \, dx$ acquire the other sign by choosing zero variations for that part of the range where $\Delta$ could have the first sign. With those possibilities, the second variation could be made negative or positive by choice of small variations: a result to be excluded. Hence $\Delta$ must always have the same sign through the range.

That the quantity $\Delta$ may be of uniform sign, whatever values the arguments may acquire, the quantities $f_{33}$ and $f_{55}$ must be of the same sign, and $f_{33}f_{55} - f_{35}^2$ (the quantity which has been denoted by $\theta$) must always be positive; for

$$f_{33} \Delta = \{ f_{33} (v' + \alpha v + \lambda w) + f_{35} (w' + \mu v + \beta w) \}^2 + (f_{33}f_{55} - f_{35}^2) (w' + \mu v + \beta w)^2,$$

$$f_{55} \Delta = (f_{33}f_{55} - f_{35}^2) (v' + \alpha v + \lambda w)^2 + \{ f_{35} (v' + \alpha v + \lambda w) + f_{55} (w' + \mu v + \beta w) \}^2.$$

The sign of $\Delta$ is then the same as the required common sign of $f_{33}$ and $f_{35}$ throughout the range. If $f_{33}f_{55} - f_{35}^2$ were negative, a variation, making either of the quantities

$$f_{33}(v' + \alpha v + \lambda w) + f_{35}(w' + \mu v + \beta w), \quad f_{35}(v' + \alpha v + \lambda w) + f_{55}(w' + \mu v + \beta w),$$

equal to zero, would make $f_{33}\Delta$ or $f_{55}\Delta$ negative; while a variation, such as $w' + \mu v + \beta w = 0$ or such as $v' + \alpha v + \lambda w = 0$, would make $f_{33}\Delta$ or $f_{55}\Delta$ positive. These opposing contingencies are to be excluded.

If $f_{33}f_{55} - f_{35}^2 = 0$, a possibility that still requires $f_{33}$ and $f_{55}$ to be of the same sign, then there are the two types of small variation which make $\Delta$ vanish; and for such variations, the second variation of the integral vanishes. It would then become necessary to consider variations of the third order and of the fourth order of the integral for those particular small variations. If the original integral is to have a maximum or a minimum, the 'third' variation for those small variations must vanish and the 'fourth' variation must then be definitely positive or definitely negative. As the possibility belongs to a special condition and the general analysis is elaborate—too elaborate to be set out, for the truncated generality of condition that remains—it will be excluded from current consideration, and can be discussed in any particular problem where it occurs. We shall therefore assume that $f_{33}f_{55} - f_{35}^2$ does not vanish along the range of the integral; and therefore, for the present purpose, $f_{33}f_{55} - f_{35}^2$ must be positive.

Thus the (*Legendre*) test has the form: *the quantities $f_{33}$ and $f_{55}$ must have the same sign throughout the range, and $f_{33}f_{55} - f_{35}^2$ must be positive throughout the range*\*: if the common sign of $f_{33}$ and $f_{55}$ is positive, a minimum for the integral is admissible through the characteristic curve: if that common sign is negative, a maximum is admissible.

### The Jacobi test: conjugates.

**149.** It remains to consider the arguments of $\Delta$, distinct from the limitations upon the coefficients $f_{33}, f_{35}, f_{55}$: because these arguments admit of wide diversity, owing to the arbitrary character of the regular functions $v$ and $w$. We now take the limits of the integral to be fixed, as having been determined in the discussion of the first variation; so that $v$ and $w$ vanish at the lower limit and vanish at the upper limit also. These arguments are

$$v' + \alpha v + \lambda w, \quad w' + \mu v + \beta w,$$

with values of $\alpha$, $\lambda$, $\mu$, $\beta$ that have been obtained: or, when these values are inserted, they are

$$\frac{p}{\eta_1 \zeta_2 - \eta_2 \zeta_1} \text{ and } \frac{q}{\eta_1 \zeta_2 - \eta_2 \zeta_1},$$

---

\* If $f_{33}f_{55} > f_{35}^2$, then $f_{33}$ and $f_{55}$ must have the same sign. The statement is made as in the text, in order to leave the actual sign of $f_{33}$ and $f_{55}$ in more direct evidence.

where

$$p = \begin{vmatrix} v', & v, & w \\ \eta_1', & \eta_1, & \zeta_1 \\ \eta_2', & \eta_2, & \zeta_2 \end{vmatrix}, \quad q = \begin{vmatrix} w', & v, & w \\ \zeta_1', & \eta_1, & \zeta_1 \\ \zeta_2', & \eta_2, & \zeta_2 \end{vmatrix},$$

and where the quantity $\eta_1\zeta_2 - \eta_2\zeta_1$ does not vanish in the range.

So long as $p$ and $q$ do not vanish together—and, of course, simultaneous zero values of $v$ and $w$ through the range are excluded for the present aim—the quantity $\Delta$ remains of uniform sign. But $\Delta$ would vanish if $p$ and $q$ could vanish simultaneously—a contingency to be excluded: and this could happen when

$$v = \rho\eta_1 + \sigma\eta_2, \quad w = \rho\zeta_1 + \sigma\zeta_2,$$

where $\rho$ and $\sigma$ may be functions of $x$, if

$$\eta_1\rho' + \eta_2\sigma' = 0, \quad \zeta_1\rho' + \zeta_2\sigma' = 0,$$

that is, if $\rho'$ and $\sigma'$ both vanish so that $\rho$ and $\sigma$ are constants. Now $v$ and $w$ are to vanish at the lower limit, while $\kappa\eta$ and $\kappa\zeta$ represent the variation to a consecutive characteristic curve. For the present purpose, we take any such consecutive curve to begin at the same point as the original curve now under examination; hence, at that lower limit, we can take $\rho\eta_1 + \sigma\eta_2 = 0$ and $\rho\zeta_1 + \sigma\zeta_2 = 0$.

If, after passing that lower limit, $\rho\eta_1 + \sigma\eta_2$ and $\rho\zeta_1 + \sigma\zeta_2$ can again vanish simultaneously within the range or (what is another mode of stating the same possibility) if the range of integration should extend as far as (or further than) the first succeeding value of $x$ where $\rho\eta_1 + \sigma\eta_2$ and $\rho\zeta_1 + \sigma\zeta_2$ vanish, then we have a possible variation $v = \rho\eta_1 + \sigma\eta_2$ and $w = \rho\zeta_1 + \sigma\zeta_2$, such that (i) $v$ and $w$ vanish at the lower limit and at the upper limit, (ii) $v$ and $w$ are variations that are not zero, and (iii) $p$ and $q$ vanish along the whole range. Then we should have $\Delta = 0$ throughout the range for a certain non-zero variation; and we could not assert the existence of a maximum or a minimum for the integral. Accordingly, the contingency, which admits this result, must be prevented: that is, it must not be possible to have a variation $\kappa(\rho\eta_1 + \sigma\eta_2)$ and $\kappa(\rho\zeta_1 + \sigma\zeta_2)$, vanishing at the beginning of the range and also at the end of the range (or earlier than the end of the range). Such a variation represents change to a consecutive curve: and the vanishing of such a variation can take place only at an intersection of the original curve and the consecutive curve; hence the range, beginning at any point of a characteristic curve, must not extend so far as the first point of intersection with a consecutive characteristic curve drawn so as to meet the original curve. Such a point of intersection (if any such point exists for one consecutive curve among all the consecutive curves that can be drawn) is called the *conjugate* of the original point which marks the lower limit of the integral; and consequently we have the third (*Jacobi*) test for the existence of a maximum or a minimum:—*the range of integration, beginning at any point on a characteristic curve, must not extend as far as the conjugate of that initial point.*

We proceed to construct the analytical expression of this limitation upon the range of integration.

*Property of a fundamental group of integral-sets: covariantive function.*

**150**. We write

$$\frac{\partial \phi}{\partial A_1}, \ \frac{\partial \phi}{\partial A_2}, \ \frac{\partial \phi}{\partial A_3}, \ \frac{\partial \phi}{\partial A_4}, = \phi_1(x), \ \phi_2(x), \ \phi_3(x), \ \phi_4(x) \Bigg\}$$

$$\frac{\partial \psi}{\partial A_1}, \ \frac{\partial \psi}{\partial A_2}, \ \frac{\partial \psi}{\partial A_3}, \ \frac{\partial \psi}{\partial A_4}, = \psi_1(x), \ \psi_2(x), \ \psi_3(x), \ \psi_4(x) \Bigg\}$$

so that $\phi_r(x)$ and $\psi_r(x)$ (for $r = 1, 2, 3, 4$) constitute an integral-set of the two equations

$$f_{33}\eta'' + f_{35}\zeta'' = (f_{22} - f_{23}')\,\eta - \qquad\qquad f_{33}'\,\eta' + (f_{24} - f_{34}')\,\zeta + (f_{25} - f_{34} - f_{35}')\,\zeta' \Bigg\}$$
$$f_{35}\eta'' + f_{55}\zeta'' = (f_{24} - f_{25}')\,\eta + (f_{34} - f_{25} - f_{35}')\,\eta' + (f_{44} - f_{45}')\,\zeta - \qquad\qquad f_{55}'\,\zeta' \Bigg\}.$$

Let

$$W = \begin{vmatrix} \phi_1(x), & \phi_2(x), & \phi_3(x), & \phi_4(x) \\ \psi_1(x), & \psi_2(x), & \psi_3(x), & \psi_4(x) \\ \phi_1'(x), & \phi_2'(x), & \phi_3'(x), & \phi_4'(x) \\ \psi_1'(x), & \psi_2'(x), & \psi_3'(x), & \psi_4'(x) \end{vmatrix} ;$$

then

$$\frac{dW}{dx} = \begin{vmatrix} \phi_1(x), & \phi_2(x), & \phi_3(x), & \phi_4(x) \\ \psi_1(x), & \psi_2(x), & \psi_3(x), & \psi_4(x) \\ \phi_1''(x), & \phi_2''(x), & \phi_3''(x), & \phi_4''(x) \\ \psi_1'(x), & \psi_2'(x), & \psi_3'(x), & \psi_4'(x) \end{vmatrix} + \begin{vmatrix} \phi_1(x), & \phi_2(x), & \phi_3(x), & \phi_4(x) \\ \psi_1(x), & \psi_2(x), & \psi_3(x), & \psi_4(x) \\ \phi_1'(x), & \phi_2'(x), & \phi_3'(x), & \phi_4'(x) \\ \psi_1''(x), & \psi_2''(x), & \psi_3''(x), & \psi_4''(x) \end{vmatrix}.$$

The two characteristic equations, on being resolved for $\phi_r''(x)$ and $\psi_r''(x)$ as values of $\eta''$ and $\zeta''$, give

$$\theta\phi_r''(x) = (-f_{33}'f_{55} + f_{35}f_{35}' - f_{34}f_{35} + f_{25}f_{35})\,\phi_r'(x)$$
$$+ \text{ linear terms in } \phi_r(x), \ \psi_r(x), \ \psi_r'(x),$$

and

$$\theta\psi_r''(x) = (-f_{33}f_{55}' + f_{35}f_{35}' - f_{25}f_{35} + f_{34}f_{35})\,\psi_r'(x)$$
$$+ \text{ linear terms in } \phi_r(x), \ \psi_r(x), \ \phi_r'(x),$$

where $\theta = f_{33}f_{55} - f_{35}^2$. When these expressions for $\phi_r''(x)$ and $\psi_r''(x)$, for $r = 1, 2, 3, 4$, are substituted in the two determinants in the foregoing value for $\frac{dW}{dx}$, we have

$$\theta\frac{dW}{dx} = (-f_{33}'f_{55} + f_{35}f_{35}' - f_{34}f_{35} + f_{25}f_{35})\,W$$
$$+ (-f_{33}f_{55}' + f_{35}f_{35}' - f_{25}f_{35} + f_{34}f_{35})\,W ;$$

and therefore
$$\theta \frac{dW}{dx} = -\theta' W,$$
that is,
$$\theta W = A,$$
where $A$ is a constant. The Legendre test is to be satisfied; thus $\theta$ is positive through the range of the independent variable. Again, the quantities $\phi_r(x)$, for $r = 1, 2, 3, 4$, are linearly independent, so that there is no identical linear relation along them; and likewise there is no identical linear relation among the quantities $\psi_r(x)$, for $r = 1, 2, 3, 4$. Now constants $k$ can always be chosen so that, if
$$\eta = k_1 \phi_1(x) + k_2 \phi_2(x) + k_3 \phi_3(x) + k_4 \phi_4(x),$$
$$\zeta = k_1 \psi_1(x) + k_2 \psi_2(x) + k_3 \psi_3(x) + k_4 \psi_4(x),$$
$$\eta' = k_1 \phi_1'(x) + k_2 \phi_2'(x) + k_3 \phi_3'(x) + k_4 \phi_4'(x),$$
$\eta$, $\zeta$, $\eta'$ vanish for any assigned value of $x$. But
$$k_1 \psi_1'(x) + k_2 \psi_2'(x) + k_3 \psi_3'(x) + k_4 \psi_4'(x),$$
being $\zeta'$, cannot vanish for that assigned value; because then the differential equations would require $\eta'' = 0$ and $\zeta'' = 0$ and, by successive differentiations, would make every derivative of $\eta$ and every derivative of $\zeta$ vanish for the assigned value. The quantities $\eta$ and $\zeta$, given as expansions in the vicinity of the assigned value, would then vanish everywhere in that vicinity— contrary to the property that the set of quantities $\phi_r(x)$ and the set of quantities $\psi_r(x)$ are, each of them, linearly independent among themselves. Hence $W$ cannot vanish within the range; and therefore *the constant $A$ is not zero.*

**151.** Further, this property does not depend upon any particular choice of integrals $\phi_r$ and $\psi_r$ (for $r = 1, 2, 3, 4$) of the subsidiary characteristic equations. If any other complete set $\eta_r$ and $\zeta_r$ (for $r = 1, 2, 3, 4$) be chosen so that
$$\left. \begin{array}{l} \eta_r = c_{r1} \phi_1(x) + c_{r2} \phi_2(x) + c_{r3} \phi_3(x) + c_{r4} \phi_4(x) \\ \zeta_r = c_{r1} \psi_1(x) + c_{r2} \psi_2(x) + c_{r3} \psi_3(x) + c_{r4} \psi_4(x) \end{array} \right\},$$
where the determinant $D(c_{11}, c_{22}, c_{33}, c_{44})$ must be different from zero, we have
$$W(\eta, \zeta) = W(\phi, \psi) D(c_{11}, c_{22}, c_{33}, c_{44}),$$
and therefore
$$\theta W(\eta, \zeta) = A',$$
where the constant $A'$ is not zero.

The determinant $W$ is, in fact, a covariant for a fundamental system of integrals of the subsidiary equations.

*Note.* We have the corollaries:

(i) The functions $\phi_1(x)$, $\phi_2(x)$, $\phi_3(x)$, $\phi_4(x)$ cannot simultaneously vanish, nor can the functions $\psi_1(x)$, $\psi_2(x)$, $\psi_3(x)$, $\psi_4(x)$ simultaneously vanish; and

(ii) The derivatives $\phi_1'(x)$, $\phi_2'(x)$, $\phi_3'(x)$, $\phi_4'(x)$ cannot simultaneously vanish, nor can the derivatives $\psi_1'(x)$, $\psi_2'(x)$, $\psi_3'(x)$, $\psi_4'(x)$ simultaneously vanish.

### Critical function for the Jacobi test.

**152.** We take one set of integrals of the characteristic equations

$$\begin{aligned}
\eta_1 &= b_1\,\phi_1(x) + b_2\,\phi_2(x) + b_3\,\phi_3(x) + b_4\,\phi_4(x) = \Sigma b_r\,\phi_r(x) \\
\zeta_1 &= b_1\,\psi_1(x) + b_2\,\psi_2(x) + b_3\,\psi_3(x) + b_4\,\psi_4(x) = \Sigma b_r\,\psi_r(x)
\end{aligned}$$

and determine the constants $b_1$, $b_2$, $b_3$, $b_4$ by the conditions that

$$\eta_1 = 0, \quad \zeta_1 = 0, \quad \eta_1' = \lambda W_0, \quad \zeta_1' = 0,$$

at $x = x_0$, where $\lambda$ is an arbitrary constant, and $W_0$, $= W(x_0)$, is the value at $x_0$ of the function $W$ (of § 150) which nowhere vanishes. Then

$$\Sigma b_r \phi_r(x_0) = 0, \quad \Sigma b_r \psi_r(x_0) = 0, \quad \Sigma b_r \phi_r'(x_0) = \lambda W_0, \quad \Sigma b_r \psi_r'(x_0) = 0 ;$$

consequently

$$\begin{aligned}
b_1 &= \lambda\,\{ \qquad k_{34}\psi_2'(x_0) + k_{42}\psi_3'(x_0) + k_{23}\psi_4'(x_0)\}, \\
-b_2 &= \lambda\,\{k_{34}\psi_1'(x_0) \qquad + k_{41}\psi_3'(x_0) + k_{13}\psi_4'(x_0)\}, \\
b_3 &= \lambda\,\{k_{24}\psi_1'(x_0) + k_{41}\psi_2'(x_0) \qquad + k_{12}\psi_4'(x_0)\}, \\
-b_4 &= \lambda\,\{k_{23}\psi_1'(x_0) + k_{31}\psi_2'(x_0) + k_{12}\psi_3'(x_0) \qquad \},
\end{aligned}$$

where

$$\left\| \begin{array}{cccc} \phi_1(x_0), & \phi_2(x_0), & \phi_3(x_0), & \phi_4(x_0) \\ \psi_1(x_0), & \psi_2(x_0), & \psi_3(x_0), & \psi_4(x_0) \end{array} \right\| = k_{12}, k_{13}, k_{14}, k_{23}, k_{24}, k_{34}.$$

We take another set of integrals of the characteristic equations

$$\begin{aligned}
\eta_2 &= c_1\phi_1(x) + c_2\phi_2(x) + c_3\phi_3(x) + c_4\phi_4(x) \\
\zeta_2 &= c_1\psi_1(x) + c_2\psi_2(x) + c_3\psi_3(x) + c_4\psi_4(x)
\end{aligned}$$

and determine the constants $c_1$, $c_2$, $c_3$, $c_4$ by the conditions that

$$\eta_1 = 0, \quad \zeta_1 = 0, \quad \eta_1' = 0, \quad \zeta_1' = -\mu W_0,$$

at $x = x_0$, where $\mu$ is an arbitrary constant. The set of integrals $\eta_2$ and $\zeta_2$ is manifestly independent of the set of integrals $\eta_1$ and $\zeta_1$. Then

$$\Sigma c_r \phi_r(x_0) = 0, \quad \Sigma c_r \psi_r(x_0) = 0, \quad \Sigma c_r \phi_r'(x_0) = 0, \quad \Sigma c_r \psi_r'(x_0) = -\mu W_0 ;$$

consequently

$$\begin{aligned}
c_1 &= \mu\,\{ \qquad k_{34}\phi_2'(x_0) + k_{42}\phi_3'(x_0) + k_{23}\phi_4'(x_0)\}, \\
-c_2 &= \mu\,\{k_{34}\phi_1'(x_0) \qquad + k_{41}\phi_3'(x_0) + k_{13}\phi_4'(x_0)\}, \\
c_3 &= \mu\,\{k_{24}\phi_1'(x_0) + k_{41}\phi_2'(x_0) \qquad + k_{12}\phi_4'(x_0)\}, \\
-c_4 &= \mu\,\{k_{23}\phi_1'(x_0) + k_{31}\phi_2'(x_0) + k_{12}\phi_3'(x_0) \qquad \}.
\end{aligned}$$

We write

$$\Delta(x_0, x) = \begin{vmatrix} \phi_1(x_0), & \phi_2(x_0), & \phi_3(x_0), & \phi_4(x_0) \\ \psi_1(x_0), & \psi_2(x_0), & \psi_3(x_0), & \psi_4(x_0) \\ \phi_1(x), & \phi_2(x), & \phi_3(x), & \phi_4(x) \\ \psi_1(x), & \psi_2(x), & \psi_3(x), & \psi_4(x) \end{vmatrix}$$

$$= k_{12}(x_0) k_{34}(x) + k_{23}(x_0) k_{14}(x) + k_{31}(x_0) k_{24}(x)$$
$$+ k_{14}(x_0) k_{23}(x) + k_{24}(x_0) k_{31}(x) + k_{34}(x_0) k_{12}(x);$$

also we write

$$W_0 = W(x_0) = k_{12} l_{34} + k_{23} l_{14} + k_{31} l_{24} + k_{14} l_{23} + k_{24} l_{31} + k_{34} l_{12},$$

where

$$\begin{Vmatrix} \phi_1'(x_0), & \phi_2'(x_0), & \phi_3'(x_0), & \phi_4'(x_0) \\ \psi_1'(x_0), & \psi_2'(x_0), & \psi_3'(x_0), & \psi_4'(x_0) \end{Vmatrix} = l_{12}, l_{13}, l_{14}, l_{23}, l_{24}, l_{34}.$$

Then we have

$$\eta_1 \zeta_2 - \zeta_1 \eta_2 = \{\Sigma b_r \phi_r(x)\} \{\Sigma c_r \psi_r(x)\} - \{\Sigma b_r \psi_r(x)\} \{\Sigma c_r \phi_r(x)\}.$$

On the right-hand side, the terms, involving $b_1$, $c_1$, $b_2$, $c_2$, $\phi_1(x)$, $\psi_1(x)$, $\phi_2(x)$, $\psi_2(x)$ alone, are

$$(b_1 c_2 - b_2 c_1) \{\phi_1(x) \psi_2(x) - \psi_1(x) \phi_2(x)\}.$$

Now

$$\phi_1(x) \psi_2(x) - \psi_1(x) \phi_2(x) = k_{12}(x);$$

and

$$-\frac{1}{\lambda\mu}(b_1 c_2 - b_2 c_1)$$
$$= \{k_{34} \psi_2'(x_0) + k_{42} \psi_3'(x_0) + k_{23} \psi_4'(x_0)\} \{k_{34} \phi_1'(x_0) + k_{41} \phi_3'(x_0) + k_{13} \phi_4'(x_0)\}$$
$$- \{k_{34} \phi_2'(x_0) + k_{42} \phi_3'(x_0) + k_{23} \phi_4'(x_0)\} \{k_{34} \psi_1'(x_0) + k_{41} \psi_3'(x_0) + k_{13} \psi_4'(x_0)\}$$
$$= k_{34}(k_{34} l_{12} + k_{31} l_{24} + k_{14} l_{23} + k_{24} l_{31} + k_{23} l_{14}) + (k_{23} k_{41} - k_{42} k_{13}) l_{34}.$$

But

$$k_{34} k_{12} + k_{31} k_{24} + k_{23} k_{14} = 0 ;$$

and therefore

$$-\frac{1}{\lambda\mu}(b_1 c_2 - b_2 c_1) = k_{34}(x_0) W_0.$$

The other terms can be arranged in similar groups, five in number; and the final form of $\eta_1 \zeta_2 - \eta_2 \zeta_1$ is

$$\eta_1 \zeta_2 - \eta_2 \zeta_1 = -\lambda\mu W_0 \{k_{12}(x) k_{34}(x_0) + \ldots + k_{34}(x) k_{12}(x_0)\}$$
$$= -\lambda\mu W_0 \Delta(x_0, x).$$

The function $\Delta(x_0, x)$ vanishes when $x = x_0$; let $x_1$ be the first value of $x$ greater than $x_0$ for which it vanishes, so that

$$\Delta(x_0, x_1) = 0.$$

Then the quantity $\eta_1 \zeta_2 - \eta_2 \zeta_1$, which vanishes at the initial value $x_0$ of $x$ and is not identically zero, afterwards first vanishes for a value $x_1$ of $x$.

**153.** Now consider a variation $v$ and $w$, such that
$$v = \rho \eta_1 + \sigma \eta_2, \quad w = \rho \zeta_1 + \sigma \zeta_2,$$
where $\rho$ and $\sigma$ are non-zero constants. At $x = x_0$, we have
$$v = 0, \quad w = 0, \quad v' = \rho \lambda, \quad w' = -\sigma \mu,$$
where $\rho \lambda$ and $\sigma \mu$ are non-zero constants; thus $v$ and $w$ constitute a non-zero variation of the characteristic curve. If it is possible to have a consecutive characteristic through the initial point $x_0$, meeting the original again in the subsequent point $x_1$ (or, if there be other points of intersection, in the first subsequent point which will be denoted by $x_1$), then at that point $x_1$ we have
$$v = 0, \quad w = 0.$$
Because $\rho$ and $\sigma$ are not zero, we have, for $x = x_1$,
$$\eta_1 \zeta_2 - \eta_2 \zeta_1 = 0,$$
and therefore
$$\Delta (x_0, x_1) = 0.$$
But the range of the characteristic curve, when it begins at $x_0$, must not extend so far as the conjugate of that initial point, that is, not so far as the limiting position of its intersection with a consecutive characteristic through $x_0$. Hence the *range of integration must not extend so far as the value $x_1$, where $x_1$ is the first root of the equation* $\Delta (x_0, x_1) = 0$ *which is greater than $x_0$.*

We thus have the analytical expression of the geometrical limit, which restricts the range along the characteristic curve to lie within the conjugate of the initial point.

*Note.* One assumption has been made tacitly—that, if $A_1$ be the conjugate of an initial point $A_0$, no point can be chosen in $A_0 A_1$ so that its conjugate shall also lie in $A_0 A_1$. The justification of the assumption will be deferred until the discussion of the variations which allow variation of $x$ as well as of $y$ and $z$.

### Summary of tests.

**154.** The results thus far obtained for special variations may be summarised as follows, when an integral $\int f(x, y, y', z, z') dx$ is to possess a maximum or a minimum:

I. (*a*) The variables $y$ and $z$, as functions of $x$, must satisfy the two characteristic equations
$$\frac{\partial f}{\partial y} - \frac{d}{dx}\left(\frac{\partial f}{\partial y'}\right) = 0, \quad \frac{\partial f}{\partial z} - \frac{d}{dx}\left(\frac{\partial f}{\partial z'}\right) = 0 ;$$
and quantities $\eta$ and $\zeta$ must be determined as the primitive of the subsidiary equations
$$\frac{\partial \Omega}{\partial \eta} - \frac{d}{dx}\left(\frac{\partial \Omega}{\partial \eta'}\right) = 0, \quad \frac{\partial \Omega}{\partial \zeta} - \frac{d}{dx}\left(\frac{\partial \Omega}{\partial \zeta'}\right) = 0,$$
where $\Omega$ is a homogeneous quadratic function of $\eta$, $\eta'$, $\zeta$, $\zeta'$.

(b) The primitive of the characteristic equations is of the form
$$y = \phi\,(x,\, A_1,\, A_2,\, A_3,\, A_4), \quad z = \psi\,(x,\, A_1,\, A_2,\, A_3,\, A_4);$$
and the primitive of the subsidiary equations is
$$\eta = a_1\,\phi_1\,(x) + a_2\,\phi_2\,(x) + a_3\,\phi_3\,(x) + a_4\,\phi_4\,(x),$$
$$\zeta = a_1\,\psi_1\,(x) + a_2\,\psi_2\,(x) + a_3\,\psi_3\,(x) + a_4\,\psi_4\,(x),$$
where $\phi_r\,(x) = \dfrac{\partial\phi}{\partial A_r}$ and $\psi_r\,(x) = \dfrac{\partial\psi}{\partial A_r}$, for $r = 1,\, 2,\, 3,\, 4$.

II. The quantity
$$\frac{\partial^2 f}{\partial y'^2}\frac{\partial^2 f}{\partial z'^2} - \left(\frac{\partial^2 f}{\partial y'\partial z'}\right)^2$$
must be positive through the range of integration, the common sign of $\dfrac{\partial^2 f}{\partial y'^2}$ and $\dfrac{\partial^2 f}{\partial z'^2}$ (persistent through the range) being positive for a minimum and negative for a maximum.

III. The range of integration, beginning at a value $x_0$, must not extend along the characteristic curve as far as the conjugate of $x_0$, this conjugate being determined by the least value of $x_1$, greater than $x_0$, which is a root of the equation
$$\Delta\,(x_0,\, x_1) = \begin{vmatrix} \phi_1\,(x_0), & \phi_2\,(x_0), & \phi_3\,(x_0), & \phi_4\,(x_0) \\ \psi_1\,(x_0), & \psi_2\,(x_0), & \psi_3\,(x_0), & \psi_4\,(x_0) \\ \phi_1\,(x_1), & \phi_2\,(x_1), & \phi_3\,(x_1), & \phi_4\,(x_1) \\ \psi_1\,(x_1), & \psi_2\,(x_1), & \psi_3\,(x_1), & \psi_4\,(x_1) \end{vmatrix} = 0.$$

# CHAPTER V.

INTEGRALS INVOLVING TWO DEPENDENT VARIABLES AND THEIR FIRST
DERIVATIVES: GENERAL WEAK VARIATIONS.

## THE METHOD OF WEIERSTRASS.

**155.** We now proceed to consider the effect of general weak variations, under which $x$, $y$, $z$ are subjected simultaneously to small arbitrary variations, so that they become $x + \kappa u$, $y + \kappa v$, $z + \kappa w$ respectively. The quantities $x$, $y$, $z$ are now made functions of a new independent variable $t$, which is supposed to increase throughout the range of integration; $u$, $v$, $w$ are arbitrary regular functions of $t$, so that they (as well as their derivatives) remain finite and continuous over the range. Denoting derivatives of $x$, $y$, $z$ by $x_1, x_2, \ldots, y_1, y_2, \ldots, z_1, z_2, \ldots$, we have

$$y' = \frac{y_1}{x_1}, \quad z' = \frac{z_1}{x_1}.$$

Let the integral $\int f(x, y, y', z, z')\, dx$, when transformed, become

$$I = \int g(x, y, z, x_1, y_1, z_1)\, dt,$$

where

$$g(x, y, z, x_1, y_1, z_1) = x_1 f(x, y, y', z, z').$$

**156.** Now this function $g$ satisfies a single permanent identity, whatever be the form of the function $f$ (and therefore of $g$), because $g$ involves six arguments, while $f$ involves only five. We have

$$\frac{\partial g}{\partial x_1} = f - \frac{y_1}{x_1} \frac{\partial f}{\partial y'} - \frac{z_1}{x_1} \frac{\partial f}{\partial z'},$$

$$\frac{\partial g}{\partial y_1} = \frac{\partial f}{\partial y'}, \quad \frac{\partial g}{\partial z_1} = \frac{\partial f}{\partial z'},$$

and therefore

$$x_1 \frac{\partial g}{\partial x_1} + y_1 \frac{\partial g}{\partial y_1} + z_1 \frac{\partial g}{\partial z_1} = x_1 f = g,$$

the identity in question.

It can be obtained also as follows, the method leading to other relations. As there is nothing to specify any requirement concerning an independent variable, we may substitute any other variable $T$ for $t$; and then, if

$$dt = \mu\, dT,$$

where $\mu$ may be variable or constant, so that

$$\frac{dx}{dT} = \mu x_1, \quad \frac{dy}{dT} = \mu y_1, \quad \frac{dz}{dT} = \mu z_1,$$

we have

$$g\left(x, y, z, \frac{dx}{dT}, \frac{dy}{dT}, \frac{dz}{dT}\right) dT = g\left(x, y, z, x_1, y_1, z_1\right) dt,$$

that is,

$$g\left(x, y, z, \mu x_1, \mu y_1, \mu z_1\right) = \mu g\left(x, y, z, x_1, y_1, z_1\right).$$

Let

$$\frac{\partial}{\partial x_1} g\left(x, y, z, x_1, y_1, z_1\right) = g_1\left(x, y, z, x_1, y_1, z_1\right),$$

$$\frac{\partial}{\partial y_1} g\left(x, y, z, x_1, y_1, z_1\right) = g_2\left(x, y, z, x_1, y_1, z_1\right),$$

$$\frac{\partial}{\partial z_1} g\left(x, y, z, x_1, y_1, z_1\right) = g_3\left(x, y, z, x_1, y_1, z_1\right);$$

then

$$g_1\left(x, y, z, \mu x_1, \mu y_1, \mu z_1\right) = g_1\left(x, y, z, x_1, y_1, z_1\right),$$

$$g_2\left(x, y, z, \mu x_1, \mu y_1, \mu z_1\right) = g_2\left(x, y, z, x_1, y_1, z_1\right),$$

$$g_3\left(x, y, z, \mu x_1, \mu y_1, \mu z_1\right) = g_3\left(x, y, z, x_1, y_1, z_1\right).$$

Various corollaries may be derived.

(i) The first derivatives of $g$, viz. $g_1$, $g_2$, $g_3$, are unaltered when $x_1$, $y_1$, $z_1$ are multiplied by any quantity $\mu$, variable or constant (but not zero). In particular, let $\mu \frac{ds}{dt} = 1$; then

$$\mu x_1 = \frac{dx}{ds} = x', \quad \mu y_1 = \frac{dy}{ds} = y', \quad \mu z_1 = \frac{dz}{ds} = z'.$$

Hence the first derivatives of $g$ are unaltered when we make the arc-length $s$ to be the independent variable, so that

$$g_1\left(x, y, z, x_1, y_1, z_1\right) = g_1\left(x, y, z, x', y', z'\right);$$

and similarly for $g_2$ and $g_3$.

(ii) In the equations for $g$, $g_1$, $g_2$, $g_3$, let $\mu = -1$; then if the simultaneous change of sign of $x_1, y_1, z_1$ does not affect any radicals, we have

$$g\left(x, y, z, -x_1, -y_1, -z_1\right) = -g\left(x, y, z, x_1, y_1, z_1\right),$$

$$g_r\left(x, y, z, -x_1, -y_1, -z_1\right) = g_r\left(x, y, z, x_1, y_1, z_1\right),$$

for $r = 1, 2, 3$.

(iii) Let $\mu = 1 + \epsilon$, where $\epsilon$ may be constant or variable; and take the expansions, in powers of $\epsilon$, in the relation

$$g\left(x, y, z, \mu x_1, \mu y_1, \mu z_1\right) = \mu g\left(x, y, z, x_1, y_1, z_1\right).$$

When the coefficients of $\epsilon$ are equated, we have

$$x_1 \frac{\partial g}{\partial x_1} + y_1 \frac{\partial g}{\partial y_1} + z_1 \frac{\partial g}{\partial z_1} = g,$$

the identity satisfied by the function $g$.

*Relations among second derivatives of the subject of integration.*

**157.** This identity (as in the earlier instances in § 51 and § 105) gives rise to certain relations among the second derivatives of $g$ with regard to its arguments, one set of them being necessary for use almost immediately. When we differentiate the identity partially with respect to $x_1$, $y_1$, $z_1$ in turn, we have

$$x_1 g_{11} + y_1 g_{13} + z_1 g_{15} = 0,$$
$$x_1 g_{13} + y_1 g_{33} + z_1 g_{35} = 0,$$
$$x_1 g_{15} + y_1 g_{35} + z_1 g_{55} = 0,$$

where differentiations with respect to $x$, $x_1$, $y$, $y_1$, $z$, $z_1$, are denoted by subscripts 0, 1, 2, 3, 4, 5, so that

$$g_{11} = \frac{\partial^2 g}{\partial x_1{}^2}, \quad g_{13} = \frac{\partial^2 g}{\partial x_1 \partial y_1}, \quad g_{15} = \frac{\partial^2 g}{\partial x_1 \partial z_1},$$

and so on.

These three equations shew that all the six derivatives can be expressed in terms of three of their number, or in terms of three independent quantities. Thus, if we write

$$g_{11} = A y_1{}^2 z_1{}^2, \quad g_{33} = B z_1{}^2 x_1{}^2, \quad g_{55} = C x_1{}^2 y_1{}^2 \left.\right\}$$
$$g_{35} = F x_1{}^2 y_1 z_1, \quad g_{51} = G x_1 y_1{}^2 z_1, \quad g_{13} = H x_1 y_1 z_1{}^2 \left.\right\},$$

we have

$$A + H + G = 0,$$
$$H + B + F = 0,$$
$$G + F + C = 0.$$

Selecting appropriate trios, we have

$$
\left.\begin{array}{l} A = B + C + 2F \\ -H = B \quad\;\; + F \\ -G = \quad\;\; C + F \end{array}\right\},\quad
\left.\begin{array}{l} B = C + A + 2G \\ -F = C \quad\;\; + G \\ -H = \quad\;\; A + G \end{array}\right\},\quad
\left.\begin{array}{l} C = A + B + 2H \\ -G = A \quad\;\; + H \\ -F = \quad\;\; B + H \end{array}\right\}.
$$

Also we have

$$
\left.\begin{array}{l} -2F = B + C - A \\ -2G = C + A - B \\ -2H = A + B - C \end{array}\right\},
$$

together with relations

$$BC - F^2 = CA - G^2 = AB - H^2$$
$$= \tfrac{1}{4}(2BC + 2CA + 2AB - A^2 - B^2 - C^2).$$

One case (§ 156) arises by taking

$$x = t, \quad x_1 = 1, \quad y_1 = y', \quad z_1 = z'.$$

The critical conditions in the Legendre test (§ 148) were that $g_{33}$ and $g_{55}$ should have the same sign while $g_{33}g_{55} - g_{35}{}^2$ should be positive: that is, the quantity

$$BC - F^2$$

must be positive. In the circumstances, it is clear that $A$, $B$, $C$ must have the same sign; but this condition is not sufficient to secure all the foregoing tests, and there is the additional condition

$$2BC + 2CA + 2AB > A^2 + B^2 + C^2.$$

We shall return, later (§ 162), to the relations among the second derivatives.

*First variation of the integral: characteristic equations: terminal conditions*

**158.** Let $J$ denote the value of the integral, when a small variation is effected upon $x$, $y$, $z$, so that

$$J = \int g\,(x + \kappa u,\ y + \kappa v,\ z + \kappa w,\ x_1 + \kappa u_1,\ y_1 + \kappa v_1,\ z_1 + \kappa w_1)\,dt$$

$$= I + \kappa I_1 + \tfrac{1}{2}\kappa^2 I_2 + K_3,$$

where $\kappa I_1$ and $\tfrac{1}{2}\kappa^2 I_2$ are the respective aggregates of terms containing the first power alone of $\kappa$ and the second power alone of $\kappa$, and where $K_3$ is the aggregate of all the terms involving the third and higher powers of $\kappa$. The 'first' variation in $J - I$, viz. $\kappa I_1$, governs the value of $J - I$, unless $I_1$ actually vanishes. If $I_1$ does not vanish, $J - I$ can be made to change its sign by changing the sign of $\kappa$; and the integral $I$ could not then possess a maximum or a minimum value. It is therefore necessary that, if $I$ is to have either kind of value, the quantity $I_1$ shall vanish, and that it shall vanish for all arbitrary small variations. We have

$$I_1 = \int \left( u\,\frac{\partial g}{\partial x} + v\,\frac{\partial g}{\partial y} + w\,\frac{\partial g}{\partial z} + u_1\,\frac{\partial g}{\partial x_1} + v_1\,\frac{\partial g}{\partial y_1} + w_1\,\frac{\partial g}{\partial z_1} \right) dt$$

$$= \left[ u\,\frac{\partial g}{\partial x_1} + v\,\frac{\partial g}{\partial y_1} + w\,\frac{\partial g}{\partial z_1} \right]$$

$$+ \int \left[ u\left\{ \frac{\partial g}{\partial x} - \frac{d}{dt}\left(\frac{\partial g}{\partial x_1}\right) \right\} + v\left\{ \frac{\partial g}{\partial y} - \frac{d}{dt}\left(\frac{\partial g}{\partial y_1}\right) \right\} + w\left\{ \frac{\partial g}{\partial z} - \frac{d}{dt}\left(\frac{\partial g}{\partial z_1}\right) \right\} \right] dt,$$

where the terms in the first line are to be taken at the limits.

Consider, first, variations which vanish at the upper limit and the lower limit alike: for them, $I_1$ consists solely of the integral in the second line in the last expression. Then, the functions $u$, $v$, $w$ are arbitrary, and they are independent of one another; and for all such functions, the integral (being the value of $I_1$ for these variations) must vanish. Hence we must have

$$\left. \begin{aligned} E_x &= \frac{\partial g}{\partial x} - \frac{d}{dt}\left(\frac{\partial g}{\partial x_1}\right) = 0 \\[4pt] E_y &= \frac{\partial g}{\partial y} - \frac{d}{dt}\left(\frac{\partial g}{\partial y_1}\right) = 0 \\[4pt] E_z &= \frac{\partial g}{\partial z} - \frac{d}{dt}\left(\frac{\partial g}{\partial z_1}\right) = 0 \end{aligned} \right\}.$$

For, if they did not vanish, we could choose $u$ to have the same sign as $E_x$ or the opposite sign, $v$ to have the same sign as $E_y$ or the opposite sign, $w$ to have the same sign as $E_z$ or the opposite sign; and the freedom of choice in each selection is unaffected by the choice in the other selections. Such choices could make $I_1$ positive or negative at will, a result that is excluded; hence the choices must not be open, though $u$, $v$, $w$ remain arbitrary: that is, $E_x$, $E_y$, $E_z$ must vanish separately.

Next, for $I_1$ consider a small variation, which is zero at the lower limit but not necessarily zero at the upper limit. The integral in $I_1$ now vanishes, owing to the equations $E_x = 0$, $E_y = 0$, $E_z = 0$. Hence, in order that $I_1$ may vanish, we must have

$$u \frac{\partial g}{\partial x_1} + v \frac{\partial g}{\partial y_1} + w \frac{\partial g}{\partial z_1} = 0$$

at the upper limit. If the upper limit be fixed by assigned data, the requirement is automatically satisfied because $u$, $v$, $w$ then vanish. If the upper limit be not thus fixed, then the relation

$$u \frac{\partial g}{\partial x_1} + v \frac{\partial g}{\partial y_1} + w \frac{\partial g}{\partial z_1} = 0$$

provides a condition or conditions at that mobile upper limit.

Similarly, by considering another variation which does not vanish at the lower limit, we must have

$$u \frac{\partial g}{\partial x_1} + v \frac{\partial g}{\partial y_1} + w \frac{\partial g}{\partial z_1} = 0$$

at the lower limit. The requirement is automatically satisfied if the lower limit be fixed by assigned data, because $u$, $v$, $w$ then vanish. If the lower limit be not thus fixed, then the relation

$$u \frac{\partial g}{\partial x_1} + v \frac{\partial g}{\partial y_1} + w \frac{\partial g}{\partial z_1} = 0$$

provides a condition or conditions at that mobile lower limit.

We thus have two kinds of requirements, necessary and sufficient to secure that $I_1$ shall vanish for all weak variations. Firstly, the equations

$$E_x = 0, \quad E_y = 0, \quad E_z = 0$$

must be satisfied everywhere in the range of integration. Secondly, the relation

$$u \frac{\partial g}{\partial x_1} + v \frac{\partial g}{\partial y_1} + w \frac{\partial g}{\partial z_1} = 0$$

must be satisfied at the upper limit and at the lower limit, separately, for all admissible variations. The second set of requirements will, directly or indirectly, determine the limits of the range of integration when they are not actually fixed by the assigned data.

Accordingly, we shall now consider the limits as known; they will no longer be subject to variation, when further investigations (in particular, concerning the transformations of $I_2$) are being pursued.

The three current equations are called the *characteristic equations*. They determine $x$, $y$, $z$ as functions of the independent parametric variable $t$; and we shall assume that these functions are analytic. The elimination of $t$, if desired, gives two relations between $x$, $y$, $z$, free from the unessential variable $t$. By either form of primitive, a curve, usually a skew curve, is determined: it is called the *characteristic curve*.

*Continuity of certain derivatives through free discontinuities on the curve.*

**159.** One property of the function $g$, the extension of the property (§§ 47, 102) possessed in the case when there is only a single dependent variable originally, can be established. It is as follows:

*In passing through a free\* point, where a discontinuity in the direction of the characteristic curve occurs, the quantities*

$$\frac{\partial g}{\partial x_1}, \quad \frac{\partial g}{\partial y_1}, \quad \frac{\partial g}{\partial z_1}$$

*are continuous in value.*

With the same notation and the same figure as before (*l.c.*) to consider the integral near such a discontinuity at a value $T$ in the range between $t_0$ and $t_1$, we take a variation of the characteristic curve, such that $u = 0$, $v = 0$, $w = 0$ along $t_0 t_2$ and also along $t_3 t_1$. The arc of the curve (though not the direction of the arc) is continuous in the range $t_2 T t_3$, so that we may assume

$$u = (t_3 - t)(t - t_2)\, \Phi(t), \quad v = (t_3 - t)(t - t_2)\, \Psi(t), \quad w = (t_3 - t)(t - t_2)\, \mathrm{X}(t),$$

where $\Phi(t)$, $\Psi(t)$, $\mathrm{X}(t)$ are regular functions of $t$, which do not vanish at $T$ and otherwise are arbitrary. In the range $T t_3$ and indefinitely near $T$, we denote the values of the derivatives of $g$ by

$$\left(\frac{\partial g}{\partial x_1}\right)_+, \quad \left(\frac{\partial g}{\partial y_1}\right)_+, \quad \left(\frac{\partial g}{\partial z_1}\right)_+,$$

in the limit as $t_3$ approaches $T$; and, similarly, in the range $t_2 T$ and indefinitely near $T$, we denote their values by

$$\left(\frac{\partial g}{\partial x_1}\right)_-, \quad \left(\frac{\partial g}{\partial y_1}\right)_-, \quad \left(\frac{\partial g}{\partial z_1}\right)_-,$$

in the limit as $t_2$ approaches $T$.

The integral $I_1$ is to vanish for all small variations, and therefore for the foregoing variation. Between $t_0$ and $t_2$, the contribution to $I_1$ is zero, because

---

\* That is, a point which is not fixed by assigned data.

$u, v, w$ are kept zero in that part of the range. Between $t_3$ and $t_1$, the contribution likewise is zero, for the same reason. Thus

$$I_1 = \int_{t_2}^{T} + \int_{T}^{t_3} \left[ u \left\{ \frac{\partial g}{\partial x} - \frac{d}{dt} \left( \frac{\partial g}{\partial x_1} \right) \right\} + v \left\{ \frac{\partial g}{\partial y} - \frac{d}{dt} \left( \frac{\partial g}{\partial y_1} \right) \right\} + w \left\{ \frac{\partial g}{\partial z} - \frac{d}{dt} \left( \frac{\partial g}{\partial z_1} \right) \right\} \right] dt$$

$$+ \left[ u \frac{\partial g}{\partial x_1} + v \frac{\partial g}{\partial y_1} + w \frac{\partial g}{\partial z_1} \right]_{t_2}^{T} + \left[ u \frac{\partial g}{\partial x_1} + v \frac{\partial g}{\partial y_1} + w \frac{\partial g}{\partial z_1} \right]_{T}^{t_3}.$$

Now $I_1$ must vanish. The integrals vanish, because the characteristic equations are satisfied everywhere along the curve. The part, outside the integral and taken at the limits $t_2$ and $T$ of the portion $t_2 T$ of the range, vanishes at $t_2$ but not at $T$; and the part, outside the integral and taken at the limits $T$ and $t_3$ of the portion $T t_3$ of the range, vanishes at $t_3$ but not at $T$. Hence we have

$$u_T \left( \frac{\partial g}{\partial x_1} \right)_{-} + v_T \left( \frac{\partial g}{\partial y_1} \right)_{-} + w_T \left( \frac{\partial g}{\partial z_1} \right)_{-} - \left\{ u_T \left( \frac{\partial g}{\partial x_1} \right)_{+} + v_T \left( \frac{\partial g}{\partial y_1} \right)_{+} + w_T \left( \frac{\partial g}{\partial z_1} \right)_{+} \right\} = 0.$$

Now $u_T, v_T, w_T$ are arbitrary functions; they are independent of one another; and, because the place $T$ is free, $u_T, v_T, w_T$ are not required (by assigned data) to vanish. Hence the equation can only be satisfied for all such arbitrary functions, if

$$\left( \frac{\partial g}{\partial x_1} \right)_{-} = \left( \frac{\partial g}{\partial x_1} \right)_{+}, \quad \left( \frac{\partial g}{\partial y_1} \right)_{-} = \left( \frac{\partial g}{\partial y_1} \right)_{+}, \quad \left( \frac{\partial g}{\partial z_1} \right)_{-} = \left( \frac{\partial g}{\partial z_1} \right)_{+},$$

being properties which constitute the continuity of the three derivatives in value, on the passage through the free point of discontinuity in the direction of the characteristic curve.

### The three characteristic equations are equivalent to two.

**160.** We note at once that, in virtue of the identity satisfied by the function $g$, the three characteristic equations in § 158 are really equivalent to only two independent equations.

When the complete derivative of the identity

$$x_1 \frac{\partial g}{\partial x_1} + y_1 \frac{\partial g}{\partial y_1} + z_1 \frac{\partial g}{\partial z_1} = g$$

with respect to $t$ is formed, we have

$$x_1 \frac{d}{dt} \left( \frac{\partial g}{\partial x_1} \right) + y_1 \frac{d}{dt} \left( \frac{\partial g}{\partial y_1} \right) + z_1 \frac{d}{dt} \left( \frac{\partial g}{\partial z_1} \right) + x_2 \frac{\partial g}{\partial x_1} + y_2 \frac{\partial g}{\partial y_1} + z_2 \frac{\partial g}{\partial z_1}$$

$$= \frac{dg}{dt}$$

$$= x_1 \frac{\partial g}{\partial x} + y_1 \frac{\partial g}{\partial y} + z_1 \frac{\partial g}{\partial z} + x_2 \frac{\partial g}{\partial x_1} + y_2 \frac{\partial g}{\partial y_1} + z_2 \frac{\partial g}{\partial z_1};$$

and therefore
$$x_1 E_x + y_1 E_y + z_1 E_z = 0,$$
a result which establishes the statement.

But, owing to their convenience in the subsequent analysis, the three equations $E_x = 0$, $E_y = 0$, $E_z = 0$ will be retained for use.

Further, when the three characteristic equations are combined with the identity satisfied by $g$, the fact that they are equivalent to two equations can also be indicated as follows. We have

$$E_x = \frac{\partial g}{\partial x} - \frac{d}{dt}\left(\frac{\partial g}{\partial x_1}\right)$$
$$= x_1 g_{01} + y_1 g_{03} + z_1 g_{05} - (x_1 g_{10} + y_1 g_{12} + z_1 g_{14} + x_2 g_{11} + y_2 g_{13} + z_2 g_{15})$$
$$= y_1 (g_{03} - g_{12}) - z_1 (g_{14} - g_{05}) - x_2 g_{11} - y_2 g_{13} - z_2 g_{15};$$

and therefore

$$x_1 E_x = x_1 y_1 (g_{03} - g_{12}) - x_1 z_1 (g_{14} - g_{05}) + x_2 (y_1 g_{13} + z_1 g_{15}) - x_1 y_2 g_{13} - x_1 z_2 g_{15}$$
$$= \{x_1 y_1 (g_{03} - g_{12}) - (x_1 y_2 - x_2 y_1) g_{13}\} - \{x_1 z_1 (g_{14} - g_{05}) - (z_1 x_2 - z_2 x_1) g_{15}\}.$$

Similarly

$$y_1 E_y = \{y_1 z_1 (g_{25} - g_{43}) - (y_1 z_2 - y_2 z_1) g_{35}\} - \{x_1 y_1 (g_{03} - g_{12}) - (x_1 y_2 - x_2 y_1) g_{13}\},$$

and

$$z_1 E_z = \{x_1 z_1 (g_{41} - g_{05}) - (z_1 x_2 - z_2 x_1) g_{15}\} - \{y_1 z_1 (g_{25} - g_{43}) - (y_1 z_2 - y_2 z_1) g_{35}\}.$$

Thus, again, we have, on addition,
$$x_1 E_x + y_1 E_y + z_1 E_z = 0.$$

Also, because $E_x = 0$, $E_y = 0$, $E_z = 0$, there is a quantity $\Gamma$ such that

$$\left.\begin{aligned}
y_1 z_1 (g_{25} - g_{43}) - (y_1 z_2 - y_2 z_1) g_{35} &= \Gamma \\
z_1 x_1 (g_{41} - g_{05}) - (z_1 x_2 - z_2 x_1) g_{51} &= \Gamma \\
x_1 y_1 (g_{03} - g_{21}) - (x_1 y_2 - x_2 y_1) g_{13} &= \Gamma
\end{aligned}\right\} :$$

relations which can be written in the form

$$z_1 x_1 \{g_{41} - g_{05} - R(x_1 y_2 - y_1 x_2) + Q(z_1 x_2 - x_1 z_2)\}$$
$$= x_1 y_1 \{g_{03} - g_{21} + P(x_1 y_2 - y_1 x_2) - R(z_1 x_2 - x_1 z_2)\}$$
$$= y_1 z_1 (g_{25} - g_{43}).$$

*Ex.* Shew that the two characteristic equations of § 134 are

$$\frac{1}{z_1 x_1}\{z_1 x_1 (g_{41} - g_{05}) - y_1 z_1 (g_{25} - g_{43}) - (z_1 x_2 - z_2 x_1) g_{51} + (y_1 z_2 - y_2 z_1) g_{35}\} = 0,$$

$$\frac{1}{x_1 y_1}\{x_1 y_1 (g_{03} - g_{21}) - y_1 z_1 (g_{25} - g_{43}) - (x_1 y_2 - x_2 y_1) g_{13} + (y_1 z_2 - y_2 z_1) g_{35}\} = 0.$$

*Primitive of the three characteristic equations: subsidiary equations.*

**161.** The three equations $E_x = 0$, $E_y = 0$, $E_z = 0$ are equivalent to two independent equations when combined with the identity satisfied by the function $g$. Their primitive is of the form

$$x = \chi\,(t + \alpha,\ \alpha_1,\ \alpha_2,\ \alpha_3,\ \alpha_4)$$
$$y = \phi\,(t + \alpha,\ \alpha_1,\ \alpha_2,\ \alpha_3,\ \alpha_4)\Bigg\},$$
$$z = \psi\,(t + \alpha,\ \alpha_1,\ \alpha_2,\ \alpha_3,\ \alpha_4)$$

where $\alpha$, $\alpha_1$, $\alpha_2$, $\alpha_3$, $\alpha_4$ are arbitrary constants, one of them obviously occurring as additive to $t$ because $t$ does not occur explicitly in the characteristic equations. The constants $\alpha_1$, $\alpha_2$, $\alpha_3$, $\alpha_4$ in the primitive are essential; but $\alpha$ is merely incidental.

These equations are satisfied identically (for all values of the arbitrary constants) when $x$, $y$, $z$ are substituted; and they remain satisfied when the arbitrary constants are changed. Let these be subjected to small variations, so as to become $\alpha + \kappa\alpha'$, $\alpha_r + \kappa a_r$ (for $r = 1, 2, 3, 4$), where $\alpha'$, $a_1$, $a_2$, $a_3$, $a_4$ are themselves arbitrary; and let the consequent small variations in $x$, $y$, $z$ change these variables to $x + \kappa\xi$, $y + \kappa\eta$, $z + \kappa\zeta$ respectively. Writing

$$\chi'(t),\ \phi'(t),\ \psi'(t),\ \text{for}\ \frac{\partial\chi}{\partial t},\ \frac{\partial\phi}{\partial t},\ \frac{\partial\psi}{\partial t};$$

$$\chi_r(t),\ \phi_r(t),\ \psi_r(t),\ \text{for}\ \frac{\partial\chi}{\partial\alpha_r},\ \frac{\partial\phi}{\partial\alpha_r},\ \frac{\partial\psi}{\partial\alpha_r}\ (r = 1, 2, 3, 4);$$

we have

$$\xi = \alpha\chi'(t) + \sum_{r=1}^{4} a_r\chi_r(t)$$
$$\eta = \alpha\phi'(t) + \sum_{r=1}^{4} a_r\phi_r(t)\Bigg\}.$$
$$\zeta = \alpha\psi'(t) + \sum_{r=1}^{4} a_r\psi_r(t)$$

Further, we take

$$\theta_r(t) = \chi'(t)\,\phi_r(t) - \phi'(t)\,\chi_r(t)$$
$$\vartheta_r(t) = \chi'(t)\,\psi_r(t) - \psi'(t)\,\chi_r(t)\Bigg\},$$

for $r = 1, 2, 3, 4$. We now form combinations of $\eta$ and $\xi$, $\zeta$ and $\xi$, such that

$$Y = x_1\eta - y_1\xi = a_1\theta_1(t) + a_2\theta_2(t) + a_3\theta_3(t) + a_4\theta_4(t)$$
$$Z = x_1\zeta - z_1\xi = a_1\vartheta_1(t) + a_2\vartheta_2(t) + a_3\vartheta_3(t) + a_4\vartheta_4(t)\Bigg\},$$

the arbitrary constant $\alpha'$ no longer occurring in the magnitudes thus selected.

With these changes effected upon the variables $x$, $y$, $z$ in the equations $E_x = 0$, $E_y = 0$, $E_z = 0$, the actual characteristic equations remain the same.

Thus the changed form of $E_x$ is still equal to $E_x$, and the same holds for $E_y$ and for $E_z$. We write

$$
\begin{aligned}
2\Omega = \ & g_{00}\xi^2 + g_{22}\eta^2 + g_{44}\zeta^2 + 2g_{02}\xi\eta + 2g_{04}\xi\zeta + 2g_{24}\eta\zeta \\
& + 2g_{01}\xi\xi' + 2g_{03}\xi\eta' + 2g_{05}\xi\zeta' \\
& + 2g_{21}\eta\xi' + 2g_{23}\eta\eta' + 2g_{25}\eta\zeta' \\
& + 2g_{41}\zeta\xi' + 2g_{43}\zeta\eta' + 2g_{45}\zeta\zeta' \\
& + g_{11}\xi'^2 + g_{33}\eta'^2 + g_{55}\zeta'^2 + 2g_{13}\xi'\eta' + 2g_{15}\xi'\zeta' + 2g_{35}\eta'\zeta' ;
\end{aligned}
$$

and we note that $t$ does not occur explicitly in $g$, so that the change of $\alpha$ into $\alpha + \kappa\alpha'$ does not affect $g$ in the small variation except in so far as it enters into the expressions for $\xi$, $\eta$, $\zeta$. Then the changes in the first derivatives of $g$ with respect to $x$, $x_1$, $y$, $y_1$, $z$, $z_1$, make them become

$$
\frac{\partial g}{\partial x} + \kappa\frac{\partial \Omega}{\partial \xi}, \quad \frac{\partial g}{\partial x_1} + \kappa\frac{\partial \Omega}{\partial \xi'},
$$

$$
\frac{\partial g}{\partial y} + \kappa\frac{\partial \Omega}{\partial \eta}, \quad \frac{\partial g}{\partial y_1} + \kappa\frac{\partial \Omega}{\partial \eta'},
$$

$$
\frac{\partial g}{\partial z} + \kappa\frac{\partial \Omega}{\partial \zeta}, \quad \frac{\partial g}{\partial z_1} + \kappa\frac{\partial \Omega}{\partial \zeta'},
$$

respectively. The three characteristic equations are unaltered by these changes: hence

$$
\left.
\begin{aligned}
E_\xi &= \frac{\partial \Omega}{\partial \xi} - \frac{d}{dt}\left(\frac{\partial \Omega}{\partial \xi'}\right) = 0 \\
E_\eta &= \frac{\partial \Omega}{\partial \eta} - \frac{d}{dt}\left(\frac{\partial \Omega}{\partial \eta'}\right) = 0 \\
E_\zeta &= \frac{\partial \Omega}{\partial \zeta} - \frac{d}{dt}\left(\frac{\partial \Omega}{\partial \zeta'}\right) = 0
\end{aligned}
\right\} .
$$

These three equations, which clearly are linear in $\xi$, $\eta$, $\zeta$ and their derivatives and are of the second order, are called the *subsidiary characteristic equations*. Their variables $\xi$, $\eta$, $\zeta$ are such as to keep $x + \kappa\xi$, $y + \kappa\eta$, $z + \kappa\zeta$ as integrals of the original characteristic equations. Moreover, we have

$$
x_1 E_x + y_1 E_y + z_1 E_z = 0 ;
$$

and therefore

$$
(x_1 + \kappa\xi_1)(E_x + \kappa E_\xi) + (y_1 + \kappa\eta_1)(E_y + \kappa E_\eta) + (z_1 + \kappa\zeta_1)(E_z + \kappa E_\zeta) = 0,
$$

where only terms of the first order in $\kappa$ are to be retained. Thus

$$
x_1 E_\xi + y_1 E_\eta + z_1 E_\zeta = 0,
$$

so that the three subsidiary equations are equivalent to only two independent equations. Accordingly, this relation must be satisfied unconditionally, whatever forms be given to the coefficients $g_{lm}$.

In place of the three equations, we shall substitute two independent equations having $Y$ and $Z$ as their dependent variables; but, in order to secure guidance as to their form, we shall effect an initial modification in the expression of the 'second' variation in our integral $I$.

*The 'second' variation: first modified form.*

**162.** Since the 'first' variation $I_1$ has been made to vanish, thereby providing the characteristic equations as well as terminal conditions which settle the limits of $I$, the governing magnitude of $J - I$ is the 'second' variation $\frac{1}{2}\kappa^2 \int \Theta\, dx$, where

$$
\begin{aligned}
\Theta =\ & g_{11}u'^2 + g_{33}v'^2 + g_{55}w'^2 + 2g_{13}u'v' + 2g_{15}u'w' + 2g_{35}v'w' \\
& + 2g_{01}uu' + 2g_{03}uv' + 2g_{05}uw' \\
& + 2g_{21}vu' + 2g_{23}vv' + 2g_{25}vw' \\
& + 2g_{41}wu' + 2g_{43}wv' + 2g_{45}ww' \\
& + g_{00}u^2 + g_{22}v^2 + g_{44}w^2 + 2g_{02}uv + 2g_{04}uw + 2g_{24}vw.
\end{aligned}
$$

It will be noticed that $\Theta$ is the same function of $u$, $u'$, $v$, $v'$, $w$, $w'$ as $2\Omega$ is (§ 161) of $\xi$, $\xi'$, $\eta$, $\eta'$, $\zeta$, $\zeta'$. In both of them, it is assumed that the values of $x$, $y$, $z$, which are determined by the characteristic equations, are substituted in the coefficients $g_{lm}$.

It has been pointed out (§ 157) that the six coefficients $g_{11}$, $g_{33}$, $g_{55}$, $g_{13}$, $g_{15}$, $g_{35}$ can be expressed in terms of three quantities initially defined by means of some selected trio among them. No preferential dependence is accorded in the foregoing analysis to a selection of two of three variables $x$, $y$, $z$ over the remaining variable. But as convenience can emerge from a choice of $y$ and $z$, because we then recur to the earlier analysis of §§ 130—154 on making $x = t$, we shall select $g_{33}$, $g_{35}$, $g_{55}$ as the three quantities of reference for the six coefficients. We take*

$$
g_{33} = x_1^2 P, \quad g_{35} = x_1^2 R, \quad g_{55} = x_1^2 Q \,;
$$

and then the other three coefficients are expressible by the relations

$$
\left.
\begin{aligned}
g_{13} &= - x_1 y_1 P - x_1 z_1 R \\
g_{15} &= \phantom{- x_1 y_1 P} - x_1 y_1 R - x_1 z_1 Q \\
g_{11} &= \phantom{-} y_1^2 P + 2 y_1 z_1 R + z_1^2 Q
\end{aligned}
\right\}.
$$

In order to obtain the first modified expression for the second variation $I_2$, we take a quantity

$$
T = A u^2 + B v^2 + C w^2 + 2 F v w + 2 G w u + 2 H u v,
$$

where $A$, $B$, $C$, $F$, $G$, $H$ are functions of $t$ at our disposal. As the limits of the integral $I$ now are fixed, the quantity $T$ vanishes at each limit, because $u$, $v$, $w$ vanish separately at each limit.

---

\* It is easy to verify that

$$
\frac{\partial^2 f}{\partial y'^2} = x_1^3 P, \quad \frac{\partial^2 f}{\partial y'\,\partial z'} = x_1^3 R, \quad \frac{\partial^2 f}{\partial z'^2} = x_1^3 Q,
$$

so that, when $x = t$, the quantities $P$, $R$, $Q$ are the critical quantities which arise (§ 148) in the discussion of the special weak variations.

With these values, the terms in $\Theta$ of the second degree in $u'$, $v'$, $w'$ are, together, equal to

$$P(x_1v' - y_1u')^2 + 2R(x_1v' - y_1u')(x_1w' - y_1u') + Q(x_1w' - z_1u')^2.$$

Accordingly, let $V$ and $W$ denote the quantities

$$V = x_1v - y_1u, \quad W = x_1w - z_1u,$$

so that

$$\begin{aligned} V' &= x_1v' - y_1u' + x_2v - y_2u \\ W' &= x_1w' - z_1u' + x_2w - z_2u \end{aligned} \Bigg\} .$$

We proceed to choose the coefficients $A$, $B$, $C$, $F$, $G$, $H$ in the quantity $T$, so that, if possible, we may have

$$\begin{aligned} \Theta + \frac{dT}{dt} = &\ PV'^2 + 2RV'W' + QW'^2 \\ &+ 2KVV' + 2LWV' + 2MVW' + 2NWW' + DV^2 + 2SVW + EW^2. \end{aligned}$$

**163.** In order that this relation may hold, the coefficients of the various combinations of $u$, $u'$, $v$, $v'$, $w$, $w'$ must be the same on the two sides.

(i) The coefficients of terms of the second degree in $u'$, $v'$, $w'$ are the same, because of the expressions for $g_{11}$, $g_{33}$, $g_{55}$, $g_{13}$, $g_{15}$, $g_{35}$.

(ii) In order that the terms in $(u, v, w \gtrless u', v', w')$ may be the same on the two sides, we must have

$$\begin{aligned}
g_{45} + C &= Qx_1x_2 + x_1{}^2N \\
g_{23} + B &= Px_1x_2 + x_1{}^2K \\
g_{43} + F &= Rx_1x_2 + x_1{}^2L \\
g_{25} + F &= Rx_1x_2 + x_1{}^2M \\
g_{41} + G &= -Qz_1x_2 - Ry_1x_2 - x_1y_1L - x_1z_1N \\
g_{05} + G &= -Qx_1z_2 - Rx_1y_2 - x_1y_1M - x_1z_1N \\
g_{21} + H &= -Py_1x_2 - Rz_1x_2 - x_1y_1K - x_1z_1M \\
g_{03} + H &= -Px_1y_2 - Rx_1z_2 - x_1y_1K - x_1z_1L \\
g_{01} + A &= \ \ Py_1y_2 + R(y_1z_2 + y_2z_1) + Qz_1z_2 \\
&\quad + y_1{}^2K + y_1z_1(L + M) + z_1{}^2N
\end{aligned} \Bigg\} .$$

(iii) In order that the terms of the second degree in $u$, $v$, $w$ may be the same on the two sides, we must have

$$\begin{aligned}
g_{44} + C' &= Qx_2{}^2 + 2x_1x_2N + x_1{}^2E \\
g_{22} + B' &= Px_2{}^2 + 2x_1x_2K + x_1{}^2D \\
g_{24} + F' &= Rx_2{}^2 + x_1x_2(L + M) + x_1{}^2S \\
g_{04} + G' &= -Qx_2z_2 - Rx_2y_2 - x_1y_2L - y_1x_2M - (x_1z_2 + z_1x_2)N - x_1y_1S - x_1z_1E \\
g_{02} + H' &= -Px_2y_2 - Rx_2z_2 - (x_1y_2 + y_1x_2)K - z_1x_2L - x_1z_2M - x_1y_1D - x_1z_1S \\
g_{00} + A' &= \ \ Py_2{}^2 + 2Ry_2z_2 + Qz_2{}^2 + 2y_1y_2K + 2z_1y_2L + 2y_1z_2M + 2z_1z_2N \\
&\quad + y_1{}^2D + 2y_1z_1S + z_1{}^2E
\end{aligned} \Bigg\} .$$

Here, there are fifteen equations in all; and they contain thirteen unknown quantities $A$, $B$, $C$, $F$, $G$, $H$; $K$, $L$, $M$, $N$, $D$, $S$, $E$. One of the two equations, involving $F$ in the set of nine, can be removed if we retain

$$g_{25} - g_{43} = x_1^2 (M - L);$$

one of the two, which involve $G$ in that set, can be removed if we retain

$$g_{05} - g_{41} = Q(z_1 x_2 - x_1 z_2) + R(y_1 x_2 - x_1 y_2) - x_1 y_1 (M - L);$$

and one of the two, which involve $H$ in that set, can be removed if we retain

$$g_{03} - g_{21} = P(y_1 x_2 - x_1 y_2) + R(z_1 x_2 - x_1 z_2) + x_1 z_1 (M - L).$$

But any two of these retained equations are satisfied in virtue of the third, because of the two fundamental and independent characteristic equations as given in § 160. Hence the total of fifteen equations must, in number, be reduced by two units in counting the tale of independent equations. Thus there are thirteen independent equations, potentially sufficient for the determination of the thirteen unknown quantities which occur linearly.

**164.** For convenience, all the fifteen will be used in the reductions, account being taken of those which ultimately are independent, viz. six out of the first set of nine, together with the one retained equation, and the six in the second set. In another aspect of the aggregate of equations, we may regard the set of nine as defining quantities $A$, $B$, $C$, $F$, $G$, $H$ and leaving one relation, free from these quantities so defined: the other set of six, together with this one relation, can (after substitution for $A$, $B$, $C$, $F$, $G$, $H$) be regarded as determining the quantities $K$, $L$, $M$, $N$, $D$, $S$, $E$.

It is, however, unnecessary to resolve the equations, and obtain the explicit expressions for all these quantities $A$, $B$, ...: provided the one aim of the investigation is the construction, explicit and complete, of the final normal form for $I_2$. This form is obtained, in terms of the quantities $P$, $R$, $Q$, and in terms of the integrals of the subsidiary characteristic equation. Moreover, the primitive of this subsidiary equation is derived from that of the characteristic equation. Thus the actual knowledge of these quantities $A$, $B$, ... is not required for the construction of the normal form; it is ancillary to the construction of an intermediate form, which itself is of transient importance.

The process adopted in the present case is, in fact, a fusion of the two processes in § 52 whereby, first, a form

$$Pw_1^2 + Gw^2 + \frac{dT}{dt}$$

was constructed for $\Theta$; and, next, a form

$$P\left(w_1 - \frac{z_1}{z}w\right)^2 + \frac{d}{dt}(T + Hw^2)$$

was deduced.

If these separate processes were desired in the present case, the first stage is attainable by keeping the expression for $T$ (on p. 236) as simple as possible; for example, we could take

$$K = 0, \quad N = 0, \quad M + L = 0,$$

initially.

The actual method adopted is more direct than the amplification of the earlier double process. Moreover, it has the added advantage that it can be extended, without further change, to the case of any number of dependent variables.

*The three subsidiary equations reducible to two, with modified variables.*

**165.** Next, we modify the form of the three subsidiary characteristic equations, of which the full initial expressions are

$$E_\xi = g_{00}\xi + g_{02}\eta + g_{04}\zeta + g_{01}\xi' + g_{03}\eta' + g_{05}\zeta'$$
$$- \frac{d}{dt}(g_{01}\xi + g_{21}\eta + g_{41}\zeta + g_{11}\xi' + g_{13}\eta' + g_{15}\zeta') = 0,$$

$$E_\eta = g_{02}\xi + g_{22}\eta + g_{42}\zeta + g_{21}\xi' + g_{23}\eta' + g_{25}\zeta'$$
$$- \frac{d}{dt}(g_{03}\xi + g_{23}\eta + g_{43}\zeta + g_{13}\xi' + g_{33}\eta' + g_{35}\zeta') = 0,$$

$$E_\zeta = g_{04}\xi + g_{24}\eta + g_{44}\zeta + g_{41}\xi' + g_{43}\eta' + g_{45}\zeta'$$
$$- \frac{d}{dt}(g_{05}\xi + g_{25}\eta + g_{45}\zeta + g_{15}\xi' + g_{35}\eta' + g_{55}\zeta') = 0.$$

In the first place, these three equations are equivalent to only two independent equations (§ 161) in virtue of the relation

$$x_1 E_\xi + y_1 E_\eta + z_1 E_\zeta = 0,$$

which now will be verified.

Partial differentiations of the identity satisfied by $g$ give relations

$$\frac{\partial g}{\partial x} = x_1 g_{01} + y_1 g_{03} + z_1 g_{05},$$

$$\frac{\partial g}{\partial y} = x_1 g_{21} + y_1 g_{23} + z_1 g_{25},$$

$$\frac{\partial g}{\partial z} = x_1 g_{41} + y_1 g_{43} + z_1 g_{45}.$$

When the first of these identities is differentiated with regard to $t$, we have

$$x_1 g_{00} + y_1 g_{02} + z_1 g_{04} + x_2 g_{01} + y_2 g_{03} + z_2 g_{05}$$
$$= x_2 g_{01} + y_2 g_{03} + z_2 g_{05} + x_1 g_{01}' + y_1 g_{03}' + z_1 g_{05}',$$

and therefore

$$x_1 (g_{00} - g_{01}') + y_1 (g_{02} - g_{03}') + z_1 (g_{04} - g_{05}') = 0.$$

Similarly from the other two identities, we have

$$x_1(g_{02} - g_{21}') + y_1(g_{22} - g_{23}') + z_1(g_{24} - g_{25}') = 0,$$

$$x_1(g_{04} - g_{05}') + y_1(g_{24} - g_{25}') + z_1(g_{44} - g_{45}') = 0.$$

In virtue of these three relations, the coefficients of $\xi$, of $\eta$, and of $\zeta$, in $x_1 E_\xi + y_1 E_\eta + z_1 E_\zeta$ vanish separately.

Again, we have obtained the relations

$$x_1 g_{11} + y_1 g_{13} + z_1 g_{15} = 0,$$

$$x_1 g_{13} + y_1 g_{33} + z_1 g_{35} = 0,$$

$$x_1 g_{15} + y_1 g_{35} + z_1 g_{55} = 0.$$

In virtue of these, the coefficients of $\xi''$, of $\eta''$, and of $\zeta''$, in $x_1 E_\xi + y_1 E_\eta + z_1 E_\zeta$ vanish separately.

Next, differentiating with regard to $t$ the first of the three relations just quoted, we have

$$x_1 g_{11}' + y_1 g_{13}' + z_1 g_{15}' + x_2 g_{11} + y_2 g_{13} + z_2 g_{15} = 0\,;$$

and therefore

$$x_1 g_{11}' + y_1(g_{13}' + g_{03} - g_{21}) + z_1(g_{15}' + g_{05} - g_{41})$$

$$= y_1(g_{03} - g_{21}) + z_1(g_{05} - g_{41}) - x_2 g_{11} - y_2 g_{13} - z_2 g_{15}$$

$$= y_1(g_{03} - g_{21}) + z_1(g_{05} - g_{41}) - x_2(y_1^2 P + 2 y_1 z_1 R + z_1^2 Q)$$

$$\qquad\qquad + y_2(x_1 y_1 P + x_1 z_1 R) + z_2(x_1 y_1 R + x_1 z_1 Q)$$

$$= 0,$$

because of one of the fundamental characteristic equations as they are stated in § 160. Similarly, from the second and the third of the foregoing relations, we have

$$x_1(g_{13}' + g_{21} - g_{03}) + y_1 g_{33}' + z_1(g_{35}' + g_{25} - g_{43}) = 0,$$

$$x_1(g_{15}' + g_{41} - g_{05}) + y_1(g_{35}' + g_{43} - g_{25}) + z_1 g_{55}' = 0.$$

In virtue of these three relations, the coefficients of $\xi'$, of $\eta'$, and of $\zeta'$, in $x_1 E_\xi + y_1 E_\eta + z_1 E_\zeta$ vanish separately.

Consequently, we have

$$x_1 E_\xi + y_1 E_\eta + z_1 E_\zeta = 0.$$

**166.** We therefore need retain only two (out of the three) subsidiary characteristic equations. We select the second and the third, and proceed to change their expressions, using the variables $Y$ and $Z$ (already introduced) instead of $\xi$, $\eta$, $\zeta$, with the definitions

$$Y = x_1 \eta - y_1 \xi, \quad Z = x_1 \zeta - z_1 \xi,$$

so that

$$\left.\begin{aligned} Y' &= x_1 \eta' - y_1 \xi' + x_2 \eta - y_2 \xi \\ Z' &= x_1 \zeta' - z_1 \xi' + x_2 \zeta - z_2 \xi \end{aligned}\right\},$$

where it will be noted that, if $u = \xi$, $v = \eta$, $w = \zeta$, then $V = Y$, $W = Z$, the quantities that occur in the modified expression for $\Theta + \dfrac{dT}{dt}$ in § 162.

Substituting the expressions given for the various coefficients $g_{lm}$ in $\Theta$, we have

$$
\begin{aligned}
g_{13}\xi' + g_{33}\eta' + g_{35}\zeta' &= -(x_1 y_1 P + x_1 z_1 R)\,\xi' + x_1{}^2 P\eta' + x_1{}^2 R\zeta' \\
&= x_1\{P(x_1\eta' - y_1\xi') + R(x_1\zeta' - z_1\xi')\} \\
&= x_1\{PY' + RZ' - P(x_2\eta - y_2\xi) - R(x_2\zeta - z_2\xi)\},
\end{aligned}
$$

and

$$
\begin{aligned}
g_{03}\xi + g_{23}\eta + g_{43}\zeta &= -(H\xi + B\eta + F\zeta) - \xi(Px_1 y_2 + Rx_1 z_2 + x_1 y_1 K + x_1 z_1 L) \\
&\quad + \eta(Px_1 x_2 + x_1{}^2 K) + \zeta(Rx_1 x_2 + x_1{}^2 L);
\end{aligned}
$$

therefore

$$
\begin{aligned}
&g_{03}\xi + g_{23}\eta + g_{43}\zeta + g_{13}\xi' + g_{33}\eta' + g_{35}\zeta' \\
&\quad = -(H\xi + B\eta + F\zeta) + x_1(PY' + RZ' + KY + LZ).
\end{aligned}
$$

Again,

$$
\begin{aligned}
g_{21}\xi' + g_{23}\eta' + g_{25}\zeta' &= -(H\xi' + B\eta' + F\zeta') - \xi'(Py_1 x_2 + Rz_1 x_2 + x_1 y_1 K + x_1 z_1 M) \\
&\quad + \eta'(Px_1 x_2 + x_1{}^2 K) + \zeta'(Rx_1 x_2 + x_1{}^2 M) \\
&= -(H\xi' + B\eta' + F\zeta') \\
&\quad + (Px_2 + x_1 K)(x_1\eta' - y_1\xi') + (Rx_2 + x_1 M)(x_1\zeta' - z_1\xi'),
\end{aligned}
$$

and

$$
\begin{aligned}
g_{02}\xi + g_{22}\eta + g_{24}\zeta &= -(H'\xi + B'\eta + C'\zeta) \\
&\quad - \xi(Px_2 y_2 + Rx_2 z_2 + x_1 y_2 K + y_1 x_2 K + z_1 x_2 L + x_1 z_2 M \\
&\qquad\qquad + x_1 y_1 D + x_1 z_1 S) \\
&\quad + \eta(Px_2{}^2 + 2x_1 x_2 K + x_1{}^2 D) + \zeta(Rx_2{}^2 + x_1 x_2 L + x_1 x_2 M + x_1{}^2 S) \\
&= -(H'\xi + B'\eta + C'\zeta) \\
&\quad + (Px_2 + x_1 K)(x_2\eta - y_2\xi) + (Rx_2 + x_1 M)(x_2\zeta - z_2\xi) \\
&\quad + (x_2 K + x_1 D)(x_1\eta - y_1\xi) + (x_2 L + x_1 S)(x_1\zeta - z_1\xi);
\end{aligned}
$$

therefore

$$
\begin{aligned}
&g_{02}\xi + g_{22}\eta + g_{24}\zeta + g_{21}\xi' + g_{23}\eta' + g_{25}\zeta' \\
&\quad = -\frac{d}{dt}(H\xi + B\eta + C\zeta) + x_2(PY' + RZ' + KY + LZ) \\
&\qquad\qquad + x_1(KY' + MZ' + DY + SZ).
\end{aligned}
$$

Substituting these values in the subsidiary equation $E_\eta = 0$, and removing a non-vanishing* factor $x_1$, we have

$$
E_Y = \frac{d}{dt}(PY' + RZ' + KY + LZ) - (KY' + MZ' + DY + SZ) = 0.
$$

---

* As $x$ varies uniformly through the range and $t$ increases throughout the range, $x_1$ can only vanish for isolated places at the utmost.

Proceeding in the same way with the subsidiary equation $E_\zeta = 0$ and again removing the non-vanishing factor $x_1$, we have

$$E_Z = \frac{d}{dt}(RY' + QZ' + MY + NZ) - (LY' + NZ' + SY + EZ) = 0.$$

We therefore take

$$E_Y = 0, \quad E_Z = 0,$$

being two independent equations, as the modified form of the subsidiary equations. Manifestly linear and of the second order, they can be regarded as determining $Y$ and $Z$ as functions of $t$; because the seven quantities $K$, $L$, $M$, $N$, $D$, $S$, $E$, do not involve $\xi$, $\eta$, $\zeta$, that is, do not involve $Y$ and $Z$.

*One selected integral of the two subsidiary equations.*

**167**. One important property, in the form of an integral of these two simultaneous equations, will prove useful in the further reduction of $\Theta + \frac{dT}{dt}$ to its final form. We write

$$2\square = PY'^2 + 2RY'Z' + QZ'^2$$
$$+ 2KYY' + 2LZY' + 2MYZ' + 2NZZ' + DY^2 + 2SYZ + EZ^2,$$

where it will be noticed that $2\square$ is the same function of $Y$ and $Z$ as $\Theta + \frac{dT}{dt}$ is of $V$ and $W$; and then the two subsidiary equations $E_Y = 0$ and $E_Z = 0$ can be written in the form

$$\frac{\partial \square}{\partial Y} - \frac{d}{dt}\left(\frac{\partial \square}{\partial Y'}\right) = 0, \quad \frac{\partial \square}{\partial Z} - \frac{d}{dt}\left(\frac{\partial \square}{\partial Z'}\right) = 0.$$

As each of these two equations is linear and of the second order, their complete primitive giving the expression of $Y$ and $Z$—to its expression we shall return later (§ 170)—is compounded of linear combinations of four linearly independent sets of integrals $Y_1$ and $Z_1$, $Y_2$ and $Z_2$, $Y_3$ and $Z_3$, $Y_4$ and $Z_4$. The property in question is that *there exist two sets of integrals $Y_1$ and $Z_1$, $Y_r$ and $Z_r$* (where $Y_r$ and $Z_r$ are either one of the other three sets, or are linear combinations of them in the form

$$Y_r = c_2 Y_2 + c_3 Y_3 + c_4 Y_4, \quad Z_r = c_2 Z_2 + c_3 Z_3 + c_4 Z_4,$$

the coefficients $c$ being constants) *such that, if $\square_1$ and $\square_r$ denote the values of $\square$ when $Y_1$ and $Z_1$, and $Y_r$ and $Z_r$, are substituted for $Y$ and $Z$, the relation*

$$Y_1 \frac{\partial \square_r}{\partial Y_r'} + Z_1 \frac{\partial \square_r}{\partial Z_r'} = Y_r \frac{\partial \square_1}{\partial Y_1'} + Z_r \frac{\partial \square_1}{\partial Z_1'}$$

*is satisfied; and the choice is not unique.*

The quantities $\square_1$ and $\square_2$ are homogeneous and quadratic in their respective arguments, and they have the same coefficients; hence the relation

$$Y_1 \frac{\partial \square_2}{\partial Y_2} + Y_1' \frac{\partial \square_2}{\partial Y_2'} + Z_1 \frac{\partial \square_2}{\partial Z_2} + Z_1' \frac{\partial \square_2}{\partial Z_2'} = Y_2 \frac{\partial \square_1}{\partial Y_1} + Y_2' \frac{\partial \square_1}{\partial Y_1'} + Z_2 \frac{\partial \square_1}{\partial Z_1} + Z_2' \frac{\partial \square_1}{\partial Z_1'}$$

holds, the two sides being equal to the same expression lineo-linear in each set of arguments. But

$$\frac{\partial \square_2}{\partial Y_2} = \frac{d}{dt}\left(\frac{\partial \square_2}{\partial Y_2'}\right), \quad \frac{\partial \square_2}{\partial Z_2} = \frac{d}{dt}\left(\frac{\partial \square_2}{\partial Z_2'}\right),$$

$$\frac{\partial \square_1}{\partial Y_1} = \frac{d}{dt}\left(\frac{\partial \square_1}{\partial Y_1'}\right), \quad \frac{\partial \square_1}{\partial Z_1} = \frac{d}{dt}\left(\frac{\partial \square_1}{\partial Z_1'}\right);$$

and so the relation can be written

$$\frac{d}{dt}\left(Y_1 \frac{\partial \square_2}{\partial Y_2'} + Z_1 \frac{\partial \square_2}{\partial Z_2'}\right) = \frac{d}{dt}\left(Y_2 \frac{\partial \square_1}{\partial Y_1'} + Z_2 \frac{\partial \square_1}{\partial Z_1'}\right).$$

Hence

$$Y_1 \frac{\partial \square_2}{\partial Y_2'} + Z_1 \frac{\partial \square_2}{\partial Z_2'} - \left(Y_2 \frac{\partial \square_1}{\partial Y_1'} + Z_2 \frac{\partial \square_1}{\partial Z_1'}\right) = k,$$

where $k$ is a constant. As $Y_1$ and $Z_1$, $Y_2$ and $Z_2$, are individual and not general integrals of their equations, $k$ is a specific (and not an arbitrary) constant.

If $k$ is zero, for a primarily selected pair of sets $Y_1$ and $Z_1$, $Y_2$ and $Z_2$, the property is established for that pair.

If $k$ is not zero, take another pair of sets $Y_1$ and $Z_1$, $Y_3$ and $Z_3$, where $Y_3$ and $Z_3$ are not a linear combination of $Y_1$ and $Z_1$, $Y_2$ and $Z_2$. Then, in the same way, we find

$$Y_1 \frac{\partial \square_3}{\partial Y_3'} + Z_1 \frac{\partial \square_3}{\partial Z_3'} - \left(Y_3 \frac{\partial \square_1}{\partial Y_1'} + Z_3 \frac{\partial \square_1}{\partial Z_1'}\right) = l,$$

where $l$ is a specific constant.

If $l$ is zero, the property is established for the pair of sets $Y_1$ and $Z_1$, $Y_3$ and $Z_3$.

If $l$ is not zero, take a set of integrals

$$Y_r = l Y_2 - k Y_3, \quad Z_r = l Z_2 - k Z_3,$$

so that $Y_r$ and $Z_r$ are not merely $Y_1$ and $Z_1$, nor merely $Y_2$ and $Z_2$, nor merely $Y_3$ and $Z_3$. Then

$$l \frac{\partial \square_2}{\partial Y_2'} - k \frac{\partial \square_3}{\partial Y_3'} = \frac{\partial \square_r}{\partial Y_r'}, \quad l \frac{\partial \square_2}{\partial Z_2'} - k \frac{\partial \square_3}{\partial Z_3'} = \frac{\partial \square_r}{\partial Z_r'},$$

owing to the linearity of the first derivatives of $\square_2$, $\square_3$, $\square_r$ in their arguments. Then we have

$$Y_1 \frac{\partial \square_r}{\partial Y_r'} + Z_1 \frac{\partial \square_r}{\partial Z_r'} = Y_r \frac{\partial \square_1}{\partial Y_1'} + Z_r \frac{\partial \square_1}{\partial Z_1'},$$

establishing the property for the pair of sets $Y_1$ and $Z_1$, $Y_r$ and $Z_r$.

Manifestly the property is not unique; because we can make $Y_r$ and $Z_r$ take the place of $Y_2$ and $Z_2$ as a set of integrals, and we still have the sets $Y_3$ and $Z_3$, $Y_4$ and $Z_4$, to associate with $Y_1$ and $Z_1$, with the new $Y_2$ and $Z_2$, and with one another.

For our immediate purpose, it is sufficient to possess the relation for two distinct sets of integrals of the equations.

*Final normal form of second variation.*

**168.** We now can proceed to the final form of the quantity $\Theta + \dfrac{dT}{dt}$. In connection with any set $Y$ and $Z$ of integrals of the subsidiary equations, we introduce quantities $\alpha$, $\beta$, $\gamma$, $\delta$ which satisfy the two relations

$$Y' + \alpha Y + \gamma Z = 0, \quad Z' + \beta Y + \delta Z = 0,$$

these relations, as they stand, being insufficient to define the four quantities. Then

$$PY' + RZ' + KY + LZ = (K - P\alpha - R\beta)\, Y + (L - P\gamma - R\delta)\, Z,$$
$$KY' + MZ' + DY + SZ = (D - K\alpha - M\beta)\, Y + (S - K\gamma - M\delta)\, Z;$$
$$RY' + QZ' + MY + NZ = (M - R\alpha - Q\beta)\, Y + (N - R\gamma - Q\delta)\, Z,$$
$$LY' + NZ' + SY + EZ = (S - L\alpha - N\beta)\, Y + (E - L\gamma - N\delta)\, Z;$$

and the subsidiary characteristic equations $E_Y = 0$ and $E_Z = 0$ are satisfied if

$$\left.\begin{aligned}
K &= P\alpha + R\beta \\
M &= R\alpha + Q\beta \\
L &= P\gamma + R\delta \\
N &= R\gamma + Q\delta \\
D &= K\alpha + M\beta = P\alpha^2 + 2R\alpha\beta + Q\beta^2 \\
S &= K\gamma + M\delta \\
  &= L\alpha + N\beta = P\alpha\gamma + R(\alpha\delta + \beta\gamma) + Q\beta\delta \\
E &= L\gamma + N\delta = P\gamma^2 + 2R\gamma\delta + Q\delta^2
\end{aligned}\right\},$$

where it will be noted that the apparently different values of $S$, viz. $K\gamma + M\delta$ and $L\alpha + N\beta$, agree.

Thus far, the relations hold for any set $Y$ and $Z$ of integrals of the subsidiary equations. But a single set is not sufficient to define the quantities $\alpha$, $\beta$, $\gamma$, $\delta$. Accordingly, we take two independent sets $Y_1$ and $Z_1$, $Y_2$ and $Z_2$, and associate both sets with these four quantities, the association not affecting the values of $K$, $L$, $M$, $N$, $D$, $S$, $E$. Then we have

$$\left.\begin{aligned} Y_1' + \alpha Y_1 + \gamma Z_1 = 0 \\ Y_2' + \alpha Y_2 + \gamma Z_2 = 0 \end{aligned}\right\}, \qquad \left.\begin{aligned} Z_1' + \beta Y_1 + \delta Z_1 = 0 \\ Z_2' + \beta Y_2 + \delta Z_2 = 0 \end{aligned}\right\},$$

so that

$$\frac{\alpha}{Z_1 Y_2' - Z_2 Y_1'} = \frac{\gamma}{Y_2 Y_1' - Y_1 Y_2'} = \frac{\beta}{Z_1 Z_2' - Z_2 Z_1'} = \frac{\delta}{Y_2 Z_1' - Y_1 Z_2'} = \frac{1}{Y_1 Z_2 - Y_2 Z_1},$$

where $Y_1 Z_2 - Y_2 Z_1$ is not zero, because the two sets of integrals $Y_1$ and $Z_1$, $Y_2$ and $Z_2$, are independent, in the sense that $Y_2$ and $Z_2$ are not merely the same constant multiples of $Y_1$ and $Z_1$.

The original characteristic equations are known to be satisfied in connection with the quantities $K$, $L$, $M$, $N$, without this assignment of values. The subsidiary equations are satisfied in connection with the additional quantities $D$, $S$, $E$, as well as $K$, $L$, $M$, $N$, without this assignment of values. All that has been done is to assign values to $K$, $L$, $M$, $N$, $D$, $S$, $E$, associated with two independent sets of integrals $Y_1$ and $Z_1$, $Y_2$ and $Z_2$. Now we have seen that any two independent sets of integrals $Y_1$ and $Z_1$, $Y_2$ and $Z_2$, are such that

$$Y_1 \frac{\partial \square_2}{\partial Y_2'} + Z_1 \frac{\partial \square_2}{\partial Z_2'} - \left( Y_2 \frac{\partial \square_1}{\partial Y_1'} + Z_2 \frac{\partial \square_1}{\partial Z_1'} \right) = k,$$

where $k$ is a specific constant, and that an appropriate choice of two sets can be made so as to secure that the specific constant is zero. Now

$$Y_1 \frac{\partial \square_2}{\partial Y_2'} + Z_1 \frac{\partial \square_2}{\partial Z_2'} - \left( Y_2 \frac{\partial \square_1}{\partial Y_1'} + Z_2 \frac{\partial \square_1}{\partial Z_1'} \right)$$

$$= Y_1 (PY_2' + RZ_2' + KY_2 + LZ_2) + Z_1 (RY_2' + QZ_2' + MY_2 + NZ_2)$$

$$\quad - Y_2 (PY_1' + RZ_1' + KY_1 + LZ_1) - Z_2 (RY_1' + QZ_1' + MY_1 + NZ_1)$$

$$= (Y_1 Z_2 - Y_2 Z_1)(L - P\gamma - R\delta) - (Y_1 Z_2 - Y_2 Z_1)(M - R\alpha - Q\beta).$$

Thus the values assigned to the various quantities, in particular to $L$ and $M$, require a zero value for the specific constant value of the left-hand side; and consequently the two sets of integrals $Y_1$ and $Z_1$, $Y_2$ and $Z_2$, of the subsidiary characteristic equations must be chosen (as they always can be chosen) so that

$$Y_1 \frac{\partial \square_2}{\partial Y_2'} + Z_1 \frac{\partial \square_2}{\partial Z_2'} = Y_2 \frac{\partial \square_1}{\partial Y_1'} + Z_2 \frac{\partial \square_1}{\partial Z_1'} .$$

With these values now made definite for $K$, $L$, $M$, $N$, $D$, $S$, $E$, we have

$$\Theta + \frac{dT}{dt} = PV'^2 + 2RV'W' + QW'^2$$

$$+ 2KVV' + 2LWV' + 2MVW' + 2NWW' + DV^2 + 2SVW + EW^2$$

$$= P(V' + \alpha V + \gamma W)^2 + Q(W' + \beta V + \delta W)^2$$

$$+ 2R(V' + \alpha V + \gamma W)(W' + \beta V + \delta W)$$

$$= \frac{1}{(Y_1 Z_2 - Y_2 Z_1)^2} \left\{ P \begin{vmatrix} V', & V, & W \\ Y_1', & Y_1, & Z_1 \\ Y_2', & Y_2, & Z_2 \end{vmatrix}^2 + Q \begin{vmatrix} W', & V, & W \\ Z_1', & Y_1, & Z_1 \\ Z_2', & Y_2, & Z_2 \end{vmatrix}^2 \right.$$

$$\left. + 2R \begin{vmatrix} V', & V, & W \\ Y_1', & Y_1, & Z_1 \\ Y_2', & Y_2, & Z_2 \end{vmatrix} \begin{vmatrix} W', & V, & W \\ Z_1', & Y_1, & Z_1 \\ Z_2', & Y_2, & Z_2 \end{vmatrix} \right\} = \mathbf{U},$$

which (with the foregoing limitation on the selection of the sets $Y_1$ and $Z_1$, $Y_2$ and $Z_2$) is the final reduced form for $\Theta + \dfrac{dT}{dt}$.

The expression for $T$ is

$$T = Au^2 + Bv^2 + Cw^2 + 2Fvw + 2Gwu + 2Huv$$

$$= - \{g_{01}u^2 + g_{23}v^2 + g_{45}w^2 + (g_{25} + g_{43})vw + (g_{41} + g_{05})wu + (g_{03} + g_{21})uv\}$$

$$+ \frac{PV + RW}{Y_1 Z_2 - Y_2 Z_1} \begin{vmatrix} x_2 v - y_2 u, & V, & W \\ Y_1', & Y_1, & Z_1 \\ Y_2', & Y_2, & Z_2 \end{vmatrix} + \frac{RV + QW}{Y_1 Z_2 - Y_2 Z_1} \begin{vmatrix} x_2 w - z_2 u, & V, & W \\ Z_1', & Y_1, & Z_1 \\ Z_2', & Y_2, & Z_2 \end{vmatrix}.$$

But $\int \frac{dT}{dt} dt = [T]$, taken at the limits. The limits have been settled by the conditions deduced from the first variation of the original integral $I$, so that $u$, $v$, $w$ (and therefore $V$ and $W$) vanish at each limit; and thus $\int \frac{dT}{dt} dt = 0$, between the limits. Thus the second variation of the integral has become equal to

$$\tfrac{1}{2} \kappa^2 \int \mathbf{U} \, dt.$$

### Primitive of the subsidiary equations.

**169.** This reduction of the second variation of $I$ manifestly depends upon a knowledge of the quantities $Y$ and $Z$; and the significance of the normal form depends upon properties of two independent sets of integrals. The discussion thus requires a knowledge of the primitive of the subsidiary characteristic equations.

Now these subsidiary equations arose from the original characteristic equations by the consideration of varied values $x + \kappa\xi$, $y + \kappa\eta$, $z + \kappa\zeta$ of $x$, $y$, $z$, these values still satisfying the characteristic equations; and the values of $\xi$, $\eta$, $\zeta$ were derived (§ 161) from those of $x$, $y$, $z$ in the primitive of the characteristic equations. By means of these quantities, two other quantities $Y$ and $Z$ were constructed, such that

$$\left. \begin{array}{l} Y = x_1\eta - y_1\xi = a_1\theta_1(t) + a_2\theta_2(t) + a_3\theta_3(t) + a_4\theta_4(t) \\ Z = x_1\zeta - z_1\xi = a_1\vartheta_1(t) + a_2\vartheta_2(t) + a_3\vartheta_3(t) + a_4\vartheta_4(t) \end{array} \right\},$$

where $a_1$, $a_2$, $a_3$, $a_4$ are independent arbitrary constants. Also, $Y$ and $Z$ satisfy the linear equations

$$\frac{\partial \square}{\partial Y} - \frac{d}{dt}\left(\frac{\partial \square}{\partial Y'}\right) = 0, \qquad \frac{\partial \square}{\partial Z} - \frac{d}{dt}\left(\frac{\partial \square}{\partial Z'}\right) = 0,$$

the primitive of which is constituted by a set of values $Y$ and $Z$, involving four arbitrary independent constants linearly. Clearly $\theta_r(t)$ and $\vartheta_r(t)$, where $r = 1, 2, 3, 4$ in turn, are a set of integrals; if, then, these four sets are linearly independent of one another, the foregoing expressions for $Y$ and $Z$ constitute the primitive of the subsidiary equations. We proceed to establish this linear independence.

**170.** The primitive of the characteristic equations is

$$
\begin{aligned}
x &= \chi\,(t + \alpha,\ \alpha_1,\ \alpha_2,\ \alpha_3,\ \alpha_4) \\
y &= \phi\,(t + \alpha,\ \alpha_1,\ \alpha_2,\ \alpha_3,\ \alpha_4) \\
z &= \psi\,(t + \alpha,\ \alpha_1,\ \alpha_2,\ \alpha_3,\ \alpha_4)
\end{aligned},
$$

where the essential constants are $\alpha_1,\ \alpha_2,\ \alpha_3,\ \alpha_4$. The constants are determinable, by the assignment of initial values to $x,\ y,\ z$ and of initial values to $x',\ y',\ z'$; but as the variable $t$ is not specific and any function of $t$ may be taken in its place, the initial conditions become precise by the assignment of an initial value $l$ to $\dfrac{dy}{dx}\left(=\dfrac{y'}{x'}\right)$, and an initial value $m$ to $\dfrac{dz}{dx}\left(=\dfrac{z'}{x'}\right)$, as well as of values $b$ and $c$ to $y$ and $z$, when $x = a$, and $t = t_0$. In this form, the primitive of the characteristic equations becomes

$$
\begin{aligned}
y &= y\,(x,\ b,\ l,\ c,\ m) \\
z &= z\,(x,\ b,\ l,\ c,\ m)
\end{aligned}.
$$

The set of essential arbitrary independent constants $\alpha_1,\ \alpha_2,\ \alpha_3,\ \alpha_4$ is expressible in terms of the arbitrary independent constants $b,\ l,\ c,\ m$, by means of four independent equations

$$
\begin{aligned}
\alpha_1 &= \omega_1\,(b,\ l,\ c,\ m) \\
\alpha_2 &= \omega_2\,(b,\ l,\ c,\ m) \\
\alpha_3 &= \omega_3\,(b,\ l,\ c,\ m) \\
\alpha_4 &= \omega_4\,(b,\ l,\ c,\ m)
\end{aligned}.
$$

As the four constants in each set are independent, these functional forms are such that neither of the Jacobians

$$
J\left(\frac{\alpha_1,\ \alpha_2,\ \alpha_3,\ \alpha_4}{b,\ l,\ c,\ m}\right),\quad
J\left(\frac{b,\ l,\ c,\ m}{\alpha_1,\ \alpha_2,\ \alpha_3,\ \alpha_4}\right),
$$

vanishes.

Now the quantities $\dfrac{dy}{dx}$ and $\dfrac{dz}{dx}$, derived from the second form of the characteristic integral, must have the same values as those derived from the first form; hence

$$
\frac{dy}{dx} = \frac{\phi'(t)}{\chi'(t)},\qquad \frac{dz}{dx} = \frac{\psi'(t)}{\chi'(t)}.
$$

Again, from the equivalence of the two forms, we have

$$
\frac{\partial y}{\partial \alpha_r} = \frac{dy}{dx}\frac{\partial x}{\partial \alpha_r} + \frac{\partial y}{\partial b}\frac{\partial b}{\partial \alpha_r} + \frac{\partial y}{\partial l}\frac{\partial l}{\partial \alpha_r} + \frac{\partial y}{\partial c}\frac{\partial c}{\partial \alpha_r} + \frac{\partial y}{\partial m}\frac{\partial m}{\partial \alpha_r},
$$

for $r = 1,\ 2,\ 3,\ 4$; and therefore

$$
\begin{aligned}
\theta_r(t) &= \chi'(t)\frac{\partial y}{\partial \alpha_r} - \phi'(t)\frac{\partial x}{\partial \alpha_r} \\
&= \chi'(t)\left\{\frac{\partial y}{\partial b}\frac{\partial b}{\partial \alpha_r} + \frac{\partial y}{\partial l}\frac{\partial l}{\partial \alpha_r} + \frac{\partial y}{\partial c}\frac{\partial c}{\partial \alpha_r} + \frac{\partial y}{\partial m}\frac{\partial m}{\partial \alpha_r}\right\}.
\end{aligned}
$$

Similarly, proceeding from the equation in $z$, we have

$$\Im_r(t) = \chi'(t)\frac{\partial z}{\partial \alpha_r} - \psi'(t)\frac{\partial x}{\partial \alpha_r}$$

$$= \chi'(t)\left\{\frac{\partial z}{\partial b}\frac{\partial b}{\partial \alpha_r} + \frac{\partial z}{\partial l}\frac{\partial l}{\partial \alpha_r} + \frac{\partial z}{\partial c}\frac{\partial c}{\partial \alpha_r} + \frac{\partial z}{\partial m}\frac{\partial m}{\partial \alpha_r}\right\}.$$

Now if there were a linear relation among the four quantities $\theta_r(t)$ with constant coefficients, and a linear relation among the four quantities $\Im_r(t)$ with the same constant coefficients, of a type

$$\sum_{r=1}^{4}\mu_r\theta_r(t) = 0, \qquad \sum_{r=1}^{4}\mu_r\Im_r(t) = 0,$$

such relations would have the forms

$$\frac{\partial y}{\partial b}\Sigma\mu_r\frac{\partial b}{\partial \alpha_r} + \frac{\partial y}{\partial l}\Sigma\mu_r\frac{\partial l}{\partial \alpha_r} + \frac{\partial y}{\partial c}\Sigma\mu_r\frac{\partial c}{\partial \alpha_r} + \frac{\partial y}{\partial m}\Sigma\mu_r\frac{\partial m}{\partial \alpha_r} = 0,$$

$$\frac{\partial z}{\partial b}\Sigma\mu_r\frac{\partial b}{\partial \alpha_r} + \frac{\partial z}{\partial l}\Sigma\mu_r\frac{\partial l}{\partial \alpha_r} + \frac{\partial z}{\partial c}\Sigma\mu_r\frac{\partial c}{\partial \alpha_r} + \frac{\partial z}{\partial m}\Sigma\mu_r\frac{\partial m}{\partial \alpha_r} = 0.$$

But these have the form of linear relations, with constant coefficients, between the set

$$\frac{\partial y}{\partial b}, \quad \frac{\partial y}{\partial l}, \quad \frac{\partial y}{\partial c}, \quad \frac{\partial y}{\partial m}$$

together, and the set

$$\frac{\partial z}{\partial b}, \quad \frac{\partial z}{\partial l}, \quad \frac{\partial z}{\partial c}, \quad \frac{\partial z}{\partial m}$$

together; and it has been proved (§ 140) that no such linear relations exist. The foregoing apparent relations must be evanescent, so that the constant coefficients must vanish; hence

$$\mu_1\frac{\partial b}{\partial \alpha_1} + \mu_2\frac{\partial b}{\partial \alpha_2} + \mu_3\frac{\partial b}{\partial \alpha_3} + \mu_4\frac{\partial b}{\partial \alpha_4} = 0,$$

$$\mu_1\frac{\partial l}{\partial \alpha_1} + \mu_2\frac{\partial l}{\partial \alpha_2} + \mu_3\frac{\partial l}{\partial \alpha_3} + \mu_4\frac{\partial l}{\partial \alpha_4} = 0,$$

$$\mu_1\frac{\partial c}{\partial \alpha_1} + \mu_2\frac{\partial c}{\partial \alpha_2} + \mu_3\frac{\partial c}{\partial \alpha_3} + \mu_4\frac{\partial c}{\partial \alpha_4} = 0,$$

$$\mu_1\frac{\partial m}{\partial \alpha_1} + \mu_2\frac{\partial m}{\partial \alpha_2} + \mu_3\frac{\partial m}{\partial \alpha_3} + \mu_4\frac{\partial m}{\partial \alpha_4} = 0.$$

The determinant of the coefficients of the constants $\mu_1$, $\mu_2$, $\mu_3$, $\mu_4$ is

$$J\left(\frac{b,\ l,\ c,\ m}{\alpha_1,\ \alpha_2,\ \alpha_3,\ \alpha_4}\right),$$

which does not vanish; and so these equations can only be satisfied if

$$\mu_1 = 0, \quad \mu_2 = 0, \quad \mu_3 = 0, \quad \mu_4 = 0.$$

Consequently, there is no linear relation between the integrals $\theta_1(t)$, $\theta_2(t)$, $\theta_3(t)$, $\theta_4(t)$ with constant coefficients, and there is no linear relation between the integrals $\vartheta_1(t)$, $\vartheta_2(t)$, $\vartheta_3(t)$, $\vartheta_4(t)$ with constant coefficients.

Hence the primitive* of the subsidiary characteristic equations is

$$Y = a_1\theta_1(t) + a_2\theta_2(t) + a_3\theta_3(t) + a_4\theta_4(t) \Big\}$$
$$Z = a_1\vartheta_1(t) + a_2\vartheta_2(t) + a_3\vartheta_3(t) + a_4\vartheta_4(t) \Big\},$$

where $a_1$, $a_2$, $a_3$, $a_4$ are independent arbitrary constants.

Clearly $\theta_1(t)$ and $\vartheta_1(t)$; $\theta_2(t)$ and $\vartheta_2(t)$; $\theta_3(t)$ and $\vartheta_3(t)$; $\theta_4(t)$ and $\vartheta_4(t)$; are four sets of integrals, each linearly independent of the others.

*Note.* In connection with the deviation from the characteristic curve for integrals involving only derivatives of the first order, it was pointed out (*Note*, § 52) that an apparently arbitrary representation of the variation was taken by including only a quantity

$$w = x_1 v - y_1 u$$

(which measures the normal deviation), and ignoring or omitting a quantity

$$x_1 u + y_1 v$$

(which, measuring the tangential slip, does not alter the shape of the curve). And it was seen that, in the reduction of the second variation, this omitted quantity cannot enter when $w$ is retained—that, in effect, $w$ gives the whole of the small variation which need be retained.

The same considerations occur in the case of skew curves. There is an apparently arbitrary representation of the full expression of the variation, by retaining only the quantities

$$Y = x_1\eta - y_1\xi, \quad Z = x_1\zeta - z_1\xi,$$

and by ignoring any tangential slip along the skew curve. The justification of the result emerges in the same way as the justification for the simpler instance. Solely for the passing purpose, we take the arc of the characteristic skew curve for the independent variable; and we resolve an arbitrary small variation $\kappa\xi$, $\kappa\eta$, $\kappa\zeta$ into three reciprocally perpendicular components, giving a displacement $\kappa\tau$ along the tangent, a displacement $\kappa\dfrac{1}{\rho}\nu$ along the principal normal, and a displacement $\kappa\dfrac{1}{\rho}\beta$ along the binormal. Then we have (with the customary notation for the direction-cosines of the principal lines of the curve at the point)

$$\kappa\xi = x'\kappa\tau + \rho x''\kappa\frac{1}{\rho}\nu + \rho\,(y'z'' - z'y'')\,\kappa\frac{1}{\rho}\beta,$$

---

* The primitive of the subsidiary characteristic equations has been deduced from the primitive (supposed known) of the original characteristic equations. If, however, only a special set of integrals of the latter were known, the subsidiary equations would have to be integrated, leading to the same primitive.

that is,

$$\xi = x'\tau + x''\nu + (y'z'' - z'y'')\,\beta\,;$$

and, similarly,

$$\eta = y'\tau + y''\nu + (z'x'' - x'z'')\,\beta,$$
$$\zeta = z'\tau + z''\nu + (x'y'' - y'x'')\,\beta.$$

Hence

$$Y = x'\eta - y'\xi = (x'y'' - y'x'')\,\nu - \beta z'',$$
$$Z = x'\zeta - z'\xi = (x'z'' - z'x'')\,\nu + \beta y''.$$

The second variation has been proved to be expressible in terms of $Y$ and $Z$ alone—that is, of $\nu$ and $\beta$ alone (were these selected, to the ignoration of $\tau$)—by means of expressions, into which $\tau$ could not then enter. The apparent ignoration of the displacement, due to a tangential slip, is justified alike in analysis and significance.

*Property of fundamental sets of integrals of the subsidiary characteristic equations.*

**171.** The subsidiary characteristic equations are

$$PY'' + RZ'' = \qquad\quad - P'\,Y' + (M - L - R')\,Z' + (D - K')\,Y + (S - L')\,Z,$$
$$RY'' + QZ'' = (L - M - R')\,Y' - \qquad\quad Q'Z' + (S - M')\,Y + (E - N')\,Z\,;$$

and they are satisfied by the four linearly independent sets of integrals $\theta_r(t)$ and $\vartheta_r(t)$, for $r = 1, 2, 3, 4$. We have

$$(PQ - R^2)\,Y'' = - \{QP' + R(L - M - R')\}\,Y' + \{Q(M - L - R') + RQ'\}\,Z'$$
$$+ \text{terms in } Y \text{ and } Z,$$
$$(PQ - R^2)\,Z'' = \{P(L - M - R') + RP'\}\,Y' - \{R(M - L - R') + PQ'\}\,Z'$$
$$+ \text{terms in } Y \text{ and } Z.$$

Denoting

$$\begin{vmatrix} \theta_1(t), & \theta_2(t), & \theta_3(t), & \theta_4(t) \\ \vartheta_1(t), & \vartheta_2(t), & \vartheta_3(t), & \vartheta_4(t) \\ \theta_1'(t), & \theta_2'(t), & \theta_3'(t), & \theta_4'(t) \\ \vartheta_1'(t), & \vartheta_2'(t), & \vartheta_3'(t), & \vartheta_4'(t) \end{vmatrix}$$

by $J$, we have

$$\frac{dJ}{dt} = \begin{vmatrix} \theta_1(t), & \theta_2(t), & \theta_3(t), & \theta_4(t) \\ \vartheta_1(t), & \vartheta_2(t), & \vartheta_3(t), & \vartheta_4(t) \\ \theta_1''(t), & \theta_2''(t), & \theta_3''(t), & \theta_4''(t) \\ \vartheta_1'(t), & \vartheta_2'(t), & \vartheta_3'(t), & \vartheta_4'(t) \end{vmatrix} + \begin{vmatrix} \theta_1(t), & \theta_2(t), & \theta_3(t), & \theta_4(t) \\ \vartheta_1(t), & \vartheta_2(t), & \vartheta_3(t), & \vartheta_4(t) \\ \theta_1'(t), & \theta_2'(t), & \theta_3'(t), & \theta_4'(t) \\ \vartheta_1''(t), & \vartheta_2''(t), & \vartheta_3''(t), & \vartheta_4''(t) \end{vmatrix}.$$

When we substitute for $\theta_r''(t)$ (that is, for $Y''$) its above value in the first determinant on the right-hand side, and for $\vartheta_r''(t)$ (that is, for $Z''$) its above value in the second determinant there, we have

$$(PQ - R^2)\,\frac{dJ}{dt} = - \{QP' + R(L - M - R')\}\,J - \{R(M - L - R') + PQ'\}\,J$$
$$= - (QP' + PQ' - 2RR')\,J.$$

Hence

$$(PQ - R^2)\, J = A',$$

where $A'$ is a constant.

The sets of integrals, $\theta_r(t)$ and $\Im_r(t)$, for $r = 1, 2, 3, 4$, are linearly independent, so that no relations of the form

$$\Sigma\mu_r\theta_r(t) = 0, \quad \Sigma\mu_r\Im_r(t) = 0,$$

where $\mu_1$, $\mu_2$, $\mu_3$, $\mu_4$ are constants, can exist; hence $J$ is not zero. Next, it will appear that, if a maximum or a minimum is to exist for our integral, the quantity $PQ - R^2$ must always be positive and may never vanish within the range. Hence $A'$, the product of two non-vanishing factors, cannot be zero. Thus

$$(PQ - R^2)\, J = A',$$

where $A'$ is a non-vanishing constant.

Clearly $A'$ will be a specific constant, depending upon the specific selection of the four linearly independent sets of integrals. Any four linearly independent sets of integrals are linearly expressible in terms of any other four linearly independent sets; thus

$$Y_m = \sum_{r=1}^{4} c_{mr}\,\theta_r(t), \quad Z_m = \sum_{r=1}^{4} c_{mr}\,\Im_r(t),$$

for $m = 1, 2, 3, 4$, where the determinant $\nabla$ of the coefficients $c_{mr}$ is not zero. The determinant $\bar{J}$ of the quantities $Y_r$ and $Z_r$ is equal to the foregoing determinant $J$ multiplied by the non-vanishing constant factor $\nabla$: that is, $\bar{J} = J\nabla$. Thus the determinant $J$ is covariantive among the four linearly independent sets of integrals.

*Corollary.* Moreover, as the quantity $J$ is not zero, the following inferences hold:

The four quantities $\theta_1(t)$, $\theta_2(t)$, $\theta_3(t)$, $\theta_4(t)$ do not have a common root in the range of the variable $t$, nor do their four derivatives; and, similarly, the four quantities $\Im_1(t)$, $\Im_2(t)$, $\Im_3(t)$, $\Im_4(t)$ do not have a common root in that range, nor do their four derivatives.

*Discussion of the second variation : the Legendre test.*

**172.** The second variation of the integral $I$ has been expressed in the form $\frac{1}{2}\kappa^2 \int \mathbf{U}\, dt$, taken between fixed limits; and, here,

$$\mathbf{U} = P\,(V' + \alpha V + \beta W)^2 + Q\,(W' + \gamma V + \delta W)^2$$
$$+ 2R\,(V' + \alpha V + \beta W)\,(W' + \gamma V + \delta W),$$

with the values of $\alpha$, $\beta$, $\gamma$, $\delta$ which have been determined. As the limits are fixed, $u, v, w$ (and therefore $V$ and $W$) vanish at each limit: subject to this property, the quantities $V$ and $W$ are arbitrary regular functions of $t$.

If the original integral $I$ is to provide a maximum, then in general we must have $\int \mathbf{U}\, dt$ negative, for all variations; and, owing to the arbitrary character of $V$ and $W$, this requirement can be met only if $\mathbf{U}$ is negative for all variations. If that integral is to provide a minimum, a similar argument leads to the conclusion that $\mathbf{U}$ must be positive for all variations. If it could happen that $\int \mathbf{U}\, dt$ is zero, without $V$ and $W$ being everywhere zero, the second variation of $I$ would vanish for such variations; and then the change of $I$ would be governed by the third variation.

We shall set aside this possibility of a conditionally vanishing second variation, as being more restricted than is desirable in the quite general discussion; if and when instances occur, they can be discussed specially, the course of discussion having been illustrated in an earlier example (p. 33). Accordingly, for our purpose, we now require that $\mathbf{U}$ shall be steadily positive (for a minimum) or steadily negative (for a maximum), so that it shall never vanish except for zero variations $u$, $v$, $w$, and therefore for zero variations $V$ and $W$.

Now so long as $V' + \alpha V + \beta W$ and $W' + \gamma V + \delta W$ do not vanish simultaneously, the conditions, which are necessary and sufficient to secure that $\mathbf{U}$ shall always have one sign, are:

(i)  the quantities $P$ and $Q$ must each have one, and only one, sign throughout the range;

(ii)  the persistent sign of $P$ and the persistent sign of $Q$ must be the same;

(iii)  the quantity $PQ - R^2$ must be positive throughout the range.

The third condition implies the second, but the second condition does not carry the third. If $PQ - R^2$ could be negative, or $P$ alone could change sign, or $Q$ alone could change sign, $\mathbf{U}$ would not have always the same sign for all variations. If $PQ - R^2$ could be zero, while the first two conditions were satisfied, variations represented by

$$P(V' + \alpha V + \beta W) + Q(W' + \gamma V + \delta W) = 0$$

would make $\mathbf{U}$ zero, a possibility to be excluded.

Manifestly, if the persistent sign of $P$ and of $Q$ be positive, a minimum is admissible; if it be negative, a maximum is admissible. Other conditions may arise from other causes: these conditions must be maintained, as necessary conditions.

These conditions are called the *Legendre* test, in extension of the fact that the corresponding condition for the simplest problem was first stated by Legendre.

*Conjugates on a characteristic curve: the Jacobi test.*

**173.** Next, we must consider the possibility that the quantities $V' + \alpha V + \beta W$ and $W' + \gamma V + \delta W$, viz.

$$\begin{vmatrix} V', & V, & W \\ Y_1', & Y_1, & Z_1 \\ Y_2', & Y_2, & Z_2 \end{vmatrix}, \qquad \begin{vmatrix} W', & V, & W \\ Z_1', & Y_1, & Z_1 \\ Z_2', & Y_2, & Z_2 \end{vmatrix},$$

may vanish simultaneously for a pair of sets of integrals of the subsidiary characteristic, such that $Y_1 Z_2 - Y_2 Z_1$ is not zero, and such that

$$Y_1 \frac{\partial \square_2}{\partial Y_2'} + Z_1 \frac{\partial \square_2}{\partial Z_2'} = Y_2 \frac{\partial \square_1}{\partial Y_1'} + Z_2 \frac{\partial \square_1}{\partial Z_1'}.$$

The only way in which this can happen is that, for any pair of sets of integrals $Y_1$ and $Z_1$, $Y_2$ and $Z_2$, which satisfy the conditions as stated, it may be possible to have variations

$$\begin{aligned} V &= \lambda Y_1 + \mu Y_2 \\ W &= \lambda Z_1 + \mu Z_2 \end{aligned} \bigg\},$$

where $\lambda$ and $\mu$ are non-zero constants. All variations $V$ and $W$ must vanish at each limit. All the requisite possibilities, that make failure admissible, will be attained, if we can have sets of integrals $\lambda Y_1 + \mu Y_2$, $\lambda Z_1 + \mu Z_2$, zero together at the lower limit and zero together at the upper limit, while $\lambda Y_1 + \mu Y_2$ and $\lambda Z_1 + \mu Z_2$ are not otherwise zero together (*a fortiori*, if they are zero at only isolated places) in the range of integration. In order to secure that **U** shall not become zero, through non-zero variations of such a type, we must secure the exclusion of the type. It therefore must be impossible to have a non-zero combination of any two conditioned pairs of sets of integrals, $\lambda Y_1 + \mu Y_2$ and $\lambda Z_1 + \mu Z_2$ being the combination, which vanishes at the lower limit, and also at the upper limit, of the range of integration. We may state the condition in a different form, by requiring that the range, beginning at a place where any combination $\lambda Y_1 + \mu Y_2$ and $\lambda Z_1 + \mu Z_2$ is made to vanish, shall not extend so far as the nearest place where some such combination again vanishes.

Such nearest place will, as before, be called the *conjugate* of the initial place. The required condition is called the *Jacobi test*, in extension of the fact that the corresponding condition for the simplest problem was first stated by Jacobi.

*Critical equation determining conjugates.*

**174.** The analytical expression of this test requires the fuller consideration of the characteristic curve. As the argument follows almost exactly the earlier argument in the case of the special variation (§ 152), it can be explained rather briefly.

We take, adjacent to the characteristic curve, another consecutive curve represented by a set of integrals $Y_1$ and $Z_1$ of the subsidiary equations, the deviation being represented by its projections $\kappa Y_1$ and $\kappa Z_1$ upon the axes of $y$ and of $z$ respectively. The conditions, assigned for the unique determination of this consecutive curve, are that, at $t = t_0$,

$$Y_1 = 0, \quad Z_1 = 0, \quad Y_1' = \rho J(t_0), \quad Z_1' = 0,$$

where $\rho$ is a non-zero constant. Then constants $b_1$, $b_2$, $b_3$, $b_4$, such that

$$\sum_{r=1}^{4} b_r \theta_r(t_0) = 0, \quad \Sigma b_r \vartheta_r(t_0) = 0, \quad \Sigma b_r \theta_r'(t_0) = \rho J(t_0), \quad \Sigma b_r \vartheta_r'(t_0) = 0,$$

give

$$Y_1 = b_1 \theta_1(t) + b_2 \theta_2(t) + b_3 \theta_3(t) + b_4 \theta_4(t) = \rho \begin{vmatrix} \theta_1(t) & \theta_2(t) & \theta_3(t) & \theta_4(t) \\ \theta_1(t_0) & \theta_2(t_0) & \theta_3(t_0) & \theta_4(t_0) \\ \vartheta_1(t_0) & \vartheta_2(t_0) & \vartheta_3(t_0) & \vartheta_4(t_0) \\ \vartheta_1'(t_0) & \vartheta_2'(t_0) & \vartheta_3'(t_0) & \vartheta_4'(t_0) \end{vmatrix},$$

$$Z_1 = b_1 \vartheta_1(t) + b_2 \vartheta_2(t) + b_3 \vartheta_3(t) + b_4 \vartheta_4(t) = \rho \begin{vmatrix} \vartheta_1(t) & \vartheta_2(t) & \vartheta_3(t) & \vartheta_4(t) \\ \theta_1(t_0) & \theta_2(t_0) & \theta_3(t_0) & \theta_4(t_0) \\ \vartheta_1(t_0) & \vartheta_2(t_0) & \vartheta_3(t_0) & \vartheta_4(t_0) \\ \vartheta_1'(t_0) & \vartheta_2'(t_0) & \vartheta_3'(t_0) & \vartheta_4'(t_0) \end{vmatrix}.$$

We take a second consecutive curve represented by a set of integrals $Y_2$ and $Z_2$ of the subsidiary equations, with $\kappa Y_2$ and $\kappa Z_2$ as the projections, upon the axes of $y$ and of $z$, of the corresponding deviation. The conditions, assigned for the unique determination of this curve, are that, at $t = t_0$,

$$Y_2 = 0, \quad Z_2 = 0, \quad Y_2' = 0, \quad Z_2' = -\sigma J(t_0),$$

where $\sigma$ is a non-zero constant. Then constants $c_1$, $c_2$, $c_3$, $c_4$, such that

$$\sum_{r=1}^{4} c_r \theta_r(t_0) = 0, \quad \Sigma c_r \vartheta_r(t_0) = 0, \quad \Sigma c_r \theta_r'(t_0) = 0, \quad \Sigma c_r \vartheta_r'(t_0) = -\sigma J(t_0),$$

give

$$Y_2 = c_1 \theta_1(t) + c_2 \theta_2(t) + c_3 \theta_3(t) + c_4 \theta_4(t) = \sigma \begin{vmatrix} \theta_1(t) & \theta_2(t) & \theta_3(t) & \theta_4(t) \\ \theta_1(t_0) & \theta_2(t_0) & \theta_3(t_0) & \theta_4(t_0) \\ \vartheta_1(t_0) & \vartheta_2(t_0) & \vartheta_3(t_0) & \vartheta_4(t_0) \\ \theta_1'(t_0) & \theta_2'(t_0) & \theta_3'(t_0) & \theta_4'(t_0) \end{vmatrix},$$

$$Z_2 = c_1 \vartheta_1(t) + c_2 \vartheta_2(t) + c_3 \vartheta_3(t) + c_4 \vartheta_4(t) = \sigma \begin{vmatrix} \vartheta_1(t) & \vartheta_2(t) & \vartheta_3(t) & \vartheta_4(t) \\ \theta_1(t_0) & \theta_2(t_0) & \theta_3(t_0) & \theta_4(t_0) \\ \vartheta_1(t_0) & \vartheta_2(t_0) & \vartheta_3(t_0) & \vartheta_4(t_0) \\ \theta_1'(t_0) & \theta_2'(t_0) & \theta_3'(t_0) & \theta_4'(t_0) \end{vmatrix}.$$

In order that these two sets of integrals $Y_1$ and $Z_1$, $Y_2$ and $Z_2$, may be suited for our purpose, they must satisfy two requirements.

In the first place, they must make the quantity

$$Y_1 \frac{\partial \square_2}{\partial Y_2'} + Z_1 \frac{\partial \square_2}{\partial Z_2'} - \left( Y_2 \frac{\partial \square_1}{\partial Y_1'} + Z_2 \frac{\partial \square_1}{\partial Z_1'} \right)$$

zero. As $Y_1$ and $Z_1$, $Y_2$ and $Z_2$, are two sets of integrals, this quantity is bound to be constant; and its constant value can be taken as its value anywhere. When $t = t_0$, we have $Y_1$, $Z_1$, $Y_2$, $Z_2$ each equal to zero: so the constant is zero. The first requirement is satisfied.

In the second place, the two sets of integrals are to be independent, in the sense that the magnitude

$$Y_1 Z_2 - Y_2 Z_1$$

must not be zero in general. It vanishes when $t = t_0$; but $V$ and $W$ vanish then also, so that (as in § 64) no infinity occurs in the expression for $\mathbf{U}$ at $t = t_0$. But after the initial value of $t$, the quantities $V$ and $W$ must not vanish simultaneously until the conjugate of $t_0$ in the range is attained. Moreover, the quantity $(Y_1 Z_2 - Y_2 Z_1)^{-2}$ occurs in the expression for $\mathbf{U}$, so that $Y_1 Z_2 - Y_2 Z_1$ ought not to vanish in the range after the initial point. Now

$$Y_1 Z_2 - Y_2 Z_1 = \{ \Sigma b_r \theta_r(t) \} \{ \Sigma c_r \mathfrak{I}_r(t) \} - \{ \Sigma c_r \theta_r(t) \} \{ \Sigma b_r \mathfrak{I}_r(t) \}$$
$$= \Sigma (b_1 c_2 - b_2 c_1) \{ \theta_1(t) \mathfrak{I}_2(t) - \mathfrak{I}_1(t) \theta_2(t) \}.$$

Writing

$$\left\| \begin{matrix} \theta_1(t), & \theta_2(t), & \theta_3(t), & \theta_4(t) \\ \mathfrak{I}_1(t), & \mathfrak{I}_2(t), & \mathfrak{I}_3(t), & \mathfrak{I}_4(t) \end{matrix} \right\| = k_{12}(t), \ k_{13}(t), \ k_{14}(t), \ k_{23}(t), \ k_{24}(t), \ k_{34}(t),$$

$$\left\| \begin{matrix} \theta_1'(t), & \theta_2'(t), & \theta_3'(t), & \theta_4'(t) \\ \mathfrak{I}_1'(t), & \mathfrak{I}_2'(t), & \mathfrak{I}_3'(t), & \mathfrak{I}_4'(t) \end{matrix} \right\| = l_{12}(t), \ l_{13}(t), \ l_{14}(t), \ l_{23}(t), \ l_{24}(t), \ l_{34}(t),$$

$$\left| \begin{matrix} \theta_1(t), & \theta_2(t), & \theta_3(t), & \theta_4(t) \\ \mathfrak{I}_1(t), & \mathfrak{I}_2(t), & \mathfrak{I}_3(t), & \mathfrak{I}_4(t) \\ \theta_1(t_0), & \theta_2(t_0), & \theta_3(t_0), & \theta_4(t_0) \\ \mathfrak{I}_1(t_0), & \mathfrak{I}_2(t_0), & \mathfrak{I}_3(t_0), & \mathfrak{I}_4(t_0) \end{matrix} \right| = \Delta(t_0, t),$$

we have

$$b_1 = \rho \{ k_{34}(t_0) \mathfrak{I}_2'(t_0) + k_{42}(t_0) \mathfrak{I}_3'(t_0) + k_{23}(t_0) \mathfrak{I}_4'(t_0) \},$$
$$- b_2 = \rho \{ k_{34}(t_0) \mathfrak{I}_1'(t_0) + k_{41}(t_0) \mathfrak{I}_3'(t_0) + k_{13}(t_0) \mathfrak{I}_4'(t_0) \},$$
$$c_1 = \sigma \{ k_{34}(t_0) \theta_2'(t_0) + k_{42}(t_0) \theta_3'(t_0) + k_{23}(t_0) \theta_4'(t_0) \},$$
$$- c_2 = \sigma \{ k_{34}(t_0) \theta_1'(t_0) + k_{41}(t_0) \theta_3'(t_0) + k_{13}(t_0) \theta_4'(t_0) \};$$

and similarly for the other constants $b$ and $c$. Then, as before (§ 152),

$$- \frac{1}{\rho\sigma} (b_1 c_2 - b_2 c_1) = \{ k_{23}(t_0) k_{41}(t_0) - k_{42}(t_0) k_{13}(t_0) \} l_{34}(t_0)$$
$$+ k_{34}(t_0) \{ k_{34}(t_0) l_{12}(t_0) + k_{31}(t_0) l_{24}(t_0) + k_{14}(t_0) l_{23}(t_0) + k_{24}(t_0) l_{31}(t_0) + k_{23}(t_0) l_{14}(t_0) \}$$
$$= k_{34}(t_0) J(t_0);$$

and similarly for the other combinations of the coefficients $b_l c_m - b_m c_l$. Thus

$$- \frac{Y_1 Z_2 - Y_2 Z_1}{\rho \sigma J (t_0)}$$

$$= k_{34} (t_0) k_{12} (t) + k_{23} (t_0) k_{14} (t) + k_{42} (t_0) k_{13} (t) + k_{12} (t_0) k_{34} (t) + k_{13} (t_0) k_{42} (t) + k_{14} (t_0) k_{23} (t)$$
$$= \Delta (t_0, t) ;$$

and therefore

$$Y_1 Z_2 - Y_2 Z_1 = - \rho \sigma J (t_0) \Delta (t_0, t).$$

Here $\rho$ and $\sigma$ are non-zero constants; $J (t_0)$ does not vanish; and $\Delta (t_0, t)$ does not vanish identically. Hence $Y_1 Z_2 - Y_2 Z_1$, though it vanishes when $t = t_0$, does not vanish identically along the range; and thus the second requirement for the two sets of integrals is satisfied.

Now consider a variation $V$ and $W$, given by

$$V = \lambda Y_1 + \mu Y_2, \quad W = \lambda Z_1 + \mu Z_2,$$

where $\lambda$ and $\mu$ are non-zero constants. At $t = t_0$, we have $V = 0$ and $W = 0$, so that the variation vanishes (as it is bound to vanish) at the lower limit. The conjugate of $t_0$ is the first place after the lower limit at which two integrals $\lambda Y_1 + \mu Y_2$ and $\lambda Z_1 + \mu Z_2$, vanishing at the lower limit, again vanish. Denoting this conjugate by $t_1$, we have

$$\lambda Y_1 (t_1) + \mu Y_2 (t_1) = 0, \quad \lambda Z_1 (t_1) + \mu Z_2 (t_1) = 0 ;$$

as $\lambda$ and $\mu$ are not zero, we must have

$$Y_1 (t_1) Z_2 (t_1) - Z_1 (t_1) Y_2 (t_1) = 0,$$

that is,

$$\Delta (t_0, t_1) = 0.$$

Thus *the conjugate of any point $t_0$ on the characteristic curve is given by the first root $t_1$, greater than $t_0$, of the equation*

$$\Delta (t_0, t_1) = 0 ;$$

*and the range of the integral, in order to satisfy the Jacobi test; must not extend as far as the conjugate of the lower limit.*

**175.** We take, in illustration, the question of a particular curve in space, viz.,

*Let it be required to find a maximum or minimum for the integral*

$$V = \int \mu \, ds,$$

*where $\mu$ is any continuous function in space, and the integral is taken from one surface—that is, from any unspecified point on that surface—to another non-intersecting surface—that is, to any unspecified point on that other surface.*

(The problem relates to the Principle of the minimum 'reduced path' of a ray in a heterogeneous medium, where $\mu$ is the index of refraction at any point in the medium : according to this Principle, the actual path of a ray between any two points of its course provides the minimum length of reduced path between them. Newton enunciated a theorem, according to which the path of the ray is that of a particle moving with a velocity

$\mu$ under a system of forces with a potential; the problem associates this theorem, so interpreted, with the Principle of least action. Further, for such a system of rays, the quantity $V$, when regarded as a function of position in space represented by the upper limit of the integral, is Hamilton's Characteristic Function*.)

(A)  The integral is

$$\int \mu \, (x_1{}^2 + y_1{}^2 + z_1{}^2)^{\frac{1}{2}} \, dt,$$

so that the function $g$ is $\mu \, (x_1{}^2 + y_1{}^2 + z_1{}^2)^{\frac{1}{2}}$, while $\mu$ is any function of $x$, $y$, $z$. When substitution is made in the characteristic equations, they become

$$\frac{\partial \mu}{\partial x} (x_1{}^2 + y_1{}^2 + z_1{}^2)^{\frac{1}{2}} - \frac{d}{dt} \left\{ \frac{\mu x_1}{(x_1{}^2 + y_1{}^2 + z_1{}^2)^{\frac{1}{2}}} \right\} = 0,$$

with two others. When we take the length of arc as the independent variable in order to simplify the equations already formed, the characteristic equations become

$$\frac{\partial \mu}{\partial x} - \frac{d}{ds} \left( \mu \frac{dx}{ds} \right) = 0, \quad \frac{\partial \mu}{\partial y} - \frac{d}{ds} \left( \mu \frac{dy}{ds} \right) = 0, \quad \frac{\partial \mu}{\partial z} - \frac{d}{ds} \left( \mu \frac{dz}{ds} \right) = 0 \ ;$$

or, in full,

$$\frac{\partial \mu}{\partial x} - \frac{d\mu}{ds} \frac{dx}{ds} - \mu \frac{d^2 x}{ds^2} = 0,$$

$$\frac{\partial \mu}{\partial y} - \frac{d\mu}{ds} \frac{dy}{ds} - \mu \frac{d^2 y}{ds^2} = 0,$$

$$\frac{\partial \mu}{\partial z} - \frac{d\mu}{ds} \frac{dz}{ds} - \mu \frac{d^2 z}{ds^2} = 0.$$

In the first place, the three equations are at once seen to be equivalent to two only, when coupled with the relation

$$\left( \frac{dx}{ds} \right)^2 + \left( \frac{dy}{ds} \right)^2 + \left( \frac{dz}{ds} \right)^2 = 1 \ ;$$

for, on multiplying them by $\dfrac{dx}{ds}, \dfrac{dy}{ds}, \dfrac{dz}{ds}$ respectively, and adding, we have the identity

$$\frac{\partial \mu}{\partial x} \frac{dx}{ds} + \frac{\partial \mu}{\partial y} \frac{dy}{ds} + \frac{\partial \mu}{\partial z} \frac{dz}{ds} - \frac{d\mu}{ds} = 0.$$

Next, eliminating $\dfrac{d\mu}{ds}$ and $\mu$ determinantally, we have

$$\begin{vmatrix} \dfrac{\partial \mu}{\partial x}, & \dfrac{dx}{ds}, & \dfrac{d^2 x}{ds^2} \\[2mm] \dfrac{\partial \mu}{\partial y}, & \dfrac{dy}{ds}, & \dfrac{d^2 y}{ds^2} \\[2mm] \dfrac{\partial \mu}{\partial z}, & \dfrac{dz}{ds}, & \dfrac{d^2 z}{ds^2} \end{vmatrix} = 0 :$$

that is, the three lines whose direction-cosines are proportional to

$$\frac{\partial \mu}{\partial x}, \ \frac{\partial \mu}{\partial y}, \ \frac{\partial \mu}{\partial z} \ ; \ \frac{dx}{ds}, \ \frac{dy}{ds}, \ \frac{dz}{ds} \ ; \ \frac{d^2 x}{ds^2}, \ \frac{d^2 y}{ds^2}, \ \frac{d^2 z}{ds^2} \ ;$$

lie in one plane. Hence the osculating plane of the curve at any point contains the normal to the level surface $\mu =$ constant at that point : or the direction of the path just outside that surface and its direction just inside are in one plane with the normal to the surface.

* For the discussion of these matters, reference may be made to R. A. Herman's *Treatise on Geometrical Optics*, ch. x, ch. xi.

Again, we have

$$N^2 = \left(\frac{\partial \mu}{\partial x}\right)^2 + \left(\frac{\partial \mu}{\partial y}\right)^2 + \left(\frac{\partial \mu}{\partial z}\right)^2 = \left(\frac{d\mu}{ds}\right)^2 + \frac{\mu^2}{\rho^2},$$

where $\rho$ is the radius of circular curvature of the curve at the point. If $\psi$ be the angle between the direction of the curve and that of the outward-drawn normal to the surface,

$$\cos \psi = \frac{1}{N}\frac{\partial \mu}{\partial x}\frac{dx}{ds} + \frac{1}{N}\frac{\partial \mu}{\partial y}\frac{dy}{ds} + \frac{1}{N}\frac{\partial \mu}{\partial z}\frac{dz}{ds}$$

$$= \frac{1}{N}\frac{d\mu}{ds};$$

or, taking account of the value of $N$,

$$\frac{\mu}{\rho}\cos \psi = \pm \frac{d\mu}{ds}\sin \psi.$$

We have $\frac{\partial \mu}{\partial x}$, $\frac{\partial \mu}{\partial y}$, $\frac{\partial \mu}{\partial z}$ as the direction-cosines of the outward-drawn normal—in the direction, that is, of increasing $x$, $y$, $z$; if, then, $RPN = \psi$ and $P'PN' = \psi + d\psi$, the angle of contingence $d\epsilon$ at $P$ is $-d\psi$. But

$$\rho = \frac{ds}{d\epsilon};$$

in the figure $\rho$ is positive, and $d\mu$ is positive: thus

$$\frac{\mu}{\rho}\cos \psi = \frac{d\mu}{ds}\sin \psi,$$

that is,

$$\mu \cos \psi \cdot d\psi = -d\mu \sin \psi,$$

and therefore

$$\mu \sin \psi = constant,$$

a first integral of the characteristic equations.

These two results are the expression of the customary laws of refraction, originally due to Snell.

(B) Specific knowledge of the form of $\mu$, expressing its value in terms of its position in space, is required before a second integral of the characteristic equations can be obtained. Moreover, in the absence of this specific knowledge, the construction of the subsidiary characteristic equations remains merely formal analysis; and the detailed form of Jacobi's range-test cannot be obtained precisely.

As regards the Legendre test, we have

$$\frac{\partial^2 g}{\partial y_1^2} = \mu\frac{x_1^2 + z_1^2}{(x_1^2 + y_1^2 + z_1^2)^{\frac{3}{2}}} = x_1^2 P,$$

$$\frac{\partial^2 g}{\partial y_1 \partial z_1} = -\mu\frac{y_1 z_1}{(x_1^2 + y_1^2 + z_1^2)^{\frac{3}{2}}} = x_1^2 R,$$

$$\frac{\partial^2 g}{\partial z_1^2} = \mu\frac{x_1^2 + y_1^2}{(x_1^2 + y_1^2 + z_1^2)^{\frac{3}{2}}} = x_1^2 Q.$$

Hence $P$ is always positive, $Q$ is always positive, and $PQ - R^2$ is always positive. Consequently, the obtained path provides a minimum value for the integral $V$, within a range that must satisfy the Jacobi test.

(C) To take a specific example, suppose that

$$\mu^2 = \mu_0^2 - (x^2 + y^2 + z^2),$$

where $\mu_0$ is a constant. (It is easy to see that no generality is provided by taking $\mu_0^2 - A^2(x^2 + y^2 + z^2)$, where $A$ is a constant, for the value of $\mu^2$.) The characteristic equations are

$$-x + \frac{dx}{ds} r \frac{dr}{ds} - (\mu_0^2 - r^2) \frac{d^2 x}{ds^2} = 0,$$

(on multiplying throughout by $\mu$, and writing $r^2$ for $x^2 + y^2 + z^2$), with two others. Multiplying this equation by $2\frac{dx}{ds}$ and integrating, we have

$$(r^2 - \mu_0^2)\left(\frac{dx}{ds}\right)^2 = x^2 - a^2 ;$$

and, similarly,

$$(r^2 - \mu_0^2)\left(\frac{dy}{ds}\right)^2 = y^2 - b^2,$$

$$(r^2 - \mu_0^2)\left(\frac{dz}{ds}\right)^2 = z^2 - c^2,$$

from the other two equations, with the relation

$$a^2 + b^2 + c^2 = \mu_0^2$$

among the constants, when account is taken of the persistent relation

$$\left(\frac{dx}{ds}\right)^2 + \left(\frac{dy}{ds}\right)^2 + \left(\frac{dz}{ds}\right)^2 = 1.$$

These three equations manifestly are satisfied by

$$x = a \sin t = \chi(t), \quad y = b \sin(t + \beta) = \phi(t), \quad z = c \sin(t + \gamma) = \psi(t) :$$

no advantage arises from taking $x = a \sin(t + \gamma)$, because the explicit absence of $t$ from the equations shews than an unessential constant (such as $\gamma$) can occur as additive to $t$.

· The quantities arising as integrals of the subsidiary equations are

$$\left.\begin{array}{l} \theta_1(t) = \chi'(t)\phi_1(t) - \phi'(t)\chi_1(t) = -b\sin t\cos(t+\beta) \\ \theta_2(t) = \chi'(t)\phi_2(t) - \phi'(t)\chi_2(t) = a\cos t\sin(t+\beta) \\ \theta_3(t) = \chi'(t)\phi_3(t) - \phi'(t)\chi_3(t) = ab\cos t\cos(t+\beta) \\ \theta_4(t) = \chi'(t)\phi_4(t) - \phi'(t)\chi_4(t) = 0 \end{array}\right\},$$

where $\chi_1(t)$, $\chi_2(t)$, $\chi_3(t)$, $\chi_4(t)$ denote $\frac{\partial\chi}{\partial a}$, $\frac{\partial\chi}{\partial b}$, $\frac{\partial\chi}{\partial\beta}$, $\frac{\partial\chi}{\partial\gamma}$, and likewise for $\phi$ and $\psi$: and

$$\left.\begin{array}{l} \vartheta_1(t) = \chi'(t)\psi_1(t) - \psi'(t)\chi_1(t) = a\dfrac{\partial c}{\partial a}\cos t\sin(t+\gamma) - c\sin t\cos(t+\gamma) \\ \qquad\qquad = -\dfrac{a^2}{c}\cos t\sin(t+\gamma) - c\sin t\cos(t+\gamma) \\ \vartheta_2(t) = \chi'(t)\psi_2(t) - \psi'(t)\chi_2(t) = a\dfrac{\partial c}{\partial b}\cos t\sin(t+\gamma) \\ \qquad\qquad = -\dfrac{ab}{c}\cos t\sin(t+\gamma) \\ \vartheta_3(t) = \chi'(t)\psi_3(t) - \psi'(t)\chi_3(t) = 0 \\ \vartheta_4(t) = \chi'(t)\psi_4(t) - \psi'(t)\chi_4(t) = ac\cos t\cos(t+\gamma) \end{array}\right\}.$$

It is clear that there are no simultaneous relations

$$A_1\theta_1(t) + A_2\theta_2(t) + A_3\theta_3(t) + A_4\theta_4(t) = 0 \atop A_1\vartheta_1(t) + A_2\vartheta_2(t) + A_3\vartheta_3(t) + A_4\vartheta_4(t) = 0 \Big\},$$

with constant coefficients $A_1$, $A_2$, $A_3$, $A_4$; and that therefore the combinations $\theta_1$ and $\vartheta_1$, $\theta_2$ and $\vartheta_2$, $\theta_3$ and $\vartheta_3$, $\theta_4$ and $\vartheta_4$, constitute four linearly independent integral-sets of the subsidiary characteristic equations.

The element of arc is given by

$$\left(\frac{ds}{dt}\right)^2 = a^2\cos^2 t + b^2\cos^2(t+\beta) + c^2\cos^2(t+\gamma),$$

so that $s$ is expressible by an elliptic integral; the explicit value of $s$ is irrelevant to the solution.

The actual path of the ray, as given by the equations

$$x = a\sin t, \quad y = b\sin(t+\beta), \quad z = c\sin(t+\gamma),$$

is also given by the equations

$$\frac{x^2}{a^2} - 2\frac{xy}{ab}\cos\beta + \frac{y^2}{b^2} = \sin^2\beta,$$

$$\frac{x^2}{a^2} - 2\frac{xz}{ac}\cos\gamma + \frac{z^2}{c^2} = \sin^2\gamma,$$

$$\frac{x}{a}\sin(\beta-\gamma) + \frac{y}{b}\sin\gamma - \frac{z}{c}\sin\beta = 0;$$

and therefore it is an ellipse, the section of either of the elliptic cylinders by the plane.

The function, which determines $t_1$ (the conjugate of the initial place $t_0$), is

$$-\Delta(t_0, t_1) = \begin{vmatrix} \theta_1(t_1), & \theta_2(t_1), & \theta_3(t_1), & 0 \\ \theta_1(t_0), & \theta_2(t_0), & \theta_3(t_0), & 0 \\ \vartheta_1(t_1), & \vartheta_2(t_1), & 0, & \vartheta_4(t_1) \\ \vartheta_1(t_0), & \vartheta_2(t_0), & 0, & \vartheta_4(t_0) \end{vmatrix}$$

$$= \{\theta_2(t_1)\theta_3(t_0) - \theta_2(t_0)\theta_3(t_1)\}\{\vartheta_1(t_1)\vartheta_4(t_0) - \vartheta_1(t_0)\vartheta_4(t_1)\}$$
$$- \{\theta_1(t_1)\theta_3(t_0) - \theta_1(t_0)\theta_3(t_1)\}\{\vartheta_2(t_1)\vartheta_4(t_0) - \vartheta_2(t_0)\vartheta_4(t_1)\}.$$

Now

$$\theta_2(t_1)\theta_3(t_0) - \theta_2(t_0)\theta_3(t_1) = a^2 b\cos t_1\cos t_0\sin(t_1-t_0),$$
$$\theta_1(t_1)\theta_3(t_0) - \theta_1(t_0)\theta_3(t_1) = -ab^2\cos(t_1+\beta)\cos(t_0+\beta)\sin(t_1-t_0),$$
$$\vartheta_1(t_1)\vartheta_4(t_0) - \vartheta_1(t_0)\vartheta_4(t_1) = -a\{a^2\cos t_1\cos t_0 + c^2\cos(t_1+\gamma)\cos(t_0+\gamma)\}\sin(t_1-t_0),$$
$$\vartheta_2(t_1)\vartheta_4(t_0) - \vartheta_2(t_0)\vartheta_4(t_1) = -a^2 b\cos t_1\cos t_0\sin(t_1-t_0);$$

and therefore

$$\Delta(t_0, t_1) = a^2 b\Delta\cos t_1\cos t_0\sin^2(t_1-t_0),$$

where

$$\Delta = a^2\cos t_1\cos t_0 + b^2\cos(t_1+\beta)\cos(t_0+\beta) + c^2\cos(t_1+\gamma)\cos(t_0+\gamma).$$

We are to have

$$\Delta(t_0, t_1) = 0.$$

The merely algebraical possibilities $t_0 = \frac{1}{2}\pi$ and $t_1 = \frac{1}{2}\pi$ are illusory, because $t$ is subject to an additive unessential constant: moreover, the conjugate $t_1$ (if there is a limit) must be related to $t_0$. Hence we must have

$$\Delta = 0;$$

and therefore, at conjugate points, the tangents to the elliptic path are perpendicular to one another.

(D) As regards the original integral

$$V = \int_{x_0, y_0, z_0}^{x, y, z} \mu \, ds,$$

we still have to consider the terms at the limits.

By our general result (§ 158), where the path is to end at a terminal surface, we have

$$\frac{\partial g}{\partial x_1} X + \frac{\partial g}{\partial y_1} Y + \frac{\partial g}{\partial z_1} Z = 0$$

as a condition to be satisfied at each extremity, that is,

$$\mu \frac{dx}{ds} X + \mu \frac{dy}{ds} Y + \mu \frac{dz}{ds} Z = 0 :$$

consequently, the curve is to cut each terminal surface orthogonally.

The foregoing integral $V$, when it is expressible solely and definitely in terms of the final position $x$, $y$, $z$ (and of the initial position $x_0$, $y_0$, $z_0$) without reference to the mode of passage from that initial position, is called the *Characteristic Function**. We then have, for variations at the upper limit,

$$dV = \mu \, ds,$$

that is,

$$\frac{\partial V}{\partial x} dx + \frac{\partial V}{\partial y} dy + \frac{\partial V}{\partial z} dz = \mu \left( \frac{dx}{ds} dx + \frac{dy}{ds} dy + \frac{dz}{ds} dz \right),$$

for all variations $dx$, $dy$, $dz$ at that upper limit, because $V$ is a function only of the variables of that limit (and of the lower limit, which does not affect the upper analytically). Hence

$$\frac{\partial V}{\partial x} = \mu \frac{dx}{ds}, \quad \frac{\partial V}{\partial y} = \mu \frac{dy}{ds}, \quad \frac{\partial V}{\partial z} = \mu \frac{dz}{ds} :$$

which, expressed in the vocabulary of geometrical optics, shews that the ray is normal to the surfaces given by the level of the characteristic function. The persistence of these level surfaces, normal to a system of rays, was first stated by Malus.

(E) Again, we have

$$\left( \frac{\partial V}{\partial x} \right)^2 + \left( \frac{\partial V}{\partial y} \right)^2 + \left( \frac{\partial V}{\partial z} \right)^2 = \mu^2,$$

from the foregoing equations. Hence, when $\mu$ is given as a function of position, the foregoing is a partial differential equation of the first order satisfied by the characteristic function $V$. Its general integral, obtainable by Jacobi's method for the integration of such equations in more than two independent variables, is of the form

$$V = f(x, y, z, a, b) + c,$$

where $a$, $b$, $c$ are arbitrary constants.

The equation is satisfied for all values of the arbitrary constants, and is therefore satisfied when they are subjected to a small variation so as to become $a + \kappa A$, $b + \kappa B$, $c + \kappa C$, where $A$, $B$, $C$ themselves are arbitrary and independent of one another. When we substitute in the differential equation, and (as usual) make $\kappa$ so small that all powers above the first are negligible in comparison with the first power, we have

$$\frac{\partial f}{\partial x} \left\{ A \frac{\partial}{\partial x} \left( \frac{\partial f}{\partial a} \right) + B \frac{\partial}{\partial x} \left( \frac{\partial f}{\partial b} \right) \right\} + \frac{\partial f}{\partial y} \left\{ A \frac{\partial}{\partial y} \left( \frac{\partial f}{\partial a} \right) + B \frac{\partial}{\partial y} \left( \frac{\partial f}{\partial b} \right) \right\} + \frac{\partial f}{\partial z} \left\{ A \frac{\partial}{\partial z} \left( \frac{\partial f}{\partial a} \right) + B \frac{\partial}{\partial z} \left( \frac{\partial f}{\partial b} \right) \right\} = 0,$$

* The title is due to Hamilton: the word 'Characteristic' has no reference to that word, as used in relation to the critical equations and the critical curve for maxima or minima in general.

or, as $A$ and $B$ are independent of one another, and are arbitrary,

$$\frac{\partial f}{\partial x}\frac{\partial}{\partial x}\left(\frac{\partial f}{\partial a}\right) + \frac{\partial f}{\partial y}\frac{\partial}{\partial y}\left(\frac{\partial f}{\partial a}\right) + \frac{\partial f}{\partial z}\frac{\partial}{\partial z}\left(\frac{\partial f}{\partial a}\right) = 0,$$

$$\frac{\partial f}{\partial x}\frac{\partial}{\partial x}\left(\frac{\partial f}{\partial b}\right) + \frac{\partial f}{\partial y}\frac{\partial}{\partial y}\left(\frac{\partial f}{\partial b}\right) + \frac{\partial f}{\partial z}\frac{\partial}{\partial z}\left(\frac{\partial f}{\partial b}\right) = 0.$$

But, along the path in question,

$$\frac{\partial f}{\partial x} = \mu\frac{dx}{ds}, \quad \frac{\partial f}{\partial y} = \mu\frac{dy}{ds}, \quad \frac{\partial f}{\partial z} = \mu\frac{dz}{ds};$$

and therefore, along that path, the equations

$$\frac{dx}{ds}\frac{\partial}{\partial x}\left(\frac{\partial f}{\partial a}\right) + \frac{dy}{ds}\frac{\partial}{\partial y}\left(\frac{\partial f}{\partial a}\right) + \frac{dz}{ds}\frac{\partial}{\partial z}\left(\frac{\partial f}{\partial a}\right) = 0,$$

$$\frac{dx}{ds}\frac{\partial}{\partial x}\left(\frac{\partial f}{\partial b}\right) + \frac{dy}{ds}\frac{\partial}{\partial y}\left(\frac{\partial f}{\partial b}\right) + \frac{dz}{ds}\frac{\partial}{\partial z}\left(\frac{\partial f}{\partial b}\right) = 0,$$

are satisfied. The first of these equations shews that the path is perpendicular to the normal to a new surface

$$\frac{\partial f}{\partial a} = a',$$

where $a'$ is a parametric constant, that is, the path touches the new surface; and the second shews that it touches a new surface

$$\frac{\partial f}{\partial b} = b',$$

where $b'$ is a parametric constant. Hence the path from any point is the intersection of those surfaces of the two families

$$\frac{\partial f}{\partial a} = a', \quad \frac{\partial f}{\partial b} = b',$$

which pass through the point.

This theorem was given first by Liouville; the method, manifestly, is based upon Jacobi's method of variation of parameters.

*Ex.* 1. Verify (by using Liouville's theorem) the result obtained in (B) for the case when

$$\mu^2 = \mu_0{}^2 - (x^2 + y^2 + z^2).$$

*Ex.* 2. Find the skew curve such that the integral

$$\int_{(x_0,\, y_0,\, z_0)}^{(x,\, y,\, z)} \{\mu_0{}^2 - 2(xy + yz + zx)\}^{\frac{1}{2}}\, ds$$

is a minimum, when taken from an unspecified point on one surface to an unspecified point on another surface.

### Exclusive complete ranges.

**176.** There are two propositions, connected with these characteristic skew curves, analogous to the two propositions proved earlier (in §§ 67, 70) for the characteristic plane curves.

According to the first of them, *the range along the skew curve, limited by conjugate points $t_0$ and $t_1$, does not include within itself a range similarly bounded by conjugate points: that is, the conjugate of any place, within such a range between $t_0$ and $t_1$, lies without that range.*

Consider the function $\Delta(t_0, t)$ of § 174, where

$$\Delta(t_0, t) = Y_1(t) Z_2(t) - Y_2(t) Z_1(t).$$

It vanishes when $t = t_0$. For increasing values of $t$ from $t_0$ onwards, it does not again vanish until $t$ reaches $t_1$, the conjugate of $t_0$. Hence, at a place $t_0 + \epsilon T_0$ just within the range from $t_0$ to $t_1$, where $\epsilon$ is small and positive and $T_0$ is positive, the sign of $\Delta(t_0, t)$ is the same as at a place $t_1 - \eta T_1$ just within the same range, where $\eta$ is small and positive and $T_1$ is positive. Now

$$\Delta(t_0, t_0 + \epsilon T_0) = \epsilon^2 T_0^2 \begin{vmatrix} \theta_1'(t_0), & \theta_2'(t_0), & \theta_3'(t_0), & \theta_4'(t_0) \\ \vartheta_1'(t_0), & \vartheta_2'(t_0), & \vartheta_3'(t_0), & \vartheta_4'(t_0) \\ \theta_1(t_0), & \theta_2(t_0), & \theta_3(t_0), & \theta_4(t_0) \\ \vartheta_1(t_0), & \vartheta_2(t_0), & \vartheta_3(t_0), & \vartheta_4(t_0) \end{vmatrix} + \dots$$

$$= \epsilon^2 T_0^2 J(t_0) + \dots,$$

the unexpressed terms being the aggregate of those which involve the third and higher powers of $\epsilon$. Hence, between the range from $t_0$ to $t_1$, the sign of $\Delta(t_0, t)$ is the same as that of $J(t_0)$: or, as $J(t)$ has an unvarying sign (§ 171) along the characteristic curve, the sign of $\Delta(t_0, t)$ has one unvarying sign whatever initial point be chosen, and $\Delta(t_0, t)$ does not vanish between $t_0$ and $t_1$. Again,

$$\Delta(t_0, t_1 - \eta T_1) = \Delta(t_0, t_1) - \eta T_1 \frac{\partial \Delta(t_0, t_1)}{\partial t_1} + \dots$$

$$= -\eta T_1 \frac{\partial \Delta(t_0, t_1)}{\partial t_1} + \dots,$$

where the unexpressed terms are the aggregate of those which involve second and higher powers of $\eta$. Thus

$$\frac{\partial \Delta(t_0, t_1)}{\partial t_1}$$

does not vanish, and it has an unvarying sign whatever initial point $t_0$ be chosen.

Next, consider the equation

$$\Delta(t_0, t_1) = 0,$$

where $t_1$ is the first value of $t$ (greater than $t_0$) at which $\Delta(t_0, t)$ vanishes. Let

$$t_1 + t_0 = 2\theta, \quad t_1 - t_0 = 2\rho;$$

the equation becomes

$$\Delta(\theta - \rho, \theta + \rho) = 0;$$

and now $2\rho$ is the smallest value of $t_1 - t_0$ satisfying the equation. But

$$\Delta(\theta - \rho, \theta + \rho) = \Delta(\theta + \rho, \theta - \rho),$$

whatever $\theta$ and $\rho$ may be; hence, coupled with a root $\rho$ of the equation, there is a root $-\rho$, because $\Delta(\theta - \rho, \theta + \rho)$ is an even function of $\rho$. Also

$$\Delta(\theta, \theta) = 0,$$

whatever $\theta$ may be; hence, for any value of $\zeta$,

$$\Delta (\theta - \zeta,\ \theta + \zeta) = \Delta (\theta + \zeta,\ \theta - \zeta)$$
$$= \zeta^2 J(\theta) + \text{even powers of } \zeta$$
$$= \zeta^2 \Phi(\theta,\ \zeta^2),$$

where $\Phi(\theta,\ 0) = J(\theta)$, a quantity of invariable sign. Thus

$$\Phi(\theta,\ \rho^2) = 0;$$

as $\zeta = \rho$ is the smallest positive root of the equation $\Delta (\theta - \rho,\ \theta + \rho) = 0$, so $\zeta = -\rho$ is the numerically smallest negative root of $\Phi(\theta,\ \rho^2) = 0$, that is, of

$$\Delta (\theta + \rho,\ \theta - \rho) = 0,$$

or

$$\Delta (t_0,\ t_1) = 0,$$

regarded as an equation in $t_0$ when $t_1$ is given. Thus there is no value $t_0 + \epsilon T_0$ (where $\epsilon$ and $T_0$ are positive), such that

$$\Delta (t_0 + \epsilon T_0,\ t_1)$$

vanishes, that is, we cannot have $\dfrac{\partial \Delta (t_0,\ t_1)}{\partial t_0}$ equal to zero.

Now take a range on the characteristic curve from $t_0 + \epsilon T_0$ to $t_1 + \epsilon \Theta$, where $\epsilon$ and $T_0$ are positive; and let it be a complete range, so that $t_1 + \epsilon \Theta$ is the conjugate of $t_0 + \epsilon T_0$. We thus have

$$\Delta (t_0 + \epsilon T_0,\ t_1 + \epsilon \Theta) = 0,$$

that is,

$$\Delta (t_0,\ t_1) + \epsilon T_0 \frac{\partial \Delta (t_0,\ t_1)}{\partial t_0} + \epsilon \Theta \frac{\partial \Delta (t_0,\ t_1)}{\partial t_1} + \ldots = 0,$$

the unspecified terms involving the second and higher powers of $E$. Hence, effectively, when $\epsilon$ is small,

$$T_0 \frac{\partial \Delta (t_0,\ t_1)}{\partial t_0} + \Theta \frac{\partial \Delta (t_0,\ t_1)}{\partial t_1} = 0.$$

Now the coefficient of $T_0$ does not vanish, and the coefficient of $\Theta$ does not vanish; hence $\epsilon T_0$ and $\epsilon \Theta$ are of the same order of magnitude.

If $T_0$ and $\Theta$ are of opposite signs, in consequence of the signs of the derivatives of $\Delta (t_0,\ t_1)$, that is, if $\Theta$ is negative, then the complete range beginning at $t_0' (= t_0 + \epsilon T_0)$ is included within the old range and ends at $t_1' (< t_1)$, the arc $t_1 t_1'$ being of the same order of magnitude as $t_0 t_0'$. At $t_0'$ conditions generally—in particular, the sign of $J(t_0')$—are the same as at $t_0$. If then a place $t_0' + \epsilon T_0'$, with both $\epsilon$ and $T_0'$ positive, be taken within $t_0' t_1'$, the foregoing analysis (with merely a change of symbols) would lead to the conjugate of $t_0' + \epsilon T_0'$, and would place it within $t_0' t_1'$, at a distance from $t_1'$ within that range of the same order of magnitude as the distance $\epsilon T_0'$: that is, a new (and more restricted) range $t_0'' t_1''$ would be included within $t_0' t_1'$. Proceeding thus with gradual steps, that shorten simultaneously at both ends the length of the range by quantities of the same order, we should come to a place $t_2$, within the middle of the range, which would have its conjugate at an arc-distance from $t_1$ of the same kind of magnitude as the arc-distance of $t_2$ from $t_0$. We could, in effect, have $t_2$ and a point in its vicinity $t_2 + \eta T_2$ as conjugates. But this is impossible: for

$$\Delta\,(t_2,\,t_2 + \eta T_2) = \eta^2\,T_2^2 J\,(t_2) + \ldots,$$

the unspecified terms being of the third and higher degrees in $\eta$; and this quantity cannot vanish.

Hence the fundamental hypothesis on which this impossible result rests —viz. that $T_0$ and $\Theta$ are of opposite signs—is untenable. We know that $\Theta$ and $T_0$ vanish together, and differ from zero simultaneously; hence they must be of the same sign. Also $\epsilon T_0$ and $\epsilon \Theta$ are of the same order of magnitude; hence the conjugate of $t_0 + \epsilon T_0$ is beyond $t_1$, at an arc-distance from $t_1$ of the same order of magnitude as the arc-distance represented by $\epsilon T_0$. When we proceed along $t_0 t_1$, from $t_0$ initially, through successive points, the conjugates of these successive points are successively moved along the characteristic curve beyond $t_1$, the respective movements of the new final point and of the new initial point being of the same order of magnitude.

We thus infer the proposition that a complete range between two conjugate points on the characteristic curve can never include within itself another complete range.

### Contiguous characteristics.

**177.** The other proposition, to which reference was made in § 176, is of use when strong variations of characteristic curves arise for consideration. It is as follows:

*Let $P_2$ be any point $(x_2, y_2, z_2)$ within a range $P_0 P_1$ on a characteristic curve, the respective values of $t$ for $P_0, P_2, P_1$ being $t_0, t_2, t_1$, such that*

$$t_0 < t_2 < t_1,$$

*so that $\Delta\,(t_0, t)$, which vanishes identically when $t = t_0$ and vanishes functionally when $t = t_1$ because $t_1$ is the conjugate of $t_0$, does not vanish when $t = t_2$. Let $Q_2$ be any point $(x_2 + \epsilon X_2,\ y_2 + \epsilon Y_2,\ z_2 + \epsilon Z_2)$ in the immediate vicinity of $P_2$; then it is possible to draw a unique characteristic curve contiguous to the curve $P_0 P_2 P_1$ such that it shall pass through $P_0$ and $Q_2$.*

A characteristic curve, contiguous to the characteristic

$$\left. \begin{aligned} x &= \chi\,(t + \alpha,\ \alpha_1,\ \alpha_2,\ \alpha_3,\ \alpha_4) \\ y &= \phi\,(t + \alpha,\ \alpha_1,\ \alpha_2,\ \alpha_3,\ \alpha_4) \\ z &= \psi\,(t + \alpha,\ \alpha_1,\ \alpha_2,\ \alpha_3,\ \alpha_4) \end{aligned} \right\},$$

will be possible, if slightly varied parameters $\alpha_1 + \kappa g_1,\ \alpha_2 + \kappa g_2,\ \alpha_3 + \kappa g_3,\ \alpha_4 + \kappa g_4$ can be found so as to satisfy the necessary conditions. Also, for the initial point on the contiguous characteristic, the initial value of $t$ will differ from $t_0$ by a small quantity, which may be denoted by $\epsilon T_0$. As the initial point is unaltered, we have

$$0 = T_0 \chi'(t_0) + g_1 \frac{\partial \chi(t_0)}{\partial \alpha_1} + g_2 \frac{\partial \chi(t_0)}{\partial \alpha_2} + g_3 \frac{\partial \chi(t_0)}{\partial \alpha_3} + g_4 \frac{\partial \chi(t_0)}{\partial \alpha_4},$$

$$0 = T_0 \phi'(t_0) + g_1 \frac{\partial \phi(t_0)}{\partial \alpha_1} + g_2 \frac{\partial \phi(t_0)}{\partial \alpha_2} + g_3 \frac{\partial \phi(t_0)}{\partial \alpha_3} + g_4 \frac{\partial \phi(t_0)}{\partial \alpha_4},$$

$$0 = T_0 \psi'(t_0) + g_1 \frac{\partial \psi(t_0)}{\partial \alpha_1} + g_2 \frac{\partial \psi(t_0)}{\partial \alpha_2} + g_3 \frac{\partial \psi(t_0)}{\partial \alpha_3} + g_4 \frac{\partial \psi(t_0)}{\partial \alpha_4}.$$

Hence, eliminating $T_0$ between the first and second of these relations, and then eliminating it between the first and third of them, we have two conditions

$$\left. \begin{array}{l} 0 = g_1 \theta_1(t_0) + g_2 \theta_2(t_0) + g_3 \theta_3(t_0) + g_4 \theta_4(t_0) \\ 0 = g_1 \vartheta_1(t_0) + g_2 \vartheta_2(t_0) + g_3 \vartheta_3(t_0) + g_4 \vartheta_4(t_0) \end{array} \right\}.$$

When these two conditions are retained, only a single one of the three relations need then be retained as well; it serves to determine $T_0$, if $g_1, g_2, g_3, g_4$ are known.

Again, the value of $t$ at $Q_2$ on the hypothetical contiguous characteristic will differ from $t_2$ by a small quantity which may be denoted by $\epsilon T_2$; then taking account of the coordinates of $P_2$ and of $Q_2$, we have

$$X_2 = T_2 \chi'(t_2) + g_1 \frac{\partial \chi(t_2)}{\partial \alpha_1} + g_2 \frac{\partial \chi(t_2)}{\partial \alpha_2} + g_3 \frac{\partial \chi(t_2)}{\partial \alpha_3} + g_4 \frac{\partial \chi(t_2)}{\partial \alpha_4},$$

$$Y_2 = T_2 \phi'(t_2) + g_1 \frac{\partial \phi(t_2)}{\partial \alpha_1} + g_2 \frac{\partial \phi(t_2)}{\partial \alpha_2} + g_3 \frac{\partial \phi(t_2)}{\partial \alpha_3} + g_4 \frac{\partial \phi(t_2)}{\partial \alpha_4},$$

$$Z_2 = T_2 \psi'(t_2) + g_1 \frac{\partial \psi(t_2)}{\partial \alpha_1} + g_2 \frac{\partial \psi(t_2)}{\partial \alpha_2} + g_3 \frac{\partial \psi(t_2)}{\partial \alpha_3} + g_4 \frac{\partial \psi(t_2)}{\partial \alpha_4}.$$

Eliminating $T_2$ between the first and second of these relations, and then eliminating it between the first and third of them, we have two conditions

$$\left. \begin{array}{l} \chi'(t_2) Y_2 - \phi'(t_2) X_2 = g_1 \theta_1(t_2) + g_2 \theta_2(t_2) + g_3 \theta_3(t_2) + g_4 \theta_4(t_2) \\ \chi'(t_2) Z_2 - \psi'(t_2) X_2 = g_1 \vartheta_1(t_2) + g_2 \vartheta_2(t_2) + g_3 \vartheta_3(t_2) + g_4 \vartheta_4(t_2) \end{array} \right\}.$$

When these two conditions are retained, only a single one of the three relations involving $T_2$ need be retained as well; it serves to determine $T_2$, if $g_1, g_2, g_3, g_4$ are known.

But there are four conditions, involving the constants $g_1, g_2, g_3, g_4$ linearly. The determinant of the coefficients of the four constants in these linear conditions is

$$\Delta(t_0, t_2),$$

which does not vanish; because $t_2 > t_0$ in the one direction, and also $< t_1$ in the other direction, $P_2$ lying within the range from $t_0$ to $t_1$. Hence the equation

$$g_1 \Delta (t_0, t_2) = \begin{vmatrix} \chi'(t_2) Y_2 - \phi'(t_2) X_2, & \theta_2(t_2), & \theta_3(t_2), & \theta_4(t_2) \\ \chi'(t_2) Z_2 - \psi'(t_2) X_2, & \vartheta_2(t_2), & \vartheta_3(t_2), & \vartheta_4(t_2) \\ 0 & , & \theta_2(t_0), & \theta_3(t_0), & \theta_4(t_0) \\ 0 & , & \vartheta_2(t_0), & \vartheta_3(t_0), & \vartheta_4(t_0) \end{vmatrix},$$

and three similar equations, determine $g_1, g_2, g_3, g_4$ uniquely as four finite constants. As their values are now known, we infer the value of $T_0$ and the value of $T_2$; and thus all the elements of the contiguous characteristic curve are known.

Hence a unique contiguous characteristic curve can be drawn as required.

*Corollary.* If we were dealing with a characteristic as given in the earlier form

$$y = \phi \, (x, A_1, A_2, A_3, A_4),$$
$$z = \psi \, (x, A_1, A_2, A_3, A_4),$$

the corresponding conditions are easily seen to be

$$0 = g_1 \frac{\partial \phi (x_0)}{\partial A_1} + g_2 \frac{\partial \phi (x_0)}{\partial A_2} + g_3 \frac{\partial \phi (x_0)}{\partial A_3} + g_4 \frac{\partial \phi (x_0)}{\partial A_4},$$

$$0 = g_1 \frac{\partial \psi (x_0)}{\partial A_1} + g_2 \frac{\partial \psi (x_0)}{\partial A_2} + g_3 \frac{\partial \psi (x_0)}{\partial A_3} + g_4 \frac{\partial \psi (x_0)}{\partial A_4},$$

$$Y_2 - \phi' \, (x_2) X_2 = g_1 \frac{\partial \phi (x_2)}{\partial A_1} + g_2 \frac{\partial \phi (x_2)}{\partial A_2} + g_3 \frac{\partial \phi (x_2)}{\partial A_3} + g_4 \frac{\partial \phi (x_2)}{\partial A_4},$$

$$Z_2 - \psi' \, (x_2) X_2 = g_1 \frac{\partial \psi (x_2)}{\partial A_1} + g_2 \frac{\partial \psi (x_2)}{\partial A_2} + g_3 \frac{\partial \psi (x_2)}{\partial A_3} + g_4 \frac{\partial \psi (x_2)}{\partial A_4};$$

and the inference, as to the existence of a unique contiguous characteristic curve, is made as before.

### Integrals involving any number of dependent variables, and their first derivatives.

**178.** The following results, analogous to those which have been obtained for integrals involving two dependent variables and their first derivatives, apply to integrals involving any number of such variables and their first derivatives. Proofs can be obtained by analysis exactly similar to that used for the simple instance, when there are two dependent variables.

An integral is given, involving $n-1$ dependent variables $y_2, \ldots, y_n$ and their first derivatives $p_2, \ldots, p_n$ with respect to the independent variable $x$, in a form

$$\int f(x, y_2, p_2, y_3, p_3, \ldots, y_n, p_n) \, dx.$$

(A)  In order that the integral may have a maximum or a minimum, the $n-1$ dependent variables must satisfy the *characteristic equations*

$$\frac{\partial f}{\partial y_r} - \frac{d}{dx}\left(\frac{\partial f}{\partial p_r}\right) = 0,$$

for $r = 2, 3, \ldots, n$.

(B)  When the limits of the integral are fixed by assigned data, those data determine the $2n - 2$ arbitrary constants that occur in the primitive of the characteristic equations. When a limit of the integral is not fixed, so that a variation at that limit represented by $\xi$, $\eta_2$, $\eta_3$, $\ldots$, $\eta_n$ must satisfy one or more equations initially prescribed, the relation

$$\xi\left(f - p_2\frac{\partial f}{\partial p_2} - \cdots - p_n\frac{\partial f}{\partial p_n}\right) + \eta_2\frac{\partial f}{\partial p_2} + \cdots + \eta_n\frac{\partial f}{\partial p_n} = 0$$

must be satisfied at each such limit.

When conditions (A) and (B) are satisfied, the first variation of the integral vanishes.

(C)  Let the primitive of the characteristic equations be

$$y_r = \phi_r(x, A_1, A_2, \ldots, A_{2n-2}),$$

where $A_1, A_2, \ldots, A_{2n-2}$ are arbitrary constants, and $r = 2, \ldots, n$. Take varied arbitrary constants $A_m + \kappa a_m$ instead of $A_m$, for $m = 1, 2, \ldots, 2n - 2$; and write

$$\eta_r = \sum_{m=1}^{2n-2} a_m \frac{\partial \phi_r(x, A_1, \ldots, A_{2n-2})}{\partial A_m}.$$

Also, let

$$2\Theta = \Sigma\Sigma \frac{\partial^2 f}{\partial y_r \partial y_s} \eta_r \eta_s$$

$$+ \Sigma\Sigma \frac{\partial^2 f}{\partial p_r \partial y_s} \eta_r' \eta_s$$

$$+ \Sigma\Sigma \frac{\partial^2 f}{\partial p_r \partial p_s} \eta_r' \eta_s',$$

where the summations are for $r$ and $s$, $= 2, \ldots, n$, independently of one another. Then the quantities $\eta_2, \ldots, \eta_n$ satisfy the equations

$$\frac{\partial \Theta}{\partial \eta_r} - \frac{d}{dx}\left(\frac{\partial \Theta}{\partial \eta_r'}\right) = 0$$

for $r = 2, \ldots, n$; and they constitute the primitive of these equations (called the *subsidiary characteristic equations*), when $a_1, \ldots, a_{2n-2}$ are taken as arbitrary constants.

(D)  Let the small variations applied to the original integral be $y_r + \kappa v_r$ in place of $y_r$, for all the variables. Then the second variation of the integral can be expressed in the form

$$\tfrac{1}{2}\kappa^2 \int \sum_{r,\,s} \sum \frac{\partial^2 f}{\partial p_r \partial p_s}\, Y_r Y_s,$$

the double summation being taken as in (C); where

$$Y_r = v_r' + \sum_{m=2}^{n} \lambda_{rm} v_m,$$

and the equation

$$\eta_{rk}' + \sum_{m=2}^{n} \lambda_{rm}\eta_{mk} = 0$$

holds for $n-1$ linearly independent integral-sets

$$\eta_{2k},\ \eta_{3k},\ \ldots,\ \eta_{nk}$$

of the subsidiary characteristic equations given by $k = 2, \ldots, n$.

(E)  In order that the second variation may have a persistent sign, either always positive (for a minimum) or always negative (for a maximum) for all weak variations, two tests must be satisfied: the extended *Legendre* test, relating to the derivatives $\dfrac{\partial^2 f}{\partial p_r \partial p_s}$, and the extended *Jacobi* test, relating to the range of the integral.

(F)  The Legendre test may be stated in various forms. The form, that seems simplest for inspection, is to require that all the inequalities

$$\frac{\partial^2 f}{\partial p_m^2}\,\frac{\partial^2 f}{\partial p_r^2} > \left(\frac{\partial^2 f}{\partial p_m \partial p_r}\right)^2$$

shall be satisfied for all combinations of $r$ and $m$, $= 2, \ldots, n$: so that $\dfrac{\partial^2 f}{\partial p_2^2}, \ \ldots, \ \dfrac{\partial^2 f}{\partial p_n^2}$ must all have the same sign, and are not to vanish anywhere in the range. When the sign is positive, the integral admits a minimum; when the sign is negative, the integral admits a maximum.

(G)  The Jacobi test limits the range of the integral so that, beginning with a lower limit $x_0$, it shall not extend as far as a (greater) upper limit $x_1$, where $x_1$ is the smallest root $x_1$, greater than $x_0$, of an equation $\Phi(x_0, x_1) = 0$; and this quantity $\Phi(x_0, x_1)$ is

$$\begin{vmatrix} \phi_{21}(x_1), & \phi_{22}(x_1), & \ldots\ldots, & \phi_{2,\,2n-2}(x_1) \\ \phi_{31}(x_1), & \phi_{32}(x_1), & \ldots\ldots, & \phi_{3,\,2n-2}(x_1) \\ \ldots\ldots\ldots\ldots & & & \ldots\ldots\ldots \\ \phi_{n1}(x_1), & \phi_{n2}(x_1), & \ldots\ldots, & \phi_{n,\,2n-2}(x_1) \\ \phi_{21}(x_0), & \phi_{22}(x_0), & \ldots\ldots, & \phi_{2,\,2n-2}(x_0) \\ \phi_{31}(x_0), & \phi_{32}(x_0), & \ldots\ldots, & \phi_{3,\,2n-2}(x_0) \\ \ldots\ldots\ldots\ldots & & & \ldots\ldots\ldots \\ \phi_{n1}(x_0), & \phi_{n2}(x_0), & \ldots\ldots, & \phi_{n,\,2n-2}(x_0) \end{vmatrix},$$

where

$$\phi_{ms} = \frac{\partial \phi_m (x, A_1, \ldots, A_{2n-2})}{\partial A_s},$$

for $m = 2, \ldots, n$, and $s = 1, \ldots, 2n - 2$.

(H) There are corresponding theorems for the integral, when it is taken in the form $\int F(y_1, y_1', y_2, y_2', \ldots, y_n, y_n') \, dt$, where the integrand $F$ satisfies the identity

$$F = y_1' \frac{\partial F}{\partial y_1'} + \ldots + y_n' \frac{\partial F}{\partial y_n'}.$$

# CHAPTER VI.

## Integrals involving two dependent variables, with derivatives of the second order: special weak variations.

### *First variation of an integral.*

**179.** We proceed to consider (but with merely brief developments) integrals which involve one independent variable, two dependent variables, and derivatives of the latter of the first order and the second order. The typical integral can be expressed by

$$\int_{x_1}^{x_2} f(x, y, y', y'', z, z', z'') \, dx,$$

with the customary limitations on the function $f$. The enquiry is as to the conditions under which this integral can assume a maximum or minimum.

In the present chapter, we shall deal almost entirely with special variations. Thus, we except $x$ from variation, save at the limits; there, variation may have to be imposed on $x$, in order to admit the possibility of mobile limits, unfixed specifically by express data. We suppose $y$ and $z$ respectively changed to $y + \kappa Y$ and $z + \kappa Z$, where $Y$ and $Z$ are regular (though arbitrary) functions of $x$, independent moreover of $\kappa$. To include the instances of mobile limits, we suppose the upper limit to become $x_2 + \kappa U_2$ and the lower to become $x_1 + \kappa U_1$, where $U_2$ and $U_1$ will govern variations at a boundary curve or on a boundary surface: these boundaries being supposed given when the extremities are not fixed. The completely varied value of the integral is

$$\int_{x_1 + \kappa U_1}^{x_2 + \kappa U_2} f(x, y + \kappa Y, y' + \kappa Y', y'' + \kappa Y'', z + \kappa Z, z' + \kappa Z', z'' + \kappa Z'') \, dx.$$

By arguments precisely analogous to those in the preceding investigations (such as in §§ 32, 82), we can express the variation of the integral in the form

$$\kappa \int_{x_1}^{x_2} \left\{ \left( Y \frac{\partial f}{\partial y} + Y' \frac{\partial f}{\partial y'} + Y'' \frac{\partial f}{\partial y''} \right) + \left( Z \frac{\partial f}{\partial z} + Z' \frac{\partial f}{\partial z'} + Z'' \frac{\partial f}{\partial z''} \right) \right\} dx$$
$$+ \kappa \left( U_2 f_2 - U_1 f_1 \right) + K_2,$$

where $f_2$ and $f_1$ are the values of $f(x, y, y', y'', z, z', z'')$ at $x_2$ and at $x_1$ respectively, and where $K_2$ denotes the aggregate of terms of the second and higher degrees in $\kappa$. Now

$$\int_{x_1}^{x_2} Y' \frac{\partial f}{\partial y'} dx = \left[ Y \frac{\partial f}{\partial y'} \right]_{x_1}^{x_2} - \int_{x_1}^{x_2} Y \frac{d}{dx} \left( \frac{\partial f}{\partial y'} \right) dx,$$

$$\int_{x_1}^{x_2} Y'' \frac{\partial f}{\partial y''} dx = \left[ Y' \frac{\partial f}{\partial y''} - Y \frac{d}{dx} \left( \frac{\partial f}{\partial y''} \right) \right]_{x_1}^{x_2} + \int_{x_1}^{x_2} Y \frac{d^2}{dx^2} \left( \frac{\partial f}{\partial y''} \right) dx ;$$

and similarly for the terms in $Z'$ and $Z''$ in the foregoing integral. Hence the full variation of the integral is

$$\kappa \int_{x_1}^{x_2} \left[ Y \left\{ \frac{\partial f}{\partial y} - \frac{d}{dx} \left( \frac{\partial f}{\partial y'} \right) + \frac{d^2}{dx^2} \left( \frac{\partial f}{\partial y''} \right) \right\} + Z \left\{ \frac{\partial f}{\partial z} - \frac{d}{dx} \left( \frac{\partial f}{\partial z'} \right) + \frac{d^2}{dx^2} \left( \frac{\partial f}{\partial z''} \right) \right\} \right] dx$$

$$+ \kappa \left[ Uf + Y \left\{ \frac{\partial f}{\partial y'} - \frac{d}{dx} \left( \frac{\partial f}{\partial y''} \right) \right\} + Y' \frac{\partial f}{\partial y''} + Z \left\{ \frac{\partial f}{\partial z'} - \frac{d}{dx} \left( \frac{\partial f}{\partial z''} \right) \right\} + Z' \frac{\partial f}{\partial z''} \right]_{x_1}^{x_2} + K_2.$$

*Characteristic equations: terminal conditions.*

**180.** In order that the integral may possess a maximum or a minimum, the first variation must vanish for all admissible variations; consequently, each portion of that first variation must vanish, when it arises from variations absolutely independent of all the others. Thus, first of all,

$$\int_{x_1}^{x_2} \left[ Y \left\{ \frac{\partial f}{\partial y} - \frac{d}{dx} \left( \frac{\partial f}{\partial y'} \right) + \frac{d^2}{dx^2} \left( \frac{\partial f}{\partial y''} \right) \right\} + Z \left\{ \frac{\partial f}{\partial z} - \frac{d}{dx} \left( \frac{\partial f}{\partial z'} \right) + \frac{d^2}{dx^2} \left( \frac{\partial f}{\partial z''} \right) \right\} \right] dx = 0 ;$$

and therefore, as $\kappa Y$ and $\kappa Z$ are independent of one another and are arbitrary, we must have

$$\frac{\partial f}{\partial y} - \frac{d}{dx} \left( \frac{\partial f}{\partial y'} \right) + \frac{d^2}{dx^2} \left( \frac{\partial f}{\partial y''} \right) = 0,$$

$$\frac{\partial f}{\partial z} - \frac{d}{dx} \left( \frac{\partial f}{\partial z'} \right) + \frac{d^2}{dx^2} \left( \frac{\partial f}{\partial z''} \right) = 0.$$

These are the *characteristic equations*, current along the whole of the skew curve which is to produce a maximum or a minimum; and they determine the nature of the *characteristic curve*.

**181.** Next, we have

$$\left[ Uf + Y \left\{ \frac{\partial f}{\partial y'} - \frac{d}{dx} \left( \frac{\partial f}{\partial y''} \right) \right\} + Y' \frac{\partial f}{\partial y''} + Z \left\{ \frac{\partial f}{\partial z'} - \frac{d}{dx} \left( \frac{\partial f}{\partial z''} \right) \right\} + Z' \frac{\partial f}{\partial z''} \right]_{x_1}^{x_2} = 0.$$

It is possible to have a variation at a mobile limit, quite independently of all variations (if any) at the other limit. Hence the relation

$$Uf + Y \left\{ \frac{\partial f}{\partial y'} - \frac{d}{dx} \left( \frac{\partial f}{\partial y''} \right) \right\} + Y' \frac{\partial f}{\partial y''} + Z \left\{ \frac{\partial f}{\partial z'} - \frac{d}{dx} \left( \frac{\partial f}{\partial z''} \right) \right\} + Z' \frac{\partial f}{\partial z''} = 0$$

is satisfied at each limit.

The quantities $Y$, $Y'$, $Z$, $Z'$ belong to a terminal variation of the characteristic curve along the $x$-ordinate alone, while the actual variation at the limit takes place along the boundary curve or on the boundary surface. If then $V$ and $W$ be quantities along this boundary as determined by the quantity $U$, the former analysis (used in § 86, ii) shews that

$$Y = V - y'U, \quad Z = W - z'U,$$
$$Y' = V' - y''U, \quad Z' = W' - z''U.$$

When these values are inserted, the condition at each limit becomes;

$$U\left[ f - y' \left\{ \frac{\partial f}{\partial y'} - \frac{d}{dx}\left( \frac{\partial f}{\partial y''} \right) \right\} - z' \left\{ \frac{\partial f}{\partial z'} - \frac{d}{dx}\left( \frac{\partial f}{\partial z''} \right) \right\} - y'' \frac{\partial f}{\partial y''} - z'' \frac{\partial f}{\partial z''} \right]$$
$$+ V\left\{ \frac{\partial f}{\partial y'} - \frac{d}{dx}\left( \frac{\partial f}{\partial y''} \right) \right\} + W\left\{ \frac{\partial f}{\partial z'} - \frac{d}{dx}\left( \frac{\partial f}{\partial z''} \right) \right\} + V' \frac{\partial f}{\partial y''} + W' \frac{\partial f}{\partial z''} = 0.$$

(i) When the limit lies on a boundary curve, the equations of which are

$$y = \theta(x), \quad z = \Im(x),$$

we have

$$V = \theta'(x) U, \quad W = \Im'(x) U.$$

Also, for this boundary,

$$y' = \theta'(x), \quad z' = \Im'(x),$$

and so

$$V' = \theta''(x) U, \quad W' = \Im''(x) U.$$

When these values are inserted in the foregoing relation, and when the terminal point is not fixed so that $U$ is not zero, we have

$$f - \{y' - \theta'(x)\} \left\{ \frac{\partial f}{\partial y'} - \frac{d}{dx}\left( \frac{\partial f}{\partial y''} \right) \right\} - \{z' - \Im'(x)\} \left\{ \frac{\partial f}{\partial z'} - \frac{d}{dx}\left( \frac{\partial f}{\partial z''} \right) \right\}$$
$$- \{y'' - \theta''(x)\} \frac{\partial f}{\partial y''} - \{z'' - \Im''(x)\} \frac{\partial f}{\partial z''} = 0.$$

Conditions of an intermediate type might be assigned. Thus, if the extreme point were fixed in position but free in direction, we should have

$$U = 0, \quad V = 0, \quad W = 0,$$

from $U = 0$, $Y = 0$, $Z = 0$, while $Y'$ and $Z'$ are arbitrary; then the resulting conditions are

$$\frac{\partial f}{\partial y''} = 0, \quad \frac{\partial f}{\partial z''} = 0,$$

which must be satisfied at the limit.

If the extreme point is fixed both in position and direction, then

$$Y = 0, \quad Z = 0, \quad Y' = 0, \quad Z' = 0,$$

at that rigorously fixed extremity.

(ii) When the limit lies freely on a boundary surface, the equation of which is

$$S(x, y, z) = 0,$$

then for any direction on the surface at a point,

$$\frac{\partial S}{\partial x} + y'\frac{\partial S}{\partial y} + z'\frac{\partial S}{\partial z} = 0.$$

Consequently, we have

$$\frac{\partial S}{\partial x} U + \frac{\partial S}{\partial y} V + \frac{\partial S}{\partial z} W = 0,$$

$$V'\frac{\partial S}{\partial y} + W'\frac{\partial S}{\partial z}$$

$$+ U\left(\frac{\partial^2 S}{\partial x^2} + y'\frac{\partial^2 S}{\partial x\partial y} + z'\frac{\partial^2 S}{\partial x\partial z}\right) + V\left(\frac{\partial^2 S}{\partial x\partial y} + y'\frac{\partial^2 S}{\partial y^2} + z'\frac{\partial^2 S}{\partial y\partial z}\right)$$

$$+ W\left(\frac{\partial^2 S}{\partial x\partial z} + y'\frac{\partial^2 S}{\partial y\partial z} + z'\frac{\partial^2 S}{\partial z^2}\right) = 0.$$

At the limit of the range, we therefore have

$$f - y'\left\{\frac{\partial f}{\partial y'} - \frac{d}{dx}\left(\frac{\partial f}{\partial y''}\right)\right\} - z'\left\{\frac{\partial f}{\partial z'} - \frac{d}{dx}\left(\frac{\partial f}{\partial z''}\right)\right\} - y''\frac{\partial f}{\partial y''} - z''\frac{\partial f}{\partial z''}$$

$$= \lambda\left(\frac{\partial^2 S}{\partial x^2} + y'\frac{\partial^2 S}{\partial x\partial y} + z'\frac{\partial^2 S}{\partial x\partial z}\right) + \mu\frac{\partial S}{\partial x},$$

$$\frac{\partial f}{\partial y'} - \frac{d}{dx}\left(\frac{\partial f}{\partial y''}\right) = \lambda\left(\frac{\partial^2 S}{\partial x\partial y} + y'\frac{\partial^2 S}{\partial y^2} + z'\frac{\partial^2 S}{\partial y\partial z}\right) + \mu\frac{\partial S}{\partial y},$$

$$\frac{\partial f}{\partial z'} - \frac{d}{dx}\left(\frac{\partial f}{\partial z''}\right) = \lambda\left(\frac{\partial^2 S}{\partial x\partial z} + y'\frac{\partial^2 S}{\partial y\partial z} + z'\frac{\partial^2 S}{\partial z^2}\right) + \mu\frac{\partial S}{\partial z},$$

$$\frac{\partial f}{\partial y''} = \lambda\frac{\partial S}{\partial y},$$

$$\frac{\partial f}{\partial z''} = \lambda\frac{\partial S}{\partial z}.$$

Similarly for conditions of an intermediate type.

### Form of the primitive of the characteristic equations.

**182.** The primitive of the characteristic equations

$$\frac{\partial f}{\partial y} - \frac{d}{dx}\left(\frac{\partial f}{\partial y'}\right) + \frac{d^2}{dx^2}\left(\frac{\partial f}{\partial y''}\right) = 0, \qquad \frac{\partial f}{\partial z} - \frac{d}{dx}\left(\frac{\partial f}{\partial z'}\right) + \frac{d^2}{dx^2}\left(\frac{\partial f}{\partial z''}\right) = 0$$

is required. When we denote derivatives of $f$ with respect to $y$, $y'$, $y''$, $z$, $z'$, $z''$ by subscripts $4, 5, 6, 7. 8, 9$ respectively, according to the conventions

$$\frac{\partial^2 f}{\partial z''^2} = f_{99}, \qquad \frac{\partial^2 f}{\partial y''\partial z''} = f_{69}, \qquad \frac{\partial^2 f}{\partial y''^2} = f_{66},$$

$$\frac{\partial^2 f}{\partial z'\partial z''} = f_{89}, \qquad \frac{\partial^2 f}{\partial y'\partial z''} = f_{59}, \qquad \frac{\partial^2 f}{\partial z'\partial y''} = f_{68}, \qquad \frac{\partial^2 f}{\partial y'\partial y''} = f_{56},$$

and so on, the characteristic equations take the form

$$f_{66}y'''' + f_{69}z'''' + f_{66}'y''' + (f_{69}' + f_{68} - f_{59})z''' + \ldots = 0,$$

$$f_{69}y'''' + f_{99}z'''' + (f_{69}' + f_{59} - f_{68})y''' + f_{99}'z''' + \ldots = 0,$$

where the unexpressed terms do not involve $y''''$, $y'''$, $z''''$, $z'''$. For reasons which will appear later, it becomes requisite (i) that $f_{66}$ and $f_{99}$ shall have one and the same uniform sign throughout the range of the variable $x$, and (ii) that the quantity $f_{66}f_{99} - f_{69}^2$ shall be uniformly positive throughout that range. Denoting this positive quantity by $\sigma$, the equations (resolved for $y''''$ and $z''''$) give

$$\sigma y'''' = -y'''\{f_{99}f_{66}' - f_{69}(f_{69}' + f_{59} - f_{68})\} - z'''\{f_{99}(f_{69}' + f_{68} - f_{59}) - f_{69}f_{99}'\} + \ldots,$$

$$\sigma z'''' = -y'''\{f_{66}(f_{69}' + f_{59} - f_{68}) - f_{69}f_{66}'\} - z'''\{f_{66}f_{99}' - f_{69}(f_{69}' + f_{68} - f_{59})\} + \ldots,$$

where the unexpressed terms do not involve $y''''$, $y'''$, $z''''$, $z'''$.

When $z$ and its derivatives are eliminated between these equations in the manner used in § 136 for a simpler case, the final equation in $y$ is of the form

$$\frac{d^8y}{dx^8} = \text{function of } \left(\frac{d^7y}{dx^7}, \frac{d^6y}{dx^6}, \ldots, \frac{dy}{dx}, y, x\right).$$

Its primitive involves eight arbitrary independent constants; and, if these are such that, when $x = a$,

$$\frac{d^ry}{dx^r} = b_r, \quad (\text{for } r = 0, 1, \ldots, 7),$$

then

$$y = b_0 + b_1(x - a) + \frac{b_2}{2!}(x - a)^2 + \ldots + \frac{b_7}{7!}(x - a)^7 + (x - a)^8 R,$$

where $R$ is a function of $b_0, \ldots, b_7$, and is a regular function of $x - a$. After this value of $y$ has been inserted in the equations that lead to the elimination of $z$, they provide a value of $z$ which is a regular function of $x - a$, involving the eight arbitrary coefficients $b_0, \ldots, b_7$.

Similarly, when the equations are combined so as to lead to the elimination of $y$, we are led to an equation of the eighth order in $z$ alone, the primitive of which involves eight arbitrary independent constants. If these were such that, when $x = a$,

$$\frac{d^rz}{dx^r} = c_r, \quad (\text{for } r = 0, 1, \ldots, 7),$$

then

$$z = c_0 + c_1(x - a) + \frac{c_2}{2!}(x - a)^2 + \ldots + \frac{c_7}{7!}(x - a)^7 + (x - a)^8 S,$$

where $S$ is a function of $c_0, \ldots, c_7$, and is a regular function of $x - a$. By using the equations that lead to the elimination of $y$, the value of $y$ is derived as a regular function of $x - a$, involving the eight arbitrary coefficients $c_0, \ldots, c_7$.

The two simultaneous values of $y$ and $z$ in the former case, and the two simultaneous values of $y$ and $z$ in the latter case, constitute separately the primitive of the characteristic equations. They are, therefore, equivalent to one another. But there is another form of the primitive, evenly balanced as between two sets of constants associated with the arbitrary values, assigned to $y$, to $z$, and to their derivatives, when $x = a$; it is as follows. We take

$$p_1 = \frac{dy}{dx}, \quad p_2 = \frac{dp_1}{dx}, \quad p_3 = \frac{dp_2}{dx},$$

$$q_1 = \frac{dz}{dx}, \quad q_2 = \frac{dq_1}{dx}, \quad q_3 = \frac{dq_2}{dx};$$

and then the second form of the characteristic equations is

$$\frac{dp_3}{dx} = P_3(x, y, z, p_1, p_2, p_3, q_1, q_2, q_3),$$

$$\frac{dq_3}{dx} = Q_3(x, y, z, p_1, p_2, p_3, q_1, q_2, q_3).$$

We thus have eight equations. There exists a primitive of these eight equations such that, when $x = a$,

$$y = a_0, \quad p_1 = a_1, \quad p_2 = a_2, \quad p_3 = a_3,$$

$$z = \alpha_0, \quad q_1 = \alpha_1, \quad q_2 = \alpha_2, \quad q_3 = \alpha_3,$$

where $a_0, a_1, a_2, a_3, \alpha_0, \alpha_1, \alpha_2, \alpha_3$ are arbitrarily assigned constants, independent of one another, provided the functions $P_3$ and $Q_3$ are regular in their immediate vicinity; and the primitive expresses the eight variables $y, p_1, p_2, p_3, z, q_1, q_2, q_3$ as regular functions of $x - a$. The expression of this primitive is

$$y = a_0 + a_1(x-a) + \frac{a_2}{2!}(x-a)^2 + \frac{a_3}{3!}(x-a)^3 + (x-a)^4 G,$$

$$z = \alpha_0 + \alpha_1(x-a) + \frac{\alpha_2}{2!}(x-a)^2 + \frac{\alpha_3}{3!}(x-a)^3 + (x-a)^4 H,$$

where $G$ and $H$ are functions of the eight constants $a_0, a_1, a_2, a_3, \alpha_0, \alpha_1, \alpha_2, \alpha_3$, and are regular functions of $x - a$; and the values of $p_1, p_2, p_3, q_1, q_2, q_3$ are derivable from those of $y$ and $z$, by the first six equations of the set of eight above. Each of the two foregoing primitives is equivalent to this new primitive. Manifestly we have

$$b_0 = a_0, \quad b_1 = a_1, \quad b_2 = a_2, \quad b_3 = a_3,$$

$$c_0 = \alpha_0, \quad c_1 = \alpha_1, \quad c_2 = \alpha_2, \quad c_3 = \alpha_3;$$

also $b_4, b_5, b_6, b_7$ are functions of $a_0, a_1, a_2, a_3, \alpha_0, \alpha_1, \alpha_2, \alpha_3$; and similarly for $c_4, c_5, c_6, c_7$.

*Properties of the primitive of the characteristic equations.*

**183.** Some properties of this primitive may be established at once.

In the first place, *there are no linear relations of the form*

$$\sum_{r=1}^{\infty} A_r \phi_r(x) = 0, \quad \sum_{r=1}^{\infty} A_r \psi_r(x) = 0,$$

*with constant coefficients* $A_1, ..., A_8$, *where*

$$\phi_1(x) = \frac{\partial y}{\partial a_0}, \quad \phi_2(x) = \frac{\partial y}{\partial a_1}, \quad \phi_3(x) = \frac{\partial y}{\partial a_2}, \quad \phi_4(x) = \frac{\partial y}{\partial a_3},$$

$$\phi_5(x) = \frac{\partial y}{\partial \alpha_0}, \quad \phi_6(x) = \frac{\partial y}{\partial \alpha_1}, \quad \phi_7(x) = \frac{\partial y}{\partial \alpha_2}, \quad \phi_8(x) = \frac{\partial y}{\partial \alpha_3},$$

$$\psi_1(x) = \frac{\partial z}{\partial a_0}, \quad \psi_2(x) = \frac{\partial z}{\partial a_1}, \quad \psi_3(x) = \frac{\partial z}{\partial a_2}, \quad \psi_4(x) = \frac{\partial z}{\partial a_3},$$

$$\psi_5(x) = \frac{\partial z}{\partial \alpha_0}, \quad \psi_6(x) = \frac{\partial z}{\partial \alpha_1}, \quad \psi_7(x) = \frac{\partial z}{\partial \alpha_2}, \quad \psi_8(x) = \frac{\partial z}{\partial \alpha_3}.$$

We have

$$\phi_1(x) = 1 + \frac{(x-a)^4}{4!}\frac{\partial b_4}{\partial a_0} + \frac{(x-a)^5}{5!}\frac{\partial b_5}{\partial a_0} + \frac{(x-a)^6}{6!}\frac{\partial b_6}{\partial a_0} + \frac{(x-a)^7}{7!}\frac{\partial b_7}{\partial a_0} + (x-a)^8\frac{\partial R}{\partial a_0},$$

$$\phi_2(x) = x-a + \frac{(x-a)^4}{4!}\frac{\partial b_4}{\partial a_1} + \frac{(x-a)^5}{5!}\frac{\partial b_5}{\partial a_1} + \frac{(x-a)^6}{6!}\frac{\partial b_6}{\partial a_1} + \frac{(x-a)^7}{7!}\frac{\partial b_7}{\partial a_1} + (x-a)^8\frac{\partial R}{\partial a_1},$$

$$\phi_3(x) = \frac{(x-a)^2}{2!} + \frac{(x-a)^4}{4!}\frac{\partial b_4}{\partial a_2} + \frac{(x-a)^5}{5!}\frac{\partial b_5}{\partial a_2} + \frac{(x-a)^6}{6!}\frac{\partial b_6}{\partial a_2} + \frac{(x-a)^7}{7!}\frac{\partial b_7}{\partial a_2} + (x-a)^8\frac{\partial R}{\partial a_2},$$

$$\phi_4(x) = \frac{(x-a)^3}{3!} + \frac{(x-a)^4}{4!}\frac{\partial b_4}{\partial a_3} + \frac{(x-a)^5}{5!}\frac{\partial b_5}{\partial a_3} + \frac{(x-a)^6}{6!}\frac{\partial b_6}{\partial a_3} + \frac{(x-a)^7}{7!}\frac{\partial b_7}{\partial a_3} + (x-a)^8\frac{\partial R}{\partial a_3},$$

$$\phi_5(x) = \frac{(x-a)^4}{4!}\frac{\partial b_4}{\partial \alpha_0} + \frac{(x-a)^5}{5!}\frac{\partial b_5}{\partial \alpha_0} + \frac{(x-a)^6}{6!}\frac{\partial b_6}{\partial \alpha_0} + \frac{(x-a)^7}{7!}\frac{\partial b_7}{\partial \alpha_0} + (x-a)^8\frac{\partial R}{\partial \alpha_0},$$

$$\phi_6(x) = \frac{(x-a)^4}{4!}\frac{\partial b_4}{\partial \alpha_1} + \frac{(x-a)^5}{5!}\frac{\partial b_5}{\partial \alpha_1} + \frac{(x-a)^6}{6!}\frac{\partial b_6}{\partial \alpha_1} + \frac{(x-a)^7}{7!}\frac{\partial b_7}{\partial \alpha_1} + (x-a)^8\frac{\partial R}{\partial \alpha_1},$$

$$\phi_7(x) = \frac{(x-a)^4}{4!}\frac{\partial b_4}{\partial \alpha_2} + \frac{(x-a)^5}{5!}\frac{\partial b_5}{\partial \alpha_2} + \frac{(x-a)^6}{6!}\frac{\partial b_6}{\partial \alpha_2} + \frac{(x-a)^7}{7!}\frac{\partial b_7}{\partial \alpha_2} + (x-a)^8\frac{\partial R}{\partial \alpha_2},$$

$$\phi_8(x) = \frac{(x-a)^4}{4!}\frac{\partial b_4}{\partial \alpha_3} + \frac{(x-a)^5}{5!}\frac{\partial b_5}{\partial \alpha_3} + \frac{(x-a)^6}{6!}\frac{\partial b_6}{\partial \alpha_3} + \frac{(x-a)^7}{7!}\frac{\partial b_7}{\partial \alpha_3} + (x-a)^8\frac{\partial R}{\partial \alpha_3}.$$

If an identical linear relation

$$\sum_{r=1}^{8} \lambda_r \phi_r(x) = 0$$

could exist, in which the coefficients $\lambda$ are constants, manifestly

$$\lambda_1 = 0, \quad \lambda_2 = 0, \quad \lambda_3 = 0, \quad \lambda_4 = 0,$$

so that the terms $(x-a)^0$, $(x-a)^1$, $(x-a)^2$, $(x-a)^3$ may disappear from the identical relation. And then

$$\lambda_5\frac{\partial b_4}{\partial \alpha_0} + \lambda_6\frac{\partial b_4}{\partial \alpha_1} + \lambda_7\frac{\partial b_4}{\partial \alpha_2} + \lambda_8\frac{\partial b_4}{\partial \alpha_3} = 0,$$

$$\lambda_5\frac{\partial b_5}{\partial \alpha_0} + \lambda_6\frac{\partial b_5}{\partial \alpha_1} + \lambda_7\frac{\partial b_5}{\partial \alpha_2} + \lambda_8\frac{\partial b_5}{\partial \alpha_3} = 0,$$

$$\lambda_5\frac{\partial b_6}{\partial \alpha_0} + \lambda_6\frac{\partial b_6}{\partial \alpha_1} + \lambda_7\frac{\partial b_6}{\partial \alpha_2} + \lambda_8\frac{\partial b_6}{\partial \alpha_3} = 0,$$

$$\lambda_5\frac{\partial b_7}{\partial \alpha_0} + \lambda_6\frac{\partial b_7}{\partial \alpha_1} + \lambda_7\frac{\partial b_7}{\partial \alpha_2} + \lambda_8\frac{\partial b_7}{\partial \alpha_3} = 0,$$

so that the terms in $(x-a)^4$, $(x-a)^5$, $(x-a)^6$, $(x-a)^7$ may disappear from that relation. The determinant of the coefficients of $\lambda_5$, $\lambda_6$, $\lambda_7$, $\lambda_8$ on the left-hand sides is

$$J\left(\frac{b_4,\ b_5,\ b_6,\ b_7}{a_0,\ a_1,\ a_2,\ a_3}\right).$$

If this could vanish, we should be able to eliminate $a_0$, $a_1$, $a_2$, $a_3$ among the relations

$$b_4,\ b_5,\ b_6,\ b_7 = \text{functions of } a_0,\ a_1,\ a_2,\ a_3,\ \alpha_0,\ \alpha_1,\ \alpha_2,\ \alpha_3,$$

and have a resulting relation

$$Q\,(a_0,\ a_1,\ a_2,\ a_3,\ b_4,\ b_5,\ b_6,\ b_7) = 0,$$

that is, a relation

$$Q\,(b_0,\ b_1,\ b_2,\ b_3,\ b_4,\ b_5,\ b_6,\ b_7) = 0,$$

contrary to the hypothesis that $b_0$, $b_1$, ..., $b_7$ are independent arbitrary constants. Thus the Jacobian $J\left(\dfrac{b_4,\ b_5,\ b_6,\ b_7}{a_0,\ a_1,\ a_2,\ a_3}\right)$ does not vanish; and so the foregoing relations can be satisfied only if

$$\lambda_5 = 0, \quad \lambda_6 = 0, \quad \lambda_7 = 0, \quad \lambda_8 = 0.$$

Hence all the constants $\lambda$ are zero. No linear relation among the functions $\phi_1(x)$, ..., $\phi_8(x)$ can exist, with constant coefficients.

Similarly, no linear relation among the functions $\psi_1(x)$, ..., $\psi_8(x)$ can exist, with constant coefficients.

The proposition is thus established.

### Subsidiary characteristic equations.

**184.** The curve—manifestly a skew curve—which represents the primitive of the characteristic equations is called the characteristic curve. The differential equations are satisfied by the primitive for all values of the arbitrary constants; and therefore they are satisfied when a small variation is effected upon these constants, the variation being as arbitrary as we please. Accordingly, we take a small variation

$$a_0 + \kappa g_1,\ \ a_1 + \kappa g_2,\ \ a_2 + \kappa g_3,\ \ a_3 + \kappa g_4,\ \text{ for } a_0,\ a_1,\ a_2,\ a_3,$$
$$\alpha_0 + \kappa g_5,\ \ \alpha_1 + \kappa g_6,\ \ \alpha_2 + \kappa g_7,\ \ \alpha_3 + \kappa g_8,\ \dots\ \alpha_0,\ \alpha_1,\ \alpha_2,\ \alpha_3,$$

in the constants, where $g_1$, ..., $g_8$ are themselves arbitrary constants. When we denote the changed values of $y$ and $z$ by $y + \kappa\eta$ and $z + \kappa\zeta$, we have

$$\eta = \sum_{r=1}^{8} g_r\,\phi_r(x), \quad \zeta = \sum_{r=1}^{8} g_r\,\psi_r(x),$$

when $\kappa$ is made very small.

But the curve given by $y + \kappa\eta$ and $z + \kappa\zeta$, as arising out of $y$ and $z$ through the small variations of the arbitrary constants, is also a characteristic curve. The differential equations satisfied by $y + \kappa\eta$ and $z + \kappa\zeta$ are the same as those satisfied by $y$ and $z$: and therefore parts of those equations, which appear as increments through the changes of $y$ and $z$ into $y + \kappa\eta$ and $z + \kappa\zeta$, must disappear. To express these increments, let

$$
\begin{aligned}
2\Omega = \ & f_{66}\eta''^2 + 2f_{69}\eta''\zeta'' + f_{99}\zeta''^2 \\
& + 2\left(f_{56}\eta'\eta'' + f_{59}\eta'\zeta'' + f_{68}\zeta'\eta'' + f_{89}\zeta'\zeta''\right) \\
& + 2\left(f_{46}\eta\eta'' + f_{49}\eta\zeta'' + f_{67}\zeta\eta'' + f_{79}\zeta\zeta''\right) \\
& + f_{55}\eta'^2 + 2f_{58}\eta'\zeta' + f_{88}\zeta'^2 \\
& + 2\left(f_{45}\eta\eta' + f_{48}\eta\zeta' + f_{57}\zeta\eta' + f_{78}\zeta\zeta'\right) \\
& + f_{44}\eta^2 + 2f_{47}\eta\zeta + f_{77}\zeta^2.
\end{aligned}
$$

The new value of $\dfrac{\partial f}{\partial y}$ is

$$
\frac{\partial f}{\partial y} + \kappa\left(\frac{\partial^2 f}{\partial y^2}\eta + \frac{\partial^2 f}{\partial y\,\partial z}\zeta + \frac{\partial^2 f}{\partial y\,\partial y'}\eta' + \frac{\partial^2 f}{\partial y\,\partial z'}\zeta' + \frac{\partial^2 f}{\partial y\,\partial y''}\eta'' + \frac{\partial^2 f}{\partial y\,\partial z''}\zeta''\right),
$$

on neglecting squares and higher powers of $\kappa$: that is, it is

$$
= \frac{\partial f}{\partial y} + \kappa\left(f_{44}\eta + f_{47}\zeta + f_{45}\eta' + f_{48}\zeta' + f_{46}\eta'' + f_{49}\zeta''\right)
$$

$$
= \frac{\partial f}{\partial y} + \kappa\frac{\partial\Omega}{\partial\eta}.
$$

The new value of $\dfrac{\partial f}{\partial y'}$ is

$$
\frac{\partial f}{\partial y'} + \kappa\frac{\partial\Omega}{\partial\eta'},
$$

and the new value of $\dfrac{\partial f}{\partial y''}$ is

$$
\frac{\partial f}{\partial y''} + \kappa\frac{\partial\Omega}{\partial\eta''}.
$$

The first of the characteristic equations now is

$$
\frac{\partial f}{\partial y} + \kappa\frac{\partial\Omega}{\partial\eta} - \frac{d}{dx}\left(\frac{\partial f}{\partial y'} + \kappa\frac{\partial\Omega}{\partial\eta'}\right) + \frac{d^2}{dx^2}\left(\frac{\partial f}{\partial y''} + \kappa\frac{\partial\Omega}{\partial\eta''}\right) = 0,
$$

and it is to be the same as the first of the characteristic equations in its original form. Hence

$$
\frac{\partial\Omega}{\partial\eta} - \frac{d}{dx}\left(\frac{\partial\Omega}{\partial\eta'}\right) + \frac{d^2}{dx^2}\left(\frac{\partial\Omega}{\partial\eta''}\right) = 0.
$$

Similarly, from the 'fact that the second of the characteristic equations is unaltered, we have

$$
\frac{\partial\Omega}{\partial\zeta} - \frac{d}{dx}\left(\frac{\partial\Omega}{\partial\zeta'}\right) + \frac{d^2}{dx^2}\left(\frac{\partial\Omega}{\partial\zeta''}\right) = 0.
$$

In these equations, the twenty-one coefficients $f_{mn}$ (for $m$, $n = 4$, 5, 6, 7, 8, 9) initially are functions of $y$, $y'$, $y''$, $z$, $z'$, $z''$. As we are deducing another characteristic from the curve initially supposed known, the values of these six arguments as functions of $x$ are substituted in the said coefficients, which thus become functions of $x$ alone. Consequently, the two equations

$$\left.\begin{array}{l} \dfrac{\partial \Omega}{\partial \eta} - \dfrac{d}{dx}\left(\dfrac{\partial \Omega}{\partial \eta'}\right) + \dfrac{d^2}{dx^2}\left(\dfrac{\partial \Omega}{\partial \eta''}\right) = 0 \\[2mm] \dfrac{\partial \Omega}{\partial \zeta} - \dfrac{d}{dx}\left(\dfrac{\partial \Omega}{\partial \zeta'}\right) + \dfrac{d^2}{dx^2}\left(\dfrac{\partial \Omega}{\partial \zeta''}\right) = 0 \end{array}\right\}$$

are linear equations in $\eta$ and $\zeta$ as the dependent variables, with coefficients that are functions of $x$. As the equations have been derived from the original characteristic equations by small variations, they are called the *subsidiary characteristic equations.*

### *Primitive of the subsidiary equations.*

**185.** When the equations are expressed more fully by substituting the explicit values of the derivatives of $\Omega$, they become

$$f_{66}\eta'''' + f_{69}\zeta'''' + f_{66}'\,\eta''' + (f_{69}' + f_{68} - f_{59})\,\zeta''' + \ldots = 0,$$
$$f_{69}\eta'''' + f_{99}\zeta'''' + (f_{69}' + f_{59} - f_{68})\,\eta''' + f_{99}'\,\zeta''' + \ldots = 0,$$

the unstated terms being linear in $\eta$, $\eta'$, $\eta''$, $\zeta$, $\zeta'$, $\zeta''$. The earlier discussion shews that the primitive of these two equations contains eight arbitrary independent constants. The equations are linear in the dependent variables; hence these arbitrary constants occur linearly in the primitive. Thus we may take the primitive in the form

$$\eta = \sum_{r=1}^{8} g_r \eta_r, \quad \zeta = \sum_{r=1}^{8} g_r \zeta_r,$$

where $\eta_r$ and $\zeta_r$ are an integral-pair for $r = 1, \ldots, 8$, and where $g_1, \ldots, g_8$ are arbitrary independent constants, provided the eight integral-pairs are linearly independent: that is, provided linear relations

$$\sum_{r=1}^{8} A_r \eta_r = 0, \quad \sum_{r=1}^{8} A_r \zeta_r = 0,$$

with constant coefficients $A$, do not exist.

But we have seen that, when $y$ and $z$ are known integrals of the characteristic equations, other known integrals of those equations can be derived by small variations of the arbitrary constants in $y$ and $z$, so as to give

$$y + \kappa \sum_{r=1}^{8} g_r \phi_r (x), \quad z + \kappa \sum_{r=1}^{8} g_r \psi_r (x).$$

Moreover, it has been proved that simultaneous linear relations

$$\sum_{r=1}^{8} \lambda_r \phi_r (x) = 0, \quad \sum_{r=1}^{8} \lambda_r \psi_r (x) = 0,$$

with constant coefficients $\lambda$, do not exist. Hence we may take

$$\eta_r = \phi_r (x), \quad \zeta_r = \psi_r (x),$$

for $r = 1, \ldots, 8$; and therefore the primitive of the subsidiary characteristic equations is given by

$$\eta = \sum_{r=1}^{8} g_r \phi_r(x), \quad \zeta = \sum_{r=1}^{8} g_r \psi_r(x),$$

where $g_1, \ldots, g_8$ are eight independent arbitrary constants.

### Invariant property of a fundamental group of integral-sets of the subsidiary equations.

**186.** Consider the determinant constructed from the eight functions $\phi_r(x)$ and the eight functions $\psi_r(x)$, and from their first three derivatives, the diagonal leading term being

$$\phi_1'''(x)\,\phi_2''(x)\,\phi_3'(x)\,\phi_4(x)\,\psi_5'''(x)\,\psi_6''(x)\,\psi_7'(x)\,\psi_8(x).$$

Let the value of the determinant be denoted by $\Psi$. Then $\dfrac{d\Psi}{dx}$ is the sum of two determinants of eight rows. The first of them has

$$\phi_1''''(x)\,\phi_2''(x)\,\phi_3'(x)\,\phi_4(x)\,\psi_5'''(x)\,\psi_6''(x)\,\psi_7'(x)\,\psi_8(x),$$

for its leading term; and the second has

$$\phi_1'''(x)\,\phi_2''(x)\,\phi_3'(x)\,\phi_4(x)\,\psi_5''''(x)\,\psi_6''(x)\,\psi_7'(x)\,\psi_8(x)$$

for its leading term. Now $\phi_r(x)$ and $\psi_r(x)$, for $r = 1, \ldots, 8$, are an integral-pair satisfying the subsidiary characteristic equations: and therefore, for each such integral-pair,

$$f_{66}\phi_r'''' + f_{69}\psi_r'''' + f_{66}'\phi_r''' + (f_{69}' + f_{68} - f_{59})\psi_r''' + \ldots = 0,$$

$$f_{69}\phi_r'''' + f_{99}\psi_r'''' + (f_{69}' + f_{59} - f_{68})\phi_r''' + f_{99}'\psi_r''' + \ldots = 0.$$

Consequently

$$\sigma\phi_r'''' + \{f_{66}'f_{99} - f_{69}(f_{69}' + f_{59} - f_{68})\}\phi_r''' + \{f_{99}(f_{69}' + f_{68} - f_{59}) - f_{69}f_{99}'\}\psi_r''' + \ldots = 0,$$

$$\sigma\psi_r'''' + \{f_{66}(f_{69}' + f_{59} - f_{68}) - f_{69}f_{66}'\}\phi_r''' + \{f_{66}f_{99}' - f_{69}(f_{69}' + f_{68} - f_{59})\}\psi_r''' + \ldots = 0,$$

where $\sigma$ denotes the quantity $f_{66}f_{99} - f_{69}^2$. When these values of $\phi_r''''$ and $\psi_r''''$ (for $r = 1, \ldots, 8$) are substituted in the two determinants that occur in the expression for $\dfrac{d\Psi}{dx}$, we have

$$\sigma\frac{d\Psi}{dx} = -\{f_{66}'f_{99} - f_{69}(f_{69}' + f_{59} - f_{68})\}\Psi - \{f_{66}f_{99}' - f_{69}(f_{69}' + f_{68} - f_{59})\}\Psi$$

$$= -\Psi\frac{d\sigma}{dx};$$

and therefore

$$\Psi\sigma = \text{constant}.$$

Now $\Psi$ cannot be zero, because linear relations

$$\sum_{r=1}^{8} A_r \phi_r(x) = 0, \quad \sum_{r=1}^{8} A_r \psi_r(x) = 0,$$

with constant coefficients $A$, do not exist. For reasons that will appear subsequently, $\sigma$ is always positive throughout the range of integration. Thus the constant, which is the value of $\Psi\sigma$, cannot be zero; let it be denoted by $C$. On the other hand, $C$ is not an arbitrary constant; for, when the characteristic curve is settled, no new arbitrary element enters into $\sigma$; and the functions $\phi_r(x)$ and $\psi_r(x)$ are specific. Thus $C$ is a specific constant, different from zero, and dependent upon the specific choice of the integral-sets $\phi_r(x)$ and $\psi_r(x)$ of the subsidiary characteristic equations. Hence we have

$$\Psi\sigma = C,$$

where $C$ is a specific constant. Consequently, the function $\Psi(x)$ does not become zero or infinite within the range of integration; it keeps the same sign throughout that range.

Moreover, this result is substantially independent of the particular choice of the fundamental system of eight integral-sets $\phi_r(x)$ and $\psi_r(x)$, of the subsidiary characteristic equations. If another fundamental system of eight integral-sets, $\chi_r(x)$ and $\omega_r(x)$, were chosen, we should have relations

$$\chi_r(x) = \sum_{s=1}^{8} k_{rs}\phi_r(x),$$

$$\omega_r(x) = \sum_{s=1}^{8} k_{rs}\psi_r(x),$$

for $r = 1, \ldots, 8$, where the determinant $|k_{rs}|$ of the constant coefficients $k$ is different from zero. Then

$$\Psi\{\chi_r(x),\, \omega_r(x)\} = |k_{rs}|\,\Psi\{\phi_r(x),\, \psi_r(x)\}$$

$$= |k_{rs}|\,\frac{C}{\sigma},$$

so that the result of changing from one fundamental system of eight integral sets to another such system is merely to change the specific constant value of the function $\Psi\sigma$.

In fact, the function $\Psi(x)$ is covariantive for all forms of primitive of the subsidiary characteristic equations.

*Particular properties of combinations of integrals of the subsidiary equations.*

**187.** A few properties of the subsidiary characteristic equations

$$\left.\begin{aligned}
\frac{\partial\Omega}{\partial\eta} - \frac{d}{dx}\left(\frac{\partial\Omega}{\partial\eta'}\right) + \frac{d^2}{dx^2}\left(\frac{\partial\Omega}{\partial\eta''}\right) = 0 \\[2mm]
\frac{\partial\Omega}{\partial\zeta} - \frac{d}{dx}\left(\frac{\partial\Omega}{\partial\zeta'}\right) + \frac{d^2}{dx^2}\left(\frac{\partial\Omega}{\partial\zeta''}\right) = 0
\end{aligned}\right\}$$

can be established at this stage, with a view to their use in the later reduction of the second variation to its normal quadratic form.

We denote by $\Omega_r$ the result of substituting $\eta_r$ and $\zeta_r$ (for $r = 1, \ldots, 8$) in place of $\eta$ and $\zeta$ in the quantity $\Omega$. Then we have, identically,

$$\eta_1 \frac{\partial \Omega_r}{\partial \eta_r} + \eta_1' \frac{\partial \Omega_r}{\partial \eta_r'} + \eta_1'' \frac{\partial \Omega_r}{\partial \eta_r''} + \zeta_1 \frac{\partial \Omega_r}{\partial \zeta_r} + \zeta_1' \frac{\partial \Omega_r}{\partial \zeta_r'} + \zeta_1'' \frac{\partial \Omega_r}{\partial \zeta_r''}$$

$$= \eta_r \frac{\partial \Omega_1}{\partial \eta_1} + \eta_r' \frac{\partial \Omega_1}{\partial \eta_1'} + \eta_r'' \frac{\partial \Omega_1}{\partial \eta_1''} + \zeta_r \frac{\partial \Omega_1}{\partial \zeta_1} + \zeta_r' \frac{\partial \Omega_1}{\partial \zeta_1'} + \zeta_r'' \frac{\partial \Omega_1}{\partial \zeta_1''},$$

because each side is equal to one and the same quantity, lineo-linear in the two sets of quantities $\eta_1$ and $\zeta_1$, $\eta_r$ and $\zeta_r$, and their derivatives. But

$$\eta_1 \frac{\partial \Omega_r}{\partial \eta_r} + \eta_1' \frac{\partial \Omega_r}{\partial \eta_r'} + \eta_1'' \frac{\partial \Omega_r}{\partial \eta_r''} = \eta_1' \frac{\partial \Omega_r}{\partial \eta_r'} + \eta_1'' \frac{\partial \Omega_r}{\partial \eta_r''} + \eta_1 \left\{ \frac{d}{dx} \left( \frac{\partial \Omega_r}{\partial \eta_r'} \right) - \frac{d^2}{dx^2} \left( \frac{\partial \Omega_r}{\partial \eta_r''} \right) \right\}$$

$$= \frac{d}{dx} \left[ \eta_1 \left\{ \frac{\partial \Omega_r}{\partial \eta_r'} - \frac{d}{dx} \left( \frac{\partial \Omega_r}{\partial \eta_r''} \right) \right\} + \eta_1' \frac{\partial \Omega_r}{\partial \eta_r''} \right],$$

with similar modifications for the other three combinations of terms. Hence

$$\eta_1 \left\{ \frac{\partial \Omega_r}{\partial \eta_r'} - \frac{d}{dx} \left( \frac{\partial \Omega_r}{\partial \eta_r''} \right) \right\} + \eta_1' \frac{\partial \Omega_r}{\partial \eta_r''} + \zeta_1 \left\{ \frac{\partial \Omega_r}{\partial \zeta_r'} - \frac{d}{dx} \left( \frac{\partial \Omega_r}{\partial \zeta_r''} \right) \right\} + \zeta_1' \frac{\partial \Omega_r}{\partial \zeta_r''}$$

$$= \eta_r \left\{ \frac{\partial \Omega_1}{\partial \eta_1'} - \frac{d}{dx} \left( \frac{\partial \Omega_1}{\partial \eta_1''} \right) \right\} + \eta_r' \frac{\partial \Omega_1}{\partial \eta_1''} + \zeta_r \left\{ \frac{\partial \Omega_1}{\partial \zeta_1'} - \frac{d}{dx} \left( \frac{\partial \Omega_1}{\partial \zeta_1''} \right) \right\} + \zeta_r' \frac{\partial \Omega_1}{\partial \zeta_1''} + C_r,$$

where $C_r$ is a constant, obviously specific and not arbitrary.

When $C_r$ is not a zero constant, we can always take another integral-set $\eta$ and $\zeta$, such that the new constant for the new functions is actually zero. Let the foregoing relation be written

$$g(1, r) = h(1, r) + C_r.$$

Take another integral-set, $\eta_s$ and $\zeta_s$, linearly independent of $\eta_1$ and $\zeta_1$, and linearly independent also of $\eta_r$ and $\zeta_r$; combining it similarly with $\eta_1$ and $\zeta_1$, we have

$$g(1, s) = h(1, s) + C_s,$$

where $C_s$ is a constant. The unfavourable circumstance would be that $C_s$ should be zero. In that event, we take

$$\eta_t = C_s \eta_r - C_r \eta_s, \quad \zeta_t = C_s \zeta_r - C_r \zeta_s,$$

so that $\eta_t$ and $\zeta_t$ is an integral-set, linearly independent of $\eta_1$ and $\zeta_1$, of $\eta_r$ and $\zeta_r$, and of $\eta_s$ and $\zeta_s$. Then, for $\eta_t$ and $\zeta_t$ thus defined,

$$g(1, t) = h(1, t),$$

a combination of $\eta_t$ and $\zeta_t$ with $\eta_1$ and $\zeta_1$, such that the specific constant $C$ is actually zero.

As any integral-set is linearly expressible (with constant coefficients) in terms of the eight sets in a fundamental system, we clearly may make $\eta_t$ and $\zeta_t$ take the place of $\eta_s$ and $\zeta_s$ in the fundamental system. Moreover, $\eta_s$ and $\zeta_s$ were taken to be any integral-set, linearly independent of $\eta_1$ and $\zeta_1$, and of $\eta_r$ and $\zeta_r$; hence, even if $C_r$ were not zero, six choices of integral-sets are possible, after $\eta_1$ and $\zeta_1$, $\eta_r$ and $\zeta_r$, have been chosen, such that the specific constant $C$ for the combination is zero. Hence we have the theorem* that *an integral-set $\eta_t$ and $\zeta_t$ can be chosen in a number of ways so that the relation*

$$\eta_1 \left\{ \frac{\partial \Omega_t}{\partial \eta_t'} - \frac{d}{dx} \left( \frac{\partial \Omega_t}{\partial \eta_t''} \right) \right\} + \eta_1' \frac{\partial \Omega_t}{\partial \eta_t''} + \zeta_1 \left\{ \frac{\partial \Omega_t}{\partial \zeta_t'} - \frac{d}{dx} \left( \frac{\partial \Omega_t}{\partial \zeta_t''} \right) \right\} + \zeta_1' \frac{\partial \Omega_t}{\partial \zeta_t''}$$

$$= \eta_t \left\{ \frac{\partial \Omega_1}{\partial \eta_1'} - \frac{d}{dx} \left( \frac{\partial \Omega_1}{\partial \eta_1''} \right) \right\} + \eta_t' \frac{\partial \Omega_1}{\partial \eta_1''} + \zeta_t \left\{ \frac{\partial \Omega_1}{\partial \zeta_1'} - \frac{d}{dx} \left( \frac{\partial \Omega_1}{\partial \zeta_1''} \right) \right\} + \zeta_t' \frac{\partial \Omega_1}{\partial \zeta_1''}.$$

*holds, where the set $\eta_t$ and $\zeta_t$ is distinct from the set $\eta_1$ and $\zeta_1$.*

### Six fundamental relations.

**188.** The foregoing analysis shews that it is certainly possible to make six selections of integral-sets $\eta_r$ and $\zeta_r$, linearly independent of $\eta_1$ and $\zeta_1$, such that the foregoing relation holds for each selection : let these sets be denoted by $t = 2, 3, 4, 5, 6, 7$, so that, for these values

$$g(1, t) = h(1, t).$$

Now take $\eta_2$ and $\zeta_2$ as an initial set, and let $m$ denote any one of the integers $3, 4, 5, 6, 7$. As before, we have

$$g(2, m) = h(2, m) + B_m,$$

where $B_m$ is a specific constant. Should $B_m$ not be zero—and we shall allow the least favourable circumstances, by assuming that no one of the constants $B_m$ is zero—we take new combinations

$$B_m \eta_n - B_n \eta_m = y_m, \quad B_m \zeta_n - B_n \zeta_m = z_m,$$

for $m = 3, 4, 5, 6$. Combining (i) $\eta_1$ and $\zeta_1$ with $y_m$ and $z_m$, and (ii) $\eta_2$ and $\zeta_2$ with $y_m$ and $z_m$, we have

$$g(1, m) = h(1, m),$$
$$g(2, m) = h(2, m),$$

for $m = 3, 4, 5, 6$.

Lastly, take $y_3$ and $z_3$ as an initial set, and let $n$ denote any one of the integers $4, 5, 6$. Then, as before, combining $y_3$ and $z_3$ with $y_n$ and $z_n$, we have

$$g(3, n) = h(3, n) + C_n,$$

where $C_n$ is a specific constant. Should $C_n$ not be zero—and again we shall allow the least favourable circumstances by assuming that no one of the constants $C_n$ is zero—we take new combinations

$$C_6 y_n - C_n y_6 = Y_n, \quad C_6 z_n - C_n z_6 = Z_n,$$

for $n = 4, 5$. Combining (i) $\eta_1$ and $\zeta_1$ with $Y_n$ and $Z_n$, (ii) $y_2$ and $z_2$ with $Y_n$ and $Z_n$, and (iii) $y_3$ and $z_3$ with $Y_n$ and $Z_n$, we have

$$g(1, n) = h(1, n),$$
$$g(2, n) = h(2, n),$$
$$g(3, n) = h(3, n),$$

for $n = 4$, and $n = 5$.

* In its present form, it is an extension of the theorem, proved in § 143 and originally due to Clebsch.

All these integral-sets $\eta_1$ and $\zeta_1$, $\eta_2$ and $\zeta_2$, $y_3$ and $z_3$, $Y_4$ and $Z_4$, $Y_5$ and $Z_5$ are linearly independent of one another. The fundamental set of integrals, for the linear expression of all particular integral-sets, can be modified so as to include, in an altered aggregate, these five sets; and this composition of the fundamental set has been made in the circumstances least favourable to our present aim.

We now take any four of the five sets, which certainly possess the specified property. Let them be $\eta_r$ and $\zeta_r$, for $r = 1, 2, 3, 4$ : we have six relations

$$\eta_m \left\{ \frac{\partial \Omega_n}{\partial \eta_n{}'} - \frac{d}{dx}\left(\frac{\partial \Omega_n}{\partial \eta_n{}''}\right)\right\} + \eta_m{}' \frac{\partial \Omega_n}{\partial \eta_n{}''} + \zeta_m \left\{ \frac{\partial \Omega_n}{\partial \zeta_n{}'} - \frac{d}{dx}\left(\frac{\partial \Omega_n}{\partial \zeta_n{}''}\right)\right\} + \zeta_m{}' \frac{\partial \Omega_n}{\partial \zeta_n{}''}$$

$$= \eta_n \left\{ \frac{\partial \Omega_m}{\partial \eta_m{}'} - \frac{d}{dx}\left(\frac{\partial \Omega_m}{\partial \eta_m{}''}\right)\right\} + \eta_n{}' \frac{\partial \Omega_m}{\partial \eta_m{}''} + \zeta_n \left\{ \frac{\partial \Omega_m}{\partial \zeta_m{}'} - \frac{d}{dx}\left(\frac{\partial \Omega_m}{\partial \zeta_m{}''}\right)\right\} + \zeta_n{}' \frac{\partial \Omega_m}{\partial \zeta_m{}''},$$

for the six arrangements out of $m, n, = 1, 2, 3, 4$. We take them in turn for the arrangements $m, n, = 1, 2 ; = 2, 3 ; = 3, 1 ; = 1, 4 ; = 2, 4 ; = 3, 4$.

First, multiply them in respective succession by

$$\begin{vmatrix} \eta_3, & \eta_4 \\ \zeta_3, & \zeta_4 \end{vmatrix}, \quad \begin{vmatrix} \eta_1, & \eta_4 \\ \zeta_1, & \zeta_4 \end{vmatrix}, \quad \begin{vmatrix} \eta_2, & \eta_4 \\ \zeta_2, & \zeta_4 \end{vmatrix}, \quad \begin{vmatrix} \eta_2, & \eta_3 \\ \zeta_2, & \zeta_3 \end{vmatrix}, \quad \begin{vmatrix} \eta_3, & \eta_1 \\ \zeta_3, & \zeta_1 \end{vmatrix}, \quad \begin{vmatrix} \eta_1, & \eta_2 \\ \zeta_1, & \zeta_2 \end{vmatrix}$$

and add. Then

$$\begin{vmatrix} \dfrac{\partial \Omega_1}{\partial \eta_1{}''}, & \eta_1{}', & \eta_1, & \zeta_1 \\[2mm] \dfrac{\partial \Omega_2}{\partial \eta_2{}''}, & \eta_2{}', & \eta_2, & \zeta_2 \\[2mm] \dfrac{\partial \Omega_3}{\partial \eta_3{}''}, & \eta_3{}', & \eta_3, & \zeta_3 \\[2mm] \dfrac{\partial \Omega_4}{\partial \eta_4{}''}, & \eta_4{}', & \eta_4, & \zeta_4 \end{vmatrix} + \begin{vmatrix} \dfrac{\partial \Omega_1}{\partial \zeta_1{}''}, & \zeta_1{}', & \eta_1, & \zeta_1 \\[2mm] \dfrac{\partial \Omega_2}{\partial \zeta_2{}''}, & \zeta_2{}', & \eta_2, & \zeta_2 \\[2mm] \dfrac{\partial \Omega_3}{\partial \zeta_3{}''}, & \zeta_3{}', & \eta_3, & \zeta_3 \\[2mm] \dfrac{\partial \Omega_4}{\partial \zeta_4{}''}, & \zeta_4{}', & \eta_4, & \zeta_4 \end{vmatrix} = 0,$$

which may be briefly stated in the form

$$\begin{vmatrix} \dfrac{\partial \Omega_1}{\partial \eta_1{}''}, & \eta_2{}', & \eta_3, & \zeta_4 \end{vmatrix} + \begin{vmatrix} \dfrac{\partial \Omega_1}{\partial \zeta_1{}''}, & \zeta_2{}', & \eta_3, & \zeta_4 \end{vmatrix} = 0.$$

Secondly, multiply them in respective succession by

$$\begin{vmatrix} \zeta_l{}', & \zeta_l \\ \zeta_r{}', & \zeta_r \end{vmatrix},$$

for $l, r, = 3, 4 ; = 1, 4 ; = 2, 4 ; = 2, 3 ; = 3, 1 ; = 1, 2$; and add. Then

$$\begin{vmatrix} \dfrac{\partial \Omega_1}{\partial \eta_1{}''}, & \eta_2{}', & \zeta_3{}', & \zeta_4 \end{vmatrix} + \begin{vmatrix} \dfrac{\partial \Omega_1}{\partial \eta_1{}'} - \dfrac{d}{dx}\left(\dfrac{\partial \Omega_1}{\partial \eta_1{}''}\right), & \eta_2, & \zeta_3{}', & \zeta_4 \end{vmatrix} = 0.$$

Thirdly, multiply them in respective succession by

$$\begin{vmatrix} \eta_l{}', & \zeta_l \\ \eta_r{}', & \zeta_r \end{vmatrix},$$

for the same combinations of $l$ and $r$; and add. Then

$$\left|\begin{array}{cccc} \dfrac{\partial\Omega_1}{\partial\zeta_1''}, & \eta_2', & \zeta_3', & \zeta_4 \end{array}\right| + \left|\begin{array}{cccc} \dfrac{\partial\Omega_1}{\partial\eta_1} - \dfrac{d}{dx}\left(\dfrac{\partial\Omega_1}{\partial\eta_1''}\right), & \eta_2', & \eta_3, & \zeta_4 \end{array}\right| = 0.$$

Fourthly, multiply them in respective succession by

$$\left|\begin{array}{cc} \zeta_l', & \eta_l \\ \zeta_r', & \eta_r \end{array}\right|,$$

for the same combinations of $l$ and $r$; and add. Then

$$\left|\begin{array}{cccc} \dfrac{\partial\Omega_1}{\partial\eta_1''}, & \eta_2', & \zeta_3', & \eta_4 \end{array}\right| + \left|\begin{array}{cccc} \dfrac{\partial\Omega_1}{\partial\zeta_1} - \dfrac{d}{dx}\left(\dfrac{\partial\Omega_1}{\partial\zeta_1''}\right), & \zeta_2, & \zeta_3', & \eta_4 \end{array}\right| = 0.$$

Fifthly, multiply them in respective succession by

$$\left|\begin{array}{cc} \eta_l', & \eta_l \\ \eta_r', & \eta_r \end{array}\right|,$$

for the same combinations of $l$ and $r$; and add. Then

$$\left|\begin{array}{cccc} \dfrac{\partial\Omega_1}{\partial\zeta_1''}, & \zeta_2', & \eta_3', & \eta_4 \end{array}\right| + \left|\begin{array}{cccc} \dfrac{\partial\Omega_1}{\partial\zeta_1} - \dfrac{d}{dx}\left(\dfrac{\partial\Omega_1}{\partial\zeta_1''}\right), & \zeta_2, & \eta_3', & \eta_4 \end{array}\right| = 0.$$

Sixthly and lastly, multiply them in respective succession by

$$\left|\begin{array}{cc} \eta_l', & \zeta_l' \\ \eta_r', & \zeta_r' \end{array}\right|,$$

for the same combinations of $l$ and $r$; and add. Then

$$\left|\begin{array}{cccc} \dfrac{\partial\Omega_1}{\partial\eta_1} - \dfrac{d}{dx}\left(\dfrac{\partial\Omega_1}{\partial\eta_1''}\right), & \eta_2', & \zeta_3', & \eta_4 \end{array}\right| + \left|\begin{array}{cccc} \dfrac{\partial\Omega_1}{\partial\zeta_1} - \dfrac{d}{dx}\left(\dfrac{\partial\Omega_1}{\partial\zeta_1''}\right), & \eta_2', & \zeta_3', & \zeta_4 \end{array}\right| = 0.$$

### First variation, by the method of Weierstrass.

**189.** The second variation of the original integral will be considered only for the special variations $y + \kappa Y$, $z + \kappa Z$, of the variables $y$ and $z$, while $x$ is left unvaried save possibly at the limits. The whole problem will not be discussed for general weak variations. We shall therefore now deal with weak variations which affect all the three variables $x$, $y$, $z$, only so far as the first variation is concerned. The investigation is desirable, not least because the customary analysis for skew curves gives no preference to one of the variables, over the other two, as the independent variable. The characteristic equations and the terminal conditions will be identified, for the special weak variations and the general weak variations: and a proposition as to the continuity of certain quantities through a free discontinuity will be obtained. But that will mark the limit of the detailed use of the Weierstrass method. (See § 206, for statements of results.)

We make $x$, $y$, $z$ to be functions of a new independent variable $t$, which can remain unspecified intrinsically; after characteristic (and other) equations have been formed, we frequently can select $s$, the length of the arc of the curve from a fixed point, as an intrinsic independent variable. Moreover, this mode of proceeding enables us to deal with strong variations where, as will be seen, the ordinary 'first' variation and the ordinary 'second' variation of the original integral do not occur. (The subject of integration does not admit of expansion in powers of $\kappa$ for strong variations, because these imply abrupt changes of the arguments.)

Derivatives of $x$, $y$, $z$ with respect to $t$ are denoted by suffix numbers; thus

$$\frac{dx}{dt} = x_1, \quad \frac{d^2x}{dt^2} = x_2,$$

and similarly for $y$ and for $z$. Hence

$$y' = \frac{y_1}{x_1}, \qquad\qquad z' = \frac{z_1}{x_1},$$

$$y'' = \frac{1}{x_1^3}(x_1 y_2 - y_1 x_2), \quad z'' = \frac{1}{x_1^3}(x_1 z_2 - z_1 x_2);$$

and the integral $\int f(x, y, y', y'', z, z', z'')\, dx$ becomes

$$\int F(x, x_1, x_2, y, y_1, y_2, z, z_1, z_2)\, dt,$$

where

$$F(x, x_1, x_2, y, y_1, y_2, z, z_1, z_2) = x_1 f(x, y, y', y'', z, z', z'').$$

As in earlier instances (§§ 38, 97, 156), the function $F$ satisfies identities, expressible in the form of partial differential equations. We have

$$\frac{\partial F}{\partial x_2} = -\frac{y_1}{x_1^2}\frac{\partial f}{\partial y''} - \frac{z_1}{x_1^2}\frac{\partial f}{\partial z''},$$

$$\frac{\partial F}{\partial y_2} = \frac{1}{x_1}\frac{\partial f}{\partial y''},$$

$$\frac{\partial F}{\partial z_2} = \frac{1}{x_1}\frac{\partial f}{\partial z''},$$

$$\frac{\partial F}{\partial x_1} = f + x_1\left\{\frac{\partial f}{\partial y'}\left(-\frac{y_1}{x_1^2}\right) + \frac{\partial f}{\partial z'}\left(-\frac{z_1}{x_1^2}\right)\right.$$
$$\left. + \frac{\partial f}{\partial y''}\left(-\frac{2y_2}{x_1^3} + 3\frac{y_1}{x_1^4}x_2\right) + \frac{\partial f}{\partial z''}\left(-\frac{2z_2}{x_1^3} + 3\frac{z_1}{x_1^4}x_2\right)\right\},$$

$$\frac{\partial F}{\partial y_1} = x_1\left\{\frac{\partial f}{\partial y'}\frac{1}{x_1} + \frac{\partial f}{\partial y''}\left(-\frac{x_2}{x_1^3}\right)\right\},$$

$$\frac{\partial F}{\partial z_1} = x_1\left\{\frac{\partial f}{\partial z'}\frac{1}{x_1} + \frac{\partial f}{\partial z''}\left(-\frac{x_2}{x_1^3}\right)\right\}.$$

Consequently, two identities are to be expected after the elimination of $f$, $\dfrac{\partial f}{\partial y'}$, $\dfrac{\partial f}{\partial z'}$, $\dfrac{\partial f}{\partial y''}$, $\dfrac{\partial f}{\partial z''}$; and they are*

$$\left.\begin{aligned} x_1 \frac{\partial F}{\partial x_2} + y_1 \frac{\partial F}{\partial y_2} + z_1 \frac{\partial F}{\partial z_2} &= 0 \\ F = x_1 \frac{\partial F}{\partial x_1} + y_1 \frac{\partial F}{\partial y_1} + z_1 \frac{\partial F}{\partial z_1} + 2x_2 \frac{\partial F}{\partial x_2} + 2y_2 \frac{\partial F}{\partial y_2} + 2z_2 \frac{\partial F}{\partial z_2} \end{aligned}\right\}.$$

Moreover, we have

$$\frac{\partial F}{\partial y_1} = \frac{\partial f}{\partial y'} - \frac{x_2}{x_1{}^2} \frac{\partial f}{\partial y''},$$

and therefore

$$\frac{\partial F}{\partial y_1} - \frac{d}{dt}\left(\frac{\partial F}{\partial y_2}\right) = \frac{\partial f}{\partial y'} - \frac{1}{x_1} \frac{d}{dt}\left(\frac{\partial f}{\partial y''}\right)$$

$$= \frac{\partial f}{\partial y'} - \frac{d}{dx}\left(\frac{\partial f}{\partial y''}\right);$$

and, similarly,

$$\frac{\partial F}{\partial z_1} - \frac{d}{dt}\left(\frac{\partial F}{\partial z_2}\right) = \frac{\partial f}{\partial z'} - \frac{d}{dx}\left(\frac{\partial f}{\partial z''}\right).$$

Further, we have (by the second identity)

$$\frac{\partial F}{\partial x_1} = f - \frac{y_1}{x_1}\frac{\partial F}{\partial y_1} - \frac{z_1}{x_1}\frac{\partial F}{\partial z_1} - 2\frac{x_2}{x_1}\frac{\partial F}{\partial x_2} - 2\frac{y_2}{x_1}\frac{\partial F}{\partial y_2} - 2\frac{z_2}{x_1}\frac{\partial F}{\partial z_2}$$

$$= f - y'\frac{\partial F}{\partial y_1} - z'\frac{\partial F}{\partial z_1} - 2y''\frac{\partial f}{\partial y''} - 2z''\frac{\partial f}{\partial z''},$$

on substituting for $\dfrac{\partial F}{\partial x_2}$, $\dfrac{\partial F}{\partial y_2}$, $\dfrac{\partial F}{\partial z_2}$; also

$$\frac{\partial F}{\partial x_2} = -y'\frac{\partial F}{\partial y_2} - z'\frac{\partial F}{\partial z_2},$$

so that

$$\frac{d}{dt}\left(\frac{\partial F}{\partial x_2}\right) = -y'\frac{d}{dt}\left(\frac{\partial F}{\partial y_2}\right) - z'\frac{d}{dt}\left(\frac{\partial F}{\partial z_2}\right) - x_1 y''\frac{\partial F}{\partial y_2} - x_1 z''\frac{\partial F}{\partial z_2}$$

$$= -y'\frac{d}{dt}\left(\frac{\partial F}{\partial y_2}\right) - z'\frac{d}{dt}\left(\frac{\partial F}{\partial z_2}\right) - y''\frac{\partial f}{\partial y''} - z''\frac{\partial f}{\partial z''}:$$

---

* If they are regarded as simultaneous partial equations of the first order for the determination of an unknown quantity $F$, their most general integral is

$$F = x_1 f\left(x, y, \frac{y_1}{x_1}, \frac{x_1 y_2 - y_1 x_2}{x_1{}^3}, z, \frac{z_1}{x_1}, \frac{x_1 z_2 - z_1 x_2}{x_1{}^3}\right)$$

$$= x_1 f\left(x, y, y', y'', z, z', z''\right),$$

where $f$ is an arbitrary function of its arguments.

consequently

$$\frac{\partial F}{\partial x_1} - \frac{d}{dt}\left(\frac{\partial F}{\partial x_2}\right) = f - y'\left\{\frac{\partial F}{\partial y_1} - \frac{d}{dt}\left(\frac{\partial F}{\partial y_2}\right)\right\} - z'\left\{\frac{\partial F}{\partial z_1} - \frac{d}{dt}\left(\frac{\partial F}{\partial z_2}\right)\right\} - y''\frac{\partial f}{\partial y''} - z''\frac{\partial f}{\partial z''}$$

$$= f - y'\left\{\frac{\partial f}{\partial y'} - \frac{d}{dx}\left(\frac{\partial f}{\partial y''}\right)\right\} - z'\left\{\frac{\partial f}{\partial z'} - \frac{d}{dx}\left(\frac{\partial f}{\partial z''}\right)\right\} - y''\frac{\partial f}{\partial y''} - z''\frac{\partial f}{\partial z''}.$$

These results will be used in connection with the terms at the limits, arising out of the conditions for the evanescence of the first variation of the integral.

*Transformation of the first variation: characteristic equations.*

**190.** The varied value of the integral $\int F dt$, when $x$, $y$, $z$ are changed into $x + \kappa X$, $y + \kappa Y$, $z + \kappa Z$, while $\kappa$ is a small arbitrary constant, is

$$\int F(x + \kappa X, \ x_1 + \kappa X_1, \ x_2 + \kappa X_2, \ y + \kappa Y, \ y_1 + \kappa Y_1, \ y_2 + \kappa Y_2,$$
$$z + \kappa Z, \ z_1 + \kappa Z_1, \ z_2 + \kappa Z_2)\, dt\,;$$

and therefore, when the function in the latter integral is expanded in powers of $\kappa$, the increment of the integral is

$$\kappa\int\left(X\frac{\partial F}{\partial x} + X_1\frac{\partial F}{\partial x_1} + X_2\frac{\partial F}{\partial x_2} + Y\frac{\partial F}{\partial y} + Y_1\frac{\partial F}{\partial y_1} + Y_2\frac{\partial F}{\partial y_2}\right.$$
$$\left. + Z\frac{\partial F}{\partial z} + Z_1\frac{\partial F}{\partial z_1} + Z_2\frac{\partial F}{\partial z_2}\right) dt + K,$$

where $K$ represents the aggregate of terms which are of the second and higher degrees in $\kappa$. (As the variable $t$ is parametric for the curve, and account is taken of variations of $x$ as well as of $y$ and of $z$, the limits in the two integrals are the same.) Now

$$\int X_1\frac{\partial F}{\partial x_1}\, dt = \left[X\frac{\partial F}{\partial x_1}\right] - \int X\frac{d}{dt}\left(\frac{\partial F}{\partial x_1}\right) dt,$$

$$\int X_2\frac{\partial F}{\partial x_2}\, dt = \left[X_1\frac{\partial F}{\partial x_2} - X\frac{d}{dt}\left(\frac{\partial F}{\partial x_2}\right)\right] + \int X\frac{d^2}{dt^2}\left(\frac{\partial F}{\partial x_2}\right) dt\,;$$

hence the part of the foregoing integral, dependent upon the quantities $X$, $X_1$, $X_2$, is equal to

$$\int X\left\{\frac{\partial F}{\partial x} - \frac{d}{dt}\left(\frac{\partial F}{\partial x_1}\right) + \frac{d^2}{dt^2}\left(\frac{\partial F}{\partial x_2}\right)\right\} dt + \left[X\left\{\frac{\partial F}{\partial x_1} - \frac{d}{dt}\left(\frac{\partial F}{\partial x_2}\right)\right\} + X_1\frac{\partial F}{\partial x_2}\right],$$

where the terms outside the integration are to be taken at the limits of the original integral. The parts of the integral, dependent upon $Y$, $Y_1$, $Y_2$ and upon $Z$, $Z_1$, $Z_2$ respectively, are to be modified similarly; and so the integral, expressing the first variation, is

$$\int (XE_x + YE_y + ZE_z)\, dt$$

$$+ \left[ X \left\{ \frac{\partial F}{\partial x_1} - \frac{d}{dt} \left( \frac{\partial F}{\partial x_2} \right) \right\} + Y \left\{ \frac{\partial F}{\partial y_1} - \frac{d}{dt} \left( \frac{\partial F}{\partial y_2} \right) \right\} + Z \left\{ \frac{\partial F}{\partial z_1} - \frac{d}{dt} \left( \frac{\partial F}{\partial z_2} \right) \right\} \right.$$

$$\left. + X_1 \frac{\partial F}{\partial x_2} + Y_1 \frac{\partial F}{\partial y_2} + Z_1 \frac{\partial F}{\partial z_2} \right],$$

where the terms in the second line are taken at the limits, and

$$
\left.
\begin{aligned}
E_x &= \frac{\partial F}{\partial x} - \frac{d}{dt} \left( \frac{\partial F}{\partial x_1} \right) + \frac{d^2}{dt^2} \left( \frac{\partial F}{\partial x_2} \right) \\
E_y &= \frac{\partial F}{\partial y} - \frac{d}{dt} \left( \frac{\partial F}{\partial y_1} \right) + \frac{d^2}{dt^2} \left( \frac{\partial F}{\partial y_2} \right) \\
E_z &= \frac{\partial F}{\partial z} - \frac{d}{dt} \left( \frac{\partial F}{\partial z_1} \right) + \frac{d^2}{dt^2} \left( \frac{\partial F}{\partial z_2} \right)
\end{aligned}
\right\}.
$$

In order that the original integral may possess a maximum or a minimum, its first variation under weak variations must vanish; and therefore the foregoing expression must vanish for all admissible variations. The part of the expression, comprised by the integral, must vanish by itself; because there are variations, zero at the limits and not zero within the range. For such variations, the terms at the limits vanish; while, subject to the acquisition of zero values at the limits, $X$, $Y$, $Z$ are arbitrary quantities within the range, varying independently of one another. In order that the integral may vanish for all such values of $X$, $Y$, $Z$, we must have

$$E_x = 0, \quad E_y = 0, \quad E_z = 0,$$

which (as usual) are the *characteristic equations*.

### *Terminal conditions.*

**191.** Next, for the terms at the limits when variations, other than those just indicated, are imposed. For these, the portion of the first variation, comprised by the integral, vanishes because the characteristic equations must be satisfied. Among the variations, the set must be included which impose no change at the lower limit; for them, the terms at the upper limit must vanish, that is, the condition

$$X_1 \frac{\partial F}{\partial x_2} + Y_1 \frac{\partial F}{\partial y_2} + Z_1 \frac{\partial F}{\partial z_2}$$

$$+ X \left\{ \frac{\partial F}{\partial x_1} - \frac{d}{dt} \left( \frac{\partial F}{\partial x_2} \right) \right\} + Y \left\{ \frac{\partial F}{\partial y_1} - \frac{d}{dt} \left( \frac{\partial F}{\partial y_2} \right) \right\} + Z \left\{ \frac{\partial F}{\partial z_1} - \frac{d}{dt} \left( \frac{\partial F}{\partial z_2} \right) \right\} = 0$$

must be satisfied for such variations as are possible at the upper limit.

We similarly infer that the same condition must be satisfied for such variations as are possible at the lower limit.

It thus appears that, in order to secure the compulsory zero value for the first variation, (i), *the three characteristic equations must be satisfied*; and (ii), *the prescribed condition must be satisfied, at each limit separately, for such variations as are possible at the respective limits.*

### *The three characteristic equations are equivalent to two.*

**192.** This investigation shews that three characteristic equations arise which must be satisfied: the former investigation (§ 180) required only two such equations. But for the form of integral now under discussion, the function $F$ satisfies two identities, whereas the function $f$ of the former integral was not thus restricted. It is easy to prove that, in virtue of these identities, a single relation connects the three equations. We have

$$\frac{dF}{dt} = x_1\frac{\partial F}{\partial x} + y_1\frac{\partial F}{\partial y} + z_1\frac{\partial F}{\partial z} + x_2\frac{\partial F}{\partial x_1} + y_2\frac{\partial F}{\partial y_1} + z_2\frac{\partial F}{\partial z_1} + x_3\frac{\partial F}{\partial x_2} + y_3\frac{\partial F}{\partial y_2} + z_3\frac{\partial F}{\partial z_2},$$

whatever be the form of a function $F$ as involving arguments $x, x_1, x_2, y, y_1, y_2, z, z_1, z_2$. Because of the second identity satisfied by the function $F$,

$$\frac{dF}{dt} = x_2\frac{\partial F}{\partial x_1} + y_2\frac{\partial F}{\partial y_1} + z_2\frac{\partial F}{\partial z_1} + x_1\frac{d}{dt}\left(\frac{\partial F}{\partial x_1}\right) + y_1\frac{d}{dt}\left(\frac{\partial F}{\partial y_1}\right) + z_1\frac{d}{dt}\left(\frac{\partial F}{\partial z_1}\right)$$

$$+ 2x_3\frac{\partial F}{\partial x_2} + 2y_3\frac{\partial F}{\partial y_2} + 2z_3\frac{\partial F}{\partial z_2} + 2x_2\frac{d}{dt}\left(\frac{\partial F}{\partial x_2}\right) + 2y_2\frac{d}{dt}\left(\frac{\partial F}{\partial y_2}\right) + 2z_2\frac{d}{dt}\left(\frac{\partial F}{\partial z_2}\right).$$

Let the two values of $\dfrac{dF}{dt}$ be equated; then

$$x_1\left\{\frac{\partial F}{\partial x} - \frac{d}{dt}\left(\frac{\partial F}{\partial x_1}\right)\right\} + y_1\left\{\frac{\partial F}{\partial y} - \frac{d}{dt}\left(\frac{\partial F}{\partial y_1}\right)\right\} + z_1\left\{\frac{\partial F}{\partial z} - \frac{d}{dt}\left(\frac{\partial F}{\partial z_1}\right)\right\}$$

$$= x_3\frac{\partial F}{\partial x_2} + 2x_2\frac{d}{dt}\left(\frac{\partial F}{\partial x_2}\right) + y_3\frac{\partial F}{\partial y_2} + 2y_2\frac{d}{dt}\left(\frac{\partial F}{\partial y_2}\right) + z_3\frac{\partial F}{\partial z_2} + 2z_2\frac{d}{dt}\left(\frac{\partial F}{\partial z_2}\right)$$

$$= \frac{d^2}{dt^2}\left(x_1\frac{\partial F}{\partial x_2} + y_1\frac{\partial F}{\partial y_2} + z_1\frac{\partial F}{\partial z_2}\right) - x_1\frac{d^2}{dt^2}\left(\frac{\partial F}{\partial x_2}\right) - y_1\frac{d^2}{dt^2}\left(\frac{\partial F}{\partial y_2}\right) - z_1\frac{d^2}{dt^2}\left(\frac{\partial F}{\partial z_2}\right),$$

and therefore, because of the first identity,

$$x_1 E_x + y_1 E_y + z_1 E_z = 0.$$

That is, the three characteristic equations are connected by one relation; and so, because of the two identities satisfied by $F$, they are equivalent to only two independent equations, the number of characteristic equations when the unrestricted function $f$ is the subject of integration.

*Comparison of terminal conditions with conditions in § 181.*

**193.** Further, the form of condition at the limits, which (§ 191) has been obtained, can be harmonised at once with the earlier form (§ 181). The new form of the condition is

$$X_1 \frac{\partial F}{\partial x_2} + Y_1 \frac{\partial F}{\partial y_2} + Z_1 \frac{\partial F}{\partial z_2}$$

$$+ X \left\{ \frac{\partial F}{\partial x_1} - \frac{d}{dt} \left( \frac{\partial F}{\partial x_2} \right) \right\} + Y \left\{ \frac{\partial F}{\partial y_1} - \frac{d}{dt} \left( \frac{\partial F}{\partial y_2} \right) \right\} + Z \left\{ \frac{\partial F}{\partial z_1} - \frac{d}{dt} \left( \frac{\partial F}{\partial z_2} \right) \right\} = 0.$$

Now

$$\frac{\partial F}{\partial x_1} - \frac{d}{dt} \left( \frac{\partial F}{\partial x_2} \right) = f - y' \left\{ \frac{\partial f}{\partial y'} - \frac{d}{dx} \left( \frac{\partial f}{\partial y''} \right) \right\} - z' \left\{ \frac{\partial f}{\partial z'} - \frac{d}{dx} \left( \frac{\partial f}{\partial z''} \right) \right\} - y'' \frac{\partial f}{\partial y''} - z'' \frac{\partial f}{\partial z''},$$

$$\frac{\partial F}{\partial y_1} - \frac{d}{dt} \left( \frac{\partial F}{\partial y_2} \right) = \frac{\partial f}{\partial y'} - \frac{d}{dx} \left( \frac{\partial f}{\partial y''} \right),$$

$$\frac{\partial F}{\partial z_1} - \frac{d}{dt} \left( \frac{\partial F}{\partial z_2} \right) = \frac{\partial f}{\partial z'} - \frac{d}{dx} \left( \frac{\partial f}{\partial z''} \right);$$

also

$$\frac{\partial F}{\partial x_2} = - \frac{y_1}{x_1} \frac{\partial F}{\partial y_2} - \frac{z_1}{x_1} \frac{\partial F}{\partial z_2} = - y' \frac{\partial F}{\partial y_2} - z' \frac{\partial F}{\partial z_2},$$

so that

$$X_1 \frac{\partial F}{\partial x_2} + Y_1 \frac{\partial F}{\partial y_2} + Z_1 \frac{\partial F}{\partial z_2} = (Y_1 - y' X_1) \frac{\partial F}{\partial y_2} + (Z_1 - z' X_1) \frac{\partial F}{\partial z_2}.$$

For comparison with the earlier form, we have

$$X = U, \quad Y = V, \quad Z = W.$$

The change in the value of $y'$ is

$$\frac{y_1 + \kappa Y_1}{x_1 + \kappa X_1} - \frac{y_1}{x_1}$$

$$= \frac{\kappa}{x_1^2} (x_1 Y_1 - y_1 X_1) + \dots$$

$$= \kappa \frac{1}{x_1} (Y_1 - y' X_1),$$

neglecting powers of $\kappa$ higher than the first: thus

$$\kappa V' = \kappa \frac{1}{x_1} (Y_1 - y' X_1),$$

so that

$$Y_1 - y' X_1 = x_1 V',$$

and therefore

$$(Y_1 - y' X_1) \frac{\partial F}{\partial y_2} = x_1 \frac{\partial F}{\partial y_2} V' = V' \frac{\partial f}{\partial y''}.$$

Similarly,

$$(Z_1 - z' X_1) \frac{\partial F}{\partial z_2} = W' \frac{\partial f}{\partial z''}.$$

When these values are substituted, the agreement between the expressions for the conditions at the limits of the integrals in the two forms is complete.

The detailed inferences from the conditions, according as an extremity of the characteristic curve (corresponding to a limit of the integral) (i) is fixed in position, with or without a fixed direction; (ii) is obliged to lie on a boundary skew curve, with or without restrictions as to direction; or (iii) is obliged to lie on a boundary surface, with or without restrictions as to direction: are, all of them, the same as in § 181.

### *Continuity of certain derivatives through a free discontinuity on the characteristic curve.*

**194.** It has been assumed that no point of discontinuity in direction, or in curvature, or in both direction and curvature (but not of actual currency), of the curve can occur in the range. Now suppose that some such place occurs; the course of the argument will be seen to allow that a finite number of such places may exist. At each of them (it will be sufficient to prove the properties at one of them), *the quantities*

$$\frac{\partial F}{\partial x_2}, \quad \frac{\partial F}{\partial x_1} - \frac{d}{dt}\left(\frac{\partial F}{\partial x_2}\right), \quad \frac{\partial F}{\partial y_2}, \quad \frac{\partial F}{\partial y_1} - \frac{d}{dt}\left(\frac{\partial F}{\partial y_2}\right), \quad \frac{\partial F}{\partial z_2}, \quad \frac{\partial F}{\partial z_1} - \frac{d}{dt}\left(\frac{\partial F}{\partial z_2}\right),$$

*are unchanged in passing through an isolated free place of discontinuity.*

Let any such place be given by the value $T$ of the independent variable $t$. Let $t'$ be any place just before $T$ and $t''$ any place just after $T$, so that

$$t' < T < t'',$$

the range from $t'$ through $T$ to $t''$ being chosen small but not infinitesimal. Consider a variation of the curve represented as before by $\kappa X$, $\kappa Y$, $\kappa Z$, such that

$X, Y, Z = 0$, separately, in all the range anterior to $t'$:

$\quad = 0$, separately, in all the range posterior to $t''$:

$$\left.\begin{array}{l} X = (t - t')^2 (t'' - t)^2 \phi(t) \\ Y = (t - t')^2 (t'' - t)^2 \psi(t) \\ Z = (t - t')^2 (t'' - t)^2 \chi(t) \end{array}\right\}, \text{ in the range from } t' \text{ to } t'':$$

where $\phi(t)$, $\psi(t)$, $\chi(t)$ are three arbitrary regular functions of $t$. Then $X, Y, Z, X_1, Y_1, Z_1$ all vanish when $t = t'$; and likewise they all vanish when $t = t''$. The coefficient of $\kappa$ in the first variation of the integral is

$$\int_{t'}^{T} (X E_x + Y E_y + Z E_z)\, dt + \int_{T}^{t''} (X E_x + Y E_y + Z E_z)\, dt$$

$$+ \left[R\right]_{t'}^{T} + \left[R\right]_{T}^{t''},$$

where

$$R = X \left\{ \frac{\partial F}{\partial x_1} - \frac{d}{dt} \left( \frac{\partial F}{\partial x_2} \right) \right\} + Y \left\{ \frac{\partial F}{\partial y_1} - \frac{d}{dt} \left( \frac{\partial F}{\partial y_2} \right) \right\} + Z \left\{ \frac{\partial F}{\partial z_1} - \frac{d}{dt} \left( \frac{\partial F}{\partial z_2} \right) \right\}$$
$$+ X_1 \frac{\partial F}{\partial x_2} + Y_1 \frac{\partial F}{\partial y_2} + Z_1 \frac{\partial F}{\partial z_2}.$$

The range of the curve from $t'$ to $T$ is regular; at every point along that range, $E_x = 0$, $E_y = 0$, $E_z = 0$. Similarly, for the regular range from $T$ on to $t''$, the same equations hold. Hence the two integrals in the foregoing expression vanish.

At $t'$, the six quantities $X$, $Y$, $Z$, $X_1$, $Y_1$, $Z_1$ vanish; hence the value of $R$ at $t'$ vanishes. Similarly, the value of $R$ at $t''$ vanishes. If we denote the value of $R$ at $T$ as belonging to the range from $t'$ to $T$ by $R_-$, and its value at the same place as belonging to the range from $T$ to $t''$ by $R_+$, the foregoing coefficient becomes

$$R_- - R_+.$$

But the first variation must vanish: hence

$$R_- = R_+.$$

Now $X$, $Y$, $Z$, and their derivatives, are regular at $T$ and suffer no discontinuity; the functions $\phi(t)$, $\psi(t)$, $\chi(t)$ are arbitrary; and the condition must be satisfied for all variations. Hence

$$\left\{ \frac{\partial F}{\partial u_1} - \frac{d}{dt} \left( \frac{\partial F}{\partial u_2} \right) \right\}_- = \left\{ \frac{\partial F}{\partial u_1} - \frac{d}{dt} \left( \frac{\partial F}{\partial u_2} \right) \right\}_+,$$
$$\left( \frac{\partial F}{\partial u_2} \right)_- = \left( \frac{\partial F}{\partial u_2} \right)_+,$$

for $u = x, y, z$, separately. Therefore the six quantities in question suffer no discontinuity in value as they pass through a free place (if any) on the characteristic curve where discontinuity of direction, or of curvature, or of direction and curvature, can occur.

But manifestly the argument would fail at a place $T$ where there could be a condensation of such points, because we could not then have the same kind of expressions for $X$, $Y$, $Z$. Thus the number of such points must, at the utmost, be finite; and each must be isolated.

Further, if $X$, $Y$, $Z$, $X_1$, $Y_1$, $Z_1$ all are necessarily zero, from assigned data, no inferences as to continuity can then be drawn: such a place is not 'free.'

### The second variation, for special weak variations.

**195.** Now that the terms of the first degree in $\kappa$ in the variation of the original integral have been made to vanish, that variation will be governed by the value of the aggregate of terms of the second degree in $\kappa$. Moreover, the terminal conditions, arising out of the evanescence of the aggregate of terms involving the first power of $\kappa$, have led to the determination of the limits of the integral, when these are not settled by precise data. For the present purpose, therefore, the limits of the integral can be taken as fixed.

Dealing solely with special variations, being those which affect $y$ and $z$ alone, we can now take the variation of the original integral as equal to

$$\tfrac{1}{2}\kappa^2 \int \Theta\, dx + K_3,$$

the integral being taken between fixed limits; $K_3$ denotes the aggregate of terms of the third and higher degrees in $\kappa$; and

$$\begin{aligned}
\Theta = \quad & f_{66} Y''^2 + 2f_{69} Y'' Z'' + f_{99} Z''^2 \\
& + 2\,(f_{56} Y'Y'' + f_{59} Y'Z'' + f_{68} Z'Y'' + f_{89} Z'Z'') \\
& + 2\,(f_{46} YY'' + f_{49} YZ'' + f_{67} ZY'' + f_{79} ZZ'') \\
& + f_{55} Y'^2 + 2f_{58} Y'Z' + f_{88} Z'^2 \\
& + 2\,(f_{45} YY' + f_{48} YZ' + f_{57} ZY' + f_{78} ZZ') \\
& + f_{44} Y^2 + 2f_{47} YZ + f_{77} Z^2,
\end{aligned}$$

where $\Theta$ is the same function of $Y$, $Z$ and their derivatives, as $2\Omega$ is of the quantities $\eta$, $\zeta$, and their derivatives, connected (§ 183) with the subsidiary characteristic equations.

To consider the significance of the second variation, as regards the existence of a maximum or minimum of the integral, it is desirable to change the expression into a simpler but equivalent normal form in modified variables, for which conditions as to a persistent positive value or a persistent negative value (that is, a non-zero value) can be enunciated. As in former similar instances, we shall seek a quantity $\square$, where

$$\begin{aligned}
\square = \; & LY'^2 + 2MY'Z' + NZ'^2 + AY^2 + 2BYZ + CZ^2 \\
& + 2DYY' + 2EYZ' + 2FZY' + 2GZZ',
\end{aligned}$$

the coefficients $A, \ldots, N$ being functions of $x$ alone, such that

$$\begin{aligned}
\Theta + \frac{d\square}{dx} = \; & f_{66}\,(Y'' + \alpha Y' + \beta Z' + \gamma Y + \delta Z)^2 \\
& + 2f_{69}\,(Y'' + \alpha Y' + \beta Z' + \gamma Y + \delta Z)\,(Z'' + \theta Y' + \phi Z' + \psi Y + \chi Z) \\
& + f_{99}\,(Z'' + \theta Y' + \phi Z' + \psi Y + \chi Z)^2,
\end{aligned}$$

which, when obtained, is the normal quadratic form of required type. The coefficients $\alpha, \ldots, \chi$ are to be functions of $x$ alone, that is, independent of $Y$, $Z$, and their derivatives.

### Relations for reduction to the normal form.

**196.** As before (§§ 147, 149), the arguments in the normal form can be expected to be such that they vanish, if $Y$ and $Z$ are made equal to $\eta$ and $\zeta$ respectively, where $\eta$ and $\zeta$ constitute an integral-set of the subsidiary equations. Noting this as a possible result to be attained, but of course without making it an initial assumption, we merely take two quantities

$\eta$ and $\zeta$ as assignable functions of $x$, the determination of their values being part of the analytical problem. In connection with these quantities $\eta$ and $\zeta$, we postulate eight other quantities $\alpha, \ldots, \chi$, and require them to satisfy two relations

$$\left.\begin{array}{l} \eta'' + \alpha\eta' + \beta\zeta' + \gamma\eta + \delta\zeta = 0 \\ \zeta'' + \theta\eta' + \phi\zeta' + \psi\eta + \chi\zeta = 0 \end{array}\right\}.$$

Then we take

$$\begin{aligned} \mathbf{Y}_2 &= f_{66}\eta'' + f_{69}\zeta'' + f_{56}\eta' + f_{68}\zeta' + f_{46}\eta + f_{67}\zeta \\ &= -L\eta' - M\zeta' - D\eta - F\zeta, \end{aligned}$$

if quantities $L, M, D, F$, being functions of $x$ alone, satisfy the relations

$$\left.\begin{array}{l} f_{56} + L = \alpha f_{66} + \theta f_{69} \\ f_{68} + M = \beta f_{66} + \phi f_{69} \\ f_{46} + D = \gamma f_{66} + \psi f_{69} \\ f_{67} + F = \delta f_{66} + \chi f_{69} \end{array}\right\}.$$

We similarly take

$$\begin{aligned} \mathbf{Z}_2 &= f_{69}\eta'' + f_{99}\zeta'' + f_{59}\eta' + f_{89}\zeta' + f_{49}\eta + f_{79}\zeta \\ &= -M\eta' - N\zeta' - E\eta - G\zeta, \end{aligned}$$

if quantities $M, N, E, G$, also being functions of $x$ alone, satisfy the relations

$$\left.\begin{array}{l} f_{59} + M = \alpha f_{69} + \theta f_{99} \\ f_{89} + N = \beta f_{69} + \phi f_{99} \\ f_{49} + E = \gamma f_{69} + \psi f_{99} \\ f_{79} + G = \delta f_{69} + \chi f_{99} \end{array}\right\}.$$

Further, we take

$$\mathbf{Y}_1 = f_{56}\eta'' + f_{59}\zeta'' + f_{55}\eta' + f_{58}\zeta' + f_{45}\eta + f_{57}\zeta;$$

and then

$$\begin{aligned} \mathbf{Y}_1 - \frac{d\mathbf{Y}_2}{dx} &= (f_{56} + L)\eta'' + (f_{59} + M)\zeta'' \\ &\quad + (f_{55} + L' + D)\eta' + (f_{58} + M' + F)\zeta' + (f_{45} + D')\eta + (f_{57} + F')\zeta \\ &= -D\eta' - E\zeta' - A\eta - B\zeta, \end{aligned}$$

if

$$\left.\begin{array}{l} f_{55} + 2D + L' = \alpha^2 f_{66} + \qquad\quad 2\alpha\theta f_{69} + \quad \theta^2 f_{99} \\ f_{58} + E + F + M' = \alpha\beta f_{66} + (\alpha\phi + \beta\theta)f_{69} + \theta\phi f_{99} \\ f_{45} + D' + A = \alpha\gamma f_{66} + (\alpha\psi + \gamma\theta)f_{69} + \theta\psi f_{99} \\ f_{57} + F' + B = \alpha\delta f_{66} + (\alpha\chi + \delta\theta)f_{69} + \theta\chi f_{99} \end{array}\right\}.$$

We similarly take

$$\mathbf{Z}_1 = f_{68}\eta'' + f_{89}\zeta'' + f_{58}\eta' + f_{88}\zeta' + f_{48}\eta + f_{78}\zeta;$$

and then

$$\begin{aligned} \mathbf{Z}_1 - \frac{d\mathbf{Z}_2}{dx} &= (f_{68} + M)\eta'' + (f_{89} + N)\zeta'' \\ &\quad + (f_{58} + M' + E)\eta' + (f_{88} + N' + G)\zeta' + (f_{48} + E')\eta + (f_{78} + G')\zeta \\ &= -F\eta' - G\zeta' - B\eta - C\zeta, \end{aligned}$$

if

$$\left.\begin{array}{l} f_{58} + E + F\ \ + M' = \alpha\beta f_{66} + (\alpha\phi + \beta\theta)\, f_{69} + \ \theta\phi f_{99} \\ f_{88} + 2G + N' = \ \beta^2 f_{66} + \ \ \ \ \ \ \ \ \ 2\beta\phi f_{69} + \ \ \phi^2 f_{99} \\ f_{48} + E'\ + B\ = \ \beta\gamma f_{66} + (\gamma\phi + \beta\psi)\, f_{69} + \phi\psi f_{99} \\ f_{78} + G'\ + C\ = \ \beta\delta f_{66} + (\delta\phi + \beta\chi)\, f_{69} + \phi\chi f_{99} \end{array}\right\}.$$

Lastly, we take

$$\mathbf{Y} = f_{46}\eta'' + f_{49}\zeta'' + f_{45}\eta' + f_{48}\zeta' + f_{44}\eta + f_{47}\zeta,$$
$$\mathbf{Z} = f_{67}\eta'' + f_{79}\zeta'' + f_{57}\eta' + f_{78}\zeta' + f_{47}\eta + f_{77}\zeta;$$

and then

$$\mathbf{Y} - \frac{d\mathbf{Y}_1}{dx} + \frac{d^2\mathbf{Y}_2}{dx^2} = (f_{46} + D)\,\eta'' + (f_{49} + E)\,\zeta''$$
$$+ (f_{45} + D' + A)\,\eta' + (f_{48} + E' + B)\,\zeta' + (f_{44} + A')\,\eta + (f_{47} + B')\,\zeta,$$

$$\mathbf{Z} - \frac{d\mathbf{Z}_1}{dx} + \frac{d^2\mathbf{Z}_2}{dx^2} = (f_{67} + F)\,\eta'' + (f_{79} + G)\,\zeta''$$
$$+ (f_{57} + F' + B)\,\eta' + (f_{78} + G' + C)\,\zeta' + (f_{47} + B')\,\eta + (f_{77} + C')\,\zeta,$$

both of which vanish if, in addition to all the foregoing relations,

$$\left.\begin{array}{l} f_{44} + A'\ = \ \gamma^2 f_{66} + \ \ \ \ \ \ \ \ \ 2\gamma\psi f_{69} + \ \psi^2 f_{99} \\ f_{47} + B'\ = \gamma\delta f_{66} + (\gamma\chi + \delta\psi)\, f_{69} + \psi\chi f_{99} \\ f_{77} + C'\ = \ \delta^2 f_{66} + \ \ \ \ \ \ \ \ \ 2\delta\chi f_{69} + \ \chi^2 f_{99} \end{array}\right\}.$$

Thus, in addition to the two initial fundamental equations

$$\left.\begin{array}{l} \eta'' + \alpha\eta' + \beta\zeta' + \gamma\eta\ + \delta\zeta = 0 \\ \zeta'' + \theta\eta' + \phi\zeta' + \psi\eta + \chi\zeta = 0 \end{array}\right\},$$

we have, in all, eighteen relations—the relation, involving the quantity $F_{58} + E + F + M'$, occurs twice. In these relations, there occur eight unknown functions $\alpha, \ldots, \chi$, and ten unknown functions $A, \ldots, N$. Together with $\eta$ and $\zeta$, these constitute twenty unknown quantities; and there are, in all, twenty equations to determine them. If then the equations are consistent with one another—their corporate consistency will be proved, with added restrictions (§ 199) on the two initial equations—they are potentially sufficient for the determination of the unknown quantities.

*First normal form of second variation.*

**197.** When regard is paid to the definition of the quantities $\mathbf{Y}_2$, $\mathbf{Y}_1$, $\mathbf{Y}$; $\mathbf{Z}_2$, $\mathbf{Z}_1$, $\mathbf{Z}$; the two relations

$$\mathbf{Y} - \frac{d\mathbf{Y}_1}{dx} + \frac{d^2\mathbf{Y}_2}{dx^2} = 0, \quad \mathbf{Z} - \frac{d\mathbf{Z}_1}{dx} + \frac{d^2\mathbf{Z}_2}{dx^2} = 0,$$

are exactly the same as the subsidiary characteristic equations

$$\frac{\partial\Omega}{\partial\eta} - \frac{d}{dx}\left(\frac{\partial\Omega}{\partial\eta'}\right) + \frac{d^2}{dx^2}\left(\frac{\partial\Omega}{\partial\eta''}\right) = 0, \quad \frac{\partial\Omega}{\partial\zeta} - \frac{d}{dx}\left(\frac{\partial\Omega}{\partial\zeta'}\right) + \frac{d^2}{dx^2}\left(\frac{\partial\Omega}{\partial\zeta''}\right) = 0.$$

Hence the quantities $\eta$ and $\zeta$ in the foregoing equations can be identified as an integral-set of the subsidiary characteristic equations.

Assuming, temporarily, that the unknown quantities $\alpha, \ldots, \chi, A, \ldots, N$ can be determined from the foregoing aggregate of relations, we have

$$\Theta + \frac{d\square}{dx} = \Delta,$$

where

$$
\begin{aligned}
\Delta =\, & f_{66} Y''^2 + 2f_{69} Y'' Z'' + f_{99} Z''^2 \\
& + 2(f_{56} + L)\, Y' Y'' + 2(f_{68} + M)\, Z' Y'' + 2(f_{59} + M)\, Y' Z'' + 2(f_{89} + N)\, Z' Z'' \\
& + (f_{55} + 2D + L')\, Y'^2 + 2(f_{58} + E + F + M')\, Y' Z' + (f_{88} + 2G + N')\, Z'^2 \\
& + 2(f_{45} + D' + A)\, Y Y' + 2(f_{48} + E' + B)\, Y Z' + 2(f_{57} + F' + B)\, Z Y' \\
& \hspace{6cm} + 2(f_{78} + G' + C)\, Z Z' \\
& + (f_{44} + A')\, Y^2 + 2(f_{47} + B')\, Y Z + (f_{77} + C')\, Z^2 \\
=\, & f_{66} (Y'' + \alpha Y' + \beta Z' + \gamma Y + \delta Z)^2 \\
& + 2f_{69} (Y'' + \alpha Y' + \beta Z' + \gamma Y + \delta Z)(Z'' + \theta Y' + \phi Z' + \psi Y + \chi Z) \\
& + f_{99} (Z'' + \theta Y' + \phi Z' + \psi Y + \chi Z)^2,
\end{aligned}
$$

(the required normal form), on using the foregoing relations.

The terminal conditions, either gave the limits as fixed by the initial data, or deduced them as fixed from initial relations less direct than specific data. Consequently, $Y, Z$ and their derivatives can now be considered (§ 195) as vanishing at each limit of the original integral; and therefore

$$\int \Delta\, dx = \int \left( \Theta + \frac{d\square}{dx} \right) dx$$

$$= \int \Theta\, dx + [\square],$$

where $\square$ is taken at the limits. But, so taken, $\square$ vanishes; hence

$$\int \Delta\, dx = \int \Theta\, dx;$$

that is, the integral $\int \Delta\, dx$ determines the second variation.

### Determination of the coefficients from four integral-sets of the subsidiary equations.

**198.** The relations, determining the quantities $\alpha, \ldots, \chi, A, \ldots, N$ in the reduction to the normal quadratic form, have been assigned in connection with two quantities $\eta$ and $\zeta$ postulated initially as two unknown quantities. It has since been proved that $\eta$ and $\zeta$ can be taken as any integral-set of the subsidiary characteristic equations

$$\frac{\partial \Omega}{\partial \eta} - \frac{d}{dx}\left( \frac{\partial \Omega}{\partial \eta'} \right) + \frac{d^2}{dx^2}\left( \frac{\partial \Omega}{\partial \eta''} \right) = 0, \quad \frac{\partial \Omega}{\partial \zeta} - \frac{d}{dx}\left( \frac{\partial \Omega}{\partial \zeta'} \right) + \frac{d^2}{dx^2}\left( \frac{\partial \Omega}{\partial \zeta''} \right) = 0.$$

We now proceed to shew that all the relations, determining the eighteen quantities, are satisfied when, in turn, we take $\eta$ and $\zeta$ to be four integral-sets $\eta_1$ and $\zeta_1$, $\eta_2$ and $\zeta_2$, $\eta_3$ and $\zeta_3$, $\eta_4$ and $\zeta_4$, which are linearly independent of one another. To secure this independence, the determinant $\omega$, where

$$\omega = \begin{vmatrix} \eta_1', & \zeta_1', & \eta_1, & \zeta_1 \\ \eta_2', & \zeta_2', & \eta_2, & \zeta_2 \\ \eta_3', & \zeta_3', & \eta_3, & \zeta_3 \\ \eta_4', & \zeta_4', & \eta_4, & \zeta_4 \end{vmatrix} = |\, \eta_1', \zeta_2', \eta_3, \zeta_4 \,|,$$

(the second line-symbol for the determinant being the notation of § 188), must not vanish; for other reasons, it must not be a constant.

With these four integral-sets, we have

$$\eta_r'' + \alpha\eta_r' + \beta\zeta_r' + \gamma\eta_r + \delta\zeta_r = 0,$$
$$\zeta_r'' + \theta\eta_r' + \phi\zeta_r' + \psi\eta_r + \chi\zeta_r = 0,$$

for $r = 1, 2, 3, 4$, so that

$$\left.\begin{aligned} \alpha\omega + |\, \eta_1'', \zeta_2', \eta_3, \zeta_4 \,| = 0 \\ \beta\omega - |\, \eta_1'', \eta_2', \eta_3, \zeta_4 \,| = 0 \\ \gamma\omega + |\, \eta_1'', \eta_2', \zeta_3', \zeta_4 \,| = 0 \\ \delta\omega - |\, \eta_1'', \eta_2', \zeta_3', \eta_4 \,| = 0 \end{aligned}\right\}, \qquad \left.\begin{aligned} \theta\omega + |\, \zeta_1'', \zeta_2', \eta_3, \zeta_4 \,| = 0 \\ \phi\omega - |\, \zeta_1'', \eta_2', \eta_3, \zeta_4 \,| = 0 \\ \psi\omega + |\, \zeta_1'', \eta_2', \zeta_3', \zeta_4 \,| = 0 \\ \chi\omega - |\, \zeta_1'', \eta_2', \zeta_3', \eta_4 \,| = 0 \end{aligned}\right\},$$

thus giving explicit values for $\alpha, \dots, \chi$. Further, we have

$$\omega' = |\, \eta_1'', \zeta_2', \eta_3, \zeta_4 \,| + |\, \eta_1', \zeta_2'', \eta_3, \zeta_4 \,| + |\, \eta_1', \zeta_2', \eta_3', \zeta_4 \,| + |\, \eta_1', \zeta_2', \eta_3, \zeta_4' \,|$$
$$= -\alpha\omega - \phi\omega + 0 + 0 = -(\alpha + \phi)\,\omega,$$

a result that will be used later.

Of the other quantities, $L, M, D, F$ are determinable by the four relations

$$\frac{\partial\Omega_r}{\partial\eta_r''} + L\eta_r' + M\zeta_r' + D\eta_r + F\zeta_r = 0,$$

for $r = 1, 2, 3, 4$, so that

$$\left.\begin{aligned} L\omega + \begin{vmatrix} \dfrac{\partial\Omega_1}{\partial\eta_1''}, & \zeta_2', & \eta_3, & \zeta_4 \end{vmatrix} = 0 \\[2mm] M\omega - \begin{vmatrix} \dfrac{\partial\Omega_1}{\partial\eta_1''}, & \eta_2', & \eta_3, & \zeta_4 \end{vmatrix} = 0 \\[2mm] D\omega + \begin{vmatrix} \dfrac{\partial\Omega_1}{\partial\eta_1''}, & \eta_2', & \zeta_3', & \zeta_4 \end{vmatrix} = 0 \\[2mm] F\omega - \begin{vmatrix} \dfrac{\partial\Omega_1}{\partial\eta_1''}, & \eta_2', & \zeta_3', & \eta_4 \end{vmatrix} = 0 \end{aligned}\right\}.$$

The quantities $M, N, E, G$ are determinable by the four relations

$$\frac{\partial\Omega_r}{\partial\zeta_r''} + M\eta_r' + N\zeta_r' + E\eta_r + G\zeta_r = 0,$$

for $r = 1, 2, 3, 4$, so that

$$
\left.\begin{aligned}
M\omega + \left| \frac{\partial\Omega_1}{\partial\zeta_1''}, \ \zeta_2', \ \eta_3, \ \zeta_4 \right| &= 0 \\[2mm]
N\omega - \left| \frac{\partial\Omega_1}{\partial\zeta_1''}, \ \eta_2', \ \eta_3, \ \zeta_4 \right| &= 0 \\[2mm]
E\omega + \left| \frac{\partial\Omega_1}{\partial\zeta_1''}, \ \eta_2', \ \zeta_3', \ \zeta_4 \right| &= 0 \\[2mm]
G\omega - \left| \frac{\partial\Omega_1}{\partial\zeta_1''}, \ \eta_2', \ \zeta_3', \ \eta_4 \right| &= 0
\end{aligned}\right\}.
$$

The quantities $D, E, A, B$ are determinable by the four relations

$$
\frac{\partial\Omega_r}{\partial\eta_r'} - \frac{d}{dx}\left(\frac{\partial\Omega_r}{\partial\eta_r''}\right) + D\eta_r' + E\zeta_r' + A\eta_r + B\zeta_r = 0,
$$

for $r = 1, 2, 3, 4$, so that

$$
\left.\begin{aligned}
D\omega + \left| \frac{\partial\Omega_1}{\partial\eta_1'} - \frac{d}{dx}\left(\frac{\partial\Omega_1}{\partial\eta_1''}\right), \ \zeta_2', \ \eta_3, \ \zeta_4 \right| &= 0 \\[2mm]
E\omega - \left| \frac{\partial\Omega_1}{\partial\eta_1'} - \frac{d}{dx}\left(\frac{\partial\Omega_1}{\partial\eta_1''}\right), \ \eta_2', \ \eta_3, \ \zeta_4 \right| &= 0 \\[2mm]
A\omega + \left| \frac{\partial\Omega_1}{\partial\eta_1'} - \frac{d}{dx}\left(\frac{\partial\Omega_1}{\partial\eta_1''}\right), \ \eta_2', \ \zeta_3', \ \zeta_4 \right| &= 0 \\[2mm]
B\omega - \left| \frac{\partial\Omega_1}{\partial\eta_1'} - \frac{d}{dx}\left(\frac{\partial\Omega_1}{\partial\eta_1''}\right), \ \eta_2', \ \zeta_3', \ \eta_4 \right| &= 0
\end{aligned}\right\}.
$$

Finally, the quantities $F, G, B, C$ are determinable by the four relations

$$
\frac{\partial\Omega_r}{\partial\zeta_r'} - \frac{d}{dx}\left(\frac{\partial\Omega_r}{\partial\zeta_r''}\right) + F\eta_r' + G\zeta_r' + B\eta_r + C\zeta_r = 0,
$$

for $r = 1, 2, 3, 4$, so that

$$
\left.\begin{aligned}
F\omega + \left| \frac{\partial\Omega_1}{\partial\zeta_1'} - \frac{d}{dx}\left(\frac{\partial\Omega_1}{\partial\zeta_1''}\right), \ \zeta_2', \ \eta_3, \ \zeta_4 \right| &= 0 \\[2mm]
G\omega - \left| \frac{\partial\Omega_1}{\partial\zeta_1'} - \frac{d}{dx}\left(\frac{\partial\Omega_1}{\partial\zeta_1''}\right), \ \eta_2', \ \eta_3, \ \zeta_4 \right| &= 0 \\[2mm]
B\omega + \left| \frac{\partial\Omega_1}{\partial\zeta_1'} - \frac{d}{dx}\left(\frac{\partial\Omega_1}{\partial\zeta_1''}\right), \ \eta_2', \ \zeta_3', \ \zeta_4 \right| &= 0 \\[2mm]
C\omega - \left| \frac{\partial\Omega_1}{\partial\zeta_1'} - \frac{d}{dx}\left(\frac{\partial\Omega_1}{\partial\zeta_1''}\right), \ \eta_2', \ \zeta_3', \ \eta_4 \right| &= 0
\end{aligned}\right\}.
$$

Among these sixteen results, there are two expressions for $M$, which are equal because of the first relation in §188; two expressions for $D$, equal because of the second relation; two expressions for $E$, equal because of the third relation; two expressions for $F$, equal because of the fourth relation; two expressions for $G$, equal because of the fifth relation; and two expressions for $B$, equal because of the sixth relation. The expressions for the ten quantities $A, \dots, N$ are thus consistent with one another.

### The values of the coefficients satisfy the relations in § 196.

**199.** But these values of $A, \dots, N$ must satisfy further groups of relations. Initially, those values were defined in connection with two quantities $\eta$ and $\zeta$, proved to be any single integral-set of the subsidiary characteristic equations. Now, they have been expressed in terms of four such integral-sets. The two aggregates must be shewn to agree.

(i) Thus we had an initial relation

$$f_{56} + L = \alpha f_{66} + \theta f_{69},$$

the first in one group of four such relations in § 196. The later value of $L$ is given by

$$- L\omega = \left| \frac{\partial \Omega_1}{\partial \eta_1''},\ \zeta_2',\ \eta_3,\ \zeta_4 \right|$$

$$= \left| f_{66}\eta_1'' + f_{69}\zeta_1'' + f_{56}\eta_1' + f_{68}\zeta_1' + f_{46}\eta_1 + f_{67}\zeta_1,\ \zeta_2',\ \eta_3,\ \zeta_4 \right|$$

$$= f_{66}\left| \eta_1'',\ \zeta_2',\ \eta_3,\ \zeta_4 \right| + f_{69}\left| \zeta_1'',\ \zeta_2',\ \eta_3,\ \zeta_4 \right| + f_{56}\left| \eta_1',\ \zeta_2',\ \eta_3,\ \zeta_4 \right|$$

$$= - f_{66}\alpha\omega - f_{69}\theta\omega + f_{56}\omega.$$

Therefore the initial relation is satisfied. Similarly for the other three members of that group.

(ii) Again, there is a group of four relations, of which the first is

$$f_{59} + M = \alpha f_{69} + \theta f_{99}.$$

This relation, by similar analysis, is seen to be satisfied, as well as the other three members of the group.

(iii) There is a set of three relations, occurring in the other three groups, and involving $L'$, $M'$, $N'$; they are

$$\left.\begin{array}{l} f_{55} + 2D + L' = \alpha^2 f_{66} + \qquad\quad 2\alpha\theta f_{69} + \quad \theta^2 f_{99} \\ f_{58} + E + F + M' = \alpha\beta f_{66} + (\alpha\phi + \beta\theta) f_{69} + \theta\phi f_{99} \\ f_{88} + 2G + N' = \beta^2 f_{66} + \qquad\quad 2\beta\phi f_{69} + \quad \phi^2 f_{99} \end{array}\right\}.$$

Differentiating the relation

$$L\omega + \left| \frac{\partial \Omega_1}{\partial \eta_1''},\ \zeta_2',\ \eta_3,\ \zeta_4 \right| = 0,$$

which has been verified in (i) above, we have

$$L'\omega + L\omega' + \left| \frac{d}{dx}\left(\frac{\partial \Omega_1}{\partial \eta_1''}\right),\ \zeta_2',\ \eta_3,\ \zeta_4 \right|$$

$$+ \left| \frac{\partial \Omega_1}{\partial \eta_1''},\ \zeta_2'',\ \eta_3,\ \zeta_4 \right| + \left| \frac{\partial \Omega_1}{\partial \eta_1''},\ \zeta_2',\ \eta_3',\ \zeta_4 \right| + \left| \frac{\partial \Omega_1}{\partial \eta_1''},\ \zeta_2',\ \eta_3,\ \zeta_4' \right| = 0.$$

Now

$$\left|\ \frac{\partial\Omega_1}{\partial\eta_1''},\ \zeta_2',\ \eta_3,\ \zeta_4'\ \right| = 0\ ;$$

$$\left|\ \frac{\partial\Omega_1}{\partial\eta_1''},\ \zeta_2',\ \eta_3',\ \zeta_4\ \right| = -\left|\ \frac{\partial\Omega_1}{\partial\eta_1''},\ \eta_2',\ \zeta_3',\ \zeta_4\ \right| = D\omega\ ;$$

$$\left|\ \frac{\partial\Omega_1}{\partial\eta_1''},\ \zeta_2'',\ \eta_3,\ \zeta_4\ \right| = -\left|\ \frac{\partial\Omega_1}{\partial\eta_1''},\ \theta\eta_2' + \phi\zeta_2' + \psi\eta_2 + \chi\zeta_2,\ \eta_3,\ \zeta_4\ \right|$$

$$= -\theta\left|\ \frac{\partial\Omega_1}{\partial\eta_1''},\ \eta_2',\ \eta_3,\ \zeta_4\ \right| - \phi\left|\ \frac{\partial\Omega_1}{\partial\eta_1''},\ \zeta_2',\ \eta_3,\ \zeta_4\ \right|$$

$$= -\theta M\omega + \phi L\omega.$$

Also

$$\left|\ \frac{d}{dx}\left(\frac{\partial\Omega_1}{\partial\eta_1''}\right),\ \zeta_2',\ \eta_3,\ \zeta_4\ \right| = \left|\ \frac{\partial\Omega_1}{\partial\eta_1},\ \zeta_2',\ \eta_3,\ \zeta_4\ \right| - \left|\ \frac{\partial\Omega_1}{\partial\eta_1} - \frac{d}{dx}\left(\frac{\partial\Omega_1}{\partial\eta_1''}\right),\ \zeta_2',\ \eta_3,\ \zeta_4\ \right|\ ;$$

here the second determinant is equal to $-D\omega$, and the first determinant is

$$\left|\ f_{56}\eta_1'' + f_{59}\zeta_1'' + f_{55}\eta_1' + f_{58}\zeta_1' + f_{45}\eta_1 + f_{57}\zeta_1,\ \zeta_2',\ \eta_3,\ \zeta_4\ \right|$$

$$= \left|\ (f_{55} - \alpha f_{56} - \theta f_{59})\,\eta_1' + (f_{58} - \beta f_{56} - \phi f_{59})\,\zeta_1' + (f_{45} - \gamma f_{56} - \psi f_{59})\,\eta_1 \right.$$
$$\left. + (f_{57} - \delta f_{56} - \chi f_{59})\,\zeta_1,\ \zeta_2',\ \eta_3,\ \zeta_4\ \right|$$

$$= (f_{55} - \alpha f_{56} - \theta f_{59})\,\omega$$

$$= (f_{55} + \alpha L + \theta M - \alpha^2 f_{66} - 2\alpha\theta f_{69} - \theta^2 f_{99})\,\omega,$$

so that

$$\left|\ \frac{d}{dx}\left(\frac{\partial\Omega_1}{\partial\eta_1''}\right),\ \zeta_2',\ \eta_3,\ \zeta_4\ \right| = (D + f_{55} + \alpha L + \theta M - \alpha^2 f_{66} - 2\alpha\theta f_{69} - \theta^2 f_{99})\,\omega.$$

Finally,

$$\omega' = -(\alpha + \phi)\,\omega.$$

Thus the differentiated relation becomes

$$L'\omega - L(\alpha + \phi)\,\omega + (D + f_{55} + \alpha L + \theta M - \alpha^2 f_{66} - 2\alpha\theta f_{69} - \theta^2 f_{99})\,\omega$$
$$- \theta M\omega + \phi L\omega + D\omega = 0,$$

that is,

$$f_{55} + 2D + L' = \alpha^2 f_{66} + 2\alpha\theta f_{69} + \theta^2 f_{99}:$$

or the required relation involving $L'$ is satisfied.

Similarly, the relation involving $M'$, and the relation involving $N'$, are satisfied by the postulated values.

(iv) There is a set of four relations, occurring in the same three groups, and involving $D'$, $E'$, $F'$, $G'$; they are

$$\left.\begin{aligned}
f_{45} + D' + A &= \alpha\gamma\,f_{66} + (\alpha\psi + \gamma\theta)\,f_{69} + \theta\psi\,f_{99} \\
f_{48} + E' + B &= \beta\gamma\,f_{66} + (\beta\psi + \gamma\phi)\,f_{69} + \phi\psi\,f_{99} \\
f_{57} + F' + B &= \alpha\delta\,f_{66} + (\alpha\chi + \delta\theta)\,f_{69} + \theta\chi\,f_{99} \\
f_{78} + G' + C &= \beta\delta\,f_{66} + (\beta\chi + \delta\phi)\,f_{69} + \phi\chi\,f_{99}
\end{aligned}\right\}.$$

Differentiating the relation

$$D\omega + \left| \frac{\partial \Omega_1}{\partial \eta_1''}, \; \eta_2', \; \zeta_3', \; \zeta_4 \right| = 0,$$

which has been verified in (i) above, we have

$$D'\omega + D\omega' + \left| \frac{d}{dx}\left(\frac{\partial \Omega_1}{\partial \eta_1''}\right), \; \eta_2', \; \zeta_3', \; \zeta_4 \right|$$

$$+ \left| \frac{\partial \Omega_1}{\partial \eta_1''}, \; \eta_2'', \; \zeta_3', \; \zeta_4 \right| + \left| \frac{\partial \Omega_1}{\partial \eta_1''}, \; \eta_2', \; \zeta_3'', \; \zeta_4 \right| + \left| \frac{\partial \Omega_1}{\partial \eta_1''}, \; \eta_2', \; \zeta_3', \; \zeta_4' \right| = 0.$$

The last determinant is zero. For the last determinant but one, we have

$$\left| \frac{\partial \Omega_1}{\partial \eta_1''}, \; \eta_2', \; \zeta_3'', \; \zeta_4 \right| = (f_{68} - \beta f_{66})\,\psi\omega - (f_{46} - \gamma f_{66})\,\phi\omega,$$

on reduction. For the last determinant but two, we have

$$\left| \frac{\partial \Omega_1}{\partial \eta_1''}, \; \eta_2'', \; \zeta_3', \; \zeta_4 \right| = (f_{56} - \theta f_{69})\,\gamma\omega - (f_{46} - \psi f_{69})\,\alpha\omega,$$

on reduction. Also

$$\left| \frac{d}{dx}\left(\frac{\partial \Omega_1}{\partial \eta_1''}\right), \; \eta_2', \; \zeta_3', \; \zeta_4 \right|$$

$$= \left| \frac{\partial \Omega_1}{\partial \eta_1}, \; \eta_2', \; \zeta_3', \; \zeta_4 \right| - \left| \frac{\partial \Omega_1}{\partial \eta_1} - \frac{d}{dx}\left(\frac{\partial \Omega_1}{\partial \eta_1''}\right), \; \eta_2', \; \zeta_3', \; \zeta_4 \right|$$

$$= \omega\,(f_{45} - \gamma f_{56} - \psi f_{59}) + A\,\omega,$$

on reduction; and

$$\omega' = -(\alpha + \phi)\,\omega.$$

Substituting, and reducing, we find (on rejecting the factor $\omega$)

$$D' + f_{45} + A - \alpha\gamma f_{66} - (\alpha\psi + \gamma\theta)f_{69} - \theta\psi f_{99} = 0,$$

which is the first of the above set of four relations.

Proceeding similarly from the relations which express $E$, $F$, $G$, we obtain the remaining three relations in the above set.

(v) Finally, we have the three remaining relations from the same three groups; they are

$$\left. \begin{aligned} f_{44} + A' &= \gamma^2 f_{66} + 2\gamma\psi f_{69} + \psi^2 f_{99} \\ f_{47} + B' &= \gamma\delta f_{66} + (\gamma\chi + \delta\psi)f_{69} + \psi\chi f_{99} \\ f_{77} + C' &= \delta^2 f_{66} + 2\delta\chi f_{69} + \chi^2 f_{99} \end{aligned} \right\}.$$

Differentiating the relation

$$A\omega + \left| \frac{\partial \Omega_1}{\partial \eta_1'} - \frac{d}{dx}\left(\frac{\partial \Omega_1}{\partial \eta_1''}\right), \; \eta_2', \; \zeta_3', \; \zeta_4 \right| = 0,$$

which has been verified in (iv) above, we have

$$A'\omega + A\omega' + \left| \frac{d}{dx}\left(\frac{\partial\Omega_1}{\partial\eta_1'}\right) - \frac{d^2}{dx^2}\left(\frac{\partial\Omega_1}{\partial\eta_1''}\right), \; \eta_2', \; \zeta_3', \; \zeta_4 \right|$$

$$+ \left| \frac{\partial\Omega_1}{\partial\eta_1'} - \frac{d}{dx}\left(\frac{\partial\Omega_1}{\partial\eta_1''}\right), \; \eta_2'', \; \zeta_3', \; \zeta_4 \right| + \left| \frac{\partial\Omega_1}{\partial\eta_1'} - \frac{d}{dx}\left(\frac{\partial\Omega_1}{\partial\eta_1''}\right), \; \eta_2', \; \zeta_3'', \; \zeta_4 \right|$$

$$+ \left| \frac{\partial\Omega_1}{\partial\eta_1'} - \frac{d}{dx}\left(\frac{\partial\Omega_1}{\partial\eta_1''}\right), \; \eta_2', \; \zeta_3', \; \zeta_4' \right| = 0,$$

the last of the four determinants being an identical zero. As to the rest of them, we have

$$\frac{\partial\Omega}{\partial\eta} - \frac{d}{dx}\left(\frac{\partial\Omega}{\partial\eta'}\right) + \frac{d^2}{dx^2}\left(\frac{\partial\Omega}{\partial\eta''}\right) = 0,$$

for all the integral-sets; and thus the first determinant

$$= \left| \frac{\partial\Omega_1}{\partial\eta_1}, \; \eta_2', \; \zeta_3', \; \zeta_4 \right|$$

$$= \left| f_{46}\eta_1'' + f_{49}\zeta_1'' + f_{44}\eta_1, \; \eta_2', \; \zeta_3', \; \zeta_4 \right|$$

$$= (f_{44} - \gamma f_{46} - \psi f_{49})\,\omega.$$

The second determinant

$$= - \left| \frac{\partial\Omega_1}{\partial\eta_1'} - \frac{d}{dx}\left(\frac{\partial\Omega_1}{\partial\eta_1''}\right), \; \alpha\eta_2' + \gamma\eta_2, \; \zeta_3', \; \zeta_4 \right| = \alpha A\omega - \gamma D\omega;$$

the third determinant

$$= - \left| \frac{\partial\Omega_1}{\partial\eta_1'} - \frac{d}{dx}\left(\frac{\partial\Omega_1}{\partial\eta_1''}\right), \; \eta_2', \; \phi\zeta_3' + \psi\eta_3, \; \zeta_4 \right| = \phi A\omega - \psi E\omega;$$

and $\omega' = -(\alpha + \phi)\,\omega$. Substituting and removing the factor $\omega$, we have

$$A' + f_{44} - \gamma^2 f_{66} - 2\gamma\psi f_{69} - \psi^2 f_{99} = 0,$$

which is the first of the above set of three relations.

Proceeding similarly from the expressions for $B$ and $C$, we obtain the other two relations in the set.

*Final normal form of the 'second' variation: the Legendre test.*

**200.** Thus the various quantities $\alpha, \ldots, \chi, A, \ldots, N$ have been expressed in terms of four integral-sets of the subsidiary characteristic equations, selected (and the selection is not unique) so as to satisfy the six relations obtained in § 188. As before, let

$$\omega = \left| \begin{array}{cccc} \eta_1', & \zeta_1', & \eta_1, & \zeta_1 \\ \eta_2', & \zeta_2', & \eta_2, & \zeta_2 \\ \eta_3', & \zeta_3', & \eta_3, & \zeta_3 \\ \eta_4', & \zeta_4', & \eta_4, & \zeta_4 \end{array} \right|.$$

This quantity $\omega$ may be made to vanish at the lower limit; but it does not vanish steadily along the range. We write

$$\mathbf{v} = \begin{vmatrix} Y'', & Y', & Z', & Y, & Z \\ \eta_1'', & \eta_1', & \zeta_1', & \eta_1, & \zeta_1 \\ \eta_2'', & \eta_2', & \zeta_2', & \eta_2, & \zeta_2 \\ \eta_3'', & \eta_3', & \zeta_3', & \eta_3, & \zeta_3 \\ \eta_4'', & \eta_4', & \zeta_4', & \eta_4, & \zeta_4 \end{vmatrix}, \qquad \mathbf{w} = \begin{vmatrix} Z'', & Y', & Z', & Y, & Z \\ \zeta_1'', & \eta_1', & \zeta_1', & \eta_1, & \zeta_1 \\ \zeta_2'', & \eta_2', & \zeta_2', & \eta_2, & \zeta_2 \\ \zeta_3'', & \eta_3', & \zeta_3', & \eta_3, & \zeta_3 \\ \zeta_4'', & \eta_4', & \zeta_4', & \eta_4, & \zeta_4 \end{vmatrix}.$$

Then the integral (§ 197), which determines the second variation, is

$$\int \frac{1}{\omega^2}(f_{66}\mathbf{v}^2 + 2f_{69}\mathbf{v}\mathbf{w} + f_{99}\mathbf{w}^2)\,dx,$$

taken between the limits which are fixed, either by means of data initially given or by inferences from initially postulated conditions.

Now that the second variation has been reduced to this normal form, the argument, leading to the requirements which can ensure a sign persistent for all admissible variations, follows the customary course. The quantity $\omega$ does not vanish along the range; it may vanish at the lower limit, but always

$$\mathbf{v} = \omega\,(Y'' + \alpha Y' + \beta Z' + \gamma Y + \delta Z), \quad \mathbf{w} = \omega\,(Z'' + \theta Y' + \phi Z' + \psi Y + \chi Z),$$

and so no analytical difficulty arises there because $\omega$ vanishes. Consequently, the critical quantity is

$$f_{66}\mathbf{v}^2 + 2f_{69}\mathbf{v}\mathbf{w} + f_{99}\mathbf{w}^2;$$

and it must be considered for all admissible non-zero small variations represented by $\kappa Y$ and $\kappa Z$. In order that the sign of this quantity may remain unaltered for all non-zero values of $\mathbf{v}$ and $\mathbf{w}$, *it is necessary that*

$$f_{66}f_{99} - f_{69}^2$$

*shall be positive** throughout the range of integration.* An additional statement made for completeness, and consistent with this condition, is that *the quantities $f_{66}$ and $f_{99}$ must everywhere in the range be different from zero and have the same sign.* If this common sign be positive, so that the second variation is positive, the original integral can possess a minimum: if this common sign be negative, the original integral can possess a maximum.

We thus have a second criterion to be satisfied. It is called the *Legendre test*, as it is the extension of the test which Legendre obtained for the simplest type of integrals capable of maxima or minima.

---

* It must never be negative. If it vanishes at isolated places without changing sign, the second variation of the integral would vanish for non-zero variations such that $f_{66}\mathbf{v} + f_{96}\mathbf{w} = 0$; the third variation of the integral for those particular non-zero variations would have to be considered. The consideration is omitted as a general issue, and can be undertaken for any individual instance in which it arises.

*Limitation of range : Jacobi test.*

**201.** To make (and keep) the criterion effective, we must secure that the quantities **v** and **w** do not simultaneously vanish for any admissible small variation $\kappa Y$ and $\kappa Z$—a requirement that leads to the third criterion, which is the extension of the Jacobi test. We proceed to its formulation.

The subject of integration, in the integral which expresses the second variation, involves the critical quantities **v** and **w**, and also the critical quantity $\omega$. Now the value of $\omega$ is

$$\begin{vmatrix} \eta_1', & \eta_2', & \eta_3', & \eta_4' \\ \zeta_1', & \zeta_2', & \zeta_3', & \zeta_4' \\ \eta_1, & \eta_2, & \eta_3, & \eta_4 \\ \zeta_1, & \zeta_2, & \zeta_3, & \zeta_4 \end{vmatrix} ;$$

and it is not to vanish within the range of integration. Consequently, we must not have any constants $a_1$, $a_2$, $a_3$, $a_4$, such that the functions

$$a_1\eta_1 + a_2\eta_2 + a_3\eta_3 + a_4\eta_4, \quad a_1\zeta_1 + a_2\zeta_2 + a_3\zeta_3 + a_4\zeta_4,$$

and their first derivatives—all being functions of $x$—vanish simultaneously within the range.

Again, we have to secure that **v** and **w**, which arise through non-zero variations $\kappa Y$ and $\kappa Z$, and which are bound to vanish at the limits of the range of integration, shall not vanish continuously through the range of integration; because, otherwise, the second variation would vanish, for such variations $\kappa Y$ and $\kappa Z$ as allow **v** and **w** to vanish. (The third variation would then have to vanish for those variations. This possibility has been set aside in earlier cases and will also now be set aside; it would be discussed in any particular instance in which it might arise.) Hence there must not be constants $a_1$, $a_2$, $a_3$, $a_4$, such that $Y$ and $Z$ have values

$$\left. \begin{array}{l} Y = a_1\eta_1 + a_2\eta_2 + a_3\eta_3 + a_4\eta_4 \\ Z = a_1\zeta_1 + a_2\zeta_2 + a_3\zeta_3 + a_4\zeta_4 \end{array} \right\} ,$$

simultaneously throughout the range. But $Y$ and $Z$ are to vanish at the lower limit, a requirement which could always be met by appropriate constants $a_1$, $a_2$, $a_3$, $a_4$. In order to leave greater freedom for these constants, this initial requirement will be met, by the assignment of initial values to the quantities $\eta_r$ and $\zeta_r$ and their derivatives, for $r = 1, 2, 3, 4$, where we take the integral-sets already (§ 198) specified for the construction of the quantities **v** and **w**.

Now $Y$ and $Z$ are to vanish at the upper limit also. Yet the foregoing expressions for $Y$ and $Z$ are to be excluded, because they would lead to non-zero variations that would allow $\mathbf{v}$ and $\mathbf{w}$ to vanish throughout the range. Also, $\omega$ is not to vanish. For these reasons combined, it must not be possible to have

$$\sum_{r=1}^{4} a_r \eta_r', \quad \sum_{r=1}^{4} a_r \zeta_r', \quad \sum_{r=1}^{4} a_r \eta_r, \quad \sum_{r=1}^{4} a_r \zeta_r,$$

vanishing simultaneously in the range. Therefore *the range must not extend so far as the first value of $x$, greater than the initial value $x_0$, for which these four quantities simultaneously vanish.* Thus we have a possible upper limit for the range, and denote it by $x_1$, when there exists such a value of $x$; and then the Jacobi test makes $x_1$ the upper limit of a range which has $x_0$ for its lower limit. The range of the integral, consequently, is not to extend so far as the conjugate of the lower limit.

We must obtain the analytical determination of the upper limit $x_1$, to be taken as the conjugate of $x_0$, the lower limit.

### *Critical equation, in connection with the Jacobi test.*

**202.** The four integral-sets, $\eta_1$ and $\zeta_1$, $\eta_2$ and $\zeta_2$, $\eta_3$ and $\zeta_3$, $\eta_4$ and $\zeta_4$, are linearly expressible (§ 184) in terms of the eight fundamental integral-sets $\phi_r(x)$ and $\psi_r(x)$, for $r = 1, \ldots, 8$. Let

$$\left.\begin{aligned}
\eta_1 &= \sum_{r=1}^{8} k_r \, \phi_r(x), \quad \zeta_1 = \sum_{r=1}^{8} k_r \, \psi_r(x) \\
\eta_2 &= \sum_{r=1}^{8} l_r \, \phi_r(x), \quad \zeta_2 = \sum_{r=1}^{8} l_r \, \psi_r(x) \\
\eta_3 &= \sum_{r=1}^{8} m_r \phi_r(x), \quad \zeta_3 = \sum_{r=1}^{8} m_r \psi_r(x) \\
\eta_4 &= \sum_{r=1}^{8} n_r \, \phi_r(x), \quad \zeta_4 = \sum_{r=1}^{8} n_r \, \psi_r(x)
\end{aligned}\right\}.$$

The quantities $\kappa\eta$ and $\kappa\zeta$, in each instance, denote a small variation from the characteristic curve under consideration, chosen so as to give a contiguous characteristic curve. For every curve, the constants are determined by initial values chosen at the lower limit. Accordingly, we choose

$$\eta_r = 0, \quad \zeta_r = 0; \quad \eta_r' = 0, \quad \zeta_r' = 0;$$

for $r = 1, 2, 3, 4$, at the lower limit $x_0$—a selection which makes each of the four contiguous curves touch the fundamental characteristic curve at that lower limit. Among these four contiguous curves, we discriminate by the assignment of other initial conditions at that place $x = x_0$, as follows:

$$\left.\begin{aligned}
\eta_1'' &= \rho, \quad \zeta_1'' = 0, \quad \eta_1''' = 0, \quad \zeta_1''' = 0 \\
\eta_2'' &= 0, \quad \zeta_2'' = \sigma, \quad \eta_2''' = 0, \quad \zeta_2''' = 0 \\
\eta_3'' &= 0, \quad \zeta_3'' = 0, \quad \eta_3''' = \tau, \quad \zeta_3''' = 0 \\
\eta_4'' &= 0, \quad \zeta_4'' = 0, \quad \eta_4''' = 0, \quad \zeta_4''' = \mu
\end{aligned}\right\},$$

where $\rho$, $\sigma$, $\tau$, $\mu$ are finite non-zero constants. Manifestly, the curves are distinct from one another; and each of them is distinct from the characteristic curve which (§ 182) would arise, if either $\rho$ or $\sigma$ or $\tau$ or $\mu$ were to vanish.

All these assigned conditions are in accord with the six relations of § 188, satisfied by the four integral-sets $\eta_r$ and $\zeta_r$ (for $r = 1, 2, 3, 4$). Those relations require that certain constants are to be zero: when the foregoing values are substituted, they do, in fact, make each of the said constants equal to zero.

The eight constants $k_1, \ldots, k_8$ in the expressions for $\eta_1$ and $\zeta_1$, are thus subject to the eight linear relations

$$\Sigma k_r \phi_r\ (x_0) = 0, \quad \Sigma k_r \psi_r\ (x_0) = 0, \quad \Sigma k_r \phi_r'\ (x_0) = 0, \quad \Sigma k_r \psi_r'\ (x_0) = 0,$$

$$\Sigma k_r \phi_r''\ (x_0) = \rho, \quad \Sigma k_r \psi_r''(x_0) = 0, \quad \Sigma k_r \phi_r'''(x_0) = 0, \quad \Sigma k_r \psi_r'''(x_0) = 0.$$

The determinant of the coefficients of the constants $k$ on the left-hand sides

$$|\ \phi_1(x_0),\ \ \psi_2(x_0),\ \ \phi_3'\ (x_0),\ \ \psi_4'\ (x_0),\ \ \phi_5''\ (x_0),\ \ \psi_6''\ (x_0),\ \ \phi_7'''(x_0),\ \ \psi_8'''(x_0)\ |$$

is the value, at $x_0$, of the quantity $\Psi$ of § 186, a quantity there proved not to vanish anywhere in the range : we denote this value by $\Psi_0$. Let $[\xi]$ denote the minor of any constituent $\xi$ in the determinantal expression for $\Psi_0$, so that, for example,

$$[\phi_1''\ (x_0)] = |\ \phi_2(x_0),\ \ \psi_3(x_0),\ \ \phi_4'\ (x_0),\ \ \psi_5'\ (x_0),\ \ \psi_6''\ (x_0),\ \ \phi_7'''(x_0),\ \ \psi_8'''(x_0)\ |,$$

$$-[\phi_2''\ (x_0)] = |\ \phi_1(x_0),\ \ \psi_3(x_0),\ \ \phi_4'\ (x_0),\ \ \psi_5'\ (x_0),\ \ \psi_6''\ (x_0),\ \ \phi_7'''(x_0),\ \ \psi_8'''(x_0)\ |,$$

and so on. Then we have

$$k_1\Psi_0 = \rho\ [\phi_1''\ (x_0)], \quad k_2\Psi_0 = \rho\ [\phi_2''\ (x_0)], \quad k_3\Psi_0 = \rho\ [\phi_3''\ (x_0)], \quad k_4\Psi_0 = \rho\ [\phi_4''\ (x_0)],$$

and so on : generally

$$k_r \Psi_0 = \rho\ [\phi_r''\ (x_0)],$$

for $r = 1, \ldots, 8$. Similarly, for the same values of $r$ in succession,

$$-\ l_r\Psi_0 = \sigma\ [\psi_r''\ (x_0)],$$

$$m_r\Psi_0 = \tau\ [\phi_r'''\ (x_0)],$$

$$-\ n_r\Psi_0 = \mu\ [\psi_r'''\ (x_0)].$$

We therefore have unique finite non-zero values for all the coefficients in the four integral-sets $\eta_1$ and $\zeta_1$, $\eta_2$ and $\zeta_2$, $\eta_3$ and $\zeta_3$, and $\eta_4$ and $\zeta_4$.

**203.** Now consider, once more, the quantity $\omega$, where

$$\omega = \begin{vmatrix} \eta_1, & \eta_2, & \eta_3, & \eta_4 \\ \zeta_1, & \zeta_2, & \zeta_3, & \zeta_4 \\ \eta_1', & \eta_2', & \eta_3', & \eta_4' \\ \zeta_1', & \zeta_2', & \zeta_3', & \zeta_4' \end{vmatrix}.$$

It vanishes at the lower limit $x_0$ of the integral; but no difficulty is thereby caused in the second variation, because

$$\frac{v}{\omega} = Y'' + \alpha Y' + \beta Z' + \gamma Y + \delta Z, \qquad \frac{w}{\omega} = Z'' + \theta Y' + \phi Z' + \psi Y + \chi Z.$$

It must not again vanish within the range of integration; and that range must extend so far as the first value $x_1$, of $x$ which is greater than $x_0$, at which the quantity $\omega$ could vanish. But

$$\omega\,(\rho\sigma\tau\mu)^{-1}\Psi_0{}^4 =$$

$$\begin{vmatrix} \Sigma\phi_r\,(x)\,[\phi_r{}''(x_0)], & \Sigma\phi_r\,(x)\,[\psi_r{}''(x_0)], & \Sigma\phi_r\,(x)\,[\phi_r{}'''(x_0)], & \Sigma\phi_r\,(x)\,[\psi_r{}'''(x_0)] \\ \Sigma\psi_r\,(x)\,[\phi_r{}''(x_0)], & \Sigma\psi_r\,(x)\,[\psi_r{}''(x_0)], & \Sigma\psi_r\,(x)\,[\phi_r{}'''(x_0)], & \Sigma\psi_r\,(x)\,[\psi_r{}'''(x_0)] \\ \Sigma\phi_r{}'(x)\,[\phi_r{}''(x_0)], & \Sigma\phi_r{}'(x)\,[\psi_r{}''(x_0)], & \Sigma\phi_r{}'(x)\,[\phi_r{}'''(x_0)], & \Sigma\phi_r{}'(x)\,[\psi_r{}'''(x_0)] \\ \Sigma\psi_r{}'(x)\,[\phi_r{}''(x_0)], & \Sigma\psi_r{}'(x)\,[\psi_r{}''(x_0)], & \Sigma\psi_r{}'(x)\,[\phi_r{}'''(x_0)], & \Sigma\psi_r{}'(x)\,[\psi_r{}'''(x_0)] \end{vmatrix}.$$

The determinant on the right-hand side can be evaluated by the repeated use of the known theorem relating to determinants made up of constituents which are the complementary minors of determinants, themselves made up of the constituents of a determinant such as $\Psi_0$. Thus, in the expansion of the foregoing determinantal expression for $\omega$, the coefficient of

$$\begin{vmatrix} \phi_1\,(x), & \phi_2\,(x), & \phi_3\,(x), & \phi_4\,(x) \\ \psi_1\,(x), & \psi_2\,(x), & \psi_3\,(x), & \psi_4\,(x) \\ \phi_1{}'(x), & \phi_2{}'(x), & \phi_3{}'(x), & \phi_4{}'(x) \\ \psi_1{}'(x), & \psi_2{}'(x), & \psi_3{}'(x), & \psi_4{}'(x) \end{vmatrix}$$

is

$$\begin{vmatrix} [\phi_1{}''(x_0)], & [\phi_2{}''(x_0)], & [\phi_3{}''(x_0)], & [\phi_4{}''(x_0)] \\ [\psi_1{}''(x_0)], & [\psi_2{}''(x_0)], & [\psi_3{}''(x_0)], & [\psi_4{}''(x_0)] \\ [\phi_1{}'''(x_0)], & [\phi_2{}'''(x_0)], & [\phi_3{}'''(x_0)], & [\phi_4{}'''(x_0)] \\ [\psi_1{}'''(x_0)], & [\psi_2{}'''(x_0)], & [\psi_3{}'''(x_0)], & [\psi_4{}'''(x_0)] \end{vmatrix}.$$

By the theorem quoted, this last determinant (made up of first minors of $\Psi_0$) is equal to

$$\Psi_0{}^3 \begin{vmatrix} \phi_5\,(x_0), & \phi_6\,(x_0), & \phi_7\,(x_0), & \phi_8\,(x_0) \\ \psi_5\,(x_0), & \psi_6\,(x_0), & \psi_7\,(x_0), & \psi_8\,(x_0) \\ \phi_5{}'(x_0), & \phi_6{}'(x_0), & \phi_7{}'(x_0), & \phi_8{}'(x_0) \\ \psi_5{}'(x_0), & \psi_6{}'(x_0), & \psi_7{}'(x_0), & \psi_8{}'(x_0) \end{vmatrix}.$$

Similarly for other terms in the expansion: the coefficient of

$$|\,\phi_p(x), \quad \psi_q(x), \quad \phi_r{}'(x), \quad \psi_s{}'(x)\,|,$$

where $p, q, r, s$, are any combination from 1, 2, 3, 4, 5, 6, 7, 8, is

$$\Psi_0{}^3\,|\,\phi_{9-s}(x_0), \quad \psi_{9-r}(x_0), \quad \phi'_{9-q}(x_0), \quad \psi'_{9-p}(x_0)\,|,$$

by the same theorem. Consequently, when all the evaluated terms are added together, we have

$$\frac{\omega\Psi_0}{\rho\sigma\tau\mu} = |\,\phi_1(x), \ \psi_2(x), \ \phi_3{}'(x), \ \psi_4{}'(x), \ \phi_5(x_0), \ \psi_6(x_0), \ \phi_7{}'(x_0), \ \psi_8{}'(x_0)\,|$$

$$= \Delta\,(x_0, x),$$

where $\Delta(x_0, x)$ denotes the determinant of eight rows, made up of $\phi(x)$, $\psi(x)$, $\phi(x_0)$, $\psi(x_0)$, and their first derivatives.

Now $\Psi_0$ does not vanish, nor become infinte. The quantities $\rho, \sigma, \tau, \mu$ are finite quantities, respectively determining the our characteristics contiguous to the characteristic under consideration. The quantity $\omega$ is not to vanish along the range; hence the function $\Delta(x_0, x)$ must not vanish along the range for any value of $x$ within the range greater than $x_0$.

The *conjugate* of $x_0$ is given by that value of $x$, say $x_1$, for which

$$\sum_{r=1}^{4} a_r \eta_r = 0, \quad \sum_{r=1}^{4} a_r \zeta_r = 0, \quad \sum_{r=1}^{4} a_r \eta_r' = 0, \quad \sum_{r=1}^{4} a_r \zeta_r' = 0,$$

without $a_1, a_2, a_3, a_4$ themselves vanishing: that is, for which $\omega$ vanishes. Hence *the conjugate, given by $x_1$, of the lower limit of the range of integration given by $x_0$, is the first root (greater than $x_0$) of the equation*

$$\Delta(x_0, x_1) = 0.$$

If the original integral is to possess a maximum or a minimum, the range of that integral, beginning at $x_0$, must not extend as far as the value $x_1$ which defines the conjugate of $x_0$ on the characteristic curve.

We thus have the third test for the possession of a maximum or minimum; after the analogy of preceding instances, it is called the *Jacobi* test.

*A range, bounded by two conjugates, does not enclose any similarly bounded range.*

**204.** In enunciating this result, one property has been tacitly assumed by anticipation: viz. a complete range along a characteristic curve, between a point and its conjugate, cannot contain within itself another complete range of the same type: in other words, *the conjugate of any point within a range lies without the range.* The establishment of this property can be effected as in the preceding cases (§§ 119, 176), by proceeding from the function $\Delta(x_0, x)$, the outline being as follows.

This function $\Delta(x_0, x)$ does not vanish anywhere that lies actually between $x_0$ and its conjugate $x_1$. Let two points be taken in the range, one near $x_0$ given by $x_0 + \lambda \xi_0$ (where $\lambda$ and $\xi_0$ are positive, and $\lambda$ is small), and one near $x_1$ given by $x_1 - \epsilon \xi_1$ (where $\epsilon$ and $\xi_1$ are positive, and $\epsilon$ is small); the two magnitudes

$$\Delta(x_0, x_0 + \lambda \xi_0), \quad \Delta(x_0, x_1 - \epsilon \xi_1)$$

have the same sign. But, approximately, for sufficiently small values of $\lambda$, we have

$$\Delta(x_0, x_0 + \lambda \xi_0) = \tfrac{1}{6} \lambda^4 \xi_0^4 \Psi_0,$$

while

$$\Delta(x_0, x_1 - \epsilon \xi_1) = \Delta(x_0, x_1) - \epsilon \xi_1 \frac{\partial \Delta(x_0, x_1)}{\partial x_1}$$

$$= -\epsilon \xi_1 \frac{\partial \Delta(x_0, x_1)}{\partial x_1},$$

so that $\dfrac{\partial \Delta\,(x_0,\,x_1)}{\partial x_1}$ has a persistent sign, opposite to the persistent sign of $\Psi$ whatever initial point be chosen. Also, as $\Delta\,(x_0,\ x_1 - \epsilon\xi_1)$ does not vanish, $\dfrac{\partial \Delta\,(x_0,\,x_1)}{\partial x_1}$ does not vanish.

When a point $x_0 + \alpha\xi$ is taken, with $\alpha$ and $\xi$ positive, and its conjugate $x_1 + \alpha\xi'$ is wanted, $\alpha$ being small, we have

$$\Delta\,(x_0 + \alpha\xi,\ x_1 + \alpha\xi') = 0;$$

so that, when $\alpha$ is sufficiently small, we have

$$\xi\,\frac{\partial \Delta\,(x_0,\,x_1)}{\partial x_0} + \xi'\,\frac{\partial \Delta\,(x_0,\,x_1)}{\partial x_1} = 0.$$

If $\xi$ and $\xi'$ have opposite signs, the new range would lie within the old; and the distances of the new ends from the old ends would be of the same order of magnitude. This diminished range would similarly lead to a more restricted range, because of the persistence of the sign of $\Psi$. Gradually, as before, we could wear down the range (on the hypothesis that $\xi$ and $\xi'$ had opposite signs) until the function $\Delta\,(x',\,x'')$ would palpably cease to vanish when $x''$ is sufficiently near $x'$. That hypothesis must therefore be abandoned. Consequently, $\xi'$ and $\xi$ have the same sign; and they are of the same order of magnitude.

Proceeding gradually from $x_0$, and taking the new conjugate for each new initial point—which conjugate must lie beyond each immediately earlier conjugate—we infer the property that the conjugate of any point within a complete range lies outside the range.

### Summary of tests.

**205.** The conditions, when satisfied, secure the existence of a maximum or a minimum for weak variations. The Euler test and the terminal conditions make the first variation vanish. The Legendre test and the Jacobi test preserve one sign for the second variation—either positive (when there is a minimum) or negative (when there is a maximum)—through all weak variations that are possible.

*General weak variations: statement of results concerning the 'second' variation.*

**206.** When general weak variations are considered, so that we have to deal with the integral

$$\int F\,(x,\ x_1,\ x_2,\ y,\ y_1,\ y_2,\ z,\ z_1,\ z_2)\,dt,$$

and the variations are $x + \kappa X,\ y + \kappa Y,\ z + \kappa Z$, the results connected with the first variation of that integral have already been obtained. Much laborious algebra is required for the reduction of its second variation to a normal form.

The results are significant, not least in connection with the discussion of a characteristic curve, of characteristics contiguous to that curve, and of a range along the curve bounded by conjugate points, with the definition of conjugates already used. The discussion follows, in more general shape, the discussion already given; and the algebraical calculations, in themselves and in their process, are the extension—an elaborate extension—of calculations of the same kind already effected for simpler cases. It may, therefore, suffice to give an outline only, arranged in successive statements. Three fundamental quantities are definitely established, in (II); the other inferences are stated, without detailed proofs being given.

(I)   The function $F$ satisfies two identities

$$x_1 \frac{\partial F}{\partial x_2} + y_1 \frac{\partial F}{\partial y_2} + z_1 \frac{\partial F}{\partial z_2} = 0,$$

$$F = x_1 \frac{\partial F}{\partial x_1} + y_1 \frac{\partial F}{\partial y_1} + z_1 \frac{\partial F}{\partial z_1} + 2x_2 \frac{\partial F}{\partial x_2} + 2y_2 \frac{\partial F}{\partial y_2} + 2z_2 \frac{\partial F}{\partial z_2}.$$

Derivatives of $F$ with regard to the variables

$$x, \quad x_1, \quad x_2, \quad y, \quad y_1, \quad y_2, \quad z, \quad z_1, \quad z_2,$$

are denoted by numerical suffixes 1, 2, 3, 4, 5, 6, 7, 8, 9, respectively; and the variable quantities

$$X, \quad X_1, \quad X_2, \quad Y, \quad Y_1, \quad Y_2, \quad Z, \quad Z_1, \quad Z_2,$$

are denoted respectively by

$$\theta_1, \quad \theta_2, \quad \theta_3, \quad \theta_4, \quad \theta_5, \quad \theta_6, \quad \theta_7, \quad \theta_8, \quad \theta_9.$$

We shall want combinations of $Y$ and $X$, of $Z$ and $X$, and the derivatives of these combinations; we write

$$\phi_4 = V \;\; = x_1 Y - y_1 X = x_1 \theta_4 - y_1 \theta_1,$$
$$\phi_5 = V' \;\; = x_1 \theta_5 - y_1 \theta_2 + (x_2 \theta_4 - y_2 \theta_1),$$
$$\phi_6 = V'' = x_1 \theta_6 - y_1 \theta_3 + 2(x_2 \theta_5 - y_2 \theta_2) + (x_3 \theta_4 - y_3 \theta_1),$$
$$\phi_7 = W \;\; = x_1 Z - z_1 X = x_1 \theta_7 - z_1 \theta_1,$$
$$\phi_8 = W' \;\; = x_1 \theta_8 - z_1 \theta_2 + (x_2 \theta_7 - z_2 \theta_1),$$
$$\phi_9 = W'' = x_1 \theta_9 - z_1 \theta_3 + 2(x_2 \theta_8 - z_2 \theta_2) + (x_3 \theta_7 - z_3 \theta_1).$$

(II)   Differentiating the first of the identities satisfied by $F$ with respect to $x_2$, to $y_2$, and to $z_2$, separately, we have

$$x_1 F_{33} + y_1 F_{36} + z_1 F_{39} = 0,$$
$$x_1 F_{36} + y_1 F_{66} + z_1 F_{69} = 0,$$
$$x_1 F_{39} + y_1 F_{69} + z_1 F_{99} = 0.$$

Consequently, there exist three quantities $g_{66}$, $g_{69}$, $g_{99}$ such that

$$F_{66} = x_1^2 g_{66}, \quad F_{69} = x_1^2 g_{69}, \quad F_{99} = x_1^2 g_{99},$$
$$F_{36} = -x_1 y_1 g_{66} - x_1 z_1 g_{69}, \quad F_{39} = -x_1 y_1 g_{69} - x_1 z_1 g_{99},$$
$$F_{33} = y_1^2 g_{66} + 2y_1 z_1 g_{69} + z_1^2 g_{99}.$$

The three quantities $g_{66}, g_{69}, g_{99}$ are fundamental; they are, save as to a power of $x_1$, equal to the quantities $f_{66}, f_{69}, f_{99}$ of §§ 182—5.

(III) The first variation of the integral has been made to vanish, thereby leading to the characteristic equations and utilising the terminal data so that the limits of the integral can be considered as fixed.

The second variation of the integral is $\frac{1}{2}\kappa^2 \int \Theta \, dt$, taken between the fixed limits, where

$$\Theta = \overset{1,\,...,\,9}{\underset{m,\,n}{\Sigma\Sigma}} F_{mn}\theta_m\theta_n,$$

the full expression of $\Theta$ being $F_{99}\theta_9{}^2 + 2F_{89}\theta_8\theta_9 + \ldots + 2F_{12}\theta_1\theta_2 + F_{11}\theta_1{}^2$. We take a quantity

$$\square = A\theta_2 + B\theta_5{}^2 + C\theta_8{}^2 + 2F\theta_5\theta_8 + 2G\theta_2\theta_8 + 2H\theta_2\theta_5$$
$$+ a\theta_1{}^2 + b\theta_4{}^2 + c\theta_7{}^2 + 2f\theta_4\theta_7 + 2g\theta_1\theta_7 + 2h\theta_1\theta_4$$
$$+ 2\theta_1(\alpha\theta_2 + \beta\theta_5 + \gamma\theta_8) + 2\theta_4(\lambda\theta_2 + \mu\theta_5 + \nu\theta_8) + 2\theta_7(\rho\theta_2 + \sigma\theta_5 + \tau\theta_8).$$

The first aim of the analysis is to modify the second variation, by expressing $\Theta + \dfrac{d\square}{dt}$ as equal to $\Phi$, where

$$\Phi = \overset{4,\,...,\,9}{\underset{m,\,n}{\Sigma\Sigma}} g_{mn}\phi_m\phi_n,$$

and, in $\Phi$, as few terms are retained as possible. For then

$$\int \left(\Theta + \frac{d\square}{dt}\right) dt = \int \Theta \, dt + [\square] = \int \Theta \, dt,$$

because the limits are now fixed and therefore every variable $\theta$ in $\square$ vanishes at each limit ($\theta_3, \theta_6, \theta_9$ are absent from $\square$): that is,

$$\int \Theta \, dt = \int \Phi \, dt.$$

Thus $\Phi$ can be regarded as an intermediate normal quadratic form for the integral.

As regards the retention of as few terms in $\Phi$ as possible, we have

$$g_{99}W''^2 + 2g_{89}W'W'' + 2g_{79}WW'' + g_{88}W'^2 + 2g_{78}WW' + g_{77}W^2$$
$$= g_{99}W''^2 + (g_{88} - g_{89}' - 2g_{79})W'^2 + (g_{77} - g_{78}' + g_{79}'')W^2$$
$$+ \frac{d}{dt}\{g_{89}W'^2 + 2g_{79}WW' + (g_{78} - g_{79}')W^2\}.$$

The term in the last line, with its sign changed, can be moved into $\dfrac{d\square}{dt}$; and then only the squared terms in the first line will be left: that is, by a change of the coefficients $g_{mn}$ which initially are at our disposal, and without loss of generality, we can choose

$$g_{89} = 0, \quad g_{79} = 0, \quad g_{78} = 0.$$

Similarly, without loss of generality, we can take

$$g_{56} = 0, \quad g_{46} = 0, \quad g_{45} = 0,$$

in the form initially postulated for $\Phi$.

We therefore can take, in umbral notation,

$$\Phi = (\gamma_9\phi_9 + \gamma_8\phi_8 + \gamma_7\phi_7 + \gamma_6\phi_6 + \gamma_5\phi_5 + \gamma_4\phi_4)^2.$$

The six umbral symbols $\gamma_4$, $\gamma_5$, $\gamma_6$, $\gamma_7$, $\gamma_8$, $\gamma_9$ combine quadratically into the significance of reality, according to the laws

$$\gamma_9^2 = g_{99}, \quad \gamma_6\gamma_9 = g_{69}, \quad \gamma_6^2 = g_{66},$$

(the three fundamental quantities already introduced);

$$\gamma_8\gamma_9 = 0, \quad \gamma_7\gamma_9 = 0, \quad \gamma_7\gamma_8 = 0, \quad \gamma_5\gamma_6 = 0, \quad \gamma_4\gamma_6 = 0, \quad \gamma_4\gamma_5 = 0,$$

(according to the foregoing initial simplifications of $\Phi$); and, for all other combinations,

$$\gamma_m\gamma_n = g_{mn},$$

these non-zero coefficients $g_{mn}$ remaining.

The consequent modifications, by the absorption of

$$- \{g_{89}W'^2 + 2g_{79}WW' + (g_{78} - g_{79}')\,W^2\}$$

into $\square$, will be supposed made, so that the coefficients in $\square$ (as yet unused) will be regarded as having absorbed the necessary changes.

(IV) Owing to the expressions for the real variables $\phi$ in terms of the real variables $\theta$, we have

$$\begin{aligned}
\sum_{r=4}^{9} \gamma_r\phi_r = \quad & \gamma_9 x_1\theta_9 + (2\gamma_9 x_2 + \gamma_8 x_1)\,\theta_8 + (\gamma_9 x_3 + \gamma_8 x_2 + \gamma_7 x_1)\,\theta_7 \\
& + \gamma_6 x_1\theta_6 + (2\gamma_6 x_2 + \gamma_5 x_1)\,\theta_5 + (\gamma_6 x_3 + \gamma_5 x_2 + \gamma_4 x_1)\,\theta_4 \\
& - (\gamma_9 z_1 + \gamma_6 y_1)\,\theta_3 \\
& - (2\gamma_9 z_2 + 2\gamma_6 y_2 + \gamma_8 z_1 + \gamma_5 y_1)\,\theta_2 \\
& - (\gamma_9 z_3 + \gamma_6 y_3 + \gamma_8 z_2 + \gamma_5 y_2 + \gamma_7 z_1 + \gamma_4 y_1)\,\theta_1.
\end{aligned}$$

Then we are to have

$$\Theta + \frac{d\square}{dt} = \Phi = \left\{ \sum_{r=4}^{9} \gamma_r\phi_r \right\}^2,$$

where, on the right-hand side, we substitute the equivalent expression which is quadratic in $\theta$.

(V) Equating the coefficients of the various combinations of the variables $\theta_9$, ..., $\theta_1$ on the two sides of the last relation in (IV), we have relations between the known coefficients $F_{mn}$, the unknown coefficients $g_{mn}$, and the coefficients in $\square$. The total number of relations in this aggregate is forty-five, being the number of coefficients $F_{mn}$.

(i) The terms, involving $(\theta_9, \theta_6, \theta_3)^2$ in $\Theta + \dfrac{d\square}{dt}$ and $\Phi$, agree without any relation, owing to the expressions for $F_{mn}$ (with $m, n = 3, 6, 9$) in terms of the three coefficients $g_{99}, g_{69}, g_{66}$.

(ii) The terms, involving $(\theta_9, \theta_6, \theta_3)(\theta_8, \theta_5, \theta_2)$ agree, if

$$\left.\begin{aligned}
F_{89} + C &= \quad 2g_{99}x_1 x_2 \\
F_{59} + F &= \quad 2g_{69}x_1 x_2 + g_{59}x_1{}^2 \\
F_{29} + G &= -2g_{99}x_1 z_2 - 2g_{69}x_1 y_2 - g_{59}x_1 y_1
\end{aligned}\right\},$$

$$\left.\begin{aligned}
F_{68} + F &= \quad 2g_{69}x_1 x_2 + g_{68}x_1{}^2 \\
F_{56} + B &= \quad 2g_{66}x_1 x_2 \\
F_{26} + H &= -2g_{69}x_1 z_2 - 2g_{66}x_1 y_2 - g_{68}x_1 z_1
\end{aligned}\right\}.$$

$$\left.\begin{aligned}
F_{38} + G &= -2g_{99}z_1 x_2 - 2g_{69}y_1 x_2 \qquad\qquad - g_{68}x_1 y_1 \\
F_{35} + H &= \qquad\qquad\; -2g_{69}z_1 x_2 - 2g_{66}y_1 x_2 - g_{59}x_1 z_1 \\
F_{23} + A &= \quad 2g_{99}z_1 z_2 + 2g_{69}(z_1 y_2 + y_1 z_2) + 2g_{66}y_1 y_2 \\
&\qquad + g_{68}z_1 y_1 + g_{59}z_1 y_1
\end{aligned}\right\}.$$

In these relations, $F$ occurs twice, $G$ twice, $H$ twice; and their repeated values must be the same. But from the fundamental identities, we have

$$x_1 F_{23} + y_1 F_{26} + z_1 F_{29} = -\frac{\partial F}{\partial x_2},$$

$$x_1 F_{23} + y_1 F_{35} + z_1 F_{38} + 2x_2 F_{33} + 2y_2 F_{36} + 2z_2 F_{39} = -\frac{\partial F}{\partial x_2},$$

which give

$$x_1 A + y_1 H + z_1 G = \frac{\partial F}{\partial x_2},$$

and

$$y_1(F_{26} + F_{35}) + z_1(F_{29} - F_{38}) = 2(x_2 F_{33} + y_2 F_{36} + z_2 F_{39}).$$

Similarly

$$x_1 H + y_1 B + z_1 F = \frac{\partial F}{\partial y_2},$$

and

$$z_1(F_{59} - F_{68}) + x_1(F_{35} - F_{26}) = 2(x_2 F_{36} + y_2 F_{66} + z_2 F_{69});$$

and

$$x_1 G + y_1 F + z_1 C = \frac{\partial F}{\partial z_2},$$

and

$$x_1(F_{38} - F_{29}) + y_1(F_{68} - F_{59}) = 2(x_2 F_{39} + y_2 F_{69} + z_2 F_{99}).$$

Of these necessary equations, the three which are free from $A, B, C, F, G, H$, are satisfied by the foregoing nine postulated equations, if only

$$F_{59} - F_{68} = (g_{59} - g_{68})x_1{}^2.$$

Therefore the whole set of equations defines $A, B, C, F, G, H, g_{59}, g_{68}$, always subject to the latter equation.

(iii) The terms, involving $(\theta_9, \theta_6, \theta_3 \big\rangle \theta_7, \theta_4, \theta_1)$ agree, if

$$F_{79} + \tau = g_{99}x_1x_3$$
$$F_{49} + \nu = g_{69}x_1x_3 + g_{59}x_1x_2 + g_{49}x_1{}^2$$
$$F_{19} + \gamma = -g_{99}x_1z_3 - g_{69}x_1y_3 - g_{59}x_1y_2 - g_{49}x_1y_1$$

$$F_{67} + \sigma = g_{69}x_1x_3 + g_{68}x_1x_2 + g_{67}x_1{}^2$$
$$F_{64} + \mu = g_{66}x_1x_3$$
$$F_{16} + \beta = -g_{69}x_1z_3 - g_{66}x_1y_3 - g_{68}x_1z_2 - g_{67}x_1z_1$$

$$F_{37} + \rho = -g_{99}z_1x_3 - g_{69}y_1x_3 - g_{68}y_1x_2 - g_{67}x_1y_1$$
$$F_{34} + \lambda = -g_{69}z_1x_3 - g_{66}y_1x_3 - g_{59}z_1x_2 - g_{49}x_1z_1$$
$$F_{13} + \alpha = g_{99}z_1z_3 + g_{69}(z_1y_3 + y_1z_3) + g_{66}y_1y_3$$
$$+ g_{59}z_1y_2 + g_{68}y_1z_2 + g_{49}y_1z_1 + g_{67}y_1z_1$$

which may be regarded as defining the nine quantities $\alpha, \ldots, \tau$, in terms of new quantities $g_{59}, g_{49}, g_{67}, g_{68}$.

(iv) The terms, involving $(\theta_8, \theta_5, \theta_2)^2$ agree, if

$$F_{88} + C' + 2\tau = 4g_{99}x_2{}^2 + g_{88}x_1{}^2$$
$$F_{85} + F' + \nu + \sigma = 4g_{69}x_2{}^2 + 2g_{59}x_1x_2 + 2g_{68}x_1x_2 + g_{58}x_1{}^2$$
$$F_{82} + G' + \gamma + \rho = -4g_{99}x_2z_2 - 4g_{69}x_2y_2$$
$$- 2g_{68}x_1y_2 - 2g_{59}y_1x_2 - g_{88}x_1z_1 - g_{58}x_1y_1$$
$$F_{55} + B' + 2\mu = 4g_{66}x_2{}^2 + g_{55}x_1{}^2$$
$$F_{52} + H' + \beta + \lambda = -4g_{69}x_2z_2 - 4g_{66}x_2y_2$$
$$- 2g_{68}z_1x_2 - 2g_{59}z_1z_2 - g_{58}x_1z_1 - g_{55}x_1y_1$$
$$F_{22} + A' + 2\alpha = 4g_{99}z_2{}^2 + 8g_{69}z_2y_2 + 4g_{66}y_2{}^2$$
$$+ 4g_{68}z_1y_2 + 4g_{59}y_1z_2 + g_{88}z_1{}^2 + 2g_{58}y_1z_1 + g_{55}y_1{}^2$$

The quantities $A, \ldots, H, \alpha, \ldots, \tau$ have been defined in the earlier group (ii) and the earlier group (iii) of relations. These six conditions therefore define $g_{55}, g_{58}, g_{88}$; and they leave three relations to be satisfied.

(v) The terms in $(\theta_8, \theta_5, \theta_2 \big\rangle \theta_7, \theta_4, \theta_1)$ agree, if

$$F_{78} + \tau' + c = 2g_{99}x_2x_3 + g_{88}x_1x_2$$
$$F_{48} + \nu' + f = 2g_{69}x_2x_3 + 2g_{59}x_2{}^2 + 2g_{49}x_1x_2 + g_{68}x_1x_3 + g_{58}x_1x_2 + g_{48}x_1{}^2$$
$$F_{18} + \gamma' + g = -2g_{99}x_2z_3 - 2g_{69}x_2y_3 - 2g_{59}x_2{}^2 - 2g_{49}y_1x_2$$
$$- g_{88}x_1z_2 - g_{68}x_1y_3 - g_{58}x_1y_2 - g_{48}x_1y_1$$
$$F_{57} + \sigma' + f = 2g_{69}x_2x_3 + 2g_{68}x_2{}^2 + 2g_{67}x_1x_2 + 2g_{59}x_1x_3 + g_{58}x_1x_2 + g_{57}x_1{}^2$$
$$F_{45} + \mu' + b = 2g_{66}x_2x_3 + g_{55}x_1x_2$$
$$F_{15} + \beta' + h = -2g_{69}x_2z_3 - 2g_{66}x_2y_3 - 2g_{68}x_2z_2 - 2g_{67}z_1x_2$$
$$- g_{59}x_1z_3 - g_{58}x_1z_2 - g_{55}x_1y_2 - g_{57}x_1z_1$$

$$F_{27} + \rho' + g = -2g_{99}z_2x_3 - 2g_{69}y_2x_3 - 2g_{68}y_2x_2 - 2g_{67}x_1y_2$$
$$\left.\begin{aligned} &\qquad\qquad\quad - g_{88}z_1x_2 - g_{59}y_1x_3 - g_{58}y_1x_2 - g_{57}x_1y_1 \\ F_{24} + \lambda' + h &= -2g_{69}z_2x_3 - 2g_{59}z_2x_2 - 2g_{49}x_1z_2 - 2g_{66}x_3y_2 \\ &\qquad\qquad\quad - g_{68}z_1x_3 - g_{58}z_1x_2 - g_{48}x_1z_1 - g_{55}y_1x_2 \\ F_{12} + \alpha' + a &= \phantom{-}2g_{99}z_2z_3 + 2g_{69}(z_2y_3 + y_2z_3) + 2g_{66}y_2y_3 \\ &\qquad\qquad\quad + g_{59}(2y_2z_2 + y_1z_3) + g_{68}(2y_2z_2 + z_1y_3) \\ &\qquad\qquad\quad + g_{88}z_1z_2 + g_{58}(z_1y_2 + y_1z_2) + g_{55}y_1y_2 \\ &\qquad\qquad\quad + 2g_{49}y_1z_2 + 2g_{67}z_1y_2 + (g_{48} + g_{57})y_1z_1 \end{aligned}\right\}.$$

(vi) The terms in $(\theta_7, \theta_4, \theta_1)^2$ agree if

$$F_{77} + c' = \phantom{-}g_{99}x_3{}^2 + g_{88}x_2{}^2 + g_{77}x_1{}^2$$
$$\left.\begin{aligned} F_{74} + f' &= \phantom{-}g_{69}x_3{}^2 + (g_{59} + g_{68})x_2x_3 + (g_{49} + g_{67})x_1x_3 + g_{58}x_2{}^2 + (g_{48} + g_{57})x_1x_2 + g_{47}x_1{}^2 \\ F_{71} + g' &= -g_{99}x_3z_3 - g_{69}x_3y_3 - g_{59}y_2x_3 - g_{49}y_1x_3 \\ &\qquad - g_{88}x_2z_2 - g_{68}x_2y_3 - g_{58}x_2y_2 - g_{48}y_1x_2 \\ &\qquad - g_{77}x_1z_1 - g_{67}x_1y_3 - g_{57}x_1y_2 - g_{47}x_1y_1 \\ F_{44} + b' &= \phantom{-}g_{66}x_3{}^2 + g_{55}x_2{}^2 + g_{44}x_1{}^2 \\ F_{14} + h' &= -g_{69}x_3z_3 - g_{66}x_3y_3 - g_{68}z_2x_3 - g_{67}z_1x_3 \\ &\qquad - g_{59}x_2z_3 - g_{58}x_2z_2 - g_{57}z_1x_2 - g_{55}x_2z_2 \\ &\qquad - g_{49}x_1z_3 - g_{48}x_1z_2 - g_{47}x_1z_1 - g_{44}x_1y_1 \\ F_{11} + a' &= \phantom{-}g_{99}z_3{}^2 + 2g_{69}y_3z_3 + g_{66}y_3{}^2 + g_{88}z_2{}^2 + 2g_{58}z_2y_2 + g_{55}y_2{}^2 + g_{77}z_1{}^2 + 2g_{47}y_1z_1 + g_{44}y_1{}^2 \\ &\qquad + 2g_{95}y_2z_3 + 2g_{94}y_1z_3 + 2g_{68}y_3z_2 + 2g_{48}y_1z_2 + 2g_{67}y_3z_1 + 2g_{57}y_2z_1 \end{aligned}\right\}.$$

(vii) Various relations are satisfied by the quantities $F_{mn}$, which arise as derivatives of the two identities satisfied by the subject of integration. Thus, from the first of these identities, we have

$$x_1F_{37} + y_1F_{67} + z_1F_{97} = 0,$$
$$x_1F_{34} + y_1F_{64} + z_1F_{94} = 0,$$
$$x_1F_{31} + y_1F_{61} + z_1F_{91} = 0,$$

and thence

$$x_1\alpha + y_1\beta + z_1\gamma = 0,$$
$$x_1\lambda + y_1\mu + z_1\nu = 0,$$
$$x_1\rho + y_1\sigma + z_1\tau = 0.$$

Again, from that same identity, we have

$$-\frac{\partial F}{\partial x_2} = x_1F_{38} + y_1F_{68} + z_1F_{98} = -x_1G - y_1F - z_1C,$$

$$-\frac{\partial F}{\partial y_2} = x_1F_{35} + y_1F_{65} + z_1F_{95} = -x_1H - y_1B - z_1F,$$

$$-\frac{\partial F}{\partial z_2} = x_1F_{32} + y_1F_{62} + z_1F_{92} = -x_1A - y_1H - z_1G;$$

$$x_1G' + y_1F' + z_1C' + x_2G + y_2F + z_2C$$
$$= x_1F_{19} + x_2F_{29} + x_3F_{39} + y_1F_{49} + y_2F_{59} + y_3F_{69} + z_1F_{79} + z_2F_{89} + z_3F_{99},$$

with two similar relations.

From the second identity, we have

$$0 = x_1 F_{22} + y_1 F_{25} + z_1 F_{28} + 2x_2 F_{23} + 2y_2 F_{26} + 2z_2 F_{29},$$
$$0 = x_1 F_{25} + y_1 F_{55} + z_1 F_{58} + 2x_2 F_{35} + 2y_2 F_{56} + 2z_2 F_{59},$$
$$0 = x_1 F_{28} + y_1 F_{85} + z_1 F_{88} + 2x_2 F_{38} + 2y_2 F_{68} + 2z_2 F_{89}.$$

When the values of $F_{19}, \ldots, F_{99}, F_{22}, \ldots, F_{89}$ are substituted here, these relations are satisfied identically.

Similarly with the other relations, that can be derived from the two identities in the same manner as those in Chapter II and Chapter III: it can be verified that each of them is identically satisfied when the values of $F_{mn}$ are substituted.

(VI) There are subsidiary characteristic equations, associated with the fundamental characteristic equations. Let $2\Omega$ denote the same function of $\xi, \xi', \xi'', \eta, \eta', \eta'', \zeta, \zeta', \zeta''$, as $\Theta$ is of $\theta_1, \ldots, \theta_9$, with the coefficients $F_{mn}$. The *subsidiary equations* are

$$\frac{\partial \Omega}{\partial \xi} - \frac{d}{dt}\left(\frac{\partial \Omega}{\partial \xi'}\right) + \frac{d^2}{dt^2}\left(\frac{\partial \Omega}{\partial \xi''}\right) = 0,$$

$$\frac{\partial \Omega}{\partial \eta} - \frac{d}{dt}\left(\frac{\partial \Omega}{\partial \eta'}\right) + \frac{d^2}{dt^2}\left(\frac{\partial \Omega}{\partial \eta''}\right) = 0,$$

$$\frac{\partial \Omega}{\partial \zeta} - \frac{d}{dt}\left(\frac{\partial \Omega}{\partial \zeta'}\right) + \frac{d^2}{dt^2}\left(\frac{\partial \Omega}{\partial \zeta''}\right) = 0,$$

which are really equivalent to two equations in virtue of the relations among the coefficients.

(VII) Corresponding to $V$ and $W$, we introduce the combinations

$$\mathbf{y} = x_1 \eta - y_1 \xi,$$
$$\mathbf{z} = x_1 \zeta - z_1 \xi.$$

We denote by $2\Upsilon$, the same function of $\mathbf{y}, \mathbf{y}_1, \mathbf{y}_2, \mathbf{z}, \mathbf{z}_1, \mathbf{z}_2$, as $\Phi$ is of $V, V', V'', W, W', W''$. Then $\mathbf{y}$ and $\mathbf{z}$ satisfy the equations

$$\frac{\partial \Upsilon}{\partial \mathbf{y}} - \frac{d}{dt}\left(\frac{\partial \Upsilon}{\partial \mathbf{y}'}\right) + \frac{d^2}{dt^2}\left(\frac{\partial \Upsilon}{\partial \mathbf{y}''}\right) = 0,$$

$$\frac{\partial \Upsilon}{\partial \mathbf{z}} - \frac{d}{dt}\left(\frac{\partial \Upsilon}{\partial \mathbf{z}'}\right) + \frac{d^2}{dt^2}\left(\frac{\partial \Upsilon}{\partial \mathbf{z}''}\right) = 0,$$

which are the modified subsidiary equations.

The primitive of the subsidiary equations, and therefore also the primitive of the modified subsidiary equations, can be derived directly from that of the fundamental characteristic equations. Every integral-set is made up of linear combinations, with constant coefficients, of eight linearly independent integral-sets which constitute a fundamental system.

There are at least five such integral-sets, any four of which (in pair-combinations) provide six relations, identical in form with the six relations of § 188: a selected four are denoted by $\mathbf{y}_1$ and $\mathbf{z}_1$, $\mathbf{y}_2$ and $\mathbf{z}_2$, $\mathbf{y}_3$ and $\mathbf{z}_3$, $\mathbf{y}_4$ and $\mathbf{z}_4$.

(VIII)  By means of the relation

$$\int \Phi\, dt = \int \left( \Phi + \frac{d\nabla}{dt} \right) dt - [\nabla]$$

$$= \int \mathbf{I}\, dt,$$

where $\nabla$ is a homogeneous quadratic expression in $V$, $V'$, $W$, $W'$, we can (by analysis exactly similar to the analysis in § 200) obtain an expression

$$\mathbf{I} = \frac{1}{\theta^2} \left( g_{99} \mathbf{w}^2 + 2 g_{69} \mathbf{w} \mathbf{v} + g_{66} \mathbf{v}^2 \right),$$

where

$$\theta = \begin{vmatrix} \mathbf{y}_1', & \mathbf{z}_1', & \mathbf{y}_1, & \mathbf{z}_1 \\ \mathbf{y}_2', & \mathbf{z}_2', & \mathbf{y}_2, & \mathbf{z}_2 \\ \mathbf{y}_3', & \mathbf{z}_3', & \mathbf{y}_3, & \mathbf{z}_3 \\ \mathbf{y}_4', & \mathbf{z}_4', & \mathbf{y}_4, & \mathbf{z}_4 \end{vmatrix},$$

$$\mathbf{v} = \begin{vmatrix} V'', & V', & W', & V, & W \\ \mathbf{y}_1'', & \mathbf{y}_1', & \mathbf{z}_1', & \mathbf{y}_1, & \mathbf{z}_1 \\ \mathbf{y}_2'', & \mathbf{y}_2', & \mathbf{z}_2', & \mathbf{y}_2, & \mathbf{z}_2 \\ \mathbf{y}_3'', & \mathbf{y}_3', & \mathbf{z}_3', & \mathbf{y}_3, & \mathbf{z}_3 \\ \mathbf{y}_4'', & \mathbf{y}_4', & \mathbf{z}_4', & \mathbf{y}_4, & \mathbf{z}_4 \end{vmatrix}, \qquad \mathbf{w} = \begin{vmatrix} W'', & V', & W', & V, & W \\ \mathbf{z}_1'', & \mathbf{y}_1', & \mathbf{z}_1', & \mathbf{y}_1, & \mathbf{z}_1 \\ \mathbf{z}_2'', & \mathbf{y}_2', & \mathbf{z}_2', & \mathbf{y}_2, & \mathbf{z}_2 \\ \mathbf{z}_3'', & \mathbf{y}_3', & \mathbf{z}_3', & \mathbf{y}_3, & \mathbf{z}_3 \\ \mathbf{z}_4'', & \mathbf{y}_4', & \mathbf{z}_4', & \mathbf{y}_4, & \mathbf{z}_4 \end{vmatrix},$$

where $\theta$ is a quantity that does not vanish within the range of integration.

This form $\int \mathbf{I}\, dt$ is the final normal form, as $\int \Phi\, dt$ was the first or intermediate normal form, of the second variation.

(IX)  The discussion of this result proceeds in the same way as the discussion in §§ 200—204.

The *Legendre test* requires that, everywhere in the range,

$$g_{99} g_{66} > g_{69}{}^2.$$

When this is satisfied, a common positive sign for $g_{66}$ and $g_{99}$ admits a minimum for the original integral, and a common negative sign for $g_{66}$ and $g_{99}$ admits a maximum.

The *Jacobi test* requires that the range of integration, which begins at an initial place $t_0$ on the characteristic curve, shall not extend as far as the place $t_1$, the conjugate of $t_0$: where $t_1$ is the first value of $t_1$ (greater than $t_0$) at which the determinant

$$| \mathbf{y}_1'(t_1), \quad \mathbf{z}_2'(t_1), \quad \mathbf{y}_3(t_1), \quad \mathbf{z}_4(t_1), \quad \mathbf{y}_5'(t_0), \quad \mathbf{z}_6'(t_0), \quad \mathbf{y}_7(t_0), \quad \mathbf{z}_8(t_0) |$$

vanishes.

(X)  A complete range on the characteristic curve, bounded by a place and its conjugate, does not contain within itself another complete range.

# CHAPTER VII.

## Ordinary integrals under strong variations, and the Weierstrass test: Solid of least resistance: Action.

### *General notion of a small variation.*

**207.** The systematic analysis dealing with irregular small variations, which can be represented by a jagged line resoluble into small straight pieces, is due to Weierstrass. He introduced it (and developed it, thereby meeting some general objections of Steiner) so as to discuss possible maxima and minima of integrals, which involve one original dependent variable and its first derivative. Jagged lines were recognised by Legendre* as possibly leading—in the case of Newton's problem of the solid resistance, he obtained jagged lines which do lead—to results, reduced below the minimum† provided by the characteristic curve under small variations that now are called 'weak.' Such lines of a zig-zag type were excluded from consideration in their analysis by Legendre and succeeding generations of writers, through a definition which was implicit at first and was made explicit only afterwards. It had been assumed, usually without any statement or only on a later definite statement kept unobtrusive, that, when the variation of the dependent variable $y$ is maintained small, the variation of its derivative $y'$ will also be small. The assumption seemed to be that the smallness of $y'$, consequent upon the smallness of $y$, was too obvious to merit even mention. There was the further assumption, also implicit and ignored in statement, that the variation of $y'$ is to be of the same small order as the small variation of $y$. All such variations were represented by $\delta y$, $\delta y'$, $\delta dy$, and so on; the mere presence of the Lagrange symbol $\delta$, in this association, apparently ensured the requirement of smallness.

Moreover, in spite of restrictions in the range of the integral to small variations $\delta y$ of the ordinate alone. circumstances compelled the consideration of small variations of $x$ at the limits of the range (as at mobile limits, in § 31), though these were not admitted elsewhere; and a compromise had to be framed, expedient for the discussion of such limits.

---

* "Mémoire sur la manière de distinguer les maxima des minima dans le calcul de variations," *Hist. de l'Acad. Roy. des Sciences* (1788), pp. 21—25.

† Legendre (*l.c.*) called a maximum or a minimum of such a type *relative or accidental*.

Manifestly the general notion of the slight deformation of a curve, into a contiguous curve, merely requires in essence that the distance, between the position of a point on the curve and the displaced position of the corresponding point on the deformed curve, must always be small, whatever point be chosen. If this requirement be met, the contiguous curve gives a small variation of the original curve, whatever be its shape in detail. We are accustomed, over a stretch of continuous curve, to very rapid small changes which, while of full importance in detail, would be deemed of slight statistical account as compared with the smoothness of an averaging curve over that stretch. Restrictions at every point, as to changes of direction, of curvature and the like, that all of these changes shall be small when the variation is small, are essential limitations upon the completeness of the variations admitted or imposed. A minimum for restricted (weak) small variations, which fails to be a minimum for unrestricted (strong) small variations, cannot be claimed as a true minimum, any more than the summit of a mountain pass could be regarded as a true minimum, because there is no higher point to which a pedestrian need rise when passing from the valley on one side of the pass to the valley on the other side, or as a true maximum, because there is no lower point to which a pedestrian need descend when passing from the ridge on one side of the pass to the ridge on the other side. A true minimum and a true maximum must possess their distinctive properties, in all circumstances of variation, and not merely for a selected group of such circumstances.

Accordingly, we proceed to the consideration of *strong* variations, indicating by the term specially the types that can be resolved into an aggregate of variations of a specifically simple character. The character of these component simple strong variations must now be explained.

### Strong variations: fundamental constituent type.

**208.** In all the preceding investigations, the small variations of the characteristic curve have been of the type styled *weak*. They are such that, if the point $x + \kappa\xi$, $y + \kappa\eta$ be the varied position of a point $x, y$ on a curve, for which $x$ and $y$ are expressed as functions of a continuously increasing parameter $t$, the quantities $\xi$ and $\eta$ are arbitrary functions of $t$, independent of $\kappa$ and limited by the requirement of being everywhere regular functions, so that they possess their due succession of derivatives which also are regular functions.

We now proceed to consider small variations which are not limited to this class. But the mere exclusion of such limitation would be mainly a negative description. For the purposes of mathematical representation, it is convenient to select particular types of small variation from among those which hitherto have been excluded from consideration. Thus we have already (§§ 47, 102) had small variations, where isolated discontinuities of direction within a range have been admissible; though the number of such discontinuities in any finite range has been limited, owing to the requirement of isolation. We have mentioned, though only by the way of illustration of possible small variations that are not of the definite weak type, the instance

$$\xi = \kappa a \sin (t\kappa^n).$$

Such a quantity $\xi$ is of course finite, continuous, and small when $\kappa$ is small ($n$ is a constant). But the first derivative of $\xi$ is of an order smaller than the order of $\xi$ when $n$ is positive. When $n$ is negative, the derivative of $\xi$ is of an order not so small as that of $\xi$; it may be finite, or it may even be large and oscillate rapidly, when $\kappa$ is small. In all such instances, the expansion of the subject of integration in powers of $\kappa$, after the small variation has been imposed, would not be effective to the end for which it was to be used. Again, it is conceivable that we could have small variations over a range continuous, not merely in arc but also in direction, yet not continuous in curvature; an integrand of the second order would suffer a discontinuity in passing through such a place, even if the place were isolated; and difficulties would arise over the curvature properties of the contiguous varied curve. Similarly, if the subject of integration involves derivatives of order higher than the second, we should be faced with the possibility of abrupt changes in the rate of curvature.

**209.** An indication of one method, of taking some of such non-regular small variations into account, is given by the consideration of an integral

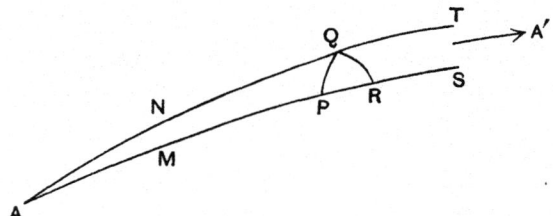

with a jagged variation. Let $AMPRS...$ be a characteristic curve, on which $A'$ (the conjugate of $A$) lies beyond $S$. Let any point $P$ in this range $AS$ be displaced to a neighbouring place $Q$. Because $P$ lies within the range bounded by $A$ and its conjugate $A'$, a consecutive characteristic $ANQT...$ can be drawn (§ 70) from $A$ to pass through $Q$; and this consecutive curve does not meet $AMPRS...$ until $A'$.

Suppose that the conditions for a minimum are required. (The alternative conditions for a maximum follow by changing, throughout the explanation, a greater inequality into a lesser inequality.) Accordingly, if the integral is to be a minimum along a characteristic curve $AMP$ from $A$ to $P$, and if we denote its value by $I_{AMP}$ when taken along that curve, we must have

$$I_{ANQ} + I_{QP} > I_{AMP},$$

where

(i)    $I_{ANQ}$ is the integral along $ANQ$ from $A$ to $Q$;

(ii)   $I_{QP}$ is the integral along $QP$, the value of the independent variable of integration along $QP$ increasing from $Q$ up to $P$; and

(iii) $I_{AMP}$ is the integral along $AMP$ from $A$ to $P$.

Now $ANQT...$ equally is a characteristic curve, determined by slightly different initial conditions. Equally it provides a minimum: so that, taking $AMPQ$ as a variation of $ANQ$, we have

$$I_{AMP} + I_{PQ} > I_{ANQ},$$

where the integrals $I_{AMP}$ and $I_{ANQ}$ are the same as before, and $I_{PQ}$ is the integral along $PQ$, the value of the independent variable of integration along $PQ$ increasing from $P$ up to $Q$. (The integral $I_{PQ}$ is not the integral $I_{QP}$ reversed: along $PQ$ in $I_{PQ}$, the independent variable of integration is increasing; and likewise, along $QP$ in $I_{QP}$, it is increasing.)

Next, consider $ANQ$ and $QR$ as a variation of the continuous length $AMPR$. Owing to the requirement of a minimum, we have

$$I_{ANQ} + I_{QR} > I_{AMPR},$$

where $I_{ANQ}$ is the same as before, $I_{QR}$ is the integral from $Q$ to $R$ along $QR$, and $I_{AMPR}$ is the integral along the characteristic arc $AMPR$. Thus

$$I_{AMP} + I_{PQ} + I_{QR} > I_{ANQ} + I_{QR} > I_{AMPR}.$$

On the left-hand side, the sum of the integrals is the whole integral, first along $AMP$ from $A$ to $P$, next along $PQ$ from $P$ to $Q$, and then along $QR$ from $Q$ to $R$. Thus the integral along $AMPQR$ is greater than the integral along $AMPR$. Hence if the minimum condition is obeyed for a broken line $AMPQ$ as a variation of a characteristic arc $AQ$, and for a broken line $ANQR$ as a variation of a characteristic arc $AR$, it is obeyed for a line $AMPQR$, with a jagged portion $PQR$, as a variation of the characteristic arc $AMPR$.

It follows that *the minimum condition will be secured for any jagged variation of a characteristic arc*, composed of portions such as $PQR$ which may be taken on either side of the arc $AMPR...$, *when a requirement is satisfied, that the minimum condition holds along any characteristic curve, separately for each broken curve, such as $ANQP$ for $AMP$, and such as $ANQR$ for $AMPR$.* In other words, *the requirement is to be satisfied, whatever point $P$ be chosen on a characteristic arc, and whatever small arc be drawn from that chosen point to any contiguous point $Q$.*

**210.** It should be noted that considerable freedom, thus far, is allowed to the form of the small arcs, such as $PQ$ and $QR$. No assumption has been made, tacitly or overtly, that they are straight lines: but we have assumed (tacitly) that such an arc $PQ$ is continuous in arc-length, and that it does not cut the characteristic curve except at $P$. The assumption of continuity of arc-length in the variation remains an assumption. The other assumption is not essential: for, if $PQ$ crossed the curve at points $U, V, \ldots$, we should consider $PQ$ resolved into portions $PU, UV, \ldots$, each similar to $PQR$ in the figure; and then we should treat each portion in the same way as $PQR$. We do not therefore need to modify the requirement of the minimum condition for a broken line such as $AQP$ for $AP$, where $P$ is any point on the line and $PQ$ is any direction through $P$.

Further, there need be no assumption as to continuity of direction, or of curvature, and so on, in the broken or jagged variation. Thus consider a curve such as $PQSTR$. We can draw characteristics $AQ$, $AS$, $AT$, consecutive to $APR$. When the minimum condition holds for a single broken line such as $AQP$ taken as a variation of $AP$, for every

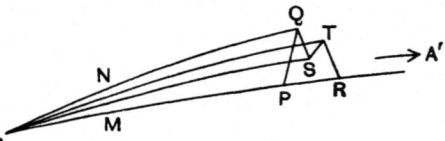

point on any characteristic curve and for every direction through the point, it holds generally. For, with the preceding notation, we have

$$I_{AMP} + I_{PQ} > I_{ANQ},$$
$$I_{ANQ} + I_{QS} > I_{AS},$$
$$I_{AS} + I_{ST} > I_{AT},$$
$$I_{AT} + I_{TR} > I_{AMPR};$$

and therefore

$$I_{AMP} + I_{PQ} + I_{QS} + I_{ST} + I_{TR} > I_{AMPR}.$$

Hence the minimum condition will hold for any accumulation of broken jagged small variations such as $PQSTR$, as it will hold for even the simplest variation such as the broken line $AQP$ for $AP$, *if the minimum condition be obeyed for each such simplest variation from every point $P$ on the general characteristic curve to any contiguous point $Q$ off the arc.*

*Note.* As already stated, all the explanations and all the results hold as regards the conditions requisite for a possible maximum, provided the greater inequalities are changed into lesser inequalities.

### *Strong variation of ordinary integral with first derivatives.*

**211.** Variations of the nature indicated are called *strong* variations. It is to be noted that they are only one type of the variations, distinct in their character from those which have hitherto been considered. The explanation just given shews that some of these strong variations can be compounded of variations of the simple broken-line type, while each individual component of the composite variation is itself a possible variant of the central curve. Further, when the condition for a minimum (or a maximum) holds at every place on that curve and for every direction at that place, the effect of the variation becomes cumulative. The condition for a minimum (or a maximum) is satisfied for a quite general composite strong variation, if it is always satisfied for the quite general ultimate constituent of fundamental type.

We therefore proceed to consider the analytical expression of this condition, initially individual for a broken line, and secure in its cumulative effect. As the introduction of strong variations, and the expression of the condition for the simplest and most frequent type of such variations, are due to Weierstrass, the analytical requirement for the fundamental constituent strong variation may fitly be called the *Weierstrass test*. It can be obtained as follows for the simplest kind of integral.

**212.** Let a point $Q$, with coordinates $x + \kappa\xi$, $y + \kappa\eta$, be the displaced position of a point $P$ on the central characteristic curve. The arc $PQ$ joining $P$ and $Q$ is not restricted, save that it shall be continuous in the intrinsic qualities (such as tangency, curvature, rate of curvature, and so on) represented by the derivatives of the dependent variable which occur in the subject of integration. If, and when, there should come a cessation of continuity in any of those properties, the arc in question would be held to cease at such a point $T$; a new consecutive characteristic $AT$ would be drawn through $T$, and a new arc would be considered to begin at $T$. Thus within the limited range $PQ$, the quantities $\xi$ and $\eta$ will be regarded as regular functions of some parameter along that arc, valid only up to $Q$. For variations beyond $Q$, new functions $\xi$ and $\eta$, similarly regular along a new limited arc, will be required, and so on: the test will be demanded in connection with each such elementary arc.

It is to be noted that all such variations are of a restricted type; thus they do not cover variations such as

$$\kappa \cos (nt\kappa^{-\frac{1}{2}}), \quad \kappa \sin (nt\kappa^{-\frac{1}{2}}),$$

to take simple functions in terms of which, by Fourier series, some less simple variations can be represented.

We shall deal, in the first instance, with the simplest kind of integral

$$I = \int_{t_0}^{t_1} F(x, y, x_1, y_1)\, dt,$$

already considered in Chapter II; and we shall assume that the same conditions, as before, attach to the integral. Also, all the tests, there found sufficient and necessary to secure a maximum or a minimum for weak variations, will be supposed to be already satisfied. We now have to consider the combination

$$I_{ANQ} + I_{QP} - I_{AMP},$$

where $AMP$ is the central characteristic; $ANQ$ is the consecutive characteristic through $Q$ a point contiguous to $P$, provided (§§ 70—72) $P$ is a point
within a range $AA'$ bounded by the initial point $A$ and its conjugate $A'$

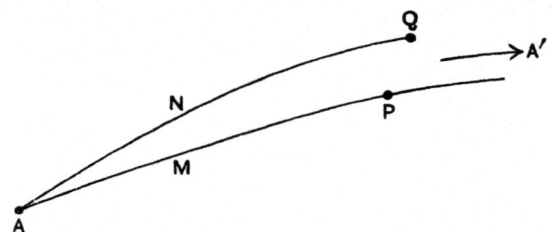

By definition, $I_{QP}$ is the integral taken from $Q$ to $P$ along any curve joining
$P$ and $Q$, continuous in arc, direction, curvature and the like properties, but
not meeting $AMP$ save at $P$.

Denoting the small weak variation from any point $M$ on $AMP$ to the corresponding point $N$ on $ANQ$ by $\kappa u$ and $\kappa v$, and regarding the integral $I_{ANQ}$
as the variation of $I_{AMP}$, we have (§§ 41, 45),

$$I_{ANQ} - I_{AMP} = \kappa \int_A^P (x_1 v - y_1 u)\, \mathfrak{C}\, dt + \left[ \kappa \left( u \frac{\partial F}{\partial x_1} + v \frac{\partial F}{\partial y_1} \right) \right]_A^P + [\kappa^2],$$

where $[\kappa^2]$ denotes the aggregate of terms of the second and higher powers
of small quantities. Now $u = 0$, $v = 0$, at $A$; and $u = \xi$, $v = \eta$, at $P$: also
everywhere along the characteristic curve, the characteristic equation $\mathfrak{C} = 0$
is satisfied. Hence, neglecting terms of the second and higher orders as compared with terms of the first order, we have

$$I_{ANQ} - I_{AMP} = \kappa \left( \xi \frac{\partial F}{\partial x_1} + \eta \frac{\partial F}{\partial y_1} \right).$$

**213.** The function $F(x, y, x_1, y_1)$, in its initial form which involves
$x, y, x_1, y_1$ *as unknown functions of t, is not a perfect differential with respect
to t*; and therefore the integral $I_{QP}$, being

$$\int_Q^P F(x, y, x_1, y_1)\, dt,$$

cannot depend solely upon the values at $P$ and at $Q$, but will depend partly
upon the path from $Q$ to $P$.

The simplest curve to select is the straight line joining $Q$ to $P$. We shall
therefore begin with the case when the arc $QP$ is a straight line. Let $\lambda$ and
$\mu$ be the direction-cosines of the line $QP$ in the direction from $Q$ to $P$; and
et $l$ denote the length of $PQ$, so that

$$\kappa\xi = -\lambda l, \quad \kappa\eta = -\mu l.$$

Also, let $S$ be any point on $QP$, and let $QS = \sigma$, so that $\sigma$ may range from $0$
to $l$ along the path $QP$ from $Q$ to $P$. The coordinates of $S$ are

$$x - \lambda(l - \sigma), \quad y - \mu(l - \sigma).$$

As the integral $\int F(x, y, x_1, y_1)\, dt$ is invariantive (§ 38), so far as concerns change of the independent variable $t$ of integration, we shall take $\sigma$ as the current independent variable of integration for $I_{QP}$. As now $x_1 = \lambda,\ y_1 = \mu$, along $QP$, we have

$$I_{QP} = \int_0^l F\{x - \lambda(l - \sigma),\ y - \mu(l - \sigma),\ \lambda,\ \mu\}\, d\sigma,$$

while $l$ is a small quantity of the same order of magnitude as $\kappa$. The function $F$ is regular over the range of integration which, moreover, is small. Hence, by the first theorem of mean value, there is a positive proper fraction $\epsilon$ such that

$$I_{QP} = lF(x - \lambda\epsilon l,\ y - \mu\epsilon l,\ \lambda,\ \mu)$$
$$= lF(x, y, \lambda, \mu) + [l^2],$$

where $[l^2]$ denotes an aggregate of quantities of the second and higher orders in $l$, that is, in $\kappa$. Consequently, up to small quantities of the first order, we can take

$$I_{QP} = lF(x, y, \lambda, \mu).$$

Thus

$$I_{ANQ} + I_{QP} - I_{AMP} = l\left\{ F(x, y, \lambda, \mu) - \lambda\frac{\partial F(x, y, x_1, y_1)}{\partial x_1} - \mu\frac{\partial F(x, y, x_1, y_1)}{\partial y_1} \right\},$$

accurate up to the first order. Now

$$F(x, y, x_1, y_1) = x_1\frac{\partial F(x, y, x_1, y_1)}{\partial x_1} + y_1\frac{\partial F(x, y, x_1, y_1)}{\partial y_1}$$

identically; and therefore

$$F(x, y, \lambda, \mu) = \lambda\frac{\partial F(x, y, \lambda, \mu)}{\partial \lambda} + \mu\frac{\partial F(x, y, \lambda, \mu)}{\partial \mu}.$$

Hence

$$I_{ANQ} + I_{QP} - I_{AMP} = lE(x, y, x_1, y_1, \lambda, \mu),$$

where

$$E(x, y, x_1, y_1, \lambda, \mu) = F(x, y, \lambda, \mu) - \lambda\frac{\partial F(x, y, x_1, y_1)}{\partial x_1} - \mu\frac{\partial F(x, y, x_1, y_1)}{\partial y_1}$$

$$= \lambda\left\{ \frac{\partial F(x, y, \lambda, \mu)}{\partial \lambda} - \frac{\partial F(x, y, x_1, y_1)}{\partial x_1} \right\} + \mu\left\{ \frac{\partial F(x, y, \lambda, \mu)}{\partial \mu} - \frac{\partial F(x, y, x_1, y_1)}{\partial y_1} \right\}.$$

This quantity $E(x, y, x_1, y_1, \lambda, \mu)$ depends, partly upon the position $x, y$ on the characteristic curve and the direction $x_1, y_1$ of the tangent to the characteristic curve at that position, and partly upon the direction $\lambda, \mu$ of the short line $PQ$ through that position: so that, if there is a variation off the curve, $PQ$ must not coincide in direction with the tangent at $P$. Further, $F(x, y, x_1, y_1)$ is homogeneous of the first degree in $x_1$ and $y_1$; and therefore $\dfrac{\partial F(x, y, x_1, y_1)}{\partial x_1}$ and $\dfrac{\partial F(x, y, x_1, y_1)}{\partial y_1}$ are homogeneous in those quantities of order zero. They are unchanged in value when $x_1$ and $y_1$ are multiplied by the same quantity; hence, when we multiply them by $\dfrac{dt}{ds}$ so that they become $\dfrac{dx}{ds}\,(=x')$ and $\dfrac{dy}{ds}\,(=y')$, where $s$ is the arc-length along the characteristic curve, the derivatives become $\dfrac{\partial}{\partial x'}\,F(x,y,x',y')$ and $\dfrac{\partial}{\partial y'}\,F(x,y,x',y')$. We now have

$$E(x,y,x_1,y_1,\lambda,\mu) = F(x,y,\lambda,\mu) - \lambda\frac{\partial}{\partial x'}\,F(x,y,x',y') - \mu\frac{\partial}{\partial y'}\,F(x,y,x',y')$$

$$= \lambda\left\{\frac{\partial}{\partial\lambda}\,F(x,y,\lambda,\mu) - \frac{\partial}{\partial x'}\,F(x,y,x',y')\right\} + \mu\left\{\frac{\partial}{\partial\mu}\,F(x,y,\lambda,\mu) - \frac{\partial}{\partial y'}\,F(x,y,x',y')\right\}.$$

### The Weierstrass E-function, and Weierstrass test.

**214.** The integral $I$ is to be a maximum or minimum for all small variations. The conditions for weak variations have been obtained and are satisfied. If the integral is to be a minimum under the strong variations composed of variations such as $ANQP$ for $AMP$, then

$$E(x,y,x_1,y_1,\lambda,\mu)$$

must be positive, for all positions $x$, $y$ on the characteristic curve, and for all directions $\lambda$, $\mu$ through every position giving a variation, that is, for all directions off the tangent; while, if the integral is to be a maximum, the same function must be negative for all positions $x$, $y$ on the characteristic curve, and for all non-tangential directions through every position.

The function $lE$ measures the excess over a minimum or the deficiency from a maximum. Conversely, if $E$ is always positive, an excess always arises, so that there is a minimum ; while, if $E$ is always negative, a deficiency always arises, so that there is a maximum. The function is often called the *excess-function*, though it is a deficiency when a maximum exists. As it is due to Weierstrass, it will be called the *Weierstrass E-function*. The test—that it is to be always positive for a minimum, and always negative for a maximum, along the characteristic curve—will be called the *Weierstrass test*. Manifestly, it is distinct in character, as in source, from the Euler test, the Legendre test, and the Jacobi test. The requirement is that *the quantity*

$$\lambda\left\{\frac{\partial}{\partial\lambda}\,F(x,y,\lambda,\mu) - \frac{\partial}{\partial x'}\,F(x,y,x',y')\right\}$$

$$+ \mu\left\{\frac{\partial}{\partial\mu}\,F(x,y,\lambda,\mu) - \frac{\partial}{\partial y'}\,F(x,y,x',y')\right\},$$

*for all places on the characteristic curve, and for every direction $\lambda$ and $\mu$ through each place distinct from the tangential direction, shall have the same sign*: the sign being positive for a minimum, and negative for a maximum.

It is a necessary test. It is a sufficient test, for the type of strong variation imposed. But that type of strong variation is only a selected type.

*Examples of the E-function as a test.*

**215.** The Weierstrass test is simple, in its application to all integrals which are of the form $\int f(x, y)\, ds$ or, what is the equivalent, are of the form $\int f(x, y)\, (x_1^2 + y_1^2)^{\frac{1}{2}}\, dt$.

For such integrals, the $E$-function is

$$\lambda \left\{ f(x, y)\, \frac{\lambda}{(\lambda^2 + \mu^2)^{\frac{1}{2}}} - f(x, y)\, \frac{x'}{(x'^2 + y'^2)^{\frac{1}{2}}} \right\}$$
$$+ \mu \left\{ f(x, y)\, \frac{\mu}{(\lambda^2 + \mu^2)^{\frac{1}{2}}} - f(x, y)\, \frac{y'}{(x'^2 + y'^2)^{\frac{1}{2}}} \right\},$$

which is equal to

$$f(x, y)\, \{1 - (\lambda x' + \mu y')\},$$

because $(\lambda^2 + \mu^2)^{\frac{1}{2}} = 1$, $(x'^2 + y'^2)^{\frac{1}{2}} = 1$. If $\chi$ be the inclination of $QP$ to the tangent to the curve, so that $\chi$ lies between $0$ and $2\pi$ for any direction that is non-tangential, we have

$$E = f(x, y)\, (1 - \cos \chi).$$

If $f(x, y)$ does not change its sign along the characteristic curve, then—as the sign of $f(x, y)$ may be taken positive—we have $E$ always positive. Therefore a minimum is admissible, and does occur, so far as concerns the requirement of strong variations of the selected type.

*Ex.* 1.  Thus the *test is satisfied for a catenoid of revolution.*  The integral is

$$2\pi \int y\, ds \; ;$$

and so, as the tests are satisfied—with a limited range—for weak variations (§ 30, Ex. 1), we still have a minimum.

*Ex.* 2.  Similarly, *the test is satisfied for a gravity brachistochrone.*  The integral effectively (§ 34, Ex. 3) is

$$\int (y + a)^{-\frac{1}{2}}\, ds,$$

where the positive sign is given to the radical.

The Weierstrass test, established (as above) for the case when $PQ$ is a straight line, can be extended to all integrals

$$\int f(x, y)\, ds, = \int f(x, y)\, (x_1^2 + y_1^2)^{\frac{1}{2}}\, dt,$$

when $P$ and $Q$ are joined by any arc (and not alone by a straight line).

As before, we have

$$I_{ANQ} - I_{AMP} = \kappa \left( \xi \frac{\partial F}{\partial x_1} + \eta \frac{\partial F}{\partial y_1} \right)$$
$$= \kappa f(x, y) \left\{ \xi \frac{x_1}{(x_1^2 + y_1^2)^{\frac{1}{2}}} - \eta \frac{y_1}{(x_1^2 + y_1^2)^{\frac{1}{2}}} \right\}$$
$$= - l f(x, y)\, (\lambda x' + \mu y')$$
$$= - l f(x, y) \cos \chi,$$

where $l$ is the length of the straight line $PQ$. For $I_{QP}$, we have

$$I_{QP} = \int f(X, Y)\, ds,$$

where $X$, $Y$ are the coordinates of any point on the arc $PQ$ as now drawn. Hence, if the total length of the arc be $L$, we have

$$I_{QP} = L f(\bar{X}, \bar{Y}),$$

where $f(\bar{X}, \bar{Y})$ represents a mean value of $f(X, Y)$ along the arc, while $\bar{X}$ differs from $x$, and $\bar{Y}$ differs from $y$, by quantities of the same order as $L$ at least. Thus

$$f(\bar{X}, \bar{Y}) = f(x, y) + [L],$$

where $L$ represents the aggregate of the terms in $L$, $L^2$, ...; and so

$$I_{QP} = L f(x, y) + L[L]$$
$$= L f(x, y) + [\kappa^2],$$

that is, to the first order of small quantities,

$$I_{QP} = L f(x, y).$$

Hence, to the first order of small quantities,

$$I_{ANQ} + I_{QP} - I_{AMP} = (L - l \cos \chi) f(x, y).$$

Now $L$ is always greater than $l$, the rectilinear distance between $P$ and $Q$; and therefore $L - l \cos \chi$ is positive for all values of $\chi$. Consequently

$$I_{ANQ} + I_{QP} - I_{AMP},$$

for the integrals in question, is always positive, whatever form the continuous arc $PQ$ may take. Hence the Weierstrass test, with the less specialised type of arc, is satisfied for integrals $\int f(x, y)\, ds$, thus admitting a minimum.

*Ex.* 3. *The Weierstrass test is satisfied for all geodesics.*

With the customary notation of differential geometry\*, and when $p$, $q$ are the parameters on a surface, any arc on a surface is

$$\int (E p_1{}^2 + 2F p_1 q_1 + G q_1{}^2)^{\frac{1}{2}}\, dt;$$

and the geodesic quality is determined by the characteristic equation†. For the construction of the Weierstrass test, we take a point $P$ given by $p$, $q$, and a neighbouring point $Q$ given by $p + \kappa \xi$, $q + \kappa \eta$; and then, as in the text, we have

$$I_{ANQ} - I_{AMP} = \kappa \left\{ \xi \frac{E p_1 + F q_1}{(E p_1{}^2 + 2F p_1 q_1 + G q_1{}^2)^{\frac{1}{2}}} + \eta \frac{F p_1 + G q_1}{(E p_1{}^2 + 2F p_1 q_1 + G q_1{}^2)^{\frac{1}{2}}} \right\}$$

$$= \kappa \left\{ \xi \left( E \frac{dp}{ds} + F \frac{dq}{ds} \right) + \eta \left( F \frac{dp}{ds} + G \frac{dq}{ds} \right) \right\}.$$

Now if $X$, $Y$, $Z$ be the space-coordinates of $P$, and $X_1$, $Y_1$, $Z_1$ those of $Q$,

$$X_1 - X = x_1 \xi + x_2 \eta + [\kappa^2], \qquad Y_1 - Y = y_1 \xi + y_2 \eta + [\kappa^2], \qquad Z_1 - Z = z_1 \xi + z_2 \eta + [\kappa^2];$$

---

\* See my *Lectures on Differential Geometry*, chap. ii.

† *Ib.* chap. v.

and therefore

$$l^2 = E\xi^2 + 2F\xi\eta + G\eta^2,$$

while

$$\frac{dx}{ds} = x_1 \frac{dp}{ds} + x_2 \frac{dq}{ds}, \quad \frac{dy}{ds} = y_1 \frac{dp}{ds} + y_2 \frac{dq}{ds}, \quad \frac{dz}{ds} = z_1 \frac{dp}{ds} + z_2 \frac{dq}{ds}.$$

Thus

$$(X_1 - X)\frac{dx}{ds} + (Y_1 - Y)\frac{dy}{ds} + (Z_1 - Z)\frac{dz}{ds} = E\xi \frac{dp}{ds} + F\left(\xi \frac{dq}{ds} + \eta \frac{dp}{ds}\right) + G\eta \frac{dq}{ds}.$$

But the left-hand side

$$= l \cos \psi,$$

where $l$ is the length $PQ$, and $\psi$ is the angle between $PQ$ and the tangent to the characteristic curve. If then, as before, $\chi$ is the angle between $QP$ and this tangent, $\chi = \pi + \psi$; and so

$$I_{ANQ} - I_{AMP} = -l \cos \chi.$$

Also $I_{QP} = l$; and therefore the Weierstrass function is

$$l(1 - \cos \chi),$$

which is always positive. The Weierstrass test is satisfied.

*Ex.* 4. Discuss the possible minimum of the integral

$$\int \left(\frac{dy}{dx}\right)^2 dx,$$

taken between two given values of $x$.

The characteristic equation is

$$\frac{d}{dx}\left(\frac{dy}{dx}\right) = 0,$$

so that the primitive is

$$y = ax + b,$$

where $a$ and $b$ are constants determined by the end-points $(x_1, y_1)$ and $(x_2, y_2)$.

The Legendre test gives $\frac{\partial^2 f}{\partial p^2}, = 2$; it is satisfied for a minimum. As the characteristic curve is a straight line, there is no finite conjugate of any initial point; so the Jacobi test is satisfied unconditionally.

(i) When the integral is taken in the Weierstrass form, it is

$$\int \frac{y_1^2}{x_1} dt.$$

Either of the (equivalent) characteristic equations leads to

$$y = ax + b.$$

For the Legendre test, we take

$$P = \frac{1}{x_1^2}\frac{\partial^2 F}{\partial y_1^2} = \frac{2}{x_1^3},$$

which is positive if $x_1$ is positive. There is no conjugate of a value $t_0$; for if

$$x = ct, \quad y = act + b,$$

with the notation of §§ 61, 63, we have

$$\phi_1(t) = 0, \quad \phi_2(t) = 0, \quad \psi_1(t) = ct, \quad \psi_2(t) = 1, \quad \phi'(t) = c, \quad \psi'(t) = ac,$$

and so

$$\chi(t) = c^2 t, \quad \omega(t) = c,$$

and the function $Z(t, t_0) = c^3(t - t_0)$, which has no zero except $t_0$. Thus for general weak variations, the integral admits a minimum.

But the subject of integration is rational (and so the Weierstrass test cannot be satisfied : see § 217). The $E$-function is

$$= \lambda \left( -\frac{\mu^2}{\lambda^2} + \frac{y'^2}{x'^2} \right) + \mu \left( 2\frac{\mu}{\lambda} - 2\frac{y'}{x'} \right)$$

$$= \frac{1}{\lambda x'^2} (\mu x' - \lambda y')^2,$$

and so changes sign with $\lambda$. The straight line does not provide a true minimum, that is, a minimum also under strong variations.

(ii) We here have taken the change of sign of $\lambda$ as dominating the $E$-function; and it is sufficient, for our purpose, to determine whether $E$ can change its sign. In this case, other curves, not regular in direction but jagged, can easily be constructed which lead to values smaller than the minimum under weak variations.

Let $AB$ be the straight line ; $PQ$ any element. Take $PT$ parallel to the axis of $x$, where $PT$ may be large compared with $PQ$, if we please ; and let $RTQ = \epsilon$, $RPQ = a$.

When for the portion $PQ$ of $AB$, we substitute the broken line $PTTQ$, along $PT$ we have $\frac{dy}{dx} = 0$ and along $TQ$ we have $\frac{dy}{dx} = \tan \epsilon$ ; thus the value of the integral for $PTTQ$ is

$= 0$, for $PT$ ; $+ TR \tan^2 \epsilon$, for $TQ$ ; $= TR \tan^2 \epsilon$

$$= QR \tan \epsilon$$

$$= dx \tan a \tan \epsilon.$$

The value of the element of the integral for $PQ$ along $AB$ is

$$dx \tan^2 a ;$$

as $\epsilon < a$, the former is less than the latter. Taking a jagged line from $A$ to $B$, made up of portions such as $PTQ$, the integral for the straight line $AB$ is diminished in the ratio $\frac{\tan \epsilon}{\tan a}$, and clearly can be made as small as we please.

*Ex.* 5. In the figure, $OB = 1$ ; $OA = a$, where $0 < a < \frac{1}{2}$ ; $OA' = 1 - a$, $BC = 1$. Along $OA'$, let

$$x = \frac{1-a}{a} t, \ y = 0, \text{ from } t = 0 \text{ to } t = a ;$$

along $A'A''$, let

$$x = 1 - t, \ y = \frac{t-a}{1-2a}, \text{ from } t = a \text{ to } t = 1 - a ;$$

along $A''C$, let

$$x = \frac{1}{a}\{(1-a)\,t - (1-2a)\}, \ \ y = 1,$$

from $t = 1 - a$ to $t = 1$.

Prove that

$$\int_0^C \left(\frac{dy}{dx}\right)^2 dx, \quad = \int_0^1 \frac{y_1^2}{x_1} dt, \text{ taken along } OA', \ A'A'', \ A''C,$$

so that there are two places of discontinuity, is equal to

$$\frac{-1}{1-2a},$$

so that there is no minimum for the integral.

(Bromwich.)

*Ex.* 6. It was proved (§ 34, Ex. 9) that the integral

$$\int (1+y^2) \left(\frac{dy}{dx}\right)^{-2} dx$$

satisfies the Euler test along a curve

$$y = \sinh (cx + a) \ ;$$

that the Legendre critical function is everywhere positive ; and that there is no conjugate (as required by the Jacobi test, if there is to be a limit to the range along the curve) of any initial point along this characteristic curve. Thus the integral satisfies the tests for a minimum, in so far as these emerge through weak variations.

It was pointed out, however, that there are variations from the characteristic curve by which the integral could be reduced from its value along that curve : such variations giving rise to a jagged curve and being, in fact, what have been called strong variations. Accordingly, we proceed to the Weierstrass test, taking the integral in the form

$$\int (1+y^2) \frac{x_1^{\ 3}}{y_1^{\ 2}} dt.$$

The Weierstrass $E$-function, for this integral, is

$$\lambda \left\{ (1+y^2)\, 3 \frac{\lambda^2}{\mu^2} - (1+y^2)\, 3 \frac{x'^2}{y'^2} \right\} + \mu \left\{ -(1+y^2)\, 2 \frac{\lambda^3}{\mu^3} + (1+y^2)\, 2 \frac{x'^3}{y'^3} \right\},$$

which is equal to

$$\frac{1}{\mu^2 y'^3}\, (1+y^2)\, (\mu x' - \lambda y')^2\, (\lambda y' + 2\mu x').$$

Now $(\mu x' - \lambda y')^2 = \sin^2 \chi$, and $\chi$ does not vanish for variations under consideration ; while the $E$-function must have one sign. But $\lambda$, $\mu$ are the direction-cosines of any arbitrary direction through a point on the curve ; for some directions $\lambda y' + 2\mu x'$ is positive, for others it is negative ; the $E$-function does not preserve a uniform sign under all directions of the strong variation imposed.

Thus the integral does not possess a minimum or a maximum. The minimum under weak variations is not a true minimum.

*Ex.* 7. It is easy to verify that, for the equivalent integrals

$$\int y \left(\frac{dy}{dx}\right)^{-2} dx, \qquad \int y \frac{x_1^{\ 3}}{y_1^{\ 2}} dt,$$

the characteristic curve is a parabola

$$4cy = (x+a)^2 \ ; \quad \text{or} \quad y = ct^2, \quad x+a = 2ct.$$

The Legendre test is everywhere satisfied except at the vertex of the parabola, and the critical magnitude is positive when the parabola lies on the positive side of the axis of $x$, while it is negative when the parabola lies on the negative side of the axis. The Jacobi test assigns no limitation to the range along the parabola for a maximum or a minimum.

For the Weierstrass $E$-function, derived (as always) from the form with a parametric variable $t$, we have

$$E = \lambda \left( 3y \frac{\lambda^2}{\mu^2} - 3y \frac{x'^2}{y'^2} \right) + \mu \left( -2y \frac{\lambda^3}{\mu^3} + 2y \frac{x'^3}{y'^3} \right)$$

$$= \frac{y}{\mu^2 y'^3} (\lambda y' - \mu x')^2 (\lambda y' + 2\mu x').$$

At any point, there clearly are directions for which $E$ is positive and directions for which $E$ is negative. Consequently, there is neither a true maximum nor a true minimum, although the necessary conditions under a weak variation are satisfied.

*Ex.* 8. *Required the minimum (if any) of the integral**

$$\int_{-1}^{1} x^2 \left(\frac{dy}{dx}\right)^2 dx,$$

with the requirements: (i), that $y$ and $\frac{dy}{dx}$ are to be regular functions of $x$ in the range; and (ii), that $y=a$ when $x=-1$ and $y=b$ when $x=1$, where $a$ and $b$ are two constants different from one another. (Otherwise stated, the requirement is to have a regular curve through $-1$, $a$, and $1$, $b$, two points with unequal ordinates.)

(I) The first integral of the characteristic equation is

$$x^2 \frac{dy}{dx} = -A,$$

so that the characteristic curve is $xy = A + Bx$: or, with the assigned conditions,

$$y = \tfrac{1}{2}(a+b) - \tfrac{1}{2}(b-a)\frac{1}{x}.$$

Manifestly the characteristic curve suffers a discontinuity at the place $x=0$. The discontinuity could be avoided only by having $a=b$: in that case, $y$ would be a mere constant.

The Legendre test is satisfied, because the critical quantity $2x^2$ is always positive; and therefore the test admits a minimum.

As regards the Jacobi test for a limitation of the range, any other characteristic curve through the point $-1$, $a$, and distinct from the central characteristic is

$$y - a = A\left(1 + \frac{1}{x}\right),$$

where $A$ is not equal to $\tfrac{1}{2}(b-a)$. The two curves do not intersect save at $x=-1$, $y=a$. Thus there is no limitation on the range.

Consequently, the three tests for weak variations are satisfied: but the requirement as to continuity in $y$ is not satisfied.

For the Weierstrass test, the integral takes the form

$$\int x^2 \frac{y_1^2}{x_1} dt.$$

Then the $E$-function is given by

$$E = x^2 \frac{\mu^2}{\lambda} - \lambda\left(-x^2 \frac{y_1^2}{x_1^2}\right) - \mu\left(x^2 \frac{2y_1}{x_1}\right)$$

$$= \frac{1}{\lambda} x^2 \left(\mu - \lambda \frac{y_1}{x_1}\right)^2,$$

which changes sign with $\lambda$, a direction-cosine of an arbitrary direction at the point $P$. Thus the $E$-function can change its sign. Accordingly the $E$-test is not satisfied: there is *no real minimum* for the integral.

* This integral is of historical interest. It was discussed in a paper by Weierstrass (quoted on p. 335), where he dealt with the assumption that a minimum exists for an integral of a rational subject of integration which can never be negative. The example shews that, under postulated conditions which merely assume that $y$ is not throughout equal to one and the same constant, no minimum exists.

A similar assumption, as to the necessary existence of a minimum for an integral

$$\iiint \left\{ \left(\frac{\partial v}{\partial x}\right)^2 + \left(\frac{\partial v}{\partial y}\right)^2 + \left(\frac{\partial v}{\partial z}\right)^2 \right\} dx\, dy\, dz,$$

is the foundation of Dirichlet's principle. It is proved hereafter (§ 385, Ex. 1) that this integral does not satisfy the $E$-test for strong variations and, consequently, does not possess a real minimum.

(II) In illustration, Weierstrass postulates* a function

$$y = \tfrac{1}{2}(a+b) + \tfrac{1}{2}(b-a)\frac{\tan^{-1}\dfrac{x}{\epsilon}}{\tan^{-1}\dfrac{1}{\epsilon}}$$

selecting for the inverse function the branch such that, as $x$ ranges from $-1$ to $+1$, while $\epsilon$ is a positive quantity,

$$-\tfrac{1}{2}\pi < \tan^{-1}\frac{x}{\epsilon} < \tfrac{1}{2}\pi.$$

This function $y$ conforms to all the specified initial conditions. Denoting the integral by $J$, we have

$$J = \left(\frac{b-a}{2\tan^{-1}\dfrac{1}{\epsilon}}\right)^2 \int_{-1}^{1} \frac{\epsilon^2 x^2}{(x^2+\epsilon^2)^2}\,dx$$

$$= \left(\frac{b-a}{2\tan^{-1}\dfrac{1}{\epsilon}}\right)^2 \epsilon \left(\tan^{-1}\frac{1}{\epsilon} - \frac{\epsilon}{1+\epsilon^2}\right).$$

No restriction has been imposed on the quantity $\epsilon$, except that it is to be positive.

When $\epsilon$ is very large and positive, $J$ tends to the value $\tfrac{1}{6}(b-a)^2$. When $\epsilon$ is very small, still being positive, $J$ tends to a value $\dfrac{(b-a)^2}{\pi}\epsilon$; that is, the integral, necessarily positive, can be made as small as we please. Thus the integral tends to a lower limit zero.

But $J$ cannot attain this lower limit. If it could, as $x$ is not zero, we should be obliged to have $\dfrac{dy}{dx}$ everywhere zero in the range: and therefore $y$ would have to be constant in the range. Now $y$ is continuous, and its values at $-1$ and $+1$ differ; the constant value is therefore impossible.

Consequently, the integral cannot have a minimum value.

(III) If $b$ and $a$ were made equal, so as to avoid the last inference, the integral of the characteristic equation becomes merely $y = a$. The integral is always equal to zero: there is no question as to the existence of a maximum or a minimum.

In the use of Dirichlet's principle, the consideration of which by Weierstrass gives rise to the particular problem, the corresponding result would imply that the small variations there introduced are, all of them, constant: the limiting conditions would require this constant to be zero: so that, in fact, there would be no variation.

*Special form of the E-function, containing a squared variable factor.*

**216.** It may be noted that the Weierstrass $E$-function took the form

$$W(1 - \cos\chi)$$

($W$ a non-vanishing factor) in some particular instances when the Weierstrass test was satisfied, that is,

$$\frac{W}{1+\cos\chi}(\mu x' - \lambda y')^2;$$

and that it has a factor $(\mu x' - \lambda y')^2$ in other particular instances where the test is not satisfied.

* *Ges. Werke*, vol. ii, p. 53.

That the property of possessing the squared factor $(\mu x' - \lambda y')^2$—which vanishes when the arbitrary direction coincides with the tangent to the characteristic curve—is general, may be seen as follows. The function is

$$E = \lambda \left\{ \frac{\partial}{\partial \lambda} F(x, y, \lambda, \mu) - \frac{\partial}{\partial x'} F(x, y, x', y') \right\}$$

$$+ \mu \left\{ \frac{\partial}{\partial \mu} F(x, y, \lambda, \mu) - \frac{\partial}{\partial y'} F(x, y, x', y') \right\}.$$

Now take two variables $p$ and $q$ such that

$$p = \theta \lambda + (1 - \theta) x', \quad q = \theta \mu + (1 - \theta) y',$$

for values of $\theta$ ranging between 0 and 1. For any function $g(p, q)$, we have

$$g(p, q) = \int \left( \frac{\partial g}{\partial p} dp + \frac{\partial g}{\partial q} dq \right),$$

and therefore

$$g(\lambda, \mu) - g(x', y') = \int_0^1 \left( \frac{\partial g}{\partial p} \frac{dp}{d\theta} + \frac{\partial g}{\partial q} \frac{dq}{d\theta} \right) d\theta.$$

First, let

$$g(p, q) = \frac{\partial}{\partial p} F(x, y, p, q),$$

so that

$$\frac{\partial g}{\partial p} = \frac{\partial^2}{\partial p^2} F(x, y, p, q) \quad = \quad q^2 P(x, y, p, q),$$

$$\frac{\partial g}{\partial q} = \frac{\partial^2}{\partial p \partial q} F(x, y, p, q) = - qp P(x, y, p, q),$$

where $P(x, y, p, q)$ is the critical quantity of § 51; then

$$\frac{\partial}{\partial \lambda} F(x, y, \lambda, \mu) - \frac{\partial}{\partial x'} F(x, y, x', y')$$

$$= \int_0^1 \left\{ (\lambda - x') q^2 - (\mu - y') qp \right\} P(x, y, p, q) \, d\theta.$$

Similarly, taking

$$g(p, q) = \frac{\partial}{\partial q} F(x, y, p, q),$$

we have

$$\frac{\partial}{\partial \mu} F(x, y, \lambda, \mu) - \frac{\partial}{\partial y'} F(x, y, x', y')$$

$$= \int_0^1 \left\{ - (\lambda - x') qp + (\mu - y') p^2 \right\} P(x, y, p, q) \, d\theta.$$

When these expressions are substituted in $E$, it takes the form

$$E = \int_0^1 [\lambda \left\{ (\lambda - x') q^2 - (\mu - y') qp \right\} + \mu \left\{ - (\lambda - x') qp + (\mu - y') p^2 \right\}] P(x, y, p, q) \, d\theta$$

$$= (\lambda y' - \mu x')^2 \int_0^1 (1 - \theta) P(x, y, p, q) \, d\theta.$$

We know that, provided the earlier conditions are satisfied, $P(x, y, x', y')$ is a regular function, of uniform sign along the curve. When $p$ and $q$ are substituted for $x'$ and $y'$, we cannot thence assert this uniformity of sign for $P(x, y, p, q)$. If it happens that $P(x, y, p, q)$ still remains uniform in sign when we substitute $p$ and $q$ for $x'$ and $y'$, then the first theorem of mean value may be applied. We note that $\int_0^1 (1 - \theta)\, d\theta = \frac{1}{2}$: some positive fractional quantity $\epsilon$, lying between 0 and 1, but not equal to zero because $P(x, y, p, q)$ is a variable quantity, exists such that

$$E = \tfrac{1}{2}(\lambda y' - \mu x')^2 P(x, y, p_\epsilon, q_\epsilon).$$

Thus $E$ has been transformed, so as to shew the existence of the square factor $(\lambda y' - \mu x')^2$.

### Modified form of the E-test.

**217.** The new form of $E$ allows another form of expression for the Weierstrass test. In the Legendre test, the function $P(x, y, x_1, y_1)$ is of uniform sign everywhere along the curve, where $x_1$ and $y_1$ are the derivatives along the curve, being $x'\left(\dfrac{ds}{dt}\right)^{-1}$ and $y'\left(\dfrac{ds}{dt}\right)^{-1}$. The $E$-function is always of one sign, if $P(x, y, p_\epsilon, q_\epsilon)$ is always of one sign, where the new function is the value of $P(x, y, x_1, y_1)$ when $p_\epsilon$ and $q_\epsilon$ are substituted for $x_1$ and $y_1$. Now $p_\epsilon$ and $q_\epsilon$ represent a pair of quantities, intermediate between $x'$ and $y'$ along the tangent, and $\lambda$ and $\mu$ definitely and arbitrarily off the tangent and not coinciding with the tangent: that is, we may take $p_\epsilon$ and $q_\epsilon$ as an arbitrary pair of quantities definitely off the tangent, arbitrary in range when—as is the fact for the first form of the $E$-function—$\lambda$ and $\mu$ represent an arbitrary direction. Hence the $E$-function does not change its sign if $P(x, y, p_\epsilon, q_\epsilon)$ does not change its sign: thus *the Weierstrass test requires the function $P(x, y, p_\epsilon, q_\epsilon)$ not to change its sign when arbitrary quantities $p_\epsilon$ and $q_\epsilon$. distinct from $x_1$ and $y_1$, are substituted for $x_1$ and $y_1$ in $P(x, y, x_1, y_1)$:* or, as an unmodified equivalent, it requires that *the integral*

$$\int_0^1 (1 - \theta)\, P(x, y, p, q)\, d\theta$$

*shall not change its sign.*

*Ex.* 1. Consider integrals of the type

$$\int f(x, y)\,(x_1{}^2 + y_1{}^2)^{\frac{1}{2}}\, dt.$$

Here

$$P(x, y, x_1, y_1) = \frac{1}{x_1{}^2}\frac{\partial^2}{\partial y_1{}^2}\{f(x, y)(x_1{}^2 + y_1{}^2)^{\frac{1}{2}}\}$$

$$= \frac{f(x, y)}{(x_1{}^2 + y_1{}^2)^{\frac{3}{2}}}.$$

Now

$$p_\epsilon{}^2 + q_\epsilon{}^2 = \epsilon^2 + (1 - \epsilon)^2 + 2\epsilon(1 - \epsilon)(\lambda x' + \mu y')$$

$$= 1 - 4\epsilon(1 - \epsilon)\cos^2 \tfrac{1}{2}\chi\,;$$

$\epsilon$ is some quantity intermediate between 0 and 1, so that $\epsilon(1-\epsilon) \lessgtr \frac{1}{4}$, and $\chi$ is not zero nor $2\pi$: thus $(p_\epsilon{}^2 + q_\epsilon{}^2)^{\frac{1}{2}}$, with the usual positive sign for the radical, is positive. Hence $P(x, y, p_\epsilon, q_\epsilon)$ is always positive for these integrals; the Weierstrass test, for a minimum in these instances, is satisfied.

*Ex.* 2. Verify that this modified form of the Weierstrass test, for the existence of a minimum, is not satisfied for the integrals

$$\int (1+y^2)\, x_1{}^3 y_1{}^{-2}\, dt, \qquad \int y y_1{}^{-2} x_1{}^3\, dt,$$

in Ex. 6 and Ex. 7 of § 215.

*Note.* The Legendre test, that $P(x, y, x_1, y_1)$ remains uniform in sign; and the Weierstrass test, that $P(x, y, p_\epsilon, q_\epsilon)$ remains uniform in sign; are distinct from one another. The Legendre test holds only *along* the curve: the Weierstrass test holds only for arbitrary directions *off* the curve; neither test can be inferred from the other.

*Neither maximum nor minimum when the parametric integrand is rational.*

**218.** It must also be noticed that, in the examples when the subject of integration $F(x, y, x_1, y_1)$ has been rational, the $E$-function has not possessed a persistent sign for all varied arguments: though, when the subject of integration has not been rational, there can be a persistent sign. The former property is typical: the examples provide illustrations of the general theorem that, *when the function $F(x, y, x_1, y_1)$ is rational in $x_1$ and $y_1$, the E-function for the integral $\int F(x, y, x_1, y_1)$ does not conserve a persistent sign for all values of its arguments.*

We have seen (§ 38) that

$$F(x, y, \mu x_1, \mu y_1) = \mu F(x, y, x_1, y_1),$$

where $\mu$ is any quantity, constant or variable, other than zero. The function $F$ either is uniform everywhere in the range of integration: or, if multiform, (such as a radical), it is restricted to a single branch among its values. Also, if

$$g(x, y, x_1, y_1) = \frac{\partial}{\partial x_1} F(x, y, x_1, y_1),$$

$$h(x, y, x_1, y_1) = \frac{\partial}{\partial y_1} F(x, y, x_1, y_1),$$

it follows that

$$g(x, y, \mu x_1, \mu y_1) = g(x, y, x_1, y_1),$$
$$h(x, y, \mu x_1, \mu y_1) = h(x, y, x_1, y_1),$$

where the functions $g$ and $h$ are of the same character as $F$. Thus for the integral $\int (1 + y^2)\, x_1{}^3 y_1{}^{-2} dt$, the subject of integration is uniform. For the length of an arc on a surface represented by $\int (E p_1{}^2 + 2 F p_1 q_1 + G q_1{}^2)^{\frac{1}{2}} dt$, the subject of integration is not uniform, and we should deal only with the positive radical; though, when the variables $p_1$ and $q_1$ pass together through zero values, we should be obliged to pass to the negative radical, if continuity alone dominated the value of the function*.

When the function is uniform, it acquires only a single value for assigned variables. When these variables change and, after all changes, return to their initial value, the uniform function returns to its same initial value. But if the function be multiform, such as a radical, then the function changes its branch, on passing through a zero where radical branches are equal.

In the foregoing relations, and now assuming that $F$ is uniform or remains a uniform branch of a multiform function, let $\mu = -1$. Then

$$F(x, y, -x_1, -y_1) = -F(x, y, x_1, y_1),$$
$$g(x, y, -x_1, -y_1) = g(x, y, x_1, y_1), \quad h(x, y, -x_1, -y_1) = h(x, y, x_1, y_1).$$

Moreover, these verify the relation

$$F(x, y, x_1, y_1) = x_1 g(x, y, x_1, y_1) + y_1 g(x, y, x_1, y_1),$$

when $x_1$ and $y_1$ are changed into $-x_1$ and $-y_1$. But (e.g.) for the arc-integral $\int (x_1{}^2 + y_1{}^2)^{\frac{1}{2}} dt$, on the assumption that $ds$ is always positive, we do not have

$$\{(-x_1)^2 + (-y_1)^2\}^{\frac{1}{2}} = -(x_1{}^2 + y_1{}^2)^{\frac{1}{2}},$$

nor

$$\frac{-x_1}{\{(-x_1)^2 + (-y_1)^2\}^{\frac{1}{2}}} = \frac{x_1}{(x_1{}^2 + y_1{}^2)^{\frac{1}{2}}}, \qquad \frac{-y_1}{\{(-x_1)^2 + (-y_1)^2\}^{\frac{1}{2}}} = \frac{y_1}{(x_1{}^2 + y_1{}^2)^{\frac{1}{2}}},$$

in dealing with the integral; so that the foregoing relations hold only for a uniform function, or for a branch of a non-uniform function throughout a field within which it is restricted to be uniform.

Now

$$E(x, y, x_1, y_1, \lambda, \mu) = \lambda \{g(x, y, \lambda, \mu) - g(x, y, x_1, y_1)\}$$
$$+ \mu \{h(x, y, \lambda, \mu) - h(x, y, x_1, y_1)\};$$

and therefore

$$E(x, y, x_1, y_1, -\lambda, -\mu) = -\lambda \{g(x, y, -\lambda, -\mu) - g(x, y, x_1, y_1)\}$$
$$- \mu \{h(x, y, -\lambda, -\mu) - h(x, y, x_1, y_1)\}$$
$$= -E(x, y, x_1, y_1, \lambda, \mu),$$

when $F$ is rational. Thus, reversion of the direction $\lambda$, $\mu$ leads, on the present hypothesis that $F$ is uniform, to a change of sign in the $E$-function; or, for a rational function $F(x, y, x_1, y_1)$, the $E$-function can change its sign, and the Weierstrass test is not satisfied.

---

* An instance is provided by the cycloid

$$x = a(t - \sin t), \quad y = a \cos t,$$

when we pass from one branch of a cycloid, through a cusp at $t = 2\pi$, to the next branch; we have $\frac{ds}{dt} = 2a \sin \frac{1}{2} t$ for the first branch, $\frac{ds}{dt} = -2a \sin \frac{1}{2} t$ for the second branch.

## *Solid of minimum resistance.*

**219.** We now proceed to consider the classical problem of the solid of least resistance to uniform motion through a fluid. Historically important as originally propounded*, it remains of scientific importance, especially in regard to submarines and to air-craft. There is difficulty in framing trustworthy laws of resistance, to be formulated as inferences from experiments. Physical conditions are gravely complicated. They differ so greatly in different circumstances that laws of resistance, normal and frictional alike, depend not merely upon varying speeds but upon varying powers of those speeds, affected (among other causes) by the disturbance of the surrounding medium due to the mode of propulsion. Thus there are laws of resistance, varying as different powers of the speed for different speeds, in the case of cruisers and destroyers, all driven by screw-propellers. There are other laws of resistance, also varying as different powers of the speed for different speeds, when the movement is effected by agencies external to the moving body, such as towage in experimental tanks and wind-pressure upon sails.

The determination of the shape of the solid is, of course, only one of the essential investigations in the full resolution of the problem.

*Ex.* 1. *Required the form of the surface of revolution such that, when it moves uniformly in a fluid in the direction of its axis, it undergoes the least possible resistance; the resisting pressure at any point on the surface being proportional to the square of the component of the velocity along the normal to the surface.*

(This problem, usually associated with the name of Newton, is the earliest modern problem to be solved by the calculus of variations: and thus is classical in the history of the subject†.)

We choose the axis of revolution for the axis of $y$. The positive direction of this axis is taken in the direction of motion; the velocity is assumed constant and equal to $v$. If $\psi$ be the inclination to $Oy$ of the outward-drawn normal, the whole resistance to the body, due to the pressure, is proportional to

$$\int 2\pi x \, ds \cdot v^2 \cos^2 \psi \cdot \cos \psi,$$

integrated along the meridian curve. Thus the whole resistance is proportional to

$$\int x \cos^3 \psi \, ds,$$

that is, to

$$\int_{t_0}^{t_1} \frac{x \, x_1^3}{x_1^2 + y_1^2} \, dt.$$

---

* In Newton's *Principia* (1687), Book ii, Prop. xxxv, Theorem xxviii, in the scholium. Newton's solution is a statement, in geometrical terms, of the first integral of the characteristic equation: no indication is given as to the construction of the result.

† Todhunter's *Researches in the calculus of variations*, (1871), chap. ix.

The subject of integration $F(x, y, x_1, y_1)$ is given by

$$F(x, y, x_1, y_1) = \frac{x \, x_1^3}{x_1^2 + y_1^2}.$$

(i)  We note at once that, for the problem, the form of function $F(x, y, x_1, y_1)$, which arises out of the hypothesis adopted for the law of resistance, is rational in $x_1$ and $y_1$. Thus (§ 217) the $E$-function cannot maintain a uniform sign; the Weierstrass test is not satisfied.  Consequently, *there is no true minimum*.  But a different law of resistance (such as the cube of the normal velocity) would lead to a different, and frequently to a non-rational, form of $F$.  Our present concern is with the classical problem, for which the $E$-function is

that is,

$$\frac{1}{x} E = \lambda \left\{ \frac{\lambda^2 (\lambda^2 + 3\mu^2)}{(\lambda^2 + \mu^2)^2} - \frac{x'^2 (x'^2 + 3y'^2)}{(x'^2 + y'^2)^2} \right\} + \mu \left\{ - \frac{2\lambda^3 \mu}{(\lambda^2 + \mu^2)^2} + \frac{2x'^3 y'}{(x^2 + y'^2)^2} \right\},$$

$$E = \frac{x}{(\lambda^2 + \mu^2)(x'^2 + y'^2)} (\mu x' - \lambda y')^2 \{(y'^2 - x'^2)\lambda + 2x'y'\mu\}$$
$$= -x \sin^2 \chi \, \cos(3\psi + \chi),$$

when we take $x' = \cos \psi$, $y' = \sin \psi$, $\lambda = \cos(\psi + \chi)$, $\mu = \sin(\psi + \chi)$.  Thus* $E$ varies in sign for varying values of $\chi$; it is positive or negative, according as $\cos(3\psi + \chi)$ is positive or negative.

(ii)  As regards the other tests arising out of conditions to be satisfied when the small variations are weak, we begin with the characteristic equation.  The subject of integration, being

$$\frac{x \, x_1^3}{x_1^2 + y_1^2},$$

does not involve $y$ explicitly.  We therefore take the characteristic equation

$$\frac{\partial F}{\partial y} - \frac{d}{dt}\left( \frac{\partial F}{\partial y_1} \right) = 0 ;$$

a first integral is given by

$$\frac{x \, x_1^3 y_1}{(x_1^2 + y_1^2)^2} = c,$$

where $c$ is an arbitrary constant.

If, at any place or at a limited number of isolated places, the characteristic curve suddenly changes its direction, the value of $c$ is unaltered in the passage through such a place (§ 47).

First, consider the possibility that $c$ should be zero, so that the curve can meet the axis where, as the surface is one of revolution, the curve will terminate.  Away from the axis, we then have either $x_1 = 0$ or $y_1 = 0$.  When $x_1 = 0$, then $x = a$, so that the curve is a straight line parallel to the axis; when $x_1$ is not zero, then $y_1$ is zero, and we have a line perpendicular to the axis. Thus we have the surface of a circular cylinder, the (zero) value of $c$ being unaltered for passage of the curve through $A$. If $AN = a$, the pressure on the cylinder is proportional to

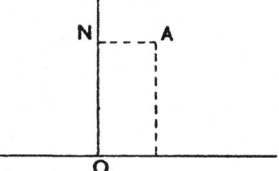

$$2\pi v^2 \int_N^A \frac{x \, x_1^3}{x_1^2 + y_1^2} \, dt ;$$

for, along the part $x_1 = 0$, the resistance in the direction of motion is zero.  Along $NA$, we have $y_1 = 0$, so that the pressure is proportional to

$$2\pi v^2 \int_N^A x x_1 \, dt$$
$$= \pi v^2 a^2,$$

as is to be expected.

_____
* This expression for $E$ was first obtained by Weierstrass.

(iii) The critical function in the Legendre test is

$$P = \frac{1}{x_1{}^2} \frac{\partial^2 F}{\partial y_1{}^2}$$

$$= \frac{2x\, x_1}{(x_1{}^2 + y_1{}^2)^3} (3y_1{}^2 - x_1{}^2).$$

In the present case when $c = 0$, the magnitude $P$ is zero over the range $x_1 = 0$. Over the range $y_1 = 0$ from $A$ to $N$,

$$P = -2\frac{x}{x_1{}^3} \; ;$$

and $x_1$ is positive along $AN$, so that $P$ is negative along $AN$. Thus along the whole surface, the magnitude $P$ is either zero or negative. The cylindrical surface does not provide a minimum for weak variations.

(iv) Next, let $c$ be different from zero. With the figure (p. 340), when $x$ and $t$ (or $x$ and $s$) increase together, we have $x_1$ positive, $y_1$ negative, $\frac{1}{2}\pi < \psi < \pi$, so we take

$$y_1 = -qx_1.$$

As the curve is

$$\frac{x\, x_1{}^3\, y_1}{(x_1{}^2 + y_1{}^2)^2} = c,$$

we have

$$x = -\frac{c}{q}(1 + q^2)^2 \; ;$$

and now

$$dy = -q\, dx,$$

so that

$$y = y_0 - c \log q + cq^2 + \tfrac{3}{4}cq^4.$$

To determine whether there is a conjugate $q_1$ (in the Jacobi sense) of an initial point $q_0$ on the curve, we construct the critical function $Z(q, q_0)$ of § 63. With the former notation, we have

$$\chi(q) = \frac{dx}{dq}\frac{\partial y}{\partial c} - \frac{dy}{dq}\frac{\partial x}{\partial c}$$

$$= \frac{dx}{dq}\left(\frac{\partial y}{\partial c} + q\frac{\partial x}{\partial c}\right)$$

$$= \frac{dx}{dq}\{\tfrac{3}{4}q^4 + q^2 - \log q - (1 + q^2)^2\}$$

$$= -\tfrac{1}{4}\frac{dx}{dq}\{(q^2 + 2)^2 + 4 \log q\},$$

and

$$\omega(q) = \frac{dx}{dq}\frac{\partial y}{\partial y_0} - \frac{dy}{dq}\frac{\partial x}{\partial y_0}$$

$$= \frac{dx}{dq} \; ;$$

and therefore

$$Z(q, q_0) = \omega(q_0)\,\chi(q) - \omega(q)\,\chi(q_0)$$

$$= \tfrac{1}{4}\frac{dx}{dq}\left(\frac{dx}{dq}\right)_{q=q_0}\{(q_0{}^2 + 2)^2 + 4 \log q_0 - (q^2 + 2)^2 - \log q\}.$$

Now with a continually increasing or a continually decreasing positive quantity $q$, the unction $(q^2 + 2)^2 + \log q$ never resumes its initial value. We therefore cannot have a zero va ue for $Z(q, q_0)$, except at $q = q_0$, from the last factor. Also

$$\frac{dx}{dq} = \frac{c}{q^2}(1 + q^2)(1 - 3q^2) \; ;$$

thus if $q$, beginning at $q_0$, can attain a value $\pm 3^{-\frac{1}{2}}$, we may have a zero for $Z(q, q_0)$: that is, as $q$ is to be positive, we must consider the value $q = \dfrac{1}{\sqrt{3}}$.

Reverting to the function $P$, we have

$$P = \frac{2x}{x_1{}^3} \frac{3q^2 - 1}{(1 + q^2)^3},$$

so that, as $x$ and $x_1$ are positive, we shall have $P$ always positive if $q > \dfrac{1}{\sqrt{3}}$. Thus both the Legendre test and the Jacobi test admit the possibility of a minimum (necessarily for weak variations alone, because only weak variations have been imposed) if $q$, beginning at a value $q_0$, does not attain a value $\dfrac{1}{\sqrt{3}}$.

(v) It therefore is desirable to trace the curve. We have

$$\frac{dx}{dq} = \frac{c}{q^2}(1 + q^2)(1 - 3q^2), \qquad \frac{dy}{dq} = -\frac{c}{q}(1 + q^2)(1 - 3q^2);$$

and therefore a cusp (or some higher singularity) may be expected at $q = \dfrac{1}{\sqrt{3}}$. We take

$$q = \frac{1}{\sqrt{3}} + \mu,$$

where $\mu$ is small in the vicinity of the place. We find, for the values of $x$ and $y$ in that vicinity,

$$-x = \frac{16c}{3\sqrt{3}}\left(1 + \tfrac{9}{4}\mu^2 + \frac{3\sqrt{3}}{2}\mu^3 + \ldots\right),$$

$$y = y_0 + c\left(\tfrac{5}{12} + \tfrac{1}{2}\log 3\right) + c\left(4\mu^2 + \frac{2}{\sqrt{3}}\mu^3 + \ldots\right),$$

shewing that the place $C$ is a cusp, the tangential direction at which is given by $\dfrac{dy}{dx} = -\dfrac{1}{\sqrt{3}}$. The two branches of the curve, on either side of the cusp, are given: $CA$, by a positive value of $\mu$; and $CB$, by a negative value of $\mu$. The angle $CTO$ is 30°.

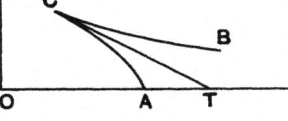

Along $CA$, $q$ $\left(\text{which} = -\dfrac{dy}{dx}\right)$ increases, so that $q_1$ is positive in the direction from $C$ to $A$; along $CB$, $q$ decreases, so that $q_1$ is negative in the direction from $C$ to $B$.

Now

$$\frac{x_1}{x} = \left(-\frac{1}{q} + \frac{4q}{1 + q^2}\right)q_1 = \frac{3q^2 - 1}{q(1 + q^2)}q_1;$$

hence

$$P = \frac{2xx_1(3y_1{}^2 - x_1{}^2)}{(x_1{}^2 + y_1{}^2)^3}$$

$$= \frac{2x(3q^2 - 1)}{(1 + q^2)^3}\frac{1}{x_1{}^3} = \frac{2q}{(1 + q^2)^2}\frac{1}{x_1{}^2 q_1}.$$

Along $CA$, $q$ is positive and $q_1$ is positive; while along $CB$, $q$ is positive and $q_1$ is negative. Thus along $CA$, the quantity $P$ is positive; the Legendre test admits a minimum for the branch $CA$. And clearly $c$ is a negative constant.

Thus, for weak variations, the characteristic curve is

$$x = \frac{a}{q}(1 + q^2)^2, \qquad y = y_0 + a\left(\log q - q^2 - \tfrac{3}{4}q^4\right),$$

on changing the negative $c$ into a positive $a$. The Legendre test is satisfied. The Jacobi test imposes no limit on a range, that begins anywhere on $CA$ and extends in the direction $C$ towards $A$. Hence we have a minimum for weak variations.

(vi) But the Weierstrass test, for the simplest strong variation, was proved to be not satisfied. What is a minimum, under weak variations, is not a minimum under the simplest strong variation. Therefore it is not a true minimum.

(vii) It was stated (§ 207) that Legendre obtained, by means of jagged lines, a curve (for the generation of the solid of revolution) which leads to an integral less than the quasi-minimum provided by the characteristic curve for weak variations.

The following is a simplified form of Legendre's construction* for the purpose. Let $PQ$ be any arc of the characteristic curve. Let $QT$ be the tangent at the lower extremity; through $P$, draw $PC$ making with the axis of $x$ the angle $\gamma$ equal to $QTO$; and through $Q$ draw the ordinate $NQC$.

Along the arc $PQ$, we have

$$p^2 < \tan^2 \gamma$$

except at the extremity $Q$; so that for the curved arc $PQ$, we have

$$I_{PQ} = \int_M^N \frac{x}{1+p^2}\,dx = \int \frac{xx_1^3}{y_1^2 + x_1^2}\,dt$$
$$> \int_M^N \frac{x}{1+\tan^2 \gamma}\,dx.$$

Along the line $PC$, we have

$$I_{PC} = \int_M^N \frac{x}{1+\tan^2 \gamma}\,dx.$$

Along the line $CQ$, we have $p = \infty$, $x_1 = 0$: thus

$$I_{CQ} = 0.$$

Consequently

$$I_{PC} + I_{CQ} < I_{PQ}:$$

the broken line $PCQ$ provides a smaller integral than the arc $PQ$.

(viii) The jagged line (mentioned by Legendre in his memoir) contains the secret of the paradox that the apparent minimum, obtained as satisfying all the tests for weak variations, can be reduced. The requirements of the jagged line lead to the Weierstrass test.

*Ex.* 2. In the historical example just considered, the resistance is taken to be due wholly to the normal pressure. But experience and observation shew that frictional resistance is often more important than the pressure resistance; thus for destroyers, the frictional resistance can be 80 per cent. of the total at a speed of 12 knots and 50 per cent. at a speed of 20 knots, while for cruisers the corresponding percentages are 90 and 80 at these respective speeds†.

The analysis must be modified when friction is taken into account. Let $C$ and $C'$ denote the absolute constants for pressure-resistance and for friction-resistance respec-

---

* *L.c.*, p. 320 (*foot-note*).

† Sir W. H. White, *British Assoc. Rep.* (1899), p. 850. For more recent experimental results mainly connected with air-craft, the *Reports and Memoranda* of the Aeronautical Research Committee may be consulted.

tively. With the preceding notation, the total resistance is

$$2\pi C v^2 \int x \cos^3 \psi \, ds + 2\pi C' v^2 \int x \sin^3 \psi \, ds$$

$$= 2\pi C v^2 \int x \frac{x_1^3 + a y_1^3}{x_1^2 + y_1^2} \, dt,$$

where $C' = aC$, and $a$ is a constant to be determined by experiment. Thus the integral $\int F dt$ must be made a minimum, where

$$F = x \frac{x_1^3 + a y_1^3}{x_1^2 + y_1^2}.$$

The subject of integration is a rational function of $x_1$ and $y_1$; and so the characteristic curve

$$\frac{\partial F}{\partial y_1} = c$$

cannot (§ 217) provide a minimum because the Weierstrass test is not satisfied. The $E$-function, occurring in the test, is

$$E = F(\lambda, \mu) - \lambda \frac{\partial F'}{\partial x_1} - \mu \frac{\partial F'}{\partial y_1};$$

or, when we take $x_1 = x' = \cos \psi$, $y_1 = y' = \sin \psi$, $\lambda = \cos(\psi + \chi)$, $\mu = \sin(\psi + \chi)$, we have

$$\frac{1}{x} E = \lambda^3 + a\mu^3$$
$$- \lambda \left(\cos^2 \psi + 2\cos^2 \psi \sin^2 \psi - 2a \cos \psi \sin^3 \psi\right)$$
$$- \mu \left(-2\cos^3 \psi \sin \psi + a \sin^2 \psi + 2a \sin^2 \psi \cos^2 \psi\right).$$

Now

$$\lambda^3 - \lambda \left(\cos^2 \psi + 2\cos^2 \psi \sin^2 \psi\right) + 2\mu \cos^3 \psi \sin \psi = -\sin^2 \chi \cos(3\psi + \chi),$$

$$\mu^3 - \mu \left(\sin^2 \psi + 2\sin^2 \psi \cos^2 \psi\right) + 2\lambda \sin^3 \psi \cos \psi = \sin^2 \chi \sin(3\psi + \chi);$$

and therefore if $a = \tan \alpha$, where $0 < \alpha < \frac{1}{2}\pi$ because $a$ is positive,

$$E = -x \sin^2 \chi \sec \alpha \cos(3\psi + \alpha + \chi).$$

Manifestly, as $\chi$ ranges from 0 to $2\pi$, the function $E$ changes its sign: the Weierstrass test is not satisfied. *There is no solid of revolution experiencing a minimum resistance, when the law of resistance assumes that resistance varies as the square of the velocity.*

*Ex.* 3. When *the resistance of the medium, through which the body moves, varies as the cube of the velocity*—a law within the range of observed results—the initial analysis is similar.

With the same notation as in the preceding example, the total resistance is

$$2\pi C v^3 \int x \frac{x_1^4 + a y_1^4}{(x_1^2 + y_1^2)^{\frac{3}{2}}} \, dt,$$

so that, in the integral $\int F dt$ to be made a minimum, we have

$$F = x \frac{x_1^4 + a y_1^4}{(x_1^2 + y_1^2)^3}.$$

Here, the function $F$ is not a rational function of $x_1$ and $y_1$; the general theorem of § 217 cannot be invoked.

As usual, the first integral of the characteristic equation is

$$\frac{\partial F}{\partial y_1} = c,$$

that is,

$$x \{a \sin^3 \psi \, (1 + 3 \cos^2 \psi) - 3 \cos^4 \psi \sin \psi\} = c,$$

which is the characteristic equation of the curve.

For the Legendre test, we have

$$P = \frac{1}{x_1^2} \frac{\partial^2 F}{\partial y_1^2}$$

$$= \frac{3x}{(x_1^2 + y_1^2)^{\frac{7}{2}}} \{ -(x_1^4 + a y_1^4) + 4 (1 + a) x_1^2 y_1^2 \},$$

so that

$$P \left(\frac{ds}{dt}\right)^3 = 3x \{4 (1 + a) \sin^2 \psi \cos^2 \psi - \cos^4 \psi - a \sin^4 \psi\}$$

$$= 3x \cos^4 \psi \{4 (1 + a) t^2 - 1 - a t^4\}$$

$$= 3x a \cos^4 \psi \, (t_1^2 - t^2) (t^2 - t_2^2),$$

where $t$ denotes $\tan \psi$, while $t_1^2$ and $t_2^2 \, (< t_1^2)$ are the two real and positive roots of the quadratic

$$4 (1 + a) z - 1 - a z^2 = 0.$$

As $P$ must be positive for a minimum under the Legendre test, $t^2$ must lie between $t_1^2$ and $t_2^2$; in particular, $t^2$ may lie in the immediate vicinity of a value $2 + \frac{2}{a}$, for which $(t_1^2 - t^2)(t^2 - t_2^2)$ acquires its positive (maximum) value.

The $E$-function, which occurs in the Weierstrass test, now is

$$\frac{1}{x} E = \lambda^4 + a \mu^4$$

$$- \lambda \, (\cos^3 \psi + 3 \cos^3 \psi \sin^2 \psi - 3a \cos \psi \sin^4 \psi)$$

$$- \mu \, (- 3 \cos^2 \psi \sin \psi + a \sin^3 \psi + 3a \sin^3 \psi \cos \psi).$$

Let

$$J (\chi) = \tfrac{3}{2} - \tfrac{3}{4} \cos 4\psi - \tfrac{3}{2} \cos (4\psi + \chi) - \cos (4\psi + 2\chi) - \tfrac{1}{2} \cos (4\psi + 3\chi),$$

$$G (\chi) = \cos 2\psi + 2 \cos (2\psi + \chi) \, ;$$

then

$$E = \tfrac{1}{2} x \, (1 - \cos \chi) \, [J (\chi) - G (\chi) + a \{J (\chi) + G (\chi)\}] = \tfrac{1}{4} x \, (1 - \cos \chi) \, W (\chi),$$

so that the sign of $E$, for non-zero variations, depends on the sign of $W (\chi)$. Now

$$W (-2\psi) = 5a - 3 - (2a + 6) \cos 2\psi - (3a + 3) \cos^2 2\psi$$

$$= (-12 + 16 a t^2 + 4 a t^4) \cos^4 \psi \, ;$$

$$W (\pi - 2\psi) = 5 - 3a + (2 + 6a) \cos 2\psi - (3 + 3a) \cos^2 2\psi$$

$$= (4 + 16 t^2 - 12 a t^4) \cos^4 \psi.$$

When $t^2$ has a value in the vicinity of $2 + \frac{2}{a}$, the Legendre test is satisfied. When $t^2 = 2 + \frac{2}{a}$,

$$-12 + 16 a t^2 + 4 a t^4 = 20 + 32a + a t^4,$$

$$4 + 16 t^2 - 12 a t^4 = -\frac{1}{a} [12 + 60a + 48a^2],$$

the former of which is certainly positive and the latter of which is certainly negative. Thus there are places on the range at, and in the vicinity of, the place $t^2 = 2 + \frac{2}{a}$, for which $E$ is certainly positive, arising out of the factor $W(-2\psi)$, and is also certainly negative, arising out of the factor $W(\pi - 2\psi)$: that is, at each such place there are directions $\chi$, some of which make $E$ positive and some of which make $E$ negative. Thus $E$ does not keep a uniform sign over the range: the Weierstrass test is not satisfied: *there is no solid of minimum resistance when the resistance varies as the cube of the velocity.*

*Ex.* 4. Prove that there is no solid of minimum pressure-resistance, and that there is no solid of minimum friction-resistance, when the resistance varies as the fifth power of the velocity.

[The $E$-function for the pressure-resistance is

$$\frac{32E}{x\,(1-\cos\chi)}=10-15\,\{\cos 2\psi+2\cos (2\psi+\chi)\}$$

$$-6\,\{3\cos 4\psi+6\cos (4\psi+\chi)+4\cos (4\psi+2\chi)+2\cos (4\psi+3\chi)\}$$

$$-\{5\cos 6\psi+2\cos (6\psi+5\chi)+4\cos (6\psi+2\chi)+6\cos (6\psi+3\chi)$$

$$+8\cos (6\psi+2\chi)+10\cos (6\psi+\chi)\}$$

$$=W(\chi);$$

here

$$W(-2\psi)=-20\,(1+\cos 2\psi)^3,\quad W(\pi)=4\,(1+\cos 2\psi)^3,$$

so that for some value of $\chi$ between $\pi$ and $2\pi-2\psi$, the $E$-function vanishes and changes its sign. The Weierstrass test is not satisfied.

Similarly for the $E$-function in the case of the solid of revolution, as regards friction-resistance.]

### *Strong variations, of more general type.*

**220.** In the strong variations, that have been discussed and have led to the Weierstrass test, the deviation from complete regularity has been resolved so that it is composed of elements, each of which is of a simple type. This

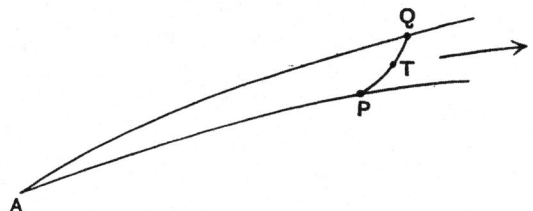

elementary deviation is geometrically represented by a continuous small arc $PQ$, passing from a point $P$ to a contiguous point $Q$ through which a consecutive characteristic curve $AQ...$ is drawn. The investigations of Weierstrass, so far as they have been made known, were limited to the case when the arc is a straight line. For a strong variation of this character, explicitly acknowledged to be only a particular kind of strong variation, the $E$-function has been constructed; and it has led to a fourth test, differing in nature from those initiated by Euler, Legendre, and Jacobi.

It has been pointed out (§ 215) that, for integrals of the form

$$\int f(x, y)\, ds, \quad \int f(x, y)\,(x_1^2 + y_1^2)^{\frac{1}{2}}\, dt,$$

while the Weierstrass test (for a minimum) is always satisfied when $PQ$ is a straight line, the test (that is, the occurrence of an increase) is equally satisfied when $PQ$ is made any continuously curved arc from $P$ to $Q$. The explanation of § 209, shewing that a broken line from $P$ to $Q$ can be resolved, so that the function need be considered for each piece alone, can manifestly be extended to the case when $PQ$ is made up of a number of pieces of continuously curved arc, each of which can be considered alone. We now shall consider the general problem when $PQ$ is a curvilinear arc; some of the analysis will be useful when we come to consider strong variations for integrals that involve derivatives of the second order.

Consider $AQ, QTP$ as a variation of $AP$, where $AP$ and $AQ$ are characteristic curves. As before, we have

$$I_{AQ} - I_{AP} = \kappa \left( \xi \frac{\partial F}{\partial x'} + \eta \frac{\partial F}{\partial y'} \right),$$

to the first order of small quantities, where $\kappa\xi$ and $\kappa\eta$ are the coordinates of $Q$ with respect to $P$. To complete the expression for the whole variation, we want $I_{QTP}$, the integral being taken along $QTP$ from $Q$ to $P$. It is therefore necessary to have the values of the variables and their derivatives at any point $T$ of $QTP$.

Let the coordinates of $T$ with respect to $P$ be $X$ and $Y$; and let the arc $PT = S$, the arc $PQ = \sigma$. Then if
$$S = \sigma\theta,$$
$\theta$ ranges from 0 to 1 as $T$ moves from $P$ to $Q$.

Now along any curve whatever, we have (with the usual notation)
$$x' = \cos\psi, \quad y' = \sin\psi, \quad \rho x'' = -\sin\psi, \quad \rho y'' = \cos\psi;$$
and therefore
$$\frac{d}{ds}(\rho x'') = -\frac{1}{\rho}x', \quad \frac{d}{ds}(\rho y'') = -\frac{1}{\rho}y',$$
so that
$$\rho^2 x''' + \rho\rho' x'' + x' = 0, \quad \rho^2 y''' + \rho\rho' y'' + y' = 0,$$
equations that are useful for the expansions of $x'$ and $y'$ (and therefore of $x$ and $y$) in terms of $s$. As the curve is arbitrarily assumed, we can imagine it defined by means of the curvature, and of initial values of $x'$ and $y'$ at $P$; and so we take
$$x' = a_1, \quad y' = b_1, \quad x'' = -\frac{b_1}{c}, \quad y'' = \frac{a_1}{c},$$

at $P$. We proceed to make $\theta$ the variable in the expansion; and we make $\sigma$ the small quantity in terms of the magnitude of which all the small quantities are expressed.

We take, at any point,
$$\rho = \sigma R,$$

so that $R$, being a function of $S$, is a function of $\sigma\theta$. Now, along $PTQ$,

$$\rho^2 \frac{d^3X}{dS^3} + \rho\, \frac{d\rho}{dS}\, \frac{d^2X}{dS^2} + \frac{dX}{dS} = 0,$$

and therefore

$$R^2 \frac{d^3X}{d\theta^3} + R\, \frac{dR}{d\theta}\, \frac{d^2X}{d\theta^2} + \frac{dX}{d\theta} = 0.$$

To express $X$ as a series in powers of $\theta$ satisfying this equation, we have, when $\theta$ is zero,

$$\frac{dX}{dS} = a_1, \quad \frac{dY}{dS} = b_1, \quad \frac{d^2X}{dS^2} = -\frac{b_1}{c}, \quad \frac{d^2Y}{dS^2} = \frac{a_1}{c};$$

or if

$$c = \sigma C,$$

so that, for the radius of curvature,

$$R = C + \theta C_1 + \tfrac{1}{2}\theta^2 C_2 + \dots,$$

the initial values are

$$\frac{dX}{d\theta} = \sigma a_1, \quad \frac{dY}{d\theta} = \sigma b_1, \quad \frac{d^2X}{d\theta^2} = -\sigma\frac{b_1}{C}, \quad \frac{d^2Y}{d\theta^2} = \sigma\frac{a_1}{C}.$$

The expansions take the form

$$\frac{dX}{d\theta} = \sigma\left(a_1 - \frac{b_1}{C}\,\theta + \frac{a_3}{2!}\,\theta^2 + \dots\right),$$

$$\frac{dY}{d\theta} = \sigma\left(b_1 + \frac{a_1}{C}\,\theta + \frac{b_3}{2!}\,\theta^2 + \dots\right),$$

and therefore, as $X$ and $Y$ vanish when $\theta = 0$,

$$X = \sigma\left(a_1\theta - \frac{b_1}{2C}\,\theta^2 + \frac{a_3}{3!}\,\theta^3 + \dots\right) = \sigma\Theta_x,$$

$$Y = \sigma\left(b_1\theta + \frac{a_1}{2C}\,\theta^2 + \frac{b_3}{3!}\,\theta^3 + \dots\right) = \sigma\Theta_y,$$

the values of the coefficients $a_3,\ b_3,\dots$ being definite: they are linear in $a_1$ and $b_1$, and they involve the coefficients in $R$.

Also at $Q$, we have $\kappa\xi$ and $\kappa\eta$ as the values of $X$ and $Y$, while $\theta$ there is unity; thus

$$\kappa\xi = \sigma\left(a_1 - \frac{b_1}{2C} + \frac{a_3}{3!} - \dots\right) = \sigma l,$$

$$\kappa\eta = \sigma\left(b_1 + \frac{a_1}{2C} + \frac{b_3}{3!} + \dots\right) = \sigma m,$$

using $l$ and $m$ to denote the two expressions. Further, if $S' = QT$ so that $S'$ is measured from $Q$ towards $P$, we have $S + S' = \sigma$, and so

$$\frac{dX}{dS'} = -\frac{dX}{\sigma\, d\theta} = -\frac{d\Theta_x}{d\theta}, \quad \frac{dY}{dS'} = -\frac{dY}{\sigma\, d\theta} = -\frac{d\Theta_y}{d\theta},$$

so that
$$\left(\frac{d\Theta_x}{d\theta}\right)^2 + \left(\frac{d\Theta_y}{d\theta}\right)^2 = 1.$$

Thus we may take
$$\frac{d\Theta_x}{d\theta} = L, \quad \frac{d\Theta_y}{d\theta} = M.$$

### Modified E-function.

**221.** With these values, we have

$$I_{QTP} = \int_Q^P F\left(x+X, y+Y, \frac{dX}{dS'}, \frac{dY}{dS'}\right) dS'$$

$$= -\sigma \int_1^0 F\left(x+X, y+Y, -\frac{d\Theta_x}{d\theta}, -\frac{d\Theta_y}{d\theta}\right) d\theta$$

$$= \sigma \int_0^1 F\left(x+\sigma\Theta_x, y+\sigma\Theta_y, -L, -M\right) d\theta$$

$$= \sigma \int_0^1 F\left(x, y, -L, -M\right) d\theta + [\sigma^2],$$

where $[\sigma^2]$ is an aggregate of quantities of the second and higher orders in $\sigma$. Now $a_1$ and $b_1$ denote any direction at $P$: for that direction, substitute the opposite direction $-a_1$, and $-b_1$. Then $-L$ and $-M$ become $L$ and $M$, while $l$ and $m$ become $-l$ and $-m$, all four expressions being linear in $a_1$ and $b_1$. Hence

$$I_{AQ} + I_{QTP} - I_{AP} = \sigma \int_0^1 F(x, y, L, M) d\theta - \sigma\left(l\frac{\partial F}{\partial x'} + m\frac{\partial F}{\partial y'}\right).$$

In the definite integral, the only quantities involving $\theta$ are $L$, $M$, quantities connected with the arc $PTQ$; let $\bar{\theta}$ be a mean value of $\theta$ (with $\bar{L}$ and $\bar{M}$ as the corresponding values of $L$ and $M$) such that $0 < \bar{\theta} < 1$, and

$$\int_0^1 F(x, y, L, M) d\theta = F(x, y, \bar{L}, \bar{M}).$$

Also

$$F(x, y, \bar{L}, \bar{M}) = \bar{L}\frac{\partial}{\partial \bar{L}} F(x, y, \bar{L}, \bar{M}) + \bar{M}\frac{\partial}{\partial \bar{M}} F(x, y, \bar{L}, \bar{M}).$$

Let $E'$ denote the combination of integrals, so that

$$E' = I_{AQ} + I_{QTP} - I_{AP};$$

then

$$E' = \sigma\left\{\bar{L}\frac{\partial}{\partial \bar{L}} F(x, y, \bar{L}, \bar{M}) - l\frac{\partial}{\partial x'} F(x, y, x', y')\right\}$$

$$+ \sigma\left\{\bar{M}\frac{\partial}{\partial \bar{M}} F(x, y, \bar{L}, \bar{M}) - m\frac{\partial}{\partial y'} F(x, y, x', y')\right\}.$$

In this expression, $\bar{L}$ and $\bar{M}$ are the direction-cosines of a tangent to $QTP$ at the point having the mean value $\bar{\theta}$; also

$$l\sigma = d\cos(\chi + \psi), \quad m\sigma = d\sin(\chi + \psi),$$

where $d$ is the length of the line $PQ$, $\psi$ is the inclination of the tangent of the characteristic curve to the axis of $x$, and $\chi$ is the angle between $PQ$ and this tangent.

*The function $E'$ must be positive* at all points of the curve for all values of $l$ and $m$ (and so of $\overline{L}$ and $\overline{M}$) *if there is to be a minimum*; it must be similarly *negative if there is to be a maximum*.

*Ex.* 1. Verify that $E'$ is positive (so that a minimum is admissible) for all integrals of the type

$$\int f(x, y) (x'^2 + y'^2)^{\frac{1}{2}} dt.$$

*Ex.* 2. The arc $PTQ$ is taken to be circular, an arbitrary radius $c$ being chosen, as in § 222 (*post*).

Prove that the variation of the integral can be expressed in the form

$$E = D \frac{\sin^{-1} \dfrac{D}{2c}}{\dfrac{D}{2c}} \left( \overline{\lambda} \frac{\partial \overline{F}}{\partial \overline{\lambda}} + \overline{\mu} \frac{\partial \overline{F}}{\partial \overline{\mu}} \right) - D \left( \frac{\partial F}{\partial x'} \cos \phi + \frac{\partial F}{\partial y'} \sin \phi \right),$$

where $\phi$ is the (arbitrary) inclination of the chord $QP$ to the axis of $x$, where $\overline{F}$ denotes $F(x, y, \overline{\lambda}, \overline{\mu})$, $\overline{\lambda} = \cos(\phi + \epsilon)$, $\overline{\mu} = \sin(\phi + \epsilon)$, and where $\epsilon$ is a mean value of $\chi$ such that

$$2F(x, y, \overline{\lambda}, \overline{\mu}) \sin^{-1} \frac{D}{2c} = \int_{-\sin^{-1}\frac{D}{2c}}^{\sin^{-1}\frac{D}{2c}} F\{x, y, \cos(\phi + \chi), \sin(\phi + \chi)\} \, d\chi.$$

*Ex.* 3. Consider the case when the arc $PQ$ is parabolic, a point $(X, Y)$ on it being given by

$$X = \lambda t - \frac{\mu}{2c} t^2, \quad Y = \mu t + \frac{\lambda}{2c} t^2.$$

At $P$, let $t = 0$; at $Q$, let $t = T$. Then we take the straight line $QP = D$, with direction-cosines $l$, $m$: so that

$$\kappa \xi = lD = \lambda T - \frac{\mu}{2c} T^2,$$

$$\kappa \eta = mD = \mu T + \frac{\lambda}{2c} T^2.$$

Thus

$$\frac{m}{l} = \frac{\mu + \dfrac{\lambda}{2c} T}{\lambda - \dfrac{\mu}{2c} T},$$

and therefore

$$\frac{T}{2c} = \frac{\lambda m - \mu l}{\lambda l + \mu m}.$$

Let $l$, $m = \cos \phi$, $\sin \phi$; $\lambda$, $\mu = \cos a$, $\sin a$; so that $\phi = QPX$, $a = X'PX$; then, if $\chi = \phi - a = QPX'$,

$$T = 2c \tan \chi.$$

Also, if $PMQ$ is a right angle,

$$D^2 = T^2 + \frac{1}{4c^2} T^4 = T^2 \sec^2 \chi,$$

But

$$\frac{dX}{dt}=\lambda-\frac{\mu}{c}\,t, \quad \frac{dY}{dt}=\mu+\frac{\lambda}{c}\,t\;;$$

and along $QP$, if the arcs $PR$, $QR$, $PQ$ be $S$, $S'$, $\sigma$, so that $S$ increases along $PR$ and $S'$ along $QR$, we have

$$\frac{dX}{dS'}=-\frac{dX}{dS}, \quad \frac{dY}{dS'}=-\frac{dY}{dS}.$$

Then

$$F\left(x+X,\,y+Y,\,\frac{dX}{dS'},\,\frac{dY}{dS'}\right)=F\left(x+X,\,y+Y,\,-\frac{dX}{dS},\,-\frac{dY}{dS}\right).$$

Change $\lambda$ and $\mu$ into $-\lambda$ and $-\mu$, and therefore $l$ and $m$ into $-l$ and $-m$; thus

$$I_{AQ}-I_{AP}=-D\left\{l\,\frac{\partial F(x,y,x',y')}{\partial x'}+m\,\frac{\partial F(x,y,x',y')}{\partial y'}\right\},$$

and

$$
\begin{aligned}
I_{QP}&=\int_Q^P F\left(x+X,\,y+Y,\,\frac{dX}{dS},\,\frac{dY}{dS}\right)dS\\
&=\int_Q^P F\left\{x+X,\,y+Y,\,\frac{\lambda c-\mu t}{(c^2+t^2)^{\frac12}},\,\frac{\mu c+\lambda t}{(c^2+t^2)^{\frac12}}\right\}dS'\\
&=\sigma F\left\{x+X,\,y+Y,\,\frac{\lambda c-\mu\epsilon T}{(c^2+\epsilon^2 T^2)^{\frac12}},\,\frac{\mu c+\lambda\epsilon T}{(c^2+\epsilon^2 T^2)^{\frac12}}\right\},
\end{aligned}
$$

where $\epsilon$ is a real constant, lying between 0 and 1, such as to give the mean value of $F$ over the arc. But

$$
\begin{aligned}
&F\left\{x+X,\,y+Y,\,\frac{\lambda c-\mu\epsilon T}{(c^2+\epsilon^2 T^2)^{\frac12}},\,\frac{\mu c+\lambda\epsilon T}{(c^2+\epsilon^2 T^2)^{\frac12}}\right\}\\
&=F\left\{x,\,y,\,\frac{\lambda c-\mu\epsilon T}{(c^2+\epsilon^2 T^2)^{\frac12}},\,\frac{\mu c+\lambda\epsilon T}{(c^2+\epsilon^2 T^2)^{\frac12}}\right\}+[\sigma],
\end{aligned}
$$

where $[\sigma]$ represents an aggregate of terms of the first and higher degrees in $\sigma$. Thus, to the first power of $\sigma$,

$$I_{QP}=\sigma F\left\{x,\,y,\,\frac{\lambda c-\mu\epsilon T}{(c^2+\epsilon^2 T^2)^{\frac12}},\,\frac{\mu c+\lambda\epsilon T}{(c^2+\epsilon^2 T^2)^{\frac12}}\right\}.$$

Now let

$$\bar{\lambda}=\frac{\lambda c-\mu\epsilon T}{(c^2+\epsilon^2 T^2)^{\frac12}}=\frac{\lambda-2\epsilon\mu\tan\chi}{(1+4\epsilon^2\tan^2\chi)^{\frac12}},$$

$$\bar{\mu}=\frac{\mu c+\lambda\epsilon T}{(c^2+\epsilon^2 T^2)^{\frac12}}=\frac{\mu+2\epsilon\lambda\tan\chi}{(1+4\epsilon^2\tan^2\chi)^{\frac12}},$$

so that $\bar{\lambda}$ and $\bar{\mu}$ are direction-cosines of the tangent to the parabolic arc at some point between $P$ and $Q$; then,

$$
\begin{aligned}
I_{QP}&=\sigma F(x,y,\bar{\lambda},\bar{\mu})\\
&=\sigma\left\{\bar{\lambda}\,\frac{\partial F(x,y,\bar{\lambda},\bar{\mu})}{\partial\bar{\lambda}}+\bar{\mu}\,\frac{\partial F(x,y,\bar{\lambda},\bar{\mu})}{\partial\bar{\mu}}\right\}.
\end{aligned}
$$

The parabolic arc $\sigma$ is longer than $D$, the straight line $PQ$; so we may take $\sigma=(1+\theta)\,D$, where $\theta$ is positive; and we have

$$
\begin{aligned}
E'=&\,\bar{\lambda}\,\frac{\partial F(x,y,\bar{\lambda},\bar{\mu})}{\partial\bar{\lambda}}+\bar{\mu}\,\frac{\partial F(x,y,\bar{\lambda},\bar{\mu})}{\partial\bar{\mu}}\\
&-\frac{1}{1+\theta}\left\{l\,\frac{\partial F(x,y,x',y')}{\partial x'}+m\,\frac{\partial F(x,y,x',y')}{\partial y'}\right\}.
\end{aligned}
$$

It is easy to verify that, when

$$F(x, y, x_1, y_1) = f(x, y) (x_1{}^2 + y_1{}^2)^{\frac{1}{2}},$$

the value of $E'$ is

$$f(x, y) \left\{ 1 - \frac{\cos \omega}{1 + \theta} \right\},$$

where $QPX = \omega$; and $E'$ thus is necessarily positive for all values of $\omega$.

*Note.* In the figure, the parabolic arc is drawn within the angle $QPX$. It can be drawn without that angle; the only difference then is that $a$ is greater than $\phi$, a change which does not affect $E'$.

### *Strong variations, when derivatives of the second order occur.*

**222.** When we deal with integrals, still involving only one dependent variable but now involving its derivative of the second order, the corresponding expression for its change under strong variations can be obtained, though the form is more complicated and less convenient than the Weierstrass $E$-function.

As in the preceding investigation (§§ 219, 220), we take the ultimate constituent of the strong variations to be composed of (i) a variation from the characteristic curve to a consecutive characteristic, and (ii) a small arbitrary arc which is continuous in direction and curvature. The integral $I_{QP}$ has to be evaluated along this small arc. The selection of a straight line, as this small arc, is precluded from consideration; for $x_1 y_2 - y_1 x_2 = 0$ along a straight line,

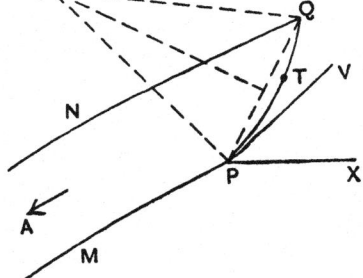

and such a relation would destroy the character of the integrand in the variation. The simplest admissible arc of non-zero curvature is a circle; and some variety in the assumption is possible through varied values of the radius. Accordingly, let $PTQ$ be a circular arc, radius $c$, and centre $O$; let $POQ = 2a$, $D = PQ$, so that

$$D = 2c \sin a.$$

The tangent to the circle at $P$ is $PV$; we take $QPX = \phi$. Then if $TOP = \theta$, so that $\theta$ varies from 0 to $2a$ as we pass from $P$ to $Q$, and if $X$ and $Y$ are the coordinates of $T$,

$$X = -c \sin (\phi - a) + c \sin (\phi - a + \theta), \quad Y = c \cos (\phi - a) - c \cos (\phi - a + \theta);$$

hence, if $S$ denote the arc $PT$, so that $S = c\theta$, we have

$$\frac{dX}{dS} = \cos (\phi - a + \theta), \qquad \frac{dY}{dS} = \sin (\phi - a + \theta),$$

$$\frac{d^2X}{dS^2} = -\frac{1}{c} \sin (\phi - a + \theta), \qquad \frac{d^2Y}{dS^2} = \frac{1}{c} \cos (\phi - a + \theta).$$

Also, along the circular arc $QT$, the direction in which the integral $I_{QP}$ is taken being from $Q$ to $P$, we have

$$\frac{dX}{dS'} = -\frac{dX}{dS}, \quad \frac{dY}{dS'} = -\frac{dY}{dS}, \quad \frac{d^2X}{dS^2} = \frac{d^2X}{dS'^2}, \quad \frac{d^2Y}{dS^2} = \frac{d^2Y}{dS'^2}.$$

Thus

$$I_{QP} = \int_Q^P G\left(x+X, \; y+Y, \; -\frac{dX}{dS}, \; -\frac{dY}{dS}, \; \frac{d^2X}{dS^2}, \; \frac{d^2Y}{dS^2}\right) dS'.$$

When we change the directions of the variations from $P$, so that $l, m$ are the direction-cosines of the line $QP$,

$$\kappa \xi = -lD, \quad \kappa \eta = -mD,$$

then

$$I_{QTP} = \int G\left(x+X, \; y+Y, \; \frac{dX}{dS}, \; \frac{dY}{dS}, \; \frac{d^2X}{dS^2}, \; \frac{d^2Y}{dS^2}\right) dS'$$

$$= 2c\alpha \, G\left(x+\bar{X}, \; y+\bar{Y}, \; \frac{\overline{dX}}{dS}, \; \frac{\overline{dY}}{dS}, \; \frac{\overline{d^2X}}{dS^2}, \; \frac{\overline{d^2Y}}{dS^2}\right),$$

where the barred symbols denote the respective values at some place between $P$ and $Q$ giving a mean value to the integral $G$; so that, if $\bar{\theta}$ be the value of $\theta$ for this mean place, $\bar{\theta} = \epsilon \cdot 2\alpha$, where $0 < \epsilon < 1$. Now $2c\alpha$ is the length of the circular arc, say $\sigma$; and

$$G\left(x+\bar{X}, \; y+\bar{Y}, \; \frac{\overline{dX}}{dS}, \; \frac{\overline{dY}}{dS}, \; \frac{\overline{d^2X}}{dS^2}, \; \frac{\overline{d^2Y}}{dS^2}\right)$$

$$= G\left(x, \; y, \; \frac{\overline{dX}}{dS}, \; \frac{\overline{dY}}{dS}, \; \frac{\overline{d^2X}}{dS^2}, \; \frac{\overline{d^2Y}}{dS^2}\right) + [\sigma],$$

where $[\sigma]$ represents an aggregate of terms of the first and higher degrees in $\sigma$. Hence, up to the first order of small quantities,

$$I_{QTP} = 2c\alpha \, G\left(x, \; y, \; \frac{\overline{dX}}{dS}, \; \frac{\overline{dY}}{dS}, \; \frac{\overline{d^2X}}{dS^2}, \; \frac{\overline{d^2Y}}{dS^2}\right).$$

Here the function $G$, on the right-hand side, is the value of the subject of integration $G(x, y, x', y', x'', y'')$, when the arc of the characteristic curve is taken to be the independent variable, and when the values of

$$\frac{\overline{dX}}{dS}, \quad \frac{\overline{dY}}{dS}, \quad \frac{\overline{d^2X}}{dS^2}, \quad \frac{\overline{d^2Y}}{dS^2},$$

at some mean place on the circle, given by

$$\cos(\phi - \alpha + \bar{\theta}), \quad \sin(\phi - \alpha + \bar{\theta}), \quad -\frac{1}{c}\sin(\phi - \alpha + \bar{\theta}), \quad \frac{1}{c}\cos(\phi - \alpha + \bar{\theta}),$$

are substituted for $x', y', x'', y''$ respectively. Thus if we write

$$\lambda = \cos(\phi - \alpha + \bar{\theta}), \quad \mu = \sin(\phi - \alpha + \bar{\theta}),$$

we have

$$I_{QTP} = 2c\alpha \, G\left(x, \; y, \; \lambda, \; \mu, \; \frac{-\mu}{c}, \; \frac{\lambda}{c}\right) = D\frac{\sin^{-1}\dfrac{D}{2c}}{\dfrac{D}{2c}}\bar{G}.$$

*The form of the E-function, when second derivatives occur.*

**223.** Next, as in § 98, we have

$$I_{ANQ} - I_{AMP} = \kappa \left[ u \left\{ \frac{\partial G}{\partial x_1} - \frac{d}{dt} \left( \frac{\partial G}{\partial x_2} \right) \right\} + u_1 \frac{\partial G}{\partial x_2} + v \left\{ \frac{\partial G}{\partial y_1} - \frac{d}{dt} \left( \frac{\partial G}{\partial y_2} \right) \right\} + v_1 \frac{\partial G}{\partial y_2} \right]_A^P,$$

taken at the limits $A$ and $P$.

At $A$, we have $u = 0$, $v = 0$; and the curve $AQ$ is consecutive to the curve $AP$, so that there

$$\frac{v_1}{u_1} = \frac{y_1}{x_1}.$$

Hence, at $A$,

$$u_1 \frac{\partial G}{\partial x_2} + v_1 \frac{\partial G}{\partial y_2} = \frac{x_1 u_1 + y_1 v_1}{x_1^2 + y_1^2} \left( x_1 \frac{\partial G}{\partial x_2} + y_1 \frac{\partial G}{\partial y_2} \right) = 0,$$

owing to the identity satisfied by $G$ everywhere. Thus the terms contributed to $I_{ANQ} - I_{AMP}$ at the lower limit vanish.

At the upper limit $P$, we have

$$u = \xi, \quad v = \eta,$$

and

$$\kappa \xi = -lD = -D \cos \phi, \quad \kappa \eta = -mD = -D \sin \phi.$$

As the integrals are invariantive (§ 104) for change of the independent variable, we make the arc-length along the characteristic curve the independent variable for the curve, so that $t = s$, $x_1 = x'$, $x_2 = x''$, $y_1 = y'$, $y_2 = y''$. Owing to the identity already quoted, which now has the form

$$x' \frac{\partial G}{\partial x''} + y' \frac{\partial G}{\partial y''} = 0,$$

there exists a quantity $K$ such that*

$$\frac{\partial G}{\partial x''} = -Ky', \quad \frac{\partial G}{\partial y''} = Kx';$$

and therefore

$$\kappa \xi' \frac{\partial G}{\partial x''} + \kappa \eta' \frac{\partial G}{\partial y''} = K \left\{ \cos \psi \frac{d}{ds} (D \sin \phi) - \sin \psi \frac{d}{ds} (D \cos \phi) \right\}$$

$$= K \left\{ \frac{dD}{ds} \sin (\phi - \psi) + D \frac{d\phi}{ds} \cos (\phi - \psi) \right\},$$

where $\psi$ is the inclination of the tangent at $P$ to the characteristic curve. Here the magnitude $D$ is an arbitrary small quantity at our choice; and $\phi$ is an arbitrary direction, capable of all variations round $P$, wherever $P$ be chosen. When any value of $D$, necessarily small, and any value of $\phi$, which has any

---

* This quantity $K$ is equal to $\frac{1}{x'^2} \frac{\partial F}{\partial q}$, when the integral is taken in the form $\int F(x, y, p, q) \, dx$; see § 101.

range, are chosen, the consecutive curve $ANQ$ is uniquely determinate (§ 123) the values of $D$ and $\phi$ entering into its determination. Thus we can take

$$\frac{dD}{ds} = 0, \quad \frac{d\phi}{ds} = 0;$$

and, with these assumptions,

$$I_{ANQ} - I_{AMP} = -D\left[ l\left\{ \frac{\partial G}{\partial x'} - \frac{d}{ds}\left( \frac{\partial G}{\partial x''} \right) \right\} + m\left\{ \frac{\partial G}{\partial y'} - \frac{d}{ds}\left( \frac{\partial G}{\partial y''} \right) \right\} \right].$$

Hence, collecting the results, and denoting by $E_c$ the change in the integral due to the strong variation of $AMP$ into $ANQP$, we have

$$E_c = I_{ANQ} + I_{QP} - I_{AMP}$$

$$= D\left[ \frac{\sin^{-1}\dfrac{D}{2c}}{\dfrac{D}{2c}} G\left( x, y, \lambda, \mu, \frac{-\mu}{c}, \frac{\lambda}{c} \right) - l\left\{ \frac{\partial G}{\partial x'} - \frac{d}{ds}\left( \frac{\partial G}{\partial x''} \right) \right\} - m\left\{ \frac{\partial G}{\partial y'} - \frac{d}{ds}\left( \frac{\partial G}{\partial y''} \right) \right\} \right],$$

which, for the present problem, is a function corresponding to the Weierstrass $E$-function.

In $E_c$, the quantity $D$ is an arbitrary length, while $l$ and $m$ represent an arbitrary direction. The radius $c$ of the circular arc is arbitrary; but the arc requires that $c \geqslant \frac{1}{2}D$. The quantities $\lambda$ and $\mu$ represent a mean direction along the arc $PTQ$; and they will not differ largely from the direction-cosines $(l, m)$ of the chord joining the arc. Finally, in the terms multiplied by $l$ and $m$, the function $G$ is the value of $G(x, y, x', y', x'', y'')$ along the characteristic curve : that is, it is

$$G\left( x, \ y, \ x', \ y', \ \frac{-y'}{\rho}, \ \frac{x'}{\rho} \right),$$

where $\rho$ is the radius of curvature of the characteristic curve at the point $P$.

*Note.* For a more general arc $PTQ$, the analysis of § 220 can be used for the construction of the corresponding $E$-function; it need not be set out in detail at this stage.

*Strong variations of integrals, with two dependent variables and their first derivatives.*

**224.** The corresponding tests, when the integrals involve two dependent variables, can be constructed in the same manner for strong variations. We consider the same simplest types of strong variation as before, used for the construction of the Weierstrass $E$-test; any non-regular variation of that type can be compounded of an aggregate of simple non-regular variations, each of which possesses the particular character chosen (§ 209). In these elementary variations we substitute, for an arc of the characteristic twisted curve, an arc of a consecutive characteristic curve—such a consecutive curve can always be drawn (§ 177)—and a small arc, continuous in direction and curvature.

The simplest test, being the Weierstrass $E$-test for curves in space, is obtained when this small arc is a straight line: the integrals, under consideration, involving only derivatives of the dependent variable of the first order. Using the same figure as before (§ 222) save that, now, the characteristic curves $AMP$ and $ANQ$ are to be twisted curves, we have to consider the integral

$$I_{ANQ} + I_{QP} - I_{AMP},$$

the subject of integration being a function $g(x, y, z, x_1, y_1, z_1)$ which satisfies the identity

$$g = x_1 \frac{\partial g}{\partial x_1} + y_1 \frac{\partial g}{\partial y_1} + z_1 \frac{\partial g}{\partial z_1}.$$

Let the small variation from $P$ to $Q$ be represented by $\kappa\xi$, $\kappa\eta$, $\kappa\zeta$, where these quantities are the coordinates of $Q$ relative to $P$; and let $l$, $m$, $n$ be the direction-cosines of the straight line $QP$, measured from $Q$ towards $P$. Then, if $D$ is the length of $PQ$, we have

$$\kappa\xi = -lD, \quad \kappa\eta = -mD, \quad \kappa\zeta = -nD.$$

We now have, as in § 158,

$$I_{ANQ} - I_{AMP} = \kappa \left[ u \frac{\partial g}{\partial x_1} + v \frac{\partial g}{\partial y_1} + w \frac{\partial g}{\partial z_1} \right]_A^P + [\kappa^2],$$

with the usual signification for $[\kappa^2]$. At $A$, the quantities $u$, $v$, $w$ vanish, because $AQ$ is a consecutive characteristic through $A$; and at $P$, they are equal to $\xi$, $\eta$, $\zeta$ respectively. Thus

$$I_{ANQ} - I_{AMP} = -D\left( l \frac{\partial g}{\partial x_1} + m \frac{\partial g}{\partial y_1} + n \frac{\partial g}{\partial z_1} \right).$$

But it has been proved (§ 156) that $\dfrac{\partial g}{\partial x_1}$, $\dfrac{\partial g}{\partial y_1}$, $\dfrac{\partial g}{\partial z_1}$ are unaltered when any other independent variable is substituted for $t$. Therefore, using the arc-length $s$ of the characteristic curve as the independent variable, and still using $g$ to denote $g(x, y, z, x', y', z')$, we have

$$I_{ANQ} - I_{AMP} = -D\left( l \frac{\partial g}{\partial x'} + m \frac{\partial g}{\partial y'} + n \frac{\partial g}{\partial z'} \right).$$

### The E-function for skew curves.

**225.** We now require the integral $I_{QP}$, taken along the straight line $QP$ from $Q$ to $P$. Let $T$ be any point on $PQ$, such that $QT = \sigma'$: then $\sigma'$ ranges from 0 to $D$. If $X$, $Y$, $Z$ be the coordinates of $T$ relative to $P$,

$$X = -l(D - \sigma'), \quad Y = -m(D - \sigma'), \quad Z = -n(D - \sigma'),$$

so that

$$\frac{dX}{d\sigma'} = l, \quad \frac{dY}{d\sigma'} = m, \quad \frac{dZ}{d\sigma'} = n.$$

Along $QP$ take the arc-length to be the independent variable—as in § 156, the integral is invariantive whatever independent variable be chosen; hence

$$I_{QP} = \int_0^D g\,(x + X,\ y + Y,\ z + Z,\ l,\ m,\ n)\,d\sigma'$$

$$= \int_0^D \{g\,(x,\ y,\ z,\ l,\ m,\ n) + [D]\}\,d\sigma',$$

where $[D]$ represents an aggregate of terms of the first and higher degrees in $D$. When only the most important term in $I_{QP}$ is retained, that is, the term involving the first power of $D$, we have

$$I_{QP} = D g\,(x,\ y,\ z,\ l,\ m,\ n)$$

$$= D \left( l\,\frac{\partial \bar{g}}{\partial l} + m\,\frac{\partial \bar{g}}{\partial m} + n\,\frac{\partial \bar{g}}{\partial n} \right),$$

where $\bar{g}$ denotes $g\,(x,\ y,\ z,\ l,\ m,\ n)$.

If now we write

$$I_{ANQ} + I_{QP} - I_{AMP} = DE\,(x,\ y,\ z,\ x',\ y',\ z',\ l,\ m,\ n) = DE,$$

where $E$ measures the variation of the integral for the strong small variation adopted, we have

$$E = l\left(\frac{\partial \bar{g}}{\partial l} - \frac{\partial g}{\partial x'}\right) + m\left(\frac{\partial \bar{g}}{\partial m} - \frac{\partial g}{\partial y'}\right) + n\left(\frac{\partial \bar{g}}{\partial n} - \frac{\partial g}{\partial z'}\right),$$

where $\bar{g}$ and $g$ denote $g\,(x,\ y,\ z,\ l,\ m,\ n)$ and $g\,(x,\ y,\ z,\ x',\ y',\ z')$ respectively, while $l,\ m,\ n$ denote any arbitrary direction at a point $x,\ y,\ z$ on the characteristic curve, and $x',\ y',\ z'$ denote the tangential direction of the curve at that point.

Now we can have a single elementary variation of the foregoing character, or any number of such single variations, or any simultaneous accumulation of them. If for each of them, taken for every point on the characteristic curve and for every direction through that point, the $E$-function is positive, it is positive for their aggregate: the value of the integral has been increased by the variation; and therefore, so far as this criterion is concerned, the integral is a minimum along the characteristic curve. If, on the other hand, the $E$-function is always negative in the like circumstances, the value of the integral has been decreased; and therefore, so far as the criterion is concerned, the integral is a maximum.

We thus obtain the Weierstrass test for integrals, involving originally two dependent variables and their first derivatives with regard to another variable initially regarded as independent:

*When the E-function is everywhere and always negative at a place on the characteristic curve, the integral can have a maximum; if the E-function similarly is positive, the integral can have a minimum.*

*Alternative integral form of the E-function.*

**226.** Another form can be given to the $E$-function, shewing its relation to the function in the Legendre test, and also the distinction between the Legendre test and the Weierstrass test.

Let

$$p = \theta l \ + (1 - \theta)\, x' = x' + \theta\, (l \ - x'),$$
$$q = \theta m + (1 - \theta)\, y' = y' + \theta\, (m - y'),$$
$$r = \theta n \ + (1 - \theta)\, z' = z' + \theta\, (n \ - z'),$$

so that $p$, $q$, $r$ constitute a combination varying, for values of $\theta$ from 0 to 1, between the combination $x'$, $y'$, $z'$ and the combination $l$, $m$, $n$; they therefore are proportional to the direction-cosines of an inclination between $x'$, $y'$, $z'$— that of the tangent to the curve—and $l$, $m$, $n$—the arbitrary direction off the tangent. Now for any function $F$ of $p$, $q$, $r$, we have

$$F\,(l,\, m,\, n) - F\,(x',\, y',\, z') = \int_{x',\,y',\,z'}^{l,\,m,\,n} \left(\frac{\partial F}{\partial p}\, dp + \frac{\partial F}{\partial q}\, dq + \frac{\partial F}{\partial r}\, dr\right)$$
$$= \int_0^1 \left\{\frac{\partial F}{\partial p}\,(l - x') + \frac{\partial F}{\partial q}\,(m - y') + \frac{\partial F}{\partial r}\,(n - z')\right\}\, d\theta \,;$$

and this relation will be used to transform the three terms in $E$.

First, we take

$$F\,(p,\, q,\, r) = \frac{\partial}{\partial p}\, g\,(x,\, y,\, z,\, p,\, q,\, r)\,;$$

the subject of integration on the right-hand side becomes

$$(l - x')\frac{\partial^2 g}{\partial p^2} + (m - y')\frac{\partial^2 g}{\partial q \partial p} + (n - z')\frac{\partial^2 g}{\partial r \partial p}\,.$$

In connection with the six second derivatives of $g$ with respect to $p$, $q$, $r$, there are six quantities $P\,(x,\, y,\, z,\, p,\, q,\, r)$, $R\,(x,\, y,\, z,\, p,\, q,\, r)$, $Q\,(x,\, y,\, z,\, p,\, q,\, r)$, say $P_\theta$, $R_\theta$, $Q_\theta$, when the foregoing values of $p$, $q$, $r$ are substituted, such that (§ 162)

$$\frac{\partial^2 g}{\partial p^2} = q^2 P_\theta + 2qr\, R_\theta + r^2 Q_\theta, \quad \frac{\partial^2 g}{\partial p \partial q} = -pq\, P_\theta - pr\, R_\theta, \quad \frac{\partial^2 g}{\partial p \partial r} = -pq\, R_\theta - pr\, Q_\theta.$$

Thus the subject of integration is equal to

$$P_\theta\,\{(l - x')\, q^2 - (m - y')\, pq\} + R_\theta\,\{2qr\,(l - x') - pr\,(m - y') - pq\,(n - z')\}$$
$$+ Q_\theta\,\{(l - x')\, r^2 - (n - z')\, pr\}\,;$$

but

$$(l - x')\, q - (m - y')\, p = ly' - mx',$$
$$(l - x')\, r - \,(n - z')\, p = lz' - nx',$$

on substitution for $p$ and $q$. Hence

$$\frac{\partial \bar{g}}{\partial l} - \frac{\partial g}{\partial x'} = (ly' - mx') \int_0^1 (P_\theta q + R_\theta r)\, d\theta + (lz' - nx') \int_0^1 (R_\theta q + Q_\theta r)\, d\theta.$$

Similarly, when we take

$$F(p,\, q,\, r) = \frac{\partial}{\partial q}\, g\, (x,\, y,\, z,\, p,\, q,\, r),$$

we find

$$\frac{\partial \bar{g}}{\partial m} - \frac{\partial g}{\partial y'} = -(ly' - mx') \int_0^1 P_\theta p\, d\theta - (lz' - nx') \int_0^1 R_\theta p\, d\theta;$$

and when we take

$$F(p,\, q,\, r) = \frac{\partial}{\partial r}\, g\, (x,\, y,\, z,\, p,\, q,\, r),$$

we find

$$\frac{\partial \bar{g}}{\partial n} - \frac{\partial g}{\partial z'} = -(ly' - mx') \int_0^1 R_\theta p\, d\theta - (lz' - nx') \int_0^1 Q_\theta p\, d\theta.$$

Hence

$$E = l \left( \frac{\partial \bar{g}}{\partial l} - \frac{\partial g}{\partial x'} \right) + m \left( \frac{\partial \bar{g}}{\partial m} - \frac{\partial g}{\partial y'} \right) + n \left( \frac{\partial \bar{g}}{\partial n} - \frac{\partial g}{\partial z'} \right)$$

$$= (ly' - mx') \int_0^1 \{l\, (P_\theta q + R_\theta r) - mP_\theta p - nR_\theta p\}\, d\theta$$

$$+ (lz' - nx') \int_0^1 \{l\, (R_\theta q + Q_\theta r) - mR_\theta p - nQ_\theta p\}\, d\theta.$$

Again,

$$lq - mp = (1 - \theta)(ly' - mx'), \quad lr - np = (1 - \theta)(lz' - nx');$$

and therefore

$$E = (ly' - mx')^2 \int_0^1 (1 - \theta) P_\theta d\theta + (lz' - nx')^2 \int_0^1 (1 - \theta) Q_\theta d\theta$$

$$+ 2 (ly' - mx')(lz' - nx') \int_0^1 (1 - \theta) R_\theta d\theta.$$

Let $P_\epsilon,\, R_\epsilon,\, Q_\epsilon$, where $\epsilon$ is a real quantity lying between 0 and 1, denote mean values in these integrals such that

$$\int_0^1 (1 - \theta) P_\theta d\theta = \tfrac{1}{2} P_\epsilon, \quad \int_0^1 (1 - \theta) R_\theta d\theta = \tfrac{1}{2} R_\epsilon, \quad \int_0^1 (1 - \theta) Q_\theta d\theta = Q_\epsilon;$$

because $P_\theta,\, R_\theta,\, Q_\theta$ are functions of $\theta$, the quantity $\epsilon$ is not zero, so that $P_\epsilon,\, R_\epsilon,\, Q_\epsilon$ are not the quantities $P,\, R,\, Q$ along the curve. Thus

$$E = \tfrac{1}{2} \{P_\epsilon\, (ly' - mx')^2 + 2R_\epsilon\, (ly' - mx')(lz' - nx') + Q_\epsilon\, (lz' - nx')^2\}.$$

One further modification may be made in this expression. Let $\chi$ be the angle between the directions $l,\, m,\, n$, and $x',\, y',\, z'$; so that $\epsilon$, as substituted for $\theta$ in the expressions for $p,\, q,\, r$, gives an inclination to the direction $x',\, y',\, z'$, lying between 0 and $\chi$. Then

$$(ly' - mx') + (mz' - ny')^2 + (nx' - lz')^2 = \sin^2 \chi,$$

so that we have

$$mz' - ny' = \sin \chi \sin \phi,$$
$$ly' - mx' = \sin \chi \cos \phi \cos \xi,$$
$$nx' - lz' = \sin \chi \cos \phi \sin \xi;$$

and now

$$E = \tfrac{1}{2} \sin^2 \chi \cos^2 \phi \, (P_\epsilon \cos^2 \xi - 2R_\epsilon \cos \xi \sin \xi + Q_\epsilon \sin^2 \xi),$$

which may be taken as the definite form of $E$.

The function $E$ vanishes if $\chi = 0$: that is, if the direction $l$, $m$, $n$ coincides with $x'$, $y'$, $z'$, so that there is no variation off the curve. It vanishes if $\phi = \tfrac{1}{2}\pi$, which can occur only in the same event. It can have a uniform sign for all values of $\xi$, that is, for all directions $l$, $m$, $n$, only if

$$P_\epsilon Q_\epsilon > R_\epsilon^2,$$

a condition which requires $P_\epsilon$ and $Q_\epsilon$ to have the same sign. Let this condition be satisfied. When the common sign of $P_\epsilon$ and $Q_\epsilon$ is positive, $E$ is positive, and a minimum is secured, so far as this test is concerned. When the common sign is negative, $E$ is negative and, correspondingly, a maximum is secured.

Hence *the E-function test requires that*

$$P_\epsilon Q_\epsilon > R_\epsilon^2:$$

*with $P_\epsilon > 0$ and $Q_\epsilon > 0$ for a minimum, and $P_\epsilon < 0$ and $Q_\epsilon < 0$ for a maximum; where $P_\epsilon$, $R_\epsilon$, $Q_\epsilon$ are the values of $P$, $R$, $Q$, when quantities $x' + \epsilon\lambda$, $y' + \epsilon\mu$, $z' + \epsilon\nu$, are substituted for $x'$, $y'$, $z'$, and $0 < \epsilon < 1$; and $\lambda$, $\mu$, $\nu$ are quantities such that $\lambda^2 + \mu^2 + \nu^2 = 4 \sin^2 \tfrac{1}{2}\chi$, where $0 < \chi < 2\pi$.*

*Note.* It is obvious that this $E$-function test is distinct from the Legendre test, arising through weak variations. The Legendre test affects the quantities $P$, $Q$, $R$ for directions *along* the curve. The $E$-function test becomes significant only for directions such that $\sin \chi$ is definitely not zero. That is, the Weierstrass $E$-function test affects the quantities $P$, $Q$, $R$ for directions *off* the curve; it is therefore distinct from the Legendre test.

*Ex.* 1.  The $E$-function test is satisfied for all integrals of the form

$$\int f(x, y, z) \, ds, \quad = \int f(x, y, z) \, (x_1^2 + y_1^2 + z_1^2)^{\tfrac{1}{2}} \, dt.$$

For

$$\frac{\partial g}{\partial x_1} = f(x, y, z) \, \frac{x_1}{(x_1^2 + y_1^2 + z_1^2)^{\tfrac{1}{2}}} = f(x, y, z) \, x',$$

and so

$$\frac{\partial \bar{g}}{\partial l} = f(x, y, z) \, l.$$

Similarly for the other derivatives of $\bar{g}$. Thus

$$E = f(x, y, z) \, \{1 - (lx' + my' + nz')\}$$
$$= f(x, y, z) \, (1 - \cos \chi),$$

which is always positive for a strong variation in a direction not coinciding with $x'$, $y'$, $z'$; that is, the Weierstrass test, so far as it is concerned, admits a minimum.

*Ex.* 2. The arc $PQ$, in the investigation in § 225, has been made a straight line.

When, for integrals $\int f(x, y, z)\, ds$, it is made any arc of continuous direction and curvature between $P$ and $Q$, and of length $\sigma$, the variation of the integral becomes

$$f(x, y, z)\, \sigma - Df(x, y, z)\, (lx' + my' + nz')$$
$$= f(x, y, z)\, (\sigma - D \cos \chi),$$

which is always positive because $\sigma > D$.

Hence, for the integrals cited, the condition for a minimum is satisfied, in the case of all strong variations that can be represented by any continuous line, however broken in direction and curvature.

*Ex.* 3. In the general investigation for any integral $\int g\, dt$, the arc $PQ$ has been taken a straight line as manifestly the simplest form.

Other forms of arcs can be chosen: either quite general in the same manner as for a rectilinear arc, already considered; or special as for a parabolic arc in § 221, Ex. 3, or a circular arc in § 222. But no form of arc other than a straight line leads to a test that admits of easy application.

Thus we might make the arc a helix with both curvatures constant; and varieties arise according to the orientation of the circular cylinder. Limiting the illustration to the case when the axis of the cylinder is parallel to the axis of $z$ in the general investigation, we take $a$ to be the radius of the cylinder and $a$ the slope of the helix. The arbitrary direction of the small variation at $P$ being $l$, $m$, $n$, we take

$$l = \cos \phi \cos \theta, \quad m = \cos \phi \sin \theta, \quad n = \sin \phi,$$

so that $lD$, $mD$, $nD$ are the coordinates $X$, $Y$, $Z$ of $Q$ with respect to $P$: then

$$X = lD = a \cos \theta, \quad Y = mD = a \sin \theta, \quad Z = nD = a\theta \tan a,$$

and, if $\sigma$ is the length of the arc $PQ$,

$$\sigma = a\theta \sec a.$$

Let $\vartheta$ be a current coordinate along the helix, ranging from $0$ to $\theta$ as we pass along the arc from $P$ to $Q$: at any point on the arc, the coordinates relative to $P$ are

$$a \cos \vartheta, \quad a \sin \vartheta, \quad a\vartheta \tan a,$$

while the direction-cosines at that point are

$$-\cos a \sin \vartheta, \quad \cos a \cos \vartheta, \quad \sin a.$$

Then, as in the text,

$$I_{QP} = \int_0^\theta g\, (x + a \cos \vartheta, \; y + a \sin \vartheta, \; z + a\vartheta \tan a, \; -\cos a \sin \vartheta, \; \cos a \cos \vartheta, \; \sin a) \frac{a\, d\vartheta}{\cos a}$$
$$= \frac{a\theta}{\cos a}\, g\, (x, y, z, \; -\cos a \sin \overline{\theta}, \; \cos a \cos \overline{\theta}, \; \sin a) + [D^2],$$

with the usual significance of $[D^2]$, while $\overline{\theta} = \epsilon\theta$, where $\epsilon$ is a real constant such that $0 < \epsilon < 1$.

But we have

$$\sin a = \frac{\tan \phi}{(\theta^2 + \tan^2 \phi)^{\frac{1}{2}}}, \quad \cos a = \frac{\theta}{(\theta^2 + \tan^2 \phi)^{\frac{1}{2}}},$$
$$a = D \cos \phi, \quad \sigma = D\, (\theta^2 \cos^2 \phi + \sin^2 \phi)^{\frac{1}{2}};$$

and so the variation of the integral is

$$D\, (\theta^2 \cos^2 \phi + \sin^2 \phi)^{\frac{1}{2}}\, g \left\{ x, y, z, \; \frac{-\theta \sin \overline{\theta}}{(\theta^2 + \tan^2 \phi)^{\frac{1}{2}}}, \; \frac{\theta \cos \overline{\theta}}{(\theta^2 + \tan^2 \phi)^{\frac{1}{2}}}, \; \frac{\tan \phi}{(\theta^2 + \tan^2 \phi)^{\frac{1}{2}}} \right\}$$
$$- D \left( \cos \phi \cos \theta \frac{\partial g}{\partial x'} + \cos \phi \sin \theta \frac{\partial g}{\partial y'} + \sin \phi \frac{\partial g}{\partial z'} \right).$$

If, as before, we write

$$\frac{\partial g}{\partial x'} = g_1\,(x,\,y,\,z,\,x',\,y',\,z'),$$

and similarly for $\frac{\partial g}{\partial y'}$, $\frac{\partial g}{\partial z'}$; then the variation of the integral can be expressed in the form

$$D\,[-\theta\sin\overline{\theta}\;g_1\,(x,\,y,\,z,\,-\theta\sin\overline{\theta},\;\theta\cos\overline{\theta},\;\tan\phi) - \cos\theta\;g_1\,(x,\,y,\,z,\,x',\,y',\,z')]\cos\phi$$

$$+\,D\,[\,\theta\cos\overline{\theta}\;g_2\,(x,\,y,\,z,\,-\theta\sin\overline{\theta},\;\theta\cos\overline{\theta},\;\tan\phi) - \sin\theta\;g_2\,(x,\,y,\,z,\,x',\,y',\,z')]\cos\phi$$

$$+\,D\,[g_3\,(x,\,y,\,z,\,-\theta\sin\overline{\theta},\;\theta\cos\overline{\theta},\;\tan\phi) - g_3\,(x,\,y,\,z,\,x',\,y',\,z')]\sin\phi,$$

where $\overline{\theta} = \epsilon\theta$, while $0 < \epsilon < 1$; and $\theta$, $\phi$, are any arbitrary angles determining the direction of the chord joining the extremities of the arbitrary helical arc.

### *Omission of E-function, when integrals involve curvatures of curves in space.*

**227.** The last example shews that the types of expression, as given, are unwieldy in general effect.

Accordingly, we shall not proceed to the corresponding expression for the *E*-function of an integral which, containing two original dependent variables *y* and *z* and their derivatives of the second order, will require consideration of the circular curvature of the curve, of its torsion, and of their variations; a linear arc *PQ* would be ineffective for the maintenance of the character of the subject of integration.

### *'Least' Action.*

**228.** Consider* the dogma often called the *Principle of Least Action*, originated by Maupertuis†.

Let *T* denote the kinetic energy of a conservative system in motion, and let *V* denote its potential energy. Throughout the motion,

$$T + V = E,$$

---

* A slight reference to the Principle has already been made, in § 175 ; see also § 34, Ex. 6.

† Its author, who announced it in 1744, declared it to be a "metaphysical principle," on which all the canons of motion are based. In the preface to his "Essai de cosmologie" (*Œuvres*, Lyon, 1756, vol. i, p. xiv), he referred to a recent investigation by Euler, containing a proof of the principle for a body moving under a central force. A further statement (*l.c.*, p. 42) is as follows :

"L'action est proportionnelle au produit de la masse par la vitesse et par l'espace. Maintenant " voici ce principe, si sage, si digne de l'Être suprême : *Lorsqu'il arrive quelque changement dans* " *la Nature, la quantité d'action employée pour ce changement est toujours la plus petite qu'il soit* " *possible.*"

Thomson and Tait's statement of the Principle (*Natural Philosophy*, vol. i, 2nd ed., 1890, § 327) is:

"Of all the different sorts of paths along which a conservative system may be guided to move ": from one configuration to another, with the sum of its potential and kinetic energies equal to "a given constant, that one for which the action is the least is such that the system will require " only to be started with the proper velocities, to move along it unguided."

For a history of the Principle, down to the date of publication of his volume, reference may be made to A. Mayer, *Geschichte des Princips der kleinsten Action* (Leipzig, 1877).

where $E$ is a constant: an equation embodying the principle of the conservation of the energy.

The detailed motion of the system is determined by the equations

$$\frac{d}{dt}\left(\frac{\partial T}{\partial \dot{\theta}}\right) - \frac{\partial T}{\partial \theta} = -\frac{\partial V}{\partial \theta},$$

taken in Lagrange's form, for each of the coordinates $\theta$ which, being independent of one another, suffice for the specification of the configuration of the system at any moment.

Let $L$ denote the Hamilton function, defined by the relation

$$L = T - V;$$

then, as $V$ is a function of the internal configuration of the system alone and is independent of the velocities of the members, the equations are

$$\frac{\partial L}{\partial \theta} - \frac{d}{dt}\left(\frac{\partial L}{\partial \dot{\theta}}\right) = 0,$$

for each of the coordinates $\theta$.

From each element of mass $m$, moving with a velocity $v$ through a rudimentary arc $dt$, the contribution to the Action is measured by

$$mv\,ds,$$
$$= mv^2\,dt.$$

Therefore the whole of the Action for the system, in any given interval beginning at any arbitrary instant denoted by $t_1$ and ending at any other arbitrary instant $t_2$, is

$$\int_{t_1}^{t_2} (\Sigma mv^2)\,dt:$$

that is, it is equal to

$$\int_{t_1}^{t_2} 2T\,dt.$$

But

$$L = T - V = 2T - E;$$

and therefore the action, in the specified interval, is

$$\int_{t_1}^{t_2} L\,dt + (t_2 - t_1)\,E.$$

### Tests of Action, for weak variations.

**229.** (A) Consider, first, a fixed interval during which the system moves from one definite configuration to another definite configuration, so that the quantity $t_2 - t_1$ is definite; then $t_2 - t_1$ (as well as $E$) is free from change. The statement, that the Action is a minimum, requires that the integral

$$\int_{t_1}^{t_2} L\,dt$$

shall be a minimum for all small variations.

(I) The Euler-Lagrange test is provided by the characteristic equations, which (as already stated) are

$$\frac{\partial L}{\partial \theta} - \frac{d}{dt}\left(\frac{\partial L}{\partial \dot{\theta}}\right) = 0,$$

for each of the coordinates of the system. The test is satisfied.

(II) The Legendre test is concerned with the second derivatives of $L$ with respect to the quantities $\dot{\theta}$. If $\dot{\theta}$ and $\dot{\phi}$ denote any pair of these quantities, the test (in whatever form it is stated) requires that, for a minimum, the relations

$$\frac{\partial^2 L}{\partial \dot{\theta}^2} > 0, \quad \frac{\partial^2 L}{\partial \dot{\phi}^2} > 0, \quad \frac{\partial^2 L}{\partial \dot{\theta}^2}\frac{\partial^2 L}{\partial \dot{\phi}^2} > \left(\frac{\partial^2 L}{\partial \dot{\theta}\partial \dot{\phi}}\right)^2,$$

shall hold for every pair of independent variables. But

$$L = T - V,$$

and $V$ does not involve any of the time-derivatives $\dot{\theta}$; hence the requirement of the Legendre test is

$$\frac{\partial^2 T}{\partial \dot{\theta}^2} > 0, \quad \frac{\partial^2 T}{\partial \dot{\phi}^2} > 0, \quad \frac{\partial^2 T}{\partial \dot{\theta}^2}\frac{\partial^2 T}{\partial \dot{\phi}^2} > \left(\frac{\partial^2 T}{\partial \dot{\theta}\partial \dot{\phi}}\right)^2,$$

for every possible variation $\dot{\theta}$ and $\dot{\phi}$.

Now the kinetic energy of the system is an essentially positive quantity in any moving system. The magnitude $2T$ is equal to

$$\Sigma m\,(\dot{x}^2 + \dot{y}^2 + \dot{z}^2),$$

summed through all the elements of the system; and it is a homogeneous quadratic function of the time-derivatives of the coordinates, so that it is expressible in the form

$$A_{11}\dot{\theta}^2 + 2A_{12}\dot{\theta}\dot{\phi} + A_{22}\dot{\phi}^2 + \ldots,$$

where the quantities $A_{11}$, $A_{12}$, $A_{22}$, ... do not involve any of the derivatives*. As this quantity is always positive.

$$A_{nn} > 0, \quad A_{mm} > 0, \quad A_{mm}A_{nn} > A_{mn}^2,$$

for all combinations $m$, $n = 1$, $2$, .... The Legendre test is satisfied.

(III) The Jacobi test requires that the range of variation (in the present instance, the time) shall not extend so far as to allow a set of the variables, agreeing in value at an initial stage with an assigned set and currently deduced from them by small variations, to agree simultaneously at a later stage with the values of that assigned set at the same later stage: all the sets of the variables satisfying the characteristic equations. More briefly stated, the Jacobi test requires that, for the admission of a minimum action,

---

* In the case of a rigid body of mass $M$, and principal moments $A$, $B$, $C$, the quantity $2T$ is (Routh, *Dynamics of a system of rigid bodies*, I, § 365)

$$M\,(\dot{x}^2 + \dot{y}^2 + \dot{z}^2) + C\,(\dot{\phi} + \dot{\psi}\cos\theta)^2 + 2\theta\dot{\psi}\,(B - A)\sin\theta\sin\phi\cos\phi$$
$$+ \theta^2\,(A\sin^2\phi + B\cos^2\phi) + \dot{\psi}^2\,(A\cos^2\phi + B\sin^2\phi)\sin^2\theta.$$

The verification of the Legendre inequalities is immediate.

its duration shall not be long enough to admit the arrival of a configuration, which is the *conjugate* of an initial configuration.

The same exclusion is made (though not expressed in the statement of the Principle) by a restriction within a range bounded by *kinetic foci*\*, in the vocabulary of kinetics. When this restriction is observed, the Jacobi test is satisfied.

(B) The establishment of the characteristic equations for a minimum, under weak variations, of the Action in a conservative field of force can be effected in a different manner, and without the time-restriction in (A).

In order that the Action

$$\int 2T dt$$

may be a minimum, under the Thomson and Tait postulation, that minimum of the integral is to be possessed subject to the condition

$$T + V = E = \text{constant.}$$

Anticipating the results of §§ 265—267, we take the characteristic equations in the form

$$\frac{d}{dt}\left[ \frac{\partial}{\partial \dot{\theta}} \{2T - \lambda (T + V - E)\} \right] - \frac{\partial}{\partial \theta} \{2T - \lambda (T + V - E)\} = 0,$$

for each of the independent coordinates $\theta$ which determine the configuration; and $\lambda$ is a multiplier which, if not a constant, is a function of $t$. Now $E$ is constant. In a conservative field, the expression for $V$ is independent of the time and the velocities; thus the typical characteristic equation is

$$\frac{d}{dt}\left\{ (2 - \lambda) \frac{\partial T}{\partial \dot{\theta}} \right\} - \frac{\partial}{\partial \theta} \{(2 - \lambda) T - \lambda V\} = 0:$$

that is,

$$\frac{d}{dt}\left\{ (2 - \lambda) \frac{\partial T}{\partial \dot{\theta}} \right\} - (2 - \lambda) \frac{\partial T}{\partial \theta} + \lambda \frac{\partial V}{\partial \theta} = 0.$$

Hence

$$(2 - \lambda)\left\{ \frac{d}{dt}\left( \frac{\partial T}{\partial \dot{\theta}} \right) - \frac{\partial T}{\partial \theta} \right\} + \lambda \frac{\partial V}{\partial \theta} - \frac{dT}{d\dot{\theta}} \frac{d\lambda}{dt} = 0.$$

The actual equations (Lagrange's) of motion are

$$\frac{d}{dt}\left( \frac{\partial T}{\partial \dot{\theta}} \right) - \frac{\partial T}{\partial \theta} = -\frac{\partial V}{\partial \theta};$$

and therefore the typical characteristic equation becomes

$$(2\lambda - 2) \frac{\partial V}{\partial \theta} = \frac{\partial T}{\partial \dot{\theta}} \frac{d\lambda}{dt}.$$

Multiply by $\dot{\theta}$, and add the resulting equations; then, as

$$\frac{dV}{dt} = \Sigma \dot{\theta} \frac{\partial V}{\partial \theta}, \quad \Sigma \dot{\theta} \frac{\partial T}{\partial \dot{\theta}} = 2T,$$

---

\* Thomson and Tait, *Natural Philosophy*, vol. i, § 357.

we have

$$(\lambda - 1)\frac{dV}{dt} = T\frac{d\lambda}{dt},$$

or, as $T + V = E = $ constant,

$$(\lambda - 1)\frac{dT}{dt} + T\frac{d\lambda}{dt} = 0,$$

and therefore

$$T(\lambda - 1) = constant.$$

Again, unless $\lambda = 1$, so that $\frac{d\lambda}{dt}$ is a zero constant, we have

$$\frac{\frac{\partial T}{\partial \dot{\theta}}}{\frac{\partial V}{\partial \theta}} = \frac{\frac{\partial T}{\partial \dot{\phi}}}{\frac{\partial V}{\partial \phi}} = \dots.$$

Now $\frac{\partial T}{\partial \dot{\theta}}, \frac{\partial T}{\partial \dot{\phi}}, \dots$ are linear in the velocities, so that these relations would not hold for a system throughout the motion, because they would provide a number of equations (not identities) among $\dot{\theta}, \dot{\phi}, \dots$. Hence we must have

$$\lambda = 1;$$

in the equation $T(\lambda - 1) = constant$, the last constant is zero. Thus the typical characteristic equation becomes

$$\frac{d}{dt}\left(\frac{\partial T}{\partial \dot{\theta}}\right) - \frac{\partial T}{\partial \theta} + \frac{\partial V}{\partial \theta} = 0,$$

satisfied for every coordinate of the system.

Hence *the characteristic equations for a minimum of the Action under weak variations are satisfied.* The other tests, arising out of the weak variations, are satisfied as before.

*Note.* It may be pointed out that, in the Thomson and Tait postulation of the Principle, the inclusion of the constancy of the whole energy is the importation of a property, which arises as a consequence of the equations of motion and which therefore is not an added condition. For, since

$$2T = \Sigma\dot{\theta}\frac{\partial T}{\partial \dot{\theta}},$$

we have

$$2\frac{dT}{dt} = \Sigma\left\{\ddot{\theta}\frac{\partial T}{\partial \dot{\theta}} + \dot{\theta}\frac{d}{dt}\left(\frac{\partial T}{\partial \dot{\theta}}\right)\right\}$$

$$= \Sigma\left\{\ddot{\theta}\frac{\partial T}{\partial \dot{\theta}} + \dot{\theta}\frac{\partial T}{\partial \theta} - \dot{\theta}\frac{\partial V}{\partial \theta}\right\},$$

because of the equations of motion. Hence

$$2\frac{dT}{dt} = \frac{dT}{dt} - \frac{dV}{dt};$$

and therefore

$$T + V = \text{constant}.$$

**230.** Still using another form of the last method and (for the sake of illustration) considering only a single particle of mass unity, which moves in the plane of $xy$ in a conservative field of force, we are to have

$$\int v \, ds$$

a minimum, subject to conditions

$$v \frac{dv}{ds} = X \frac{dx}{ds} + Y \frac{dy}{ds}, \quad \left(\frac{dx}{ds}\right)^2 + \left(\frac{dy}{ds}\right)^2 = 1.$$

Accordingly, with $s$ as the variable of reference, the subject of integration is $v$; and the equations of condition are

$$vv' - Xx' - Yy' = 0, \quad x'^2 + y'^2 - 1 = 0.$$

Therefore, for $v, x, y$ as three dependent variables, the characteristic equations for a minimum or a maximum are (§ 273)

$$\frac{d}{ds}(-\lambda v) - (1 - \lambda v') = 0,$$

$$\frac{d}{ds}(\lambda X - 2\mu x') - \lambda \left(x' \frac{\partial X}{\partial x} + y' \frac{\partial Y}{\partial x}\right) = 0,$$

$$\frac{d}{ds}(\lambda Y - 2\mu y') - \lambda \left(x' \frac{\partial X}{\partial y} + y' \frac{\partial Y}{\partial y}\right) = 0.$$

Also, if $V$ be the potential of the field of force,

$$X = -\frac{\partial V}{\partial x}, \quad Y = -\frac{\partial V}{\partial y},$$

so that

$$\frac{\partial X}{\partial y} = \frac{\partial Y}{\partial x}.$$

From the first equation, we have

$$v\lambda' = -1.$$

The second equation is

$$X\lambda' + \lambda \left(\frac{\partial X}{\partial x} x' + \frac{\partial X}{\partial y} y' - x' \frac{\partial X}{\partial x} - y' \frac{\partial Y}{\partial x}\right) - 2\mu x'' - 2\mu' x' = 0,$$

that is,

$$X\lambda' - 2\mu x'' - 2\mu' x' = 0;$$

and, similarly, the third equation is

$$Y\lambda' - 2\mu y'' - 2\mu' y' = 0.$$

Multiplying these relations by $x', y'$, and adding, we have

$$\lambda' vv' - 2\mu' = 0,$$

that is,

$$2\mu' + v' = 0;$$

so that we can satisfy the equations by taking $2\mu + v = 0$. (We only require some multiplier $\mu$ and some multiplier $\lambda$, to enable us to form the original characteristic equation.)  Again, multiplying by $x''$, $y''$, and adding, we have

$$\lambda'\,(Xx'' + Yy'') - 2\mu\,(x''^2 + y''^2) = 0,$$

or, substituting for $\lambda'$ and $2\mu$,

$$\frac{v^2}{\rho} = X\rho x'' + Y\rho y'',$$

shewing that, in any smooth groove along the contemplated path in the given field of force, the normal pressure is zero: that is, the path is free and not constrained.  Thus, along the free path, the Action satisfies the characteristic equations.

The other tests for weak variations apply as before.  The Action is a minimum under weak variations.

*Ex.  Consider the Action in the path of a heavy unit of mass moving freely.*

Let the particle be projected from a point $B$ in any direction, with a velocity due to the depth below a horizontal line $OMy$: it will describe a parabolic path $BVP...$, the directrix of the parabola being $Oy$.

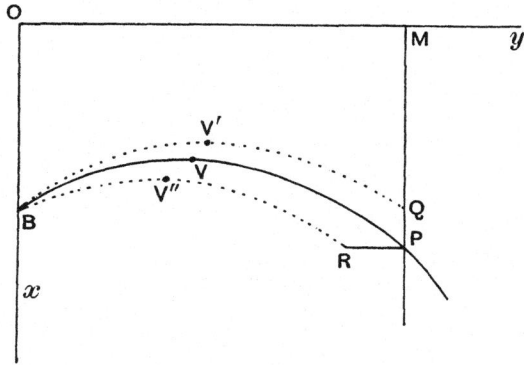

Consider two constrained paths from $B$ to $P$.  Let one of them be given by a contiguous parabola $BV'Q$, and a vertical element $QP$.  Let the other be given by another contiguous parabola $BV''R$, and a horizontal element $RP$.  (The vertices $V''$, $V$, $V'$ of all parabolas, having $OMy$ as directrix, lie on an ellipse of which $OB$ is the minor axis.)  Let $OB = a$, $OM = k$.

(i)  Let the latus rectum of $BVP$ be $4l$; the equation of the parabola, with $Ox$ and $Oy$ as axes, is

$$(y - c)^2 = 4l\,(x - l),$$

with the condition

$$c^2 = 4l\,(a - l).$$

As $l \leqslant a$, we take $l = a\cos^2 a$; then (assuming, merely for illustration, that $V$ lies between $B$ and $P$) we have $c = a\sin 2a$.  Any point on the curve is

$$x - l = l\tan^2\theta, \quad y - c = 2l\tan\theta,$$

so that $\theta$ ranges in increase from the value $-a$.  The velocity $v$ at any point is given by

$$v^2 = 2gx = 2ga\cos^2 a\sec^2\theta.$$

The element of arc $ds$ is equal to $\{(dx)^2 + (dy)^2\}^{\frac{1}{2}}$, that is, to

$$2a\cos^2 a\sec^3\theta\,d\theta.$$

Let $A$ denote the Action in the path from $B$ to $P$; then

$$A = \int_{-a}^{\phi} v \frac{ds}{d\theta} d\theta$$

$$= (8ga^3)^{\frac{1}{2}} (\tan \phi + \tfrac{1}{3} \tan^3 \phi + \tan a + \tfrac{1}{3} \tan^3 a) \cos^3 a.$$

(ii) Consider the constrained guiding path $BV'QP$. For this path, let $a + \epsilon_1$ and $\phi + \eta_1$ be the new values of $a$ and $\phi$, where $\epsilon_1$ is positive for $BV'Q$ as drawn. Because $k$, that is, $a (\sin 2a + 2 \cos^2 a \tan \phi)$, is the same for $BV'Q$ and $BVP$,

$$\eta_1 = - \epsilon_1 \frac{\cos \phi \cos (2a + \phi)}{\cos^2 a}.$$

Let $A'$ be the action for the portion $BV'Q$ of the constrained path, so that $A'$ is the same function of $a + \epsilon_1$ and $\phi + \eta_1$, as $A$ is of $a$ and $\phi$. On reduction, we find

$$\frac{A' - A}{(2ga^3)^{\frac{1}{2}}} = - \epsilon_1 \frac{\cos a \sin \phi}{\cos^3 \phi} \sin (2a + 2\phi).$$

The portion $\overrightarrow{PQ}$ is the variation of $PM\ (= X)$, that is, of $a \cos^2 a \sec^2 \phi$; thus

$$dX = - a\epsilon_1 \sin (2a + 2\phi) \sec^2 \phi.$$

The element of Action along $\overrightarrow{QP}$ is

$$(2ga)^{\frac{1}{2}} \cos a \sec \phi \cdot \overrightarrow{QP}$$

$$= - (2ga)^{\frac{1}{2}} \cos a \sec \phi \cdot dX$$

$$= \epsilon_1 (2ga^3)^{\frac{1}{2}} \cos a \sin (2a + 2\phi) \sec^3 \phi.$$

If $\overline{A}'$ denotes the whole Action for the constrained guiding path $BV'QP$, we have

$$\frac{\overline{A}' - A}{(2ga^3)^{\frac{1}{2}}} = \epsilon_1 \cos a \sin (2a + 2\phi) \sec^3 \phi\, (1 - \sin \phi),$$

which, as $\epsilon_1$ is positive, is a positive quantity provided $2a + 2\phi < \pi$.

(iii) Consider, in the same way, the constrained guiding path $BV''RP$. Let $a - \epsilon_2$ and $\phi - \eta_2$ be the new values of $a$ and $\phi$, where $\epsilon_2$ is positive for $BV''R$ as drawn. Because $MP\ (= a \cos^2 a \sec^2 \phi)$ is the same for $BV''R$ as for $BVP$,

$$\eta_2 \tan \phi = \epsilon_2 \tan a.$$

Let $A''$ be the action for the portion $BV''R$ of the constrained path. As before, we find

$$\frac{A'' - A}{(8ga^3)^{\frac{1}{2}}} = - \epsilon_2 \frac{\cos a \sin (2a + 2\phi)}{\sin 2\phi}.$$

The portion $\overrightarrow{PR}$ is the variation of $OM\ (= Y)$, that is, of $2 (\sin 2a + 2 \cos^2 a \tan \phi)$; thus

$$dY = - 2a\epsilon_2 \frac{\sin (2a + 2\phi)}{\sin 2\phi}.$$

The element of Action along $\overrightarrow{RP}$ is

$$(2ga)^{\frac{1}{2}} \cos a \sec \phi \cdot \overrightarrow{RP}$$

$$= (2ga)^{\frac{1}{2}} \cos a \sec \phi\, (- dY)$$

$$= \epsilon_2 (8ga^3)^{\frac{1}{2}} \frac{\cos a \sin (2a + 2\phi)}{\cos \phi \sin 2\phi}.$$

If $\overline{A}''$ denotes the whole Action for the path $BV''RP$, we have

$$\frac{\overline{A}'' - A}{(8ga^3)^{\frac{1}{2}}} = \epsilon_2 \frac{\cos a \sin(2a + 2\phi)}{\sin 2\phi} (\sec\phi - 1)$$

$$= \tfrac{1}{2}\epsilon_2 \cos a \sec\phi \tan\tfrac{1}{2}\phi \sin(2a + 2\phi).$$

In order that the configuration may be possible, $\phi$ must be positive : also, $\sigma_2$ is positive. Hence $\overline{A}'' - A$ is a positive quantity provided $2a + 2\phi < \pi$.

(iv) To find the conjugate of the initial point $B$, or the kinetic focus of the configuration which places the particle initially at $B$, we take a contiguous parabola through $B$ for which the value of $a$ is $a + \epsilon$. The conjugate of $B$ is the place where these two parabolas next meet. Let this place $B'$ be given by a value $\vartheta$ of $\theta$ for the parabola $BVP$, and by a value $\vartheta + \eta$ for the contiguous parabola. Because the value of $y$ at $B'$ is the same, we have, from (ii),

$$\eta = -\epsilon \frac{\cos\vartheta \cos(2a + \vartheta)}{\cos^2 a};$$

and because the value of $x$ at $B'$ is the same, we have, from (iii),

$$\eta = \epsilon \frac{\tan a}{\tan\vartheta}.$$

Hence

$$\frac{\tan a}{\tan\vartheta} + \frac{\cos\vartheta \cos(2a + \vartheta)}{\cos^2 a} = 0,$$

leading to

$$\sin(2a + 2\vartheta) = 0.$$

At the initial place $B$, the value of $\vartheta$ is $-a$. The next place where this equation is satisfied, after $\vartheta = -a$, is given by $2a + 2\vartheta = \pi$.

As $\dfrac{dy}{dx} = \cot\theta$, we have

$$\left(\frac{dy}{dx}\right)_B \left(\frac{dy}{dx}\right)_{B'} = -\cot a \cot\vartheta = -1;$$

the tangents at $B$ and $B'$ are perpendicular, $BB'$ is a focal chord, and $B'$ is found as the point of contact of the other tangent to the parabola from $T$, where $T$ is the point in which the directrix is cut by the tangent at $B$.

For a minimum Action, the range must not (under the Jacobi test) extend as far as $B'$. Hence $2a + 2\phi$ satisfies the proviso in (ii) and in (iii), when the range ends before the conjugate or the kinetic focus.

### *General weak variations : tests for minimum Action.*

**231.** All the preceding variations (except in the examples) have been of the weak type, and have been 'special': that is, no general weak variation was admitted during the course of the characteristic changes in the configuration. Moreover, in the first instance, even the total time was restricted.

In order to take account of general weak variations—the restriction in § 229, as to the constancy of the total time, is an unessential superfluity— and also in order to take later account of strong variations, we select a new current independent variable. We choose it free from any intrinsic association so direct, as the length of arc in the passage of any unit mass from one position to another, or the amount of time in the passage from one configuration to another. Thus all the coordinates, sufficient and necessary to define, not only the geometry of the configuration in any phase, but also the time which has elapsed since motion from the initially considered phase began, are now taken to be functions of this selected independent variable $u$. We shall assume that, throughout the continuous changes in successive phases of the configuration, $u$ is a regularly increasing variable.

Thus $\theta$, $\phi$, ..., $t$, in the general problem, are now to be regarded as functions of $u$. We denote their derivatives with respect to $u$ by $\theta'$, $\phi'$, ..., $t'$, so that

$$\dot{\theta} = \frac{\theta'}{t'}, \quad \dot{\phi} = \frac{\phi'}{t'}, \quad \ldots.$$

Restricting the problem to the case, where $2T$ does not explicitly involve $t$, so that $2T$ can be a general function of $\theta$, $\phi$, ..., but not of $t$, and is a homogeneous quadratic function of $\dot{\theta}$, $\dot{\phi}$, ..., we have

$$2T = K\,(\theta, \phi, \ldots, \dot{\theta}, \dot{\phi}, \ldots)$$
$$= K\left(\theta, \phi, \ldots, \frac{\theta'}{t'}, \frac{\phi'}{t'}, \ldots\right),$$

so that the Action-integral is

$$A = \int 2T\,dt.$$

We have, throughout the whole motion of the system,

$$T + V = E,$$

where $E$ is a persistent constant for the system. Hence, with the introduction of the Hamilton function $L$, defined by the relation

$$T - V = L,$$

we have

$$2T = L + E,$$

so that

$$A = \int (L + E)\,dt.$$

Now $L$ is a function of $\dot{\theta}$, $\dot{\phi}$, ..., inasmuch as these occur in $T$ and do not occur in $V$; and where they occur in $T$, they make it a quadratic function of $\dot{\theta}$, $\dot{\phi}$, ..., so that $L$ involves the new derivatives in the combinations $\frac{\theta'}{t'}$, $\frac{\phi'}{t'}$, ....
We write

$$L + E = K\,(\theta, \phi, \ldots, \dot{\theta}, \dot{\phi}, \ldots),$$

and

$$F\,(\theta, \phi, \ldots, \theta', \phi', \ldots, t') = t'K\left(\theta, \phi, \ldots, \frac{\theta'}{t'}, \frac{\phi'}{t'}, \ldots\right);$$

and, with this value of $F$, we have

$$A = \int F\,(\theta, \phi, \ldots, \theta', \phi', \ldots, t')\,du.$$

Moreover, the function $F$ is homogeneous and of the first order in the variables $\theta'$, $\phi'$, ..., $t'$, because $K$ is homogeneous of zero order in them; hence $F$ satisfies the identity

$$F = t' \frac{\partial F}{\partial t'} + \theta' \frac{\partial F}{\partial \theta'} + \phi' \frac{\partial F}{\partial \phi'} + \dots.$$

**232.** The problem now is the same, mathematically, as before.

We begin with weak variations. The characteristic equations are

$$\frac{\partial F}{\partial t} - \frac{d}{du}\left(\frac{\partial F}{\partial t'}\right) = 0,$$

$$\frac{\partial F}{\partial \theta} - \frac{d}{du}\left(\frac{\partial F}{\partial \theta'}\right) = 0,$$

$$\frac{\partial F}{\partial \phi} - \frac{d}{du}\left(\frac{\partial F}{\partial \phi'}\right) = 0,$$

and so on, for each of the variables $t$, $\theta$, $\phi$, ....

(i) In the first of these equations, we have $\dfrac{\partial F}{\partial t} = 0$, because $t$ does not occur explicitly in $F$. The equation gives

$$\frac{\partial F}{\partial t'} = \underline{\text{constant.}}$$

Now

$$F = (L + E)\, t';$$

and therefore

$$\begin{aligned}
\frac{\partial F}{\partial t'} &= L + E + t'\frac{\partial L}{\partial t'} \\
&= L + E + t'\left(-\frac{\theta'}{t'^2}\frac{\partial L}{\partial \dot\theta} - \frac{\phi'}{t'^2}\frac{\partial L}{\partial \dot\phi} - \dots\right) \\
&= L + E - \left(\dot\theta\frac{\partial L}{\partial \dot\theta} + \dot\phi\frac{\partial L}{\partial \dot\phi} + \dots\right) \\
&= L + E - \left(\dot\theta\frac{\partial T}{\partial \dot\theta} + \dot\phi\frac{\partial T}{\partial \dot\phi} + \dots\right) \\
&= L + E - 2T = 0.
\end{aligned}$$

Hence the constant is zero, owing to the constancy of the energy throughout the motion. The first characteristic equation is satisfied from this cause; it imposes no further condition.

(ii) Taking the second equation as typical of the rest, we have

$$\frac{\partial F}{\partial \theta} = t' \frac{\partial L}{\partial \theta},$$

$$\frac{\partial F}{\partial \theta'} = t' \frac{\partial L}{\partial \theta'} = \frac{\partial L}{\partial \dot\theta},$$

so that the equation is

$$t' \frac{\partial L}{\partial \theta} - \frac{d}{du}\left(\frac{\partial L}{\partial \dot\theta}\right) = 0,$$

that is,

$$\frac{\partial L}{\partial \theta} - \frac{d}{dt}\left(\frac{\partial L}{\partial \dot\theta}\right) = 0.$$

Similarly for all the other equations. These are the customary equations of motion, which therefore have their equivalent in the characteristic equations.

(iii) For the extended Legendre test, we need the quantities corresponding to the quantities $P$, $R$, $Q$ of § 178. Now, from the identity

$$F = t' \frac{\partial F}{\partial t'} + \theta' \frac{\partial F}{\partial \theta'} + \phi' \frac{\partial F}{\partial \phi'} + \cdots,$$

we have (also as identities)

$$0 = t' \frac{\partial^2 F}{\partial t'^2} + \theta' \frac{\partial^2 F}{\partial t' \partial \theta'} + \phi' \frac{\partial^2 F}{\partial t' \partial \phi'} + \cdots,$$

$$0 = t' \frac{\partial^2 F}{\partial \theta' \partial t'} + \theta' \frac{\partial^2 F}{\partial \theta'^2} + \phi' \frac{\partial^2 F}{\partial \theta' \partial \phi'} + \cdots,$$

$$0 = t' \frac{\partial^2 F}{\partial \phi' \partial t'} + \theta' \frac{\partial^2 F}{\partial \theta' \partial \phi'} + \phi' \frac{\partial^2 F}{\partial \phi'^2} + \cdots,$$

$$\cdots\cdots\cdots\cdots\cdots\cdots\cdots\cdots\cdots\cdots\cdots$$

To give effect to these relations, we introduce quantities $P_{11}$, $P_{12}$, $P_{22}$, ... such that

$$\frac{\partial^2 F}{\partial \theta'^2} = t'^2 P_{11}, \qquad \frac{\partial^2 F}{\partial \theta' \partial \phi'} = t'^2 P_{12}, \qquad \frac{\partial^2 F}{\partial \phi'^2} = t'^2 P_{22}, \ldots;$$

and then the identities are satisfied by taking

$$-\frac{\partial^2 F}{\partial \theta' \partial t'} = t' (\theta' P_{11} + \phi' P_{12} + \cdots),$$

$$-\frac{\partial^2 F}{\partial \phi' \partial t'} = t' (\theta' P_{12} + \phi' P_{22} + \cdots),$$

$$\cdots\cdots\cdots\cdots\cdots\cdots\cdots\cdots\cdots\cdots$$

$$\frac{\partial^2 F}{\partial t'^2} = \theta'^2 P_{11} + 2\theta' \phi' P_{12} + \phi'^2 P_{22} + \cdots.$$

The extended Legendre test requires that, for a minimum, all the relations

$$P_{11} > 0, \quad P_{22} > 0, \ldots,$$

$$P_{11} P_{22} > P_{12}^2, \ldots,$$

shall be satisfied. Now

$$F = (L + E) t' = 2T t' = \frac{1}{t'} (A_{11} \theta'^2 + 2A_{12} \theta' \phi' + A_{22} \phi'^2 + \cdots),$$

so that

$$\frac{\partial^2 F}{\partial \theta'^2} = 2t' \frac{\partial^2 T}{\partial \theta'^2} = \frac{1}{t'} A_{11},$$

$$\frac{\partial^2 F}{\partial \theta' \partial \phi'} = \frac{1}{t'} A_{12},$$

$$\frac{\partial^2 F}{\partial \phi'^2} = \frac{1}{t'} A_{22},$$

and so on. As $t'$ is steadily positive, when $t$ is measured with regular increase, and $u$ is a variable increasing throughout the range, it follows that the Legendre conditions become

$$A_{11} > 0, \quad A_{22} > 0, \quad A_{11} A_{22} > A_{12}, \dots,$$

which are known to be satisfied; and therefore the Legendre test, that the Action is a minimum, is satisfied.

(iv)  To apply the extended Jacobi test, the Jacobi expression (§ 178) for the second variation is needed. Let an arbitrary small variation of the characteristic configuration be represented by $t + \kappa T$, $\theta + \kappa \Theta$, $\phi + \kappa \Phi$, ..., where $T, \Theta, \Phi, \dots$ are arbitrary continuous functions of $u$. We form quantities

$$\bar{\Theta} = t'\Theta - \theta'T, \quad \bar{\Phi} = t'\Phi - \phi'T, \dots;$$

and, when there are $p$ coordinates $\theta, \phi, \dots, \psi$, we construct expressions

$$V_1 = \bar{\Theta}' + \lambda_{11}\bar{\Theta} + \lambda_{12}\bar{\Phi} + \dots + \lambda_{1p}\bar{\Psi},$$
$$V_2 = \bar{\Phi}' + \lambda_{21}\bar{\Theta} + \lambda_{22}\bar{\Phi} + \dots + \lambda_{2p}\bar{\Psi},$$
$$\dots\dots\dots\dots\dots\dots\dots\dots\dots\dots\dots\dots$$
$$V_p = \bar{\Psi}' + \lambda_{p1}\bar{\Theta} + \lambda_{p2}\bar{\Phi} + \dots + \lambda_{pp}\bar{\Psi}.$$

The Jacobi form of the second variation is

$$\tfrac{1}{2}\kappa^2 \int^{\mu, \, \nu = p}_{\mu, \, \nu = 1} \Sigma\Sigma \, P_{\mu\nu} V_\mu V_\nu \, du;$$

and the range of the integral must not extend so far as to allow

$$V_1, \, V_2, \, \dots, \, V_p,$$

simultaneously to vanish. The coefficients $\lambda$ are determined by means of $p$ linearly independent sets of integrals $\bar{\Theta}_r, \bar{\Phi}_r, \dots, \bar{\Psi}_r$, (for $r = 1, \dots, p$) of the $p$ subsidiary characteristic equations, being given by relations

$$\bar{\Theta}_r' + \lambda_{11}\bar{\Theta}_r + \lambda_{12}\bar{\Phi}_r + \dots + \lambda_{1p}\bar{\Psi}_r = 0,$$

$(r = 1, \dots, p)$ for $\lambda_{11}, \dots, \lambda_{1p}$: by relations

$$\bar{\Phi}_r' + \lambda_{21}\bar{\Theta}_r + \lambda_{22}\bar{\Phi}_r + \dots + \lambda_{2p}\bar{\Psi}_r = 0,$$

$(r = 1, \dots, p)$ for $\lambda_{21}, \dots, \lambda_{2p}$: and so on. The exclusion of the possibility of variations

$$\bar{\Theta} = \alpha_1 \bar{\Theta}_1 + \alpha_2 \bar{\Theta}_2 + \dots + \alpha_p \bar{\Theta}_p,$$
$$\bar{\Phi} = \alpha_1 \bar{\Phi}_1 + \alpha_2 \bar{\Phi}_2 + \dots + \alpha_p \bar{\Phi}_p,$$
$$\dots\dots\dots\dots\dots\dots\dots\dots\dots\dots\dots$$
$$\bar{\Psi} = \alpha_1 \bar{\Psi}_1 + \alpha_2 \bar{\Psi}_2 + \dots + \alpha_p \bar{\Psi}_p,$$

where $\alpha_1, \dots, \alpha_p$ are non-zero constants, is thus required.

Now, for the integral, the quantities $\overline{\Theta}$, $\overline{\Phi}$, ..., $\overline{\Psi}$ are to be zero at each limit. It is always possible, by choice of the constants in the primitive, to secure that $\Theta_1$, $\overline{\Theta}_2$, ..., $\overline{\Theta}_p$, shall vanish at the lowest limit; hence the upper limit must not extend as far as a value $u_1$ of $u$, after the value $u_0$ at the lower limit, such that the determinant

$$\begin{vmatrix} \overline{\Theta}_1, & \overline{\Theta}_2, & ..., & \overline{\Theta}_p \\ \overline{\Phi}_1, & \overline{\Phi}_2, & ..., & \overline{\Phi}_p \\ & ............... & \\ \overline{\Psi}_1, & \overline{\Psi}_2, & ..., & \overline{\Psi}_p \end{vmatrix}$$

shall vanish there. By analysis similar to that in § 152, we obtain a requirement that the range of the integral, beginning at $u_0$, must not extend as far as the conjugate (in the Jacobi sense) of $u_0$, being the first value of $u_1$ greater than $u_0$ arising as a root of

$$\Delta\,(u_0,\ u_1) = 0,$$

a determinant of $2p$ rows, being the only factor of the foregoing determinant which can vanish.

The dynamical interpretation follows at once. At the lower limit, we have $T = 0$, $\Theta = 0$, ..., $\Psi = 0$. But these equations are not satisfied away from the lower limit: that is, we take a configuration at any phase (coordinates and time included) and make a slight variation under 'guidance' (to use the Thomson and Tait phrase). If it can happen in the sequel that the quantities $\overline{\Theta}$, $\overline{\Phi}$, ..., $\overline{\Psi}$, descriptive of this variation and vanishing at the initial phase shall again vanish, then we should have

$$\frac{T}{t'} = \frac{\Theta}{\theta'} = \frac{\Phi}{\phi'} = ... = \frac{\Psi}{\psi'}:$$

that is, the varied configuration would essentially have become the same as the original configuration, because all the variations between the two phases are zero. The two configurations are then reciprocal kinetic foci. Thus the Jacobi test, interpreted in the vocabulary of dynamics, precludes the range of change in the configuration from extending as far as the kinetic focus, complementary to the initial configuration, when the Action is to be a minimum.

The test is assumed to be satisfied, for the estimate of a minimum. Within the range allowed by the test, the Action is a minimum.

Thus the three tests, which are necessary for the possession of a minimum if it is to be maintained under weak variations, are satisfied.

## *Strong variations imposed upon the Action.*

**233.** We now proceed to consider strong variations. As the only derivatives which occur in the integral are of the first order, we can include the class of strong variations compounded of individual constituents of the simple type of strong variation, which can be generalised from the simple Weierstrass type already (§ 209) considered. For this purpose, we take a small abrupt variation of all the variables $\theta$, $\phi$, ..., $\psi$, $t$, so that they become

$$\theta + \kappa\bar{\theta}, \ \phi + \kappa\bar{\phi}, \ \dots, \ \psi + \kappa\bar{\psi}, \ t + \kappa\bar{t}.$$

We then take, as a variation of the characteristic configuration between the initial stage and the stage represented by $\theta$, $\phi$, ..., $\psi$, $t$, the whole modification made up of

(i) the stages of a characteristic configuration (continuous, that is to say, in variables satisfying the characteristic equations) between the initial stage, and the stage represented by

$$\theta + \kappa\bar{\theta}, \ \phi + \kappa\bar{\phi}, \ \dots, \ \psi + \kappa\bar{\psi}, t + \kappa\bar{t}:$$

and

(ii) the abrupt non-characteristic change, from this configuration

$$\theta + \kappa\bar{\theta}, \ \phi + \kappa\bar{\phi}, \ \dots, \ \psi + \kappa\bar{\psi}, \ t + \kappa\bar{t},$$

back to the configuration represented by $\theta$, $\phi$, ..., $\psi$, $t$.

(I) In (i), we have a characteristic variation of the integral $\int F du$; and its value is

$$\kappa \left( \bar{\theta} \frac{\partial F}{\partial \theta'} + \bar{\phi} \frac{\partial F}{\partial \phi'} + \dots + \bar{\psi} \frac{\partial F}{\partial \psi'} + \bar{t} \frac{\partial F}{\partial t'} \right),$$

obtained exactly as in the first variation of the integral. For, over the modified characteristic configuration, the integral is

$$\int_{u_0}^{u} \bar{F} du,$$

where $\bar{F}$ is the value of $F$ under the modified configuration. If, in the current variations, we take $\theta + \kappa\Theta$, $\phi + \kappa\Phi$, ..., $\psi + \kappa\Psi$, $t + \kappa T$, as the changed values of $\theta, \phi, \dots, \psi, t$, so that $\Theta, \Phi, \dots, \Psi, T$ vanish at $u_0$ and are equal to $\bar{\theta}, \bar{\phi}, \dots, \bar{\psi}, \bar{t}$ at $u$, then

$$\int_{u_0}^{u} \bar{F} du = \int_{u_0}^{u} F du + \kappa \left( \bar{\theta} \frac{\partial F}{\partial \theta'} + \bar{\phi} \frac{\partial F}{\partial \phi'} + \dots + \bar{\psi} \frac{\partial F}{\partial \psi'} + \bar{t} \frac{\partial F}{\partial t'} \right) + K_2,$$

omitting the integral which vanishes, because the characteristic equations are satisfied everywhere along the characteristic configuration. (The modified configuration is taken also to be characteristic. As in § 209, we thus are able to build up a composite strong variation from the simple strong elements selected; and we are able to resolve a composite strong variation into these simple strong elements.) When only terms of the first order are retained, and $\kappa$ is small enough to render $K_2$ relatively negligible, the increment of the Action in the modified configuration, over the Action in the unvaried configuration, is

$$\int_{u_0}^{u} \bar{F} du - \int_{u_0}^{u} F du,$$

that is, it is the magnitude already stated.

(II) In (ii), to estimate the Action between the phases

$$\theta + \kappa\bar{\theta}, \quad \phi + \kappa\bar{\phi}, \quad \ldots, \quad \psi + \kappa\bar{\psi}, \quad t + \kappa\bar{t}, \quad \text{and} \quad \theta, \quad \phi, \quad \ldots, \quad \psi, \quad t,$$

along a non-characteristic direct change from the phase, represented by the former set of variables, to the phase, represented by the latter set of variables, it is convenient to represent the analysis in the vocabulary of multi-dimensional geometry. We imagine the variables $\theta$, $\phi$, $\ldots$, $\psi$, $t$, multiplied by invariable constants $c_1$, $c_2$, $\ldots$, $c_p$, $c$, so that all the products are of the dimensions of a line: thus $c$ is a velocity, $c_1$ is a line when $\theta$ is an angular polar coordinate, $c_2$ is unity when $\phi$ is a linear Cartesian coordinate. Hence the displacement between the two configurations can be represented by the aggregate of small coordinates

$$\kappa c_1 \bar{\theta}, \quad \kappa c_2 \bar{\phi}, \quad \ldots, \quad \kappa c_p \bar{\psi}, \quad \kappa c\bar{t}.$$

In this representation, let $\sigma$ denote the small arc between the two points which represent the respective configurations at the beginning and at the end of the change; then

$$\sigma^2 = \kappa^2 (c_1^2 \bar{\theta}^2 + c_2^2 \bar{\phi}^2 + \ldots + c_p^2 \bar{\psi}^2 + c^2 \bar{t}^2).$$

Further, let $\lambda_1$, $\lambda_2$, $\ldots$, $\lambda_p$, $\lambda$, be quantities such that

$$\kappa c_1 \bar{\theta} = -\lambda_1 \sigma, \quad \kappa c_2 \bar{\phi} = -\lambda_2 \sigma, \quad \ldots, \quad \kappa c_p \bar{\psi} = -\lambda_p \sigma, \quad \kappa c\bar{t} = -\lambda\sigma,$$

so that

$$\lambda_1^2 + \lambda_2^2 + \ldots + \lambda_p^2 + \lambda^2 = 1 ;$$

then $\lambda_1$, $\lambda_2$, $\ldots$, $\lambda_p$, $\lambda$, can be regarded as the generalised direction-cosines of the displacement in the multi-dimensional representation, corresponding to the passage, in considering the Action in (ii), from the configuration

$$\theta + \kappa\bar{\theta}, \quad \phi + \kappa\bar{\phi}, \quad \ldots, \quad \psi + \kappa\bar{\psi}, \quad t + \kappa\bar{t},$$

back to the configuration $\theta$, $\phi$, $\ldots$, $\psi$, $t$. Any phase of the configuration, intermediate between the extremes during the passage, can be represented by

$$\theta + \left(1 - \frac{s}{\sigma}\right)\kappa\bar{\theta}, \quad \phi + \left(1 - \frac{s}{\sigma}\right)\kappa\bar{\phi}, \quad \ldots, \quad \psi + \left(1 - \frac{s}{\sigma}\right)\kappa\bar{\psi}, \quad t + \left(1 - \frac{s}{\sigma}\right)\kappa\bar{t} ;$$

where $s$ ranges from 0, at the beginning of the element of Action, to $\sigma$, at the end of that element of Action. That is, we can take the coordinates of the configuration at any stage in the passage to be

$$\theta - \frac{\lambda_1}{c_1}(\sigma - s), \quad \phi - \frac{\lambda_2}{c_2}(\sigma - s), \quad \ldots, \quad \psi - \frac{\lambda_p}{c_p}(\sigma - s), \quad t - \frac{\lambda}{c}(\sigma - s).$$

(III) In conformity with this mode of representation, we represent an element in the continuous deformation of the configuration by

$$c_1 d\theta, \quad c_2 d\phi, \quad \ldots, \quad c_p d\psi, \quad c\, dt, \quad dS,$$

so that

$$c_1^2 \theta'^2 + c_2^2 \phi'^2 + \ldots + c_p^2 \psi'^2 + c^2 t'^2 = \left(\frac{dS}{du}\right)^2 = S'^2.$$

We also take generalised direction-cosines $l_1$, $l_2$, ..., $l_p$, $l$, such that

$$c_1\theta' = l_1 S', \quad c_2\phi' = l_2 S', \quad ..., \quad c_p\psi' = l_p S', \quad ct' = lS',$$

$$l_1^2 + l_2^2 + ... + l_p^2 + l^2 = 1.$$

In this representation, we have

$$F = \frac{1}{t'}(A_{11}\theta'^2 + 2A_{12}\theta'\phi' + A_{22}\phi'^2 + ... + A_{pp}\psi'^2),$$

and so

$$\frac{\partial F}{\partial \theta'} = 2\left(A_{11}\frac{\theta'}{t'} + A_{12}\frac{\phi'}{t'} + ... + A_{1p}\frac{\psi'}{t'}\right) = 2\frac{c}{l}\left(A_{11}\frac{l_1}{c_1} + A_{12}\frac{l_2}{c_2} + ... + A_{1p}\frac{l_p}{c_p}\right),$$

$$\frac{\partial F}{\partial \phi'} = 2\left(A_{12}\frac{\theta'}{t'} + A_{22}\frac{\phi'}{t'} + ... + A_{2p}\frac{\psi'}{t'}\right) = 2\frac{c}{l}\left(A_{12}\frac{l_1}{c_1} + A_{22}\frac{l_2}{c_2} + ... + A_{2p}\frac{l_p}{c_p}\right),$$

$$\cdots\cdots\cdots\cdots\cdots\cdots\cdots\cdots\cdots\cdots\cdots\cdots\cdots\cdots\cdots$$

$$\frac{\partial F}{\partial \psi'} = 2\left(A_{1p}\frac{\theta'}{t'} + A_{2p}\frac{\phi'}{t'} + ... + A_{pp}\frac{\psi'}{t'}\right) = 2\frac{c}{l}\left(A_{1p}\frac{l_1}{c_1} + A_{2p}\frac{l_2}{c_2} + ... + A_{pp}\frac{l_p}{c_p}\right),$$

$$\frac{\partial F}{\partial t'} = -\frac{1}{t'^2}(A_{11}\theta'^2 + 2A_{12}\theta'\phi' + A_{22}\phi'^2 + ... + A_{pp}\psi'^2)$$

$$= -\frac{c^2}{l^2}\left(A_{11}\frac{l_1^2}{c_1^2} + 2A_{12}\frac{l_1 l_2}{c_1 c_2} + A_{22}\frac{l_2^2}{c_2^2} + ... + A_{pp}\frac{l_p^2}{c_p^2}\right).$$

Then the element of the variation of the Action, contributed by the characteristic portion (i) of the variation, is equal to

$$\kappa\left(\bar{\theta}\frac{\partial F}{\partial \theta'} + \bar{\phi}\frac{\partial F}{\partial \phi'} + ... + \bar{\psi}\frac{\partial F}{\partial \psi'} + \bar{t}\frac{\partial F}{\partial t'}\right),$$

that is,

$$= -\sigma\left(\frac{\lambda_1}{c_1}\frac{\partial F}{\partial \theta'} + \frac{\lambda_2}{c_2}\frac{\partial F}{\partial \phi'} + ... + \frac{\lambda_p}{c_p}\frac{\partial F}{\partial \psi'} + \frac{\lambda}{c}\frac{\partial F}{\partial t'}\right) = -\sigma\Omega\frac{c}{l},$$

where

$$\Omega = \frac{A_{11}}{c_1^2}\left(2\lambda_1 l_1 - \frac{\lambda}{l}l_1^2\right) + 2\frac{A_{12}}{c_1 c_2}\left(\lambda_1 l_2 + \lambda_2 l_1 - \frac{\lambda}{l}l_1 l_2\right) + ...$$

$$+ \frac{A_{22}}{c_2^2}\left(2\lambda_2 l_2 - \frac{\lambda}{l}l_2^2\right) + ... + \frac{A_{pp}}{c_p^2}\left(2\lambda_p l_p - \frac{\lambda}{l}l_p^2\right).$$

(IV) For the element of the variation of the Action, contributed by the non-characteristic portion (ii) of the variation, let $\bar{F}$ denote the value of $F$ at any stage intermediate between the configurations

$$\theta + \kappa\bar{\theta}, \quad \phi + \kappa\bar{\phi}, \quad ..., \quad \psi + \kappa\bar{\psi}, \quad t + \kappa\bar{t}, \quad \text{and} \quad \theta, \quad \phi, \quad ..., \quad t,$$

say between $\bar{u}$ and $u$, beginning at $\bar{u}$ and ending at $u$. Then the Action during that passage is

$$\int_{\bar{u}}^{u} \bar{F}\,du$$

$$= \int_0^\sigma \bar{F}\frac{du}{ds}\,ds.$$

Now, denoting the coefficients $A_{mn}$ in the passage by $\bar{A}_{mn}$ (for all the values $m, n = 1, \ldots, p$), we have

$$\bar{A}_{mn} = A_{mn} + [\sigma],$$

where $[\sigma]$ represents an aggregate of terms of the first and higher degrees in the small quantity $\sigma$. Also

$$F = \frac{1}{\dfrac{dt}{du}}\left\{\bar{A}_{11}\left(\frac{d\theta}{du}\right)^2 + 2\bar{A}_{12}\frac{d\theta}{du}\frac{d\phi}{du} + \bar{A}_{22}\left(\frac{d\phi}{du}\right)^2 + \ldots + \bar{A}_{pp}\left(\frac{d\psi}{du}\right)^2\right\}$$

$$= \frac{ds}{du}\frac{1}{\dfrac{dt}{ds}}\left\{\bar{A}_{11}\left(\frac{d\theta}{ds}\right)^2 + 2\bar{A}_{12}\frac{d\theta}{ds}\frac{d\phi}{ds} + \bar{A}_{22}\left(\frac{d\phi}{ds}\right)^2 + \ldots + \bar{A}_{pp}\left(\frac{d\psi}{ds}\right)^2\right\};$$

and therefore

$$F\frac{du}{ds} = \frac{1}{\dfrac{dt}{ds}}\left\{\bar{A}_{11}\left(\frac{d\theta}{ds}\right)^2 + 2\bar{A}_{12}\frac{d\theta}{ds}\frac{d\phi}{ds} + \bar{A}_{22}\left(\frac{d\phi}{ds}\right)^2 + \ldots + \bar{A}_{pp}\left(\frac{d\psi}{ds}\right)^2\right\}.$$

Along the passage, the current values of the respective variables are

$$\theta - \frac{\lambda_1}{c_1}(\sigma - s), \quad \phi - \frac{\lambda_2}{c_2}(\sigma - s), \quad \ldots, \quad \psi - \frac{\lambda_p}{c_p}(\sigma - s), \quad t - \frac{\lambda}{c}(\sigma - s),$$

in which the quantities $\theta, \phi, \psi, \ldots, t$, are the values of these variables at the end of the passage. Hence, in the range for $F$,

$$\frac{dt}{ds} = \frac{\lambda}{c}, \quad \frac{d\theta}{ds} = \frac{\lambda_1}{c_1}, \quad \frac{d\phi}{ds} = \frac{\lambda_2}{c_2}, \quad \ldots, \quad \frac{d\psi}{ds} = \frac{\lambda_p}{c_p}.$$

Consequently,

$$F\frac{du}{ds} = \frac{c}{\lambda}\left(\bar{A}_{11}\frac{\lambda_1^2}{c_1^2} + 2\bar{A}_{12}\frac{\lambda_1\lambda_2}{c_1 c_2} + \bar{A}_{22}\frac{\lambda_2^2}{c_2^2} + \ldots + \bar{A}_{pp}\frac{\lambda_p^2}{c_p^2}\right)$$

$$= \frac{c}{\lambda}\left(A_{11}\frac{\lambda_1^2}{c_1^2} + 2A_{12}\frac{\lambda_1\lambda_2}{c_1 c_2} + A_{22}\frac{\lambda_2^2}{c_2^2} + \ldots + A_{pp}\frac{\lambda_p^2}{c_p^2}\right) + [\sigma],$$

where $[\sigma]$ now represents an aggregate of terms of the first and higher degrees in $\sigma$, arising out of all the coefficients $A_{mn}$. Hence the element of the variation of the Action, during the non-characteristic portion (ii) of the course, is

$$\sigma\frac{c}{\lambda}\left(A_{11}\frac{\lambda_1^2}{c_1^2} + 2A_{12}\frac{\lambda_1\lambda_2}{c_1 c_2} + A_{22}\frac{\lambda_2^2}{c_2^2} + \ldots + A_{pp}\frac{\lambda_p^2}{c_p^2}\right) + [\sigma^2],$$

where $[\sigma^2]$ is an aggregate of terms of the second and higher degrees in $\sigma$.

### Expression for the strong variation of the Action.

Let $A$ denote the Action in the characteristic configuration, from the initial phase up to the phase $\theta, \phi, \ldots, \psi, t$; and let $\bar{A}$ denote the Action in the modified configuration, represented by the characteristic passage (i) followed by the non-characteristic passage (ii). We retain terms of the first order in $\sigma$ and keep $\sigma$ small, so that the aggregate $[\sigma^2]$ is negligible. Then

$$\bar{A} - A = -\sigma\Omega\frac{c}{l} + \sigma\frac{c}{\lambda}\left\{A_{11}\frac{\lambda_1^2}{c_1^2} + 2A_{12}\frac{\lambda_1\lambda_2}{c_1 c_2} + A_{22}\frac{\lambda_2^2}{c_2^2} + \ldots + A_{pp}\frac{\lambda_p^2}{c_p^2}\right\}$$

$$= \sigma\frac{c}{l^2}\frac{\square}{\lambda},$$

where

$$\square = \frac{A_{11}}{c_1^2}(\lambda_1 l - \lambda l_1)^2 + 2\frac{A_{12}}{c_1 c_2}(\lambda_1 l - \lambda l_1)(\lambda_2 l - \lambda l_2) + \ldots$$

$$+ \frac{A_{22}}{c_2^2}(\lambda_2 l - \lambda l_2)^2 + \ldots + \frac{A_{pp}}{c_p^2}(\lambda_p l - \lambda l_p)^2.$$

*The Action is not a minimum.*

**234.** The quantities $A_{mn}$ satisfy the relations

$$A_{mm} > 0, \quad A_{nn} > 0, \quad A_{mm}A_{nn} > A_{mn}^2,$$

for all the combinations $m$, $n$, $= 1, 2, \ldots, p$. Also

$$\lambda_1^2 + \lambda_2^2 + \ldots + \lambda_p^2 + \lambda^2 = 1, \quad l_1^2 + l_2^2 + \ldots + l_p^2 + l^2 = 1,$$

where $l_1, l_2, \ldots, l_p, l$, represent the generalised direction-cosines for passage through the characteristic configurations at any phase, and $\lambda_1, \lambda_2, \ldots, \lambda_p, \lambda$ represent the generalised direction-cosines for the strong variation of the configuration at that phase. Hence, unless we have the relations

$$\frac{\lambda_1}{l_1} = \frac{\lambda_2}{l_2} = \ldots = \frac{\lambda_p}{l_p} = \frac{\lambda}{l},$$

each fraction then being necessarily equal to $+1$ or $-1$: that is, unless there is no external variation of the configuration; the quantity $\square$ is essentially positive. Also, the constant $c$ may be taken positive, without affecting the explanation or the issue: it is used merely to substitute a line-variable $ct$ in place of the time-variable.

The quantities $\lambda_1, \lambda_2, \ldots, \lambda$, are subject to the single relation

$$\lambda_1^2 + \lambda_2^2 + \ldots + \lambda_p^2 + \lambda^2 = 1;$$

and they are otherwise arbitrary. Hence all but one of them may be taken arbitrarily within a range that leaves all of them real; in particular, we may take $\lambda$ arbitrarily within a range

$$-1 \leqslant \lambda \leqslant 1.$$

Now

$$\bar{A} - A = \frac{1}{\lambda}\frac{\sigma c\square}{l^2}.$$

Consequently, $\bar{A} - A$ is positive for a positive value of $\lambda$, and is negative for a negative value of $\lambda$. We thus can choose some strong variations for which $\bar{A} - A$ is positive; and we can choose other strong variations for which $\bar{A} - A$ is negative.

We therefore infer that *the Action, in passing from one configuration to another, is not a true minimum.* It satisfies the conditions for the possession of a minimum, when only weak variations are imposed. It does not satisfy the requirement for a minimum, under the simplest (Weierstrass) type of strong variation.

### General representation (including the time) for configurations.

**235.** It is desirable to emphasise one quality of the strong variation, as it arises here for consideration. It is also desirable to indicate one essential distinction between the weak variations and the strong variations.

In the earlier investigation, and specially in the example in § 230, one simple type of strong variation was allowed: it was unnecessary to obtrude the fact that it is not a weak variation. In the figure there given, $BV'QP$ is a strong variation of $BVP$, made in the plane of $xy$; and $BV''RP$ is another strong variation of the same characteristic curve, also made in that plane. The test for the possession of a minimum is satisfied; it has been proved that the increment is positive. It will now be noted that the principal quality of the test, as there used, lies in its nature as really a Weierstrass $E$-test. The Legendre test is not even mentioned. The Jacobi test enters, only late and incidentally. For this particular pair of strong variations, the minimum under weak variations remains a minimum.

In any investigation, when account is taken of the mobility of the configuration within the range of sensitive experience, and when therefore, in the present instance, account is taken of all the variables, so as to include those of position and of time, the changes of all the variables must be considered. In fact, the interval between two phases of the system is to be measured, not solely by the space intermediate between those phases, nor solely by the time intermediate between those phases. Account of both these changes, different in nature, must be taken in an estimate of the interval. (One special method has been adopted in the preceding discussion, for the simultaneous geometrical representation of the changes and for illustration of the analysis. But the geometrical mode of representation is unessential; when this representation is omitted, the analysis persists unaltered and is essential.) Thus, usually, there is no coordinate in any diagram to represent the variable denoting the time. If, in the analysis of § 232, we write

$$ct = \omega,$$

the time-variable could be represented in a geometrical diagram by means of the coordinate $\omega$. The customary diagram, which ignores the time in the final representation of the phases because their analytical equations usually result from the elimination of the time, would come by taking

$$c = 0, \quad \omega = 0,$$

and leaving the unrepresented time as a variable.

*Ex. Consider the effect of one form of strong variation upon the Action of a unit mass under gravity* (in the Example in § 230).

(i) In the preceding analysis in § 233, we take

$$p=2, \quad \theta=x, \quad \phi=y, \quad ct=\omega ;$$
$$c_1=1, \quad c_2=1 ; \quad A_{11}=1, \quad A_{12}=0, \quad A_{22}=1.$$

Let the unit mass be projected horizontally with a velocity $V$. Take the axis of $y$ vertically downwards, the axes of $x$ and $\omega$ in a horizontal plane and perpendicular to one another. At any moment and in any position, the configuration is given by

$$x = Vt, \quad y = \tfrac{1}{2}gt^2, \quad \omega = ct.$$

In the figure, the axes $OX$, $O\Omega$ are horizontal, $OY$ is vertically downwards, $O$ being the point of projection.

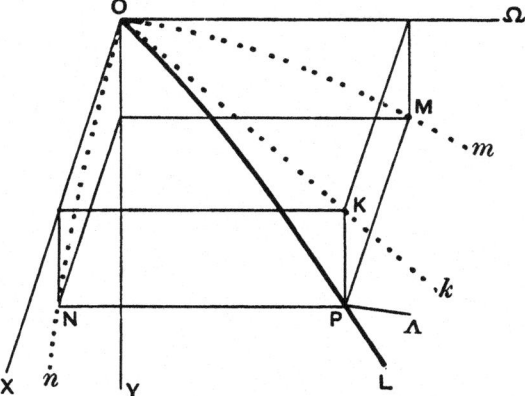

The curve $ONn$ is the parabola

$$x^2 = \frac{2V^2}{g} y, \quad \Omega = 0,$$

in the plane of $XY$, the usual plane representation, without reference to the time: here the curve is the projection of the trajectory on the plane $\Omega = 0$.

The curve $OKk$ is the straight line

$$x = \frac{V}{c} \omega, \quad Y = 0,$$

not usually represented in a diagram: here the curve is the projection of the trajectory on the plane $Y = 0$.

The curve $OMm$ is the parabola

$$\omega^2 = \frac{2c^2}{g} y, \quad X = 0,$$

not usually represented in a diagram: here the curve is the projection of the trajectory on the plane $X = 0$.

The *actual trajectory* is represented by the parabola $OPL$, being the curve common to

$$x^2 = \frac{2V^2}{g} y, \quad x = \frac{V}{c} \omega, \quad \omega^2 = \frac{2c^2}{g} y.$$

(ii) We take a place $P$ on the trajectory, $PL$ being the tangent to the characteristic path $OPL$. The direction-cosines of $PL$ are

$$l = \cos a, \quad l_1 = \cos \theta \sin a, \quad l_2 = \sin \theta \sin a,$$

where $\theta$ is the inclination of the tangent $Nn$ to $OX$, and $a$ is the inclination of $PL$ to $O\Omega$.

At $P$, we take an arbitrary direction $P\Lambda$. This direction lies off the tangent $PL$, and it does not necessarily lie in the plane $OKPL$ of the trajectory; we denote the direction-cosines of $P\Lambda$ by

$$\lambda = \cos \beta, \quad \lambda_1 = \cos \phi \sin \beta, \quad \lambda_2 = \sin \phi \sin \beta,$$

where $\lambda$ is the inclination of $P\Lambda$ to $O\Omega$; thus, as $P\Lambda$ is an arbitrary direction at $P$, the quantity $\lambda$ is such that

$$-1 \leqslant \lambda \leqslant 1.$$

If $\chi$ be the angle between $PL$ and $P\Lambda$, we have

$$\square = \cos^2 a + \cos^2 \beta - 2 \cos a \cos \beta \cos \chi,$$

a positive quantity unless either $\chi = 0$ and $a = \beta$, or $\chi = \pi$ and $a + \beta = \pi$, either of which implies that there is no variation off $PL$. Thus

$$\bar{A} - A = \sigma c \, \frac{\cos^2 a + \cos^2 \beta - 2 \cos a \cos \beta \cos \chi}{\cos^2 a \cos \beta},$$

which is positive if $-\frac{1}{2}\pi < \beta < \frac{1}{2}\pi$, and is negative if $\frac{1}{2}\pi < \beta < \frac{3}{2}\pi$.

Thus the variation of the Action is not uniformly an increase under strong variations. The Action is not a minimum.

*Note.* In the variations considered in § 230, the consideration was limited to variations in the plane of $XY$; and the time was omitted from representation. The variations there considered were the projections on that plane; the foregoing analysis is not applicable to that restricted case, for then $\cos a = 0$, $\cos \beta = 0$.

## *General review of all the tests.*

**236.** At this stage of the whole subject, it may be convenient to summarise the tests which apply to single integrals, involving one dependent variable or more than one dependent variable, as well as derivatives of the first and higher orders of the dependent variable or variables.

The tests are of two kinds, qualitative, and quantitative. The qualitative tests must be satisfied for every phase of the configuration depending upon the variables which occur in the integral. The quantitative test is concerned with the total range of the integral; it is applied to the varying configuration which, in every phase, satisfies the qualitative tests, and it is constructed in a form adapted to the configuration; but it is discrete, not continuous, in character.

The analysis has shewn that, when a test holds for a special weak variation, the same test (though necessarily in an extended form) holds for general weak variations; hence there is no need to enumerate the tests separately for special variations and for general variations.

The analysis for strong variations has been expounded for general strong variations. It is possible to deduce, from special variations, some partial results (sometimes with imperfect conditions, sometimes with additional unessential conditions, arising from the limitations) for special strong variations, though ancillary explanations become necessary. Consequently, there is ultimate simplicity in dealing directly with strong variations.

(A) The qualitative tests are three in number: the Euler (or Euler-Lagrange) test, the Legendre test, and the Weierstrass test.

The Euler test determines the generic character of the relation, between the dependent variable (or variables) and the independent variable. In order to use the test, integration is required; and the arbitrary constants arising in the complete process are determined, either by given data, or by analysis conforming to postulated conditions. The equations in the Euler test are the characteristic equations; and the relation (or relations) which they determine constitute the characteristic curve (or configuration). Moreover, the Euler test arises through weak variations only.

The Legendre test must be satisfied at every place on the characteristic curve (or at every phase of the characteristic configuration). It is constructed directly from the integral, without any reference to the characteristic curve. The requirement merely is that it shall be satisfied for the characteristic curve (or configuration); but often, indeed usually, it happens that the test, so constructed, is of a form which definitely would be satisfied, or definitely would not be satisfied, for all possible curves (or configurations) and not merely for characteristics. Moreover, like the Euler test, the Legendre test occurs through weak variations only.

The Weierstrass test must be satisfied at every place on the characteristic curve (or at every phase of the characteristic configuration). By its nature, it is significant only for the most arbitrary and general deviations from the characteristic curve (or configuration): it has no significance for internal changes along the curve (or in the configuration), which do not affect the external conformation. It must be satisfied, everywhere on the curve (or configuration), for every deviation from the characteristic conformation at every place (or in every phase). Unlike the Euler test and the Legendre test, the Weierstrass test occurs only in connection with strong variations.

(B) There is only one quantitative test: the Jacobi test. It is concerned solely with the extent of the range of integration; and it does no more than impose a limit (should a limit be found necessary) upon that range. Should the range be given as falling short of that limit, the Jacobi test is satisfied. The range must certainly not extend beyond the limit. If the range extends up to (but not beyond) the limit, special investigation is needed for a discrimination, as to the existence or the non-existence of the maximum or the minimum which may be admissible under the other tests. Like the Euler test and the Legendre test, but unlike the Weierstrass test, the Jacobi test occurs through weak variations only : though, if the range tends towards the limit at which the Jacobi test is critical, difficulty arises in the construction of the Weierstrass test, and additional special investigation then is necessary.

(C)  For weak variations, the Euler test, the Legendre test, and the Jacobi test, are necessary and are sufficient to secure a minimum or a maximum under these variations.

For strong variations, the Weierstrass test is necessary and is sufficient to confirm a minimum or a maximum allowed by the weak variations, in so far as strong variations of the limited type considered by Weierstrass are concerned. Sometimes the test is satisfied for strong variations of a wider range than those considered by Weierstrass. It is not inconceivable that other types of strong variations could be devised which, for some integrals at least, would preclude the existence of a maximum or a minimum. In any case, where the Weierstrass test, arising out of the type of strong variation in which it is constructed, is not satisfied, a true minimum or a true maximum does not exist.

# CHAPTER VIII.

*Relative (or limited) maxima and minima.*

**237.** All the problems, which can be resolved by the processes in the preceding investigations, propound demands for a maximum or a minimum that may be described as *free*: that is, there is no rationing of other magnitudes which are intrinsically connected with a solution of the problem. Thus a curve has been required (§ 30, Ex. 1) such that, by its revolution round a fixed line, it generates a surface of minimum area: the curve is not restricted, so that it shall have an assigned length; the requirement to pass through two points, though it affects the position of the curve and may have an influence upon its shape, does not make any claim upon its intrinsic character or properties.

(*a*) Occasionally it happens that a restriction is imposed which would seem to transfer the problem to another category.

Yet such transference may not be necessary in a particular instance. For example, let it be required to find a curve of given length passing from any mobile point $B$ on a line $AB$ to any mobile point $C$ on a line $AC$, so that the area between the curve $BC$ and the lines $AB$ and $AC$ is a maximum.

The maximum can hardly be described as a *free* maximum; but the preceding analysis can be made to apply to this problem. Let $l$ denote the given length; and take the arc of the curve as the independent variable (an assumption that is not always convenient analytically), so that $s$ ranges from 0 to $l$. Then we have to make

$$\tfrac{1}{2}\int r^2 d\theta$$

a maximum, that is,

$$\tfrac{1}{2}\int_0^l r \left\{1 - \left(\frac{dr}{ds}\right)^2\right\}^{\frac{1}{2}} ds$$

a maximum. Proceeding in the usual way, we find that the characteristic equation (from the vanishing of the first variation) makes the curve a circle: and that, by using the terminal conditions which make the first variation vanish, the circle is orthogonal at $B$ to $AB$ and at $C$ to $AC$, so that $A$ is its centre.

Now suppose it is required to find a curve which, revolving round the axis, shall generate a solid, the superficies to be a given constant $2\pi A$ and the volume to be a maximum: a similar problem. The preceding analysis, without

25—2

other developments, is inapplicable. For the volume is $\pi \int y^2 dx$, while the area is $2\pi \int y\, ds$. When we take $dS$ to be an element of superficies, and make $S$ the current variable for the integral to range from $0$ to $A$, the volume is

$$\tfrac{1}{2} \int y\, \frac{dx}{ds}\, dS;$$

but neither $y$ nor $\dfrac{dx}{ds}$, at any point of the curve, can be regarded generally as a function of $S$.

Accordingly, further developments, certainly to include such an instance as the latter and (being general) incidentally to cover the former, will be requisite.

(b) Conditional restrictions, of another kind, either may be imposed from the beginning or may emerge in the analysis from a particular postulation. For example, it may be required to find a curve of given length, joining two points in space and having constant non-zero circular curvature, such that its projection on a fixed plane is a minimum. The requirement, as to given length, is a restriction of the earlier type in (a). The requirement as to constant circular curvature is a restriction of an entirely different type which cannot be merged into the analysis, in the same kind of way as the preceding restriction might (but not necessarily could) be merged. On this score also, further (and other) developments are necessary.

(c) The earlier restriction arose in problems as old as the subject itself, and was contemplated by the Greeks (if not earlier) in a geometrical form. Not infrequently such problems are styled *isoperimetric*\*. The other (and later) restriction arose only after the development of the infinitesimal calculus.

The term *relative* will be used to denote problems of maxima and minima in which permanent intrinsic limitations occur, (i) such as a demand that one (or more than one) concurrent integral shall have an assigned value; or (ii) such that relations among dependent variables (whether involving derivatives or not) are assigned. Sometimes the term *limited* is used.

### Isoperimetric problems : simplest type.

**238.** We proceed to construct the tests for relative maxima and minima, first of all taking the earlier type of restriction that has been indicated. Stated in analytical terms, the problem consists in determining the tests for possible maxima and minima of an integral

$$\int V\, dx,$$

---

\* The word does not occur in classical Greek: its earliest occurrence is with Bishop Synesius early in the fifth century A.D.

when another integral acquires a constant value, say

$$\int U\, dx = \text{constant,}$$

where $U$ contains some or all of the unknown functions that occur in $V$. Equally possible is a requirement that there shall be several such integrals $\int U\, dx$, each acquiring its assigned constant value. The course of the analysis will shew that the results for the latter category of problems can be stated at once, as soon as those for the simplest category have been obtained.

It is not uncustomary to declare that, adopting the usual results for the determination of limited maxima or minima of a function $F$ when its variables are tied by explicit equations $G_1 = 0$, $G_2 = 0$, ..., (so that maxima and minima of

$$F + \lambda_1 G_1 + \lambda_2 G_2 + \ldots,$$

where $\lambda_1$, $\lambda_2$, ... are constants, are sought unconditionally, these further constants being determined with the help of the conditional equations), we must, for the solution of our problem, make the integral

$$\int (V + \lambda U)\, dx$$

(where $\lambda$ is an arbitrary constant) a maximum or a minimum.

The demand is, of course, justifiable, so far as it is a requirement which must be met; though there is nothing to imply that it will suffice to provide all the extra conditions required by the restriction. The (unstated) argument is irrelevant: the algebraical rules do not apply, as the aim is entirely different. In the problem quoted, the various unknown quantities are determined as constants: in the problem now propounded for solution, the unknown quantities are to be determined as functions of the independent variable or variables. Further, while the specified linear association of the constant integral, with the integral to be maximised or minimised, does undoubtedly provide a demand that must be satisfied, no regard is paid (if this course of argument is adopted) to other possible combinations, such as the product, of the two integrals. The crude product of the integrals leads to analysis that is at once useless, and either inconvenient or impossible; nor does advantage come from expressing the product as a double integral

$$\iint U\left(x, y, \frac{dy}{dx}, \ldots\right) V\left(X, Y, \frac{dY}{dX}, \ldots\right) \frac{\partial (x, X)}{\partial (u, v)}\, du\, dv,$$

by making $x$ and $X$ simultaneous functions of two variables $u$ and $v$.

It is safer (and more comprehensive) in the result, to avoid these initial artifices that seem to simplify the problem. We merely adopt the direct process of submitting the variable integral to the most general variations possible; and we then select, for the variable integral, only those variations which are such that, under them, the constant integral retains its constant value. The tests for maxima and minima, for the surviving variations thus limited, are then to be constructed.

## Statement of the problem.

Accordingly, taking the simplest instance where there occur one dependent variable and only its first derivative with regard to one independent variable, let it be required to find the maximum or minimum of an integral $I$, where

$$I = \int g\left(x, y, \frac{dy}{dx}\right) dx,$$

subject to the condition that an integral $J$, where

$$J = \int \phi\left(x, y, \frac{dy}{dx}\right) dx,$$

is to have a constant assigned value, the limits and the range for the two integrals being the same.

Instead of developing the analysis for the problem as thus stated, we shall proceed to the more general form of the problem in which $x$ and $y$ are functions of a new independent variable $t$. Using $x', x'', \ldots, y', y'', \ldots$ to denote derivatives of $x$ and $y$, we have

$$I = \int G(x, y, x', y')\, dt, \quad J = \int \Phi(x, y, x', y')\, dt,$$

where

$$G = x' g\left(x, y, \frac{y'}{x'}\right), \quad \Phi = x' \phi\left(x, y, \frac{y'}{x'}\right).$$

As in § 37, the functions $G$ and $\Phi$ satisfy the respective identities

$$G = x' \frac{\partial G}{\partial x'} + y' \frac{\partial G}{\partial y'}, \quad \Phi = x' \frac{\partial \Phi}{\partial x'} + y' \frac{\partial \Phi}{\partial y'}.$$

A return to the earlier form can always be made by taking $x = t$, so that $x' = 1, y' = \frac{dy}{dx}$; then $G$ turns into $g$, $\Phi$ into $\phi$, while the identities only have significance in assigning values to the unwanted and disappearing values of $\frac{\partial G}{\partial x'}$ and $\frac{\partial \Phi}{\partial x'}$.

Also, the limits of the integrals $\int G\, dt$ and $\int \Phi\, dt$ are fixed: when the extremities of the characteristic curve are partially free, the possibility is taken into account by the small variations of $x$ and of $y$, without small variations of $t$.

Finally, *the course of integration* (that is, for the same values of $x$ and $y$; and the same change of $t$, by way of increase or of decrease) *must be the same for the two integrals*: otherwise, only their net values would be considered.

*Selection of weak variations that leave the constant integral*
*unchanged in value.*

**239.** We begin with weak variations. We have to make the integral $I$ a maximum or a minimum, while the integral $J$ is to remain constant in value, under arbitrary small weak variations. But we shall not now be able to deal with either demand alone. Undoubtedly, arbitrary small variations of the single type $\kappa u$ and $\kappa v$ can be imposed so as to make the first variation of $I$ vanish; the inferred characteristic equation would lead to a determination of $x$ and $y$ as functions of $t$. Such functions of $t$ usually would not provide, for those variations, a vanishing first variation of $J$. Similarly, we might secure the constant value for $J$, by a corresponding differential equation determining $x$ and $y$ as functions of $t$, when we use small variations of the single type $\kappa u$ and $\kappa v$; but then, for such functions, the first variation of $I$ usually would not vanish.

We have seen, however, that, when we seek to maximise or minimise a given integral which is not subject to concurrent conditions, an arbitrary small composite variation (§ 73) of the type

$$\sum_{r=1}^{n} \kappa_r u_r, \quad \sum_{r=1}^{n} \kappa_r v_r,$$

would leave unaffected all the results derived from the imposition of a single variation. Accordingly, we impose an arbitrary small variation of this composite type; and then proceed to seek the limitations upon the composite variation, due to the necessity of keeping $J$ constant. Let

$$w_r = x' v_r - y' u_r,$$

$$\mathfrak{E}_1 = \frac{\partial^2 G}{\partial y \partial x'} - \frac{\partial^2 G}{\partial x \partial y'} + \frac{x' y'' - y' x''}{x' y'} \frac{\partial^2 G}{\partial x' \partial y'},$$

$$\mathfrak{E}_0 = \frac{\partial^2 \Phi}{\partial y \partial x'} - \frac{\partial^2 \Phi}{\partial x \partial y'} + \frac{x' y'' - y' x''}{x' y'} \frac{\partial^2 \Phi}{\partial x' \partial y'}.$$

Let **I** denote the varied value of $I$; there is no change in the constant value of $J$. Then by analysis, similar to that in §§ 40—45, and now applied to $I$ and to $J$ respectively, we have

$$\mathbf{I} - I = \int_{t_0}^{t_1} \mathfrak{E}_1 \left( \Sigma \kappa_r w_r \right) dt + S_1 + S_2,$$

$$0 = \int_{t_0}^{t_1} \mathfrak{E}_0 \left( \Sigma \kappa_r w_r \right) dt + T_1 + T_2,$$

where $S_1$ and $T_1$ denote terms of the first order taken at the limits; and $S_2$ and $T_2$ represent respective aggregates, each of the second and higher degrees in the arbitrary small constants $\kappa_1, \ldots, \kappa_n$. Moreover, $\mathfrak{E}_1$, $S_1$, and $S_2$, are linear in the derivatives of $G$; while $\mathfrak{E}_0$, $T_1$, and $T_2$, are linear in the derivatives of $\Phi$.

Now when the constants $\kappa_1, \ldots, \kappa_n$ are sufficiently small, the aggregates $S_2$ and $T_2$ are negligible, in comparison with the aggregates of terms of the first order in these constants. Hence the second equation can only be satisfied if the aggregate of terms of the first order vanishes by itself; that is, if the equation

$$\kappa_1 J_1 + \kappa_2 J_2 + \ldots + \kappa_n J_n + T_1 = 0$$

is satisfied for all admissible variations where, for $r = 1, \ldots, n$,

$$J_r = \int_{t_0}^{t_1} \mathfrak{E}_0 w_r \, dt.$$

Among these variations, consider the complete set which are such as to vanish at the limits, so that $T_1 = 0$ for that set. We then have

$$\kappa_1 J_1 + \kappa_2 J_2 + \ldots + \kappa_n J_n = 0 \; ;$$

and therefore

$$\kappa_n = - \kappa_1 \frac{J_1}{J_n} - \kappa_2 \frac{J_2}{J_n} - \ldots - \kappa_{n-1} \frac{J_{n-1}}{J_n},$$

unless $J_n$ vanishes. But $u_n$ and $v_n$ are arbitrary continuous functions of $t$, independent of one another: and so $w_n$ is an arbitrary continuous function of $t$. If, therefore, $J_n$ were to vanish for such a quantity $w_n$, we should be obliged to have $\mathfrak{E}_0 = 0$, an equation for the determination of $x$ and $y$ as functions of $t$; and the arbitrary constants in the primitive of that equation would be determinable by assigned values of $x$ and $y$ at the limits $t_1$ and $t_0$. No disposable element, not even an arbitrary constant, would then remain, to annihilate the first variation in $\mathbf{I} - I$ for the set of small variations in question, or even to secure the assigned constant value of $J$. Meanwhile, the condition

$$\int_{t_0}^{t_1} \mathfrak{E}_1 \left( \Sigma \kappa_r w_r \right) dt = 0,$$

attached to the first variation of $I$ when the integral is to possess a maximum or a minimum, still has to be satisfied.

When the relation $\sum_{r=1}^{n} \kappa_r J_r = 0$ is satisfied for special forms of $J$ without imposing a restrictive relation among the constants $\kappa_1, \ldots, \kappa_n$, the value of $J$ still has to remain constant. But now, in the modified expression for the change in $J$ which must vanish because of the persistent constant value, the terms of the second degree in the small constants $\kappa$ dominate the rest. Making these constants so small that the aggregate of terms of the third and higher degrees is negligible in comparison with the aggregate of terms of the second degree, the constant value of $J$ requires that a quadratic relation

$$(* \, \Diamond \, \kappa_1, \ldots, \kappa_n)^2 = 0$$

shall be satisfied, in place of the former evanescent linear relation $\sum_{r=1}^{n} \kappa_r J_r = 0$. Save for particular types of the already special form of $J$, this quadratic relation would provide values of $\kappa_n$, in terms of the other constants $\kappa$; and the analysis could still proceed.

As the integral $J$ must be of very special form in order that the equation $\mathfrak{E}_0 = 0$ may be satisfied, the discussion of the problem will be left for investigation when that form arises. We shall, in consequence, assume that $J_n$ does not vanish for all admissible variations—shall assume, in fact, that $\mathfrak{E}_0 = 0$ is not satisfied. And, by a similar argument, we infer that, now, no one of the other integrals $J_r$ vanishes for all admissible variations.

### Vanishing of the first variation of I.

**240.** In order that the integral $I$ may acquire a maximum or a minimum value, its first variation must vanish for all variations. Hence, for the selected set of variations which make $T_1$ vanish at the limits, we have

$$\kappa_1 I_1 + \kappa_2 I_2 + \ldots + \kappa_r I_r = 0,$$

where, for $r = 1, \ldots, n$,

$$I_r = \int_{t_0}^{t_1} \mathfrak{E}_1 w_r \, dt.$$

On the substitution of the value of $\kappa_n$ deduced from the relation $\sum\limits_{r=1}^{n} \kappa_r J_r = 0$, the last requirement becomes

$$\kappa_1 \left( I_1 - I_n \frac{J_1}{J_n} \right) + \kappa_2 \left( I_2 - I_n \frac{J_2}{J_n} \right) + \ldots + \kappa_{n-1} \left( I_{n-1} - I_n \frac{J_{n-1}}{J_n} \right) = 0.$$

The constants $\kappa$ in this relation are arbitrary. As $\kappa_n$ has been removed, they are independent of one another. Hence, making them all vanish save one, and selecting that one from $\kappa_1, \ldots, \kappa_{n-1}$ in turn, we have

$$I_r - I_n \frac{J_r}{J_n} = 0,$$

for $r = 1, \ldots, n-1$; that is, we have

$$\frac{I_1}{J_1} = \frac{I_2}{J_2} = \ldots = \frac{I_n}{J_n}.$$

In these relations, $I_1$ and $J_1$ depend on $u_1$ and $v_1$ which are arbitrary functions occurring in $I_1$ and $J_1$ alone, and in no other integral $I_r$ or $J_r$. Likewise, $I_2$ and $J_2$ depend on $u_2$ and $v_2$, also arbitrary functions independent of $u_1$ and $v_1$ and occurring in $I_2$ and $J_2$ alone; and so on, for all the pairs of integrals $I_r$ and $J_r$. Moreover, the limits $t_0$ and $t_1$ are fixed. Hence each of the foregoing fractions is a constant. Also, each integral involves two independent arbitrary functions $u$ and $v$; therefore that constant must be arbitrary and not specific, until it is determined by other considerations. Denoting it by $\lambda$, we have

$$I_1 - \lambda J_1 = 0, \quad I_2 - \lambda J_2 = 0, \quad \ldots, \quad I_n - \lambda J_n = 0;$$

and therefore, for $r = 1, \ldots, n$,

$$\int_{t_0}^{t_1} (\mathfrak{E}_1 - \lambda \mathfrak{E}_0) w_r \, dt = 0,$$

while $u_r$ and $v_r$, occurring in $w_r$, are arbitrary continuous functions of $t$.

This last relation, typical of the $n$ relations for $r = 1, ..., n$, can be satisfied, when account is taken of the arbitrary and continuous character of $w_r$, only if

$$\mathfrak{E}_1 - \lambda\mathfrak{E}_0 = 0,$$

which is independent of $w_r$. The same condition could be deduced, as necessary in order that any one of the relations shall be satisfied. When it is satisfied, each of the $n$ relations is satisfied.

**241.** Next, consider the other variations (if any) which are not selected so as to vanish at the limits. Then, because the first variation of $I$ vanishes, we have

$$\kappa_1 I_1 + ... + \kappa_n I_n + S_1 = 0;$$

and, because $J$ is constant, we have

$$\kappa_1 J_1 + ... + \kappa_n J_n + T_1 = 0.$$

Consequently, we have the equation

$$S_1 - \lambda T_1 = 0$$

at the limits because, owing to the condition

$$\mathfrak{E}_1 - \lambda\mathfrak{E}_0 = 0,$$

each of the quantities $I_r - \lambda J_r$ vanishes.

It therefore follows that, because the first variation of $I$ is to vanish for a maximum or a minimum, and because the first variation of $J$ is to vanish owing to its constant value, we have the condition

$$\mathfrak{E}_1 - \lambda\mathfrak{E}_0 = 0$$

everywhere in the range of integration, and we have the terminal requirement

$$S_1 - \lambda T_1 = 0,$$

$\lambda$ being an arbitrary constant—the condition and the requirement, when combined, being sufficient and necessary to secure zero values for the first variation of $I$ and of $J$.

### Characteristic equation : terminal conditions.

**242.** We now combine the functions $G$ and $\Phi$ in the form

$$F(x, y, x', y') = G(x, y, x', y') - \lambda\Phi(x, y, x', y');$$

then, if $\mathfrak{E}$ denote the former combination (§ 44) of the derivatives of $F$, viz.

$$\mathfrak{E} = \frac{\partial^2 F}{\partial y \partial x'} - \frac{\partial^2 F}{\partial x \partial y'} + \frac{x'y'' - y'x''}{x'y'}\frac{\partial^2 F}{\partial x' \partial y'},$$

we have

$$\mathfrak{E} = \mathfrak{E}_1 - \lambda\mathfrak{E}_0 = 0,$$

which is the characteristic equation for the new function $F$. Also, the terminal requirement

$$S_1 - \lambda T_1 = 0$$

becomes

$$\Sigma \kappa_r \left( u_r \frac{\partial F}{\partial x'} + v_r \frac{\partial F}{\partial y'} \right) = 0.$$

Again, we have

$$\mathbf{I} - I = \kappa_1 I_1 + \dots + \kappa_n I_n + S_1 + S_2,$$
$$0 = \kappa_1 J_1 + \dots + \kappa_n J_n + T_1 + T_2.$$

Thus, when the first variation is made to vanish,

$$\mathbf{I} - I = S_2 - \lambda T_2,$$

where the new right-hand side is the same linear function of the derivatives of $F$, as $S_2$ is of the derivatives of $G$.

To the consideration of $S_2 - \lambda T_2$ we shall proceed later (§ 246). Meanwhile, it is to be noted that, by the introduction of the function $F$ as the linear combination $G - \lambda \Phi$, we have reduced the first variation of $I$ to zero, and we have obtained an expression $S_2 - \lambda T_2$ for the whole residual variation of $I$ after its first variation has vanished. In doing so, we have taken complete account of the constancy of $J$ in value, subject (of course) to the full consideration of the residual variation $S_2 - \lambda T_2$.

When we change $G - \lambda \Phi$ into $F$, the terminal requirement $S_1 - \lambda T_1 = 0$ becomes the same as before (§ 73), viz.

$$u \frac{\partial F}{\partial x'} + v \frac{\partial F}{\partial y'} = 0,$$

to be satisfied at each limit.

When a limit is fixed, $u = 0$ and $v = 0$; the requirement is satisfied without leaving any condition.

When a limit lies on an assigned curve $\chi(x, y) = 0$, then

$$u \frac{\partial \chi}{\partial x} + v \frac{\partial \chi}{\partial y} = 0$$

at that limit; and therefore the condition

$$\frac{\partial \chi}{\partial y} \frac{\partial F}{\partial x'} - \frac{\partial \chi}{\partial x} \frac{\partial F}{\partial y'} = 0$$

must be satisfied at the limit.

*Characteristic equation and terminal conditions for the initial*
*forms of I and J.*

**243.** When we wish to return to the forms in which

$$I = \int g \left( x, y, \frac{dy}{dx} \right) dx = \int g(x, y, p) \, dx,$$

$$J = \int \phi \left( x, y, \frac{dy}{dx} \right) dx = \int \phi(x, y, p) \, dx,$$

we write
$$f(x, y, p) = g(x, y, p) - \lambda\phi(x, y, p),$$
while
$$F(x, y, x', y') = x'f\left(x, y, \frac{y'}{x'}\right).$$

The characteristic equation is

$$\frac{\partial f}{\partial y} - \frac{d}{dx}\left(\frac{\partial f}{\partial p}\right) = 0.$$

When a limit is not fixed but is merely required to lie upon a given curve
$$\chi(x, y) = 0,$$
then, at that limit, the relation

$$\frac{\partial \chi}{\partial y}\left(f - p\,\frac{\partial f}{\partial p}\right) - \frac{\partial \chi}{\partial x}\,\frac{\partial f}{\partial p} = 0$$

must be satisfied.

These results can, of course, be derived without the use of the more general forms of $I$ and of $J$; the discussion would follow the course of the analysis in Chapter I, with the modifications necessary for the simultaneous consideration of the two integrals.

### Continuity of certain functions at an isolated free discontinuity on the characteristic curve.

**244.** It may happen, owing to the character of a solution of the characteristic equation (as in § 30, Ex. 1; and § 47, Ex. 2), or owing to assigned data, that there is discontinuity in the direction of the characteristic curve at a place or at a finite number of places. As before (§ 47), we have the result that *the values of* $\dfrac{\partial F}{\partial x'}$ *and* $\dfrac{\partial F}{\partial y'}$ *are continuous in passage through a free\* place of discontinuity of direction of the characteristic curve.*

With the same notation as before (§ 47), and with the same explanations, the vanishing of the first variation in $\mathbf{I} - I$, when a single discontinuity at a place $T$ exists, requires that the equation

$$\left\{\int_{t_2}^{T} + \int_{T}^{t_3}\right\}(\Sigma\kappa_r w_r)\,\mathfrak{E}_1\,dt + \Sigma\kappa_r\left[u_r\frac{\partial G}{\partial x'} + v_r\frac{\partial G}{\partial y'}\right]_{t_2}^{T} + \Sigma\kappa_r\left[u_r\frac{\partial G}{\partial x'} + v_r\frac{\partial G}{\partial y'}\right]_{T}^{t_3} = 0$$

shall be satisfied. The constancy of the integral $J$, requiring its first variation to vanish by itself, demands that the equation

$$\left\{\int_{t_2}^{T} + \int_{T}^{t_3}\right\}(\Sigma\kappa_r w_r)\,\mathfrak{E}_0\,dt + \Sigma\kappa_r\left[u_r\frac{\partial \Phi}{\partial x'} + v_r\frac{\partial \Phi}{\partial y'}\right]_{t_2}^{T} + \Sigma\kappa_r\left[u_r\frac{\partial \Phi}{\partial x'} + v_r\frac{\partial \Phi}{\partial y'}\right]_{T}^{t_3} = 0$$

---

\* A *free* place is one where variations $u$ and $v$ are possible; at a *fixed* place, every variation represented by $u_r$ and $v_r$ is zero.

shall be satisfied. Multiply the latter equation by the constant $\lambda$ already introduced, and subtract the product from the former equation. Noting that the equation

$$\mathfrak{E}_1 - \lambda \mathfrak{E}_0 = 0$$

is satisfied everywhere along the curve (and therefore necessarily from $t_2$ to $T$ and from $T$ to $t_3$, the ranges being unaffected in their course by the discontinuity at $T$), and substituting $F$ for $G - \lambda \Phi$, we have

$$\Sigma \kappa_r \left[ u_r \frac{\partial F}{\partial x'} + v_r \frac{\partial F}{\partial y'} \right]_{t_2}^{T} + \Sigma \kappa_r \left[ u_r \frac{\partial F}{\partial x'} + v_r \frac{\partial F}{\partial y'} \right]_{T}^{t_3} = 0,$$

that is,

$$\left\{ \left( \frac{\partial F}{\partial x'} \right)_+ - \left( \frac{\partial F}{\partial x'} \right)_- \right\} \Sigma \kappa_r u_r + \left\{ \left( \frac{\partial F}{\partial y'} \right)_+ - \left( \frac{\partial F}{\partial y'} \right)_- \right\} \Sigma \kappa_r v_r = 0.$$

Now at a place that is not free, no variation is possible, and therefore $u_r = 0$ and $v_r = 0$ (for $r = 1, \ldots, n$) at such a place. The equation is then satisfied automatically: no descriptive terminal condition emerges.

At a free place capable of variation, the quantities $\Sigma \kappa_r u_r$ and $\Sigma \kappa_r v_r$ are variable, arbitrary, and independent of one another. Then the equation is satisfied, only if the relations

$$\left( \frac{\partial F}{\partial x'} \right)_+ = \left( \frac{\partial F}{\partial x'} \right)_-, \quad \left( \frac{\partial F}{\partial y'} \right)_+ = \left( \frac{\partial F}{\partial y'} \right)_-,$$

hold: that is, if $\frac{\partial F}{\partial x'}$ and $\frac{\partial F}{\partial y'}$ are continuous through the place of discontinuity.

The same holds at each isolated place of discontinuity. Thus there may be *a finite number of such places in a finite range*. But an unlimited number of such places cannot exist in a finite range; for then there would be concentration of such places in the immediate vicinity of at least one point, and such a point would not be an isolated place of discontinuity.

*Comparison with, and divergence from, the results in §§ 37—49.*

**245.** It thus appears that, so far as concerns the necessary vanishing of the first variation of the integral $I$, limited by the condition that $J$ must remain constant, all the requirements are met and all the conditions are satisfied if, for the significance of the first variation, the earlier analysis of Chapter II is applied to the integral

$$\int_{t_0}^{t_1} (G - \lambda \Phi) \, dt$$

without any limitations, that is, to the integral

$$\int_{t_0}^{t_1} F \, dt.$$

In particular, the characteristic equation is

$$\mathfrak{E}_1 - \lambda \mathfrak{E}_0 = 0.$$

Now

$$G = x' \frac{\partial G}{\partial x'} + y' \frac{\partial G}{\partial y'}, \quad \Phi = x' \frac{\partial \Phi}{\partial x'} + y' \frac{\partial \Phi}{\partial y'};$$

and therefore

$$F = x' \frac{\partial F}{\partial x'} + y' \frac{\partial F}{\partial y'}.$$

Hence

$$x' \frac{\partial F}{\partial x} + y' \frac{\partial F}{\partial y} + x'' \frac{\partial F}{\partial x'} + y'' \frac{\partial F}{\partial y'} = \frac{dF}{dt}$$

$$= x'' \frac{\partial F}{\partial x'} + y'' \frac{\partial F}{\partial y'} + x' \frac{d}{dt}\left(\frac{\partial F}{\partial x'}\right) + y' \frac{d}{dt}\left(\frac{\partial F}{\partial y'}\right),$$

so that

$$x' \left\{ \frac{\partial F}{\partial x} - \frac{d}{dt}\left(\frac{\partial F}{\partial x'}\right) \right\} + y' \left\{ \frac{\partial F}{\partial y} - \frac{d}{dt}\left(\frac{\partial F}{\partial y'}\right) \right\} = 0.$$

But

$$\frac{\partial F}{\partial x} - \frac{d}{dt}\left(\frac{\partial F}{\partial x'}\right)$$

$$= x' \frac{\partial^2 F}{\partial x \partial x'} + y' \frac{\partial^2 F}{\partial x \partial y'} - \left\{ x' \frac{\partial^2 F}{\partial x \partial x'} + y' \frac{\partial^2 F}{\partial y \partial x'} + x'' \frac{\partial^2 F}{\partial x'^2} + y'' \frac{\partial^2 F}{\partial x' \partial y'} \right\}$$

$$= - y' \left\{ \frac{\partial^2 F}{\partial y \partial x'} - \frac{\partial^2 F}{\partial x \partial y'} + \frac{x' y'' - y' x''}{x' y'} \frac{\partial^2 F}{\partial x' \partial y'} \right\}$$

$$= - y' (\mathfrak{E}_1 - \lambda \mathfrak{E}_0) = 0;$$

and similarly,

$$\frac{\partial F}{\partial y} - \frac{d}{dt}\left(\frac{\partial F}{\partial y'}\right) = x' (\mathfrak{E}_1 - \lambda \mathfrak{E}_0) = 0;$$

that is, in virtue of the single equation $\mathfrak{E}_1 - \lambda \mathfrak{E}_0 = 0$, we have the two equations

$$\frac{\partial F}{\partial x} - \frac{d}{dt}\left(\frac{\partial F}{\partial x'}\right) = 0, \quad \frac{\partial F}{\partial y} - \frac{d}{dt}\left(\frac{\partial F}{\partial y'}\right) = 0,$$

the primitive of which gives the characteristic curve when $t$ is eliminated. All the calculations in that earlier analysis, which deal with the two characteristic equations and with relations between the second derivatives of $F$, apply at this stage. For now there are the two characteristic equations, which have just been formulated and are satisfied in virtue of the single characteristic equation $\mathfrak{E}_1 - \lambda \mathfrak{E}_0 = 0$; and there is the permanent identity

$$F = x' \frac{\partial F}{\partial x'} + y' \frac{\partial F}{\partial y'}.$$

But, as regards the subsidiary characteristic equations, there is a difference from the formal analysis. The function $F$, being equal to $G - \lambda \Phi$, contains an arbitrary constant $\lambda$ which has no analogue in the former analysis. The primitive of the characteristic equations, which already contain $\lambda$, will contain two arbitrary constants of integration $A_1$ and $A_2$; therefore it is of the form

$$x = \chi(t, A_1, A_2, \lambda), \quad y = \psi(t, A_1, A_2, \lambda).$$

Thus, in all, there are three arbitrary constants. As potential data for the determination of these constants, we may have (as before) the initial values of $x$, $y$, and $\dfrac{\dot{y}}{\dot{x}}$ when $t = t_0$, say

$$a = \chi(t_0, A_1, A_2, \lambda), \quad b = \psi(t_0, A_1, A_2, \lambda), \quad m = \frac{\psi'(t_0, A_1, A_2, \lambda)}{\chi'(t_0, A_1, A_2, \lambda)};$$

and, in addition, there is the given constant value $J$ of the integral

$$\int \Phi(x, y, x', y')\, dt.$$

We may substitute, as potential data, the values of $x$ and $y$ for the initial value $t_0$ of $t$, the values of $x$ and $y$ for the final value $t_1$ of $t$, and the constant value of $J$, making five relations to determine $t_0$ and $t_1$ (being unessential constants), the constant $\lambda$, and the essential constants $A_1$, $A_2$, of integration.

When the subsidiary characteristic equations are constructed by the former process (§ 61) and their primitive is derived, a consecutive characteristic curve is used, deduced by making arbitrary small variations in the constants $A_1$, $A_2$, $\lambda$ which are bound up together in the analysis. As there is a parameter $\lambda$, which did not occur before, there will be a difference in the calculations connected with the subsidiary equations.

### 'Second' variation of the integral : normal form.

**246.** The first variation of $\mathbf{I} - I$ has been made to vanish. It now is necessary to consider the total residual variation

$$S_2 - \lambda T_2$$

of the quantity $\mathbf{I} - I$.

Complete account has been taken of the constancy in value of the integral $J$, when framing the expression for $S_2 - \lambda T_2$. This expression, being linear in the derivatives of $G$ and the derivatives of $\Phi$, can therefore be made to involve the derivatives of $F$ alone. Let

$$S_2 - \lambda T_2 = K_2 + K_3,$$

where $K_2$ denotes the aggregate of terms of the second degree in $\kappa_1, \ldots, \kappa_n$, and $K_3$ denotes the aggregate of terms of degree higher than two in those small arbitrary constants. Then the sign of $S_2 - \lambda T_2$ is governed by the sign of $K_2$. If this sign is a permanent positive for all admissible variations $\kappa u$ and $\kappa v$, a limited minimum for $I$ is admissible. If the sign is a permanent negative for those variations, a limited maximum is admissible for $I$. For brevity, we write

$$\kappa u = \sum_{r=1}^{n} \kappa_r u_r, \quad \kappa v = \sum_{r=1}^{n} \kappa_r v_r,$$

$$\kappa w = \kappa \left( x'v - y'u \right) = \sum_{r=1}^{n} \kappa_r \left( x'v_r - y'u_r \right) = \sum_{r=1}^{n} \kappa_r w_r.$$

With the former notation (§ 50), we now have

$$K_2 = \tfrac{1}{2}\kappa^2 \int \Theta \, dt,$$

where

$$\Theta = u'^2 \frac{\partial^2 F}{\partial x'^2} + 2u'v' \frac{\partial^2 F}{\partial x' \partial y'} + v'^2 \frac{\partial^2 F}{\partial y'^2}$$

$$+ 2uu' \frac{\partial^2 F}{\partial x \partial x'} + 2uv' \frac{\partial^2 F}{\partial x \partial y'} + 2vu' \frac{\partial^2 F}{\partial y \partial x'} + 2vv' \frac{\partial^2 F}{\partial y \partial y'}$$

$$+ u^2 \frac{\partial^2 F}{\partial x^2} + 2uv \frac{\partial^2 F}{\partial x \partial y} + v^2 \frac{\partial^2 F}{\partial y^2}.$$

Using the former analysis (§ 51), we take

$$\frac{1}{y'^2} \frac{\partial^2 F}{\partial x'^2} = - \frac{1}{x'y'} \frac{\partial^2 F}{\partial x' \partial y'} = \frac{1}{x'^2} \frac{\partial^2 F}{\partial y'^2} = P,$$

$$\frac{\partial^2 F}{\partial x \partial x'} = L + y'y'' P, \quad \frac{\partial^2 F}{\partial x \partial y'} = M - x'y'' P,$$

$$\frac{\partial^2 F}{\partial y \partial y'} = N + x'x'' P, \quad \frac{\partial^2 F}{\partial y \partial x'} = M - y'x'' P,$$

the two definitions of $M$ being consistent because

$$\mathfrak{E} = \mathfrak{E}_1 - \lambda \mathfrak{E}_0 = 0 \, ;$$

also,

$$\frac{\partial^2 F}{\partial x^2} = y'^2 \Gamma + y''^2 P + \frac{dL}{dt},$$

$$\frac{\partial^2 F}{\partial x \partial y} = - x'y' \Gamma - x''y'' P + \frac{dM}{dt},$$

$$\frac{\partial^2 F}{\partial y^2} = x'^2 \Gamma + x''^2 P + \frac{dN}{dt},$$

(where $\Gamma$ now takes the place of the former $G$ in § 51). Then we have

$$\Theta = Pw'^2 + \Gamma w^2 + \frac{d}{dt} \left( Lu^2 + 2Muv + Nv^2 \right)$$

$$= Pw'^2 - 2Hww' + \left( \Gamma - H' \right) w^2 + \frac{d}{dt} \left( Lu^2 + 2Muv + Nv^2 + Hw^2 \right),$$

where $H$ is a quantity (chosen free from $u$, $v$, $w$) at our disposal. The vanishing of the first variation in $\mathbf{I} - I$ has fixed the limits, when these are not fixed by initially postulated data; hence

$$\int_{t_0}^{t_1} \frac{d}{dt} (Lu^2 + 2Muv + Nv^2 + Hw^2)\, dt = \left[ Lu^2 + 2Muv + Nv^2 + Hw^2 \right]_{t_0}^{t_1}$$
$$= 0,$$

because $\kappa u = 0$ and $\kappa v = 0$ at each limit. Thus

$$\int \Theta\, dt = \int \{ Pw'^2 - 2Hww' + (\Gamma - H')\, w^2 \}\, dt,$$

where $H$ is an arbitrary function at our disposal.

Let $H$ be chosen so as to make the subject of integration in the last integral a perfect square: then

$$H^2 = P(\Gamma - H');$$

or, if

$$H = P \frac{z'}{z},$$

where $z$ is taken as a new unknown quantity instead of $H$, the variable $z$ must satisfy the equation

$$Pz'' + P'z' - \Gamma z = 0.$$

For any non-zero value of $z$ satisfying this new equation, we now have

$$\Theta = \int_{t_0}^{t_1} P \left( w' - \frac{z'}{z} w \right)^2 dt.$$

### Subsidiary equation, and its primitive.

**247.** Before considering the significance of this value for $\Theta$, the primitive of the $z$-equation must be obtained. The equation itself can be deduced in another manner which allows the primitive to be constructed concurrently.

The primitive of the characteristic equations

$$\frac{\partial F}{\partial x} - \frac{d}{dt} \left( \frac{\partial F}{\partial x'} \right) = 0, \quad \frac{\partial F}{\partial y} - \frac{d}{dt} \left( \frac{\partial F}{\partial y'} \right) = 0,$$

has been expressed in the form

$$x = \chi(t, A_1, A_2, \lambda), \quad y = \psi(t, A_1, A_2, \lambda),$$

where $A_1$, $A_2$, $\lambda$ are arbitrary constants; of these, $\lambda$ enters during the formation of the equations, and $A_1$ and $A_2$ during their integration. The equations are satisfied for all values of the arbitrary constants, and therefore will be satisfied when $A_1$ and $A_2$ are changed into $A_1 + \kappa a_1$ and $A_2 + \kappa a_2$ respectively, where $a_1$ and $a_2$ are arbitrary. But these changes mean a new characteristic curve; they give new values for the integrals denoted (§ 239) by $I_1, \ldots, I_n$, and $J_1, \ldots, J_n$. Therefore a new value for $\lambda$ will be needed; let it be denoted by $\lambda + \kappa \mu$, where $\mu$ can be expected to depend upon $a_1$ and $a_2$. Let the values of $x$ and $y$, consequent upon these changes, be denoted by $x + \kappa\xi$ and $y + \kappa\eta$. When the corresponding changed value of $t$ is denoted by $t + \kappa T$, where $T$ is a continuous function of $t$ acquiring a value $T_0$ at the lower limit $t_0$, we have

$$\xi = T\chi'(t) + a_1\chi_1(t) + a_2\chi_2(t) + \mu\chi_3(t),$$
$$\eta = T\psi'(t) + a_1\psi_1(t) + a_2\psi_2(t) + \mu\psi_3(t),$$

where

$$\chi_1(t) = \frac{\partial\chi}{\partial A_1}, \quad \chi_2(t) = \frac{\partial\chi}{\partial A_2}, \quad \chi_3(t) = \frac{\partial\chi}{\partial\lambda},$$

$$\psi_1(t) = \frac{\partial\psi}{\partial A_1}, \quad \psi_2(t) = \frac{\partial\psi}{\partial A_2}, \quad \psi_3(t) = \frac{\partial\psi}{\partial\lambda}.$$

Therefore, when we take

$$\theta_r(t) = \chi'(t)\psi_r(t) - \psi'(t)\chi_r(t),$$

for $r = 1, 2, 3$, and a quantity

$$\zeta = x'\eta - y'\xi,$$

with the same significance as before in §§ 58, 60, we have

$$\zeta = a_1\theta_1(t) + a_2\theta_2(t) + \mu\theta_3(t).$$

The first variation of $J$ vanishes for all variations in the aggregate represented by $\kappa u\, (= \Sigma\kappa_r u_r)$ and $\kappa v\, (= \Sigma\kappa_r v_r)$, so that

$$\int_{t_0}^{t_1} \mathfrak{E}_0(x'v - y'u)\, dt = 0.$$

Clearly $\xi$ is a possible value of $u$, and $\eta$ is a possible value of $v$; hence

$$\int_{t_0}^{t_1} \mathfrak{E}_0\zeta\, dt = 0.$$

Let

$$\int_{t_0}^{t_1} \mathfrak{E}_0\theta_r(t)\, dt = \gamma_r(t_1) - \gamma_r(t_0),$$

for $r = 1, 2, 3$. Then the relation* between $a_1$, $a_2$, $\mu$ is

$$a_1\{\gamma_1(t_1) - \gamma_1(t_0)\} + a_2\{\gamma_2(t_1) - \gamma_2(t_0)\} + \mu\{\gamma_3(t_1) - \gamma_3(t_0)\} = 0.$$

Now take

$$\vartheta_1(t) = \theta_1(t) - \frac{\gamma_1(t_1) - \gamma_1(t_0)}{\gamma_3(t_1) - \gamma_3(t_0)}\theta_3(t),$$

$$\vartheta_2(t) = \theta_2(t) - \frac{\gamma_2(t_1) - \gamma_2(t_0)}{\gamma_3(t_1) - \gamma_3(t_0)}\theta_3(t);$$

the final form of $\zeta$ is

$$\zeta = a_1\vartheta_1(t) + a_2\vartheta_2(t).$$

**248.** The preceding analysis gives the effect, upon the primitive of the characteristic equations, of varying the parameters $a_1$, $a_2$, $\mu$. Now consider its effect upon the characteristic equations themselves. It will produce an

---

* Not all the quantities $\gamma_r(t_1) - \gamma_r(t_0)$ can vanish simultaneously; otherwise, we should have the eventuality excluded in § 239. In the text, we assume that $\gamma_3(t_1) - \gamma_3(t_0)$ is a non-vanishing quantity.

increment in each equation, yet the equations must be unaltered: hence each increment must be zero. To find these increments, let $2\Omega$ denote the same function of $\xi$, $\eta$, and their derivatives, as $\Theta$ is of $u$, $v$, and their derivatives, so that

$$2\Omega = \xi'^2 \frac{\partial^2 F}{\partial x'^2} + 2\xi'\eta' \frac{\partial^2 F}{\partial x'\partial y'} + \eta'^2 \frac{\partial^2 F}{\partial y'^2}$$

$$+ 2\xi\xi' \frac{\partial^2 F}{\partial x \partial x'} + 2\xi\eta' \frac{\partial^2 F}{\partial x \partial y'} + 2\eta\xi' \frac{\partial^2 F}{\partial y \partial x'} + 2\eta\eta' \frac{\partial^2 F}{\partial y \partial y'}$$

$$+ \xi^2 \frac{\partial^2 F}{\partial x^2} + 2\xi\eta \frac{\partial^2 F}{\partial x \partial y} + \eta^2 \frac{\partial^2 F}{\partial y^2}.$$

Owing to the small variations, when we include only the first power of $\kappa$,

$$\frac{\partial F}{\partial x} \text{ changes into } \frac{\partial F}{\partial x} + \kappa \frac{\partial \Omega}{\partial \xi},$$

$$\frac{\partial F}{\partial x'} \cdots\cdots\cdots\cdots \frac{\partial F}{\partial x'} + \kappa \frac{\partial \Omega}{\partial \xi'},$$

$$\frac{\partial F}{\partial y} \cdots\cdots\cdots\cdots \frac{\partial F}{\partial y} + \kappa \frac{\partial \Omega}{\partial \eta},$$

$$\frac{\partial F}{\partial y'} \cdots\cdots\cdots\cdots \frac{\partial F}{\partial y'} + \kappa \frac{\partial \Omega}{\partial \eta'}.$$

Consequently, on removing the factor $\kappa$, the vanishing of the increments requires that

$$\frac{\partial \Omega}{\partial \xi} - \frac{d}{dt}\left(\frac{\partial \Omega}{\partial \xi'}\right) = 0, \quad \frac{\partial \Omega}{\partial \eta} - \frac{d}{dt}\left(\frac{\partial \Omega}{\partial \eta'}\right) = 0,$$

the coefficients of $\xi$, $\eta$, $\xi'$, $\eta'$ being known functions of $x$, because we are presumed to know the primitive of the characteristic equations.

These new equations are the *subsidiary characteristic equations*. From them, we can deduce the single equation satisfied by $\zeta$. We have

$$\frac{\partial \Omega}{\partial \xi'} = \xi' \frac{\partial^2 F}{\partial x'^2} + \eta' \frac{\partial^2 F}{\partial x'\partial y'} + \xi \frac{\partial^2 F}{\partial x \partial x'} + \eta \frac{\partial^2 F}{\partial y \partial x'}$$

$$= P(y'^2\xi' - x'y'\eta') + L\xi + M\eta + (y'y''\xi - y'x''\eta)P$$

$$= -y'P\zeta' + L\xi + M\eta,$$

$$\frac{\partial \Omega}{\partial \xi} = \xi' \frac{\partial^2 F}{\partial x \partial x'} + \eta' \frac{\partial^2 F}{\partial x \partial y'} + \xi \frac{\partial^2 F}{\partial x^2} + \eta \frac{\partial^2 F}{\partial x \partial y}$$

$$= \xi'(L + y'y''P) + \eta'(M - x'y''P) + \xi\left(y'^2\Gamma + y''^2P + \frac{dL}{dt}\right)$$

$$+ \eta\left(-x'y'\Gamma - x''y''P + \frac{dM}{dt}\right)$$

$$= \frac{d}{dt}(L\xi + M\eta) - y''P\zeta' - y'\Gamma\zeta;$$

and therefore

$$\frac{\partial \Omega}{\partial \xi} - \frac{d}{dt}\left(\frac{d\Omega}{d\xi}\right) = \quad y'\left\{\frac{d}{dt}(P\zeta') - \Gamma\zeta\right\}.$$

Similarly,

$$\frac{\partial \Omega}{\partial \eta} - \frac{d}{dt}\left(\frac{\partial \Omega}{\partial \eta'}\right) = -x'\left\{\frac{d}{dt}(P\zeta') - \Gamma\zeta\right\}.$$

Hence the two subsidiary equations are satisfied in virtue of the single equation

$$\frac{d}{dt}(P\zeta') - \Gamma\zeta = 0,$$

that is,

$$P\zeta'' + P'\zeta' - \Gamma\zeta = 0.$$

The former equation for $z$ in § 246 is precisely this equation; hence, taking $z = \zeta$, as given by the primitive of this linear equation of the second order, we have

$$\Theta = \int_{t_0}^{t_1} P\left(w' - \frac{\zeta'}{\zeta}w\right)^2 dt.$$

*Lemma to establish $\zeta$ as the primitive of the subsidiary equations.*

**249.** The primitive of the subsidiary equation has incidentally been constructed in form; we have proved that

$$\zeta = a_1 \mathfrak{S}_1(t) + a_2 \mathfrak{S}_2(t),$$

where $a_1$ and $a_2$ are arbitrary constants. This relation will be the primitive in fact, if $\mathfrak{S}_1(t)$ and $\mathfrak{S}_2(t)$ are linearly independent. The linear independence of $\mathfrak{S}_1(t)$ and $\mathfrak{S}_2(t)$ must therefore be established.

The primitive of the two fundamental equations

$$\frac{\partial F}{\partial x} - \frac{d}{dt}\left(\frac{\partial F}{\partial x'}\right) = 0, \quad \frac{\partial F}{\partial y} - \frac{d}{dt}\left(\frac{\partial F}{\partial y'}\right) = 0,$$

is

$$x = \chi(t, A_1, A_2, \lambda), \quad y = \psi(t, A_1, A_2, \lambda);$$

and the characteristic curve is obtained by the elimination of $t$ between these equations. Now

$$G(x, y, x', y') = x'g(x, y, p), \quad \Phi(x, y, x', y') = x'\phi(x, y, p),$$

so that

$$F = G - \lambda\Phi = x'(g - \lambda\phi) = x'f(x, y, p),$$

where $p = \dfrac{y'}{x'} = \dfrac{dy}{dx}$; and the single fundamental equation, in virtue of which (taken in conjunction with the identity satisfied by $F$) the two fundamental equations are satisfied, is

$$\frac{\partial f}{\partial y} - \frac{d}{dx}\left(\frac{\partial f}{\partial p}\right) = 0.$$

In this equation, the constant $\lambda$ remains in the coefficients of $f$. It is not an arbitrary constant arising through the primitive, which is uniquely determined as a uniform function of $x$, by the assignment (without regard to $\lambda$) of independent arbitrary values $b$ and $m$ to $y$ and $p$ for an initial value $a$ of $x$. This primitive may be taken to be

$$y = b + m(x-a) + (x-a)^2 R(x-a, b, m, \lambda) = h(x, b, m, \lambda),$$

where $R$ is a regular function of $x-a$. It must agree with the equation of the characteristic curve obtained by eliminating $t$ between

$$x = \chi(t, A_1, A_2, \lambda), \quad y = \psi(t, A_1, A_2, \lambda);$$

and therefore we must have

$$b = b(A_1, A_2, \lambda), \quad m = m(A_1, A_2, \lambda),$$

as relations between the two independent parameters $A_1$ and $A_2$ of integration for the one form, and the two independent parameters $b$ and $m$ of integration for the other form. Consequently

$$\frac{\partial b}{\partial A_r} \frac{\partial m}{\partial A_{r'}} - \frac{\partial b}{\partial A_{r'}} \frac{\partial m}{\partial A_r}$$

does not vanish, because the two parameters in each pair are independent.

Now

$$\frac{\partial y}{\partial b} = 1 + (x-a)^2 \frac{\partial R}{\partial b},$$

$$\frac{\partial y}{\partial m} = x - a + (x-a)^2 \frac{\partial R}{\partial m},$$

$$\frac{\partial y}{\partial \lambda} = (x-a)^2 \frac{\partial R}{\partial \lambda},$$

from the first form; manifestly no linear relation

$$C_1 \frac{\partial y}{\partial b} + C_2 \frac{\partial y}{\partial m} + C_3 \frac{\partial y}{\partial \lambda} = 0$$

can subsist, when the coefficients $C$ are non-zero constants.

When we substitute $x = \chi(t, A_1, A_2, \lambda)$ in the primitive $y = h(x, b, m, \lambda)$, the last equation must become $y = \psi(t, A_1, A_2, \lambda)$ when the values of $b$ and $m$ in terms of $A_1, A_2, \lambda$ are substituted. Hence

$$\frac{\partial \psi}{\partial A_1} = \frac{\partial y}{\partial A_1} = \frac{\partial h}{\partial x} \frac{\partial x}{\partial A_1} + \frac{\partial h}{\partial b} \frac{\partial b}{\partial A_1} + \frac{\partial h}{\partial m} \frac{\partial m}{\partial A_1},$$

$$\frac{\partial \psi}{\partial A_2} = \frac{\partial y}{\partial A_2} = \frac{\partial h}{\partial x} \frac{\partial x}{\partial A_2} + \frac{\partial h}{\partial b} \frac{\partial b}{\partial A_2} + \frac{\partial h}{\partial m} \frac{\partial m}{\partial A_2},$$

$$\frac{\partial \psi}{\partial \lambda} = \frac{\partial y}{\partial \lambda} = \frac{\partial h}{\partial x} \frac{\partial x}{\partial \lambda} + \frac{\partial h}{\partial b} \frac{\partial b}{\partial \lambda} + \frac{\partial h}{\partial m} \frac{\partial m}{\partial \lambda} + \frac{\partial h}{\partial \lambda}.$$

Therefore, as $\dfrac{dh}{dx} = \dfrac{dy}{dx} = \dfrac{\psi'(t)}{\chi'(t)}$ , we have

$$\theta_1(t) = \chi'(t)\left(\frac{\partial h}{\partial b}\frac{\partial b}{\partial A_1} + \frac{\partial h}{\partial m}\frac{\partial m}{\partial A_1}\right),$$

$$\theta_2(t) = \chi'(t)\left(\frac{\partial h}{\partial b}\frac{\partial b}{\partial A_2} + \frac{\partial h}{\partial m}\frac{\partial m}{\partial A_2}\right),$$

$$\theta_3(t) = \chi'(t)\left(\frac{\partial h}{\partial b}\frac{\partial b}{\partial \lambda} + \frac{\partial h}{\partial m}\frac{\partial m}{\partial \lambda} + \frac{\partial h}{\partial \lambda}\right).$$

If a relation

$$D_1\theta_1(t) + D_2\theta_2(t) + D_3\theta_3(t) = 0$$

could exist with constant coefficients $D_1$, $D_2$, $D_3$, we should have, on disregarding the non-vanishing factor $\chi'(t)$,

$$\frac{\partial h}{\partial b}\left(D_1\frac{\partial b}{\partial A_1} + D_2\frac{\partial b}{\partial A_2} + D_3\frac{\partial b}{\partial \lambda}\right) + \frac{\partial h}{\partial m}\left(D_1\frac{\partial m}{\partial A_1} + D_2\frac{\partial m}{\partial A_2} + D_3\frac{\partial m}{\partial \lambda}\right) + \frac{\partial h}{\partial \lambda}D_3 = 0.$$

Here, the coefficients of $\dfrac{\partial h}{\partial b}$, $\dfrac{\partial h}{\partial m}$, $\dfrac{\partial h}{\partial \lambda}$ —that is, of $\dfrac{\partial y}{\partial b}$, $\dfrac{\partial y}{\partial m}$, $\dfrac{\partial y}{\partial \lambda}$ —are constants ; and it has just been proved that no relation between these three quantities can exist with non-zero constant coefficients. Consequently

$$D_1\frac{\partial b}{\partial A_1} + D_2\frac{\partial b}{\partial A_2} + D_3\frac{\partial b}{\partial \lambda} = 0,$$

$$D_1\frac{\partial m}{\partial A_1} + D_2\frac{\partial m}{\partial A_2} + D_3\frac{\partial m}{\partial \lambda} = 0,$$

$$D_3\frac{\partial h}{\partial \lambda} = 0.$$

The Jacobian $\dfrac{\partial(b,\ m)}{\partial(A_1,\ A_2)}$ does not vanish ; and therefore

$$D_1 = 0, \quad D_2 = 0, \quad D_3 = 0,$$

so that no relation, linear and homogeneous in $\theta_1(t)$, $\theta_2(t)$, $\theta_3(t)$, exists with constant coefficients.

But if $\vartheta_1(t)$ and $\vartheta_2(t)$ are not linearly independent of one another, so that we could have

$$\rho_1\vartheta_1(t) + \rho_2\vartheta_2(t) = 0,$$

where $\rho_1$ and $\rho_2$ are constants, the relation would be a linear homogeneous relation between $\theta_1(t)$, $\theta_2(t)$, $\theta_3(t)$, with constant coefficients, such as has been proved not to exist. Hence $\vartheta_1(t)$ and $\vartheta_2(t)$ are linearly independent. Consequently

$$\zeta = a_1\vartheta_1(t) + a_2\vartheta_2(t),$$

where $a_1$ and $a_2$ are arbitrary independent constants, is the primitive of the subsidiary characteristic equation

$$P\zeta'' + P'\zeta' - \Gamma\zeta = 0.$$

*Discussion of the second variation: the Legendre test.*

**250.** The sign of the second variation in $\mathbf{I} - I$ must be persistent for all admissible small variations, represented by $\kappa w$. Thus the sign of

$$\int_{t_0}^{t_1} P \left( w' - \frac{\zeta'}{\zeta} w \right)^2 dt$$

must be uniformly positive or uniformly negative. (If the quantity could be zero, either through isolated zeros of $P$ without change of sign in $P$ when $t$ passes through any zero, or through the vanishing of $w' - \frac{\zeta'}{\zeta} w$, the second variation would vanish. The third variation would then have to vanish, either for general values of $w$ in the first alternative, or for the particular values of $w$ in the second alternative. Should that requirement be satisfied, then, further, the fourth variation, either for general variations or for the particular small variations $\kappa w$ in question, would have to possess a uniform sign. When the eventuality arises in any individual instance, it can be discussed specially for that instance.)

By the same argument as before (§ 56), the sign of $P$ must be everywhere positive in the range—then admitting the possibility of a minimum for $I$: or it must be everywhere negative in the range—then admitting the possibility of a maximum. We thus again have the *Legendre* test: the quantity

$$\frac{\partial^2 G}{\partial y'^2} - \lambda \frac{\partial^2 \Phi}{\partial y'^2},$$

or the quantity

$$\frac{\partial^2 g}{\partial p^2} - \lambda \frac{\partial^2 \phi}{\partial p^2},$$

must be uniformly positive or uniformly negative in the range.

*The Jacobi test: its analytical expression.*

**251.** We shall suppose the Legendre test satisfied. As $\vartheta_1(t)$ and $\vartheta_2(t)$ are integrals of the subsidiary equation

$$P\zeta'' + P'\zeta' - \Gamma\zeta = 0,$$

we have, as usual,

$$P \{\vartheta_1(t) \vartheta_2'(t) - \vartheta_2(t) \vartheta_1'(t)\} = \text{constant}.$$

Now $P$ never vanishes in the range; and, because $\vartheta_1(t)$ and $\vartheta_2(t)$ are linearly independent, the quantity, of which $P$ is the coefficient, does not vanish there; hence the new constant is not zero. But the new constant is not arbitrary, for $P$, $\vartheta_1(t)$, $\vartheta_2(t)$, contain no arbitrary elements so far as the last integration is concerned. Consequently, the constant is specific also so far as the integration is concerned, though of course it may involve $\lambda$. Denoting it by $K$, we have

$$\vartheta_1(t)\,\vartheta_2{}'(t) - \vartheta_2(t)\,\vartheta_1{}'(t) = \frac{K}{P},$$

where $P$ does not vanish in the range, and $K$ is a non-zero constant.

Next, consider the quantity $w' - \frac{\zeta'}{\zeta}\,w$ which occurs in the subject of integration. Among the non-zero variations represented by $\kappa w$, none may exist which can be such that $w = c\zeta$ along the range, $c$ being any constant. If such a variation $\kappa w$ were possible, the second variation of $I$ would vanish. The third and the fourth variations would then have to be considered for such variations—the third being required to vanish and the fourth not to vanish. We omit these exceptional possibilities as arising only for particular instances, which can be discussed if they occur. The exclusion of a variation $w = c\zeta$ is secured as follows. At the lower limit $t_0$ of the integral, we have $u = 0$ and $v = 0$, so that $w$ is zero there; consequently, a relation $w = c\zeta$ would require that $\zeta$ should vanish there—a demand satisfied by drawing, through the point on the characteristic curve corresponding to the lower limit, a consecutive characteristic. But $w$ is to vanish at the upper limit also. If it were possible to have $w = c\zeta$, the quantity $\zeta$ would have to vanish at that upper limit. Let $t_1$ be the first value of $t$, greater than $t_0$, at which $\zeta$ again vanishes. We restrict the range of the integral, which begins at $t_0$, so that it shall not extend as far as $t_1$. In these circumstances, there cannot be a variation such as $w = c\zeta$, which shall vanish at both limits of the range. For all other non-zero variations $w' - \frac{\zeta'}{\zeta}\,w$ is different from zero; consequently, with this limitation of range, the second variation does not vanish.

A place $t_1$, thus obtained in connection with an initial place $t_0$, is called the *conjugate* of the initial place: and so we again have the *Jacobi* test. It requires that *the range of integration must not extend as far along the characteristic curve as the conjugate of the lower limit.*

**252.** The analytical determination of $t_1$ is immediate. As we take a consecutive characteristic through the initial point $t_0$ in order to obtain the conjugate of $t_0$, we have $\xi = 0$, $\eta = 0$, and therefore $\zeta = 0$ initially: that is,

$$a_1\vartheta_1(t_0) + a_2\vartheta_2(t_0) = 0.$$

Taking $a_2 = -\vartheta_1(t_0)$, $a_1 = \vartheta_2(t_0)$, an assumption that merely modifies the scale of $\kappa\zeta$, we have

$$\zeta = \vartheta_1(t)\,\vartheta_2(t_0) - \vartheta_2(t)\,\vartheta_1(t_0)$$
$$= \Theta(t_0, t),$$

so that $\zeta$, or $\Theta(t_0, t)$, manifestly vanishes when $t = t_0$. Now the range is not to extend as far as the first value of $t$, greater than $t_0$, at which $\zeta$ vanishes:

that is, it *must not extend as far as the smallest root $t_1$ of the equation*

$$\Theta(t_0, t_1) = 0$$

*which is greater than $t_0$.*

This is the analytical equation defining $t_1$ in terms of $t_0$; it can be exhibited in another form. We have

$$\vartheta_1(t_1)\,\vartheta_2(t_0) - \vartheta_2(t_1)\,\vartheta_1(t_0)$$
$$= \{\theta_1(t_1) - \sigma_1\theta_3(t_1)\}\,\{\theta_2(t_0) - \sigma_2\theta_3(t_0)\} - \{\theta_2(t_1) - \sigma_2\theta_3(t_1)\}\,\{\theta_1(t_0) - \sigma_1\theta_3(t_0)\},$$

where

$$\sigma_1 = \frac{\gamma_1(t_1) - \gamma_1(t_0)}{\gamma_3(t_1) - \gamma_3(t_0)}, \qquad \sigma_2 = \frac{\gamma_2(t_1) - \gamma_2(t_0)}{\gamma_3(t_1) - \gamma_3(t_0)}.$$

The left-hand side is zero. When we multiply both sides by the factor $\gamma_3(t_1) - \gamma_3(t_0)$, the critical equation becomes

$$\begin{vmatrix} \theta_1(t_1), & \theta_2(t_1), & \theta_3(t_1) \\ \theta_1(t_0), & \theta_2(t_0), & \theta_3(t_0) \\ \gamma_1(t_1) - \gamma_1(t_0), & \gamma_2(t_1) - \gamma_2(t_0), & \gamma_3(t_1) - \gamma_3(t_0) \end{vmatrix} = 0,$$

which, except for a non-vanishing factor $\gamma_3(t_1) - \gamma_3(t_0)$, is the same as the equation $\Theta(t_0, t_1) = 0$.

### *A range, bounded by conjugates, contains no similar range.*

**253.** One further proposition must be established, so as to secure the effectiveness of the Jacobi test: *a complete range on the characteristic curve, bounded by a point and its conjugate, cannot contain within itself another complete range.* Take a complete range bounded by $t_0$ and $t_1$, where

$$\Theta(t_0, t_1) = 0;$$

then

$$\frac{\vartheta_1(t_1)}{\vartheta_1(t_0)} = \frac{\vartheta_2(t_1)}{\vartheta_2(t_0)} = \sigma.$$

Now $\vartheta_1(t_1)$ and $\vartheta_2(t_1)$ cannot vanish simultaneously, because

$$P\{\vartheta_1(t)\,\vartheta_2'(t) - \vartheta_2(t)\,\vartheta_1'(t)\} = K;$$

and therefore $\sigma$ is a non-vanishing quantity. Let $t_0 + \epsilon T_0$, where $\epsilon$ and $T_0$ are positive and $\epsilon$ is small, denote a place, within the complete range and near $t_0$. Then

$$\Theta(t_1, t_0 + \epsilon T_0) = \vartheta_1(t_1)\,\vartheta_2(t_0 + \epsilon T_0) - \vartheta_2(t_1)\,\vartheta_1(t_0 + \epsilon T_0)$$
$$= \sigma\{\vartheta_1(t_0)\,\vartheta_2(t_0 + \epsilon T_0) - \vartheta_2(t_0)\,\vartheta_1(t_0 + \epsilon T_0)\}$$
$$= -\sigma\Theta(t_0 + \epsilon T_0, t_0),$$

which is not zero because $t_1$ is the first value of $t$ greater than $t_0$, where $\Theta(t, t_0)$ vanishes. Thus $t_1$ is not the conjugate of $t_0 + \epsilon T_0$: let the conjugate be $t_1 + \epsilon T_1$, so that we must determine the sign of $T_1$ as given by the equation

$$\Theta(t_1 + \epsilon T_1, t_0 + \epsilon T_0) = 0.$$

The relation between $T_1$ and $T_0$ effectively is

$$T_1 \frac{\partial \Theta (t_1, t_0)}{\partial t_1} + T_0 \frac{\partial \Theta (t_1, t_0)}{\partial t_0} = 0.$$

Now

$$\frac{\partial \Theta (t_1, t_0)}{\partial t_1} = \vartheta_1{}' (t_1) \, \vartheta_2 (t_0) - \vartheta_2{}' (t_1) \, \vartheta_1 (t_0)$$

$$= \frac{1}{\sigma} \{ \vartheta_1{}' (t_1) \, \vartheta_2 (t_1) - \vartheta_2{}' (t_1) \, \vartheta_1 (t_1) \} = - \frac{1}{\sigma} \frac{K}{P_1},$$

and

$$\frac{\partial \Theta (t_1, t_0)}{\partial t_0} = \vartheta_1 (t_1) \, \vartheta_2{}' (t_0) - \vartheta_2 (t_1) \, \vartheta_1{}' (t_0)$$

$$= \sigma \{ \vartheta_1 (t_0) \, \vartheta_2{}' (t_0) - \vartheta_2 (t_0) \, \vartheta_1{}' (t_0) \} = \sigma \frac{K}{P_0};$$

and therefore

$$T_1 = T_0 \, \sigma^2 \frac{P_1}{P_0},$$

where $P_1$ and $P_0$ are the values of $P$ at $t_1$ and at $t_0$. Now $P$ does not change its sign along the range, so that $P_1/P_0$ is positive. Thus $T_1$ is positive; and it is of the same order of magnitude as $T_0$. Consequently, the conjugate of $t_0 + \epsilon T_0$ within the range lies outside the range beyond $t_1$, the distances between the two upper conjugates and the two lower conjugates being of the same order of magnitude.

Passing along the range from point to point, we pass from conjugate to conjugate, each conjugate being beyond its immediate predecessor. The proposition is established.

*A consecutive characteristic can be drawn through a point contiguous to a place within a complete range.*

**254.** One other result is proved at this stage; it will be useful when strong variations are to be discussed. It is to the effect that, *if $Q$ be a point in the near vicinity of a point $P$, which lies within a complete range $AB$ on a characteristic curve, a consecutive characteristic can be drawn from $A$ so as to pass through $Q$.*

For the purpose, it will be enough to shew that the constants, necessary and sufficient to define a consecutive characteristic, can be determined as unique finite quantities. Let such constants, if they exist for a consecutive curve, be denoted by $A_1 + \kappa b_1$, $A_2 + \kappa b_2$, $\lambda + \kappa v$, where $\kappa \xi$, $\kappa \eta$ are the co-ordinates of $Q$ relative to $P$. The analysis of § 247 shews that such constants $b_1$, $b_2$, $v$, are connected by the necessary relation

$$b_1 \{ \gamma_1 (t_1) - \gamma_1 (t_0) \} + b_2 \{ \gamma_2 (t_1) - \gamma_2 (t_0) \} + v \{ \gamma_3 (t_1) - \gamma_3 (t_0) \} = 0.$$

For such a consecutive curve, if it is possible, let $t_0 + \kappa T_0$ be the value of $t$ at $A$; let $t'$, where $t' < t_1$, be the value of $t$ at $P$; and, for the conceivable consecutive curve, let $t' + \kappa T'$ be the value of $t$ at $Q$. Then, because $A$ is common to the two curves, we have

$$0 = T_0 \, \chi'(t_0) + b_1 \, \chi_1(t_0) + b_2 \, \chi_2(t_0) + \nu \chi_3(t_0),$$
$$0 = T_0 \, \psi'(t_0) + b_1 \, \psi_1(t_0) + b_2 \, \psi_2(t_0) + \nu \psi_3(t_0),$$

where a factor $\kappa$ has been removed and, owing to their smallness as still involving the first and higher powers of $\kappa$, the other terms are ignored as negligible. Eliminating $T_0$, we have

$$0 = b_1 \theta_1(t_0) + b_2 \theta_2(t_0) + \nu \theta_3(t_0).$$

Again, at $Q$, we have (with a similar ignoration of small terms of higher orders)

$$\xi = T' \chi'(t') + b_1 \, \chi_1(t') + b_2 \, \chi_2(t') + \nu \chi_3(t'),$$
$$\eta = T' \psi'(t') + b_1 \psi_1(t') + b_2 \psi_2(t') + \nu \psi_3(t') ;$$

and therefore, on the elimination of $T$,

$$\eta \chi'(t') - \xi \psi'(t') = b_1 \theta_1(t') + b_2 \theta_2(t') + \nu \theta_3(t').$$

We thus have three equations, each of the first degree, to determine $b_1$, $b_2$, $\nu$. These three quantities will be unique, definite, and finite, if the determinant of their coefficients is not zero, that is, if

$$\begin{vmatrix} \theta_1(t') & , & \theta_2(t') & , & \theta_3(t') \\ \theta_1(t_0) & , & \theta_2(t_0) & , & \theta_3(t_0) \\ \gamma_1(t_1) - \gamma_1(t_0), & \gamma_2(t_1) - \gamma_2(t_0), & \gamma_3(t_1) - \gamma_3(t_0) \end{vmatrix}$$

does not vanish. Now this determinant

$$= \{\gamma_3(t_1) - \gamma_3(t_0)\} \, \{\vartheta_1(t') \vartheta_2(t_0) - \vartheta_2(t') \vartheta_1(t_0)\}$$
$$= \{\gamma_3(t_1) - \gamma_3(t_0)\} \, \Theta(t_0, t') :$$

the quantity $\gamma_3(t_1) - \gamma_3(t_0)$ does not vanish; the quantity $\Theta(t_0, t')$ does not vanish because $t' < t_1$, the first value of $t$ after $t_0$ at which $\Theta(t_0, t)$ vanishes; and therefore the determinant does not vanish. Also, we assume that $P$ is not in the immediate vicinity of $B$ within the range $AB$, so that $\Theta(t_0, t')$ is not very small.

Consequently $b_1$, $b_2$, $\nu$ are unique, definite, and finite; and therefore the consecutive characteristic curve can be drawn as required.

*Ex.* Prove that, if $P_1$ and $P_2$ are two points in the complete range $AB$, and if $Q_1$ and $Q_2$ are two points in the immediate vicinity of $P_1$ and $P_2$ respectively, a consecutive characteristic can be drawn through $Q_1$ and $Q_2$, whether $Q_1$ and $Q_2$ be on the same side or on opposite sides of the curve.

## EXAMPLES.

**255.** We now pass to some examples of the foregoing general results.

*Ex.* 1. *Required a curve of given length which shall enclose a maximum area under assigned conditions.*

We have indicated (§ 237) that the curve must be a circle or a circular arc, so far as its nature is concerned : the inference was derived by very special analysis. Proceeding by the general method, we take

$$A = \tfrac{1}{2} \int_{t_0}^{t_1} (xy_1 - yx_1)\, dt, \quad L = \int_{t_0}^{t_1} (x_1{}^2 + y_1{}^2)^{\frac{1}{2}}\, dt,$$

where $A$ has to be maximised, and $L$ denotes the given length. Here

$$F(x, y, x_1, y_1) = \tfrac{1}{2}(xy_1 - yx_1) - \lambda (x_1{}^2 + y_1{}^2)^{\frac{1}{2}},$$

for which form of $F$ the identity $F = x_1 \dfrac{\partial F}{\partial x_1} + y_1 \dfrac{\partial F}{\partial y_1}$ can be verified at once. The characteristic equations are

$$y_1 + \lambda \frac{d}{dt} \left\{ \frac{x_1}{(x_1{}^2 + y_1{}^2)^{\frac{1}{2}}} \right\} = 0, \quad - x_1 + \lambda \frac{d}{dt} \left\{ \frac{y_1}{(x_1{}^2 + y_1{}^2)^{\frac{1}{2}}} \right\} = 0,$$

which, with the arc-length $s$ as the independent variable now that the characteristic equations have been formed, become

$$\frac{dy}{ds} + \lambda \frac{d^2x}{ds^2} = 0, \quad - \frac{dx}{ds} + \lambda \frac{d^2y}{ds^2} = 0.$$

(It follows at once that the radius of curvature is constant, being equal to the real constant $\lambda$, and therefore that the curve is a circle.) Writing $z = x + iy$, where $i$ denotes $\sqrt{-1}$, we have

$$\frac{dz}{ds} = \frac{dx}{ds} + i \frac{dy}{ds} = e^{i\psi},$$

with the customary significance of $\psi$ for the curve. Also

$$\lambda \frac{d^2z}{ds^2} - i \frac{dz}{ds} = 0,$$

so that

$$\lambda \frac{d\psi}{ds} - 1 = 0.$$

Let $s = \lambda t$, so that $\psi = t + a$, where $a$ is an unessential constant that could be merged in $t$ ; then

$$\frac{dx}{dt} = \lambda \frac{dx}{ds} = \lambda \cos (t + a), \quad \frac{dy}{dt} = \lambda \frac{dy}{ds} = \lambda \sin (t + a).$$

Consequently the primitive of the characteristic equations is

$$x = A_1 + \lambda \sin (t + a), \quad y = A_2 - \lambda \cos (t + a),$$

where $A_1$ and $A_2$ are the arbitrary constants of integration, and $\lambda$ is the arbitrary constant arising out of the general theory. The characteristic curve is a circle of radius $\lambda$.

For the Legendre test, we have

$$P = - \frac{1}{x_1 y_1} \frac{\partial^2 F}{\partial x_1 \partial y_1} = - \frac{\lambda}{(x_1{}^2 + y_1{}^2)^{\frac{3}{2}}}.$$

The radical, which arises out of the arc-length, is always taken positive ; and $\dfrac{ds}{dt} = \lambda$, so that $\lambda$ is positive. Hence $P$ is negative. The area $A$ is a maximum, subject (it may be) to other conditions.

For the Jacobi test, we have $\chi\,(t)=x$, $\psi\,(t)=y$, so that

$$\chi'\,(t)=\lambda\cos(t+a),\quad \chi_1\,(t)=1,\quad \chi_2\,(t)=0,\quad \chi_3\,(t)=\ \ \sin(t+a),$$
$$\psi'\,(t)=\lambda\sin(t+a),\quad \psi_1\,(t)=0,\quad \psi_2\,(t)=1,\quad \psi_3\,(t)=-\cos(t+a)\ ;$$

and therefore

$$\theta_1\,(t)=-\lambda\sin(t+a),\quad \theta_2\,(t)=\lambda\cos(t+a),\quad \theta_3\,(t)=-\lambda\ :$$

the forms of which verify the general proposition that $\theta_1\,(t)$, $\theta_2\,(t)$, $\theta_3\,(t)$, are linearly independent. Also, in this case, as $\Phi=(x_1{}^2+y_1{}^2)^{\frac{1}{2}}$, we have (§ 239)

$$\mathfrak{G}_0=\frac{x_1y_2-y_1x_2}{x_1y_1}\frac{\partial^2\Phi}{\partial x_1\,\partial y_1}=\frac{1}{\lambda},$$

so that $\mathfrak{G}_0$ does not vanish. Thus

$$\gamma_1\,(t_1)-\gamma_1\,(t_0)=-\int_{t_0}^{t_1}\sin(t+a)\,dt=\cos(t_1+a)-\cos(t_0+a),$$

$$\gamma_2\,(t_1)-\gamma_2\,(t_0)=\ \ \int_{t_0}^{t_1}\cos(t+a)\,dt=\sin(t_1+a)-\sin(t_0+a),$$

$$\gamma_3\,(t_1)-\gamma_3\,(t_0)=-\int_{t_0}^{t_1}dt\qquad\ \ =-t_1+t_0.$$

The critical equation for the determination of $t_1$ is

$$\begin{vmatrix} -\lambda\sin(t_1+a) & , & \lambda\cos(t_1+a) & , & -\lambda \\ -\lambda\sin(t_0+a) & , & \lambda\cos(t_0+a) & , & -\lambda \\ \cos(t_1+a)-\cos(t_0+a), & \sin(t_1+a)-\sin(t_0+a), & -t_1+t_0 \end{vmatrix}=0\ :$$

that is,

$$\lambda^2\{(t_1-t_0)\sin(t_1-t_0)-2+2\cos(t_1-t_0)\}=0.$$

Now $t_1$ is to be greater than $t_0$. The equation is satisfied by

$$\sin\tfrac{1}{2}\,(t_1-t_0)=0,$$

the lowest value of $t_1$ from which is $t_0+2\pi$, thus yielding no conjugate within the complete circumference if there be a complete circumference. The equation is also satisfied by

$$t_1-t_0-2\tan\tfrac{1}{2}\,(t_1-t_0)=0\ ;$$

the lowest value of $t_1$, greater than $t_0$, satisfying this relation is greater than $t_0+2\pi$.

Thus there is no conjugate of the initial value of $t_0$; the Jacobi test, in the present instance, imposes no limit on the range along a circumference.

*Ex.* 2. The preceding result is of a general character. The Euler test, through the characteristic equations, makes the free curve a circle: the Legendre test is satisfied: and the Jacobi test imposes no limitation upon the range. Various problems arise through the variety of conditions associated with the isoperimetric maximum *: some of these may be taken in brief outline.

(i) *No associated conditions.*

The requirement of constant length $L$ gives

$$L=\int_{t_0}^{t_0+2\pi}(x_1{}^2+y_1{}^2)^{\frac{1}{2}}\,dt=2\pi\lambda\ ;$$

the circle, of this radius $\lambda$, can be placed anywhere.

* The earliest record of the problem, as propounded, is associated with the name of Zenodorus (*fl. ca.* 150 B.C.).

(ii) *The curve is to pass through three given points, being the angular points of a triangle ABC.*

The curve must consist of three arcs of circles on $BC$, $CA$, $AB$ respectively. As the points $A$, $B$, $C$ are fixed and not free (§ 242), the quantities $u$ and $v$ vanish at each of the points : the conditions, arising from the first variation, are satisfied without leaving any residual condition.

As no individual arc $AB$, $BC$, $CA$ is given but only the sum of the whole three arcs, the integrals

$$\int G\,dt, \quad \int \Phi\,dt,$$

are combined once for all. Thus there is only a single constant of the type $\lambda$, or the three arcs of circles have the same radius $\lambda*$, which is determined by the equation

$$\sin^{-1}\frac{a}{2\lambda} + \sin^{-1}\frac{b}{2\lambda} + \sin^{-1}\frac{c}{2\lambda} = \frac{L}{2\lambda},$$

where $a$, $b$, $c$ are the sides of the triangle. Obviously, from the geometry and this equation alike, $L \geqslant a+b+c$.

(iii) *The curve is required to pass through three points A, B, C, in a straight line, B being the intermediate point.*

The curves consist of two circular arcs $AB$ and $BC$ on one side of the line, and of a circular arc on the other side of the line, the common radius $\lambda$ being given by

$$\sin^{-1}\frac{a}{2\lambda} + \sin^{-1}\frac{c}{2\lambda} + \sin^{-1}\frac{a+c}{2\lambda} = \frac{L}{2\lambda}.$$

(iv) The discussion of the limitation upon $L$, so that the circular arcs (as drawn upon $AB$, $BC$, $CA$) shall not intersect except at $A$, $B$, $C$, in (ii) and in (iii) above, is left as an exercise.

*Ex.* 3. In a triangle $ABC$, it is required to draw an arc from any point on $BC$ to any point on $CA$, another arc from any other point on $CA$ to any point on $AB$, and a third arc from any other point on $AB$ to any new point on $BC$, so that the area enclosed by the three arcs and the intercepted portions of the sides shall be the greatest possible consistent with a given perimeter $L$, less than the perimeter of the triangle and not less than the circumference of the inscribed circle.

Prove that the free curves are arcs of circles having a common radius $\lambda$; that an arc touches the side of the triangle at an extremity; that, if

$$L = \sigma \frac{4\pi\Delta}{a+b+c},$$

where $\Delta$ is the area of the triangle, then

$$1 \leqslant \sigma \leqslant \frac{1}{\pi}\cot\tfrac{1}{2}A \cot\tfrac{1}{2}B \cot\tfrac{1}{2}C;$$

and find the value of $\lambda$.

*Ex.* 4. Determine the curve of shortest length which shall enclose an area of given extent,

(a) when no conditions are attached to the curve:

(b) when the curve is to be drawn from any point on one line $OA$ to any point on a perpendicular line $OB$, the intercepted parts of $OA$ and $OB$ being portions of the whole perimeter of the boundary.

---

* This result, together with the solution of some other isoperimetric problems, was given by Steiner (*Ges. Werke*, ii, 75—91). All his investigations are the geometrical equivalent of a discussion of first variations of the original integrals.

*Ex. 5.  Given a volume of homogeneous attracting gravitating matter, find the shape of the solid of revolution constituting the volume so that the attraction at a point on the axis of revolution shall be a maximum.*

We take the axis of $x$ for the axis of revolution, and the origin on the surface of the attracted mass at the end of the axis. When Cartesian coordinates are used, the whole attraction of the body of revolution is

$$2\pi \int_{t_0}^{t_1} \left\{ 1 - \frac{x}{(x^2+y^2)^{\frac{1}{2}}} \right\} x_1 dt,$$

the absolute measure of attraction being taken to be unity. This quantity is to be a maximum, subject to the condition

$$\pi \int_{t_0}^{t_1} y^2 x_1 \, dt = \text{constant}.$$

We therefore take

$$F = 2\pi \left\{ 1 - \frac{x}{(x^2+y^2)^{\frac{1}{2}}} \right\} x_1 - \pi \lambda y^2 x_1.$$

First of all, we note that

$$\frac{\partial^2 F}{\partial x_1 \, \partial y_1} = 0,$$

so that the Legendre test cannot be satisfied. There is no maximum or minimum, even for weak variations.

We note that $\dot{y}_1$ does not occur in $F$; hence the characteristic equation, $Y = 0$, gives

$$\frac{\partial F}{\partial y} = 0,$$

that is,

$$\left\{ 2\pi \frac{xy}{(x^2+y^2)^{\frac{3}{2}}} - 2\pi \lambda y \right\} x_1 = 0.$$

Now $x_1$ is not zero: the implication would be that the matter is a (geometrical) plane disc, which has no volume. Nor can $y$ be zero: the implication would be that the matter is a (geometrical) straight line, which has no volume. Hence the characteristic equation gives

$$x = \lambda (x^2+y^2)^{\frac{3}{2}},$$

or, when we take $\lambda = \dfrac{1}{a^2}$ and change to polar coordinates,

$$r^2 = a^2 \cos \theta.$$

The curve is symmetrical about the axis of $x$, as would be expected. The volume of the solid of revolution is

$$\tfrac{4}{15} \pi a^3,$$

so that $a$ is known.

The solid is ovate in form, somewhat like a sphere increasingly flattened towards the origin-end of the axis. The attraction (with the adopted absolute unit) is

$$\tfrac{4}{5} \pi a.$$

The attraction of the same mass, concentrated into a sphere, is greater at a point on the surface than at any other point outside or at any point inside; the attraction is continuous in magnitude for change of position of the attracted unit of mass, but suffers a discontinuity of rate as the position of that unit passes from within the mass to an outside position. The ratio of the greatest attraction of the sphere to the attraction of the solid of revolution given by the foregoing analysis is $\tfrac{1}{3} \sqrt[3]{25}$, a number slightly less than unity.

The solid provides a greatest value, not a mathematical maximum.

The discussion of the second variation of $F$ is unnecessary.

*Ex.* 6. *Find a curve in a plane such that, when the plane is rotated about the axis of x, the superficies of the generated solid is a given quantity and its volume is a maximum.*

(I)  Here we are to have the integral

$$\int y^2 x_1 \, dt$$

a maximum, while the integral

$$2\pi \int y \, (x_1{}^2 + y_1{}^2)^{\frac{1}{2}} \, dt$$

is a given quantity.  Consequently, for this problem,

$$F = y^2 x_1 - \lambda y \, (x_1{}^2 + y_1{}^2)^{\frac{1}{2}}.$$

As $F$ does not explicitly involve $x$, the characteristic equation

$$\frac{\partial F}{\partial x} - \frac{d}{dt}\left(\frac{\partial F}{\partial x_1}\right) = 0$$

has a first integral

$$y^2 - \lambda y \, \frac{x_1}{(x_1{}^2 + y_1{}^2)^{\frac{1}{2}}} = C,$$

where $C$ is an arbitrary constant of integration.

Before proceeding with the further integration, we note the Legendre test.  The critical quantity $P$ is

$$P = - \frac{1}{x_1 y_1} \frac{\partial^2 F}{\partial x_1 \partial y_1} = - \frac{\lambda y}{(x_1{}^2 + y_1{}^2)^{\frac{3}{2}}}.$$

For the moment, we shall assume $\lambda$ positive (an assumption which will turn out to be justified).  Then $P$ is always negative, so long as $y$ does not vanish.  Also, $P$ is nowhere infinite along a range of the characteristic curve free from nodes, cusps, or other singularities (at which, of course, $x_1$ and $y_1$ vanish simultaneously).  We shall assume that the Legendre test is satisfied, with these restrictions; the volume, consequently, may be a maximum.

To obtain a descriptive character of the curve, draw the ordinate $PM$ at any point $P$; and let the normal at $P$ to the curve cut the axis of $x$ in $G$.  Produce $PG$ (if production is necessary) to $N$, so that $PN = \lambda$; draw $NM'$ perpendicular to $Ox$, and along $NM'$ take $M'S = M'N$.  Then, selecting the positive sign of the square root occurring in the expression for the superficies, we have

$$\frac{x_1}{(x_1{}^2 + y_1{}^2)^{\frac{1}{2}}} = \cos GPM\,;$$

and therefore

$$C = y \, (y - \lambda \cos GPM) = - PM \cdot SM'.$$

Hence we have the construction : take a central conic, having its major axis equal to $\lambda$ (thus we have $\lambda$ a positive quantity), and the square of its minor axis equal to $-C$ (so that, if $C$ is negative, the conic is an ellipse : if positive, the conic is a hyperbola) : then the locus of $P$, that is, the characteristic curve, is the roulette of a focus as the conic rolls along the axis $Ox$.

Returning to the integration, we write $\lambda = 2a$, $C$ positive and $= b^2$, so that

$$y^2 + 2ay \cos \psi = b^2,$$

where $\dfrac{dy}{dx} = \tan \psi$.  Then, if $b = a \tan \alpha$, so that $2\alpha$ is the angle between the asymptotes of the hyperbola,

$$y = -a \cos \psi + a \sec \alpha \, (1 - \cos^2 \alpha \sin^2 \psi)^{\frac{1}{2}}.$$

Hence
$$\frac{dx}{d\psi} = a \cos \psi + a \sec a \left\{ \frac{\sin^2 a}{(1 - \cos^2 a \sin^2 \psi)^{\frac{1}{2}}} - (1 - \cos^2 a \sin^2 \psi)^{\frac{1}{2}} \right\}.$$

In the notation of Jacobian elliptic functions, let
$$\psi = \operatorname{am} t, \quad k = \cos a \, ;$$
the primitive of the characteristic equation is
$$\left. \begin{aligned} x &= c + a \operatorname{sn} t - a \left\{ E(t) - t \sin^2 a \right\} \sec a \\ y &= -a \operatorname{cn} t + a \sec a \operatorname{dn} t \end{aligned} \right\},$$
where $a \,(= \tfrac{1}{2}\lambda)$, $c$, $a$ (or $k$), are the arbitrary constants.

The determination of the possible conjugate of an initial place $t_0$ is left as an exercise in elliptic functions.

(II)　Various conditions, specifying curves, may be associated with the generic conditions.

(a)　Let the extremities be two fixed points on the same side of the axis $Ox$, neither of them lying on it; and let their coordinates be $(x_0, y_0)$, $(x_1, y_1)$. If $t_0$ and $t_1$ be the terminal values of $t$,
$$y_1 = -a \operatorname{cn} t_1 + a \sec a \operatorname{dn} t_1,$$
$$y_0 = -a \operatorname{cn} t_0 + a \sec a \operatorname{dn} t_0,$$
$$x_1 - x_0 = a (\operatorname{sn} t_1 - \operatorname{sn} t_0) - a \sec a \left\{ E(t_1) - E(t_0) - (t_1 - t_0) \sin^2 a \right\},$$
$$S = 2\pi a^2 \int_{t_0}^{t_1} (\operatorname{dn} t - \cos a \operatorname{cn} t)^2 \frac{\sec a}{\operatorname{dn} t} \, dt,$$
where $S$ is the given superficies; the constants to be determined are $a$ and $\lambda$ (which are essential), together with $t_1$ and $t_0$.

(b)　At an extremity, the terminal condition
$$u \frac{\partial F}{\partial x'} + v \frac{\partial F}{\partial y'} = 0,$$
that is,
$$Cu - v\lambda y \sin \psi = 0,$$
must be satisfied.

No residual condition ensues if the extremity, as in (a), be rightly fixed.

(i)　If the extremity, though not fixed, is bound to lie upon the axis $Ox$, then $v = 0$; and $u$ is arbitrary there. Thus $C = 0$; and therefore the curve generating the solid is
$$y (y - \lambda \cos \psi) = 0.$$
This last equation can be satisfied completely by the single equation
$$y = 0,$$
which gives zero superficies and therefore is to be excluded as a solution by itself. Or it can be satisfied completely by the single equation
$$y - \lambda \cos \psi = 0,$$
which represents a circle of radius $\lambda$ having its centre on the axis of $x$; and then
$$S = 4\pi \lambda^2,$$
so that a semi-circle, with its centre on the axis $Ox$ and radius $\lambda$, provides a solution.

(ii)　If an extremity is perfectly free, so that both $u$ and $v$ are arbitrary there, we have
$$C = 0, \quad y \sin \psi = 0,$$

at that extremity. From the first condition, it follows that

$$y(y - \lambda \cos \psi) = 0$$

is the equation of the characteristic curve; so we take

$$y - \lambda \cos \psi = 0$$

as the significant part of the equation, and it leads to a circle

$$(x - c)^2 + y^2 = \lambda^2,$$

where $c$ is arbitrary. But there is a second condition at the free extremity, viz.

$$y \sin \psi = 0,$$

so that, if the extremity does not lie on the axis of $x$, we have

$$\psi = 0.$$

Again, we have a circle with its centre on the axis of $x$.

We thus infer that, among the surfaces of revolution with assigned superficies, a sphere has the greatest volume.

(iii) If both the extremities are bound to lie on the axis, as at assigned points $A$ and $B$, we still have

$$y(y - \lambda \cos \psi) = 0.$$

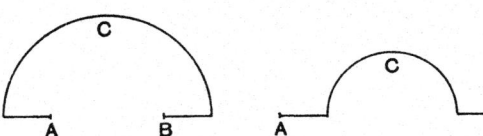

We take part of the axis $Ox$ and the semi-circle $y - \lambda \cos \psi = 0$ as the curve, choosing a configuration as indicated according to the magnitude of $S$.

*Ex.* 7. Prove that, for the surface, engendered by the revolution (round the axis of $x$) of the general curve

$$y^2 - \lambda y \frac{x}{(x_1^2 + y_1^2)^{\frac{1}{2}}} = C$$

in the preceding example, the mean measure of curvature (the sum of the two principal curvatures) is constant. (See § 289 *post*.)

*Ex.* 8. Prove that, among all curves of given length joining two points in a plane, the curve, which has its centroid with respect to the axis of $x$ nearest that axis, is a catenary.

(The two points are to be taken on the same side of the axis, and the length too small to let a joining curve meet the axis.)

*Ex.* 9. A curve of given length joins two points in a plane on the same side of the axis of $x$, and is revolved round that axis. Prove that the smallest surface engendered occurs, when the curve is a catenary with its convexity towards the axis; and that the largest surface engendered occurs, when the curve is a catenary with its concavity towards the axis.

Does any limitation upon the length exist, in either instance, for the existence of a maximum and a minimum under weak variations?

*Ex.* 10. A closed curve of given length passes through two points in a plane on the same side of the axis of $x$, and is revolved round that axis which it does not cut. Prove that, when the volume engendered is a maximum, the curve is composed of two arcs of the elastica, of which the equations are

$$x - x_0 = c\{E(u)\sec a - u\cos a\}, \quad y^2 = c^2(1 + \sin^2 a\,\mathrm{cn}\,u);$$

and that one of the arcs is convex and the other concave to the axis.

*Ex.* 11. Points $P$ and $Q$ are to be taken on the sides $AD$ and $BC$ of a rectangle $ABCD$, so that the sum of $AP$, $BQ$, and a curve joining $PQ$, is an assigned length. Prove that, when the area $APQBA$ is a maximum, $PQ$ is a semi-circle with its boundary diameter parallel to $AB$.

### *Strong variations for limited maxima and minima.*

**256.** In the investigations in limited problems from § 239 onwards, we have considered criteria which arise from weak variations. We now proceed to consider the effect of strong variations, restricting the discussion to the effect of strong variations of the simpler constituent types introduced in § 208.

We adopt the same method as before. The variations of general type are imposed upon the integral $I$, which has to be maximised or minimised, and upon the integral $J$, which has to be maintained at a constant value. Those variations are selected, which permit the conservation of the constant value for $J$; the due criteria are framed from the effect of these selected variations upon the value of $I$, as to whether it is a definite increase (implying a minimum for $I$) or a definite decrease (implying a maximum).

Accordingly, we take a range of the characteristic curve, limited by the conjugate (if there be a conjugate, under the Jacobi test) of the initial point. Within this range $AMP...$ (as in § 209), we take any point $P$; and at that point $P$, we draw a short arc $PQ$ in any direction off the characteristic curve. This arc $PQ$ can be taken of as general a form, as in the investigations concerning unrestricted maxima and minima: for the present purpose, it will be sufficient to take a small strong variation which can be represented by a small straight line $PQ$ in a diagram. Through $Q$, and through the initial point $A$, we draw a consecutive characteristic: such a curve can be drawn, because $Q$ is contiguous to a point $P$ within the range of the curve bounded by conjugates. We then consider, for $I$ and for $J$ separately, the values of the integrals along $ANQ$ and $QP$, and their respective increments for this path from $A$ to $P$ over their values for the characteristic path $AMP$.

The variation at any point in $AP$ is the customary weak variation, represented by $\kappa u$ and $\kappa v$, passing to the consecutive characteristic. At $P$, this variation is taken to be $\kappa\xi$ and $\kappa\eta$; so that, between $A$ and $P$, we have $u$ ranging continuously from 0 to $\xi$, and $v$ ranging continuously from 0 to $\eta$. The integral in the direction $QP$ is required. We therefore denote the direction-cosines of $QP$, from $Q$ to $P$, by $l$, $m$; hence, if $\sigma$ denote the length of the small straight line $PQ$ (where $\sigma$ is taken positive), we have

$$\kappa\xi = -l\sigma, \quad \kappa\eta = -m\sigma.$$

With the notation of § 239, we have

$$I_{AQ} - I_{AP} = \left[ \kappa \left( u \frac{\partial G}{\partial x_1} + v \frac{\partial G}{\partial y_1} \right) \right]_{0,0}^{\xi,\eta} + \kappa \int_{t_0}^{t} \mathfrak{E}_1 w \, dt + [\kappa^2]$$

$$= - \sigma \left( l \frac{\partial G}{\partial x_1} + m \frac{\partial G}{\partial y_1} \right) + \mu \int_{t_0}^{t} \mathfrak{E}_1 w \, dt + [\kappa^2],$$

where $t_0$ and $t$ are the values of $t$ at $A$ and at $P$, and $w = x_1 v - y_1 u$; and $[\kappa^2]$ denotes the aggregate of terms of the second and higher degrees in $\kappa$.

Again, if $X$ and $Y$ denote any point in $PQ$, we have

$$X = x - l(\sigma - t), \quad Y = y - m(\sigma - t),$$

where $t$ varies from 0 at $Q$ to $\sigma$ at $P$. Hence, for the integral taken from $Q$ to $P$, we have

$$\frac{dX}{dt} = l, \quad \frac{dY}{dt} = m;$$

and therefore

$$I_{QP} = \int_0^\sigma G\left( X, Y, \frac{dX}{dt}, \frac{dY}{dt} \right) dt$$

$$= \int_0^\sigma G \{ x - l(\sigma - t), y - m(\sigma - t), l, m \} \, dt$$

$$= \sigma G(x, y, l, m) + [\sigma^2],$$

where $[\sigma^2]$ is the aggregate of quantities of the second and higher degrees in $\sigma$.

When the terms of the second and higher orders are merged into a single aggregate $S_2$, we have

$$I_{AQ} + I_{QP} - I_{AP} = \sigma \left\{ G(x, y, l, m) - l \frac{\partial G}{\partial x_1} - m \frac{\partial G}{\partial y_1} \right\} + \kappa \int_{t_0}^{t} \mathfrak{E}_1 w \, dt + S_2,$$

as the expression for the increment of the integral $I$ when the path of integration is varied to $ANQP$ from $AMP$.

Proceeding in the same way with the integral $J$, we have

$$J_{AQ} + J_{QP} - J_{AP} = \sigma \left\{ \Phi(x, y, l, m) - l \frac{\partial \Phi}{\partial x_1} - m \frac{\partial \Phi}{\partial y_1} \right\} + \kappa \int_{t_0}^{t} \mathfrak{E}_0 w \, dt + T_2,$$

where $T_2$ is the aggregate of quantities of the second and higher orders. The whole integral $J$ is to be unaltered in value; also, for the remainder of the integral from $P$ along the characteristic path to the final upper limit $B$, there is no variation. Thus

$$J_{AQ} + J_{QP} + J_{PB} = J_{AP} + J_{PB},$$

and therefore

$$J_{AQ} + J_{QP} - J_{AP} = 0.$$

Hence, owing to the conservation of the value of $J$, the equation

$$\sigma \left\{ \Phi(x, y, l, m) - l \frac{\partial \Phi}{\partial x_1} - m \frac{\partial \Phi}{\partial y_1} \right\} + \kappa \int_{t_0}^{t} \mathfrak{E}_0 w \, dt + T_2 = 0$$

must be obeyed by all the small variations imposed. Consequently, these are the variations under which the integral $I$ can change.

Now along the characteristic curve, which is being tested, the equation

$$\mathfrak{E}_1 - \lambda \mathfrak{E}_0 = 0$$

is satisfied everywhere. Hence

$$I_{AQ} + I_{QP} - I_{AP} = \sigma \left\{ G\left(x, y, l, m\right) - l\frac{\partial G}{\partial x_1} - m\frac{\partial G}{\partial y_1} \right\} + \lambda\kappa \int_{t_0}^{t_1} \mathfrak{E}_1 w\, dt + S_2$$

$$= \sigma \left\{ F\left(x, y, l, m\right) - l\frac{\partial F}{\partial x_1} - m\frac{\partial F}{\partial y_1} \right\} + S_2 - \lambda T_2,$$

on taking account of the equation which limits the variations and again introducing the function $F, = G - \lambda\Phi$. We write

$$E_{IJ} = F\left(x, y, l, m\right) - l\frac{\partial F}{\partial x_1} - m\frac{\partial F}{\partial y_1};$$

and now

$$I_{AQ} + I_{QP} - I_{AP} = \sigma E_{IJ} + R_2,$$

where $R_2$ is the aggregate of terms $S_2 - \lambda T_2$, of the second and higher orders in the small quantity $\sigma$.

Clearly $R_2$ is negligible in comparison with the magnitude $\sigma E_{IJ}$, which is only of the first order in $\sigma$. Hence, effectively, the magnitude of the increment $I_{AQ} + I_{QP} - I_{AP}$ is measured by $\sigma E_{IJ}$; and $\sigma$ is a positive quantity. If, for every place $x$, $y$, successively and arbitrarily chosen on the characteristic curve, and for every direction $l$ and $m$ at that place, the function $E_{IJ}$ is positive, the foregoing increment is an increase: that is, the integral $I$ has a minimum value for strong variations, subject to the conservation of its constant value by the integral $J$. If, on the other hand, for the same range of variability at and off the characteristic curve, the function $E_{IJ}$ is negative, the foregoing increment is a decrease: that is, the integral $I$ has a maximum value for strong variations, subject to the conservation of its constant value by the integral $J$.

We thus have the $E$-function, and the Weierstrass test: the requirement is that the *E-function shall preserve an invariable sign, at all places on the characteristic curve and for all directions off the curve at every place.* When the sign is positive, the integral $I$ has a minimum; when the sign is negative, the integral $I$ has a maximum; in either event, for strong variations, while the integral $J$ maintains its constant assigned value.

*Note.* It will suffice now to refer to the general aggregate of tests, stated in § 236, for weak variations and strong variations combined. They need not be recapitulated in the present connection.

*Ex.* 1. Consider the $E$-test for the classical problem in Ex. 1 of § 255, requiring *a maximum area for a contour of assigned length.*

With the notation adopted, we have

$$G = \tfrac{1}{2}\left(xy_1 - yx_1\right), \quad \Phi = \left(x_1^2 + y_1^2\right)^{\frac{1}{2}},$$

$$F = \tfrac{1}{2}\left(xy_1 - yx_1\right) - \lambda\left(x_1^2 + y_1^2\right)^{\frac{1}{2}};$$

and the quantity $\lambda$ is positive, being the radius of the circle or of the circular arcs in the various cases. Here

$$F(x, y, l, m) = \tfrac{1}{2}(xm - yl) - \lambda(l^2 + m^2)^{\frac{1}{2}} = \tfrac{1}{2}(xm - yl) - \lambda,$$

$$\frac{\partial F}{\partial x_1} = -\tfrac{1}{2}y - \lambda\frac{x_1}{(x_1{}^2 + y_1{}^2)^{\frac{1}{2}}}, \quad \frac{\partial F}{\partial y_1} = \tfrac{1}{2}x - \lambda\frac{y_1}{(x_1{}^2 + y_1{}^2)^{\frac{1}{2}}};$$

and therefore

$$E_{IJ} = \tfrac{1}{2}(xm - yl) - \lambda - \left[ l\left\{ -\tfrac{1}{2}y - \lambda\frac{x_1}{(x_1{}^2 + y_1{}^2)^{\frac{1}{2}}} \right\} + m\left\{ \tfrac{1}{2}x - \lambda\frac{y_1}{(x_1{}^2 + y_1{}^2)^{\frac{1}{2}}} \right\} \right]$$

$$= -\lambda + \lambda\frac{lx_1 + my_1}{(x_1{}^2 + y_1{}^2)^{\frac{1}{2}}}$$

$$= -\lambda(1 - \cos\chi),$$

where $\chi$ is the angle between the direction $PQ$ off the characteristic curve and the tangent to the curve. For every such direction, $0 < \chi < 2\pi$; consequently $E_{IJ}$ is always negative.

Thus the area, provided by the characteristic curve under the assigned conditions, is a maximum for strong variations. It has been proved to satisfy the tests for a maximum under weak variations. It therefore is a true maximum.

*Ex.* 2. *Find, upon a sphere, a closed curve of assigned length less than the circumference of a great circle, such that the enclosed portion of the spherical surface (to be less than half the whole superficies) shall be as large as possible.*

When the radius of the sphere is taken as unity, any point on the sphere is

$$x = \sin\theta\cos\phi, \quad y = \sin\theta\sin\phi, \quad z = \cos\theta.$$

The curve is to be a closed curve: so any point within it can be taken as the pole of reference on the sphere and, therefore, as an origin for spherical polar coordinates $\theta$ and $\phi$. To determine a closed curve, we must potentially know $\theta$ as a function of $\phi$, or $\phi$ as a function of $\theta$, along its course; and $\phi$ will range from 0 to $2\pi$.

An element of area is $\sin\theta\,d\theta\,d\phi$; hence we can take *

$$A = \int_0^{2\pi} (1 - \cos\theta)\,d\phi,$$

where $\theta$ is a function of $\phi$ to be determined, explicitly or implicitly. Thus

$$G = (1 - \cos\theta)\phi_1.$$

The length of the whole arc of the curve is

$$L = \int_0^{2\pi} \left\{ \left(\frac{d\theta}{d\phi}\right)^2 + \sin^2\theta \right\}^{\frac{1}{2}} d\phi,$$

the range of the integral being the same as for the superficial area $A$. Thus

$$\Phi = \{\theta_1{}^2 + \phi_1{}^2\sin^2\theta\}^{\frac{1}{2}}.$$

Consequently, the function $F$ in the text is

$$F(\theta, \phi, \theta_1, \phi_1) = (1 - \cos\theta)\phi_1 - \lambda(\theta_1{}^2 + \phi_1{}^2\sin^2\theta)^{\frac{1}{2}}.$$

(I) As $\phi$ does not occur explicitly in $F$, but only its derivative $\phi_1$, we choose the characteristic equation

$$\frac{\partial F}{\partial\phi} - \frac{d}{dt}\left(\frac{\partial F}{\partial\phi_1}\right) = 0.$$

* We might equally take

$$A = \int \phi\sin\theta\,d\theta;$$

but there then is a further analytical difficulty of determining the limits of $\theta$, for the purpose of defining the integral. Much of the succeeding analysis will, of course, be applicable to this alternative form.

A first integral is

$$\frac{\partial F}{\partial \phi_1} = \text{constant} = 1 - a,$$

where $a$ is arbitrary; and therefore

$$a - \cos\theta - \lambda\,\frac{\phi_1\sin^2\theta}{(\theta_1{}^2 + \phi_1{}^2\sin^2\theta)^{\frac{1}{2}}} = 0.$$

We now make $\theta$ temporarily the independent variable, for the construction of the primitive of the characteristic equation; and we write $\phi' = \dfrac{d\phi}{d\theta} = \dfrac{\phi_1}{\theta_1}$. Thus

$$\phi'^2\sin^2\theta = \frac{(a - \cos\theta)^2}{\lambda^2 - a^2 + 2a\cos\theta - (\lambda^2 + 1)\cos^2\theta}.$$

Let

$$\lambda = \tan\gamma, \quad a = \cos a\,\sec\gamma;$$

the equation becomes

$$\phi'^2\sin^2\theta = -\frac{(\cos a - \cos\gamma\cos\theta)^2}{\cos^2\theta - 2\cos a\cos\gamma\cos\theta + \cos^2\gamma - \sin^2 a},$$

leading to the required primitive

$$\phi - \beta = \cos^{-1}\frac{\cos\gamma - \cos a\cos\theta}{\sin a\sin\theta},$$

in which $\beta$ and $a$ are the arbitrary constants of integration, while $\gamma\,(=\tan^{-1}\lambda)$ is the constant introduced by the general theory.

This primitive is

$$\sin a\cos\beta\sin\theta\cos\phi + \sin a\sin\beta\sin\theta\sin\phi + \cos a\cos\theta = \cos\gamma,$$

that is,

$$x\sin a\cos\beta + y\sin a\sin\beta + z\cos a = \cos\gamma.$$

The curve is a small circle of angular radius $\gamma$, with its own pole at an angular distance $a$ from the pole of reference: a result also to be inferred from the formula

$$\cos\gamma = \cos a\cos\theta + \sin a\sin\theta\cos(\phi - \beta),$$

of the spherical triangle.

(II) For the Legendre test, we have

$$-\frac{1}{\theta_1\phi_1}\frac{\partial^2 F}{\partial\theta_1\,\partial\phi_1} = -\frac{\lambda\sin^2\theta}{(\theta_1{}^2 + \phi_1{}^2\sin^2\theta)^{\frac{3}{2}}},$$

a quantity which always is negative. Hence the test admits the characteristic curve (being a small circle) as providing a maximum.

(III) For the Jacobi test, we substitute for the analysis (in the present instance) the more easily used property that a conjugate is provided by drawing a consecutive characteristic curve with the imposed limitations, that is, another small circle through the point with the same angular radius. Now all small circles, of the same size on a sphere, have their space-centres on a concentric sphere which is touched by their planes: hence, if $PC$ be the radius of the plane of the small circle which is being tested, the radius $PC'$ of a consecutive characteristic small circle in its own plane will be a consecutive generator of the cone $PCC'$, enveloping the concentric sphere. The small 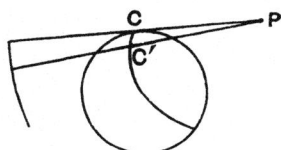 circle, in the plane $PC'$ perpendicular to the plane of the paper and beginning at $P$ in that plane, meets the small circle, in the plane $PC$ also perpendicular to the plane of the paper and also beginning at $P$ in its own plane, only at $P$ again, as the circumference is being completed.

Thus the conjugate of $P$ under the Jacobi test lies only at the end of a complete tour of the circumference. There is no conjugate to limit the range, within the assigned conditions.

But any such circle so drawn through $P$ is not unique. The small circle, which is the section of the given sphere by any tangent plane of the enveloping cone $PCC'$, will provide an adequate admissible minimum. The lack of uniqueness is, in fact, analogous to the lack of uniqueness on the part of a geodesic semi-circle on a sphere between two diametrically opposite points.

*Note.* The analysis, in the construction of the Jacobi test, can be simplified by the introduction of the angle $\omega, = CAP$. We have, in the figure

$$CA = CP = \gamma, \quad OC = a, \quad OP = \theta;$$
$$POA = t = \phi - \beta.$$

Then

$$\sin \gamma \sin \omega = \sin \theta \sin (\phi - \beta),$$
$$\cos \theta = \cos a \cos \gamma - \sin a \sin \gamma \cos \omega,$$
$$\cos \gamma = \cos a \cos \theta + \sin a \sin \theta \cos (\phi - \beta),$$
$$\frac{\sin \theta \, d\phi}{\cos a \sin \theta - \sin a \cos \theta \cos (\phi - \beta)} = \frac{-d\theta}{\sin a \sin (\phi - \beta)} = d\omega,$$
$$\sin \gamma \, d\omega = (d\theta^2 + \sin^2 \theta \, d\phi^2)^{\frac{1}{2}}.$$

The construction of the formula for the test is left as an exercise.

(IV) For the Weierstrass test, we have

$$G = (1 - \cos \theta) \phi_1, \quad \Phi = (\theta_1{}^2 + \phi_1{}^2 \sin^2 \theta)^{\frac{1}{2}}.$$

The variation is affected by changing $\theta$ and $\phi$ into $\theta + \kappa \xi$ and $\phi + \kappa \eta$; and

$$\kappa \xi = -l\sigma, \quad \kappa \eta = -m\sigma.$$

A place on the small line $PQ$ is given by

$$\bar{\theta} = \theta - l(\sigma - t), \quad \bar{\phi} = \phi - m(\sigma - t);$$

and

$$F(\theta, \phi, l, m) = (1 - \cos \theta) m - \lambda (l^2 + m^2 \sin^2 \theta)^{\frac{1}{2}}.$$

Also

$$\frac{\partial F}{\partial \theta_1} = -\frac{\lambda \theta_1}{(\theta_1{}^2 + \phi_1{}^2 \sin^2 \theta)^{\frac{1}{2}}},$$
$$\frac{\partial F}{\partial \phi_1} = (1 - \cos \theta) - \frac{\lambda \phi_1 \sin^2 \theta}{(\theta_1{}^2 + \phi_1{}^2 \sin^2 \theta)^{\frac{1}{2}}};$$

and therefore

$$E_{IJ} = (1 - \cos \theta) m - \lambda (l^2 + m^2 \sin^2 \theta)^{\frac{1}{2}} - l \left\{ -\frac{\lambda \theta_1}{(\theta_1{}^2 + \phi_1{}^2 \sin^2 \theta)^{\frac{1}{2}}} \right\}$$
$$- m \left\{ (1 - \cos \theta) - \frac{\lambda \phi_1 \sin^2 \theta}{(\theta_1{}^2 + \phi_1{}^2 \sin^2 \theta)^{\frac{1}{2}}} \right\}$$
$$= -\lambda (l^2 + m^2 \sin^2 \theta)^{\frac{1}{2}} \left\{ 1 - \frac{l\theta_1 + m\phi_1 \sin^2 \theta}{(l^2 + m^2 \sin^2 \theta)^{\frac{1}{2}} (\theta_1{}^2 + \phi_1{}^2 \sin^2 \theta)^{\frac{1}{2}}} \right\}.$$

Now

$$(l^2 + m^2 \sin^2 \theta)(\theta_1{}^2 + \phi_1{}^2 \sin^2 \theta)^{\frac{1}{2}} - (l\theta_1 + m\phi_1 \sin^2 \theta)^2 = (l\phi_1 - m\theta_1)^2 \sin^2 \theta$$
$$> 0,$$

for all variations $l\phi_1 - m\theta_1 \gtrless 0$, that is, off the small circle. Hence $E_{IJ}$ is negative for all values of $\theta$, and for all directions $l, m_i$ at $\theta$ which are off the curve.

(V) Thus the small circle provides a true maximum. For complete and unique definition, all that is required is merely specification of its position ; the angular magnitude is settled by the datum of initially assigned length.

*Ex.* 3. A closed curve of given length is to be drawn upon the surface of a right cone, so as to enclose the greatest amount of surface of the cone. Prove that, when the cone is developed, the curve is a circle.

*Ex.* 4. A curve of given length is to be drawn from one point on the equator of a sphere to another point on the equator, so that the area of the sphere included between the curve and the equator is as great as possible. Shew that, when the length is less than half the equator, the required curve is a small circle having its pole in the equator.

*Ex.* 5. A fixed curve is given on a surface. Another curve of given length is to be drawn on the surface, so that the area enclosed between the two curves is a maximum. Shew that

(i) when a developable, touching the surface along the required curve, is developed, the trace of the curve becomes circular : and

(ii) if the curve is drawn from a free point of the given curve to another free point, the two curves cut orthogonally at each point.

(Moigno-Lindelöf.)

*Isoperimetrical problems, with derivatives of order higher than the first.*

**257.** We pass to the consideration of integrals, involving one dependent variable and its derivatives of order higher than the first, and of integrals involving more than one dependent variable, when such integrals are to be maximised or minimised subject to the possession of constant values by one or more coexisting integrals. The method is the same as that adopted for integrals involving only the first derivative of one dependent variable. We adopt what may be called a distributed variation of the character previously indicated (§§ 73, 239) ; and we use it, first of all, to satisfy the demands of the constant integral or integrals, before maximising or minimising the integral capable of varied values.

Accordingly, we proceed to find the conditions for a maximum or a minimum of an integral

$$I = \int H\left(x,\, x_1,\, x_2,\, y,\, y_1,\, y_2\right) dt,$$

subject to the requirement that a coexisting integral

$$J = \int \Psi\left(x,\, x_1,\, x_2,\, y,\, y_1,\, y_2\right) dt$$

shall acquire a constant value for all variations. The functions $H$ and $\Psi$ (as in § 97) satisfy the identities

$$0 = x_1 \frac{\partial H}{\partial x_2} + y_1 \frac{\partial H}{\partial y_2}, \quad H = x_1 \frac{\partial H}{\partial x_1} + y_1 \frac{\partial H}{\partial y_1} + 2 x_2 \frac{\partial H}{\partial x_2} + 2 y_2 \frac{\partial H}{\partial y_2},$$

$$0 = x_1 \frac{\partial \Psi}{\partial x_2} + y_1 \frac{\partial \Psi}{\partial y_2}, \quad \Psi = x_1 \frac{\partial \Psi}{\partial x_1} + y_1 \frac{\partial \Psi}{\partial y_1} + 2 x_2 \frac{\partial \Psi}{\partial x_2} + 2 y_2 \frac{\partial \Psi}{\partial y_2}.$$

We take composite variations for $x$ and for $y$ in the form

$$\kappa u + \kappa' u' + \kappa'' u'' + \ldots + \kappa^{(n)} u^{(n)},$$
$$\kappa v + \kappa' v' + \kappa'' v'' + \ldots + \kappa^{(n)} v^{(n)},$$

where $\kappa$, $\kappa'$, $\kappa''$, ... are arbitrary small constants, and where $u$, $u'$, $u''$, ..., $v$, $v'$, $v''$, ... are arbitrary regular functions of $t$. Let $\mathbf{I}$ denote the varied value of $I$; then, as before, we find

$$\mathbf{I} - I = \Sigma\kappa \int \left[ u \left\{ \frac{\partial H}{\partial x} - \frac{d}{dt}\left(\frac{\partial H}{\partial x_1}\right) + \frac{d^2}{dt^2}\left(\frac{\partial H}{\partial x_2}\right) \right\} + v \left\{ \frac{\partial H}{\partial y} - \frac{d}{dt}\left(\frac{\partial H}{\partial y_1}\right) + \frac{d^2}{dt^2}\left(\frac{\partial H}{\partial y_2}\right) \right\} \right] dt$$
$$+ \Sigma\kappa \left[ u \left\{ \frac{\partial H}{\partial x_1} - \frac{d}{dt}\left(\frac{\partial H}{\partial x_2}\right) \right\} + u_1 \frac{\partial H}{\partial x_2} + v \left\{ \frac{\partial H}{\partial y_1} - \frac{d}{dt}\left(\frac{\partial H}{\partial y_2}\right) \right\} + v_1 \frac{\partial H}{\partial y_2} \right] + R_2,$$

where $R_2$ is the aggregate of terms of the second and higher degrees in the small arbitrary constants $\kappa$, and the summation is taken over the set of simultaneous variations $\kappa u$ and $\kappa v$, $\kappa' u'$ and $\kappa' v'$, ..., $\kappa^{(n)} u^{(n)}$ and $\kappa^{(n)} v^{(n)}$.

Also, from the constancy of $J$, we have

$$0 = \Sigma\kappa \int \left[ u \left\{ \frac{\partial \Psi}{\partial x} - \frac{d}{dt}\left(\frac{\partial \Psi}{\partial x_1}\right) + \frac{d^2}{dt^2}\left(\frac{\partial \Psi}{\partial x_2}\right) \right\} + v \left\{ \frac{\partial \Psi}{\partial y} - \frac{d}{dt}\left(\frac{\partial \Psi}{\partial y_1}\right) + \frac{d^2}{dt^2}\left(\frac{\partial \Psi}{\partial y_2}\right) \right\} \right] dt$$
$$+ \Sigma\kappa \left[ u \left\{ \frac{\partial \Psi}{\partial x_1} - \frac{d}{dt}\left(\frac{\partial \Psi}{\partial x_2}\right) \right\} + u_1 \frac{\partial \Psi}{\partial x_2} + v \left\{ \frac{\partial \Psi}{\partial y_1} - \frac{d}{dt}\left(\frac{\partial \Psi}{\partial y_2}\right) \right\} + v_1 \frac{\partial \Psi}{\partial y_2} \right] + S_2,$$

where $S_2$ is an aggregate similar to $R_2$, and the summation is for the same set of variations. Further, we assume the quantities $\kappa$, $\kappa'$, $\kappa''$, ... to be so small, that terms of any degree in them are negligible in comparison with terms of lower degree.

*Characteristic equations.*

**258.** Consider, first, the small variations of $x$ and $y$ which vanish at each limit. Thus, if we write

$$\int \left[ u^{(r)} \left\{ \frac{\partial H}{\partial x} - \frac{d}{dt}\left(\frac{\partial H}{\partial x_1}\right) + \frac{d^2}{dt^2}\left(\frac{\partial H}{\partial x_2}\right) \right\} + v^{(r)} \left\{ \frac{\partial H}{\partial y} - \frac{d}{dt}\left(\frac{\partial H}{\partial y_1}\right) + \frac{d^2}{dt^2}\left(\frac{\partial H}{\partial y_2}\right) \right\} \right] dt = H^{(r)},$$

and use $\Psi^{(r)}$ to denote the corresponding integral derived from $\Psi$, we have, for such variations,

$$\mathbf{I} - I = \Sigma\kappa^{(r)} H^{(r)} + R_2,$$
$$0 = \Sigma\kappa^{(r)} \Psi^{(r)} + S_2,$$

where $R_2$ and $S_2$ are negligible compared with $\Sigma\kappa H$ and $\Sigma\kappa \Psi$ respectively, unless these two sums vanish. Now for a maximum or a minimum of $I$, its first variation must vanish. The whole variation of the constant integral $J$ must vanish for all variations, and therefore its first variation vanishes. Thus

$$\kappa\Psi^{(0)} + \kappa'\Psi' + \kappa''\Psi'' + \ldots = 0,$$
$$\kappa H^{(0)} + \kappa' H' + \kappa'' H'' + \ldots = 0,$$

the former of which (associated with $J$) is a permanent requirement, and the latter (associated with $I$) a conditional requirement. Thus the latter becomes

$$\kappa' \left\{ H' - \frac{\Psi'}{\Psi^{(0)}} H^{(0)} \right\} + \kappa'' \left\{ H'' - \frac{\Psi''}{\Psi^{(0)}} H^{(0)} \right\} + \ldots = 0.$$

This new equation, linear in $\kappa', \kappa'', \ldots$, which are arbitrary constants, does not involve $\kappa$ and therefore does not depend upon $\Sigma \kappa^{(r)} \Psi^{(r)} = 0$; consequently, it can be satisfied only if

$$\frac{H^{(0)}}{\Psi^{(0)}} = \frac{H'}{\Psi'} = \frac{H''}{\Psi''} = \ldots = \lambda,$$

where the common value $\lambda$ is a constant, because the limits of the integrals are fixed. Moreover $H^{(0)}$ and $\Psi^{(0)}$ involve two quantities $u^{(0)}$ and $v^{(0)}$, which are arbitrary regular functions of $t$, independent of one another; hence $\lambda$ is an arbitrary constant.

Now let

$$F(x, x_1, x_2, y, y_1, y_2) = H(x, x_1, x_2, y, y_1, y_2) - \lambda \Psi(x, x_1, x_2, y, y_1, y_2),$$

so that $F$ satisfies the two identities of § 97, exactly similar to those satisfied by $H$ and by $\Psi$. Then, taking account of the constancy of $J$, and combining the two sets of integrals in the form

$$H^{(0)} - \lambda \Psi^{(0)} = 0, \quad H' - \lambda \Psi' = 0, \ldots,$$

we have

$$\int \left[ u^{(r)} \left\{ \frac{\partial F}{\partial x} - \frac{d}{dt} \left( \frac{\partial F}{\partial x_1} \right) + \frac{d^2}{dt^2} \left( \frac{\partial F}{\partial x_2} \right) \right\} + v^{(r)} \left\{ \frac{\partial F}{\partial y} - \frac{d}{dt} \left( \frac{\partial F}{\partial y_1} \right) + \frac{d^2}{dt^2} \left( \frac{\partial F}{\partial y_2} \right) \right\} \right] dt = 0,$$

for all the sets $u^{(r)}$ and $v^{(r)}$. But $u^{(r)}$ and $v^{(r)}$ are arbitrary regular functions of $t$, independent of one another; and therefore, as before, the relation can be satisfied only if

$$\frac{\partial F}{\partial x} - \frac{d}{dt} \left( \frac{\partial F}{\partial x_1} \right) + \frac{d^2}{dt^2} \left( \frac{\partial F}{\partial x_2} \right) = 0,$$

$$\frac{\partial F}{\partial y} - \frac{d}{dt} \left( \frac{\partial F}{\partial y_1} \right) + \frac{d^2}{dt^2} \left( \frac{\partial F}{\partial y_2} \right) = 0.$$

Further, when these characteristic equations hold, all the relations

$$H^{(r)} - \lambda \Psi^{(r)} = 0$$

are satisfied.

### Terminal conditions.

**259.** Next, consider variations which do not vanish at the limits. As the characteristic equations are satisfied, the terms of the first variation in $\mathbf{I} - I$ have become

$$\Sigma \kappa u \left\{ \frac{\partial F}{\partial x_1} - \frac{d}{dt} \left( \frac{\partial F}{\partial x_2} \right) \right\} + \Sigma \kappa u_1 \frac{\partial F}{\partial x_2} + \Sigma \kappa v \left\{ \frac{\partial F}{\partial y_1} - \frac{d}{dt} \left( \frac{\partial F}{\partial y_2} \right) \right\} + \Sigma \kappa v_1 \frac{\partial F}{\partial y_2} = 0;$$

or, as $\Sigma\kappa u$ and $\Sigma\kappa v$ represent the full small variations of $x$ and of $y$ respectively, the relation

$$U\left\{\frac{\partial F}{\partial x_1} - \frac{d}{dt}\left(\frac{\partial F}{\partial x_2}\right)\right\} + U_1\frac{\partial F}{\partial x_2} + V\left\{\frac{\partial F}{\partial y_1} - \frac{d}{dt}\left(\frac{\partial F}{\partial y_2}\right)\right\} + V_1\frac{\partial F}{\partial y_2} = 0$$

must be satisfied at each extremity of the integral for any small variation $kU$ and $kV$, where $k$ is an arbitrary small constant quantity.

### Second variation.

**260.** The complete first variation in $\mathbf{I} - I$ thus vanishes. The whole residual variation is

$$R_2 - \lambda S_2,$$

which is the same as the residual variation for the function

$$G = H - \lambda\Psi,$$

such as was considered in §§ 105–110.

In fact, when we take this function $G$, thus composed from $H$ and $\Psi$ in such a way as to include the requirement that, so far as variations are concerned, $J$ shall be constant, all the subsequent analysis follows exactly in every detail all the analysis just quoted, leading to the Legendre test, to the Jacobi test, the general characteristic, consecutive characteristics, and conjugate points limiting complete ranges, as well as the proposition (§ 102) that

$$\frac{\partial G}{\partial x_2}, \quad \frac{\partial G}{\partial x_1} - \frac{d}{dt}\left(\frac{\partial G}{\partial x_2}\right), \quad \frac{\partial G}{\partial y_2}, \quad \frac{\partial G}{\partial y_1} - \frac{d}{dt}\left(\frac{\partial G}{\partial y_2}\right),$$

are continuous in value at places of discontinuous direction or curvature on the characteristic curve—the number of such places (if any) in any finite range being limited.

### Isoperimetrical problems, with several dependent variables.

**261.** When it is required to find the maximum or the minimum of an integral the integrand of which involves several dependent variables and their derivatives, while concurrent integrals (one or more) involving those dependent variables as well as some or all of their derivatives are to maintain assigned values, a precisely similar argument leads to an analogous result. It therefore seems unnecessary to repeat the argument in detail, where the only modifications, that arise, affect the dependent variables simultaneously without any change except as regards their number.

The arbitrary small variations of $x, y, z, \ldots$ are distributed as in §§ 73, 239. When these are denoted by $\Sigma\kappa_r u_r, \Sigma\kappa_r v_r, \Sigma\kappa_r w_r, \ldots$, we have relations

$$\sum_{r=1}\int\left[\kappa_r\left\{\frac{\partial\phi_n}{\partial x} - \frac{d}{dt}\left(\frac{\partial\phi_n}{\partial x_1}\right) + \ldots\right\}u_r + \kappa_r\left\{\frac{\partial\phi_n}{\partial y} - \frac{d}{dt}\left(\frac{\partial\phi_n}{\partial y_1}\right) + \ldots\right\}v_r\right.$$
$$\left. + \kappa_r\left\{\frac{\partial\phi_n}{\partial z} - \frac{d}{dt}\left(\frac{\partial\phi_n}{\partial z_1}\right) + \ldots\right\}w_r + \ldots\right]dt = 0,$$

arising from the constancy of the integrals

$$J_n = \int \phi_n (x, x_1, \ldots, y, y_1, \ldots, z, z_1, \ldots, \ldots)\, dt,$$

for $n = 1, \ldots$. Thus we have relations, as persistent requirements,

$$\kappa_1 J_{11} + \kappa_2 J_{12} + \kappa_3 J_{13} + \ldots = 0,$$
$$\kappa_1 J_{21} + \kappa_2 J_{22} + \kappa_3 J_{23} + \ldots = 0,$$

and so on, there being one such relation corresponding to each constant integral. Also, from the vanishing first variation of the integral

$$\int \phi (x, x_1, \ldots, y, y_1, \ldots, z, z_1, \ldots, \ldots)\, dt,$$

we have a condition (arising from variations that vanish at the limit)

$$\kappa_1 I_1 + \kappa_2 I_2 + \kappa_3 I_3 + \ldots = 0,$$

where

$$I_r = \int \left[ \left\{ \frac{\partial \phi}{\partial x} - \frac{d}{dt}\left(\frac{\partial \phi}{\partial x_1}\right) + \ldots \right\} u_r + \left\{ \frac{\partial \phi}{\partial y} - \frac{d}{dt}\left(\frac{\partial \phi}{\partial y_1}\right) + \ldots \right\} v_r \right. $$
$$\left. + \left\{ \frac{\partial \phi}{\partial z} - \frac{d}{dt}\left(\frac{\partial \phi}{\partial z_1}\right) + \ldots \right\} w_r + \ldots \right] dt.$$

Let there be $m$ constant integrals $J$; then there exist, as satisfying the $m$ requirements from the integrals $J$ and the single condition from the integral $I$, $m$ constants $\lambda_1, \ldots, \lambda_m$, such that

$$I_1 = \lambda_1 J_{11} + \lambda_2 J_{21} + \ldots + \lambda_m J_{m1},$$
$$I_2 = \lambda_1 J_{12} + \lambda_2 J_{22} + \ldots + \lambda_m J_{m2},$$
$$\ldots\ldots\ldots\ldots\ldots\ldots\ldots\ldots\ldots\ldots\ldots\ldots\ldots$$
$$I_r = \lambda_1 J_{1r} + \lambda_2 J_{2r} + \ldots + \lambda_m J_{mr},$$

and so on—there being as many such relations (not fewer than $m$) as there are dependent variables. Moreover, each of the integrals $I_l$ involves sets of quantities $u_l, v_l, w_l, \ldots$; and likewise for each of the integrals $J_{pq}$. Each such set of quantities, by itself, involves a number of quantities equal to the number of dependent variables. Therefore each of the $m$ constants $\lambda$, deduced by taking any $m$ (say the first $m$) of these relations, is arbitrary.

Thus, when we take a modified integral

$$\int (\phi - \lambda_1 \phi_1 - \lambda_2 \phi_2 - \ldots - \lambda_m \phi_m)\, dt,$$

with constant values for $\lambda$, all the aggregate of integral terms in the first variation of the new integral vanish. Then, in order that the remainder of the first variation may vanish, we must have the quantity

$$u \frac{\partial}{\partial x_1}(\phi - \lambda_1 \phi_1 - \ldots - \lambda_m \phi_m) + v \frac{\partial}{\partial y_1}(\phi - \lambda_1 \phi_1 - \ldots - \lambda_m \phi_m)$$
$$+ w \frac{\partial}{\partial z_1}(\phi - \lambda_1 \phi_1 - \ldots - \lambda_m \phi_m) + \ldots$$

$$- u' \frac{\partial}{\partial x_2} (\phi - \lambda_1 \phi_1 - \dots - \lambda_m \phi_m) - v' \frac{\partial}{\partial y_2} (\phi - \lambda_1 \phi_1 - \dots - \lambda_m \phi_m)$$

$$- w' \frac{\partial}{\partial z_2} (\phi - \lambda_1 \phi_1 - \dots - \lambda_m \phi_m) + \dots,$$

where $\kappa u = \Sigma \kappa_r u_r$, $\kappa v = \Sigma \kappa_r v_r$, $\kappa w = \Sigma \kappa_r w_r$, ..., equal to zero, both at the upper limit and at the lower limit of the integral.

Accordingly, we use a new function $F$, where

$$F = \phi - \lambda_1 \phi_1 - \dots - \lambda_m \phi_m,$$

full regard having been paid to the constancy of the $m$ integrals, both in the construction of the $m$ constants and in all the constructed equations. The net result is that the whole of the analysis relating to the general problem of the limited variations is, so far as concerns all essential equations, the same as it would be for an unconditioned function $F$, once it has been constructed in the form just stated. That is, the characteristic equations and the subsidiary equations, the Legendre test, and the form of the second variation, are all expressible in terms of the function $F$ alone. The differences, for the purpose of proceeding to the detailed completion, are similar in kind to those already seen in the earlier sections of this chapter. The $m$ arbitrary constants, used in the formation of $F$, occur in the primitive of the characteristic equations as well as the arbitrary constants of integration; and the constant values of the $m$ given integrals provide $m$ additional conditions for the determination of the $m$ additional constants. Further differences, similar also to those already found, occur in the construction of the primitive of the subsidiary equations and in the determination of a complete range of integration bound by conjugate values of the independent variable. The stages in the calculations are of the same character for the wider problems, as they are for the simplest problem that already has been discussed. The detailed development of the analysis naturally is more elaborate, but only because of the more extensive demands of the more general problems.

*Ex.* 1. *A curve of given length is to join two points, not lying in one plane with a given straight line; and its moment about the line is to be a maximum or minimum. Find the curve.*

Let cylindrical coordinates be chosen, with the given line $Ox$ as axis, so that any point is $x$, $r \cos \theta$, $r \sin \theta$. The element of arc is

$$(x'^2 + r'^2 + r^2 \theta'^2)^{\frac{1}{2}} dt \, ;$$

hence the foregoing general theory requires us to apply the criteria to the integral

$$\int (r - \lambda) (x'^2 + r'^2 + r^2 \theta'^2)^{\frac{1}{2}} dt.$$

(A) As neither $x$ nor $\theta$ occurs explicitly in the subject of integration, first integrals of two of the characteristic equations are

$$\frac{\partial F}{\partial x'} = a,$$

$$\frac{\partial F}{\partial \theta'} = c,$$

where $a$ and $c$ are arbitrary constants. Moreover, by an earlier theorem (§ 160), it is sufficient to take two of the three characteristic equations. When substitution takes place, and we introduce the arc-length as the independent variable now that the characteristic equations have been formed, the foregoing equations can be transformed to become

$$(r-\lambda)\frac{dx}{ds}=a, \quad (r-\lambda)\,r^2\frac{d\theta}{ds}=c.$$

We always have

$$\left(\frac{dx}{ds}\right)^2+\left(\frac{dr}{ds}\right)^2+r^2\left(\frac{d\theta}{ds}\right)^2=1\,;$$

hence

$$ds=\frac{r\,(r-\lambda)}{\{r^2\,(r-\lambda)^2+a^2r^2+c^2\}^{\frac{1}{2}}}\,dr.$$

Let

$$r^2\,(r-\lambda)^2+a^2r^2+c^2=(r-e_1)\,(r-e_2)\,(r-e_3)\,(r-e_4),$$

$$k^2=\frac{(e_1-e_4)\,(e_2-e_3)}{(e_1-e_3)\,(e_2-e_4)},$$

$$r=\frac{e_2\,(e_1-e_3)-e_1\,(e_2-e_3)\sin^2\theta}{e_1-e_3-(e_2-e_3)\sin^2\theta}\,;$$

then

$$ds=2\frac{r\,(r-\lambda)}{\{(e_1-e_3)\,(e_2-e_4)\}^{\frac{1}{2}}}\frac{d\theta}{(1-k^2\sin^2\theta)^{\frac{1}{2}}}.$$

Using the notation of Jacobian elliptic functions, let

$$\theta=\operatorname{am}\gamma t,$$

where

$$\gamma=\{(e_1-e_3)\,(e_2-e_4)\}^{\frac{1}{2}}\,;$$

we have

$$ds=2r\,(r-\lambda)\,dt,$$

$$dx=2ar\,dt,$$

$$d\theta=2\frac{c}{r}\,dt,$$

$$r=\frac{e_2\,(e_1-e_3)-e_1\,(e_2-e_3)\operatorname{sn}^2\gamma t}{e_1-e_3-(e_2-e_3)\operatorname{sn}^2\gamma t}$$

As regards the arbitrary constants, and the data serving to determine the constants, we have

$$x=a+2a\int_{t_0}^{t}r\,dt,$$

$$\theta=\beta+2c\int_{t_0}^{t}\frac{dt}{r}\,,$$

$$l=2\int_{t_0}^{t_1}r\,(r-\lambda)\,dt,$$

where $t_0$ and $t_1$ are the limits of $t$ for the integral; $l$ is the given length of the curve; $a$, $\beta$, $a$, $c$, are the parameters in the integration of the primitive; and $\lambda$ is the arbitrary constant of the general theory. Let $x_0$ and $\theta_0$ be the values of $x$ and of $\theta$ when $t=t_0$, where ultimately $x_0$ and $\theta_0$ are unessential constants; and let $x_1$ and $\theta_1$ be their values when $t=t_1$: then

$$a=x_0, \quad \beta=\theta_0.$$

But $x_1-x_0$ and $\theta_1-\theta_0$ are known quantities, because the curve must pass through fixed points; consequently,

$$x_1-x_0=2a\int_{t_0}^{t_1}r\,dt, \quad \theta_1-\theta_0=2c\int_{t_0}^{t_1}\frac{dt}{r}\,.$$

Also, $r_0$ and $r_1$ are known constants; hence

$$r_0 = \frac{e_2(e_1 - e_3) - e_1(e_2 - e_3)\operatorname{sn}^2 \gamma t_0}{e_1 - e_3 - (e_2 - e_3)\operatorname{sn}^2 \gamma t_0},$$

$$r_1 = \frac{e_2(e_1 - e_3) - e_1(e_2 - e_3)\operatorname{sn}^2 \gamma t_1}{e_1 - e_3 - (e_2 - e_3)\operatorname{sn}^2 \gamma t_1}.$$

These four equations, together with

$$l = 2 \int_{t_0}^{t_1} r(r - \lambda)\,dt,$$

are potentially sufficient for the determination of $t_0$, $t_1$, $a$, $c$, $\lambda$.

(B) The construction of the Jacobi test, which leads to the equation for the determination of the conjugate of the initial lower limit, is manifestly an elaborate exercise in elliptic functions.

When the two terminal points lie in a plane through the axis of $x$, the relation

$$\theta_1 - \theta_0 = 0$$

holds. Thus $c = 0$; that is, $\theta$ is a constant, and the curve is plane. The integral equations are expressible by means of hyperbolic functions: and the curve is a catenary. (The analysis is then the same as in Ex. 1, § 30.)

(C) For the Legendre test in general, we have

$$\frac{\partial^2 F}{\partial r'^2} = \frac{x'^2 + r^2\theta'^2}{(x'^2 + r'^2 + r^2\theta'^2)^{\frac{3}{2}}}(r - \lambda),$$

$$\frac{\partial^2 F}{\partial \theta'^2} = \frac{r^2(x'^2 + r'^2)}{(x'^2 + r'^2 + r^2\theta'^2)^{\frac{3}{2}}}(r - \lambda),$$

$$\frac{\partial^2 F}{\partial r'\partial \theta'} = \frac{r^2\theta'r'}{(x'^2 + r'^2 + r^2\theta'^2)^{\frac{3}{2}}}(r - \lambda).$$

Thus the first part of the test, viz. that

$$\frac{\partial^2 F}{\partial r'^2}\frac{\partial^2 F}{\partial \theta'^2} - \left(\frac{\partial^2 F}{\partial r'\partial \theta'}\right)^2, \quad = \frac{r^2 x'^2}{(x'^2 + r'^2 + r^2\theta'^2)^{\frac{1}{2}}}(r - \lambda)^2,$$

should always be positive, is satisfied. Further, $r$ cannot be equal to $\lambda$, for $\dfrac{dx}{ds}$ is finite and $a$ clearly could not be zero: hence $r$ is either greater than $\lambda$ or less than $\lambda$.

When $r > \lambda$, both $\dfrac{\partial^2 F}{\partial r'^2}$ and $\dfrac{\partial^2 F}{\partial \theta'^2}$ are positive; the curve is convex to the axis of moment, and it provides a minimum moment for the given length.

When $r < \lambda$, both these derivatives are negative; the curve is concave to the axis of moment, and it provides a maximum moment for the given length.

(D) The Weierstrass test is constructed as in § 256.

We take a strong elementary variation of an arc of the characteristic curve from an initial point $A$ up to any point $P$ within a Jacobi range, by drawing a consecutive characteristic curve through $A$ and through a point $Q$ contiguous to $P$, where $PQ$ is a small straight line in an arbitrary direction at $P$. For the displacement from $P$ to $Q$, we change $x$, $r$, $\theta$, into $x + \kappa\xi$, $r + \kappa\rho$, $\theta + \kappa\vartheta$; and we take

$$\kappa\xi = -l\sigma, \quad \kappa\rho = -m\sigma, \quad r\kappa\vartheta = -n\sigma,$$

so that $l$, $m$, $n$, are the direction-cosines of the small line $\overrightarrow{QP}$, the length of which is $\sigma$. Any position on $PQ$ is given by

$$X = x - l\,(\sigma - t),$$
$$R = r - m\,(\sigma - t),$$
$$\Theta = \theta - \frac{n}{r}(\sigma - t),$$

where $t$ ranges from 0 at $Q$ to $\sigma$ at $P$.

Writing the function $F$ in the form

$$F = F\,(x,\ r,\ \theta,\ x',\ r',\ \theta'),$$

so that

we have

$$F(x,\ r,\ \theta,\ x',\ r',\ \theta') = (r - \lambda)\,(x'^2 + r'^2 + r^2\theta'^2)^{\frac{1}{2}},$$

$$I_{QP} = \sigma F\left(x,\ r,\ \theta,\ l,\ m,\ \frac{n}{r}\right)$$
$$= \sigma\,(r - \lambda)\,(l^2 + m^2 + n^2)^{\frac{1}{2}}$$
$$= \sigma\,(r - \lambda),$$

when $\sigma$ is taken small enough to render terms of the second and higher orders negligible.

Again,

$$I_{AQ} - I_{AP} = \kappa\xi\frac{\partial F}{\partial x'} + \kappa\rho\frac{\partial F}{\partial r'} + \kappa\vartheta\frac{\partial F}{\partial \theta'}$$
$$= -\sigma\left(l\frac{\partial F}{\partial x'} + m\frac{\partial F}{\partial r'} + \frac{n}{r}\frac{\partial F}{\partial \theta'}\right)$$
$$= -\sigma\,(r - \lambda)\frac{lx' + mr + nr\theta'}{(x'^2 + r'^2 + r^2\theta'^2)^{\frac{1}{2}}},$$

with the same assumption concerning $\sigma$.

The $E$-function, for the Weierstrass test, is

$$E_{IJ} = (r - \lambda)\left\{1 - \frac{lx' + mr' + nr\theta'}{(x'^2 + r'^2 + r^2\theta'^2)^{\frac{1}{2}}}\right\};$$

and

$$(x'^2 + r'^2 + r^2\theta'^2)\,(l^2 + m^2 + n^2) - (lx' + mr' + nr\theta')^2$$
$$= (lr' - mx')^2 + (mr\theta' - nr')^2 + (nx' - lr\theta')^2$$
$$> 0,$$

when the relations $l : m : n = x' : r' : r\theta'$ do not hold : that is, when we take a variation off the characteristic curve.

Thus $E_{IJ}$ is positive when $r > \lambda$, and it is negative when $r < \lambda$.

Hence all the tests combine to admit a maximum when $r < \lambda$, and to admit a minimum when $r > \lambda$.

*Relative maxima and minima, limited by equations of non-integral type :*
*general statement.*

**262.** We have indicated, in § 237 (*b*), another form of limitation upon the free variation of the quantities, which occur in an integral that is to be maximised or minimised. It arises when conditions (one or more in number) are propounded, requiring persistent relations among dependent variables and (possibly) their derivatives. But such relations are not expressed (as in the immediately preceding sections) through the assignment of constant values to integrals which cannot be explicitly evaluated: they take the form of equations free from quadratures.

($\alpha$) Thus it may be required to find the shortest line drawn on a given surface

$$S(x, y, z) = 0,$$

between two points on the surface: that is, we require a minimum for the integral

$$\int (x_1{}^2 + y_1{}^2 + z_1{}^2)^{\frac{1}{2}} dt,$$

when the variables $x$, $y$, $z$ satisfy the equation $S = 0$. It is always theoretically possible to suppose this equation resolved so as to express $z$ in terms of $x$ and $y$, and thus to change the integral so that it involves only two unconnected variables $x$ and $y$. The result of the modified problem is, however, to give the projection of the required curve on the plane $z = 0$, and thus to leave the curve as an intersection of $S = 0$ with a cylinder on this projection, having its axis parallel to the axis of $z$: not a completely satisfactory form of result. Moreover, the algebra required to attain the result may be cumbrous; it can either be difficult within reason or even be impossible. We therefore shall find it convenient to have another method of proceeding.

($\beta$) Again, it may be required to find a curve of constant circular curvature, while the curve (joining two points in space, or passing from one surface to another surface) may be required to possess some maximum or minimum property, such as that of minimum length. The analytical solution of the problem requires a minimum value of

$$\int (x_1{}^2 + y_1{}^2 + z_1{}^2)^{\frac{1}{2}} dt,$$

where the variables $x$, $y$, $z$ must satisfy the relation

$$\frac{1}{c^2} (x_1{}^2 + y_1{}^2 + z_1{}^2)^3 = (y_1 z_2 - y_2 z_1)^2 + (z_1 x_2 - x_1 z_2)^2 + (x_1 y_2 - y_1 x_2)^2.$$

Or, if the arc be taken as the independent variable, the problem requires a minimum value of $\int ds$, subject to the relation

$$x''^2 + y''^2 + z''^2 = \frac{1}{c^2}.$$

The earlier method does not apply; another method of proceeding must be devised.

($\gamma$) As a last illustration of these types of relative maxima and minima, a given problem can often be reduced in apparent complexity of statement,

and can be modified into another apparently equivalent problem, when the
new problem is subjected to conditions of a type not initially contemplated.
It is the method usually adopted in the investigations of Clebsch and others,
to which reference has been made in § 6.

Thus it may be required to find a maximum or a minimum value of an
integral

$$\int f\left(x,\ y,\ \frac{dy}{dx},\ \frac{d^2y}{dx^2}\right) dx,$$

subject to no conditions other than those of a descriptive character at the
limits of the integral. The problem can be made the same (though the consequent analysis will not be the same) as that of finding a maximum or a
minimum value of the integral

$$\int f\left(x,\ y,\ p,\ \frac{dp}{dx}\right) dx,$$

where the former dependent variable $y$ and a new dependent variable $p$ are
connected by the relation

$$\frac{dy}{dx} - p = 0,$$

which cannot be expressed in a form free from derivatives. Once more, we
find it necessary to have a new method of proceeding.

($\delta$) Moreover, there may be a combination of different types of persistent
relations, on the one hand requiring that certain given concurrent integrals
shall be constant; and, on the other hand, postulating persistent equations
free from quadrature. The two kinds of relations may be simultaneously
imposed upon all the variables concerned, thus affecting the complete freedom
of their variation when a maximum or a minimum of a given integral is to
be secured.

**263.** One precaution must be observed. In dealing with integrals which
have to be made a maximum or a minimum, we have omitted those which
can be evaluated explicitly so that they become free from quadratures. The
specific exclusion was made (§§ 13, 16) for the simplest case; and, subsequently
in other cases, it has been tacitly assumed.

In a corresponding fashion, certain types of concurrent conditions can be
ignored. Thus with a plane problem involving derivatives of any order, we
should omit those for which a concurrent condition such as

$$\phi\left(x, y, \frac{dy}{dx}\right) = 0$$

is assigned. This relation is a differential equation, the primitive of which
leads to the determination of $y$: when this value of $y$ is substituted in the
integral of the problem, the analysis becomes trivial and does not require the

processes of the calculus of variations. Similarly, we should omit a problem for which a concurrent condition such as

$$G\left(x, x_1, x_2, y, y_1, y_2\right) = 0$$

is postulated, when identities

$$G = x_1 \frac{\partial G}{\partial x_1} + y_1 \frac{\partial G}{\partial y_1} + 2x_2 \frac{\partial G}{\partial x_2} + 2y_2 \frac{\partial G}{\partial y_2}, \quad 0 = x_1 \frac{\partial G}{\partial x_2} + y_1 \frac{\partial G}{\partial y_2},$$

are satisfied; for then

$$G = x_1 g\left(x, y, \frac{dy}{dx}, \frac{d^2 y}{dx^2}\right),$$

and the condition $G = 0$ is essentially therefore a given differential equation, determining $y$ without any reference to the problem.

## Method of proceeding.

**264.** Finally, it is not uncustomary to declare (taking the first type of example merely for explanatory purposes here) that the integral

$$\int \left\{ \left(x_1{}^2 + y_1{}^2 + z_1{}^2\right)^{\frac{1}{2}} - \lambda S\left(x, y, z\right) \right\} dt$$

must satisfy the conditions of a maximum or minimum, $\lambda$ being (at least, initially) any arbitrary variable quantity. The demand, of course, is one that must be met. But there is no proof—and, in that unexpanded statement, there is no justification for an assumption—that this particular association of the equation $S = 0$ with the original integral embodies all the influence of the external condition upon the problem. Moreover, it seems desirable to connect the quantity $\lambda$ organically with the range of variations imposed.

Accordingly, we proceed without any initial postulation of a special form of association of the external condition (or conditions) with the integral that is to be maximised or minimised. From among the complete range of weak variations possible for an unlimited problem, we select the aggregate of such variations as satisfy the imposed conditions; and then the integral, which is to have a maximum or a minimum, is subjected to the variations in that selected aggregate.

The necessary analysis will be set forth for one simple (but representative) type of problem. Afterwards, briefer statements will be given for other types, without full proof.

## Simplest type of problem for limited maximum or minimum.

**265.** We therefore take, as a typical problem, the requirement that an integral

$$\int G\left(x, x_1, y, y_1, z, z_1\right) dt,$$

where $G$ satisfies the identity

$$G = x_1 \frac{\partial G}{\partial x_1} + y_1 \frac{\partial G}{\partial y_1} + z_1 \frac{\partial G}{\partial z_1},$$

shall assume a maximum or minimum value, subject to the demand that the relation

$$\phi (x, x_1, y, y_1, z, z_1) = 0$$

shall be permanently satisfied. Also we shall suppose that the relation itself is such that the function $\phi$ satisfies an identity

$$x_1 \frac{\partial \phi}{\partial x_1} + y_1 \frac{\partial \phi}{\partial y_1} + z_1 \frac{\partial \phi}{\partial z_1} = n\phi,$$

where $n$ is a constant. (The simplest case occurs when $n = 1$, so that the identities satisfied by $\phi$ and by $G$ are the same; but there is no need for limitation to this simplest value of $n$, because ultimately the relation is always used in the form $\phi = 0$.) The identities secure that

$$G = x_1 f \left( x, y, \frac{dy}{dx}, z, \frac{dz}{dx} \right), \qquad \phi = x_1{}^n g \left( x, y, \frac{dy}{dx}, z, \frac{dz}{dx} \right);$$

and the equation $g = 0$ is assumed to have no integral equivalent in the form of a single equation free from quadrature.

*First variation: its transformation, by means of the imposed condition.*

**266.** Small variations $\kappa X$, $\kappa Y$, $\kappa Z$, are imposed upon $x$, $y$, $z$, where $\kappa$ is small; $X$, $Y$, $Z$, are arbitrary regular functions of $t$, independent of one another until conditions are applied. The variation $\mathbf{I} - I$ of the integral is transformed, as before, to

$$\kappa \int \left[ X \left\{ \frac{\partial G}{\partial x} - \frac{d}{dt} \left( \frac{\partial G}{\partial x_1} \right) \right\} + Y \left\{ \frac{\partial G}{\partial y} - \frac{d}{dt} \left( \frac{\partial G}{\partial y_1} \right) \right\} + Z \left\{ \frac{\partial G}{\partial z} - \frac{d}{dt} \left( \frac{\partial G}{\partial z_1} \right) \right\} \right] dt$$
$$+ \kappa \left[ X \frac{\partial G}{\partial x_1} + Y \frac{\partial G}{\partial y_1} + Z \frac{\partial G}{\partial z_1} \right] + R_2,$$

where the bracketed terms in the second line are taken at the limits of the integral, and $R_2$ is the aggregate of terms involving the second and higher powers of $\kappa$. Now the equation $\phi = 0$ is to be persistently satisfied, and therefore must be satisfied by the varied values of $x$, $y$, $z$; consequently

$$\kappa \left( X \frac{\partial \phi}{\partial x} + X_1 \frac{\partial \phi}{\partial x_1} + Y \frac{\partial \phi}{\partial y} + Y_1 \frac{\partial \phi}{\partial y_1} + Z \frac{\partial \phi}{\partial z} + Z_1 \frac{\partial \phi}{\partial z_1} \right) + S_2 = 0,$$

where $S_2$ is the aggregate of terms involving the second and higher powers of $\kappa$. This last equation is a limitation (and is the only limitation) upon the complete independence of the otherwise arbitrary functions $X$, $Y$, $Z$; and it is only for quantities $X$, $Y$, $Z$, satisfying the last equation, that we have to consider the value of $\mathbf{I} - I$. Hence, when $\kappa$ is very small, we have

$$X \frac{\partial \phi}{\partial x} + X_1 \frac{\partial \phi}{\partial x_1} + Y \frac{\partial \phi}{\partial y} + Y_1 \frac{\partial \phi}{\partial y_1} + Z \frac{\partial \phi}{\partial z} + Z_1 \frac{\partial \phi}{\partial z_1} = 0.$$

If it were possible to use the equation, so as to express $X$ in terms of $Y$ and $Z$, the modification of $\mathbf{I} - I$ could at once be effected. But the possibility is not immediate, because both $X$ and $X_1$ are involved; the modification can, however, be effected indirectly as follows. Let $\lambda$ denote any function of $t$, at our disposal for a subsequent purpose. We have

$$\lambda \left( X_1 \frac{\partial \phi}{\partial x_1} + Y_1 \frac{\partial \phi}{\partial y_1} + Z_1 \frac{\partial \phi}{\partial z_1} \right)$$

$$= \frac{d}{dt} \left\{ \lambda \left( X \frac{\partial \phi}{\partial x_1} + Y \frac{\partial \phi}{\partial y_1} + Z \frac{\partial \phi}{\partial z_1} \right) \right\} - \left\{ X \frac{d}{dt} \left( \lambda \frac{\partial \phi}{\partial x_1} \right) + Y \frac{d}{dt} \left( \lambda \frac{\partial \phi}{\partial y_1} \right) + Z \frac{d}{dt} \left( \lambda \frac{\partial \phi}{\partial z_1} \right) \right\};$$

and therefore

$$\lambda \left( X \frac{\partial \phi}{\partial x} + X_1 \frac{\partial \phi}{\partial x_1} + Y \frac{\partial \phi}{\partial y} + Y_1 \frac{\partial \phi}{\partial y_1} + Z \frac{\partial \phi}{\partial z} + Z_1 \frac{\partial \phi}{\partial z_1} \right)$$

$$= X \left\{ \lambda \frac{\partial \phi}{\partial x} - \frac{d}{dt} \left( \lambda \frac{\partial \phi}{\partial x_1} \right) \right\} + Y \left\{ \lambda \frac{\partial \phi}{\partial y} - \frac{d}{dt} \left( \lambda \frac{\partial \phi}{\partial y_1} \right) \right\} + Z \left\{ \lambda \frac{\partial \phi}{\partial z} - \frac{d}{dt} \left( \lambda \frac{\partial \phi}{\partial z_1} \right) \right\}$$

$$+ \frac{d}{dt} \left\{ \lambda \left( X \frac{\partial \phi}{\partial x_1} + Y \frac{\partial \phi}{\partial y_1} + Z \frac{\partial \phi}{\partial z_1} \right) \right\}.$$

Having regard to the first variation in $\mathbf{I} - I$, and to the last relation as affecting the persistent equation between $X, X_1, Y, Y_1, Z, Z_1$, we now assume that $\lambda$, which was specially introduced for such subsequent determination as should be useful, satisfies an equation

$$\frac{\partial G}{\partial x} - \lambda \frac{\partial \phi}{\partial x} - \frac{d}{dt} \left\{ \frac{\partial G}{\partial x_1} - \lambda \frac{\partial \phi}{\partial x_1} \right\} = 0.$$

Then

$$X \left\{ \frac{\partial G}{\partial x} - \frac{d}{dt} \left( \frac{\partial G}{\partial x_1} \right) \right\} + Y \left\{ \frac{\partial G}{\partial y} - \frac{d}{dt} \left( \frac{\partial G}{\partial y_1} \right) \right\} + Z \left\{ \frac{\partial G}{\partial z} - \frac{d}{dt} \left( \frac{\partial G}{\partial z_1} \right) \right\}$$

$$= X \left\{ \lambda \frac{\partial \phi}{\partial x} - \frac{d}{dt} \left( \lambda \frac{\partial \phi}{\partial x_1} \right) \right\} + Y \left\{ \frac{\partial G}{\partial y} - \frac{d}{dt} \left( \frac{\partial G}{\partial y_1} \right) \right\} + Z \left\{ \frac{\partial G}{\partial z} - \frac{d}{dt} \left( \frac{\partial G}{\partial z_1} \right) \right\}$$

$$= Y \left\{ \frac{\partial G}{\partial y} - \lambda \frac{\partial \phi}{\partial y} - \frac{d}{dt} \left( \frac{\partial G}{\partial y_1} - \lambda \frac{\partial \phi}{\partial y_1} \right) \right\} + Z \left\{ \frac{\partial G}{\partial z} - \lambda \frac{\partial \phi}{\partial z} - \frac{d}{dt} \left( \frac{\partial G}{\partial z_1} - \lambda \frac{\partial \phi}{\partial z_1} \right) \right\}$$

$$- \frac{d}{dt} \left\{ \lambda \left( X \frac{\partial \phi}{\partial x_1} + Y \frac{\partial \phi}{\partial y_1} + Z \frac{\partial \phi}{\partial z_1} \right) \right\}.$$

Thus the first variation—the coefficient of $\kappa$ in the full variation—in $\mathbf{I} - I$ is

$$= \kappa \int \left[ Y \left\{ \frac{\partial G}{\partial y} - \lambda \frac{\partial \phi}{\partial y} - \frac{d}{dt} \left( \frac{\partial G}{\partial y_1} - \lambda \frac{\partial \phi}{\partial y_1} \right) \right\} + Z \left\{ \frac{\partial G}{\partial z} - \lambda \frac{\partial \phi}{\partial z} - \frac{d}{dt} \left( \frac{\partial G}{\partial z_1} - \lambda \frac{\partial \phi}{\partial z_1} \right) \right\} \right] dt$$

$$+ \kappa \left[ X \left( \frac{\partial G}{\partial x_1} - \lambda \frac{\partial \phi}{\partial x_1} \right) + Y \left( \frac{\partial G}{\partial y_1} - \lambda \frac{\partial \phi}{\partial y_1} \right) + Z \left( \frac{\partial G}{\partial z_1} - \lambda \frac{\partial \phi}{\partial z_1} \right) \right].$$

In order that $I$ may possess a maximum or a minimum, its first variation must vanish, by the customary argument; and therefore the last expression must vanish for all variations.

*Characteristic equations: terminal conditions.*

**267.** Consider, first, all those variations which vanish at each limit. When these are imposed, we must have

$$\int \left[ Y \left\{ \frac{\partial G}{\partial y} - \lambda \frac{\partial \phi}{\partial y} - \frac{d}{dt} \left( \frac{\partial G}{\partial y_1} - \lambda \frac{\partial \phi}{\partial y_1} \right) \right\} + Z \left\{ \frac{\partial G}{\partial z} - \lambda \frac{\partial \phi}{\partial z} - \frac{d}{dt} \left( \frac{\partial G}{\partial z_1} - \lambda \frac{\partial \phi}{\partial z_1} \right) \right\} \right] dt = 0.$$

This integral does not involve $X$. The quantities $Y$ and $Z$ are arbitrary; and they are independent of one another, because the permanent relation (the only relation) between $X$, $Y$, $Z$, imposes no restriction upon $Y$ and $Z$, and determines $X$ after $Y$ and $Z$ have been chosen. Hence the foregoing integral can vanish for all such functions $Y$ and $Z$, arbitrary and independent, only if

$$\frac{\partial G}{\partial y} - \lambda \frac{\partial \phi}{\partial y} - \frac{d}{dt} \left( \frac{\partial G}{\partial y_1} - \lambda \frac{\partial \phi}{\partial y_1} \right) = 0,$$

$$\frac{\partial G}{\partial z} - \lambda \frac{\partial \phi}{\partial z} - \frac{d}{dt} \left( \frac{\partial G}{\partial z_1} - \lambda \frac{\partial \phi}{\partial z_1} \right) = 0.$$

Conversely, when these equations are satisfied, the foregoing integral in the first variation (§ 266) vanishes for all variations, whatever may happen at the limits of the integral.

Consider, next, variations which do not necessarily vanish at the limits. The portion of the first variation depending upon the integral vanishes because the equations, which have just been established, are supposed to be satisfied. The remainder must therefore vanish by itself, since the whole first variation vanishes. Hence

$$X \left( \frac{\partial G}{\partial x_1} - \lambda \frac{\partial \phi}{\partial x_1} \right) + Y \left( \frac{\partial G}{\partial y_1} - \lambda \frac{\partial \phi}{\partial y_1} \right) + Z \left( \frac{\partial G}{\partial z_1} - \lambda \frac{\partial \phi}{\partial z_1} \right) = 0,$$

at each limit. In these relations, $X$, $Y$, $Z$, must, on the one hand, be consistent with the conditions at the limit; on the other hand, they must belong to the range of variation required by the permanent condition $\phi = 0$.

Consequently, as necessary and sufficient conditions that the first variation in $\mathbf{I} - I$ shall vanish for all weak variations allowed by the permanent condition, we have the set of *characteristic equations*

$$\left. \begin{array}{l} \dfrac{\partial G}{\partial x} - \lambda \dfrac{\partial \phi}{\partial x} - \dfrac{d}{dt} \left( \dfrac{\partial G}{\partial x_1} - \lambda \dfrac{\partial \phi}{\partial x_1} \right) = 0 \\[2ex] \dfrac{\partial G}{\partial y} - \lambda \dfrac{\partial \phi}{\partial y} - \dfrac{d}{dt} \left( \dfrac{\partial G}{\partial y_1} - \lambda \dfrac{\partial \phi}{\partial y_1} \right) = 0 \\[2ex] \dfrac{\partial G}{\partial z} - \lambda \dfrac{\partial \phi}{\partial z} - \dfrac{d}{dt} \left( \dfrac{\partial G}{\partial z_1} - \lambda \dfrac{\partial \phi}{\partial z_1} \right) = 0 \end{array} \right\},$$

which must be satisfied throughout the range of the integral; and there is
the equation $\phi = 0$. Further, at each limit, the relation

$$X\left(\frac{\partial G}{\partial x_1} - \lambda\frac{\partial \phi}{\partial x_1}\right) + Y\left(\frac{\partial G}{\partial y_1} - \lambda\frac{\partial \phi}{\partial y_1}\right) + Z\left(\frac{\partial G}{\partial z_1} - \lambda\frac{\partial \phi}{\partial z_1}\right) = 0$$

must be satisfied, $\kappa X$, $\kappa Y$, $\kappa Z$, being small variations conforming to conditions
at the limit and therefore admitted by the relation $\phi = 0$.

### *The three characteristic equations reducible to two, in general.*

**268**. It must at once be noted that, in association with the identity satisfied by the function $G$ and with the relation

$$x_1\frac{\partial \phi}{\partial x_1} + y_1\frac{\partial \phi}{\partial y_1} + z_1\frac{\partial \phi}{\partial z_1} = 0,$$

(being the modified form of the condition $\phi = 0$, derived through the identity
satisfied by $\phi$), the three characteristic equations are equivalent to only two
independent equations.

For, differentiating the identity satisfied by $G$, we have

$$x_2\frac{\partial G}{\partial x_1} + y_2\frac{\partial G}{\partial y_1} + z_2\frac{\partial G}{\partial z_1} + x_1\frac{d}{dt}\left(\frac{\partial G}{\partial x_1}\right) + y_1\frac{d}{dt}\left(\frac{\partial G}{\partial y_1}\right) + z_1\frac{d}{dt}\left(\frac{\partial G}{\partial z_1}\right)$$

$$= \frac{dG}{dt}$$

$$= x_1\frac{\partial G}{\partial x} + y_1\frac{\partial G}{\partial y} + z_1\frac{\partial G}{\partial z} + x_2\frac{\partial G}{\partial x_1} + y_2\frac{\partial G}{\partial y_1} + z_2\frac{\partial G}{\partial z_1},$$

so that

$$x_1\left\{\frac{\partial G}{\partial x} - \frac{d}{dt}\left(\frac{\partial G}{\partial x_1}\right)\right\} + y_1\left\{\frac{\partial G}{\partial y} - \frac{d}{dt}\left(\frac{\partial G}{\partial y_1}\right)\right\} + z_1\left\{\frac{\partial G}{\partial z} - \frac{d}{dt}\left(\frac{\partial G}{\partial z_1}\right)\right\} = 0.$$

Again, from $\phi = 0$, we have

$$x_1\frac{\partial \phi}{\partial x} + y_1\frac{\partial \phi}{\partial y} + z_1\frac{\partial \phi}{\partial z} + x_2\frac{\partial \phi}{\partial x_1} + y_2\frac{\partial \phi}{\partial y_1} + z_2\frac{\partial \phi}{\partial z_1} = 0.$$

From the relation

$$x_1\frac{\partial \phi}{\partial x_1} + y_1\frac{\partial \phi}{\partial y_1} + z_1\frac{\partial \phi}{\partial z_1} = 0,$$

we have

$$x_1\frac{d}{dt}\left(\frac{\partial \phi}{\partial x_1}\right) + y_1\frac{d}{dt}\left(\frac{\partial \phi}{\partial y_1}\right) + z_1\frac{d}{dt}\left(\frac{\partial \phi}{\partial z_1}\right) + x_2\frac{\partial \phi}{\partial x_1} + y_2\frac{\partial \phi}{\partial y_1} + z_2\frac{\partial \phi}{\partial z_1} = 0.$$

Now

$$x_1\frac{d}{dt}\left(\lambda\frac{\partial \phi}{\partial x_1}\right) + y_1\frac{d}{dt}\left(\lambda\frac{\partial \phi}{\partial y_1}\right) + z_1\frac{d}{dt}\left(\lambda\frac{\partial \phi}{\partial z_1}\right)$$

$$= \frac{d\lambda}{dt}\left(x_1\frac{\partial \phi}{\partial x_1} + y_1\frac{\partial \phi}{\partial y_1} + z_1\frac{\partial \phi}{\partial z_1}\right)$$

$$+ \lambda\left\{x_1\frac{d}{dt}\left(\frac{\partial \phi}{\partial x_1}\right) + y_1\frac{d}{dt}\left(\frac{\partial \phi}{\partial y_1}\right) + z_1\frac{d}{dt}\left(\frac{\partial \phi}{\partial z_1}\right)\right\}$$

$$= -\lambda \left( x_2 \frac{\partial \phi}{\partial x_1} + y_2 \frac{\partial \phi}{\partial y_1} + z_2 \frac{\partial \phi}{\partial z_1} \right)$$

$$= x_1 \lambda \frac{\partial \phi}{\partial x} + y_1 \lambda \frac{\partial \phi}{\partial y} + z_1 \lambda \frac{\partial \phi}{\partial z}.$$

Hence

$$x_1 \left\{ \lambda \frac{\partial \phi}{\partial x} - \frac{d}{dt}\left( \lambda \frac{\partial \phi}{\partial x_1} \right) \right\} + y_1 \left\{ \lambda \frac{\partial \phi}{\partial y} - \frac{d}{dt}\left( \lambda \frac{\partial \phi}{\partial y_1} \right) \right\} + z_1 \left\{ \lambda \frac{\partial \phi}{\partial z} - \frac{d}{dt}\left( \lambda \frac{\partial \phi}{\partial z_1} \right) \right\} = 0.$$

Let the three characteristic equations be denoted by $E_x = 0$, $E_y = 0$, $E_z = 0$: multiply them by $x_1$, $y_1$, $z_1$, and add. From the two equations just established, it follows that

$$x_1 E_x + y_1 E_y + z_1 E_z = 0,$$

which shews that *the three characteristic equations, coupled with the identities, are equivalent to two independent equations.*

*Ex.* Shew that, if the function $\phi$ does not satisfy an identity of the form

$$x_1 \frac{\partial \phi}{\partial x_1} + y_1 \frac{\partial \phi}{\partial y_1} + z_1 \frac{\partial \phi}{\partial z_1} = n\phi,$$

where $n$ is a constant, so that the quantity $x_1 \frac{\partial \phi}{\partial x_1} + y_1 \frac{\partial \phi}{\partial y_1} + z_1 \frac{\partial \phi}{\partial z_1}$ does not vanish, then the three characteristic equations are not equivalent to two only, in virtue of the identity satisfied by $G$ and of the persistent condition $\phi = 0$.

Prove also that, in this event, the equations give

$$\lambda \left( x_1 \frac{\partial \phi}{\partial x_1} + y_1 \frac{\partial \phi}{\partial y_1} + z_1 \frac{\partial \phi}{\partial z_1} \right) = c,$$

where $c$ is a non-vanishing constant, arbitrary so far as the characteristic equations are concerned.


### Form of results for special weak variations: examples.

**269.** Either by analysis similar to that in § 133, or by making $x$ the independent variable, the results as regards a maximum or a minimum for an integral

$$\int f(x, y, z, y', z')\, dx,$$

where $y'$ and $z'$ denote the first derivatives of $y$ and $z$, subject to

$$g(x, y, z, y', z') = 0,$$

are established as follows: the two characteristic equations are

$$\left.\begin{aligned}
\frac{\partial f}{\partial y} - \lambda \frac{\partial g}{\partial y} - \frac{d}{dx}\left( \frac{\partial f}{\partial y'} - \lambda \frac{\partial g}{\partial y'} \right) &= 0 \\[2mm]
\frac{\partial f}{\partial z} - \lambda \frac{\partial g}{\partial z} - \frac{d}{dx}\left( \frac{\partial f}{\partial z'} - \lambda \frac{\partial g}{\partial z'} \right) &= 0
\end{aligned}\right\},$$

being independent of one another; and the terminal condition, to be satisfied at each limit, is

$$\xi \left\{ f - y' \left( \frac{\partial f}{\partial y'} - \lambda \frac{\partial g}{\partial y'} \right) - z' \left( \frac{\partial f}{\partial z'} - \lambda \frac{\partial g}{\partial z'} \right) \right\} + \eta \left( \frac{\partial f}{\partial y'} - \lambda \frac{\partial g}{\partial y'} \right) + \zeta \left( \frac{\partial f}{\partial z'} - \lambda \frac{\partial g}{\partial z'} \right) = 0,$$

where $\xi$, $\eta$, $\zeta$, must accord with the variations (if any) possible at the limit, and must accord also with the permanent condition $g = 0$. Also, $\lambda$ is now a function of $x$, the independent variable in the present instance.

As examples, we shall consider some results relating to geodesics.

*Ex.* 1. The characteristic equations of *a geodesic on a surface* lead at once to its fundamental property.

(i) The integral to be minimised is

$$\int (x_1{}^2 + y_1{}^2 + z_1{}^2)^{\frac{1}{2}} \, dt,$$

subject to the relation

$$S(x, y, z) = 0,$$

the equation of the surface. The fundamental equations are

$$-\lambda \frac{\partial S}{\partial x} - \frac{d}{dt} \left\{ \frac{x_1}{(x_1{}^2 + y_1{}^2 + z_1{}^2)^{\frac{1}{2}}} \right\} = 0,$$

with two others : that is, introducing the length of an arc as a variable,

$$\left. \begin{array}{l} \lambda \dfrac{\partial S}{\partial x} + \dfrac{ds}{dt} \dfrac{d^2x}{ds^2} = 0 \\[2mm] \lambda \dfrac{\partial S}{\partial y} + \dfrac{ds}{dt} \dfrac{d^2y}{ds^2} = 0 \\[2mm] \lambda \dfrac{\partial S}{\partial z} + \dfrac{ds}{dt} \dfrac{d^2z}{ds^2} = 0 \end{array} \right\}.$$

with

These three equations, on multiplication by $\dfrac{dx}{ds}$, $\dfrac{dy}{ds}$, $\dfrac{dz}{ds}$, and subsequent addition, are at once seen to be equivalent (§ 268) to two independent equations.

From the equations, we have

$$\frac{1}{\dfrac{\partial S}{\partial x}} \frac{d^2x}{ds^2} = \frac{1}{\dfrac{\partial S}{\partial y}} \frac{d^2y}{ds^2} = \frac{1}{\dfrac{\partial S}{\partial z}} \frac{d^2z}{ds^2} :$$

hence the principal normal to the geodesic coincides with the normal to the surface. These two relations, together with $S = 0$, are potentially sufficient to determine $x$, $y$, $z$ as functions of $s$.

Moreover, the independent variable $t$ is not specific and is mainly at our choice. Thus, for the simple instance of special weak variations, we can take $t = x$. Again, especially after the characteristic equations have been formed, we can ·take $t = s$, neither variable having had its origin of measurement settled as yet. With the latter choice,

$$\lambda \frac{\partial S}{\partial x} + \frac{d^2x}{ds^2} = 0, \quad \lambda \frac{\partial S}{\partial y} + \frac{d^2y}{ds^2} = 0, \quad \lambda \frac{\partial S}{\partial z} + \frac{d^2z}{ds^2} = 0.$$

When the equations have been integrated so as to express $x$, $y$, $z$, in terms of $s$, any one of the last three equations can be used to determine the value of $\lambda$. No arbitrary element arises in the solution through the determination of $\lambda$, which is merely subsidiary to the analysis required for the problem.

(ii) As an illustration, let
$$S\,(x,\,y,\,z) = x^2 + y^2 + z^2 - a^2 = 0.$$
The equations are
$$2\lambda x + \frac{d^2x}{ds^2} = 0, \qquad 2\lambda y + \frac{d^2y}{ds^2} = 0, \qquad 2\lambda z + \frac{d^2z}{ds^2} = 0.$$

For the spherical surface, we have
$$x\,\frac{dx}{ds} + y\,\frac{dy}{ds} + z\,\frac{dz}{ds} = 0,$$
and therefore
$$x\,\frac{d^2x}{ds^2} + y\,\frac{d^2y}{ds^2} + z\,\frac{d^2z}{ds^2} = -\left\{\left(\frac{dx}{ds}\right)^2 + \left(\frac{dy}{ds}\right)^2 + \left(\frac{dz}{ds}\right)^2\right\} = -1.$$

Hence, multiplying the three equations by $x$, $y$, $z$, and adding, we have
$$2\lambda a^2 - 1 = 0\,;$$
and therefore the equations are
$$x + a^2\frac{d^2x}{ds^2} = 0, \qquad y + a^2\frac{d^2y}{ds^2} = 0, \qquad z + a^2\frac{d^2z}{ds^2} = 0.$$
Consequently,
$$x = c_1\cos\left(\frac{s}{a} + a_1\right), \qquad y = c_2\cos\left(\frac{s}{a} + a_2\right), \qquad z = c_3\cos\left(\frac{s}{a} + a_3\right),$$

where $c_1$, $c_2$, $c_3$, $a_1$, $a_2$, $a_3$ are arbitrary (but not independent) constants. As $x$, $y$, $z$, must satisfy the equation $x^2 + y^2 + z^2 = a^2$, we have
$$\Sigma c_r^2\cos^2 a_r = a^2, \qquad \Sigma c_r^2\sin^2 a_r = a^2, \qquad \Sigma c_r^2\sin a_r\cos a_r = 0,$$
so
$$\Sigma c_r^2 = 2a^2, \qquad \Sigma c_r^2\cos 2a_r = 0, \qquad \Sigma c_r^2\sin 2a_r = 0.$$

These relations can be used, in a variety of ways, to reduce the arbitrary constants to two independent constants : thus
$$c_1{}^2 = 2a^2\,\frac{1 + \cos(2a_2 - 2a_3)}{1 + \cos(2a_2 - 2a_3) - \cos(2a_1 - 2a_2) - \cos(2a_3 - 2a_1)},$$
with similar expressions for $c_2{}^2$ and $c_3{}^2$.

Again,
$$\begin{vmatrix} x, & c_1\cos a_1, & c_1\sin a_1 \\ y, & c_2\cos a_2, & c_2\sin a_2 \\ z, & c_3\cos a_3, & c_3\sin a_3 \end{vmatrix} = 0,$$

shewing that the curve lies in a plane through the origin, that is, through the centre of the sphere. It therefore lies along a great circle.

(iii) Wherever conveniently possible, the expression of the coordinates of a point on a surface in terms of two independent parameters is desirable. Thus, for the foregoing sphere, we may take
$$x = a\sin\theta\cos\phi, \quad y = a\sin\theta\sin\phi, \quad z = a\cos\theta\,;$$
then
$$ds = a\,(\theta'^2 + \phi'^2\sin^2\theta)^{\frac{1}{2}}\,dt.$$

The permanent equation is always satisfied by these values ; of that equation, no further account need be taken, and the quantity $\lambda$ now is unnecessary. It is the integral $\int\{\theta'^2 + \phi'^2\sin^2\theta\}^{\frac{1}{2}}\,dt$ which has to be discussed.

(iv) Quite generally, we can dispense with the method in the text and avoid the introduction of the subsidiary quantity $\lambda$, by using the Gauss general representation* of surfaces, whenever the coordinates of a point on a surface can be expressed in terms of two independent parameters $p$ and $q$.

For the purpose, we take
$$E = x_1^2 + y_1^2 + z_1^2, \quad F = x_1 x_2 + y_1 y_2 + z_1 z_2, \quad G = x_2^2 + y_2^2 + z_2^2,$$
where $x_1 = \dfrac{\partial x}{\partial p}$, $x_2 = \dfrac{\partial x}{\partial q}$, and similarly for $y$ and $z$; so that
$$EG - F^2, = (y_1 z_2 - y_2 z_1)^2 + (z_1 x_2 - x_1 z_2)^2 + (x_1 y_2 - y_1 x_2)^2,$$
is a positive quantity. Then the integral to be minimised is
$$\int (E p'^2 + 2 F p' q' + G q'^2)^{\frac{1}{2}} \, dt,$$
where $p$ and $q$ are taken as functions of a new variable $t$ along the line. It is
$$\int \left\{ E + 2F \frac{dq}{dp} + G \left( \frac{dq}{dp} \right)^2 \right\}^{\frac{1}{2}} dp,$$
when one of the parameters is taken as the independent variable for the integral.

For the first form of the integral, the characteristic equations, when $\Delta$ is used to denote $(E p'^2 + 2 F p' q' + G q'^2)^{\frac{1}{2}}$, are
$$\left. \begin{array}{l} \dfrac{\partial \Delta}{\partial p} - \dfrac{d}{dt} \left( \dfrac{E p' + F q'}{\Delta} \right) = 0 \\[2mm] \dfrac{\partial \Delta}{\partial q} - \dfrac{d}{dt} \left( \dfrac{F p' + G q'}{\Delta} \right) = 0 \end{array} \right\} ;$$
or, if the arc-variable $s$ be now introduced,
$$\left. \begin{array}{l} \dfrac{\partial E}{\partial p} \left( \dfrac{dp}{ds} \right)^2 + 2 \dfrac{\partial F}{\partial p} \dfrac{dp}{ds} \dfrac{dq}{ds} + \dfrac{\partial G}{\partial p} \left( \dfrac{dq}{ds} \right)^2 = \dfrac{d}{ds} \left( E \dfrac{dp}{ds} + F \dfrac{dq}{ds} \right) \\[2mm] \dfrac{\partial E}{\partial q} \left( \dfrac{dp}{ds} \right)^2 + 2 \dfrac{\partial F}{\partial q} \dfrac{dp}{ds} \dfrac{dq}{ds} + \dfrac{\partial G}{\partial q} \left( \dfrac{dq}{ds} \right)^2 = \dfrac{d}{ds} \left( F \dfrac{dp}{ds} + G \dfrac{dq}{ds} \right) \end{array} \right\}.$$

For the second form of the integral, the single characteristic equation is
$$\frac{1}{2} \frac{\dfrac{\partial E}{\partial q} + 2 \dfrac{\partial F}{\partial q} \dfrac{dq}{dp} + \dfrac{\partial G}{\partial q} \left( \dfrac{dq}{dp} \right)^2}{\left\{ E + 2F \dfrac{dq}{dp} + G \left( \dfrac{dq}{dp} \right)^2 \right\}^{\frac{1}{2}}} - \frac{d}{dp} \frac{F + G \dfrac{dq}{dp}}{\left\{ E + 2F \dfrac{dq}{dp} + G \left( \dfrac{dq}{dp} \right)^2 \right\}^{\frac{1}{2}}} = 0,$$
which, when expanded, is
$$(EG - F^2) \frac{d^2 q}{dp^2} = A \left( \frac{dq}{dp} \right)^3 + B \left( \frac{dq}{dp} \right)^2 + C \frac{dq}{dp} + D,$$
where
$$\left. \begin{array}{l} A = \quad G \dfrac{\partial F}{\partial q} - \tfrac{1}{2} F \dfrac{\partial G}{\partial q} - \tfrac{1}{2} G \dfrac{\partial G}{\partial p} \\[2mm] B = \quad G \dfrac{\partial E}{\partial q} - \tfrac{1}{2} E \dfrac{\partial G}{\partial q} - \tfrac{1}{2} F \dfrac{\partial G}{\partial p} + F \dfrac{\partial F}{\partial q} \\[2mm] C = - E \dfrac{\partial G}{\partial p} + \tfrac{1}{2} G \dfrac{\partial E}{\partial p} + \tfrac{1}{2} F \dfrac{\partial E}{\partial q} - F \dfrac{\partial F}{\partial p} \\[2mm] D = - E \dfrac{\partial F}{\partial p} + \tfrac{1}{2} F \dfrac{\partial E}{\partial p} + \tfrac{1}{2} E \dfrac{\partial E}{\partial q} \end{array} \right\}.$$

* *Disquisitiones generales circa superficies curvas*, Gauss, *Ges. Werke*, t. iv, pp. 217—258. An account of geodesics will be found in any treatise on geometry; the author's *Lectures on Differential Geometry* (chap. v) may be consulted.

The Legendre test-quantity, for the first form of the integral, is

$$\frac{EG - F^2}{(Ep'^2 + 2Fp'q' + Gq'^2)^{\frac{3}{2}}};$$

for the second form of the integral, it is

$$\frac{EG - F^2}{\left\{E + 2F\frac{dq}{dp} + G\left(\frac{dq}{dp}\right)^2\right\}^{\frac{3}{2}}}.$$

Both of these expressions are positive, because the positive sign is taken for the arc, and $EG - F^2$ is positive. Thus the integral can admit a minimum for weak variations.

The possible limitation of range, arising out of the Jacobi test, is particular to each surface, and possibly to a group or to groups of geodesics on the surface. It must be determined in each instance specially ; and the determination usually depends on a knowledge of the characteristic primitive.

*Ex.* 2. Shew that the Weierstrass test for a minimum is satisfied for all geodesics.

*Ex.* 3. *Find the geodesics upon the paraboloid*

$$\frac{y^2}{b} + \frac{z^2}{c} = 4x.$$

It will be assumed that $b$ is not equal to $c$; the case of the paraboloid of revolution, when $b = c$, has already been discussed (§ 34, Ex. 8). For greater precision, we shall assume $b > c > 0$.

The coordinates $x$, $y$, $z$, of any point on the surface can be expressed in terms of the parameters $p$ and $q$ of the confocal paraboloids through the point. Thus $0, p, q$ are the roots of the cubic

$$\frac{y^2}{b+\mu} + \frac{z^2}{c+\mu} = 4(x + \mu),$$

so that we have

$$b > c > 0 > -c > p > -b > q.$$

As usual,

$$\frac{y^2}{b+\mu} + \frac{z^2}{c+\mu} - 4(x + \mu) = -4\mu \frac{(\mu - p)(\mu - q)}{(b+\mu)(c+\mu)},$$

for all values of $\mu$ ; and therefore

$$y^2 = -4b\frac{(b+p)(b+q)}{b-c}, \quad z^2 = -4c\frac{(c+p)(c+q)}{c-b},$$
$$x = -b - c - p - q.$$

We have, with the foregoing definitions of $E$, $F$, $G$,

$$E = \frac{p(p-q)}{(b+p)(c+p)} = (p-q)P,$$

$$G = \frac{q(q-p)}{(b+q)(c+q)} = (q-p)Q,$$

$$F = 0.$$

Thus taking the second form in Ex. 1, (iv), we have to minimise the integral

$$\int\left[(p-q)\left\{P - Q\left(\frac{dq}{dp}\right)^2\right\}\right]^{\frac{1}{2}} dp.$$

The characteristic equation, on reduction, becomes

$$2(p-q)PQ\frac{d^2q}{dp^2} + \left(Q\frac{dq}{dp} - P\right)\left\{P - Q\left(\frac{dq}{dp}\right)^2\right\} - (p-q)\frac{dq}{dp}\left(Q\frac{dP}{dp} - P\frac{dQ}{dq}\frac{dq}{dp}\right) = 0.$$

Let a new variable $\theta$ be chosen such that

$$\frac{Q}{q+\theta}\left(\frac{dq}{dp}\right)^2 = \frac{P}{\theta+p};$$

taking logarithmic derivatives, we have

$$2\frac{\dfrac{d^2q}{dp^2}}{\dfrac{dq}{dp}} + \frac{1}{Q}\frac{dQ}{dq}\frac{dq}{dp} - \frac{1}{P}\frac{dP}{dp} - \frac{1}{q+\theta}\left(\frac{dq}{dp}+\frac{d\theta}{dp}\right) + \frac{1}{p+\theta}\left(1+\frac{d\theta}{dp}\right)=0.$$

Transforming the developed characteristic equation by using this value of $\dfrac{d^2q}{dp^2}$, we find

$$\frac{d\theta}{dp}\left(\frac{1}{p+\theta}-\frac{1}{q+\theta}\right)=0,$$

and therefore

$$\frac{d\theta}{dp}=0:$$

that is, a first integral of the characteristic equation is

$$\frac{Q}{q+\theta}\left(\frac{dq}{dp}\right)^2 = \frac{P}{\theta+p},$$

where $\theta$ is an arbitrary constant. Hence the geodesics are given by the equation

$$\frac{q\,dq}{\{q\,(b+q)(c+q)\,(\theta+q)\}^{\frac{1}{2}}} = \pm\frac{p\,dp}{\{p\,(b+p)\,(c+p)\,(\theta+p)\}^{\frac{1}{2}}}.$$

By the substitutions

$$q=\frac{bc\,\mathrm{sn}^2\,v}{c\,\mathrm{cn}^2\,v-b},\qquad p=\frac{bc\,\mathrm{sn}^2\,u}{c\,\mathrm{cn}^2\,u-b},$$

where the modulus of the elliptic functions is given by

$$k^2=\frac{c\,(b-\theta)}{\theta\,(b-c)},$$

this differential equation becomes

$$\frac{\mathrm{sn}^2\,u}{b-c+c\,\mathrm{sn}^2\,u}\,du = \pm\frac{\mathrm{sn}^2\,v}{b-c+c\,\mathrm{sn}^2\,v}\,dv.$$

The determination of the doubtful sign is made by means of the initial direction of the geodesic through an assigned initial point. The further analysis is an exercise in elliptic functions.

*Ex.* 4. A geodesic is drawn on the spheroid

$$\frac{x^2+y^2}{a^2}+\frac{z^2}{c^2}=1,$$

where $c^2=a^2\,(1-e^2)$, and any point on the surface is represented by

$$x=a\sin\theta\cos\phi,\quad y=a\sin\theta\sin\phi,\quad z=c\cos\theta.$$

Prove that the geodesic undulates between two parallels of latitude; that its equation can be made to have the form

$$d\phi = \frac{(1-e^2\sin^2\theta)^{\frac{1}{2}}\sin a}{(\sin^2\theta-\sin^2 a)^{\frac{1}{2}}\sin\theta}\,d\theta,$$

where $a$ is the smallest value of $\theta$ in its course; and that the difference in longitude, between a place of highest latitude and the place of next lowest latitude, is less than $\pi$ by

$$2\frac{(1-e^2\sin^2 a)^{\frac{1}{2}}}{\sin a}\int_0^\kappa \frac{\mathrm{dn}\,u-\mathrm{dn}^2\,u}{1+\cot^2 a\,\mathrm{sn}^2\,u}\,du,$$

where $(1-e^2\sin^2 a)^{-\frac{1}{2}}e\cos a$ is the modulus of the elliptic functions.

*Ex.* 5. The coordinates of a point on the quadric $\dfrac{x^2}{a} + \dfrac{y^2}{b} + \dfrac{z^2}{c} = 1$ are expressed in terms of the parameters $p$ and $q$ of the confocal quadrics through the point (that is, having $a+p$, $b+p$, $c+p$, and $a+q$, $b+q$, $c+q$, for the squares of their semi-axes): prove that the equation of a geodesic can be expressed in the form

$$\frac{q\,dq}{\{q\,(a+q)\,(b+q)\,(c+q)\,(\theta+q)\}^{\frac{1}{2}}} \pm \frac{p\,dp}{\{p\,(a+p)\,(b+p)\,(c+p)\,(\theta+p)\}^{\frac{1}{2}}} = 0,$$

$\theta$ being an arbitrary constant, and that

$$-2\frac{ds}{pq} = \frac{dq}{\{q\,(a+q)\,(b+q)\,(c+q)\,(\theta+q)\}^{\frac{1}{2}}} \pm \frac{dp}{\{p\,(a+p)\,(b+p)\,(c+p)\,(\theta+p)\}^{\frac{1}{2}}},$$

the upper signs being taken together, and the lower signs being taken together.

### Statement of results, when second derivatives occur with two dependent variables.

**270.** The corresponding problem, when two dependent variables and their derivatives up to the second order (inclusive) occur, may be stated as follows:

It is required to find the maximum or minimum of an integral

$$\int F\,(x,\,x_1,\,x_2,\,y,\,y_1,\,y_2,\,z,\,z_1,\,z_2)\,dt,$$

subject to a persistent relation

$$\Phi\,(x,\,x_1,\,x_2,\,y,\,y_1,\,y_2,\,z,\,z_1,\,z_2) = 0,$$

where the function $F$ satisfies the two identities

$$F = x_1\frac{\partial F}{\partial x_1} + y_1\frac{\partial F}{\partial y_1} + z_1\frac{\partial F}{\partial z_1} + 2x_2\frac{\partial F}{\partial x_2} + 2y_2\frac{\partial F}{\partial y_2} + 2z_2\frac{\partial F}{\partial z_2}, \quad 0 = x_1\frac{\partial F}{\partial x_2} + y_1\frac{\partial F}{\partial y_2} + z_1\frac{\partial F}{\partial z_2},$$

and $\Phi$ satisfies* the two exactly similar identities

$$0 = x_1\frac{\partial \Phi}{\partial x_1} + y_1\frac{\partial \Phi}{\partial y_1} + z_1\frac{\partial \Phi}{\partial z_1} + 2x_2\frac{\partial \Phi}{\partial x_2} + 2y_2\frac{\partial \Phi}{\partial y_2} + 2z_2\frac{\partial \Phi}{\partial z_2}, \quad 0 = x_1\frac{\partial \Phi}{\partial x_2} + y_1\frac{\partial \Phi}{\partial y_2} + z_1\frac{\partial \Phi}{\partial z_2}.$$

The argument (in brief outline) leading to the establishment of the three characteristic equations may be taken thus:

The variation of the integral, being $\mathbf{I} - I$, is equal to

$$\kappa \int \Sigma \left( \frac{\partial F}{\partial u}\,U + \frac{\partial F}{\partial u_1}\,U_1 + \frac{\partial F}{\partial u_2}\,U_2 \right) dt + R_2,$$

where $R_2$ is the aggregate of terms of the second and higher degrees in $\kappa$; the summation is for $u = x, y, z$, in turn, with $U = X, Y, Z$, respectively; and the small variations $\kappa X, \kappa Y, \kappa Z$, are subject to the permanent condition

$$\kappa \Sigma \left( \frac{\partial \Phi}{\partial u}\,U + \frac{\partial \Phi}{\partial u_1}\,U_1 + \frac{\partial \Phi}{\partial u_2}\,U_2 \right) + S_2$$

---

* For our purpose, the left-hand side of the first identity might be $n\Phi$, where $n$ need not be unity.

Here, $S_2$ is the aggregate of the corresponding terms of the second and higher degrees in $\kappa$; and the summation has the same significance as in the expression for $\mathbf{I} - I$.

We have, for each of these quantities $U = X$, $Y$, $Z$,

$$\int \left( \frac{\partial F}{\partial u} U + \frac{\partial F}{\partial u_1} U_1 + \frac{\partial F}{\partial u_2} U_2 \right) dt$$

$$= \int U \left\{ \frac{\partial F}{\partial u} - \frac{d}{dt} \left( \frac{\partial F}{\partial u_1} \right) + \frac{d^2}{dt^2} \left( \frac{\partial F}{\partial u_2} \right) \right\} dt + \left[ U \left\{ \frac{\partial F}{\partial u_1} - \frac{d}{dt} \left( \frac{\partial F}{\partial u_2} \right) \right\} + U_1 \frac{\partial F}{\partial u_2} \right],$$

the last set of terms being taken at the limits of the integral. The equation, derived from $\phi = 0$ and connecting the variations $\kappa X$, $\kappa Y$, $\kappa Z$, by a permanent condition, is modified into a form, which enables us to remove one of these variations—taken to be $\kappa X$—with its derivatives, from the integral in the first variation in $\mathbf{I} - I$. The identities

$$\lambda \frac{\partial \Phi}{\partial u_1} U_1 = - U \frac{d}{dt} \left( \lambda \frac{\partial \Phi}{\partial u_1} \right) + \frac{d}{dt} \left( \lambda \frac{\partial \Phi}{\partial u_1} U \right),$$

$$\lambda \frac{\partial \Phi}{\partial u_2} U_2 = U \frac{d^2}{dt^2} \left( \lambda \frac{\partial \Phi}{\partial u_2} \right) + \frac{d}{dt} \left\{ \lambda \frac{\partial \Phi}{\partial u_2} U_1 - \frac{d}{dt} \left( \lambda \frac{\partial \Phi}{\partial u_2} \right) U \right\},$$

can be verified at once, for any quantity $\lambda$ which is a function of $t$. Thus

$$\lambda \left( \frac{\partial \Phi}{\partial u} U + \frac{\partial \Phi}{\partial u_1} U_1 + \frac{\partial \Phi}{\partial u_2} U_2 \right)$$

$$= U \left\{ \lambda \frac{\partial \Phi}{\partial u} - \frac{d}{dt} \left( \lambda \frac{\partial \Phi}{\partial u_1} \right) + \frac{d^2}{dt^2} \left( \lambda \frac{\partial \Phi}{\partial u_2} \right) \right\}$$

$$+ \frac{d}{dt} \left[ \left\{ \lambda \frac{\partial \Phi}{\partial u_1} - \frac{d}{dt} \left( \lambda \frac{\partial \Phi}{\partial u_2} \right) \right\} U + \lambda \frac{\partial \Phi}{\partial u_2} U_1 \right].$$

Let the quantity $\lambda$, hitherto unspecified and at our disposal, be chosen so as to satisfy the equation

$$\frac{\partial F}{\partial x} - \lambda \frac{\partial \Phi}{\partial x} - \frac{d}{dt} \left( \frac{\partial F}{\partial x_1} - \lambda \frac{\partial \Phi}{\partial x_1} \right) + \frac{d^2}{dt^2} \left( \frac{\partial F}{\partial x_2} - \lambda \frac{\partial \Phi}{\partial x_2} \right) = 0.$$

By the use of this equation, the first variation in $\mathbf{I} - I$ can be modified, so that $X$, $X_1$, $X_2$, no longer occur in the integral; and the first variation then becomes equal to

$$\int \left[ Y \left\{ \left( \frac{\partial F}{\partial y} - \lambda \frac{\partial \Phi}{\partial y} \right) - \frac{d}{dt} \left( \frac{\partial F}{\partial y_1} - \lambda \frac{\partial \Phi}{\partial y_1} \right) + \frac{d^2}{dt^2} \left( \frac{\partial F}{\partial y_2} - \lambda \frac{\partial \Phi}{\partial y_2} \right) \right\} \right.$$

$$\left. + Z \left\{ \left( \frac{\partial F}{\partial z} - \lambda \frac{\partial \Phi}{\partial z} \right) - \frac{d}{dt} \left( \frac{\partial F}{\partial z_1} - \lambda \frac{\partial \Phi}{\partial z_1} \right) + \frac{d^2}{dt^2} \left( \frac{\partial F}{\partial z_2} - \lambda \frac{\partial \Phi}{\partial z_2} \right) \right\} \right] dt$$

$$+ \Sigma \left[ U \left\{ \frac{\partial F}{\partial u_1} - \lambda \frac{\partial \Phi}{\partial u_1} - \frac{d}{dt} \left( \frac{\partial F}{\partial u_2} - \lambda \frac{\partial \Phi}{\partial u_2} \right) \right\} + U_1 \left( \frac{\partial F}{\partial u_2} - \lambda \frac{\partial \Phi}{\partial u_2} \right) \right],$$

where the terms in the last line are taken at the limits.

For a maximum or a minimum, the first variation in $\mathbf{I} - I$ necessarily vanishes for all variations. As before, we first make the integral vanish alone, by choosing variations that are zero at each limit. The quantities $Y$ and $Z$ are arbitrary functions of $t$, independent of one another; and therefore the integral can vanish for all such arbitrary functions, only if the coefficients of $Y$ and $Z$ vanish everywhere. We therefore have the *three characteristic equations*

$$
\left.
\begin{aligned}
\frac{\partial F}{\partial x} - \lambda \frac{\partial \Phi}{\partial x} - \frac{d}{dt}\left(\frac{\partial F}{\partial x_1} - \lambda \frac{\partial \Phi}{\partial x_1}\right) + \frac{d^2}{dt^2}\left(\frac{\partial F}{\partial x_2} - \lambda \frac{\partial \Phi}{\partial x_2}\right) = 0 \\
\frac{\partial F}{\partial y} - \lambda \frac{\partial \Phi}{\partial y} - \frac{d}{dt}\left(\frac{\partial F}{\partial y_1} - \lambda \frac{\partial \Phi}{\partial y_1}\right) + \frac{d^2}{dt^2}\left(\frac{\partial F}{\partial y_2} - \lambda \frac{\partial \Phi}{\partial y_2}\right) = 0 \\
\frac{\partial F}{\partial z} - \lambda \frac{\partial \Phi}{\partial z} - \frac{d}{dt}\left(\frac{\partial F}{\partial z_1} - \lambda \frac{\partial \Phi}{\partial z_1}\right) + \frac{d^2}{dt^2}\left(\frac{\partial F}{\partial z_2} - \lambda \frac{\partial \Phi}{\partial z_2}\right) = 0
\end{aligned}
\right\},
$$

*holding everywhere through the range of integration.*

Because the portion of the vanishing first variation, which consists of the integral alone, has been made to vanish, the remainder also vanishes. Thus *the relation*

$$
\Sigma X \left\{\frac{\partial F}{\partial x_1} - \lambda \frac{\partial \Phi}{\partial x_1} - \frac{d}{dt}\left(\frac{\partial F}{\partial x_2} - \lambda \frac{\partial \Phi}{\partial x_2}\right)\right\} + \Sigma X_1 \left(\frac{\partial F}{\partial x_2} - \lambda \frac{\partial \Phi}{\partial x_2}\right) = 0,
$$

*where the summation is for the three quantities $X$, $Y$, $Z$, must hold at each limit.*

### The three characteristic equations usually reduce to two, because of identities.

**271.** The three characteristic equations are not independent of one another. In virtue of the identities satisfied by $F$ and by $\Phi$, and of the equation $\Phi = 0$, there is a linear relation between them

$$
x_1 E_x + y_1 E_y + z_1 E_z = 0.
$$

Hence, when the identities are retained, the three characteristic equations are satisfied by means of only two independent equations.

To establish the relation, we use the identities and the equation $\Phi = 0$. We have

$$
F = \Sigma x_1 \frac{\partial F}{\partial x_1} + 2\Sigma x_2 \frac{\partial F}{\partial x_2},
$$

where the summation is over the three variables $x, y, z$. Hence, differentiating with respect to $t$,

$$
\Sigma x_1 \frac{\partial F}{\partial x} + \Sigma x_2 \frac{\partial F}{\partial x_1} + \Sigma x_3 \frac{\partial F}{\partial x_2}
$$

$$
= \Sigma x_1 \frac{d}{dt}\left(\frac{\partial F}{\partial x_1}\right) + \Sigma x_2 \frac{\partial F}{\partial x_1} + 2\Sigma x_2 \frac{d}{dt}\left(\frac{\partial F}{\partial x_2}\right) + 2\Sigma x_3 \frac{\partial F}{\partial x_2};
$$

and therefore

$$\Sigma x_1 \left\{ \frac{\partial F}{\partial x} - \frac{d}{dt} \left( \frac{\partial F}{\partial x_1} \right) \right\} = 2\Sigma x_2 \frac{d}{dt} \left( \frac{\partial F}{\partial x_2} \right) + \Sigma x_3 \frac{\partial F}{\partial x_2}.$$

Consequently

$$\Sigma x_1 \left\{ \frac{\partial F}{\partial x} - \frac{d}{dt} \left( \frac{\partial F}{\partial x_1} \right) + \frac{d^2}{dt^2} \left( \frac{\partial F}{\partial x_2} \right) \right\} = \Sigma \left\{ x_1 \frac{d^2}{dt^2} \left( \frac{\partial F}{\partial x_2} \right) + 2x_2 \frac{d}{dt} \left( \frac{\partial F}{\partial x_2} \right) + x_3 \frac{\partial F}{\partial x_2} \right\}$$

$$= \frac{d^2}{dt^2} \left\{ \Sigma x_1 \frac{\partial F}{\partial x_2} \right\}$$

$$= 0,$$

because of the second identity satisfied by $F$.

Again, differentiating the permanent relation $\Phi = 0$, we have

$$\Sigma x_1 \frac{\partial \Phi}{\partial x} + \Sigma x_2 \frac{\partial \Phi}{\partial x_1} + \Sigma x_3 \frac{\partial \Phi}{\partial x_2} = 0 ;$$

and therefore

$$\Sigma x_1 \lambda \frac{\partial \Phi}{\partial x} + \Sigma x_2 \lambda \frac{\partial \Phi}{\partial x_1} + \Sigma x_3 \lambda \frac{\partial \Phi}{\partial x_2} = 0.$$

From the identity

$$\Sigma x_1 \frac{\partial \Phi}{\partial x_1} + 2\Sigma x_2 \frac{\partial \Phi}{\partial x_2} = 0,$$

multiplied by $\lambda$ and then differentiated with respect to $t$, we have

$$\Sigma x_1 \frac{d}{dt} \left( \lambda \frac{\partial \Phi}{\partial x_1} \right) + \Sigma x_2 \lambda \frac{\partial \Phi}{\partial x_1} + 2\Sigma x_2 \frac{d}{dt} \left( \lambda \frac{\partial \Phi}{\partial x_2} \right) + 2\Sigma x_3 \lambda \frac{\partial \Phi}{\partial x_2} = 0 ;$$

or, on combining these derived equations,

$$\Sigma x_1 \left\{ \lambda \frac{\partial \Phi}{\partial x} - \frac{d}{dt} \left( \lambda \frac{\partial \Phi}{\partial x_1} \right) \right\} = 2\Sigma x_2 \frac{d}{dt} \left( \lambda \frac{\partial \Phi}{\partial x_2} \right) + \Sigma x_3 \lambda \frac{\partial \Phi}{\partial x_2}.$$

Consequently,

$$\Sigma x_1 \left\{ \lambda \frac{\partial \Phi}{\partial x} - \frac{d}{dt} \left( \lambda \frac{\partial \Phi}{\partial x_1} \right) + \frac{d^2}{dt^2} \left( \lambda \frac{\partial \Phi}{\partial x_2} \right) \right\}$$

$$= \Sigma \left\{ x_1 \frac{d^2}{dt^2} \left( \lambda \frac{\partial \Phi}{\partial x_2} \right) + 2x_2 \frac{d}{dt} \left( \lambda \frac{\partial \Phi}{\partial x_2} \right) + x_3 \lambda \frac{\partial \Phi}{\partial x_2} \right\}$$

$$= \frac{d^2}{dt^2} \left\{ \Sigma \left( x_1 \lambda \frac{\partial \Phi}{\partial x_2} \right) \right\}$$

$$= 0,$$

because of the identity $\Sigma x_1 \frac{\partial \Phi}{\partial x_2} = 0.$

If, then, the three fundamental equations be denoted by $E_x = 0$, $E_y = 0$, $E_z = 0$, we have

$$x_1 E_x + y_1 E_y + z_1 E_z = 0.$$

That is, the three equations are equivalent to only two fundamental equations, in virtue of (i) the relation $\Phi = 0$, and (ii) the identities satisfied by $F$ and by $\Phi$.

*Ex.* Prove that, when the function $\Phi$ is not such as to satisfy the two identities stated, the three characteristic equations are independent of one another. In that event, shew that, if

$$\rho = \Sigma x_1 \frac{\partial \Phi}{\partial x_1} + 2\Sigma x_2 \frac{\partial \Phi}{\partial x_2}, \qquad \sigma = \Sigma x_1 \frac{\partial \Phi}{\partial x_2},$$

the quantity $\lambda$ satisfies the equation

$$\frac{d}{dt}(\lambda \rho) = \frac{d^2}{dt^2}(\lambda \sigma).$$

*Results in forms adapted for special weak variations: characteristic equations, and terminal conditions.*

**272.** When the identities are satisfied, the equations can be transformed so as to make one of the three variables $x$, $y$, $z$—say $x$—the independent variable. Writing $y' = \frac{y_1}{x_1} = \frac{dy}{dx}$, and so for other derivatives with respect to $x$, we find, from the theory of partial differential equations of the first order, that the most general integral of the two identities (now regarded as equations) satisfied by $F$ is ·

$$F = x_1 f(x, y, y', y'', z, z', z''),$$

and that the most general integral of the two identities (similarly regarded as equations) satisfied by $\Phi$ is

$$\Phi = x_1 \phi(x, y, y', y'', z, z', z''),$$

where there are no limitations upon the form of $f$ or upon the form of $\phi$. The integral to be maximised or minimised is

$$\int f(x, y, y', y'', z, z', z'') \, dx;$$

the permanent equation of condition is

$$\phi(x, y, y', y'', z, z', z'') = 0;$$

and $\lambda$, an unknown function of $t$, becomes an unknown function of $x$.

There now are *two characteristic equations* to be satisfied throughout the range of integration. These two equations are

$$\left.\begin{array}{l} \dfrac{\partial f}{\partial y} - \lambda \dfrac{\partial \phi}{\partial y} - \dfrac{d}{dx}\left(\dfrac{\partial f}{\partial y'} - \lambda \dfrac{\partial \phi}{\partial y'}\right) + \dfrac{d^2}{dx^2}\left(\dfrac{\partial f}{\partial y''} - \lambda \dfrac{\partial \phi}{\partial y''}\right) = 0 \\[2mm] \dfrac{\partial f}{\partial z} - \lambda \dfrac{\partial \phi}{\partial z} - \dfrac{d}{dx}\left(\dfrac{\partial f}{\partial z'} - \lambda \dfrac{\partial \phi}{\partial z'}\right) + \dfrac{d^2}{dx^2}\left(\dfrac{\partial f}{\partial z''} - \lambda \dfrac{\partial \phi}{\partial z''}\right) = 0 \end{array}\right\}.$$

To express *the conditions at the limits* of the integral, let $\kappa\xi$, $\kappa\eta$, $\kappa\zeta$ denote a small variation at a limit, consistent with the permanent relation $\phi = 0$, and

29—2

consistent also with the assigned data (if any), but arbitrary within these restrictions. Then, if

$$A = \frac{\partial f}{\partial y'} - \lambda \frac{\partial \phi}{\partial y'} - \frac{d}{dx}\left(\frac{\partial f}{\partial y''} - \lambda \frac{\partial \phi}{\partial y''}\right),$$

$$B = \frac{\partial f}{\partial z'} - \lambda \frac{\partial \phi}{\partial z'} - \frac{d}{dx}\left(\frac{\partial f}{\partial z''} - \lambda \frac{\partial \phi}{\partial z''}\right),$$

$$C = \frac{\partial f}{\partial y''} - \lambda \frac{\partial \phi}{\partial y''},$$

$$D = \frac{\partial f}{\partial z''} - \lambda \frac{\partial \phi}{\partial z''},$$

the relation

$$\xi(f - y'A - z'B - y''C - z''D) + \eta A + \zeta B + \eta'C + \zeta'D = 0$$

must be satisfied at each limit.

Varying inferences can be drawn from the relation, according as a limit is

(i) fixed in position and direction :

(ii) fixed in position and free in direction :

(iii) required to lie freely on a twisted curve :

(iv) required to lie freely on a surface :

possibly subject to the condition $\phi(x, y, y', y'', z, z', z'') = 0$. They are similar to the respective inferences drawn in § 181.

Finally, the condition $\phi = 0$ may, in fact, have a form that is simpler than the form postulated generally. Thus it may be explicitly free from $y''$ and $z''$; in this case, the data in (i) must conform to the postulated condition.

*Ex. Required the shortest curve of given constant circular curvature to join two points in space.*

It seems preferable not to take the arc-variable as the independent variable, at least until the critical equations have been formed. After that stage in the analysis, the arc-variable can be so assigned for purposes of analytical description.

For the constant circular curvature, we have

$$a = \frac{1}{\rho^2} = \left(\frac{d^2x}{ds^2}\right)^2 + \left(\frac{d^2y}{ds^2}\right)^2 + \left(\frac{d^2z}{ds^2}\right)^2$$

$$= \frac{1}{\mathbf{e}^3}(\mathbf{eg} - \mathbf{f}^2),$$

where, if $x_1$ and $x_2$ denote $\frac{dx}{dt}$ and $\frac{d^2x}{dt^2}$, and similarly for derivatives of $y$ and $z$,

$$\mathbf{e} = x_1^2 + y_1^2 + z_1^2, \quad \mathbf{f} = x_1 x_2 + y_1 y_2 + z_1 z_2, \quad \mathbf{g} = x_2^2 + y_2^2 + z_2^2.$$

When $s$ is taken later to be the independent variable, $\mathbf{e}$ becomes unity always, $\mathbf{f}$ becomes zero always, and $\mathbf{g}$ (in the problem) acquires the constant value $a$.

The permanent equation $\phi = 0$ can be taken

$$\Phi = \mathbf{eg} - \mathbf{f}^2 - a\mathbf{e}^3 = 0.$$

Here

$$\frac{\partial \Phi}{\partial x_1} = 2x_1 \mathbf{g} - 2x_2 \mathbf{f} - 6a\mathbf{e}^2 x_1, \qquad \frac{\partial \Phi}{\partial x_2} = 2x_2 \mathbf{e} - 2x_1 \mathbf{f},$$

$$\frac{\partial \Phi}{\partial y_1} = 2y_1 \mathbf{g} - 2y_2 \mathbf{f} - 6a\mathbf{e}^2 y_1, \qquad \frac{\partial \Phi}{\partial y_2} = 2y_2 \mathbf{e} - 2y_1 \mathbf{f},$$

$$\frac{\partial \Phi}{\partial z_1} = 2z_1 \mathbf{g} - 2z_2 \mathbf{f} - 6a\mathbf{e}^2 z_1 , \qquad \frac{\partial \Phi}{\partial z_2} = 2z_2 \mathbf{e} - 2z_1 \mathbf{f},$$

so that

$$x_1 \frac{\partial \Phi}{\partial x_2} + y_1 \frac{\partial \Phi}{\partial y_2} + z_1 \frac{\partial \Phi}{\partial z_2} = 0,$$

$$x_1 \frac{\partial \Phi}{\partial x_1} + y_1 \frac{\partial \Phi}{\partial y_1} + z_1 \frac{\partial \Phi}{\partial z_1} + 2x_2 \frac{\partial \Phi}{\partial x_2} + 2y_2 \frac{\partial \Phi}{\partial y_2} + 2z_2 \frac{\partial \Phi}{\partial z_2} = 0 ;$$

and therefore the requisite identities are satisfied by the permanent equation.

The problem is to minimise the integral $\int (x_1^2 + y_1^2 + z_1^2)^{\frac{1}{2}} dt$, subject to the permanent relation $\Phi = 0$. As, both in the integral and in the relation, $x$, $y$, $z$, do not occur but only their derivatives, we take

$$u = x_1, \quad v = y_1, \quad w = z_1 ;$$

and now we have to minimise the integral

$$\int (u^2 + v^2 + w^2)^{\frac{1}{2}} dt, \quad = \int F dt,$$

subject to the condition

$$\Phi = \mathbf{e}\mathbf{g} - \mathbf{f}^2 - a\mathbf{e}^3 = 0,$$

where, for the moment,

$$\mathbf{e} = u^2 + v^2 + w^2, \quad \mathbf{f} = uu_1 + vv_1 + ww_1, \quad \mathbf{g} = u_1^2 + v_1^2 + w_1^2.$$

For this analytic form of the problem, the characteristic equations are

$$\frac{\partial F}{\partial u} - \lambda \frac{\partial \Phi}{\partial u} + \frac{d}{dt}\left(\lambda \frac{\partial \Phi}{\partial u_1}\right) = 0,$$

$$\frac{\partial F}{\partial v} - \lambda \frac{\partial \Phi}{\partial v} + \frac{d}{dt}\left(\lambda \frac{\partial \Phi}{\partial v_1}\right) = 0,$$

$$\frac{\partial F}{\partial w} - \lambda \frac{\partial \Phi}{\partial w} + \frac{d}{dt}\left(\lambda \frac{\partial \Phi}{\partial w_1}\right) = 0,$$

the function $F$ not involving $u_1$, $v_1$, $w_1$.

Now that the characteristic equations have been formed, we particularise the variable $t$ in them, so that it becomes the arc-element of the characteristic curve; and we pass back to $x$, $y$, $z$, so that these become the dependent variables which are subject to relations

$$x'^2 + y'^2 + z'^2 = 1, \quad x''^2 + y''^2 + z''^2 = a = \frac{1}{a^2},$$

where $a$ is the value of the constant radius of curvature. Thus the characteristic equations become

$$\frac{dx}{ds} + \lambda \, 4a \, \frac{dx}{ds} + \frac{d}{ds}\left(2\lambda \frac{d^2 x}{ds^2}\right) = 0,$$

together with the two others; that is, they become

$$x' (1 + 4a\lambda) + 2\lambda' x'' + 2\lambda x''' = 0,$$

$$y' (1 + 4a\lambda) + 2\lambda' y'' + 2\lambda y''' = 0,$$

$$z' (1 + 4a\lambda) + 2\lambda' z'' + 2\lambda z''' = 0.$$

Multiplying by $x'$, $y'$, $z'$, adding, and noting that, as $x'x'' + y'y'' + z'z'' = 0$, also

$$x'x''' + y'y''' + z'z''' = -(x''^2 + y''^2 + z''^2) = -a,$$

we have

$$1 + 2a\lambda = 0,$$

so that $\lambda$ is constant, and therefore $\lambda' = 0$. Thus

$$x' + a^2 x''' = 0, \quad y' + a^2 y''' = 0, \quad z' + a^2 z''' = 0;$$

and therefore

$$x' = B_1 \cos \frac{s}{a} + C_1 \sin \frac{s}{a}, \quad y' = B_2 \cos \frac{s}{a} + C_2 \sin \frac{s}{a}, \quad z' = B_3 \cos \frac{s}{a} + C_3 \sin \frac{s}{a}.$$

The terms at the limits give

$$\left[ \lambda \left( U \frac{\partial \Phi}{\partial u_1} + V \frac{\partial \Phi}{\partial v_1} + W \frac{\partial \Phi}{\partial w_1} \right) \right] = 0,$$

that is,

$$[X'x'' + Y'y'' + Z'z''] = 0.$$

But, from the equations

$$x'^2 + y'^2 + z'^2 = 1, \quad x''^2 + y''^2 + z''^2 = a,$$

we have, everywhere,

$$x'X' + y'Y' + z'Z' = 0, \quad x''X'' + y''Y'' + z''Z'' = 0.$$

By the equations, the first of these gives

$$x'''X' + y'''Y' + z'''Z' = 0;$$

and therefore we have, everywhere along the curve,

$$\frac{d}{ds}(X'x'' + Y'y'' + Z'z'') = 0,$$

that is,

$$[X'x'' + Y'y'' + Z'z''] = 0,$$

at the limits taken together. No limitation is thus imposed on the constants.

Again, the values of $x'$, $y'$, $z'$, must always satisfy the equation

$$x'^2 + y'^2 + z'^2 = 1,$$

so that we must have

$$B_1^2 + B_2^2 + B_3^2 = 1, \quad C_1^2 + C_2^2 + C_3^2 = 1, \quad B_1 C_1 + B_2 C_2 + B_3 C_3 = 0.$$

In virtue of these relations, they satisfy the equation

$$x''^2 + y''^2 + z''^2 = \frac{1}{a^2}.$$

Further, we have

$$x = A_1 + aB_1 \sin \frac{s}{a} - aC_1 \cos \frac{s}{a} \left.\begin{array}{l} \\ \\ \end{array}\right\}$$
$$y = A_2 + aB_2 \sin \frac{s}{a} - aC_2 \cos \frac{s}{a} \quad;$$
$$z = A_3 + aB_3 \sin \frac{s}{a} - aC_3 \cos \frac{s}{a}$$

and therefore

$$\begin{vmatrix} x - A_1, & B_1, & C_1 \\ y - A_2, & B_2, & C_2 \\ z - A_3, & B_3, & C_3 \end{vmatrix} = 0,$$

$$(x - A_1)^2 + (y - A_2)^2 + (z - A_3)^2 = a^2;$$

that is, the required curve is a circle of radius $a$.

Finally, let one extremity be taken at the origin and the other at a distance $d$ along the axis of $x$; and let the length of the arc be $l$. The terminal conditions become

$$0 = A_1 - aC_1, \quad d = A_1 + aB_1 \sin \frac{l}{a} - aC_1 \cos \frac{l}{a},$$

$$0 = A_2 - aC_2, \quad 0 = A_2 + aB_2 \sin \frac{l}{a} - aC_2 \cos \frac{l}{a},$$

$$0 = A_3 - aC_3, \quad 0 = A_3 + aB_3 \sin \frac{l}{a} - aC_3 \cos \frac{l}{a};$$

these lead to the relation

$$d = 2a \sin \frac{l}{2a},$$

in conformity with the preceding result, that the curve is a circle of radius $a$.

### Postulation of more than one permanent non-integral relation.

**273.** Sometimes more than a single permanent condition will be postulated, when there are several dependent simultaneous variables. If any such permanent condition could be integrated by itself, its primitive could be used to reduce the number of dependent variables by one unit: the problem would remain with a diminished number of variables. Accordingly, it can be assumed that all conditions (if any) of this type have been used; and the problem therefore is to maximise or minimise an integral

$$\int F(x, x_1, \ldots; y, y_1, \ldots; z, z_1, \ldots; u, u_1, \ldots; \ldots) \, dt,$$

when the variables are subject to relations

$$\phi_n(x, x_1, \ldots; y, y_1, \ldots; z, z_1, \ldots; u, u_1, \ldots; \ldots) = 0,$$

for $n = 1, 2, \ldots, r$.

The same kind of argument, as that which has been used in §§ 266, 270, leads to the conclusion that, if the first variation of $F$ is to vanish for all arbitrary small weak variations obeying the equations $\phi_1 = 0, \ldots, \phi_r = 0$, we introduce $r$ unknown quantities $\mu_1, \mu_2, \ldots, \mu_r$, and construct the characteristic equations in the form

$$\frac{\partial F}{\partial \theta} - \Sigma \mu_r \frac{\partial \phi_r}{\partial \theta} - \frac{d}{dt}\left(\frac{\partial F}{\partial \theta'} - \Sigma \mu_r \frac{\partial \phi_r}{\partial \theta'}\right) + \frac{d^2}{dt^2}\left(\frac{\partial F}{\partial \theta''} - \Sigma \mu_r \frac{\partial \phi_r}{\partial \theta''}\right) - \ldots = 0,$$

where $\theta$ denotes, in turn, each of the dependent variables $x, y, z, u, \ldots$ in the original integral.

It is unnecessary to pursue the general stage further, so as to include derivatives of order higher than the first. As has been indicated in § 262 ($\gamma$), the problem can be transformed so that the integral shall no longer involve derivatives of order higher than the first, by the introduction of suitable external conditions $\phi_r = 0$, each such condition itself containing no derivative of order higher than the first.

*Ex. A particle is moving in a smooth plane vertical groove under gravity in a resisting medium, the groove having the form of a catenary; find the equation satisfied by the law of resistance, when the brachistochrone is a catenary.*

Take the arc to be the independent variable; the axis of $y$ to be vertically downward; and $\phi(v)$ to be the law of resistance, where $v$ is the velocity. The permanent equations are

$$x'^2 + y'^2 = 1, \quad vv' - gy' + \phi(v) = 0,$$

while we have to minimise the integral

$$\int \frac{ds}{v}.$$

The characteristic equations are

$$-\frac{1}{v^2} - \mu_1\left(v' + \frac{d\phi}{dv}\right) + \frac{d}{ds}(\mu_1 v) = 0,$$

$$-\frac{d}{ds}(2\mu_2 x') = 0,$$

$$\frac{d}{ds}(\mu_1 g - 2\mu_2 y') = 0.$$

Thus there are five equations, together with the equation of the catenary, to determine $x$, $y$, $v$, $\phi(v)$, and the two multipliers $\mu_1$, $\mu_2$.

At once, we have

$$2\mu_2 x' = a,$$

$$2\mu_2 y' = \mu_1 g + b,$$

where $a$ and $b$ are constants; and the first equation gives

$$\frac{1}{v^2} + \mu_1 \frac{d\phi}{dv} = v\mu_1' = \frac{v}{g}\frac{d}{ds}(a\tan\psi).$$

For a catenary, we have

$$s = c\tan\psi,$$

so that the last result is

$$\frac{1}{v^2} + \mu_1 \frac{d\phi}{dv} = \frac{va}{gc} = \frac{v}{\rho^3},$$

where $\rho$ is a constant. Also

$$\mu_1 = -\frac{b}{g} + \frac{s}{\rho^3},$$

so that, if

$$\frac{v^3 - \rho^3}{v^2 \frac{\partial\phi}{\partial v}} + \frac{b}{g}\rho^3 = G(v),$$

we have

$$s = G(v).$$

But

$$v\frac{dv}{ds} + \phi(v) = gy'$$

$$= g\frac{s}{(s^2 + c^2)^{\frac{1}{2}}};$$

hence

$$\frac{v}{G'(v)} + \phi(v) = g\frac{G}{(G^2 + c^2)^{\frac{1}{2}}},$$

a differential equation of the second order satisfied by $\phi(v)$.

# CHAPTER IX.

## Double integrals with derivatives of the first order: weak variations: minimal surfaces.

(i) The discussion of double integrals (and of multiple integrals) in the problems of the calculus of variations is much less advanced towards completion than is the discussion of single integrals. One main reason is that the central critical equation is a partial differential equation, of the second order in even the simplest case. Consequently, difficulties arise through lack of effective methods for the construction of a primitive.

(ii) The most interesting special problem, which depends upon double integrals, is that of determining the surface of least superficial extent limited by an assigned boundary curve. The problem was initiated by Lagrange*, who may be regarded as the pioneer in the subject; he obtained the partial differential equation of the second order characteristic of such surfaces, interpretable as the property that the radii of curvature are equal and opposite. A primitive of the equation was obtained in a substantially general form by Monge†; the effectively useful form of this primitive, now generally used, is due to Weierstrass‡. Further investigations are specially due to Schwarz§ who, in particular, obtained the equations of a minimal surface (§ 304, *post*) determined by the most general conditions, that are contemplated in the Cauchy existence-theorem concerning primitives of partial equations of the second order. The geometry of minimal surfaces is amply discussed ‖ by Darboux.

(iii) For the general problem of double integrals involving derivatives of the first order, reference may be made to two papers by Kobb¶, who has extended, to the treatment of such integrals, the Weierstrass method for single integrals, so that all the variables may receive simultaneous variations, whether these are weak, or are strong variations of the simplest type as considered by Weierstrass. The method here adopted for the reduction of the second variation differs from Kobb's; as indicated in § 164 for a simpler problem, it is a modification of the method developed successively by Legendre, Hesse, and Clebsch, for single integrals. Moreover, the construction of the Weierstrass $E$-test, connected with the simplest type of strong variation, is less purely analytical than Kobb's.

(iv) The extension, to double integrals, of the Jacobi test—that is, the limitation of the range within which a surface may provide a maximum or a minimum for weak variations—appears to have received no detailed attention for the general problem, and very little attention for the special problem of minimal surfaces. As regards the latter, a test was outlined in a paper** by the author; the result there obtained has been made specifically general for minimal surfaces (§§ 315, 316, *post*), and is applied to two particular instances.

---

* See references, p. 4, *foot-note*.
† *Applications de l'analyse à la géométrie*, p. 211.
‡ *Berl. Monatsb.* (1866), pp. 612—625, 855—856.
§ *Gesammelte Abhandlungen*, vol. i, *passim*.
‖ *Théorie générale des surfaces*, vol. i, book iii.
¶ *Acta Math.*, t. xvi (1892—3), pp. 65—140; *ib.* t. xvii (1893), pp. 321—343.
** *Ann. di Mat.*, Ser. iii, t. xxi (1912—3), pp. 121—142.

*Double integrals of the simplest type : special weak variations.*

**274.** We now pass to the consideration of the conditions under which multiple integrals can possess maxima and minima. We shall deal first with double integrals, so that there are two independent variables, to be denoted by $x$ and $y$; and we shall assume that there is a single dependent variable, to be denoted by $z$. The simplest type of such integral will involve the partial derivatives of $z$ of the first order alone, denoted by $p$ and $q$. The integral $I$ thus is of the form

$$I = \iint f(x, y, z, p, q)\, dx\, dy,$$

with duly specified limits of integration, initially free from variation.

*First variation.*

In the first place, we shall deal only with maxima and minima under weak variations. In order to obtain some guiding results, as in Chapter I for single integrals, we shall consider only special weak variations. Thus we must consider the effect, made upon the value of $I$ by substituting a varied value $z + \kappa Z$ for $z$, where $\kappa$ is a small arbitrary constant of the same kind as has already been considered, and where $Z$ is a uniform regular (but otherwise arbitrary) function of $x$ and $y$ that is independent of $\kappa$. The function $f(x, y, z, p, q)$, sometimes briefly denoted by $f$, will be supposed regular when its arguments are continuous, so that it is expansible in powers of $\kappa$, after the variation is imposed.

By way of notation, we denote partial derivatives of $f$ with regard to $x, y, z, p, q$, by suffixes 1, 2, 3, 6, 9, respectively*, so that

$$\frac{\partial f}{\partial x} = f_1, \quad \frac{\partial f}{\partial y} = f_2, \quad \frac{\partial f}{\partial z} = f_3, \quad \frac{\partial f}{\partial p} = f_6, \quad \frac{\partial f}{\partial q} = f_9,$$

$$\frac{\partial^2 f}{\partial z \partial p} = f_{36}, \quad \frac{\partial^2 f}{\partial z \partial q} = f_{39}, \quad \frac{\partial^2 f}{\partial p^2} = f_{66},$$

and so on; differentiation with regard to two variables is represented by combination of the two associated integers, and repeated differentiation with regard to a single variable is represented by a repetition of the associated integer. Also, we denote $\dfrac{\partial Z}{\partial x}$ and $\dfrac{\partial Z}{\partial y}$ by $P$ and $Q$, so that $P$ and $Q$ also are regular functions of $x$ and $y$. If $J$ represent the varied value of $I$, then (with an assumption of unchanged limits)

$$J - I = \iint \{ f(x, y, z + \kappa Z, p + \kappa P, q + \kappa Q) - f(x, y, z, p, q) \}\, dx\, dy$$

$$= \kappa \iint (Z f_3 + P f_6 + Q f_9)\, dx\, dy + R_2 = \kappa \mathbf{A} + R_2,$$

---

* A reason, for the apparently arbitrary assignment of these integers, will appear (§ 290) when we come to consider weak variations in general.

where

$$\mathbf{A} = \iint (Zf_3 + Pf_6 + Qf_9)\,dx\,dy,$$

and $R_2$ is the aggregate of all the terms involving $\kappa^2$, $\kappa^3$, ... in the expansion of $J - I$. The quantity $\kappa\mathbf{A}$, as usual, is called the *first variation* of $I$.

### Lemma in double integration.

**275.** We proceed to modify the expression for the double integral $\mathbf{A}$, which extends over an area (simply or multiply connected); the boundary or boundaries of the said area may be fixed, or they may admit of mobility limited by assigned conditions. Assuming, for the immediate present, that the boundaries are fixed, we suppose (merely by way of example) that the area is bounded outwardly by a curve such as $AHDE$ and inwardly by a curve such as $BGCF$. With the customary convention that the positive direction of a boundary is such that the included area lies to the left when looking downwards at the enclosed area as depicted, the positive direction of the outer boundary line is $AHDE$ and of the inner boundary line is $BFCG$. Moreover, we assume the limits in $\mathbf{A}$ to be taken so that the whole of the upper limit is external to the whole of the associated lower limit.

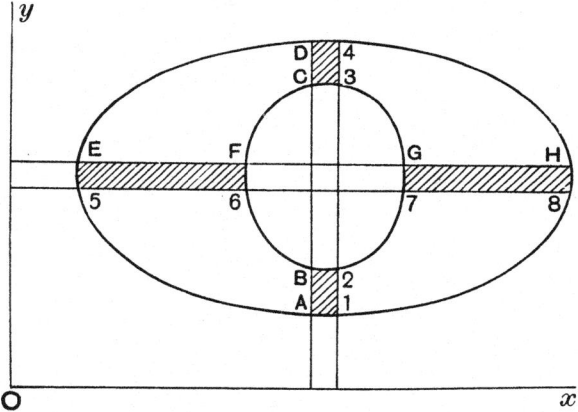

A lemma in integration will lead to the desired simplification of the double integral. It is as follows:

Let $\Phi$ and $\Psi$ denote two functions of the variables $x$ and $y$, completely regular over a bounded area and along its boundary. Then

$$\iint \left( \frac{\partial\Phi}{\partial x} + \frac{\partial\Psi}{\partial y} \right) dx\,dy = \int (\Phi\,dy - \Psi\,dx),$$

the double integral being taken positively for both variables over the whole of the included area, and the single integral being taken positively along the whole boundary of that area.

We have

$$\iint \frac{\partial \Phi}{\partial x} \, dx \, dy = \int \{(\Phi_8 - \Phi_7) + (\Phi_6 - \Phi_5)\} \, dy,$$

because integration along $EFGH$, with the same breadth $dy$ of strip, (in the positive direction of $y$ in the area) extends over the shaded area $EF65E$ with the upper limit at $F$ and the lower at $E$, and over the shaded area $GH87G$ with the upper limit at $H$ and the lower limit at $G$. For the positive direction of description of the whole boundary, this positive element $dy$ is such that

$$(dy)_8 = dy, \quad (dy)_7 = -dy, \quad (dy)_6 = dy, \quad (dy)_5 = -dy.$$

Thus the foregoing element $\{(\Phi_8 - \Phi_7) + (\Phi_6 - \Phi_5)\} \, dy$ is equal to

$$(\Phi \, dy)_8 + (\Phi \, dy)_7 + (\Phi \, dy)_6 + (\Phi \, dy)_5 :$$

that is, it is equal to the portion of $\int \Phi \, dy$ for that strip $dy$, when this integral is taken positively round the boundary. Hence, summing for all the strips, with increasing $y$ (and therefore positive $dy$ for the area), we have

$$\iint \frac{\partial \Phi}{\partial x} \, dx \, dy = \int \Phi \, dy,$$

the single integral being taken positively round the whole boundary.

Again,

$$\iint \frac{\partial \Psi}{\partial y} \, dx \, dy = \int \{(\Psi_4 - \Psi_3) + (\Psi_2 - \Psi_1)\} \, dx,$$

because integration along $ABCD$, with the same breadth $dx$ of strip, (in the positive direction of $x$ in the area) extends over the shaded area $AB21A$ with the upper limit at $B$ and the lower at $A$, and over the shaded area $CD43C$ with the upper limit at $D$ and the lower limit at $C$. For the positive direction of description of the whole boundary, this positive element $dx$ is such that

$$(dx)_4 = -dx, \quad (dx)_3 = dx, \quad (dx)_2 = -dx, \quad (dx)_1 = dx.$$

Thus the foregoing element $\{(\Psi_4 - \Psi_3) + (\Psi_2 - \Psi_1)\} \, dx$ is equal to

$$-(\Psi \, dx)_4 - (\Psi \, dx)_3 - (\Psi \, dx)_2 - (\Psi \, dx)_1 :$$

that is, is equal to the portion of $-\int \Psi \, dx$ for that strip $dx$, when this integral is taken positively round the boundary. Hence, summing for all the strips, with increasing $x$ (and therefore positive $dx$ for the area), we have

$$\iint \frac{\partial \Psi}{\partial y} \, dx \, dy = -\int \Psi \, dx,$$

the single integral being taken positively round the whole boundary.

Combining the two results, we obtain the enunciated lemma.

*Note.* We adopt the convention that, in a double integral $\iint f \, dx \, dy$, in a triple integral $\iiint f \, dx \, dy \, dz$, and so on, the integration is to be effected, first for the element $dx$ which occurs first in the sequence $dx\,dy$, $dx\,dy\,dz$, ... proceeding from left to right, between an upper limit for $x$ as an assigned function of the remaining variable or variables, and a lower limit for $x$ as another assigned function of the remaining variable or variables. In integrations subsequent to the first, should there be more than one, the same convention for the order of successive operations is adopted.

Also, in order to change the variables of integration from one set to another, we use the theorem*

$$\iint f\,dx\,dy = \iint f J\left(\frac{x,\,y}{u,\,v}\right) du\,dv,$$

where

$$J\left(\frac{x,\,y}{u,\,v}\right) = \frac{\partial x}{\partial u}\frac{\partial y}{\partial v} - \frac{\partial x}{\partial v}\frac{\partial y}{\partial u};$$

and similarly for triple integrals.

Accordingly, for the integral $\iint \dfrac{\partial\Phi}{\partial x}\,dx\,dy$, we take it to be the integral, with respect to $y$, of $\displaystyle\int \frac{\partial\Phi}{\partial x}\,dx$ between limits: that is, $\displaystyle\int \Phi\,dy$. For the integral $\iint \dfrac{\partial\Psi}{\partial y}\,dx\,dy$, which is

$$\iint \frac{\partial\Psi}{\partial y}\,J\left(\frac{x,\,y}{y,\,x}\right) dy\,dx, \quad = -\iint \frac{\partial\Psi}{\partial y}\,dy\,dx,$$

we take it (from the latter expression for the value) to be the integral, with respect to $x$, of $-\displaystyle\int \frac{\partial\Psi}{\partial y}\,dy$ between limits: that is, $-\displaystyle\int \Psi\,dx$.

The conclusion, as to the positive direction round the whole boundary, is deduced as in the foregoing proof of the lemma.


*Modification of expression for first variation.*

**276.** In order to apply the lemma so as to modify the expression for **A**, we write

$$\Phi = Zf_6, \quad \Psi = Zf_9;$$

and then

$$\iint \left( Pf_6 + Z\frac{\partial f_6}{\partial x} + Qf_9 + Z\frac{\partial f_9}{\partial y} \right) dx\,dy = \int Z\,(f_6\,dy - f_9\,dx),$$

the single integral being taken positively round the whole boundary of the included area: thus

$$\iint (Pf_6 + Qf_9)\,dx\,dy = -\iint Z\left(\frac{\partial f_6}{\partial x} + \frac{\partial f_9}{\partial y}\right) dx\,dy + \int Z\,(f_6\,dy - f_9\,dx).$$

* See Picard, *Traité d'Analyse*, t. i, ch. iv. The transformation for double integrals appears to have been effected first by Euler; for triple integrals, first by Lagrange; for integrals of general multiplicity, first by Jacobi.

Consequently,

$$\mathbf{A} = \iint Z \left( f_3 - \frac{\partial f_6}{\partial x} - \frac{\partial f_9}{\partial y} \right) dx\, dy + \int Z (f_6\, dy - f_9\, dx),$$

where the double integral is still taken positively along the axis of $x$ and along the axis of $y$ for the whole of the bounded area, and the single integral is taken positively along the whole boundary of the included area.

But as stated already for the present problem (which is the simplest as regards boundary conditions), the limits are free from arbitrary small variations. Therefore, for this problem at present, we have

$$Z = 0$$

everywhere along the boundary. Hence the first variation becomes

$$\mathbf{A} = \iint Z \left( f_3 - \frac{\partial f_6}{\partial x} - \frac{\partial f_9}{\partial y} \right) dx\, dy.$$

### Characteristic equation : characteristic surface.

**277.** In order that the integral $I$ may possess a maximum or a minimum, the quantity $J - I$ must possess a sign that is unchangeable for all the admissible variations. The magnitude of $R_2$, owing to the smallness of $\kappa$, will be negligible as compared with the magnitude of the first variation, unless the latter should actually vanish and be not merely small.

Consider, first, a small variation which is not zero over the surface, but is zero at the boundary (that is, at the limits of integration), so that the term $\int Z (f_6\, dy - f_9\, dx)$ does not survive for that variation. If the quantity $f_3 - \dfrac{\partial f_6}{\partial x} - \dfrac{\partial f_9}{\partial y}$ does not vanish everywhere, then ($Z$ being at our choice, because all variations are admissible) we can choose $Z$ positive where the quantity is positive and $Z$ negative where the quantity is negative, so that $\iint Z \left( f_3 - \dfrac{\partial f_6}{\partial x} - \dfrac{\partial f_9}{\partial y} \right) dx\, dy$ is positive; and we can choose $Z$ negative where the quantity is positive and $Z$ positive where the quantity is negative, and make the same integral to be negative. Moreover, $\kappa$ can be taken positive or negative at will. On the supposition that has been adopted, we can make the first variation positive or negative at will : we could therefore, on that supposition, make $J - I$ positive or negative at will : the integral could not have a maximum or a minimum. The supposition must therefore be excluded for our purpose: that is, we must have

$$f_3 - \frac{\partial f_6}{\partial x} - \frac{\partial f_9}{\partial y} = 0,$$

or

$$f_3 - (f_{16} + f_{29}) - (p f_{36} + q f_{39}) - (r f_{66} + 2s f_{69} + t f_{99}) = 0,$$

a condition to be satisfied, everywhere in the range, by the function $f$.

It is the *characteristic equation*, being a partial differential equation of the second order to be satisfied by the variable $z$ as a function of $x$ and $y$. The surface, thus determined, is called the *characteristic surface*.

Both forms of the characteristic equation will be used. The earlier form,

$$E = f_3 - \frac{\partial f_6}{\partial x} - \frac{\partial f_9}{\partial y} = 0,$$

will be the more immediately useful, when (§ 283) the expression for the second variation of $I$ is to be moulded into the critical shape. The later form,

$$E = f_3 - (f_{16} + f_{29}) - (pf_{36} + qf_{39}) - (rf_{66} + 2sf_{69} + tf_{99}) = 0,$$

is the more significant as regards the actual surface. It is a partial equation of the second order, linear in the derivatives of that order, the coefficients of these second derivatives $r$, $s$, $t$, being functions of $x$, $y$, $z$, $p$, $q$. Thus the equation is of the form

$$Rr + 2Ss + Tt = U,$$

which has formed the subject of research from the days of Euler onwards. The character of the primitive depends upon the nature of the quantities $f_{66}$, $f_{69}$, $f_{99}$; and these quantities prove to be critical also in other tests subsequently to be applied.

### First variation, with mobile limits under special variations.

**278.** In the preceding discussion, simplicity was obtained by an assumption that the boundaries of the double integral are actually determinate, so that small variations at the boundaries are inadmissible. We now proceed to take the alternative hypothesis into account, so that small variations may be made at the boundary; but these small variations must be consistent with the conservation of the boundary, while the current special variations admit change of the ordinate alone. The terminal variations, and the current variations at the limits, must ultimately be brought into relation with one another.

As before, we take $\kappa Z$ to be the special variation, current in the field of integration. On the boundary of the field, we take $\kappa\xi$, $\kappa\eta$, $\kappa\zeta$, to denote the variation of a point on any portion of the boundary, so that a displaced position $x + \kappa\xi$, $y + \kappa\eta$, $z + \kappa\zeta$, must lie on an assigned boundary, whether it be a curve or a surface. We shall use $\xi_0$, $\eta_0$, $\zeta_0$, as connected with the lower limits of our integral, $\xi'$, $\eta'$, $\zeta'$, as connected with its upper limits. Then the varied value of our integral is

$$\iint_{x_0 + \kappa\xi_0,\ y_0 + \kappa\eta_0}^{x' + \kappa\xi',\ y' + \kappa\eta'} f(x,\, y,\, z + \kappa Z,\, p + \kappa P,\, q + \kappa Q)\, dx\, dy.$$

Writing $F = f(x, y, z + \kappa Z, p + \kappa P, q + \kappa Q)$ temporarily, we have

$$\iint_{x_0 + \kappa \xi_0,\ y_0 + \kappa \eta_0}^{x' + \kappa \xi',\ y' + \kappa \eta'} F \, dx \, dy$$

$$= \int_{y_0 + \kappa \eta_0}^{y' + \kappa \eta'} \left\{ \left( \int_{x'}^{x' + \kappa \xi'} + \int_{x_0}^{x'} - \int_{x_0}^{x_0 + \kappa \xi_0} \right) F \, dx \right\} dy$$

$$= \int_{y_0 + \kappa \eta_0}^{y' + \kappa \eta'} \left\{ \kappa \xi' F' - \kappa \xi_0 F_0 + \int_{x_0}^{x'} F \, dx \right\} dy,$$

where $F'$ denotes the mean value of $F$ over the range $x'$ to $x' + \kappa \xi'$, and $F_0$ denotes the mean value of $F$ over the range $x_0$ to $x_0 + \kappa \xi_0$. Now $F'$ differs from $f(x, y, z, p, q)$ at the corresponding place on the upper limit by quantities of the first and higher degrees in $\kappa$, and $F'$ is multiplied by $\kappa \xi'$. Hence, retaining explicitly the terms of the first degree in $\kappa$, and merging the rest in a complete remainder $K_2$ which shall include all terms of the second and higher degrees in $\kappa$, we have

$$\kappa \xi' F' = \kappa \xi' f' + \dots,$$

the unspecified terms being merged in $K_2$ as stated, while $f'$ denotes the value of $f$ at the upper limit. Similarly,

$$\kappa \xi_0 F_0 = \kappa \xi_0 f_0 + \dots,$$

the unspecified terms being merged in the complete remainder $K_2$, while $f_0$ denotes the value of $f$ at the lower limit.

Again,

$$\int_{y_0 + \kappa \eta_0}^{y' + \kappa \eta'} U \, dy = \int_{y'}^{y' + \kappa \eta'} U \, dy + \int_{y_0}^{y'} U \, dy - \int_{y_0}^{y_0 + \kappa \eta_0} U \, dy$$

$$= \left( \int_{y'}^{y' + \kappa \eta'} + \int_{y_0}^{y'} - \int_{y_0}^{y_0 + \kappa \eta_0} \right) U \, dy.$$

Now the quantities

$$\int_{y'}^{y' + \kappa \eta'} (\kappa \xi' f' - \kappa \xi_0 f_0) \, dy, \quad \int_{y_0}^{y_0 + \kappa \eta_0} (\kappa \xi' f' - \kappa \xi_0 f_0) \, dy,$$

are each of them an aggregate of quantities of the second and higher degrees in $\kappa$, and are therefore to be merged in the complete remainder $K_2$; and consequently

$$\int_{y_0 + \kappa \eta_0}^{y' + \kappa \eta'} (\kappa \xi' F' - \kappa \xi_0 F_0) \, dy = \kappa \int_{y_0}^{y'} (\xi' F' - \xi_0 F_0) \, dy,$$

to the first power of $\kappa$. Again, we have

$$\int_{y_0 + \kappa \eta_0}^{y' + \kappa \eta'} \left( \int_{x_0}^{x'} F \, dx \right) dy$$

$$= - \int_{x_0}^{x'} \left\{ \int_{y_0 + \kappa \eta_0}^{y' + \kappa \eta'} F \, dy \right\} dx,$$

as in § 275, *Note*. But

$$\int_{y_0 + \kappa \eta_0}^{y' + \kappa \eta'} F \, dy = \left( \int_{y'}^{y' + \kappa \eta'} + \int_{y_0}^{y'} - \int_{y_0}^{y_0 + \kappa \eta_0} \right) F \, dy.$$

As before,

$$\int_{y'}^{y'+\kappa\eta'} F\,dy = \kappa\eta' F'',$$

where $F''$ is the mean value of $F$ over the range $y'$ to $y' + \kappa\eta'$: or, again denoting by $f'$ the value of $f$ at the upper limit, we have

$$\int_{y'}^{y'+\kappa\eta'} F\,dy = \kappa\eta' f' + \ldots,$$

the unspecified terms being merged in $K_2$. Similarly,

$$\int_{y_0}^{y_0+\kappa\eta_0} F\,dy = \kappa\eta_0 f_0 + \ldots,$$

with the same mergence of unspecified terms in $K_2$.

Thus, finally, the value of the varied integral is

$$\kappa \int_{y_0}^{y'} (\xi' f' - \xi_0 f_0)\,dy - \kappa \int_{x_0}^{x'} (\eta' f' - \eta_0 f_0)\,dx + \iint_{x_0, y_0}^{x', y'} F\,dx\,dy + K_2.$$

The original integral is

$$\iint_{x_0, y_0}^{x', y'} f\,dx\,dy\,;$$

and therefore the whole variation in value of the integral is equal to

$$\kappa \int [f(\xi\,dy - \eta\,dx)] + \iint_{x_0, y_0}^{x', y'} (F - f)\,dx\,dy + K_2.$$

But the integral

$$\iint (F - f)\,dx\,dy$$

is the quantity which has been considered in the preceding discussion (§§ 274, 276): it has been proved equal to

$$\kappa \iint Z \left( f_3 - \frac{\partial f_6}{\partial x} - \frac{\partial f_9}{\partial y} \right) dx\,dy + \kappa \int Z (f_6\,dy - f_9\,dx) + \ldots,$$

with unspecified terms of the second and higher degrees in $\kappa$. The double integral extends, as before, over the whole of the bounded area; the single integral is taken along the whole of the boundary (that is, all the boundary representing the complete upper limits and the complete lower limits of the area, over which the double integral extends).

Accordingly, denoting by $R_2$ the aggregate of all the terms which include the second and higher powers of $\kappa$, we have the full variation of our original integral in the form

$$\kappa \iint Z \left( f_3 - \frac{\partial f_6}{\partial x} - \frac{\partial f_9}{\partial y} \right) dx\,dy$$

$$+ \kappa \int \{(Zf_6 + \xi f)\,dy - (Zf_9 + \eta f)\,dx\} + R_2,$$

where the single integral is taken positively round the whole boundary of the area over which the double integral extends; while $Z$, in that single integral, arises out of the variation of the ordinate at a place on the boundary, consistent with the boundary limitations.

*Characteristic equation: boundary conditions.*

**279.** As always, the possession of a maximum or a minimum requires that the first variation shall vanish—that is, that the quantity multiplied by $\kappa$ in the preceding expression shall vanish—for all possible admissible weak variations.

Among these weak variations, consider the aggregate that leave the boundary everywhere unvaried. For these, we have $Z = 0$ on the boundary, $\xi = 0$, $\eta = 0$. The double integral must then vanish for all the arbitrary variations $\kappa Z$ in the field. By precisely the same argument as before, this result requires that the equation

$$E = f_3 - \frac{\partial f_6}{\partial x} - \frac{\partial f_9}{\partial y} = 0$$

shall be satisfied everywhere within the range of integration. We thus obtain again (as is to be expected) *the characteristic equation.*

For all other weak variations, that is, for all weak variations which are admissible along the boundary, we must have

$$\int \{(Zf_6 + \xi f)\, dy - (Zf_9 + \eta f)\, dx\} = 0,$$

taken over the whole boundary of the surface over which the double integral extends. When this whole boundary consists of two or more portions, each portion being a curve complete in itself and detached topographically from the rest of the boundary, the preceding condition must be satisfied over each portion. For we can take an admissible weak variation over any such portion and, temporarily, leave all the remainder of the boundary unvaried; the condition over all that portion is all that remains for such a variation, and it must be satisfied. Similarly for every portion. Hence the condition

$$\int \{(Zf_6 + \xi f)\, dy - (Zf_9 + \eta f)\, dx\}$$

must be satisfied over each portion of the boundary, which is complete in itself and is topographically separate from all the other portions of the boundary.

**280.** More precise inferences from the condition can only be drawn, after consideration of each such portion of the boundary. When the boundary is not rigidly and definitely assigned by specific data, it usually is associated with conditions relating to a curve or curves, or to a surface or surfaces, in space. Variation along any such curve or surface is limited by the geometry of such curve or such surface: while the weak variation $\kappa Z$, for the original integral, gives variation solely along the ordinate of the surface which is the subject of investigation. Hence $\kappa Z$ must be brought into relation with the variation along the terminal curve or the terminal surface at each complete and separate portion of the boundary. Thus our original integral might be required to have one given curve as one limit, and another curve as the other limit; we must bring the variation along the ordinate into relation with the possible variations along these curves. Or our original integral might be required to pass from one given surface (the range on that surface not being specified) to another given surface (the range on that other surface also not being specified). In neither illustration would the quantity $\kappa Z$ be the displacement along the $z$-coordinate of the limiting curve or of the limiting surface respectively.

Accordingly, at the boundary, let the point $P(x, y, z)$ be displaced to $x + \kappa\xi$, $y + \kappa\eta$, $z + \kappa\zeta$, also on the boundary curve or boundary surface: and consider the consecutive characteristic surface through this point. The tangent plane to the latter characteristic is given by the equation

$$z' - \kappa\zeta = p_1(x' - \kappa\xi) + q_1(y' - \kappa\eta),$$

the point $x$, $y$, $z$, being taken as origin for our purpose: here $p_1$ and $q_1$, belonging to the consecutive surface, differ from $p$ and $q$ by quantities of the first and higher powers of $\kappa$. The intercept from $P$ made along the $z$-ordinate by this plane is $\kappa Z$, and along that ordinate $x' = 0$ and $y' = 0$; thus

$$\kappa Z = \text{value of } z' \text{ when } x' = 0 \text{ and } y' = 0$$
$$= \kappa(\zeta - p\xi - q\eta) + \dots,$$

or, to the first power of $\kappa$,

$$Z = \zeta - p\xi - q\eta.$$

Hence our condition comes to be

$$\int [\{(\zeta - p\xi - q\eta) f_6 + \xi f\} \, dy - \{(\zeta - p\xi - q\eta) f_9 + \eta f\} \, dx] = 0,$$

for each complete and separate portion of the boundary: and this condition must be satisfied for all possible variations consistent with the conservation of the boundary. Moreover, we can keep these variations continuously changing; and we can make them zero at any place or over any range along the boundary. Hence the quantity

$$\{(\zeta - p\xi - q\eta) f_6 + \xi f\} \, dy - \{(\zeta - p\xi - q\eta) f_9 + \eta f\} \, dx$$

must vanish everywhere along the boundary; for if it is positive over any range or stretch, we can make it zero and have the rest of the variation zero, and the integral would be positive, instead of zero; and similarly, if the quantity is negative.

**281.** Summing up the results, we have the following conditions as necessary and sufficient to secure the vanishing of the first variation of the original integral:

(i) the characteristic equation

$$E = 0$$

must be satisfied throughout the range of integration: and

(ii) the equation

$$\{(\zeta - p\xi - q\eta)f_6 + \xi f\}\, dy - \{(\zeta - p\xi - q\eta)f_9 + \eta f\}\, dx = 0$$

must be satisfied everywhere along each complete and separate portion of the boundary, where $\kappa\xi$, $\kappa\eta$, $\kappa\zeta$, are the actual displacements at a point on that portion of the boundary, either when that portion is a given curve or when it is compelled to lie upon a given surface.

If the boundary is a definite given curve, we have

$$\phi\,(x,\, y,\, z) = 0, \quad \psi\,(x,\, y,\, z) = 0,$$

and therefore, at a contiguous point $x + \kappa\xi$, $y + \kappa\eta$, $z + \kappa\zeta$, on such a boundary, we have

$$\frac{\partial\phi}{\partial x}\,\xi + \frac{\partial\phi}{\partial y}\,\eta + \frac{\partial\phi}{\partial z}\,\zeta = 0, \quad \frac{\partial\psi}{\partial x}\,\xi + \frac{\partial\psi}{\partial y}\,\eta + \frac{\partial\psi}{\partial z}\,\zeta = 0,$$

for non-zero values of $\xi$, $\eta$, $\zeta$; thus the relation

$$\left\{ J\left(\frac{\phi,\ \psi}{x,\ y}\right) - pJ\left(\frac{\phi,\ \psi}{y,\ z}\right) - qJ\left(\frac{\phi,\ \psi}{z,\ x}\right) \right\} \left( \frac{\partial f}{\partial p}\,dy - \frac{\partial f}{\partial q}\,dx \right)$$
$$+ \left\{ J\left(\frac{\phi,\ \psi}{y,\ z}\right) dy - J\left(\frac{\phi,\ \psi}{z,\ x}\right) dx \right\} f = 0$$

is satisfied everywhere along that limit. Similarly for each separate and complete portion of the boundary.

If the boundary is to lie on a given surface

$$F\,(x,\, y,\, z) = 0,$$

so that, for a contiguous point $x + \kappa\xi$, $y + \kappa\eta$, $z + \kappa\zeta$, we have

$$\xi\frac{\partial F}{\partial x} + \eta\frac{\partial F}{\partial y} + \zeta\frac{\partial F}{\partial z} = 0,$$

the terminal conditions are

$$\frac{f\,dy - p\left(\dfrac{\partial f}{\partial p}\,dy - \dfrac{\partial f}{\partial q}\,dx\right)}{\dfrac{\partial F}{\partial x}} = \frac{-f\,dx - q\left(\dfrac{\partial f}{\partial p}\,dy - \dfrac{\partial f}{\partial q}\,dx\right)}{\dfrac{\partial F}{\partial y}} = \frac{\dfrac{\partial f}{\partial p}\,dy - \dfrac{\partial f}{\partial q}\,dx}{\dfrac{\partial F}{\partial z}}$$

along that limit. Similarly at each separate and complete portion of a boundary lying on an assigned surface.

*Ex.* 1. The intercept on the normal to a surface between the surface and the plane $z = 0$ is denoted by $\nu$. Obtain the characteristic equation of a surface, for which the integral

$$\iint \nu \, dS$$

could be a maximum or a minimum for weak variations, in the form

$$2(r + t) = 1 - p^2 - q^2.$$

*Ex.* 2. Let the intercept between the origin, and the intersection of the axis of $z$ with the tangent plane to a surface, be denoted by $\zeta$. Shew that the characteristic equation of a surface, for which the integral

$$\iint \zeta \, dS$$

could be a maximum or a minimum, is

$$(z - px - qy)\{(1 + q^2) r - 2pqs + (1 + p^2) t\}$$
$$= 2(1 + p^2 + q^2)\{1 + p(p + xr + ys) + q(q + xs + yt)\}.$$

## *Cauchy's primitive of the characteristic equation.*

**282.** The characteristic partial differential equation is of the second order. Cauchy's general theory * establishes the existence of a unique uniform integral $z$ of that equation, determined by two conditions:

(i)  along any assigned curve (restricted not to belong to either of two special families) the quantity $z$ assumes an assigned arbitrary value, which is a regular function of $x$ and $y$; and

(ii)  along the same curve, one of the derivatives of $z$ assumes a similar assigned arbitrary value, which is a regular function of $x$ and $y$.

The source of the restriction on the assignment of the initial curve appears at the earliest stage in Cauchy's proof. Let the curve, projected orthogonally on the plane $z = 0$, be

$$\alpha(x, y) = \text{constant}.$$

Along this curve we have

$$\theta \frac{\partial \alpha}{\partial x} + \frac{\partial \alpha}{\partial y} = 0,$$

where $\theta$ denotes the value of $\dfrac{dx}{dy}$ at the point $(x, y)$ on the projected curve. Let the assigned value of $z$ be $F(x, y)$, and the assigned value of $p$ be $G(x, y)$. We have, always,

$$dz = p \, dx + q \, dy;$$

and therefore, along the curve,

$$\frac{\partial F}{\partial x} dx + \frac{\partial F}{\partial y} dy = G \, dx + q \, dy,$$

---

\* See the author's *Theory of Differential Equations*, vol. vi, chaps. xii, xx.

that is,

$$q = \frac{\partial F}{\partial y} + \left( \frac{\partial F}{\partial x} - G \right) \theta = H(x, y),$$

so that $H(x, y)$ is a known function along the curve. Also, always,

$$r\,dx + s\,dy = dp, \quad s\,dx + t\,dy = dq\,;$$

and therefore, along the curve,

$$r\,dx + s\,dy = \frac{\partial G}{\partial x}\,dx + \frac{\partial G}{\partial y}\,dy,$$

$$s\,dx + t\,dy = \frac{\partial H}{\partial x}\,dx + \frac{\partial H}{\partial y}\,dy,$$

that is,

$$s = -r\theta + \frac{\partial G}{\partial y} + \theta\,\frac{\partial G}{\partial x},$$

$$t = r\theta^2 + \frac{\partial H}{\partial y} + \left( \frac{\partial H}{\partial x} - \frac{\partial G}{\partial y} \right) \theta - \frac{\partial G}{\partial x}\,\theta^2.$$

The differential equation (§ 277) is

$$r f_{66} + 2s f_{69} + t f_{99} = f_3 - f_{16} - f_{29} - p f_{36} - q f_{39}.$$

When the values of $s$ and $t$ are substituted, the equation determines the value of $r$ along the curve, unless

$$f_{66} - 2f_{69}\theta + f_{99}\theta^2$$

vanishes. When $r$ is thus known, the values of $s$ and $t$ are known. The values of the derivatives of the third order along the curve are similarly deducible when, for the purpose, the equation is differentiated with respect to $x$ and $y$; and similarly for the derivatives of all orders. The integral is then obtainable in a series that can be proved to converge uniformly and absolutely; and so the unique integral exists as determined by the assigned conditions. The conditions are therefore effective, unless

$$f_{66} \left( \frac{\partial \alpha}{\partial x} \right)^2 + 2f_{69}\,\frac{\partial \alpha}{\partial x}\,\frac{\partial \alpha}{\partial y} + f_{99} \left( \frac{\partial \alpha}{\partial y} \right)^2 = 0,$$

that is, unless the conditions are assigned along a curve

$$f_{66}\,dy^2 - 2f_{69}\,dx\,dy + f_{99}\,dx^2 = 0.$$

It will appear, in the sequel, that $f_{66}$ and $f_{99}$ have the same sign and also that $f_{66}f_{99} > f_{69}^2$, the variables $x$ and $y$ having real values, if the surface is to provide a maximum or a minimum. Thus the critical curves are imaginary on such a surface; and there are two conjugate sets. Therefore, when the data consist of conditions, stated in terms of real quantities along real curves on the surface, the restriction, required by the existence-theorem, is actually observed in the problem. But the restriction is not necessarily observed in connection with every partial differential equation of the second order

$$E(x, y, z, p, q, r, s, t) = 0,$$

whatever its source; and the equation of the excluded critical curves is

$$\frac{\partial E}{\partial r}\,dy^2 - \frac{\partial E}{\partial s}\,dy\,dx + \frac{\partial E}{\partial t}\,dx^2 = 0.$$

These critical curves are called* *characteristics* by Cauchy, a term applied to all such curves in his method of establishing the existence of integrals of partial differential equations of all orders when there are two independent variables.

A simple geometrical interpretation can be given to the result. It appears, from the data, that the value of $z$ is given for every point upon the projected curve; that is, the surface is to pass through a given curve in space. (As we are dealing with maxima and minima, we shall assume that the curve is real.) It also appears, as a deduction from the data, that the values of $p$ and $q$ are known for every point upon that given curve: in other words, the direction-cosines of the normal to the characteristic surface are known along the curve or (what effectively is the same property) the tangent plane to the characteristic surface is known at all points along the given curve. But a mathematically continuous succession of planes through all the points on a continuous curve generates a developable surface through that curve; and therefore we can declare that *a characteristic surface is uniquely determinate by the condition that, along an assigned curve, it shall touch a given developable surface through the curve.*

It will appear that the Cauchy critical curves for a minimal surface are the nul-lines of the surface. Hence, if we desire to extend the analytical range of the data, the restrictive condition is that the analytical curve, along which the data for a minimal surface are given, is not to be a nul-line on the surface. A mode of determining the arbitrary functions in the primitive of the characteristic equation by means of the assigned data will be given later (§ 304), in a theorem due to Schwarz.

*Second variation under special variations: normal form.*

**283.** When the characteristic equation is established, and when the conditions (if any) dependent on the boundary restrictions are satisfied, the first variation of $I$ vanishes; and the magnitude of $J - I$ then depends mainly upon the terms in $R_2$ which involve only the second power of $\kappa$, when $\kappa$ is sufficiently small. Now

$$J - I = \tfrac{1}{2}\kappa^2 \iint \Theta\,dx\,dy + R_3,$$

where $R_3$ represents the aggregate of terms involving $\kappa^3$, $\kappa^4$, ... in the expansion of $J - I$. Thus the so-called 'second' variation, being the term in $\kappa^2$, depends upon the double integral, of which the integrand $\Theta$ is

$$\Theta = f_{66}P^2 + 2f_{69}PQ + f_{99}Q^2 + 2f_{36}PZ + 2f_{39}QZ + f_{33}Z^2.$$

---

* This notion of characteristics was introduced by Monge, *Application de l'analyse à la géométrie*, p. 33 (in the Liouville edition, 1850).

We proceed to modify the expression for $\Theta$. The coefficients $f_{mn}$ can be considered as functions of $x$ and $y$ alone, when the value of $z$, (with the derived values of $p$ and $q$) derived from the primitive of the characteristic equation, has been substituted in them. Introducing three disposable functions of $x$ and $y$ by $A$, $B$, $\mu$, and denoting the first derivatives of $\mu$ by $\mu_1$ and $\mu_2$, we determine these functions by the requirement that the quantity

$$\Theta + \frac{\partial}{\partial x}(AZ^2) + \frac{\partial}{\partial y}(BZ^2)$$

shall be expressible* in the form

$$f_{66}\left(P - \frac{\mu_1}{\mu}Z\right)^2 + 2f_{69}\left(P - \frac{\mu_1}{\mu}Z\right)\left(Q - \frac{\mu_2}{\mu}Z\right) + f_{99}\left(Q - \frac{\mu_2}{\mu}Z\right)^2.$$

This requirement in form will be met, provided the functions satisfy the relations

$$\left.\begin{array}{rl} A + f_{36} =& -f_{66}\dfrac{\mu_1}{\mu} - f_{69}\dfrac{\mu_2}{\mu} \\[2mm] B + f_{39} =& \qquad -f_{69}\dfrac{\mu_1}{\mu} - f_{99}\dfrac{\mu_2}{\mu} \\[2mm] \dfrac{\partial A}{\partial x} + \dfrac{\partial B}{\partial y} + f_{33} =& f_{66}\dfrac{\mu_1^2}{\mu^2} + 2f_{69}\dfrac{\mu_1\mu_2}{\mu^2} + f_{99}\dfrac{\mu_2^2}{\mu^2} \end{array}\right\}.$$

Manifestly $A$ and $B$ would be known, as soon as $\mu$ is known. The elimination of $A$ and $B$, between these three relations, leads to the equation

$$f_{33} - \frac{\partial f_{36}}{\partial x} - \frac{\partial f_{39}}{\partial y} - \frac{\partial}{\partial x}\left(f_{66}\frac{\mu_1}{\mu} + f_{69}\frac{\mu_2}{\mu}\right) - \frac{\partial}{\partial y}\left(f_{69}\frac{\mu_1}{\mu} + f_{99}\frac{\mu_2}{\mu}\right)$$

$$= f_{66}\frac{\mu_1^2}{\mu^2} + 2f_{69}\frac{\mu_1\mu_2}{\mu^2} + f_{99}\frac{\mu_2^2}{\mu^2}$$

for the determination of $\mu$, an equation which can be written in the form

$$f_{33}\mu + f_{36}\mu_1 + f_{39}\mu_2 - \frac{\partial}{\partial x}(f_{36}\mu + f_{66}\mu_1 + f_{69}\mu_2) - \frac{\partial}{\partial y}(f_{39}\mu + f_{69}\mu_1 + f_{99}\mu_2) = 0.$$

This equation, partial and of the second order, is linear. If

$$2\Omega = f_{33}\mu^2 + 2f_{36}\mu\mu_1 + 2f_{39}\mu\mu_2 + f_{66}\mu_1^2 + 2f_{69}\mu_1\mu_2 + f_{99}\mu_2^2,$$

(where it should be noted that $2\Omega$ is the same function of $\mu$, as $\Theta$ is of $Z$), the equation can be written

$$\frac{\partial\Omega}{\partial\mu} - \frac{\partial}{\partial x}\left(\frac{\partial\Omega}{\partial\mu_1}\right) - \frac{\partial}{\partial y}\left(\frac{\partial\Omega}{\partial\mu_2}\right) = 0.$$

* It is easy to see that no generality is gained by taking

$$\frac{\partial}{\partial x}(AZ^2 + \theta ZQ) + \frac{\partial}{\partial y}(BZ^2 - \theta ZP)$$

as the terms additive to $\Theta$, where $\theta$ is any function of $x$ and $y$.

*Subsidiary characteristic equation, and its primitive.*

**284.** Now the characteristic equation is

$$E = \frac{\partial f}{\partial z} - \frac{\partial}{\partial x}\left(\frac{\partial f}{\partial p}\right) - \frac{\partial}{\partial y}\left(\frac{\partial f}{\partial q}\right) = 0:$$

or, as it was written,

$$f_3 - \frac{\partial f_6}{\partial x} - \frac{\partial f_9}{\partial y} = 0.$$

We denote by $z$ the primitive of this equation, taking the most general primitive possible which involves two arbitrary functions. The characteristic equation is still satisfied, when two modified arbitrary functions take the place of the two which actually occur. Let the two arbitrary functions be denoted by $\phi$ and $\psi$, with arguments $\alpha$ and $\beta$ respectively such that [*]

$$f_{66}\left(\frac{\partial \alpha}{\partial x}\right)^2 + 2f_{69}\frac{\partial \alpha}{\partial x}\frac{\partial \alpha}{\partial y} + f_{99}\left(\frac{\partial \alpha}{\partial y}\right)^2 = 0,$$

$$f_{66}\left(\frac{\partial \beta}{\partial x}\right)^2 + 2f_{69}\frac{\partial \beta}{\partial x}\frac{\partial \beta}{\partial y} + f_{99}\left(\frac{\partial \beta}{\partial y}\right)^2 = 0;$$

then $z$ is expressible in a form

$$z = F(x, y, \phi, \phi', \ldots, \psi, \psi', \ldots).$$

When arbitrary small variations are effected upon the arbitrary functions $\phi$ and $\psi$, so that they become $\phi + \kappa\Phi$ and $\psi + \kappa\Psi$ with the customary significance for $\kappa$, let $z$ become $z + \kappa\zeta + \ldots$, the unexpressed terms being of the second and higher orders in $\kappa$. The value of $\zeta$ is given by

$$\zeta = \Phi\frac{\partial F}{\partial \phi} + \Phi'\frac{\partial F}{\partial \phi'} + \ldots + \Psi\frac{\partial F}{\partial \psi} + \Psi'\frac{\partial F}{\partial \psi'} + \ldots.$$

Manifestly $\Phi$ and $\Psi$ can be taken, in their turn, as quite arbitrary functions, so that the most general variation of $z$ can be secured; and still the characteristic equation is satisfied by this modified value $z + \kappa\zeta$ of the dependent variable.

But the effect of changing $z$ into $z + \kappa\zeta$ is to make the following changes, viz.

$f_3$ into $f_3 + \kappa(f_{33}\zeta + f_{36}\zeta_1 + f_{39}\zeta_2) +$ terms of second and higher orders,

$f_6$ into $f_6 + \kappa(f_{36}\zeta + f_{66}\zeta_1 + f_{69}\zeta_2) + \ldots\ldots\ldots\ldots\ldots\ldots\ldots\ldots$,

$f_9$ into $f_9 + \kappa(f_{39}\zeta + f_{69}\zeta_1 + f_{99}\zeta_2) + \ldots\ldots\ldots\ldots\ldots\ldots\ldots\ldots$.

As the characteristic equation is still satisfied, we have

$$\kappa\left\{(f_{33}\zeta + f_{36}\zeta_1 + f_{39}\zeta_2) - \frac{\partial}{\partial x}(f_{36}\zeta + f_{66}\zeta_1 + f_{69}\zeta_2) - \frac{\partial}{\partial y}(f_{39}\zeta + f_{69}\zeta_1 + f_{99}\zeta_2)\right\} + \ldots = 0;$$

[*] See my *Theory of Differential Equations*, vol. vi, § 186.

that is, $\zeta$ satisfies the equation

$$f_{33}\zeta + f_{36}\zeta_1 + f_{39}\zeta_2 - \frac{\partial}{\partial x}\left(f_{36}\zeta + f_{66}\zeta_1 + f_{69}\zeta_2\right) - \frac{\partial}{\partial y}\left(f_{39}\zeta + f_{69}\zeta_1 + f_{99}\zeta_2\right) = 0,$$

an equation of the second order and linear. But in the expression

$$\zeta = \Phi\frac{\partial F}{\partial \phi} + \Phi'\frac{\partial F}{\partial \phi'} + \cdots + \Psi\frac{\partial F}{\partial \psi} + \Psi'\frac{\partial F}{\partial \psi'} + \cdots,$$

$\Phi$ and $\Psi$ are two independent arbitrary functions, with the same arguments as $\phi$ and $\psi$ respectively, while the terms of the second order in the derivatives of $\zeta$ are

$$f_{66}\zeta_{11} + 2f_{69}\zeta_{12} + f_{99}\zeta_{22},$$

the same as those terms in the characteristic equation. Hence the stated expression for $\zeta$ is the primitive of the equation of the second order satisfied by $\zeta$.

As this equation in $\zeta$ is derived by small variations from the characteristic equation, it may be called the *subsidiary characteristic equation*. Its primitive is known (and has been given) when the primitive of the characteristic equation is known. But if (not the primitive, but only) a special solution of the characteristic equation is known, and if we desire to settle whether this special solution will provide a maximum or a minimum for our integral, then the primitive of the subsidiary characteristic equation must be obtained by processes of integration, the special solution first being substituted in the coefficients in that subsidiary equation.

**285.** The equation, with $\mu$ as the dependent variable, is the same as the equation with $\zeta$ as the dependent variable; and we may now suppose that we have the primitive of the $\zeta$-equation. We thus shall obtain the most general possible value of $\mu$, by taking

$$\mu = \zeta.$$

Then we have

$$\Theta + \frac{\partial}{\partial x}(AZ^2) + \frac{\partial}{\partial y}(BZ^2) = \square,$$

where

$$\square = f_{66}\left(P - \frac{\zeta_1}{\zeta}Z\right)^2 + 2f_{69}\left(P - \frac{\zeta_1}{\zeta}Z\right)\left(Q - \frac{\zeta_2}{\zeta}Z\right) + f_{99}\left(Q - \frac{\zeta_2}{\zeta}Z\right)^2$$

and

$$\left. \begin{aligned} -A &= f_{36} + f_{66}\frac{\zeta_1}{\zeta} + f_{69}\frac{\zeta_2}{\zeta} \\ -B &= f_{39} + f_{69}\frac{\zeta_1}{\zeta} + f_{99}\frac{\zeta_2}{\zeta} \end{aligned} \right\}.$$

Thus

$$\int\!\!\int \Theta\, dx\, dy = \int\!\!\int\left\{\square - \frac{\partial}{\partial x}(AZ^2) - \frac{\partial}{\partial y}(BZ^2)\right\} dx\, dy$$

$$= \int\!\!\int \square\, dx\, dy - \int Z^2(A\, dy - B\, dx),$$

where the single integral is taken positively round the whole of the boundary of the area over which the double integral extends.

If the original integral is to possess a maximum or a minimum, the integral $\iint \Theta \, dx \, dy$ must possess a sign which is uniform for all small variations that can be imposed.

The variations along the boundary of the area do not substantially constrain the range of variations within the area; at the utmost, they are only the forms of these surface variations at the boundary limits when these last are mobile. Now the terminal conditions connected with the vanishing of the first variation are satisfied, either (i) owing to data initially assigned as fixing the limits, in which case $Z$ is everywhere zero along the boundary: or (ii) by the determination of those limits with the range of original data. In the latter case, we may consider that the limits have been determined in connection with the vanishing of the first variation. Hence, when the second variation is considered, these limits are to be regarded as known and fixed; consequently, they admit no small variations of the characteristic surface along their length. Thus along the boundary we may, in considering $\iint \Theta \, dx \, dy$, assume $Z$ to be zero along the whole of the boundary; and then the sign of the second variation depends upon the sign of $\iint \square \, dx \, dy$, with the foregoing value of $\square$. This sign is to be uniform for all possible values of the function $Z$, taken to be any arbitrary regular function of $x$ and $y$, which now is merely required to vanish along the boundary.

*Discussion of the normal form of second variation: the Legendre test.*

**286.** Consider, first, those variations $Z$ of the foregoing type, such that the quantities

$$P - \frac{\zeta_1}{\zeta} Z, \quad Q - \frac{\zeta_2}{\zeta} Z,$$

do not vanish together. Then $\square$, which is equal to

$$\frac{1}{f_{66}} \left[ \left\{ f_{66} \left( P - \frac{\zeta_1}{\zeta} Z \right) + f_{69} \left( Q - \frac{\zeta_2}{\zeta} Z \right) \right\}^2 + (f_{66} f_{99} - f_{69}{}^2) \left( Q - \frac{\zeta_2}{\zeta} Z \right)^2 \right]$$

and to

$$\frac{1}{f_{99}} \left[ \left\{ f_{99} \left( Q - \frac{\zeta_2}{\zeta} Z \right) + f_{69} \left( P - \frac{\zeta_1}{\zeta} Z \right) \right\}^2 + (f_{66} f_{99} - f_{69}{}^2) \left( P - \frac{\zeta_1}{\zeta} Z \right)^2 \right],$$

will be uniform in sign, if

$f_{66}$ is uniform in sign throughout the range,

$f_{99}$ is uniform in sign throughout the range (these signs of $f_{66}$ and $f_{99}$ being the same), and

$f_{66} f_{99} - f_{69}{}^2$ is positive throughout the range.

Suppose it possible that $f_{66}f_{99} - f_{69}{}^2$ could be negative or that it could vanish. In the former case, we could make $\square$ of a sign opposite to $f_{66}$ and $f_{99}$, by choosing a variation

$$f_{66}\left(P - \frac{\zeta_1}{\zeta}Z\right) + f_{69}\left(Q - \frac{\zeta_2}{\zeta}Z\right) = 0,$$

or a variation

$$f_{99}\left(Q - \frac{\zeta_2}{\zeta}Z\right) + f_{69}\left(P - \frac{\zeta_1}{\zeta}Z\right) = 0,$$

while neither $P - \dfrac{\zeta_1}{\zeta}Z$ nor $Q - \dfrac{\zeta_2}{\zeta}Z$ vanishes; and we could make it of the same sign as $f_{66}$ and $f_{99}$, by choosing either $P - \dfrac{\zeta_1}{\zeta}Z$ or $Q - \dfrac{\zeta_2}{\zeta}Z$ (but not both of these quantities) equal to zero. In the latter case, we could make $\square$ vanish for all variations, such that

$$f_{66}\left(P - \frac{\zeta_1}{\zeta}Z\right) + f_{69}\left(Q - \frac{\zeta_2}{\zeta}Z\right) = 0.$$

The first possibility must be excluded rigidly: that is, the quantity $f_{66}f_{99} - f_{69}{}^2$ must never be negative. The second possibility is more limited in significance: under it, $\square$ could vanish: and so, for this variation $Z$, the second variation of the integral could vanish. We shall, temporarily, exclude this latter possibility, just as (temporarily) the possibility that $P - \dfrac{\zeta_1}{\zeta}Z$ and $Q - \dfrac{\zeta_2}{\zeta}Z$ may simultaneously vanish has been excluded.

When these conditions are satisfied, $\square$ and consequently $\iint \square\, dx\, dy$ are uniform in sign for all admissible variations. The sign is positive when the common sign of $f_{66}$ and of $f_{99}$ is positive—that is, a minimum is admissible for the integral $I$, if other conditions are satisfied. The sign of the second variation is negative when the common sign of $f_{66}$ and of $f_{99}$ is negative—that is, a maximum is admissible for the integral $I$ if other conditions are satisfied.

This collective test—that $f_{66}$ and $f_{99}$ may nowhere vanish in the range and everywhere must have the same sign, and $f_{66}f_{99} - f_{69}{}^2$ must everywhere be positive in the range—corresponds to the second (Legendre) test in the case of integrals of the first order when there is only one independent variable. By an extension of the former usage, and although Legendre's researches did not cover integrals of the kind now under discussion, it is convenient to call this test the *Legendre test*.

We shall now assume that the Legendre test is satisfied, noting the possibility that $f_{66}f_{99} - f_{69}{}^2$ may be zero—a possibility which, if existing, must be further considered (§ 288).

*The Jacobi test, in descriptive form.*

**287.** The quantity $\square$, and therefore the second variation of $I$, is now of uniform sign for all possible variations, unless the range of integration is such that non-zero variations can occur which make

$$P - \frac{\zeta_1}{\zeta} Z = 0, \quad Q - \frac{\zeta_2}{\zeta} Z = 0,$$

simultaneously over the whole range. This occurrence is possible only if

$$Z = C\zeta,$$

where $C$ is a pure constant.

The most general value of $\zeta$ is supposed to be known; it is given by the primitive of the subsidiary characteristic equation. The quantity $Z$ is to vanish along the whole boundary of the region of integration. We may imagine such a region with an 'initial' boundary—it might be a point, or a closed oval curve, in the simplest cases. Along such an initial boundary, the variations are to be zero, and therefore $Z$ is zero along that boundary. Consequently, we choose the arbitrary elements in $\zeta$, so that the quantity $\zeta$ vanishes along that boundary but does not vanish everywhere on the characteristic surface. Now consider this selected quantity $\zeta$, which is a function of $x$ and $y$; it may never again vanish within the range of the surface, after passing from the initial boundary. We therefore cannot have $Z = C\zeta$, because $Z$ is to vanish at what may be termed the 'final' boundary—which, in the simplest instances, usually is a closed oval curve. In such an event, the non-zero variations, which could annihilate $\square$, do not exist. We should not then be forced to impose a limit upon the range of the integral to secure a non-vanishing second variation, that is, to secure the chance that the integral $I$ may possess a maximum or a minimum.

But if the region of integration is such, that the quantity $\zeta$, which vanishes along the initial boundary, can again vanish at a final boundary or within the region, then over that region, ranging between the initial boundary and a boundary constituted by this new curve on the characteristic surface where $\zeta$ again vanishes, we can have a non-zero variation of the characteristic surface represented by $Z = C\zeta$. For such a variation, $\square$ vanishes: and therefore, for such a variation, the second variation of $I$ vanishes.

In this case, $J - I$ is equal to the quantity denoted by $R_3$; but it is equal to $R_3$, only when this particular variation $Z = C\zeta$ is used. It then would become necessary to examine the term in $R_3$ which involves $\kappa^3$ alone: if that term does not vanish, a change in the sign of the arbitrary constant $\kappa$ changes the sign of the governing term in $R_3$: and the integral $I$ would not possess either a maximum or a minimum. If however the term in $R_3$, which involves $\kappa^3$, does vanish for the variation $Z = C\zeta$, we must consider the term in $R_3$ which involves $\kappa^4$ alone; but we need only consider it for the variation $Z = C\zeta$. A positive sign would, so far, imply a minimum for $I$: a negative sign would, so far, imply a maximum for $I$. And in the rare event of the term in $\kappa^4$ also vanishing for the particular variation, we should have to consider higher powers of $\kappa$ in $R_3$: the possibilities become so extremely special, as to require examination solely in the rare instances where they occur.

The net general result is that the region of integration, initially bounded, must not extend so far as to allow the quantity $\zeta$, a primitive of the subsidiary characteristic equation chosen so as to vanish over that initial boundary, to vanish again. Now $z$, the primitive of the characteristic equation, determines the characteristic surface: and $z + \kappa\zeta$, another primitive of that equation derived from $z$ by a small variation of the arbitrary elements, determines another characteristic surface. The two surfaces intersect in what has been called the initial boundary, because $\zeta$ has been made to vanish along that curve. The limiting position, as $\kappa$ tends to zero, of the first curve of succeeding intersection of the two surfaces marks the range where $\zeta$ again first vanishes. This limit of the second intersection of the two consecutive characteristic surfaces is called the *conjugate* of the first intersection. We therefore infer that, in order to secure a maximum or secure a minimum for the integral, *the region of integration bounded by an initial curve must not extend so far as the conjugate of that initial curve.*

As with the preceding test, so here, and even although Jacobi's researches did not include integrals of the kind now under discussion, it is convenient to call the criterion just established the *Jacobi* test.

**288.** The special case of partial exception (§ 286) to the Legendre test, represented by the relation

$$f_{66}f_{99} - f_{69}^2 = 0,$$

must be mentioned: but it will not be discussed in full. When that relation is satisfied, the second variation of the integral $I$ vanishes for a small variation of the characteristic surface given by

$$\theta\left(P - \frac{\zeta_1}{\zeta}Z\right) + Q - \frac{\zeta_2}{\zeta}Z = 0,$$

where $f_{69} = \theta f_{99}, f_{66} = \theta f_{69}$. If the integral of the equation

$$dx = \theta\, dy,$$

where $\theta$ now is a function of $x$ and $y$ only, be

$$g(x, y) = c,$$

the most general value of $Z$ satisfying the foregoing relation is

$$Z = \zeta\Phi\{g(x, y)\},$$

where $\Phi$ is an arbitrary function. For such a variation $\kappa Z$ of the characteristic surface, the second variation of $I$ vanishes; and then $J - I$ depends upon the value of $R_3$ for that value of $Z$.

The discussion of $R_3$ now follows, generally, the same lines as the former discussion when the region of integration extends as far as a conjugate curve. The instances, when the relation $f_{66}f_{99} = f_{69}{}^2$ holds, are so special that the investigation is simpler to carry out in each particular instance as it arises. Analytically, the case belongs to that class of linear partial differential equations of the second order (that is, linear in the derivatives of the second order), when in Ampère's method* of integration the arguments are one and the same, and when in Cauchy's method† of integration as developed by Darboux the two 'characteristic curves' are one and the same.

**289.** We now consider, summarily, one important example, the simplest form of minimal surfaces.

*Ex. 1. Obtain the differential equation of minimal surfaces, that is, of surfaces, whose total area between two boundary non-intersecting curves is a minimum.*

(The differential equation is characteristic of the surface at every free point, and has no functional dependence upon boundary conditions which are stated solely to make the definition of minimal surfaces clearer.)

With the notation adopted, the area is equal to the integral

$$\iint (1 + p^2 + q^2)^{\frac{1}{2}} \, dx \, dy,$$

taken between the indicated limits. Here $f = (1 + p^2 + q^2)^{\frac{1}{2}}$, so that neither $x$, $y$, nor $z$, occurs explicitly in $f$; and thus the characteristic equation is

$$rf_{66} + 2sf_{69} + tf_{99} = 0,$$

that is, on substitution,

$$(1 + q^2) r - 2pqs + (1 + p^2) t = 0,$$

the required equation‡.

The equation expresses the property, that the principal radii of curvature are equal and opposite, or that (what is called) the mean curvature of the surface is zero. For a plane catenary, the radius of curvature is equal and opposite to the intercept of the normal between the curve and the directrix; hence the surface, obtained by rotating a catenary about its directrix, is minimal (§ 30, Ex. 1).

The integration of the general equation will be effected later (§ 301). Meanwhile, we may note that the equation of this catenoid of revolution should satisfy this characteristic equation. When we take the directrix to be the axis of $z$, the parallel through the vertex to be the plane $z = 0$, and adjust the scale, the equation is

$$(x^2 + y^2)^{\frac{1}{2}} = \rho = \cosh z.$$

*. See my *Theory of Differential Equations*, vol. vi, ch. xvii.

† *Ib.* vol. vi, §§ 256 *sqq.*

‡ It was first obtained by Lagrange, *Misc. Taur.* vol. ii (1760–1), p. 190; *Œuvres de Lagrange*, t. i, p. 356. It constitutes the earliest analytical result in the theory of minimal surfaces, of the history of which Darboux (*Théorie générale des surfaces*, t. i, pp. 267—280) gives an outline.

For this equation, the values of the derivatives given by

$$x = p \cosh z \sinh z, \quad y = q \cosh z \sinh z,$$
$$1 = r \cosh z \sinh z + p^2 \cosh 2z,$$
$$0 = s \cosh z \sinh z + pq \cosh 2z,$$
$$1 = t \cosh z \sinh z + q^2 \cosh 2z;$$

satisfy the relation

$$(1 + q^2) r - 2pqs + (1 + p^2) t = 0.$$

We may also note that the Legendre test is satisfied. For we have

$$f_{66} = \frac{\partial^2 f}{\partial p^2} = \frac{1 + q^2}{(1 + p^2 + q^2)^{\frac{3}{2}}}, \quad f_{99} = \frac{\partial^2 f}{\partial q^2} = \frac{1 + p^2}{(1 + p^2 + q^2)^{\frac{3}{2}}},$$

$$f_{69} = \frac{\partial^2 f}{\partial p \, \partial q} = \frac{-pq}{(1 + p^2 + q^2)^{\frac{3}{2}}};$$

hence $f_{66}$, $f_{99}$, $f_{66}f_{99} - f_{69}^2$ are always positive. Thus, so far as concerns this test, the surface is minimal.

*Ex. 2. Find the conjugate of a parallel of latitude upon a catenoid of revolution (that is, the range within which the catenoid provides definitely the minimal surface joining the parallel and that conjugate curve).*

As in the last example, we represent the catenoid by the equation

$$\cosh z = (x^2 + y^2)^{\frac{1}{2}} = \rho.$$

This equation is a particular solution of the characteristic equation. So far from being a primitive, it contains no arbitrary element; consequently, no solution (least of all, the primitive) of the subsidiary characteristic equation can be deduced from it. This subsidiary equation (§ 284), in the present instance where the function $f$ in the original integral does not involve $z$ explicitly, becomes

$$\frac{\partial}{\partial x} (f_{66} \zeta_1 + f_{69} \zeta_2) + \frac{\partial}{\partial y} (f_{69} \zeta_1 + f_{99} \zeta_2) = 0.$$

Manifestly, a consecutive catenoid can be drawn through the assigned parallel, though it is not the only minimal surface nor (in any respect) a general minimal surface through that parallel. As the immediate aim is an illustration of the general analysis, we shall consider the intersection of the two catenoids. Symmetry suggests that this intersection, being the conjugate, will be another parallel, so that we shall assume $\zeta$ to be a function of $\rho$ only.

With this assumption, we have

$$\zeta_1 = \frac{x}{\rho} \frac{\partial \zeta}{\partial \rho}, \quad \zeta_2 = \frac{y}{\rho} \frac{\partial \zeta}{\partial \rho},$$

$$f_{66} \zeta_1 + f_{69} \zeta_2 = \frac{1}{(1 + p^2 + q^2)^{\frac{3}{2}}} \{\zeta_1 + q (q\zeta_1 - p\zeta_2)\}$$

$$= \frac{\zeta_1}{(1 + p^2 + q^2)^{\frac{3}{2}}}$$

$$= \frac{x}{\rho^4} (\rho^2 - 1)^{\frac{3}{2}} \frac{\partial \zeta}{\partial \rho};$$

and

$$f_{69} \zeta_1 + f_{99} \zeta_2 = \frac{y}{\rho^4} (\rho^2 - 1)^{\frac{3}{2}} \frac{\partial \zeta}{\partial \rho}.$$

Thus the subsidiary equation becomes

$$\frac{\partial}{\partial x} \left\{ \frac{x}{\rho^4} (\rho^2 - 1)^{\frac{3}{2}} \frac{\partial \zeta}{\partial \rho} \right\} + \frac{\partial}{\partial y} \left\{ \frac{y}{\rho^4} (\rho^2 - 1)^{\frac{3}{2}} \frac{\partial \zeta}{\partial \rho} \right\} = 0,$$

that is,

$$\frac{2}{\rho^4}(\rho^2-1)^{\frac{3}{2}}\frac{\partial\zeta}{\partial\rho}+x\,\frac{x}{\rho}\,\frac{\partial}{\partial\rho}\left\{\frac{1}{\rho^4}(\rho^2-1)^{\frac{3}{2}}\frac{\partial\zeta}{\partial\rho}\right\}+y\,\frac{y}{\rho}\,\frac{\partial}{\partial\rho}\left\{\frac{1}{\rho^4}(\rho^2-1)^{\frac{3}{2}}\frac{\partial\zeta}{\partial\rho}\right\}=0,$$

and therefore

$$\frac{2}{\rho^4}(\rho^2-1)^{\frac{3}{2}}\frac{\partial\zeta}{\partial\rho}+\rho\,\frac{\partial}{\partial\rho}\left\{\frac{1}{\rho^4}(\rho^2-1)^{\frac{3}{2}}\frac{\partial\zeta}{\partial\rho}\right\}=0,$$

an ordinary equation. A first integral is

$$\frac{1}{\rho^2}(\rho^2-1)^{\frac{3}{2}}\frac{\partial\zeta}{\partial\rho}=a.$$

After the change of the independent variable to $z$, the primitive is found to be

$$\zeta=a\,(z-\coth z)+b,$$

where $a$ and $b$ are arbitrary constants. When $z_1$ is the value of $z$ for the initial parallel, the quantity

$$a\,\{z-\coth z-(z_1-\coth z_1)\}$$

is a value of $\zeta$ which vanishes on that parallel. We take $z_1$ as positive: the only other value of $z$, for which $\zeta$ vanishes, is (as in § 30, Ex. 1) negative, say $-z_2$. Then

$$\coth z_1+\coth z_2=z_1+z_2,$$

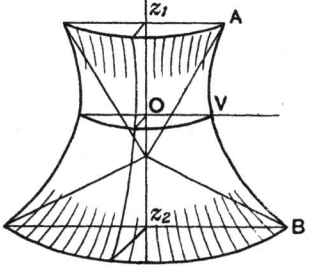

where $z_2$ is the depth, below the vertex, of the parallel which is conjugate to the parallel at a height $z_1$ above the vertex.

The geometrical construction is as before. In a meridian section, the tangents at $A$ and $B$, points on that meridian in the conjugate parallels, meet at a point in the directrix.

*Ex.* 3. Shew that the only minimal surface, the equation of which is expressible in the form

$$z=\phi(x)+\psi(y),$$

is

$$e^{az}=\cos ax\sec ay,$$

where $a$ is any constant.

---

### *General weak variations: transformation of the integral for completely independent variables.*

**290.** Without further discussion of special variations at this stage, we pass to the consideration of general weak variations which can be imposed simultaneously on all the variables $x, y, z$, throughout the range of integration. For this purpose, we make each of these three variables functions of two new independent variables $u$ and $v$, which (subject to the sole condition of independence) are at our choice. We therefore now treat $x, y, z$, as three dependent variables, functions of $u$ and $v$; and the current variables $u$ and $v$ do not occur explicitly in the integral which is to be maximised or minimised.

For brevity, we shall use a suffix numerical notation in two modes. According to one of these modes, derivatives of $x, y, z$, with regard to $u$ and $v$ will be indicated; according to the other, derivatives of a quantity $\Omega$ with regard to its arguments (which do not explicitly involve $u$ and $v$) will be indicated.

Thus, for the former, we write

$$x, \quad \frac{\partial x}{\partial u}, \quad \frac{\partial x}{\partial v}, \quad \frac{\partial^2 x}{\partial u^2}, \quad \frac{\partial^2 x}{\partial u \partial v}, \quad \frac{\partial^2 x}{\partial v^2} = x, \quad x_1, \quad x_2, \quad x_{11}, \quad x_{12}, \quad x_{22};$$

$$y, \quad \frac{\partial y}{\partial u}, \quad \frac{\partial y}{\partial v}, \quad \frac{\partial^2 y}{\partial u^2}, \quad \frac{\partial^2 y}{\partial u \partial v}, \quad \frac{\partial^2 y}{\partial v^2} = y, \quad y_1, \quad y_2, \quad y_{11}, \quad y_{12}, \quad y_{22};$$

$$z, \quad \frac{\partial z}{\partial u}, \quad \frac{\partial z}{\partial v}, \quad \frac{\partial^2 z}{\partial u^2}, \quad \frac{\partial^2 z}{\partial u \partial v}, \quad \frac{\partial^2 z}{\partial v^2} = z, \quad z_1, \quad z_2, \quad z_{11}, \quad z_{12}, \quad z_{22}.$$

In connection with these derivatives, we shall have to deal with combinations $E$, $G$, $F$, such that

$$x_1^2 + y_1^2 + z_1^2 = E, \quad x_2^2 + y_2^2 + z_2^2 = G,$$

$$x_1 x_2 + y_1 y_2 + z_1 z_2 = F.$$

We also shall denote quantities $l$, $m$, $n$, $V$, by the definitions

$$l = y_1 z_2 - z_1 y_2, \qquad m = z_1 x_2 - x_1 z_2, \qquad n = x_1 y_2 - y_1 x_2,$$

$$V^2 = EG - F^2,$$

so that

$$l^2 + m^2 + n^2 = V^2;$$

and we write

$$l x_{11} + m y_{11} + n z_{11} = VL,$$

$$l x_{12} + m y_{12} + n z_{12} = VM,$$

$$l x_{22} + m y_{22} + n z_{22} = VN.$$

Thus $E$, $F$, $G$, $L$, $M$, $N$ are the 'fundamental magnitudes' of a surface[*] which, when known, determine the surface uniquely save as to its orientation and position; and $l/V$, $m/V$, $n/V$, are the direction-cosines of the positive direction of the normal, $V$ being given the positive radical sign when $u$ and $v$ are real[†].

By the other use of a numerical suffix notation, we associate numbers 1, 4, 7, with $x$, $x_1$, $x_2$; numbers 2, 5, 8, with $y$, $y_1$, $y_2$; and numbers 3, 6, 9, with $z$, $z_1$, $z_2$. When a quantity $\Omega$ denotes a function in which $x$, $y$, $z$, $x_1$, $y_1$, $z_1$, $x_2$, $y_2$, $z_2$ occur (but not $u$, not $v$, and no higher derivatives of $x$ or $y$ or $z$), we express first and second derivatives of $\Omega$ with regard to its arguments according to the scheme

$$\frac{\partial \Omega}{\partial z} = \Omega_3, \quad \frac{\partial \Omega}{\partial z_1} = \Omega_6, \quad \frac{\partial \Omega}{\partial z_2} = \Omega_9,$$

$$\frac{\partial^2 \Omega}{\partial z_1^2} = \Omega_{66}, \quad \frac{\partial^2 \Omega}{\partial z_1 \partial z_2} = \Omega_{69}, \quad \frac{\partial^2 \Omega}{\partial z_2^2} = \Omega_{99},$$

and so on.

[*] See my *Lectures on Differential Geometry*, chapter ii.

[†] It may be convenient, hereafter, to take $u$ and $v$ as complex (conjugate) variables, when occasion arises. Unless such an exceptional assumption is specifically indicated, $u$ and $v$ will be assumed real.

As the original independent variables are $x$ and $y$, we shall be able to return, from the general forms now to be considered to the special forms already considered, by taking $u = x$, $v = y$; and then

$$x_1 = 1, \quad x_2 = 0; \quad y_1 = 0, \quad y_2 = 1;$$

$$z_1 = p, \quad z_2 = q.$$

To express $p$ and $q$ in terms of the new derivatives, we have

$$z_1 du + z_2 dv = dz$$

$$= p\,dx + q\,dy = p\,(x_1 du + x_2 dv) + q\,(y_1 du + y_2 dv),$$

and therefore

$$\frac{p}{z_1 y_2 - y_1 z_2} = \frac{q}{x_1 z_2 - z_1 x_2} = \frac{1}{x_1 y_2 - y_1 x_2}.$$

Also, by the theorem relating to the change of current variables in a double integral, we have

$$\iint \Theta\, dx\, dy = \iint \Theta\, (x_1 y_2 - y_1 x_2)\, du\, dv.$$

Our original integral was

$$\iint f(x, y, z, p, q)\, dx\, dy.$$

When the changes of the independent variables are effected, it becomes

$$\iint \Omega\, (x, y, z, x_1, y_1, z_1, x_2, y_2, z_2)\, du\, dv,$$

where

$$\Omega\, (x, y, z, x_1, y_1, z_1, x_2, y_2, z_2) = (x_1 y_2 - y_1 x_2) f(x, y, z, p, q).$$

Now $f(x, y, z, p, q)$ is a function of five arguments, as general as can occur in regard to those arguments; while $\Omega$, though a function of nine arguments, is bound to $f$ by this relation. Thus $\Omega$ cannot be a general function of its nine arguments; we may expect that $\Omega$ will satisfy certain equations of condition, which will be satisfied identically owing to its lack of functional generality in its nine arguments.

### Properties of the integrand in the transformed integral.

**291.** To establish the identities satisfied by $\Omega$, we may proceed from the foregoing relation. We have, on taking partial derivatives with respect to $x_1, y_1, z_1, x_2, y_2, z_2$, the six equations

$$\Omega_4 = \phantom{-} y_2 f - y_2 p \frac{\partial f}{\partial p} + x_2 p \frac{\partial f}{\partial q}$$

$$\Omega_5 = -x_2 f - y_2 q \frac{\partial f}{\partial p} + x_2 q \frac{\partial f}{\partial q}$$

$$\Omega_6 = \phantom{-x_2 f - {}} y_2 \frac{\partial f}{\partial p} - x_2 \frac{\partial f}{\partial q}$$

$$\Omega_7 = -y_1 f + y_1 p \frac{\partial f}{\partial p} - x_1 p \frac{\partial f}{\partial q}$$

$$\Omega_8 = \phantom{-} x_1 f + y_1 q \frac{\partial f}{\partial p} - x_1 q \frac{\partial f}{\partial q}$$

$$\Omega_9 = \phantom{-x_1 f + {}} -y_1 \frac{\partial f}{\partial p} + x_1 \frac{\partial f}{\partial q}$$

together with
$$\Omega = (x_1 y_2 - y_1 x_2) f,$$
seven equations in $\Omega$ and its first derivatives, while they involve the three quantities $f, \dfrac{\partial f}{\partial p}, \dfrac{\partial f}{\partial q}$. We therefore can expect four relations free from these three quantities; and we find

$$\left.\begin{array}{l} x_1 \Omega_4 + y_1 \Omega_5 + z_1 \Omega_6 = \Omega \\ x_2 \Omega_4 + y_2 \Omega_5 + z_2 \Omega_6 = 0 \\ x_1 \Omega_7 + y_1 \Omega_8 + z_1 \Omega_9 = \dot{0} \\ x_2 \Omega_7 + y_2 \Omega_8 + z_2 \Omega_9 = \Omega \end{array}\right\},$$

which are four fundamental identities satisfied by every function $\Omega$ of this type.

These identities can be established by a different process, which leads also to relations used in the discussion of strong variations. In changing to $u$ and $v$ as new independent variables, no special choice is dictated; if therefore other such variables $u'$ and $v'$ are chosen, and if the new derivatives are denoted by $\theta_r'$ instead of $\theta_r$ (for $\theta = x, y, z$, and for $r = 1, 2$), we have

$$\iint \Omega (x, y, z, x_1, y_1, z_1, x_2, y_2, z_2)\, du\, dv = \iint \Omega (x, y, z, x_1', y_1', z_1', x_2', y_2', z_2')\, du'\, dv'.$$

Let
$$du' = \alpha\, du + \beta\, dv, \quad dv' = \gamma\, du + \delta\, dv,$$
where $\alpha\delta - \beta\gamma$ is not zero; then
$$x_1 = \alpha x_1' + \gamma x_2', \quad x_2 = \beta x_1' + \delta x_2',$$
with similar relations concerning $y_1$ and $y_2$, $z_1$ and $z_2$; and so

$$\Omega (x, y, z, \alpha x_1' + \gamma x_2', \alpha y_1' + \gamma y_2', \alpha z_1' + \gamma z_2', \beta x_1' + \delta x_2', \beta y_1' + \delta y_2', \beta z_1' + \delta z_2')$$
$$= (\alpha\delta - \beta\gamma)\, \Omega (x, y, z, x_1', y_1', z_1', x_2', y_2', z_2').$$

Several inferences can be drawn from this relation as follows:

(i) Let $\alpha = 1, \beta = 0, \gamma = 0, \delta = -1$; then
$$\Omega (x, y, z, x_1, y_1, z_1, -x_2, -y_2, -z_2) = -\Omega (x, y, z, x_1, y_1, z_1, x_2, y_2, z_2):$$

(ii) Let $\alpha = -1, \beta = 0, \gamma = 0, \delta = 1$; then
$$\Omega (x, y, z, -x_1, -y_1, -z_1, x_2, y_2, z_2) = -\Omega (x, y, z, x_1, y_1, z_1, x_2, y_2, z_2):$$

(iii) Let $\alpha = -1, \beta = 0, \gamma = 0, \delta = -1$; then
$$\Omega (x, y, z, -x_1, -y_1, -z_1, -x_2, -y_2, -z_2) = \Omega (x, y, z, x_1, y_1, z_1, x_2, y_2, z_2).$$

(iv) We have
$$\alpha \frac{\partial \Omega}{\partial x_1} + \beta \frac{\partial \Omega}{\partial x_2} = (\alpha\delta - \beta\gamma) \frac{\partial \Omega}{\partial x_1'},$$

and so, taking the three preceding substitutions in turn, we have

$$\Omega_4(x, y, z, \quad x_1, \quad y_1, \quad z_1, -x_2, -y_2, -z_2) = -\Omega_4(x, y, z, x_1, y_1, z_1, x_2, y_2, z_2),$$

$$\Omega_4(x, y, z, -x_1, -y_1, -z_1, \quad x_2, \quad y_2, \quad z_2) = \Omega_4(x, y, z, x_1, y_1, z_1, x_2, y_2, z_2),$$

$$\Omega_4(x, y, z, -x_1, -y_1, -z_1, -x_2, -y_2, -z_2) = -\Omega_4(x, y, z, x_1, y_1, z_1, x_2, y_2, z_2);$$

and exactly the same relations hold for $\Omega_5$, $\Omega_6$, as for $\Omega_4$.

Also

$$\gamma \frac{\partial \Omega}{\partial x_1} + \delta \frac{\partial \Omega}{\partial x_2} = (\alpha\delta - \beta\gamma) \frac{\partial \Omega}{\partial x_2'},$$

and so

$$\Omega_7(x, y, z, \quad x_1, \quad y_1, \quad z_1, -x_2, -y_2, -z_2) = \Omega_7(x, y, z, x_1, y_1, z_1, x_2, y_2, z_2),$$

$$\Omega_7(x, y, z, -x_1, -y_1, -z_1, \quad x_2, \quad y_2, \quad z_2) = -\Omega_7(x, y, z, x_1, y_1, z_1, x_2, y_2, z_2),$$

$$\Omega_7(x, y, z, -x_1, -y_1, -z_1, -x_2, -y_2, -z_2) = -\Omega_7(x, y, z, x_1, y_1, z_1, x_2, y_2, z_2);$$

and exactly the same relations hold for $\Omega_8$, $\Omega_9$, as for $\Omega_7$.

(v) Let $\alpha = 1 + \epsilon$, $\beta = \zeta$, $\gamma = \theta$, $\delta = 1 + \eta$, where $\epsilon$, $\zeta$, $\theta$, $\eta$, are arbitrary independent small quantities. Expanding and comparing the coefficients of $\epsilon$, $\zeta$, $\theta$, $\eta$, respectively on the two sides, we find

$$\left.\begin{aligned}
x_1 \Omega_4 + y_1 \Omega_5 + z_1 \Omega_6 &= \Omega \\
x_2 \Omega_4 + y_2 \Omega_5 + z_2 \Omega_6 &= 0 \\
x_1 \Omega_7 + y_1 \Omega_8 + z_1 \Omega_9 &= 0 \\
x_2 \Omega_7 + y_2 \Omega_8 + z_2 \Omega_9 &= \Omega
\end{aligned}\right\},$$

which are the four fundamental identities satisfied by $\Omega$.

*Ex.* Denoting second derivatives of $z$ with respect to $x$ and $y$ by $r$, $s$, $t$, as usual, prove that

$$n^3 r = l\,(x_{11}y_2^2 - 2x_{12}y_1 y_2 + x_{22}y_1^2)$$
$$+ m\,(y_{11}y_2^2 - 2y_{12}y_1 y_2 + y_{22}y_1^2)$$
$$+ n\,(z_{11}y_2^2 - 2z_{12}y_1 y_2 + z_{22}y_1^2),$$

$$n^3 s = l\,\{-x_{11}x_2 y_2 + x_{12}(x_1 y_2 + y_1 x_2) - x_{22}x_1 y_1\}$$
$$+ m\,\{-y_{11}x_2 y_2 + y_{12}(x_1 y_2 + y_1 x_2) - y_{22}x_1 y_1\}$$
$$+ n\,\{-z_{11}x_2 y_2 + z_{12}(x_1 y_2 + y_1 x_2) - z_{22}x_1 y_1\},$$

$$n^3 t = l\,(x_{11}x_2^2 - 2x_{12}x_1 x_2 + x_{22}x_1^2)$$
$$+ m\,(y_{11}x_2^2 - 2y_{12}x_1 x_2 + y_{22}x_1^2)$$
$$+ n\,(z_{11}x_2^2 - 2z_{12}x_1 x_2 + z_{22}x_1^2),$$

$$n^4\,(rt - s^2) = (LN - M^2)\,V^2, \quad n^3\{(1+q^2)\,r - 2pqs + (1+p^2)\,t\} = (GL - 2FM + EN)\,V,$$

with the notation of § 290. Shew that the measures of curvature of a surface are given by

$$\frac{rt - s^2}{(1 + p^2 + q^2)^{\frac{1}{2}}} = \frac{LN - M^2}{EG - F^2},$$

being the Gauss measure, and by

$$\frac{(1 + q^2)\,r - 2pqs + (1 + p^2)\,t}{(1 + p^2 + q^2)^{\frac{3}{2}}} = \frac{GL - 2FM + EN}{EG - F^2},$$

being the measure of mean curvature.

*Relations between second derivatives of $\Omega$.*

**292.** Some deductions, derivable from these identities and relating to the second differential coefficients of $\Omega$, will be required immediately in connection with the characteristic equations.

(*a*) Differentiating the first two of the four identities with respect to $x_1, y_1, z_1$, in turn, we have

$$x_1\Omega_{44} + y_1\Omega_{45} + z_1\Omega_{46} = 0, \quad x_2\Omega_{44} + y_2\Omega_{45} + z_2\Omega_{46} = 0;$$
$$x_1\Omega_{45} + y_1\Omega_{55} + z_1\Omega_{56} = 0, \quad x_2\Omega_{45} + y_2\Omega_{55} + z_2\Omega_{56} = 0;$$
$$x_1\Omega_{46} + y_1\Omega_{56} + z_1\Omega_{66} = 0, \quad x_2\Omega_{46} + y_2\Omega_{56} + z_2\Omega_{66} = 0.$$

Hence

$$\frac{\Omega_{44}}{l^2} = \frac{\Omega_{45}}{lm} = \frac{\Omega_{46}}{ln} = \frac{\Omega_{55}}{m^2} = \frac{\Omega_{56}}{mn} = \frac{\Omega_{66}}{n^2} = P,$$

where $P$ is a new quantity.

(*b*) Differentiating the last two of the four identities with respect to $x_2, y_2, z_2$, in turn, we similarly have

$$\frac{\Omega_{77}}{l^2} = \frac{\Omega_{78}}{lm} = \frac{\Omega_{79}}{ln} = \frac{\Omega_{88}}{m^2} = \frac{\Omega_{89}}{mn} = \frac{\Omega_{99}}{n^2} = Q,$$

where $Q$ is a new quantity.

(*c*) Differentiating the first two of the identities with respect to $x_2$, and the last two with respect to $x_1$, we have

$$x_1\Omega_{47} + y_1\Omega_{57} + z_1\Omega_{67} = \quad \Omega_7, \quad x_2\Omega_{47} + y_2\Omega_{57} + z_2\Omega_{67} = -\Omega_4,$$
$$x_1\Omega_{47} + y_1\Omega_{48} + z_1\Omega_{49} = -\Omega_7, \quad x_2\Omega_{47} + y_2\Omega_{48} + z_2\Omega_{49} = \quad \Omega_4;$$

and therefore

$$2x_1\Omega_{47} + y_1(\Omega_{48} + \Omega_{57}) + z_1(\Omega_{49} + \Omega_{67}) = 0,$$
$$2x_2\Omega_{47} + y_2(\Omega_{48} + \Omega_{57}) + z_2(\Omega_{49} + \Omega_{67}) = 0.$$

Differentiating the first two with respect to $y_2$ and the last two with respect to $y_1$, we similarly have

$$x_1(\Omega_{48} + \Omega_{57}) + 2y_1\Omega_{58} + z_1(\Omega_{59} + \Omega_{68}) = 0,$$
$$x_2(\Omega_{48} + \Omega_{57}) + 2y_2\Omega_{58} + z_2(\Omega_{59} + \Omega_{68}) = 0.$$

Differentiating similarly with respect to $z_2$ and to $z_1$, we have

$$x_1(\Omega_{49} + \Omega_{67}) + y_1(\Omega_{59} + \Omega_{68}) + 2z_1\Omega_{69} = 0,$$
$$x_2(\Omega_{49} + \Omega_{67}) + y_2(\Omega_{59} + \Omega_{68}) + 2z_2\Omega_{69} = 0.$$

Hence

$$\frac{2\Omega_{47}}{l^2} = \frac{\Omega_{48} + \Omega_{57}}{lm} = \frac{\Omega_{49} + \Omega_{67}}{ln} = \frac{2\Omega_{58}}{m^2} = \frac{\Omega_{59} + \Omega_{68}}{mn} = \frac{2\Omega_{69}}{n^2}$$
$$= 2R,$$

where $R$ is a new quantity.

*Ex.* 1. Establish the connection of these quantities $P$, $R$, $Q$, with the second derivatives of $f(x, y, z, p, q)$ with respect to $p$ and $q$, in the forms

$$P = \frac{1}{n^3}\left(y_2{}^2\,\frac{\partial^2 f}{\partial p^2} - 2x_2 y_2\,\frac{\partial^2 f}{\partial p\,\partial q} + x_2{}^2\,\frac{\partial^2 f}{\partial q^2}\right),$$

$$R = \frac{1}{n^3}\left\{-y_1 y_2\,\frac{\partial^2 f}{\partial p^2} + (x_1 y_2 + y_1 x_2)\,\frac{\partial^2 f}{\partial p\,\partial q} - x_1 x_2\,\frac{\partial^2 f}{\partial q^2}\right\},$$

$$Q = \frac{1}{n^3}\left(y_1{}^2\,\frac{\partial^2 f}{\partial p^2} - 2x_1 y_1\,\frac{\partial^2 f}{\partial p\,\partial q} + x_1{}^2\,\frac{\partial^2 f}{\partial q^2}\right);$$

and shew that

$$PQ - R^2 = \frac{1}{n^4}\left\{\frac{\partial^2 f}{\partial p^2}\,\frac{\partial^2 f}{\partial q^2} - \left(\frac{\partial^2 f}{\partial p\,\partial q}\right)^2\right\}.$$

*Ex.* 2. Prove the relations:

$$\Omega_{48} = f - p\,\frac{\partial f}{\partial p} - q\,\frac{\partial f}{\partial q} + lmR, \qquad \Omega_{57} = -f + p\,\frac{\partial f}{\partial p} + q\,\frac{\partial f}{\partial q} + lmR;$$

$$\Omega_{68} = \frac{\partial f}{\partial p} + mnR \qquad , \qquad \Omega_{59} = -\frac{\partial f}{\partial p} + mnR;$$

$$\Omega_{49} = \frac{\partial f}{\partial q} + lnR \qquad , \qquad \Omega_{67} = -\frac{\partial f}{\partial q} + lnR.$$

*Ex.* 3. Prove the following relations:

$$\Omega_4 = z_2\,\frac{\partial f}{\partial q} - y_2\left(p\,\frac{\partial f}{\partial p} + q\,\frac{\partial f}{\partial q} - f\right)$$

$$\Omega_5 = -z_2\,\frac{\partial f}{\partial p} + x_2\left(p\,\frac{\partial f}{\partial p} + q\,\frac{\partial f}{\partial q} - f\right)$$

$$\Omega_6 = y_2\,\frac{\partial f}{\partial p} - x_2\,\frac{\partial f}{\partial q}$$

$$\Omega_7 = y_1\left(p\,\frac{\partial f}{\partial p} + q\,\frac{\partial f}{\partial q} - f\right) - z_1\,\frac{\partial f}{\partial q}$$

$$\Omega_8 = -x_1\left(p\,\frac{\partial f}{\partial p} + q\,\frac{\partial f}{\partial q} - f\right) + z_1\,\frac{\partial f}{\partial p}$$

$$\Omega_9 = x_1\,\frac{\partial f}{\partial q} - y_1\,\frac{\partial f}{\partial p}$$

$$\Omega = -l\,\frac{\partial f}{\partial p} - m\,\frac{\partial f}{\partial q} - n\left(p\,\frac{\partial f}{\partial p} + q\,\frac{\partial f}{\partial q} - f\right).$$

*Note.* For other relations, affecting the derivatives of $\Omega$, and useful in the reduction of the second variation of $\Omega$ to an intermediate normal form, reference may be made to the first of Kobb's memoirs already (p. 457) quoted.

## *First variation of the integral* $\iint \Omega\,du\,dv$.

**293.** Small variations are now to be imposed upon $x, y, z$, simultaneously. To express these variations, let the variables become $x + \kappa X$, $y + \kappa Y$, $z + \kappa Z$, respectively; as the variations still are to be weak, the quantities $X$, $Y$, $Z$, are taken to be regular functions of $u$ and $v$ which, otherwise, are arbitrary, are independent of one another, and (as usual) are independent of $\kappa$. With the same notation for derivatives of $X$, $Y$, $Z$, as for derivatives of $x, y, z$, the increment of the integral has the form

$$\iint \Omega\,(x + \kappa X,\ y + \kappa Y,\ z + \kappa Z,\ x_1 + \kappa X_1,\ y_1 + \kappa Y_1,\ z_1 + \kappa Z_1,$$
$$x_2 + \kappa X_2,\ y_2 + \kappa Y_2,\ z_2 + \kappa Z_2)\,du\,dv$$

$$-\iint \Omega\,(x, y, z, x_1, y_1, z_1, x_2, y_2, z_2)\,du\,dv.$$

This increment is equal to

$$\kappa \iint \Psi \, du \, dv + K_2,$$

where $K_2$ denotes the aggregate of terms involving the second and higher powers of $\kappa$ after expansion; where

$$\Psi = X \frac{\partial \Omega}{\partial x} + Y \frac{\partial \Omega}{\partial y} + Z \frac{\partial \Omega}{\partial z} + X_1 \frac{\partial \Omega}{\partial x_1} + Y_1 \frac{\partial \Omega}{\partial y_1} + Z_1 \frac{\partial \Omega}{\partial z_1} + X_2 \frac{\partial \Omega}{\partial x_2} + Y_2 \frac{\partial \Omega}{\partial y_2} + Z_2 \frac{\partial \Omega}{\partial z_2};$$

and where the double integral is taken over the whole field of variation of $u$ and $v$, taken for increase of $u$ and of $v$, the field being completely bounded in such manner as may be assigned.

We adopt the same convention as before (§ 275) for the positive direction of the boundary (or the distinct portions of the boundary) of the field of variation of $u$ and $v$. Hence, as

$$\frac{\partial}{\partial u} \left( X \frac{\partial \Omega}{\partial x_1} \right) = X_1 \frac{\partial \Omega}{\partial x_1} + X \frac{\partial}{\partial u} \left( \frac{\partial \Omega}{\partial x_1} \right), \quad \frac{\partial}{\partial v} \left( X \frac{\partial \Omega}{\partial x_2} \right) = X_2 \frac{\partial \Omega}{\partial x_2} + X \frac{\partial}{\partial v} \left( \frac{\partial \Omega}{\partial x_2} \right),$$

we have (§ 275)

$$\iint X_1 \frac{\partial \Omega}{\partial x_1} \, du \, dv = \int X \frac{\partial \Omega}{\partial x_1} \, dv - \iint X \frac{\partial}{\partial u} \left( \frac{\partial \Omega}{\partial x_1} \right) du \, dv,$$

$$\iint X_2 \frac{\partial \Omega}{\partial x_2} \, du \, dv = -\int X \frac{\partial \Omega}{\partial x_2} \, du - \iint X \frac{\partial}{\partial v} \left( \frac{\partial \Omega}{\partial x_2} \right) du \, dv,$$

the line-integral being taken in the positive direction along each portion of the boundary. Consequently,

$$\iint \left( X \frac{\partial \Omega}{\partial x} + X_1 \frac{\partial \Omega}{\partial x_1} + X_2 \frac{\partial \Omega}{\partial x_2} \right) du \, dv$$
$$= \iint X \left\{ \frac{\partial \Omega}{\partial x} - \frac{\partial}{\partial u} \left( \frac{\partial \Omega}{\partial x_1} \right) - \frac{\partial}{\partial v} \left( \frac{\partial \Omega}{\partial x_2} \right) \right\} du \, dv + \int X \left( \frac{\partial \Omega}{\partial x_1} \, dv - \frac{\partial \Omega}{\partial x_2} \, du \right).$$

Similarly,

$$\iint \left( Y \frac{\partial \Omega}{\partial y} + Y_1 \frac{\partial \Omega}{\partial y_1} + Y_2 \frac{\partial \Omega}{\partial y_2} \right) du \, dv$$
$$= \iint Y \left\{ \frac{\partial \Omega}{\partial y} - \frac{\partial}{\partial u} \left( \frac{\partial \Omega}{\partial y_1} \right) - \frac{\partial}{\partial v} \left( \frac{\partial \Omega}{\partial y_2} \right) \right\} du \, dv + \int Y \left( \frac{\partial \Omega}{\partial y_1} \, dv - \frac{\partial \Omega}{\partial y_2} \, du \right),$$

$$\iint \left( Z \frac{\partial \Omega}{\partial z} + Z_1 \frac{\partial \Omega}{\partial z_1} + Z_2 \frac{\partial \Omega}{\partial z_2} \right) du \, dv$$
$$= \iint Z \left\{ \frac{\partial \Omega}{\partial z} - \frac{\partial}{\partial u} \left( \frac{\partial \Omega}{\partial z_1} \right) - \frac{\partial}{\partial v} \left( \frac{\partial \Omega}{\partial z_2} \right) \right\} du \, dv + \int Z \left( \frac{\partial \Omega}{\partial z_1} \, dv - \frac{\partial \Omega}{\partial z_2} \, du \right).$$

Let

$$\mathbf{E}_x = \frac{\partial \Omega}{\partial x} - \frac{\partial}{\partial u} \left( \frac{\partial \Omega}{\partial x_1} \right) - \frac{\partial}{\partial v} \left( \frac{\partial \Omega}{\partial x_2} \right) \Bigg)$$

$$\mathbf{E}_y = \frac{\partial \Omega}{\partial y} - \frac{\partial}{\partial u} \left( \frac{\partial \Omega}{\partial y_1} \right) - \frac{\partial}{\partial v} \left( \frac{\partial \Omega}{\partial y_2} \right) \Bigg\};$$

$$\mathbf{E}_z = \frac{\partial \Omega}{\partial z} - \frac{\partial}{\partial u} \left( \frac{\partial \Omega}{\partial z_1} \right) - \frac{\partial}{\partial v} \left( \frac{\partial \Omega}{\partial z_2} \right) \Bigg)$$

then

$$\iint \Psi\, du\, dv = \iint (X\mathbf{E}_x + Y\mathbf{E}_y + Z\mathbf{E}_z)\, du\, dv$$

$$+ \int \left\{ \left( X\frac{\partial \Omega}{\partial x_1} + Y\frac{\partial \Omega}{\partial y_1} + Z\frac{\partial \Omega}{\partial z_1} \right) dv - \left( X\frac{\partial \Omega}{\partial x_2} + Y\frac{\partial \Omega}{\partial y_2} + Z\frac{\partial \Omega}{\partial z_2} \right) du \right\},$$

where the double integral is taken positively over the range of the $u$-$v$ field of integration, and the single integral is taken positively along the whole boundary of that field.

*Vanishing of the first variation : characteristic equations :*
*boundary condition.*

**294.** In order that the original integral may possess a maximum or a minimum, the first variation (that is, the coefficient of $\kappa$ in the expansion) of the integral must vanish for all admissible variations. Thus we must have

$$\iint \Psi\, du\, dv = 0,$$

for all variations $X$, $Y$, $Z$, regular but otherwise arbitrary through the field, and for all variations admissible along the boundary.

Consider, first of all, those variations which vanish along the boundary. For all of them, we must have

$$\iint (X\mathbf{E}_x + Y\mathbf{E}_y + Z\mathbf{E}_z)\, du\, dv = 0,$$

where $X$, $Y$, $Z$, are regular, arbitrary except at the boundary where they vanish, and independent of one another. For variations such that $Y$ and $Z$ (but not $X$) vanish, there remains only the single term $X\mathbf{E}_x$ in the subject of integration; if $\mathbf{E}_x$ is not zero everywhere, we could choose $X$ of the same sign as $\mathbf{E}_x$ and make the integral positive, and we could choose $X$ of the sign opposite to that of $\mathbf{E}_x$ and make the integral negative. Both possibilities contravene the requirement to vanish; and they arise solely on the hypothesis that $\mathbf{E}_x$ is not zero. Hence we must have

$$\mathbf{E}_x = 0.$$

Similarly, we must have

$$\mathbf{E}_y = 0, \quad \mathbf{E}_z = 0,$$

arising from the selection of other variations. Hence, in all, we have

$$\mathbf{E}_x = 0, \quad \mathbf{E}_y = 0, \quad \mathbf{E}_z = 0.$$

In virtue of them, the double integral always vanishes for all variations $X$, $Y$, $Z$. They are the *characteristic equations*.

Consider, next, variations which need not vanish everywhere along the boundary. In order that the first variation of the original integral may vanish, there still remains the requirement that the line-integral shall vanish; hence we must have

$$\iint \left\{ \left( X \frac{\partial \Omega}{\partial x_1} + Y \frac{\partial \Omega}{\partial y_1} + Z \frac{\partial \Omega}{\partial z_1} \right) dv - \left( X \frac{\partial \Omega}{\partial x_2} + Y \frac{\partial \Omega}{\partial y_2} + Z \frac{\partial \Omega}{\partial z_2} \right) du \right\} = 0,$$

over the whole boundary.

When the boundary consists of several portions, each complete in itself and detached from all the others, the foregoing relation must be satisfied along each such portion by itself. For we then can choose a variation vanishing along the whole of every portion of the boundary, except along any one selected portion: the foregoing integral receives only zero contributions from the portions other than the selected portion, and it vanishes as a whole; hence the contribution from that portion vanishes. In other words, the line-integral vanishes along that portion by itself.

A similar choice of other variations leads to the corresponding result for other portions of the boundary, in each instance separately. Hence, as stated, the line-integral must vanish for each portion of the boundary, which is detached and is complete in itself.

Again, the requirement must be satisfied for all classes of small variations consistent with the terminal conditions. These conditions, when they do not exclude all variations—that is, when they do not settle the boundary rigidly in every portion—allow us to choose continuous values of $X$, $Y$, $Z$, which may be zero over any part of a boundary rim, however small the parts or the intervals between different parts, and are not zero in the intervals; and the requirement still must be satisfied. Hence, at every place along each part of the boundary, the condition

$$\left( X \frac{\partial \Omega}{\partial x_1} + Y \frac{\partial \Omega}{\partial y_1} + Z \frac{\partial \Omega}{\partial z_1} \right) dv - \left( X \frac{\partial \Omega}{\partial x_2} + Y \frac{\partial \Omega}{\partial y_2} + Z \frac{\partial \Omega}{\partial z_2} \right) du = 0$$

must be satisfied; it is taken to be the first general form of the *boundary condition*.

**295.** The inferences, in further detail, depend upon assigned data or conditions attaching to each particular portion or to any portions of the boundary. Thus if any portion be rigidly fixed at every point, we should have $X = 0$, $Y = 0$, $Z = 0$, at every point on such portion; and the requirement would be satisfied without any residuary condition. Again, a portion of a boundary might be given as a specific curve in space; along that curve, we should have definite equations such as

$$\phi (x, y, z) = 0, \quad \psi (x, y, z) = 0,$$

and $X$, $Y$, $Z$, would be subject to the relations

$$\frac{\partial \phi}{\partial x} X + \frac{\partial \phi}{\partial y} Y + \frac{\partial \phi}{\partial z} Z = 0, \quad \frac{\partial \psi}{\partial x} X + \frac{\partial \psi}{\partial y} Y + \frac{\partial \psi}{\partial z} Z = 0;$$

and then we should have

$$\begin{vmatrix} \dfrac{\partial\Omega}{\partial x_1}, & \dfrac{\partial\phi}{\partial x}, & \dfrac{\partial\psi}{\partial x} \\[2ex] \dfrac{\partial\Omega}{\partial y_1}, & \dfrac{\partial\phi}{\partial y}, & \dfrac{\partial\psi}{\partial y} \\[2ex] \dfrac{\partial\Omega}{\partial z_1}, & \dfrac{\partial\phi}{\partial z}, & \dfrac{\partial\psi}{\partial z} \end{vmatrix} dv - \begin{vmatrix} \dfrac{\partial\Omega}{\partial x_2}, & \dfrac{\partial\phi}{\partial x}, & \dfrac{\partial\psi}{\partial x} \\[2ex] \dfrac{\partial\Omega}{\partial y_2}, & \dfrac{\partial\phi}{\partial y}, & \dfrac{\partial\psi}{\partial y} \\[2ex] \dfrac{\partial\Omega}{\partial z_2}, & \dfrac{\partial\phi}{\partial z}, & \dfrac{\partial\psi}{\partial z} \end{vmatrix} du = 0,$$

everywhere along that portion of the boundary. Or a terminal condition may be assigned which requires a portion of the boundary to lie on a surface

$$F(x, y, z) = 0,$$

in which case the only condition imposed on $X$, $Y$, $Z$, along that portion is

$$\frac{\partial F}{\partial x} X + \frac{\partial F}{\partial y} Y + \frac{\partial F}{\partial z} Z = 0;$$

and we should have

$$\begin{vmatrix} \dfrac{\partial\Omega}{\partial x_1}, & \dfrac{\partial\Omega}{\partial y_1}, & \dfrac{\partial\Omega}{\partial z_1} \\[2ex] \dfrac{\partial F}{\partial x}, & \dfrac{\partial F}{\partial y}, & \dfrac{\partial F}{\partial z} \end{vmatrix} dv - \begin{vmatrix} \dfrac{\partial\Omega}{\partial x_2}, & \dfrac{\partial\Omega}{\partial y_2}, & \dfrac{\partial\Omega}{\partial z_2} \\[2ex] \dfrac{\partial F}{\partial x}, & \dfrac{\partial F}{\partial y}, & \dfrac{\partial F}{\partial z} \end{vmatrix} du = 0,$$

(the same determinants being chosen from the two arrays), everywhere along that portion of the boundary.

In each instance, either we have a definite limit, assigned precisely from the beginning: or we deduce, from assigned conditions, relations that make the boundary precise. When, in the latter event, the relations are satisfied, variations at the boundary are no longer of effect: the boundary has been made free from variation, by satisfying the assigned conditions.

*Agreement of boundary condition with condition for special weak variation.*

**296.** We can bring the boundary condition for the general weak variation, as just given, into agreement with the terminal conditions for the special weak variation as given in § 280.

From the first two of the four identities satisfied by $\Omega$, we have

$$(x_1 y_2 - y_1 x_2)\,\Omega_4 = (\Omega - z_1\Omega_6)\,y_2 + z_2\Omega_6 y_1 = y_2\Omega - (x_1 y_2 - y_1 x_2)\,p\Omega_6.$$

Also

$$\Omega = (x_1 y_2 - y_1 x_2)\,f.$$

Hence

$$\Omega_4 = \quad y_2 f - p\Omega_6.$$

Similarly,

$$\Omega_5 = -\,x_2 f - q\Omega_6,$$
$$\Omega_7 = -\,y_1 f - p\Omega_9,$$
$$\Omega_8 = \quad x_1 f - q\Omega_9.$$

Consequently,

$$X\Omega_4 + Y\Omega_5 + Z\Omega_6 = (Z - pX - qY)\,\Omega_6 + y_2 Xf - x_2 Yf,$$

$$X\Omega_7 + Y\Omega_8 + Z\Omega_9 = (Z - pX - qY)\,\Omega_9 - y_1 Xf + x_1 Yf;$$

and therefore

$$(X\Omega_4 + Y\Omega_5 + Z\Omega_6)\,dv - (X\Omega_7 + Y\Omega_8 + Z\Omega_9)\,du$$

$$= (Z - pX - qY)(\Omega_6 dv - \Omega_9 du) + Xf(y_2 dv + y_1 du) - Yf(x_2 dv + x_1 du)$$

$$= (Z - pX - qY)\left\{\frac{\partial f}{\partial p}(y_2 dv + y_1 du) - \frac{\partial f}{\partial q}(x_2 dv + x_1 du)\right\} + Xf\,dy - Yf\,dx$$

$$= \left\{(Z - pX - qY)\frac{\partial f}{\partial p} + Xf\right\}dy - \left\{(Z - pX - qY)\frac{\partial f}{\partial q} + Yf\right\}dx;$$

so that the two forms are in exact agreement.

The further inferences have already (§ 281) been given for each form.

### *Continuity of certain derivatives through a free discontinuity on the characteristic surface.*

**297.** In the preceding investigation, an assumption has been made implicitly that, while the surface is continuous, there are neither lines nor positions where the derivatives of $x, y, z$, with regard to $u$ and to $v$, can suffer discontinuity. We can prove, in the same way as for places of discontinuous direction upon plane curves (§§ 47, 102) and upon skew curves (§§ 159, 194), that, *when there are isolated free lines of discontinuity of direction upon the characteristic surface, the three quantities*

$$\frac{\partial \Omega}{\partial x_1}\,dv - \frac{\partial \Omega}{\partial x_2}\,du, \quad \frac{\partial \Omega}{\partial y_1}\,dv - \frac{\partial \Omega}{\partial y_2}\,du, \quad \frac{\partial \Omega}{\partial z_1}\,dv - \frac{\partial \Omega}{\partial z_2}\,du,$$

*are continuous in value during passage across such lines.*

Let $CD$ be part of a free line of discontinuous direction; $C''D''$ be part of a contiguous line on the surface on one side of $CD$, and $C'D'$ be part of another contiguous line on the other side of $CD$. Consider variations of the surface, which are zero everywhere along $C''D''$, and over all parts of the surface between $C''D''$ and all the boundary beyond $C''D''$ on the side further from $CD$; and also are zero everywhere along $C'D'$, and over all parts of the surface between $C'D'$ and all the boundary within $C'D'$ on the side further from $CD$. Such variations can be represented by $\kappa X, \kappa Y, \kappa Z$, where

$$X = (u'' - U)(v'' - V)(U - u')(V - v')f(u, v),$$

$$Y = (u'' - U)(v'' - V)(U - u')(V - v')g(u, v),$$

$$Z = (u'' - U)(v'' - V)(U - u')(V - v')h(u, v),$$

where the variables $(U, V)$ denote any point on $CD$; the variables $(u'', v'')$ the contiguous point on $C''D''$: the variables $(u', v')$ the contiguous point on $C'D'$: and $f(u, v)$, $g(u, v)$, $h(u, v)$, are independent arbitrary regular functions of $u$ and $v$.

For every weak variation, the whole first variation of our integral, which of course must vanish for a maximum or a minimum of that integral, is

$$\kappa \iint (X\mathbf{E}_x + Y\mathbf{E}_y + Z\mathbf{E}_z)\, du\, dv$$

$$+ \kappa \int \left\{ \left( X\frac{\partial\Omega}{\partial x_1} + Y\frac{\partial\Omega}{\partial y_1} + Z\frac{\partial\Omega}{\partial z_1} \right) dv - \left( X\frac{\partial\Omega}{\partial x_2} + Y\frac{\partial\Omega}{\partial y_2} + Z\frac{\partial\Omega}{\partial z_2} \right) du \right\}.$$

The double integral vanishes for the portions of the surface (i) between the whole inner boundary and $C'D'$, and (ii) between $C''D''$ and the whole outer boundary, because $X = 0$, $Y = 0$, $Z = 0$, over those portions. It vanishes for the portions of the surface (iii) between $C'D'$ and $CD$, and (iv) between $CD$ and $C''D''$, because the equations

$$\mathbf{E}_x = 0, \quad \mathbf{E}_y = 0, \quad \mathbf{E}_z = 0,$$

are satisfied, everywhere over these portions. Hence the whole double integral vanishes.

Let $\left( \dfrac{\partial\Omega}{\partial t_m} \right)^{+}$, for $t = x, y,$ and $z$, and for $m = 1$ and $2$, denote the value of $\dfrac{\partial\Omega}{\partial t_m}$ at a place on $CD$ for the portion of the surface on the same side of $CD$ as $C''D''$; and let $\left( \dfrac{\partial\Omega}{\partial t_m} \right)^{-}$, also for $t = x, y,$ and $z$, and for $m = 1$ and $2$, denote the value of $\dfrac{\partial\Omega}{\partial t_m}$ at that place on $CD$ for the portion of the surface on the same side of $CD$ as $C'D'$. The boundary terms in the foregoing expression of the first variation become

$$\kappa \left\{ B_{(C''D'')} - B^{+}{}_{(CD)} + B^{-}{}_{(CD)} - B_{(C'D')} \right\},$$

where $B_{(C''D'')}$ denotes the integral

$$\int \left\{ \left( X\frac{\partial\Omega}{\partial x_1} + Y\frac{\partial\Omega}{\partial y_1} + Z\frac{\partial\Omega}{\partial z_1} \right) dv - \left( X\frac{\partial\Omega}{\partial x_2} + Y\frac{\partial\Omega}{\partial y_2} + Z\frac{\partial\Omega}{\partial z_2} \right) du \right\}$$

taken along $C''D''$: $B^{+}{}_{(CD)}$ denotes the same integral taken along $CD$, for $CD$ as the portion of the boundary of the surface between $CD$ and $C''D''$: $B^{-}{}_{CD)}$ denotes the same integral along $CD$, for $CD$ as the portion of the boundary of the surface between $CD$ and $C'D'$: and $B_{(C'D')}$ denotes the same integral taken along $C'D'$. Now, along $C'D'$ and along $C''D''$, for the particular set of variations chosen,

$$X = 0, \quad Y = 0, \quad Z = 0;$$

and therefore

$$B_{(C''D'')} = 0, \quad B_{(C'D')} = 0.$$

Also,

$$B^+{}_{(CD)} = \int \Big[ \Big\{ X \Big( \frac{\partial \Omega}{\partial x_1} \Big)^+ + Y \Big( \frac{\partial \Omega}{\partial y_1} \Big)^+ + Z \Big( \frac{\partial \Omega}{\partial z_1} \Big)^+ \Big\} \, dv$$

$$- \Big\{ X \Big( \frac{\partial \Omega}{\partial x_2} \Big)^+ + Y \Big( \frac{\partial \Omega}{\partial y_2} \Big)^+ + Z \Big( \frac{\partial \Omega}{\partial z_2} \Big)^+ \Big\} \, du \Big],$$

$$B^-{}_{(CD)} = \int \Big[ \Big\{ X \Big( \frac{\partial \Omega}{\partial x_1} \Big)^- + Y \Big( \frac{\partial \Omega}{\partial y_1} \Big)^- + Z \Big( \frac{\partial \Omega}{\partial z_1} \Big)^- \Big\} \, dv$$

$$- \Big\{ X \Big( \frac{\partial \Omega}{\partial x_2} \Big)^- + Y \Big( \frac{\partial \Omega}{\partial y_2} \Big)^- + Z \Big( \frac{\partial \Omega}{\partial z_2} \Big)^- \Big\} \, du \Big];$$

and, in order that the boundary condition may be satisfied,

$$B^+{}_{(CD)} = B^-{}_{(CD)}.$$

But in $X$, $Y$, $Z$, which are the same along $CD$ on the side $C''D''$, as along $CD$ on the side $C'D'$, the functions $f(u, v)$, $g(u, v)$, $h(u, v)$, are arbitrary and independent of one another. Taking as a first choice, $g(u, v) = 0$ and $h(u, v) = 0$, and keeping $f(u, v)$ still quite arbitrary, we have

$$\int X \Big\{ \Big( \frac{\partial \Omega}{\partial x_1} \Big)^+ dv - \Big( \frac{\partial \Omega}{\partial x_2} \Big)^+ du \Big\} = \int X \Big\{ \Big( \frac{\partial \Omega}{\partial x_1} \Big)^- dv - \Big( \frac{\partial \Omega}{\partial x_2} \Big)^- du \Big\},$$

where $X$ involves an arbitrary function $f(u, v)$ as a factor, entirely at our disposal. Hence, at every point of $CD$, we must have

$$\Big( \frac{\partial \Omega}{\partial x_1} \Big)^+ dv - \Big( \frac{\partial \Omega}{\partial x_2} \Big)^+ du = \Big( \frac{\partial \Omega}{\partial x_1} \Big)^- dv - \Big( \frac{\partial \Omega}{\partial x_2} \Big)^- du.$$

Similarly, owing to the completely independent arbitrary characters of $g(u, v)$ in $Y$ and $h(u, v)$ in $Z$, we have

$$\Big( \frac{\partial \Omega}{\partial y_1} \Big)^+ dv - \Big( \frac{\partial \Omega}{\partial y_2} \Big)^+ du = \Big( \frac{\partial \Omega}{\partial y_1} \Big)^- dv - \Big( \frac{\partial \Omega}{\partial y_2} \Big)^- du,$$

$$\Big( \frac{\partial \Omega}{\partial z_1} \Big)^+ dv - \Big( \frac{\partial \Omega}{\partial z_2} \Big)^+ du = \Big( \frac{\partial \Omega}{\partial z_1} \Big)^- dv - \Big( \frac{\partial \Omega}{\partial z_2} \Big)^- du.$$

In other words, the three quantities

$$\frac{\partial \Omega}{\partial x_1} dv - \frac{\partial \Omega}{\partial x_2} du, \quad \frac{\partial \Omega}{\partial y_1} dv - \frac{\partial \Omega}{\partial y_2} du, \quad \frac{\partial \Omega}{\partial z_1} dv - \frac{\partial \Omega}{\partial z_2} du$$

are continuous at a free line of discontinuity of direction on the characteristic surface.

Had the line of discontinuity been fixed and not free, $X$, $Y$, $Z$ would vanish; the inferences could not then be drawn.

*The three characteristic equations : equivalent to a single independent equation.*

**298.**  We must now consider the three characteristic equations

$$\mathbf{E}_x = 0, \quad \mathbf{E}_y = 0, \quad \mathbf{E}_z = 0.$$

These equations potentially serve to determine the three variables $x, y, z$, as functions of the two independent variables $u$ and $v$, in forms

$$x = x(u, v), \quad y = y(u, v), \quad z = z(u, v).$$

When $u$ and $v$ are eliminated between these three integrated equations, which constitute the primitive of the characteristic equations, a single equation

$$S(x, y, z) = 0$$

will result.  This equation represents a surface, which is called *the characteristic surface.*

Instead, however, of seeking this single equation $S = 0$, advantage often arises from the retention of the three equations which express $x, y, z$, as functions of $u$ and $v$; and then $u$ and $v$ are parametric variables on the surface. As indicated initially, there is wide choice of the independent variables, as far as the statement of the problem is concerned.  Nothing, by way of limitation on the choice, has been imposed by the analysis.  Consequently, when the characteristic equations have been formed, we are still free to choose, within a very wide range, these independent variables as current parameters of organic curves on the surface (should choice lead to suitable forms), such as lines of curvature (always real), or asymptotic lines (sometimes real, sometimes imaginary and conjugate), or nul-lines (always conjugate imaginary lines on real surfaces).

As regards these characteristic equations, there is the property of fundamental importance that, *in virtue of the four identities satisfied by the function* $\Omega(x, y, z, x_1, y_1, z_1, x_2, y_2, z_2)$, *the three characteristic equations* $\mathbf{E}_x = 0$, $\mathbf{E}_y = 0$, $\mathbf{E}_z = 0$, *are equivalent to a single equation*

$$\mathbf{E} = 0,$$

*which is independent of the identities* *.

---

* It may be noted, in elucidation of this result, that, when the four identities are regarded as four simultaneous partial differential equations of the first order, involving the one dependent variable $\Omega$ and the nine independent variables $x, \ldots, z_2$, the most general primitive of those four equations is

$$\Omega = (x_1 y_2 - y_1 x_2) f\left(x, y, z, \frac{z_1 y_2 - y_1 z_2}{x_1 y_2 - y_1 x_2}, \frac{x_1 z_2 - z_1 x_2}{x_1 y_2 - y_1 x_2}\right),$$

where $f$ denotes an absolutely arbitrary function of its five arguments.

The remaining equation $\mathbf{E} = 0$ is then effectively the characteristic equation (§§ 277, 278) which is satisfied by the function $f$.

To modify the equations $\mathbf{E}_x$, $\mathbf{E}_y$, $\mathbf{E}_z$, the four identities satisfied by $\Omega$ are used. We have

$$\frac{\partial \Omega}{\partial x} = x_1 \Omega_{14} + y_1 \Omega_{15} + z_1 \Omega_{16},$$

from the first identity. When derivatives of $\dfrac{\partial \Omega}{\partial x_1}$ and $\dfrac{\partial \Omega}{\partial x_2}$ are fully expressed, we have

$$\frac{\partial}{\partial u}\left(\frac{\partial \Omega}{\partial x_1}\right) = x_1 \Omega_{14} + y_1 \Omega_{24} + z_1 \Omega_{34} + x_{11} \Omega_{44} + y_{11} \Omega_{45} + z_{11} \Omega_{46} + x_{12} \Omega_{47} + y_{12} \Omega_{48} + z_{12} \Omega_{49},$$

$$\frac{\partial}{\partial v}\left(\frac{\partial \Omega}{\partial x_2}\right) = x_2 \Omega_{17} + y_2 \Omega_{27} + z_2 \Omega_{37} + x_{12} \Omega_{47} + y_{12} \Omega_{57} + z_{12} \Omega_{67} + x_{22} \Omega_{77} + y_{22} \Omega_{78} + z_{22} \Omega_{79}.$$

Hence

$$-\mathbf{E}_x = -\frac{\partial \Omega}{\partial x} + \frac{\partial}{\partial u}\left(\frac{\partial \Omega}{\partial x_1}\right) + \frac{\partial}{\partial v}\left(\frac{\partial \Omega}{\partial x_2}\right)$$

$$= y_1 (\Omega_{24} - \Omega_{15}) + z_1 (\Omega_{34} - \Omega_{16}) + x_2 \Omega_{17} + y_2 \Omega_{27} + z_2 \Omega_{37} + l\Theta,$$

on using the relations of § 292: where

$$\Theta = P(lx_{11} + my_{11} + nz_{11}) + 2R(lx_{12} + my_{12} + nz_{12}) + Q(lx_{22} + my_{22} + nz_{22}).$$

In the same way, we find

$$-\mathbf{E}_y = x_1 (\Omega_{15} - \Omega_{24}) + z_1 (\Omega_{35} - \Omega_{26}) + x_2 \Omega_{18} + y_2 \Omega_{28} + z_2 \Omega_{38} + m\Theta,$$

$$-\mathbf{E}_z = x_1 (\Omega_{16} - \Omega_{34}) + y_1 (\Omega_{26} - \Omega_{35}) + x_2 \Omega_{19} + y_2 \Omega_{29} + z_2 \Omega_{39} + n\Theta.$$

But from the third identity, when differentiated with respect to $x$, $y$, $z$, in turn, we have

$$x_1 \Omega_{17} + y_1 \Omega_{18} + z_1 \Omega_{19} = 0,$$

$$x_1 \Omega_{27} + y_1 \Omega_{28} + z_1 \Omega_{29} = 0,$$

$$x_1 \Omega_{37} + y_1 \Omega_{38} + z_1 \Omega_{39} = 0.$$

Also $x_1 l + y_1 m + z_1 n = 0$; hence

$$x_1 \mathbf{E}_x + y_1 \mathbf{E}_y + z_1 \mathbf{E}_z = 0.$$

This result has been obtained by using the first identity, to give the expressions for $\dfrac{\partial \Omega}{\partial x}$, $\dfrac{\partial \Omega}{\partial y}$, $\dfrac{\partial \Omega}{\partial z}$ in $\mathbf{E}_x$, $\mathbf{E}_y$, $\mathbf{E}_z$. When we use the fourth identity in the same manner as the first, and when the results of differentiating the second identity with respect to $x$, $y$, $z$, in turn are used, we have, similarly,

$$x_2 \mathbf{E}_x + y_2 \mathbf{E}_y + z_2 \mathbf{E}_z = 0.$$

Consequently, there exists a quantity $\mathbf{E}$, such that

$$\mathbf{E}_x = l\mathbf{E}, \quad \mathbf{E}_y = m\mathbf{E}, \quad \mathbf{E}_z = n\mathbf{E};$$

and the three characteristic equations $\mathbf{E}_x = 0$, $\mathbf{E}_y = 0$, $\mathbf{E}_z = 0$, in conjunction with the four identities satisfied by $\Omega$, are equivalent to a single characteristic equation

$$\mathbf{E} = 0.$$

**299.** This single characteristic equation $\mathbf{E} = 0$ can be at once identified with the single equation that arises (§ 277) from special variations.

We have
$$\Omega = nf,$$
and therefore
$$\frac{\partial \Omega}{\partial z} = n \frac{\partial f}{\partial z}.$$
Also
$$\frac{\partial \Omega}{\partial z_1} = \Omega_6 = y_2 \frac{\partial f}{\partial p} - x_2 \frac{\partial f}{\partial q}, \quad \frac{\partial \Omega}{\partial z_2} = \Omega_9 = -y_1 \frac{\partial f}{\partial p} + x_1 \frac{\partial f}{\partial q},$$
by the formulæ in Ex. 3, § 292; hence
$$\frac{\partial}{\partial u}\left(\frac{\partial \Omega}{\partial z_1}\right) + \frac{\partial}{\partial v}\left(\frac{\partial \Omega}{\partial z_2}\right) = \left\{y_2 \frac{\partial}{\partial u}\left(\frac{\partial f}{\partial p}\right) - y_1 \frac{\partial}{\partial v}\left(\frac{\partial f}{\partial p}\right)\right\} + \left\{x_1 \frac{\partial}{\partial v}\left(\frac{\partial f}{\partial q}\right) - x_2 \frac{\partial}{\partial u}\left(\frac{\partial f}{\partial q}\right)\right\}.$$
But, for any function $\theta$ of $u$ and $v$,
$$\frac{\partial \theta}{\partial u} = x_1 \frac{\partial \theta}{\partial x} + y_1 \frac{\partial \theta}{\partial y}, \quad \frac{\partial \theta}{\partial v} = x_2 \frac{\partial \theta}{\partial x} + y_2 \frac{\partial \theta}{\partial y};$$
and therefore
$$y_2 \frac{\partial \theta}{\partial u} - y_1 \frac{\partial \theta}{\partial v} = n \frac{\partial \theta}{\partial x}, \quad -x_2 \frac{\partial \theta}{\partial u} + x_1 \frac{\partial \theta}{\partial v} = n \frac{\partial \theta}{\partial y}.$$
Consequently,
$$\frac{\partial}{\partial u}\left(\frac{\partial \Omega}{\partial z_1}\right) + \frac{\partial}{\partial v}\left(\frac{\partial \Omega}{\partial z_2}\right) = n\left\{\frac{\partial}{\partial x}\left(\frac{\partial f}{\partial p}\right) + \frac{\partial}{\partial y}\left(\frac{\partial f}{\partial q}\right)\right\}.$$
Hence
$$\mathbf{E}_z = \frac{\partial \Omega}{\partial z} - \frac{\partial}{\partial u}\left(\frac{\partial \Omega}{\partial z_1}\right) - \frac{\partial}{\partial v}\left(\frac{\partial \Omega}{\partial z_2}\right)$$
$$= n\left\{\frac{\partial f}{\partial z} - \frac{\partial}{\partial x}\left(\frac{\partial f}{\partial p}\right) - \frac{\partial}{\partial y}\left(\frac{\partial f}{\partial q}\right)\right\};$$
and therefore the quantity $\mathbf{E}$ of § 298 is given by
$$\mathbf{E} = \frac{\partial f}{\partial z} - \frac{\partial}{\partial x}\left(\frac{\partial f}{\partial p}\right) - \frac{\partial}{\partial y}\left(\frac{\partial f}{\partial q}\right).$$

This relation establishes the identification of the characteristic equations, which arise from the special weak variation, and from the general weak variation, respectively, so far as concerns functional significance.

**300.** The three characteristic equations
$$\mathbf{E}_x = 0, \quad \mathbf{E}_y = 0, \quad \mathbf{E}_z = 0,$$
are equations of the second order, as containing derivatives of $x$, $y$, $z$, of the second order with respect to $u$ and $v$. They are linear in those second derivatives, which occur in the combination
$$P(lx_{11} + my_{11} + nz_{11}) + 2R(lx_{12} + my_{12} + nz_{12}) + Q(lx_{22} + my_{22} + nz_{22}),$$
common to all the equations.

The single characteristic equation

$$\mathbf{E} = 0$$

is, naturally, of the same order as $\mathbf{E}_x = 0$, $\mathbf{E}_y = 0$, $\mathbf{E}_z = 0$. It is linear in the stated combination of the derivatives of $x$, $y$, $z$, of the second order; and it becomes the single characteristic equation, which arises in the case when special variations are applied to the integral.

The primitive of the characteristic equations in the case of the most general weak variations expresses $x$, $y$, $z$, in terms of two independent arbitrary functions of parameters. This same primitive is the not infrequent form of primitive of the single characteristic equation in the case of the special weak variations.

Manifestly, the explicit completion of this stage of the problem requires the construction of the primitive of the characteristic equation, or of the three functionally related characteristic equations. Thus we are led into the details of the theory and the method, devised by Ampère for the integration of partial differential equations of the second order.

In order to consider later (§ 305) the second variation of the original integral, we shall assume that the primitive of the characteristic equation or equations has been obtained.

### Minimal Surfaces: primitive of the characteristic equations.

**301.** We return to the problem of *minimal surfaces* (§ 289, Ex. 1)—that is, the determination of the surface of least area, bounded by two given skew curves which do not cross one another.

The integral, which expresses the area, is

$$\iint (1 + p^2 + q^2)^{\frac{1}{2}} \, dx \, dy.$$

When a change to new independent variables $u$ and $v$ is made, we have

$$(1 + p^2 + q^2)^{\frac{1}{2}} = \frac{1}{x_1 y_2 - y_1 x_2} \{ (x_1^2 + y_1^2 + z_1^2)(x_2^2 + y_2^2 + z_2^2) - (x_1 x_2 + y_1 y_2 + z_1 z_2)^2 \}^{\frac{1}{2}}$$

$$= \frac{1}{x_1 y_2 - y_1 x_2} (EG - F^2)^{\frac{1}{2}},$$

so that the integral expression for the area becomes $\iint (EG - F^2)^{\frac{1}{2}} \, du \, dv$,

$= \iint V du \, dv$. The first of the characteristic equations is

$$\frac{\partial V}{\partial x} - \frac{\partial}{\partial u} \left( \frac{\partial V}{\partial x_1} \right) - \frac{\partial}{\partial v} \left( \frac{\partial V}{\partial x_2} \right) = 0.$$

The independent variables $u$ and $v$ are completely at our choice. After the characteristic equations are formed, we select these variables to be such that

$$x_1^2 + y_1^2 + z_1^2 = 0, \quad x_2^2 + y_2^2 + z_2^2 = 0,$$

so that, as $x$, $y$, $z$ are real, $u$ and $v$ will be conjugate complex quantities*. Thus $E = 0$ and $G = 0$ permanently, together with all fully derived relations

$$\frac{\partial E}{\partial u} = 0, \quad \frac{\partial E}{\partial v} = 0, \quad \frac{\partial G}{\partial u} = 0, \quad \frac{\partial G}{\partial v} = 0;$$

and $V$ becomes $iF$. Hence

$$\frac{\partial}{\partial u}\left(\frac{\partial V}{\partial x_1}\right) = \frac{\partial}{\partial u}\left\{\frac{1}{2V}\left(E\frac{\partial G}{\partial x_1} + G\frac{\partial E}{\partial x_1} - 2F\frac{\partial F}{\partial x_1}\right)\right\} = i\frac{\partial}{\partial u}\left(\frac{\partial F}{\partial x_1}\right);$$

and, similarly, $\dfrac{\partial}{\partial v}\left(\dfrac{\partial V}{\partial x_2}\right) = i\dfrac{\partial}{\partial v}\left(\dfrac{\partial F}{\partial x_2}\right)$. The first characteristic equation becomes

$$i\frac{\partial F}{\partial x} - i\frac{\partial}{\partial u}\left(\frac{\partial F}{\partial x_1}\right) - i\frac{\partial}{\partial v}\left(\frac{\partial F}{\partial x_2}\right) = 0;$$

and so for the other two. Consequently, the characteristic equations are

$$\frac{\partial F}{\partial x} - \frac{\partial}{\partial u}\left(\frac{\partial F}{\partial x_1}\right) - \frac{\partial}{\partial v}\left(\frac{\partial F}{\partial x_2}\right) = 0,$$

$$\frac{\partial F}{\partial y} - \frac{\partial}{\partial u}\left(\frac{\partial F}{\partial y_1}\right) - \frac{\partial}{\partial v}\left(\frac{\partial F}{\partial y_2}\right) = 0,$$

$$\frac{\partial F}{\partial z} - \frac{\partial}{\partial u}\left(\frac{\partial F}{\partial z_1}\right) - \frac{\partial}{\partial v}\left(\frac{\partial F}{\partial z_2}\right) = 0:$$

that is,

$$x_{12} = 0, \quad y_{12} = 0, \quad z_{12} = 0.$$

These three equations are equivalent (as, after § 298, they should be equivalent) to a single independent equation: for, from $E = 0$ and $G = 0$, we have

$$x_1 x_{12} + y_1 y_{12} + z_1 z_{12} = 0, \quad x_2 x_{12} + y_2 y_{12} + z_2 z_{12} = 0.$$

Hence $x$, $y$, and $z$, are, each of them, of the form

$x$, $y$, $z$, = arbitrary function of $u$ + arbitrary function of $v$.

Consequently $x_1, y_1, z_1$, are functions of $u$ alone; and $x_2, y_2, z_2$, are functions of $v$ alone; each of the two sets of functions being arbitrary.

But, by definition,

$$x_1{}^2 + y_1{}^2 + z_1{}^2 = 0;$$

and so we take

$$x_1 = \tfrac{1}{2}(P^2 - Q^2), \quad y_1 = \frac{i}{2}(P^2 + Q^2), \quad z_1 = 2PQ,$$

where $P$ and $Q$ are arbitrary functions of $u$ alone. Now any arbitrary function of $u$ can be substituted for $u$. When $Q/P$ is not constant, as is usually the case, let

$$Q = uP, \quad U = P^2,$$

* When any parameters $p'$ and $q'$ are taken for the expression of $x$ and $y$, so that

$$e = \left(\frac{\partial x}{\partial p'}\right)^2 + \left(\frac{\partial y}{\partial p'}\right)^2 + \left(\frac{\partial z}{\partial p'}\right)^2, \qquad g = \left(\frac{\partial x}{\partial q'}\right)^2 + \left(\frac{\partial y}{\partial q'}\right)^2 + \left(\frac{\partial z}{\partial q'}\right)^2,$$

$$f = \frac{\partial x}{\partial p'}\frac{\partial x}{\partial q'} + \frac{\partial y}{\partial p'}\frac{\partial y}{\partial q'} + \frac{\partial z}{\partial p'}\frac{\partial z}{\partial q'},$$

then $u = constant$ and $v = constant$ are the integrals of the equations

$$e\,dp' + \{f - i\,(eg - f^2)^{\frac{1}{2}}\}\,dq' = 0,$$

$$e\,dp' + \{f + i\,(eg - f^2)^{\frac{1}{2}}\}\,dq' = 0,$$

respectively. In the vocabulary of differential geometry, $u$ and $v$ are the parameters of the nul-lines on the surface.

so that $U$ is a new arbitrary function of the newly defined variable $u$; and we now have

$$x_1 = \tfrac{1}{2}(1 - u^2) U, \quad y_1 = \frac{i}{2}(1 + u^2) U, \quad z_1 = uU.$$

Similarly for $x_2, y_2, z_2$, derivable from $x_1, y_1, z_1$, by interchanging the complex variables $u$ and $v$ with all consequent interchanges—that is, the arbitrary function $U$ of $u$ into a conjugate arbitrary function $V$ of the conjugate complex variable $v$, and $i$ into $-i$. Thus

$$x_2 = \tfrac{1}{2}(1 - v^2) V, \quad y_2 = -\frac{i}{2}(1 + v^2) V, \quad z_2 = vV.$$

We have, always,

$$dx = x_1 du + x_2 dv, \quad dy = y_1 du + y_2 dv, \quad dz = z_1 du + z_2 dv.$$

Substituting the values of $x_1, y_1, z_1, x_2, y_2, z_2$, that have been obtained, and writing

$$U = f'''(u), \quad V = g'''(v),$$

merely to facilitate the expression of quadratures, $g(v)$ and $f(u)$ being conjugate functions, we have

$$\left. \begin{aligned} x - a &= \tfrac{1}{2}(1 - u^2) f''(u) + uf'(u) - f(u) + \tfrac{1}{2}(1 - v^2) g''(v) + vg'(v) - g(v) \\ y - b &= i\{\tfrac{1}{2}(1 + u^2) f''(u) - uf'(u) + f(u)\} - i\{\tfrac{1}{2}(1 + v^2) g''(v) - vg'(v) + g(v)\} \\ z - c &= uf''(u) - f'(u) + vg''(v) - g'(v) \end{aligned} \right\},$$

where $a, b, c$ are arbitrary constants, significant only as regards the coordinate position of the surface, while $f(u)$ and $g(v)$ are two arbitrary (and conjugate) functions of the two complex (and conjugate) variables $u$ and $v$.

These equations constitute Weierstrass's form of the *primitive** of the *characteristic equation or equations of minimal surfaces*.

*Ex.* Prove that

$$\frac{p}{u + v} = \frac{q}{i(v - u)} = \frac{1}{1 - uv}$$

for all surfaces; and obtain the values of $r, s, t$.

*Note.* In the preceding investigation, the analysis requires supplementing, should the ratio $Q/P$ be constant, because a variable $u$ cannot then be taken as equal to that ratio. Similarly, for the same limiting eventuality in connection with the function $v$. Two cases thus arise: (i), when one, but not both, of the variables cannot thus be determined; (ii), when neither of the variables can thus be determined.

In the first case, let $Q/P$ be constant, but not the corresponding ratio for $v$; then

$$x - A = \tfrac{1}{2}(1 - a^2) U + F(v),$$
$$y - B = \frac{i}{2}(1 + a^2) U + G(v),$$
$$z - C = aU + H(v),$$

---

* A primitive was obtained first by Monge, afterwards by Ampère. For these earlier forms, and their merger into the Weierstrass form, see the author's *Theory of Differential Equations*, vol. vi, § 249.

where $a$ is a constant (not necessarily real), $U$ is any function of $u$, and $F(v)$, $G(v)$, $H(v)$, are any functions of $v$ such that $F'^2 + G'^2 + H'^2 = 0$. The surface is a cylinder: and is, therefore, deformable into a plane, without stretching or tearing.

In the second case, where both ratios are constant,

$$x - A = \tfrac{1}{2}(1 - a^2) U + \tfrac{1}{2}(1 - c^2) V,$$

$$y - B = \frac{i}{2}(1 + a^2) U - \frac{i}{2}(1 + c^2) V,$$

$$z - C = \qquad\quad aU + cV,$$

where $a$ and $c$ are constants which can be taken conjugate, and where $U$ and $V$ are any functions of $u$ and of $v$ respectively. Then the surface has

$$(a + c)(x - A) + i(c - a)(y - B) + (ac - 1)(z - C) = 0,$$

for its equation: that is, the surface is a plane.

*Boundary condition for minimal surfaces: general form.*

**302.** The condition along a curve, which forms any portion of the boundary of the surface making the integral a minimum, is that the relation

$$\left( X \frac{\partial \Omega}{\partial x_1} + Y \frac{\partial \Omega}{\partial y_1} + Z \frac{\partial \Omega}{\partial z_1} \right) dv - \left( X \frac{\partial \Omega}{\partial x_2} + Y \frac{\partial \Omega}{\partial y_2} + Z \frac{\partial \Omega}{\partial z_2} \right) du = 0$$

shall be satisfied in general (§ 295). For the particular variables used to represent the surface, this condition becomes

$$(X x_2 + Y y_2 + Z z_2) dv - (X x_1 + Y y_1 + Z z_1) du = 0.$$

Now, from the primitive of the characteristic equations, we have

$$x_2 dv - x_1 du = -\tfrac{1}{2}(1 - u^2) f'''(u) du + \tfrac{1}{2}(1 - v^2) g'''(v) dv,$$

with corresponding expressions for $y_2 dv - y_1 du$, $z_2 dv - z_1 du$, which occur in the condition. When we use Bonnet's minimal surface adjoint[*] to the actual surface, a point $(x_0, y_0, z_0)$ on the adjoint surface is associated with a point $(x, y, z)$ on the actual surface, by means of the equations

$$\tfrac{1}{2}(x - i x_0) = \tfrac{1}{2}(1 - u^2) f''(u) + u f'(u) - f(u),$$

$$\tfrac{1}{2}(x + i x_0) = \tfrac{1}{2}(1 - v^2) g''(v) + v g'(v) - g(v).$$

Then

$$i dx_0 = \tfrac{1}{2}(1 - v^2) g'''(v) dv - \tfrac{1}{2}(1 - u^2) f'''(u) du.$$

Similarly for the other terms in the boundary condition, which thus becomes

$$X dx_0 + Y dy_0 + Z dz_0 = 0.$$

As $\kappa X, \kappa Y, \kappa Z$, now denote a displacement along the boundary of the minimal surface, so that $\kappa X = dx$, $\kappa Y = dy$, $\kappa Z = dz$, this condition is

$$dx\, dx_0 + dy\, dy_0 + dz\, dz_0 = 0,$$

---

[*] See the author's *Lectures on Differential Geometry*, § 185: for Bonnet's memoir, see *Comptes Rendus*, t. xxxvii (1853), pp. 529—532.

a relation that is always satisfied* for a surface and its adjoint. Thus the terminal condition is associated with the adjoint surface.

The precise determination of a minimal surface can be effected, among other ways; (i), by the requirement of passing through two assigned curves in space; or (ii), by the requirement of passing through one assigned curve in space and touching, along the curve, any assigned surface through it. The latter mode, in particular, is the realisation of Cauchy's theorem concerning the existence of a primitive of the characteristic equation: it was adopted by Schwarz†, whose investigation will be given almost immediately (§ 304). The development of the analysis, however, soon ceases to belong specially to the calculus of variations: it becomes a part of the theory of analytic functions and a part of the geometric theory of minimal surfaces‡.

*The equations of the characteristic provide an actual minimum for weak variations, subject to a conjugate limit.*

**303.** One passing remark may be made. Without waiting for the establishment of the canonical form in general and for the inferred construction of the test, which is the extension of the Legendre test as applied to a double integral of the first order, it is easy to see that the second variation is always positive (and that the surface is therefore a minimum) within a range limited by the conjugate (if any) of a boundary curve.

The general weak variation is given by $x + \kappa X$, $y + \kappa Y$, $z + \kappa Z$, in place of $x, y, z$. The first variation is

$$i\kappa \iint (x_1 X_2 + x_2 X_1 + y_1 Y_2 + y_2 Y_1 + z_1 Z_2 + z_2 Z_1)\, du\, dv;$$

and has been made to vanish under the preceding requirements. The second variation, constituting (in this case) the whole of the remainder of the variation of the original integral, is

$$i\kappa^2 \int (X_1 X_2 + Y_1 Y_2 + Z_1 Z_2)\, du\, dv.$$

To have the most general weak variation possible, we take

$$f(u) + \kappa \theta(u), \quad g(v) + \kappa \Im(v),$$

as variations of $f(u)$ and $g(v)$, where $\theta(u)$ and $\Im(v)$ are arbitrary conjugate functions of the conjugate variables; then

$$X_1 = \tfrac{1}{2}(1 - u^2)\,\theta'''(u), \quad Y_1 = \tfrac{1}{2}i(1 + u^2)\,\theta'''(u), \quad Z_1 = u\theta'''(u),$$
$$X_2 = \tfrac{1}{2}(1 - v^2)\,\Im'''(v), \quad Y_2 = -\tfrac{1}{2}i(1 + v^2)\,\Im'''(v), \quad Z_2 = v\Im'''(v),$$

---

* *l.c.*, p. 299.

† *l.c.*, p. 301. Schwarz's memoir occurs in *Crelle*, vol. LXXX (1875), pp. 280—300; see p. 291, for the actual expression obtained in § 304 (*post*).

‡ For Darboux's exposition, see his *Théorie générale des surfaces*, vol. i, book iii.

so that

$$X_1 X_2 + Y_1 Y_2 + Z_1 Z_2 = \tfrac{1}{2}(1+uv)^2\, \theta'''(u)\, \vartheta'''(v).$$

As $u$ and $v$ are conjugate, let them be expressed in the form

$$u = \alpha + i\beta, \quad v = \alpha - i\beta,$$

where $\alpha$ and $\beta$ are real. Then, in the integral $\iint \Omega\, du\, dv$,

$$du\, dv = \frac{\partial\,(u,\,v)}{\partial\,(\alpha,\,\beta)}\, d\alpha\, d\beta = -\,2i\, d\alpha\, d\beta\,;$$

consequently, the second variation is

$$\kappa^2 \iint (1+uv)^2 \theta'''(u)\, \vartheta'''(v)\, d\alpha\, d\beta.$$

The quantities $\theta'''(u)$ and $\vartheta'''(v)$ are conjugate functions of $\alpha + i\beta$ and $\alpha - i\beta$; and therefore $\theta'''(u)\, \vartheta'''(v)$ is of the form $A^2 + B^2$, where $A$ and $B$ are real functions of $\alpha$ and $\beta$; and $uv = \alpha^2 + \beta^2$.

Thus the second variation is always positive so long as $\theta'''(u)$ and $\vartheta'''(v)$ do not vanish, that is, so long as the conjugate of the initial curve is not attained. Hence, within that range, the surface is a minimum.

### Determination of minimal surface by assigned conditions : Schwarz's theorem.

**304.** We have seen (§ 282) that any characteristic surface, and therefore a minimal surface, is uniquely determinate under two conditions: that it shall pass through a given regular curve in space; and that, along the curve, it shall touch a given developable surface (or, what is the equivalent, shall have its tangent plane given at every point on the curve). The precise determination of any surface requires a knowledge of the arbitrary functions $f$ and $g$ in the primitive, that is, of the arbitrary elements in the primitive; and it is attained as follows.

The general critical curves in the Cauchy existence-theorem (§ 282) are

$$f_{66}\, dy^2 - 2f_{69}\, dx\, dy + f_{99}\, dx^2 = 0.$$

For minimal surfaces, this equation is

$$(1+q^2)\, dy^2 + 2pq\, dx\, dy + (1+p^2)\, dx^2 = 0,$$

that is,

$$dx^2 + dy^2 + dz^2 = 0\,;$$

and therefore the curves are the nul-lines of the surface. Thus the arguments of the arbitrary functions in the primitive are $u$ and $v$. Moreover, the data serving to determine these arbitrary functions must be given along a curve, which does not belong to either of the families $u =$ constant, $v =$ constant.

The primitive of the differential equation of the surface is of the form

$$x = A\,(u) + A_0\,(v), \quad y = B\,(u) + B_0\,(v), \quad z = C\,(u) + C_0\,(v),$$

where $A(u)$ and $A_0(v)$ are conjugate: likewise $B(u)$ and $B_0(v)$: likewise $C(u)$ and $C_0(v)$. As

$$x_1{}^2 + y_1{}^2 + z_1{}^2 = 0, \quad x_1 x_2 + y_1 y_2 + z_1 z_2 = F, \quad x_2{}^2 + y_2{}^2 + z_2{}^2 = 0,$$

while

$$l, m, n = \left\| \begin{array}{ccc} x_1, & y_1, & z_1 \\ x_2, & y_2, & z_2 \end{array} \right\|,$$

we have

$$\left.\begin{array}{l} mz_1 - ny_1 = -x_1 F \\ mz_2 - ny_2 = x_2 F \end{array}\right\}, \quad \left.\begin{array}{l} nx_1 - lz_1 = -y_1 F \\ nx_2 - lz_2 = y_2 F \end{array}\right\}, \quad \left.\begin{array}{l} ly_1 - mx_1 = -z_1 F \\ ly_2 - mx_2 = z_2 F \end{array}\right\},$$

while the direction-cosines of the normal to the surface are

$$\mathbf{X} = \frac{l}{iF}, \quad \mathbf{Y} = \frac{m}{iF}, \quad \mathbf{Z} = \frac{n}{iF}.$$

In association with $x, y, z$, we take $x_0, y_0, z_0$, respectively, where [*]

$$x_0 = iA(u) - iA_0(v), \quad y_0 = iB(u) - iB_0(v), \quad z_0 = iC(u) - iC_0(v)$$

Then

$$\begin{aligned} dx_0 &= i \frac{\partial A}{\partial u} du - i \frac{\partial A_0}{\partial v} dv \\ &= i x_1 du - i x_2 dv \\ &= (\mathbf{Y} z_1 - \mathbf{Z} y_1) du + (\mathbf{Y} z_2 - \mathbf{Z} y_2) dv \\ &= \mathbf{Y} dz - \mathbf{Z} dy. \end{aligned}$$

But

$$x - ix_0 = 2A(u), \quad x + ix_0 = 2A_0(v);$$

therefore

$$x - i \int (\mathbf{Y} dz - \mathbf{Z} dy) = 2A(u), \quad x + i \int (\mathbf{Y} dz - \mathbf{Z} dy) = 2A_0(v).$$

Similarly,

$$y - i \int (\mathbf{Z} dx - \mathbf{X} dz) = 2B(u), \quad y + i \int (\mathbf{Z} dx - \mathbf{X} dz) = 2B_0(v),$$

$$z - i \int (\mathbf{X} dy - \mathbf{Y} dx) = 2C(u), \quad z + i \int (\mathbf{X} dy - \mathbf{Y} dx) = 2C_0(v).$$

If, for any configuration, the values of $x, y, z, \mathbf{X}, \mathbf{Y}, \mathbf{Z}$, can be expressed in terms of a parameter, these relations serve to determine the forms of the functions $A, A_0, B, B_0, C, C_0$. In particular, consider the initial curve in Cauchy's existence-theorem concerning the primitive. The quantities $x, y, z, \mathbf{X}, \mathbf{Y}, \mathbf{Z}$, in the conditions governing the theorem, are expressible in terms of some parameter $\mu$ along the initial curve; and the curve itself will be given by a relation

$$\chi(u, v) = 0,$$

* These quantities were introduced as connected with Bonnet's surface, adjoint to a given surface: see § 302.

which is not an equation involving $u$ alone or involving $v$ alone, so that along the curve $u$ and $v$ will be expressible in terms of a single quantity which may be taken to be the parameter $\mu$. Thus we shall have, along the curve,

$$A(u) = \mathbf{A}(\mu), \quad B(u) = \mathbf{B}(\mu), \quad C(u) = \mathbf{C}(\mu),$$
$$A_0(v) = \mathbf{A}_0(\mu), \quad B_0(v) = \mathbf{B}_0(\mu), \quad C_0(v) = \mathbf{C}_0(\mu).$$

When substitution takes place for $x$, $y$, $z$, $\mathbf{Y}$, $\mathbf{Z}$, in terms of $\mu$, in the expressions

$$x - i \int (\mathbf{Y}\,dz - \mathbf{Z}\,dy), \quad x + i \int (\mathbf{Y}\,dz - \mathbf{Z}\,dy),$$

we have $\mathbf{A}(\mu)$, and $\mathbf{A}_0(\mu)$; and similarly for $\mathbf{B}(\mu)$, $\mathbf{B}_0(\mu)$, $\mathbf{C}(\mu)$, $\mathbf{C}_0(\mu)$: that is, the forms of the functions $\mathbf{A}$, $\mathbf{A}_0$, $\mathbf{B}$, $\mathbf{B}_0$, $\mathbf{C}$, $\mathbf{C}_0$ are known.

Now take two conjugate variables $\rho$ and $\sigma$, and write

$$2\mathbf{A}(\rho) = x(\rho) - i \int^\rho (\mathbf{Y}\,dz - \mathbf{Z}\,dy), \quad 2\mathbf{A}_0(\sigma) = x(\sigma) + i \int^\sigma (\mathbf{Y}\,dz - \mathbf{Z}\,dy),$$

so that

$$2\mathbf{A}(\rho) + 2\mathbf{A}_0(\sigma) = x(\rho) + x(\sigma) - i \int_\sigma^\rho (\mathbf{Y}\,dz - \mathbf{Z}\,dy).$$

Also, because any function of $u$ can be substituted for $u$ and any function of $v$ for $v$, without affecting the primitive in character, we determine the functions and the variables, so that

$$A(u) = \mathbf{A}(\rho), \quad B(u) = \mathbf{B}(\rho), \quad C(u) = \mathbf{C}(\rho),$$
$$A_0(v) = \mathbf{A}_0(\sigma), \quad B_0(v) = \mathbf{B}_0(\sigma), \quad C_0(v) = \mathbf{C}_0(\sigma),$$

a determination which, owing to the curve-relations of the type $A(u) = \mathbf{A}(\mu)$, $A_0(v) = \mathbf{A}_0(\mu)$, is immediate. But $x = A(u) + A_0(v)$, and so for $y$ and for $z$: thus we finally have

$$2x = x(\rho) + x(\sigma) - i \int_\sigma^\rho (\mathbf{Y}\,dz - \mathbf{Z}\,dy) \Big)$$

$$2y = y(\rho) + y(\sigma) - i \int_\sigma^\rho (\mathbf{Z}\,dx - \mathbf{X}\,dz) \Big\rangle :$$

$$2z = z(\rho) + z(\sigma) - i \int_\sigma^\rho (\mathbf{X}\,dy - \mathbf{Y}\,dx) \Big)$$

equations of the minimal surface which satisfy the Cauchy conditions along the initial curve.

As regards the determination of $u$ as a function of $\rho$ only, and the determination of $v$ as a function of $\sigma$ only, we have

$$x_1 = \frac{\partial A}{\partial u}, \quad y_1 = \frac{\partial B}{\partial u}, \quad z_1 = \frac{\partial C}{\partial u},$$

so that

$$x_1 + iy_1 = \tfrac{1}{2}(1 - u^2) f'''(u) - \tfrac{1}{2}(1 + u^2) f'''(u) = -u^2 f'''(u),$$
$$z_1 = u f'''(u);$$

and therefore
$$- uz_1 = x_1 + iy_1.$$
But
$$x_1 = \frac{\partial x}{\partial \rho}\frac{d\rho}{du}, \quad y_1 = \frac{\partial y}{\partial \rho}\frac{d\rho}{du}, \quad z_1 = \frac{\partial z}{\partial \rho}\frac{d\rho}{du};$$

hence $u$ is given by the equation
$$- u\frac{\partial z}{\partial \rho} = \frac{\partial x}{\partial \rho} + i\frac{\partial y}{\partial \rho}.$$

Similarly, $v$ is given by the equation
$$- v\frac{\partial z}{\partial \sigma} = \frac{\partial x}{\partial \sigma} - i\frac{\partial y}{\partial \sigma}.$$

On substitution of the values of $x$, $y$, $z$, in terms of $\rho$ and $\sigma$, these two equations express $u$ as a function of $\rho$ and $v$ as a function of $\sigma$: and conversely. The arbitrary functions are then obtainable from a comparison of the forms
$$uf'''(u) = \frac{\partial z}{\partial \rho}\frac{d\rho}{du}, \quad vg'''(v) = \frac{\partial z}{\partial \sigma}\frac{d\sigma}{dv}.$$

This theorem, which leads from the data in the initial conditions to the values of $x$, $y$, $z$, constituting the required appropriate primitive, is due to Schwarz*.

*Ex.* 1. *Required the minimal surface, which touches a right circular cone of semi-vertical angle a along a circular section.*

Let the circular section be of radius $c$. When the vertex of the cone is taken as origin and its axis is taken as axis of $z$, we have
$$x = c\,\cos\mu \quad, \quad y = c\,\sin\mu \quad, \quad z = c\cot a,$$
$$\mathbf{X} = -\cos\mu\cos a, \quad \mathbf{Y} = -\sin\mu\cos a, \quad \mathbf{Z} = \sin a,$$

along the section. Thus
$$\mathbf{Y}\,dz - \mathbf{Z}\,dy = -c\sin a\cos\mu\,d\mu, \quad \mathbf{Z}\,dx - \mathbf{X}\,dz = -c\sin a\sin\mu\,d\mu, \quad \mathbf{X}\,dy - \mathbf{Y}\,dx = -c\cos a\,d\mu;$$
and therefore
$$2A(\mu) = c(\cos\mu + i\sin a\sin\mu), \quad 2A_0(\mu) = c(\cos\mu - i\sin a\sin\mu),$$
$$2B(\mu) = c(\sin\mu - i\cos a\cos\mu), \quad 2B_0(\mu) = c(\sin\mu + i\cos a\cos\mu),$$
$$2C(\mu) = c(\cot a + i\mu\cos a), \quad 2C_0(\mu) = c(\cot a - i\mu\cos a).$$

Consequently
$$2x = c(\cos\rho + \cos\sigma) + ic\sin a(\sin\rho - \sin\sigma),$$
$$2y = c(\sin\rho + \sin\sigma) - ic\sin a(\cos\rho - \cos\sigma),$$
$$2z = 2c\cot a + ic(\rho - \sigma)\cos a.$$

We have
$$2\frac{dz}{d\rho} = ic\cos a, \quad 2\frac{dx}{d\rho} = -c\sin\rho + ic\sin a\cos\rho, \quad 2\frac{dy}{d\rho} = c\cos\rho + ic\sin a\sin\rho;$$
and therefore, as
$$- u\frac{dz}{d\rho} = \frac{dx}{d\rho} + i\frac{dy}{d\rho},$$
we have
$$- u = \frac{1 + \sin a}{\cos a}e^{\rho i}.$$

* *l.c.*, p. 502.

Similarly

$$-v = \frac{1+\sin a}{\cos a} e^{-\sigma i}.$$

Hence also

$$-\frac{1}{u} = \frac{1-\sin a}{\cos a} e^{-\rho i}, \quad -\frac{1}{v} = \frac{1-\sin a}{\cos a} e^{\sigma i}.$$

Thus

$$\left.\begin{aligned}
\frac{2x}{c \cos a} &= -\tfrac{1}{2}\left(u + \frac{1}{u}\right) - \tfrac{1}{2}\left(v + \frac{1}{v}\right) \\
\frac{2y}{c \cos a} &= \frac{i}{2}\left(u - \frac{1}{u}\right) - \frac{i}{2}\left(v - \frac{1}{v}\right) \\
\frac{2(z - \cot a)}{c \cos a} &= \log u + \log v - 2\log\left(\frac{1+\sin a}{\cos a}\right)
\end{aligned}\right\},$$

shewing that the surface is a catenoid of parameter $c \cos a$.

*Ex.* 2. A minimal surface is drawn through a helix of pitch $\tan^{-1} c$ on a circular cylinder of radius unity, having its axis along the axis of $z$; and the minimal surface touches the cylinder along the helix. Obtain its equation in the form

$$z = c\tan^{-1}\frac{y}{x} + c\tan^{-1}\left\{c\left(\frac{x^2+y^2-1}{x^2+y^2+c^2}\right)^{\frac{1}{2}}\right\} + c\tanh^{-1}\left(\frac{x^2+y^2-1}{x^2+y^2+c^2}\right)^{\frac{1}{2}}.$$

*Ex.* 3. A minimal surface touches the quadric

$$\frac{x^2}{a} + \frac{y^2}{b} + \frac{z^2}{c} = 1$$

along a line of curvature, determined by the intersection with the confocal quadric

$$\frac{x^2}{a+p} + \frac{y^2}{b+p} + \frac{z^2}{c+p} = 1.$$

Obtain its equation in the form

$$2x = \{\mathbf{a}(a+\rho)\}^{\frac{1}{2}} + \{\mathbf{a}(a+\sigma)\}^{\frac{1}{2}} - \tfrac{1}{2}\left\{\frac{a(b+p)(c+p)}{p(b-a)(c-a)}\right\}^{\frac{1}{2}} \int_\sigma^\rho \left\{\frac{\mu}{(b+\mu)(c+\mu)}\right\}^{\frac{1}{2}} d\mu,$$

$$2y = \{\mathbf{b}(b+\rho)\}^{\frac{1}{2}} + \{\mathbf{b}(b+\sigma)\}^{\frac{1}{2}} - \tfrac{1}{2}\left\{\frac{b(c+p)(a+p)}{p(c-b)(a-b)}\right\}^{\frac{1}{2}} \int_\sigma^\rho \left\{\frac{\mu}{(c+\mu)(a+\mu)}\right\}^{\frac{1}{2}} d\mu,$$

$$2z = \{\mathbf{c}(c+\rho)\}^{\frac{1}{2}} + \{\mathbf{c}(c+\sigma)\}^{\frac{1}{2}} - \tfrac{1}{2}\left\{\frac{c(a+p)(b+p)}{p(a-c)(b-c)}\right\}^{\frac{1}{2}} \int_\sigma^\rho \left\{\frac{\mu}{(a+\mu)(b+\mu)}\right\}^{\frac{1}{2}} d\mu,$$

where

$$\mathbf{a} = \frac{a(a+p)}{(a-b)(a-c)}, \quad \mathbf{b} = \frac{b(b+p)}{(b-c)(b-a)}, \quad \mathbf{c} = \frac{c(c+p)}{(c-a)(c-b)}.$$

*Ex.* 4. Find the equation of a minimal surface, touching a paraboloid

$$\frac{y^2}{l} + \frac{z^2}{l'} = 4x$$

along a line of curvature.

## *The second variation of the original integral.*

**305.** The first variation in the complete variation of the original integral has been made to vanish. In consequence, we have the three characteristic equations, proved to be equivalent to one in essence; but all three are retained for the sake of analytical convenience. Also, boundary conditions have arisen: when these have not been satisfied automatically by assigned data, they have led to relations which enable us to regard the boundary as definitely fixed. Accordingly, we turn to the consideration of the 'second' variation, that is, to the coefficient of $\kappa^2$, in the complete variation. This second variation is the quantity

$$\tfrac{1}{2}\kappa^2 \iint \Theta \, du \, dv,$$

where

$$
\begin{aligned}
\Theta = {}& \Omega_{77}X_2{}^2 + 2\Omega_{78}X_2Y_2 + 2\Omega_{79}X_2Z_2 + \Omega_{88}Y_2{}^2 + 2\Omega_{89}Y_2Z_2 + \Omega_{99}Z_2{}^2 \\
& + 2\left\{(\Omega_{47}X_1X_2 + \Omega_{57}Y_1X_2 + \Omega_{67}Z_1X_2) + (\Omega_{48}X_1Y_2 + \Omega_{58}Y_1Y_2 + \Omega_{68}Y_1Z_2)\right. \\
& \qquad \left. + (\Omega_{49}X_1Z_2 + \Omega_{59}Y_1Z_2 + \Omega_{69}Z_1Z_2)\right\} \\
& + 2\left\{(\Omega_{17}XX_2 + \Omega_{27}YX_2 + \Omega_{37}ZX_2) + (\Omega_{18}XY_2 + \Omega_{28}YY_2 + \Omega_{38}ZY_2)\right. \\
& \qquad \left. + (\Omega_{19}XZ_2 + \Omega_{29}YZ_2 + \Omega_{39}ZZ_2)\right\} \\
& + \Omega_{44}X_1{}^2 + 2\Omega_{45}X_1Y_1 + 2\Omega_{46}X_1Z_1 + \Omega_{55}Y_1{}^2 + 2\Omega_{56}Y_1Z_1 + \Omega_{66}Z_1{}^2 \\
& + 2\left\{(\Omega_{14}XX_1 + \Omega_{24}YX_1 + \Omega_{34}ZX_1) + (\Omega_{15}XY_1 + \Omega_{25}YY_1 + \Omega_{35}ZY_1)\right. \\
& \qquad \left. + (\Omega_{16}XZ_1 + \Omega_{26}YZ_1 + \Omega_{36}ZZ_1)\right\} \\
& + \Omega_{11}X^2 + 2\Omega_{12}XY + 2\Omega_{13}XZ + \Omega_{22}Y^2 + 2\Omega_{23}YZ + \Omega_{33}Z^2.
\end{aligned}
$$

Reduction of this expression for the second variation to a normal quadratic form requires the addition, to $\Theta$, of groups of terms obviously analogous to the corresponding groups added in the various instances that have been considered when there is only a single independent variable. That same reduction will require also the addition, to $\Theta$, of a group of terms (§ 308) to which there is no analogy in those instances. There will also be a further reduction, so that the final normal quadratic form will be expressed by means of certain integrals of the subsidiary characteristic equations which may be conveniently considered at once.

*Subsidiary characteristic equations.*

**306.** The central characteristic equations are

$$\frac{\partial \Omega}{\partial x} - \frac{\partial}{\partial u}\left(\frac{\partial \Omega}{\partial x_1}\right) - \frac{\partial}{\partial v}\left(\frac{\partial \Omega}{\partial x_2}\right) = 0,$$

$$\frac{\partial \Omega}{\partial y} - \frac{\partial}{\partial u}\left(\frac{\partial \Omega}{\partial y_1}\right) - \frac{\partial}{\partial v}\left(\frac{\partial \Omega}{\partial y_2}\right) = 0,$$

$$\frac{\partial \Omega}{\partial z} - \frac{\partial}{\partial u}\left(\frac{\partial \Omega}{\partial z_1}\right) - \frac{\partial}{\partial v}\left(\frac{\partial \Omega}{\partial z_2}\right) = 0,$$

the three being satisfied in virtue of a single equation, when account is taken of the four identities satisfied by $\Omega$. In its most general shape, the primitive of these equations or of the single equation, all being of the second order, is given by three equations expressing $x$, $y$, $z$, in terms of $u$ and $v$, the independent variables. The corresponding surface is the characteristic surface, which is to provide a maximum or a minimum. In the expression of the primitive, two independent arbitrary functions occur; and the characteristic differential equations are satisfied identically, when the primitive relations are substituted, whatever be the arbitrary functions.

Now let small arbitrary weak variations be imposed upon these arbitrary functions, independent of one another; and let the consequent values of the variables be represented by $x + \kappa\xi$, $y + \kappa\eta$, $z + \kappa\zeta$. The characteristic equations are still satisfied and are to remain unaltered. Hence the first increment of each of the equations must vanish.

These increments are expressible at once by means of the derivatives of a function $2\Upsilon$, *where $2\Upsilon$ is the same function of $\xi$, $\eta$, $\zeta$, and their derivatives, as the function $\Theta$ in § 305 is of $X$, $Y$, $Z$, and their derivatives.* Thus $\dfrac{\partial\Omega}{\partial x}$ becomes, on the imposition of the specified small variations,

$$\frac{\partial\Omega}{\partial x} + \kappa\left(\Omega_{11}\xi + \Omega_{12}\eta + \Omega_{13}\zeta + \Omega_{14}\xi_1 + \Omega_{15}\eta_1 + \Omega_{16}\zeta_1 + \Omega_{17}\xi_2 + \Omega_{18}\eta_2 + \Omega_{19}\zeta_2\right),$$

where squares and higher powers of $\kappa$ are neglected: that is, it becomes

$$\frac{\partial\Omega}{\partial x} + \kappa\frac{\partial\Upsilon}{\partial\xi}.$$

Similarly for the small variations of all the first derivatives of $\Omega$ which occur in the characteristic equations. Hence as the total first increment of each of the equations must vanish, because each equation is unaltered, we have

$$\left.\begin{aligned}
\mathbf{E}_\xi &= \frac{\partial\Upsilon}{\partial\xi} - \frac{\partial}{\partial u}\left(\frac{\partial\Upsilon}{\partial\xi_1}\right) - \frac{\partial}{\partial v}\left(\frac{\partial\Upsilon}{\partial\xi_2}\right) = 0 \\
\mathbf{E}_\eta &= \frac{\partial\Upsilon}{\partial\eta} - \frac{\partial}{\partial u}\left(\frac{\partial\Upsilon}{\partial\eta_1}\right) - \frac{\partial}{\partial v}\left(\frac{\partial\Upsilon}{\partial\eta_2}\right) = 0 \\
\mathbf{E}_\zeta &= \frac{\partial\Upsilon}{\partial\zeta} - \frac{\partial}{\partial u}\left(\frac{\partial\Upsilon}{\partial\zeta_1}\right) - \frac{\partial}{\partial v}\left(\frac{\partial\Upsilon}{\partial\zeta_2}\right) = 0
\end{aligned}\right\},$$

which are the *subsidiary characteristic equations.*

**307.** These three equations are linear, in the variables $\xi$, $\eta$, $\zeta$, and in their derivatives up to the second order. The coefficients are the second derivatives of $\Omega$ with respect to its original arguments; the values of $x$, $y$, $z$, are supposed to be substituted, so that these coefficients come to be functions of $u$ and $v$.

The three equations are not independent; the relations among the coefficients lead to the relations

$$x_1\mathbf{E}_\xi + y_1\mathbf{E}_\eta + z_1\mathbf{E}_\zeta = 0, \quad x_2\mathbf{E}_\xi + y_2\mathbf{E}_\eta + z_2\mathbf{E}_\zeta = 0,$$

so that we have

$$\mathbf{E}_\xi = l\mathbf{E}', \quad \mathbf{E}_\eta = m\mathbf{E}', \quad \mathbf{E}_\zeta = n\mathbf{E}'.$$

We thus can take, in their place, a single equation $\mathbf{E}' = 0$. In this single equation, we take a new variable $\theta$, which is

$$\theta = l\xi + m\eta + n\zeta;$$

and this equation $\mathbf{E}'$ is linear, in $\theta$ and in its derivatives up to the second order.

If the primitive of the characteristic equations is

$$x = F(\phi, \phi', \ldots, \psi, \psi', \ldots), \quad y = G(\phi, \phi', \ldots, \psi, \psi', \ldots), \quad z = H(\phi, \phi', \ldots, \psi, \psi', \ldots),$$

where $\phi$ and $\psi$ are arbitrary functions, the primitive of the subsidiary equations is

$$\xi = \Phi \frac{\partial F}{\partial \phi} + \Phi' \frac{\partial F}{\partial \phi'} + \ldots + \Psi \frac{\partial F}{\partial \psi} + \Psi' \frac{\partial F}{\partial \psi'} + \ldots,$$

with similar expressions for $\eta$ and $\zeta$, where $\Phi$ and $\Psi$ are the new arbitrary independent functions necessary for a complete primitive. But there are limitations of a general type upon this primitive. As $u$ and $v$ are the parameters of the nul-lines upon the surface (where any function of $u$ may be substituted for $u$, and likewise any function of $v$ for $v$), we have the relations

$$x_1{}^2 + y_1{}^2 + z_1{}^2 = 0, \quad x_2{}^2 + y_2{}^2 + z_2{}^2 = 0 \,;$$

and therefore* the characteristic small variation must satisfy the equations

$$x_1 \xi_1 + y_1 \eta_1 + z_1 \zeta_1 = 0, \quad x_2 \xi_2 + y_2 \eta_2 + z_2 \zeta_2 = 0.$$

An illustration of their restrictive influence will be given later (Ex. 1, § 314): and an alternative method for the construction of $\xi$, $\eta$, $\zeta$, will be explained.

### *Modifications in the form of the second variation.*

**308.** We now proceed to the modification of the quantity $\Theta$, the subject of integration in the second variation. Already certain relations affecting the coefficients $\Omega_{rs}$ (for $r$, $s$, = 4, 5, 6, 7, 8, 9) have been obtained, viz.

$$\frac{\Omega_{44}}{l^2} = \frac{\Omega_{45}}{lm} = \frac{\Omega_{46}}{ln} = \frac{\Omega_{55}}{m^2} = \frac{\Omega_{56}}{mn} = \frac{\Omega_{66}}{n^2} = P,$$

$$\frac{\Omega_{77}}{l^2} = \frac{\Omega_{78}}{lm} = \frac{\Omega_{79}}{ln} = \frac{\Omega_{88}}{m^2} = \frac{\Omega_{89}}{mn} = \frac{\Omega_{99}}{n^2} = Q,$$

$$\frac{\Omega_{47}}{l^2} = \tfrac{1}{2}\frac{\Omega_{48} + \Omega_{57}}{lm} = \tfrac{1}{2}\frac{\Omega_{49} + \Omega_{67}}{ln} = \frac{\Omega_{58}}{m^2} = \tfrac{1}{2}\frac{\Omega_{59} + \Omega_{68}}{mn} = \frac{\Omega_{69}}{n^2} = R.$$

These relations, however, express only the sums $\Omega_{48} + \Omega_{57}$, $\Omega_{49} + \Omega_{67}$, $\Omega_{59} + \Omega_{68}$; we therefore take

$$\Omega_{59} = mnR - \alpha, \quad \Omega_{67} = nlR - \beta, \quad \Omega_{48} = lmR - \gamma,$$

$$\Omega_{68} = mnR + \alpha, \quad \Omega_{49} = nlR + \beta, \quad \Omega_{57} = lmR + \gamma,$$

---

* In forming these equations, we neglect the terms $\kappa^2 (\xi_1{}^2 + \eta_1{}^2 + \zeta_1{}^2)$ and $\kappa^2 (\xi_2{}^2 + \eta_2{}^2 + \zeta_2{}^2)$, for two reasons which ultimately are one and the same. Throughout, we take $\kappa$ so small that quantities involving the second power of $\kappa$ are to be neglected as compared with non-vanishing quantities involving the first power. Also, when terms involving $\kappa^2$ arise out of the terms originally involving the first power of $\kappa$, they form a portion of the infinitesimal change of the second order. As it is a fundamental proposition in Lie's theory of continuous groups, that effectively such a group is completely determined by the aggregate of its infinitesimal variations of the first order, we need not consider terms in $\kappa^2$ and in higher powers.

Other examples of the proposition occur in the theory of invariants and covariants of homogeneous algebraic forms, of quadratic differential forms, and of differential invariants and covariants on surfaces and in space.

where $P$, $Q$, $R$, $\alpha$, $\beta$, $\gamma$ are (in all) six quantities in terms of which the twenty-one coefficients $\Omega_{rs}$ are expressed*. When these values of the said coefficients are substituted in the corresponding terms in $\Theta$, and the resulting terms are gathered together, we find that

(i)   the aggregate of terms in $(X_1, Y_1, Z_1)^2$ is
$$P(lX_1 + mY_1 + nZ_1)^2:$$

(ii)  the aggregate of terms in $(X_2, Y_2, Z_2)^2$ is
$$Q(lX_2 + mY_2 + nZ_2)^2:$$

(iii) the aggregate of terms in $(X_1, Y_1, Z_1 \lozenge X_2, Y_2, Z_2)$ is
$$2R(lX_1 + mY_1 + nZ_1)(lX_2 + mY_2 + nZ_2)$$
$$+ 2\alpha(Z_1 Y_2 - Y_1 Z_2) + 2\beta(X_1 Z_2 - Z_1 X_2) + 2\gamma(Y_1 X_2 - X_1 Y_2):$$

it being remembered that all the aggregates belong to the double integrand.

We modify the terms in the second line in (iii). Using the lemma of § 275, we have

$$\iint \rho Z_1 Y_2 \, du\, dv = \left[\int Z\rho\, Y_2 dv\right] - \iint Z \frac{\partial}{\partial u}(\rho Y_2)\ du\, dv,$$

$$\iint \rho Y_1 Z_2 \, du\, dv = -\left[\int Z\rho\, Y_1 du\right] - \iint Z \frac{\partial}{\partial v}(\rho Y_1)\ du\, dv,$$

$$\iint \rho' Z_1 Y_2 \, du\, dv = -\left[\int Y\rho'\, Z_1 du\right] - \iint Y \frac{\partial}{\partial v}(\rho' Z_1)\ du\, dv,$$

$$\iint \rho' Y_1 Z_2 \, du\, dv = \left[\int Y\rho'\, Z_2 dv\right] - \iint Y \frac{\partial}{\partial u}(\rho' Z_2)\ du\, dv;$$

and therefore

$$\iint (\rho + \rho')(Z_1 Y_2 - Y_1 Z_2)\, du\, dv = \int \{\rho Z(Y_2 dv + Y_1 du) - \rho' Y(Z_1 du + Z_2 dv)\}$$
$$+ \iint (\rho_2 Y_1 Z - \rho_1 Y_2 Z + \rho_1' YZ_2 - \rho_2' YZ_1)\, du\, dv,$$

where the single integrals are taken positively round the whole of the boundary of the field of integration, $\rho$ and $\rho'$ being any functions of $u$ and $v$. Similarly,

$$\iint (\sigma + \sigma')(X_1 Z_2 - Z_1 X_2)\, du\, dv = \int \{\sigma X(Z_2 dv + Z_1 du) - \sigma' Z(X_1 du + X_2 dv)\}$$
$$+ \iint (\sigma_2 Z_1 X - \sigma_1 Z_2 X + \sigma_1' ZX_2 - \sigma_2' ZX_1)\, du\, dv,$$

$$\iint (\tau + \tau')(Y_1 X_2 - X_1 Y_2)\, du\, dv = \int \{\tau Y(X_2 dv + X_1 du) - \tau' X(Y_1 du + Y_2 dv)\}$$
$$+ \iint (\tau_2 X_1 Y - \tau_1 X_2 Y + \tau_1' XY_2 - \tau_2' XY_1)\, du\, dv,$$

---

* In terms of the derivatives of the function $f(x, y, z, p, q)$, we have
$$\alpha = \frac{\partial f}{\partial p}, \quad \beta = \frac{\partial f}{\partial q}, \quad \gamma = p\frac{\partial f}{\partial p} + q\frac{\partial f}{\partial q} - f;$$
see the results stated in § 292, Ex. 2.

where $\sigma$ and $\sigma'$; $\tau$ and $\tau'$; are any functions of $u$ and $v$. Now let

$$\rho + \rho' = \alpha, \quad \sigma + \sigma' = \beta, \quad \tau + \tau' = \gamma,$$

leaving the complete determination of the six quantities $\rho, \rho', \sigma, \sigma', \tau, \tau'$ open, in case such determination should be required. The expressions for

$$\iint \{\alpha (Z_1 Y_2 - Y_1 Z_2) + \beta (X_1 Z_2 - Z_1 X_2) + \gamma (Y_1 X_2 - X_1 Y_2)\} \, du \, dv,$$

thus obtained, will now be supposed to be substituted in $\iint \Theta \, du \, dv$, yielding some terms in the form of a boundary-integral and adding some terms to the field-integral.

**309.** Next, owing to the forms of the aggregate of terms in $(X_1, Y_1, Z_1)^2$, in $(X_2, Y_2, Z_2)^2$, and in the unchanged remainder of the aggregate of terms in $(X_1, Y_1, Z_1 \lozenge X_2, Y_2, Z_2)$, we introduce a quantity $W$, where $W$ is defined as

$$W = lX + mY + nZ.$$

Manifestly $\kappa W \div V$ is the *displacement*, along the normal to the characteristic surface, due to the small variation $\kappa X, \kappa Y, \kappa Z$. Then

$$W_1 = lX_1 + mY_1 + nZ_1 + l_1 X + m_1 Y + n_1 Z,$$
$$W_2 = lX_2 + mY_2 + nZ_2 + l_2 X + m_2 Y + n_2 Z.$$

We now proceed to the combination

$$\Theta + \frac{\partial}{\partial u} (AX^2 + BY^2 + CZ^2 + 2a\,YZ + 2b\,ZX + 2c\,XY)$$

$$+ \frac{\partial}{\partial v} (FX^2 + GY^2 + HZ^2 + 2f\,YZ + 2g\,ZX + 2h\,XY),$$

so that, when the double integral of this expression is formed, we obtain

$$\iint \Theta \, du\, dv + \int (AX^2 + BY^2 + CZ^2 + 2a\,YZ + 2b\,ZX + 2c\,XY)\, dv$$

$$- \int (FX^2 + GY^2 + HZ^2 + 2f\,YZ + 2g\,ZX + 2h\,XY)\, du,$$

where the single integrals are taken round the boundary in the positive direction. But $\iint \Theta \, du \, dv$ can be expressed in two parts. One is the single integral

$$\int \{(X_1 du + X_2 dv)(\tau Y - \sigma' Z) + (Y_1 du + Y_2 dv)(\rho Z - \tau' X)$$
$$+ (Z_1 du + Z_2 dv)(\sigma X - \rho' Y)\}$$

taken positively round the boundary. The other part of $\iint \Theta \, du \, dv$ is $\iint \Phi \, du \, dv$, where $\Phi$ is an aggregate of terms, homogeneous and quadratic in the arguments $X, Y, Z, X_1, Y_1, Z_1, X_2, Y_2, Z_2$, such that its terms involving only $X_1$, $Y_1$, $Z_1$, $X_2$, $Y_2$, $Z_2$, are

$$P (lX_1 + mY_1 + nZ_1)^2 + Q (lX_2 + mY_2 + nZ_2)^2$$
$$+ 2R (lX_1 + mY_1 + nZ_1) (lX_2 + mY_2 + nZ_2).$$

On the analogy of the results of earlier investigations, we postulate a possible trinomial form

$$\Phi = P (W_1 - \lambda W)^2 + 2R (W_1 - \lambda W)(W_2 - \mu W) + Q (W_2 - \mu W)^2;$$

and we consider the relations which would have to be satisfied, if this form be possible. If these relations can be established, the quantity

$$\iint \Theta\, du\, dv - \iint \Phi\, du\, dv$$

will be equal to the sum of the single integrals

$$\int (FX^2 + GY^2 + HZ^2 + 2fYZ + 2gZX + 2hXY)\, du$$

$$- \int (AX^2 + BY^2 + CZ^2 + 2aYZ + 2bZX + 2cXY)\, dv$$

$$+ \int \{(X_1 du + X_2 dv)(\sigma'Z - \tau Y) + (Y_1 du + Y_2 dv)(\tau'X - \rho Z)$$
$$+ (Z_1 du + Z_2 dv)(\rho'Y - \sigma X)\},$$

all taken positively round the whole boundary of the field of integration. But the conditions derived from the vanishing of the first variation fix the boundary, if it is mobile. We therefore can take $X = 0$, $Y = 0$, $Z = 0$, everywhere along the boundary; thus all these single integrals vanish. Hence

$$\iint \Theta\, du\, dv = \iint \Phi\, du\, dv;$$

and the second variation will thus depend upon the integral $\iint \Phi\, du\, dv$.

### Relations for the transformation.

**310.** The relations, necessary and sufficient to secure that $\Phi$ (which is the aggregate of the terms, that are quadratic in $X$, $Y$, $Z$, $X_1$, $Y_1$, $Z_1$, $X_2$, $Y_2$, $Z_2$, in $\Theta$ after the modification made in § 308) can acquire the desired trinomial form, arise from the comparison of those quadratic terms.

We have already seen that the terms, quadratic in $X_1$, $Y_1$, $Z_1$, $X_2$, $Y_2$, $Z_2$, agree without any further relations.

In order that the terms in $(X, Y, Z)(X_1, Y_1, Z_1, X_2, Y_2, Z_2)$ may agree, the following relations, for the respective terms as set out, must be satisfied; for the terms in

$$\begin{aligned}
XX_1: \quad & \Omega_{14} + A && = Pl\ (l_1 - l\lambda) + Rl\ (l_2 - l\mu) \\
XY_1: \quad & \Omega_{15} + c - \tau_2' && = Pm\ (l_1 - l\lambda) + Rm\ (l_2 - l\mu) \\
XZ_1: \quad & \Omega_{16} + b + \sigma_2 && = Pn\ (l_1 - l\lambda) + Rn\ (l_2 - l\mu)
\end{aligned}\right\},$$

$$
\begin{aligned}
XX_2: & \quad \Omega_{17} + F && = Rl\,(l_1 - l\lambda) + Ql\,(l_2 - l\mu) \\
XY_2: & \quad \Omega_{18} + h + \tau_1' = Rm\,(l_1 - l\lambda) + Qm\,(l_2 - l\mu) \\
XZ_2: & \quad \Omega_{19} + g - \sigma_1 = Rn\,(l_1 - l\lambda) + Qn\,(l_2 - l\mu)
\end{aligned}
\Bigg\},
$$

$$
\begin{aligned}
YX_1: & \quad \Omega_{24} + c + \tau_2 && = Pl\,(m_1 - m\lambda) + Rl\,(m_2 - m\mu) \\
YY_1: & \quad \Omega_{25} + B && = Pm\,(m_1 - m\lambda) + Rm\,(m_2 - m\mu) \\
YZ_1: & \quad \Omega_{26} + a - \rho_2' = Pn\,(m_1 - m\lambda) + Rn\,(m_2 - m\mu)
\end{aligned}
\Bigg\},
$$

$$
\begin{aligned}
YX_2: & \quad \Omega_{27} + h - \tau_1 = Rl\,(m_1 - m\lambda) + Ql\,(m_2 - m\mu) \\
YY_2: & \quad \Omega_{28} + G && = Rm\,(m_1 - m\lambda) + Qm\,(m_2 - m\mu) \\
YZ_2: & \quad \Omega_{29} + f + \rho_1' = Rn\,(m_1 - m\lambda) + Qn\,(m_2 - m\mu)
\end{aligned}
\Bigg\},
$$

$$
\begin{aligned}
ZX_1: & \quad \Omega_{34} + b - \sigma_2' = Pl\,(n_1 - n\lambda) + Rl\,(n_2 - n\mu) \\
ZY_1: & \quad \Omega_{35} + a + \rho_2 = Pm\,(n_1 - n\lambda) + Rm\,(n_2 - n\mu) \\
ZZ_1: & \quad \Omega_{36} + C && = Pn\,(n_1 - n\lambda) + Rn\,(n_2 - n\mu)
\end{aligned}
\Bigg\},
$$

$$
\begin{aligned}
ZX_2: & \quad \Omega_{37} + g + \sigma_1' = Rl\,(n_1 - n\lambda) + Ql\,(n_2 - n\mu) \\
ZY_2: & \quad \Omega_{38} + f - \rho_1 = Rm\,(n_1 - n\lambda) + Qm\,(n_2 - n\mu) \\
ZZ_2: & \quad \Omega_{39} + H && = Rn\,(n_1 - n\lambda) + Qn\,(n_2 - n\mu)
\end{aligned}
\Bigg\}.
$$

Finally, in order that the terms in $(X, Y, Z)^2$ may agree, the following relations must be satisfied; for the terms in

$$
\begin{aligned}
X^2 &: \ \Omega_{11} + A_1 + F_2 = P\,(l_1 - l\lambda)^2 + 2R\,(l_1 - l\lambda)\,(l_2 - l\mu) + Q\,(l_2 - l\mu)^2 \\
XY &: \ \Omega_{12} + c_1 + h_2 = P\,(l_1 - l\lambda)\,(m_1 - m\lambda) + Q\,(l_2 - l\mu)\,(m_2 - m\mu) \\
& \qquad\qquad\qquad + R\,\{(m_1 - m\lambda)\,(l_2 - l\mu) + (l_1 - l\lambda)\,(m_2 - m\mu)\} \\
XZ &: \ \Omega_{13} + b_1 + g_2 = P\,(l_1 - l\lambda)\,(n_1 - n\lambda) + Q\,(l_2 - l\mu)\,(n_2 - n\mu) \\
& \qquad\qquad\qquad + R\,\{(n_1 - n\lambda)\,(l_2 - l\mu) + (l_1 - l\lambda)\,(n_2 - n\mu)\} \\
Y^2 &: \ \Omega_{22} + B_1 + G_2 = P\,(m_1 - m\lambda)^2 + 2R\,(m_1 - m\lambda)\,(m_2 - m\mu) + Q\,(m_2 - m\mu)^2 \\
YZ &: \ \Omega_{23} + a_1 + f_2 = P\,(m_1 - m\lambda)\,(n_1 - n\lambda) + Q\,(m_2 - m\mu)\,(n_2 - n\mu) \\
& \qquad\qquad\qquad + R\,\{(n_1 - n\lambda)\,(m_2 - m\mu) + (m_1 - m\lambda)\,(n_2 - n\mu)\} \\
Z^2 &: \ \Omega_{33} + C_1 + H_2 = P\,(n_1 - n\lambda)^2 + 2R\,(n_1 - n\lambda)\,(n_2 - n\mu) + Q\,(n_2 - n\mu)^2
\end{aligned}
\right\}
$$

### Use of the subsidiary equations, in the transformation of the normal form.

**311.** Next, having regard to the form of $\Phi$, which has $W_1 - \lambda W$ and $W_2 - \mu W$ for its arguments, and bearing in remembrance the fact that, in former instances (§§ 110, 168) the arguments vanished when the arbitrary small variation was made equal to a characteristic small variation, we proceed to verify that all the conditions and equations are satisfied, if

$$
\lambda = \frac{\theta_1}{\theta}, \quad \mu = \frac{\theta_2}{\theta},
$$

where $\theta$, the quantity introduced in § 307, denotes $l\xi + m\eta + n\zeta$, while the quantities $\xi$, $\eta$, $\zeta$, constitute a solution of the subsidiary characteristic equations

$$\frac{\partial \Upsilon}{\partial \xi} - \frac{\partial}{\partial u}\left(\frac{\partial \Upsilon}{\partial \xi_1}\right) - \frac{\partial}{\partial v}\left(\frac{\partial \Upsilon}{\partial \xi_2}\right) = 0$$

$$\frac{\partial \Upsilon}{\partial \eta} - \frac{\partial}{\partial u}\left(\frac{\partial \Upsilon}{\partial \eta_1}\right) - \frac{\partial}{\partial v}\left(\frac{\partial \Upsilon}{\partial \eta_2}\right) = 0$$

$$\frac{\partial \Upsilon}{\partial \zeta} - \frac{\partial}{\partial u}\left(\frac{\partial \Upsilon}{\partial \zeta_1}\right) - \frac{\partial}{\partial v}\left(\frac{\partial \Upsilon}{\partial \zeta_2}\right) = 0$$

Thus $\theta$ is the normal displacement (§ 309) to the consecutive characteristic.

With the foregoing value of $\theta$, we have

$$\theta_1 = l\xi_1 + m\eta_1 + n\zeta_1 + l_1\xi + m_1\eta + n_1\zeta, \quad \theta_2 = l\xi_2 + m\eta_2 + n\zeta_2 + l_2\xi + m_2\eta + n_2\zeta.$$

Hence, as $2\Upsilon$ is the same function of $\xi$, $\eta$, $\zeta$, $\xi_1$, $\eta_1$, $\zeta_1$, $\xi_2$, $\eta_2$, $\zeta_2$, as $\Theta$ (in its initial form) is of $X$, $Y$, $Z$, $X_1$, $Y_1$, $Z_1$, $X_2$, $Y_2$, $Z_2$, we have

$$\begin{aligned}
\frac{\partial \Upsilon}{\partial \zeta_2} &= \Omega_{99}\zeta_2 + \Omega_{69}\zeta_1 + \Omega_{39}\zeta + \Omega_{89}\eta_2 + \Omega_{59}\eta_1 + \Omega_{29}\eta + \Omega_{79}\xi_2 + \Omega_{49}\xi_1 + \Omega_{19}\xi \\
&= n^2 Q\zeta_2 + n^2 R\zeta_1 + \{Qn(n_2 - n\mu) + Rn(n_1 - n\lambda)\}\zeta - H\zeta \\
&\quad + mnQ\eta_2 + mnR\eta_1 + \{Qn(m_2 - m\mu) + Rn(m_1 - m\lambda)\}\eta - f\eta - \alpha\eta_1 - \rho_1'\eta \\
&\quad + lnQ\xi_2 + lnR\xi_1 + \{Qn(l_2 - l\mu) + Rn(l_1 - l\lambda)\}\xi - g\xi + \beta\xi_1 + \sigma_1\xi \\
&= Qn(\theta_2 - \mu\theta) + Rn(\theta_1 - \lambda\theta) - (g\xi + f\eta + H\zeta) - (\alpha\eta_1 + \rho_1'\eta - \beta\xi_1 - \sigma_1\xi) \\
&= -(g\xi + f\eta + H\zeta) - (\alpha\eta_1 + \rho_1'\eta - \beta\xi_1 - \sigma_1\xi),
\end{aligned}$$

with the assumption postulated as to the values of $\lambda$ and $\mu$. Similarly,

$$\begin{aligned}
\frac{\partial \Upsilon}{\partial \zeta_1} &= \Omega_{96}\zeta_2 + \Omega_{66}\zeta_1 + \Omega_{36}\zeta + \Omega_{86}\eta_2 + \Omega_{56}\eta_1 + \Omega_{26}\eta + \Omega_{76}\xi_2 + \Omega_{46}\xi_1 + \Omega_{16}\xi \\
&= -(b\xi + a\eta + c\zeta) + (\alpha\eta_2 + \rho_2'\eta - \beta\xi_2 - \sigma_2\xi),
\end{aligned}$$

after substitution for the coefficients $\Omega$ and reduction. Again,

$$\begin{aligned}
\frac{\partial \Upsilon}{\partial \zeta} &= \Omega_{93}\zeta_2 + \Omega_{63}\zeta_1 + \Omega_{33}\zeta + \Omega_{83}\eta_2 + \Omega_{53}\eta_1 + \Omega_{23}\eta + \Omega_{73}\xi_2 + \Omega_{43}\xi_1 + \Omega_{13}\xi \\
&= \zeta_2\{Qn(n_2 - n\mu) + Rn(n_1 - n\lambda)\} \\
&\quad + \zeta_1\{Rn(n_2 - n\mu) + Pn(n_1 - n\lambda)\} - H\zeta_2 - C\zeta_1 \\
&\quad + \zeta\{Q(n_2 - n\mu)^2 + 2R(n_2 - n\mu)(n_1 - n\lambda) + P(n_1 - n\lambda)^2\} - H_2\zeta - C_1\zeta \\
&\quad + \eta_2\{Qm(n_2 - n\mu) + Rm(n_1 - n\lambda)\} \\
&\quad + \eta_1\{Rm(n_2 - n\mu) + Pm(n_1 - n\lambda)\} - f\eta_2 - a\eta_1 + \rho_1\eta_2 - \rho_2\eta_1 \\
&\quad + \eta\{Q(n_2 - n\mu)(m_2 - m\mu) + P(n_1 - n\lambda)(m_1 - m\lambda) \\
&\quad\quad + R(n_2 - n\mu)(m_1 - m\lambda) + R(m_2 - m\mu)(n_1 - n\lambda)\} - f_2\eta - a_1\eta \\
&\quad + \xi_2\{Ql(n_2 - n\mu) + Rl(n_1 - n\lambda)\} \\
&\quad + \xi_1\{Rl(n_2 - n\mu) + Pl(n_1 - n\lambda)\} - g\xi_2 - b\xi_1 - \sigma_1'\xi_2 + \sigma_2'\xi_1 \\
&\quad + \xi\{Q(n_2 - n\mu)(l_2 - l\mu) + R(n_1 - n\lambda)(l_1 - l\lambda) \\
&\quad\quad + R(n_2 - n\mu)(l_1 - l\lambda) + R(l_2 - l\mu)(n_1 - n\lambda)\} - g_2\xi - b_1\xi
\end{aligned}$$

33—2

$$= (n_2 - n\mu) \{Q(\theta_2 - \mu\theta) + R(\theta_1 - \lambda\theta)\}$$
$$+ (n_1 - n\lambda) \{R(\theta_2 - \mu\theta) + P(\theta_1 - \lambda\theta)\}$$
$$- \frac{\partial}{\partial u}(b\xi + a\eta + C\zeta) - \frac{\partial}{\partial v}(g\xi + f\eta + H\zeta) + \rho_1\eta_2 - \rho_2\eta_1 - \sigma_1'\xi_2 + \sigma_2'\xi_1$$
$$= - \frac{\partial}{\partial u}(b\xi + a\eta + C\zeta) - \frac{\partial}{\partial v}(g\xi + f\eta + H\zeta) + \rho_1\eta_2 - \rho_2\eta_1 - \sigma_1'\xi_2 + \sigma_2'\xi_1.$$

Hence

$$\frac{\partial \mathbf{T}}{\partial \zeta} - \frac{\partial}{\partial u}\left(\frac{\partial \mathbf{T}}{\partial \zeta_1}\right) - \frac{\partial}{\partial v}\left(\frac{\partial \mathbf{T}}{\partial \zeta_2}\right)$$
$$= \rho_1\eta_2 - \rho_2\eta_1 - \sigma_1'\xi_2 + \sigma_2'\xi_1 - \frac{\partial}{\partial u}(\alpha\eta_2 + \rho_2'\eta - \beta\xi_2 - \sigma_2\xi)$$
$$+ \frac{\partial}{\partial v}(\alpha\eta_1 + \rho_1'\eta - \beta\xi_1 - \sigma_1\xi)$$
$$= 0,$$

because $\rho + \rho' = \alpha$, $\sigma + \sigma' = \beta$. The first of the subsidiary characteristic equations is thus satisfied, and the postulated assumptions are consequently so far verified.

Similarly, the second and the third of the subsidiary equations are satisfied, in connection with the relations and the postulated assumptions as regards $\lambda$ and $\mu$.

Hence all the requirements of the relations are satisfied by means of the values of $\lambda$ and $\mu$, expressed in terms of $\theta$.

**312.** It thus becomes unnecessary to resolve the grave array of equations of transformation in § 310, so far as concerns the construction of a normal form of the second variation, which is the main aim of this section of the investigation.

The same kind of analytical position was attained (§ 163) in the construction of the normal form of the second variation of a single integral, which involves two original dependent variables and their first derivatives. There, it was pointed out that, by the immediate use of the integrals of the subsidiary equations, the two distinct processes in the second chapter could be fused into one process, in so far as the main purpose is the actual construction of the reduced normal form of the second variation.

The same fusion of two separate processes * can be effected in the present case and, indeed, in all the corresponding problems that arise for reduction of second variations to normal forms. All the remarks in § 164, concerning the simpler instance, apply here, *mutatis mutandis*.

---

* In the first of the two memoirs by Kobb already (p. 457) quoted, these separate processes are, in the main, effected: the modification, here made in § 308, is left over, in the memoirs, for use in the second stage of the reduction.

*Discussion of the normal form : the Legendre test.*

**313.** Thus the second variation of the original integral becomes

$$\tfrac{1}{2}\kappa^2 \iint \left\{ P\left( W_1 - \frac{\theta_1}{\theta} W \right)^2 + 2R\left( W_1 - \frac{\theta_1}{\theta} W \right)\left( W_2 - \frac{\theta_2}{\theta} W \right) + Q\left( W_2 - \frac{\theta_2}{\theta} W \right)^2 \right\} du\, dv;$$

and this quantity must have a persistent sign for all non-zero variations. The first variation has been made to vanish and, in doing so, either has fulfilled stated terminal requirements or has settled the elements at the boundary by means of terminal conditions. Hence, at the boundary, all non-zero variations must yield a vanishing $W$, while, at any place not on the boundary, $W$ measures the normal displacement from the characteristic surface resulting from the small variation $\kappa X, \kappa Y, \kappa Z$.

The persistence of sign in the value of the foregoing expression depends upon the coefficients $P, R, Q$, and upon the arguments $W_1 - \frac{\theta_1}{\theta} W$, $W_2 - \frac{\theta_2}{\theta} W$. Whatever range of values these arguments may have, always provided they do not vanish, the expression cannot maintain one sign unless

$$PQ > R^2.$$

If $PQ < R^2$, the expression could be made positive at will, and could be made negative at will—a combined possibility which is fatal to the possession of a maximum or a minimum.

If $PQ = R^2$, all the variations such that

$$P\left( W_1 - \frac{\theta_1}{\theta} W \right) + R\left( W_2 - \frac{\theta_2}{\theta} W \right) = 0$$

would make the second variation vanish. The third variation in the complete variation of the original integral would then have to vanish for that particular range of quantities $W$; and the fourth variation, for the same range of quantities $W$, would then have to be examined, in order to determine whether it has a permanent sign. As in other instances, the investigation is left for any special case, if and when it occurs.

Accordingly, we are left with the sole possibility

$$PQ > R^2.$$

The fulfilment of this condition over the whole surface cannot be effected, unless $P$ and $Q$ nowhere vanish and everywhere have the same sign. If the common sign be positive, it allows the expression to be positive and only positive; if that sign be negative, it allows the expression to be negative and only negative. We thus have the second test—the extension of the *Legendre* test—which must be satisfied for the possession of a maximum or a minimum: *the quantity*

$$PQ - R^2$$

*is not to vanish, and it must be positive everywhere along the surface. If $P$ and $Q$ be positive, a minimum is admissible: if $P$ and $Q$ be negative, a maximum is admissible.*

*The Jacobi test.*

**314.** In order that the second variation may have a persistent sign for all non-zero variations, it is necessary that the arguments in $\iint \Phi \, du \, dv$ shall not vanish; consequently, it must be impossible to have non-zero variations such that

$$W_1 - \frac{\theta_1}{\theta} W = 0, \quad W_2 - \frac{\theta_2}{\theta} W = 0,$$

that is, such that

$$W = \nu\theta,$$

where $\nu$ is a constant. Now $W$ is the normal displacement due to an arbitrary small variation, which now is required to vanish everywhere on the boundary; and $\theta$ is the normal displacement due to a small variation from one characteristic surface to another, or—as it may be stated—due to a small characteristic variation. If it were possible to have a relation $W = \nu\theta$, the possibility would require that $\theta$ should vanish everywhere on the whole boundary. This requirement could be met by drawing a consecutive characteristic through the lower limit, that is, through the initial curve which constitutes the boundary of the surface, and by assuming that the field of integration extends so far as to reach the first intersection of the two characteristic surfaces after the initial curve. That first intersection may be called the *conjugate* of the initial curve. We are precluded from the relation $W = \nu\theta$, and therefore from circumstances that make this relation possible; that is, *the field of integration, limited at one boundary by the initial curve which is common to the central characteristic and a consecutive characteristic, must not extend so far as the conjugate of that initial curve.*

We thus have a third test—the extension of the *Jacobi* test—which must be satisfied for the possession of a maximum or a minimum: it provides the limitation (if any) upon the range of integration with an assigned initial boundary. Further, let $\theta$ be determined—its general value is the value of $l\xi + m\eta + n\zeta$, where $\xi, \eta, \zeta$, have their primitive values as given in §§ 307, 308 —so as to vanish along the initial curve; the next nearest curve, along which $\theta$ everywhere vanishes, provides the conjugate of that initial curve and therefore provides the upper range to which the range of integration is forbidden to extend.

*Ex.* 1. We have seen that every minimal surface is given by the equations

$$x = \tfrac{1}{2}(1 - u^2)f''(u) + uf'(u) - f(u) + \tfrac{1}{2}(1 - v^2)g''(v) + vg'(v) - g(v),$$
$$y = i\{\tfrac{1}{2}(1 + u^2)f''(u) - uf'(u) + f(u)\} - i\{\tfrac{1}{2}(1 + v^2)g''(v) - vg'(v) + g(v)\},$$
$$z = uf''(u) - f'(u) + vg''(v) - g'(v),$$

where $f(u)$ and $g(v)$ are conjugate arbitrary functions in the conjugate variables $u$ and $v$ because $x, y, z$, are supposed real.

Various surfaces occur, according to the form adopted for $f(u)$ and $g(v)$. When

$$f'''(u) = \frac{c}{u^2}, \quad g'''(v) = \frac{c}{v^2},$$

and constants additive to $x$, $y$, $z$, are neglected as affecting merely position relative to coordinate axes, we find

$$\frac{x}{c} = -\tfrac{1}{2} \left( u + \frac{1}{u} \right) - \tfrac{1}{2} \left( v + \frac{1}{v} \right),$$

$$\frac{y}{c} = \tfrac{1}{2} i \left( u - \frac{1}{u} \right) - \tfrac{1}{2} i \left( v - \frac{1}{v} \right),$$

$$\frac{z}{c} = \log u + \log v \; ;$$

and therefore

$$(x^2 + y^2)^{\frac{1}{2}} = 2c \cosh \frac{x}{2c},$$

so that the surface is a catenoid*.

The *Legendre* test does not apply in the exact form as given in the text; for we have there assumed that the independent variables are real. But it has already been applied (§ 303) to all minimal surfaces, referred to $u$ and $v$ as variables ; and it is known to be satisfied always.

For the *Jacobi* test, we need the values of $\xi$, $\eta$, $\zeta$, $l$, $m$, $n$. By direct substitution, we find

$$\left.\begin{aligned}
l &= y_1 z_2 - z_1 y_2 = \frac{ic^2}{2u^2 v^2} (u + v)(1 + uv) \\
m &= z_1 x_2 - x_1 z_2 = -\frac{c^2}{2u^2 v^2} (v - u)(1 + uv) \\
n &= x_1 y_2 - y_1 x_2 = -\frac{ic^2}{2u^2 v^2} (1 - uv)(1 + uv)
\end{aligned}\right\}.$$

If $\mathbf{X}$, $\mathbf{Y}$, $\mathbf{Z}$, denote the direction-cosines to the normal to the surface, we have

$$\mathbf{X} = \frac{u + v}{1 + uv}, \quad \mathbf{Y} = i\,\frac{v - u}{1 + uv}, \quad \mathbf{Z} = \frac{uv - 1}{1 + uv}.$$

(It is easy to verify that these expressions for $\mathbf{X}$, $\mathbf{Y}$, $\mathbf{Z}$, hold for all minimal surfaces and not merely for the catenoid.)

As we are dealing with the catenoid only as an example, we shall use the Jacobi test to find the conjugate of a latitudinal section given by $z = $ constant, say by

$$uv = a^2,$$

where, of course, $a$ is a real constant. Having regard to the values of $x$, $y$, $z$, we may assume the variations $\kappa\xi$, $\kappa\eta$, $\kappa\zeta$, to be such that

$$\xi = au + \frac{\beta}{u} + av + \frac{\beta}{v} + \rho,$$

$$\eta = \gamma u + \frac{\delta}{u} + \gamma v + \frac{\delta}{v} + \sigma,$$

$$\zeta = \epsilon \log u + \epsilon \log v + \tau,$$

which satisfy the subsidiary characteristic equations, $a$, $\beta$, $\gamma$, $\delta$, $\epsilon$, $\rho$, $\sigma$, $\tau$, being constants.

These quantities $\xi$, $\eta$, $\zeta$, have also (§ 307) to satisfy the relations

$$x_1\xi_1 + y_1\eta_1 + z_1\zeta_1 = 0, \quad x_2\xi_2 + y_2\eta_2 + z_2\zeta_2 = 0.$$

* A different method, by proceeding from the Cauchy primitive of the characteristic equation, leads to this surface; see § 304, Ex. 1.

Both relations are satisfied identically, if

$$a - i\gamma = 0,$$
$$\beta + i\delta = 0,$$
$$a + \beta + i\gamma - i\delta - 2\epsilon = 0;$$

and therefore we take

$$\gamma = -ia, \quad \delta = i\beta, \quad \epsilon = a + \beta,$$

so that, now,

$$\xi = \left(a + \frac{\beta}{uv}\right)(u + v) + \rho,$$

$$\eta = -i\left(a + \frac{\beta}{uv}\right)(u - v) + \sigma,$$

$$\zeta = (a + \beta)\log uv + \tau.$$

Along the initial latitudinal curve $uv = a^2$, we must have $\xi = 0$, $\eta = 0$, $\zeta = 0$; hence

$$a + \frac{\beta}{a^2} = 0, \quad \rho = 0, \quad \sigma = 0, \quad \tau + (a + \beta)\log a^2 = 0.$$

We satisfy all requirements by taking

$$a = \tfrac{1}{2}\frac{\mu}{a}, \quad \beta = -\tfrac{1}{2}\mu a, \quad \tau = -\tfrac{1}{2}\mu\left(\frac{1}{a} - a\right)\log a^2;$$

and so

$$\xi = \tfrac{1}{2}\mu\frac{uv - a^2}{auv}(u + v),$$

$$\eta = -\tfrac{1}{2}\mu\frac{uv - a^2}{auv}(u - v),$$

$$\zeta = \tfrac{1}{2}\mu\left(\frac{1}{a} - a\right)\log\frac{uv}{a^2}.$$

The critical quantity $\theta$ is given by

$$\theta = l\xi + m\eta + n\zeta$$
$$= \frac{i\mu c^2}{4au^2v^2}(1 + uv)\left\{4(uv - a^2) + (a^2 - 1)(uv - 1)\log\frac{uv}{a^2}\right\}.$$

It manifestly vanishes when $uv = a^2$, which is the initial curve. The quantity $1 + uv$ is essentially positive; hence when $\theta$ again vanishes, it is for a value $r^2$ of $uv$, where

$$4(r^2 - a^2) + (a^2 - 1)(r^2 - 1)\log\frac{r^2}{a^2} = 0,$$

that is, for a value of $r$ such that

$$\frac{2}{r^2 - 1} - \log r = \frac{2}{a^2 - 1} - \log a.$$

This equation means that the tangents to the meridian catenary at the latitudinal parallels $uv = a^2$ and $uv = r^2$ meet in the directrix: and thus the conjugate curve is found, in agreement with the former result (§ 289, Ex. 2).

*Ex.* 2. Defining the line of curvature on a surface as a line along which consecutive normals to the surface intersect, prove that the lines of curvature on the general minimal surface are given by the equation

$$f'''(u)\,du^2 - g'''(v)\,dv^2 = 0;$$

and verify the known property that they are orthogonal curves.

*Ex.* 3. A minimal surface is given by taking

$$f'''(u) = \frac{a}{u^3}(1 + u^2), \quad g'''(v) = \frac{a}{v^3}(1 + v^2).$$

Prove that the surface is periodic; that it has only one side; and that the sections of the surface by the planes

$$y = m\pi a,$$

where $m$ is an integer, lie on the parabolic cylinder

$$z^2 = 8a\,(x + a).$$

*General expression for small variation to a consecutive minimal surface.*

**315.** The determination of the effective range of a minimal surface requires a knowledge of the quantity

$$\theta = l\xi + m\eta + n\zeta.$$

The primitive of the subsidiary characteristic equations has been stated: but it contains two arbitrary functions, and the whole expression of the primitive has to be made subject to the conditions

$$x_1\xi_1 + y_1\eta_1 + z_1\zeta_1 = 0, \quad x_2\xi_2 + y_2\eta_2 + z_2\zeta_2 = 0,$$

so that the new arbitrary functions must be limited in terms of those which have occurred in the characteristic function; and they must satisfy the requirements along the initial curve.

Instead of determining the new arbitrary functions in this manner, we may proceed to deduce a characteristic small variation from Schwarz's expression of the primitive satisfying initial conditions (§ 304). A consecutive characteristic surface through the initial curve is given by making a small arbitrary continuous variation in the directions of the normals to the surface along that curve. Denoting the direction-cosines of the normal at any point by $\mathbf{X}$, $\mathbf{Y}$, $\mathbf{Z}$, we effect this variation by changing them into $\mathbf{X} + \kappa\mathbf{X}'$, $\mathbf{Y} + \kappa\mathbf{Y}'$, $\mathbf{Z} + \kappa\mathbf{Z}'$. Any displacement $dx, dy, dz$ along the initial curve on the characteristic surface satisfies the equation

$$\mathbf{X}\,dx + \mathbf{Y}\,dy + \mathbf{Z}\,dz = 0.$$

As that initial curve lies also on the consecutive characteristic (which, of course, is drawn through the curve), we have

$$(\mathbf{X} + \kappa\mathbf{X}')\,dx + (\mathbf{Y} + \kappa\mathbf{Y}')\,dy + (\mathbf{Z} + \kappa\mathbf{Z}')\,dz = 0;$$

and therefore

$$\mathbf{X}'dx + \mathbf{Y}'dy + \mathbf{Z}'dz = 0.$$

Along the initial curve, $x, y, z$, are functions of a parameter $\mu$, as used in § 304; hence

$$\mathbf{X}'\frac{dx}{d\mu} + \mathbf{Y}'\frac{dy}{d\mu} + \mathbf{Z}'\frac{dz}{d\mu} = 0.$$

Now we can take

$$\mathbf{X} = \sin\omega\cos\chi, \quad \mathbf{Y} = \sin\omega\sin\chi, \quad \mathbf{Z} = \cos\omega,$$

so that, as $\mathbf{X}$, $\mathbf{Y}$, $\mathbf{Z}$, are known along the curve, because

$$\frac{\mathbf{X}}{y_1 z_2 - z_1 y_2} = \frac{\mathbf{Y}}{z_1 x_2 - x_1 z_2} = \frac{\mathbf{Z}}{x_1 y_2 - y_1 x_2} = \frac{1}{V},$$

$\omega$ and $\chi$ are also known along the curve. When small variations are effected in the direction of a normal to the characteristic surface, so as to lead to the consecutive characteristic, let $\omega$ and $\chi$ become $\omega + \kappa\omega'$ and $\chi + \kappa\chi'$, where $\omega'$ and $\chi'$ are finite quantities, being regular functions of $\mu$ when they are variable. Thus

$$\mathbf{X}' = \omega' \cos \omega \cos \chi - \chi' \sin \omega \sin \chi,$$
$$\mathbf{Y}' = \omega' \cos \omega \sin \chi + \chi' \sin \omega \cos \chi,$$
$$\mathbf{Z}' = - \omega' \sin \omega ;$$

and therefore, writing

$$U = \frac{dx}{d\mu} \cos \omega \cos \chi + \frac{dy}{d\mu} \cos \omega \sin \chi - \frac{dz}{d\mu} \sin \omega,$$
$$V = \frac{dx}{d\mu} \sin \omega \sin \chi - \frac{dy}{d\mu} \sin \omega \cos \chi = \mathbf{Y} \frac{dx}{d\mu} - \mathbf{X} \frac{dy}{d\mu},$$

we obtain

$$U\omega' - V\chi' = 0,$$

as the condition for the consecutive characteristic. Hence the small variations in the direction-cosines of the normals are given by

$$\begin{aligned} \mathbf{X}' &= \omega' \left( \cos \omega \cos \chi - \frac{U}{V} \sin \omega \sin \chi \right) \\ \mathbf{Y}' &= \omega' \left( \cos \omega \sin \chi + \frac{U}{V} \sin \omega \cos \chi \right) \\ \mathbf{Z}' &= - \omega' \sin \omega \end{aligned} \Bigg\}' ,$$

where $\omega'$ is arbitrarily at our disposal, constant or variable : though, if variable, $\omega'$ must be a regular function of $\mu$. On substitution, we have

$$\begin{aligned} V\mathbf{X}' &= \omega' \left( \mathbf{Y} \frac{dz}{d\mu} - \mathbf{Z} \frac{dy}{d\mu} \right) \sin \omega \\ V\mathbf{Y}' &= \omega' \left( \mathbf{Z} \frac{dx}{d\mu} - \mathbf{X} \frac{dz}{d\mu} \right) \sin \omega \\ V\mathbf{Z}' &= \omega' \left( \mathbf{X} \frac{dy}{d\mu} - \mathbf{Y} \frac{dx}{d\mu} \right) \sin \omega \end{aligned} \Bigg\} .$$

Let $\epsilon$ denote the small angle between the normals to consecutive characteristics with direction-cosines $\mathbf{X}$, $\mathbf{Y}$, $\mathbf{Z}$, and $\mathbf{X} + \kappa\mathbf{X}'$, $\mathbf{Y} + \kappa\mathbf{Y}'$, $\mathbf{Z} + \kappa\mathbf{Z}'$, so that

$$\epsilon^2 = \kappa^2 \{ (\mathbf{YZ}' - \mathbf{ZY}')^2 + (\mathbf{ZX}' - \mathbf{XZ}')^2 + (\mathbf{XY}' - \mathbf{YX}')^2 \}.$$

We have

$$\begin{aligned} \mathbf{YZ}' - \mathbf{ZY}' &= \frac{\omega' \sin \omega}{V} \left( \mathbf{XY} \frac{dy}{d\mu} - \mathbf{Y}^2 \frac{dx}{d\mu} - \mathbf{Z}^2 \frac{dx}{d\mu} + \mathbf{XZ} \frac{dz}{d\mu} \right) \\ &= \frac{\omega' \sin \omega}{V} \left\{ \mathbf{X} \left( \mathbf{X} \frac{dx}{d\mu} + \mathbf{Y} \frac{dy}{d\mu} + \mathbf{Z} \frac{dz}{d\mu} \right) - (\mathbf{X}^2 + \mathbf{Y}^2 + \mathbf{Z}^2) \frac{dx}{d\mu} \right\} \\ &= - \frac{\omega' \sin \omega}{V} \frac{dx}{d\mu}, \end{aligned}$$

and similarly for the others. Hence

$$\epsilon = -\kappa \frac{\omega' \sin \omega}{V} \frac{ds}{d\mu}.$$

Substituting for $\omega' \sin \omega$, we find

$$\kappa \mathbf{X}' = -\epsilon \frac{d\mu}{ds} \left( \mathbf{Y} \frac{dz}{d\mu} - \mathbf{Z} \frac{dy}{d\mu} \right)$$
$$\kappa \mathbf{Y}' = -\epsilon \frac{d\mu}{ds} \left( \mathbf{Z} \frac{dx}{d\mu} - \mathbf{X} \frac{dz}{d\mu} \right) \Bigg\},$$
$$\kappa \mathbf{Z}' = -\epsilon \frac{d\mu}{ds} \left( \mathbf{X} \frac{dy}{d\mu} - \mathbf{Y} \frac{dx}{d\mu} \right)$$

where $ds$ here denotes the element of arc along the initial curve with the current parameter $\mu$, and $\epsilon$ is the small angle between the normals to the central characteristic surface and the consecutive characteristic surface at the point $x$, $y$, $z$, on the common initial curve.

The equations can also be expressed in the form

$$\kappa \left( \mathbf{Y}' \frac{dz}{d\mu} - \mathbf{Z}' \frac{dy}{d\mu} \right) = \epsilon \mathbf{X} \frac{ds}{d\mu}$$
$$\kappa \left( \mathbf{Z}' \frac{dx}{d\mu} - \mathbf{X}' \frac{dz}{d\mu} \right) = \epsilon \mathbf{Y} \frac{ds}{d\mu} \Bigg\},$$
$$\kappa \left( \mathbf{X}' \frac{dy}{d\mu} - \mathbf{Y}' \frac{dx}{d\mu} \right) = \epsilon \mathbf{Z} \frac{ds}{d\mu}$$

valid along the initial curve.

### Equations of the consecutive characteristic.

**316.** We have

$$2x = x(\rho) + x(\sigma) - i \int_\sigma^\rho (\mathbf{Y} \, dz - \mathbf{Z} \, dy),$$

$$2y = y(\rho) + y(\sigma) - i \int_\sigma^\rho (\mathbf{Z} \, dx - \mathbf{X} \, dz),$$

$$2z = z(\rho) + z(\sigma) - i \int_\sigma^\rho (\mathbf{X} \, dy - \mathbf{Y} \, dx).$$

The effect of a small variation is to change $\mathbf{X}$, $\mathbf{Y}$, $\mathbf{Z}$, into $\mathbf{X} + \kappa\mathbf{X}'$, $\mathbf{Y} + \kappa\mathbf{Y}'$, $\mathbf{Z} + \kappa\mathbf{Z}'$. But the consecutive characteristic surface passes through the same initial curve; and therefore $x(\rho)$, $x(\sigma)$, $y(\rho)$, $y(\sigma)$, $z(\rho)$, $z(\sigma)$, are left unchanged by the inclination of the normals to the surfaces at any point on that curve. The changes of $x$, $y$, $z$, consequent on the variation, are denoted by $\kappa\xi$, $\kappa\eta$, $\kappa\zeta$; hence

$$2\kappa\xi = -i\kappa \int_\sigma^\rho (\mathbf{Y}' dz - \mathbf{Z}' dy),$$

$$2\kappa\eta = -i\kappa \int_\sigma^\rho (\mathbf{Z}' dx - \mathbf{X}' dz),$$

$$2\kappa\zeta = -i\kappa \int_\sigma^\rho (\mathbf{X}' dy - \mathbf{Y}' dx).$$

Now along the initial curve,

$$\kappa \left( \mathbf{Y}' \frac{dz}{d\mu} - \mathbf{Z}' \frac{dy}{d\mu} \right) = \epsilon \mathbf{X} \frac{ds}{d\mu} ;$$

and therefore

$$\kappa \int (\mathbf{Y}' dz - \mathbf{Z}' dy) = \int \epsilon \mathbf{X} ds$$

$$= \epsilon \int \mathbf{X} ds,$$

if $\epsilon$ be a constant. When the values of $\mathbf{X}$ and $ds$ in terms of $\mu$ are substituted, let $\int \mathbf{X} ds = L(\mu)$; then

$$2\kappa\xi = -i\epsilon \{ L(\rho) - L(\sigma) \}.$$

Similarly for $\eta$ and $\zeta$. We therefore have the result:

*Let*

$$\int \mathbf{X} ds = L(\mu), \quad \int \mathbf{Y} ds = M(\mu), \quad \int \mathbf{Z} ds = N(\mu),$$

*these integrals being taken along the initial curve, and let $\epsilon$ be a constant; then the components $\kappa\xi$, $\kappa\eta$, $\kappa\zeta$, of a small characteristic variation are given by*

$$2\kappa\xi = -i\epsilon \{ L(\rho) - L(\sigma) \}, \quad 2\kappa\eta = -i\epsilon \{ M(\rho) - M(\sigma) \}, \quad 2\kappa\zeta = -i\epsilon \{ N(\rho) - N(\sigma) \},$$

*$\epsilon$ being the small constant angle between the normals to the characteristic surface and the consecutive characteristic.*

Ex. 1. *Find the conjugate of a circle of latitude on a catenoid.*

The circle of latitude is a line of curvature on the catenoid; as $\epsilon$ is constant, that circle is also a line of curvature on the consecutive surface (by a theorem due to Joachimstahl).

With the notation of Ex. 1, § 304, we have, along the circle,

$$x = c \cos\mu \quad , \quad y = c \sin\mu \quad , \quad z = c \cot a,$$
$$\mathbf{X} = -\cos\mu \cos a, \quad \mathbf{Y} = -\sin\mu \cos a, \quad \mathbf{Z} = \sin a.$$

Also, along the circle, $ds = c \, d\mu$: hence

$$\int \mathbf{X} ds = -\int c \cos\mu \cos a \, d\mu = -c \sin\mu \cos a,$$

$$\int \mathbf{Y} ds = -\int c \sin\mu \cos a \, d\mu = \quad c \cos\mu \cos a,$$

$$\int \mathbf{Z} ds = \quad \int c \sin a \, d\mu \quad = \quad c\mu \sin a ;$$

and therefore

$$2\kappa\xi = i c \epsilon \cos a (\sin\rho - \sin\sigma), \quad 2\kappa\eta = -i c \epsilon \cos a (\cos\rho - \cos\sigma), \quad 2\kappa\zeta = -i c \epsilon \sin a (\rho - \sigma).$$

Now

$$-u = \frac{1 + \sin a}{\cos a} e^{\rho i}, \quad -v = \frac{1 + \sin a}{\cos a} e^{-\sigma i},$$

or, writing

$$a = \frac{1 + \sin a}{\cos a}, \quad \frac{1}{a} = \frac{1 - \sin a}{\cos a},$$

we have

$$-u = a e^{\rho i}, \quad -v = a e^{-\sigma i}.$$

Hence

$$
\begin{aligned}
2\kappa\xi &= \frac{c\epsilon \cos a}{2a\,uv}(u+v)(a^2-uv) \\
2\kappa\eta &= \frac{ic\epsilon \cos a}{2a\,uv}(v-u)(a^2-uv) \\
2\kappa\zeta &= -\tfrac{1}{2}c\epsilon \cos a\left(a-\frac{1}{a}\right)\log\frac{uv}{a^2}
\end{aligned}\;\Bigg\}.
$$

The critical quantity, which leads to the determination of the conjugate of the initial curve, is

$$l\xi + m\eta + n\zeta,$$

that is, on multiplication by $(l^2+m^2+n^2)^{-\frac{1}{2}}$,

$$\mathbf{X}\xi + \mathbf{Y}\eta + \mathbf{Z}\zeta.$$

But

$$\mathbf{X} = \frac{u+v}{1+uv}, \quad \mathbf{Y} = i\frac{v-u}{1+uv}, \quad \mathbf{Z} = \frac{uv-1}{uv+1};$$

and therefore

$$\frac{4\kappa a}{c\epsilon \cos a}(\mathbf{X}\xi + \mathbf{Y}\eta + \mathbf{Z}\zeta) = -\left\{4(uv-a^2)+(a^2-1)(uv-1)\log\frac{uv}{a^2}\right\},$$

which can be taken as the critical magnitude.

This magnitude manifestly vanishes when $uv=a^2$, that is, along the initial circle of latitude. The only other range, where it can vanish, is given by a value $uv=r^2$, that is, another circle of latitude; and the value of $r$ is given by the equation

$$2(r^2-a^2)+(a^2-1)(r^2-1)\log\frac{r}{a}=0,$$

that is, by the equation

$$\frac{2}{r^2-1}-\log r = \frac{2}{a^2-1}-\log a,$$

where $r$ and $a$ are positive quantities. The equation is the customary expression of the property that the tangents to the meridian at the points, where it is met by the conjugate parallels of latitude, intersect in the directrix.

*Ex.* 2. Consider Enneper's surface[*], for which
the parametric equations are

$$f(u)=u^3, \quad g=v^3;$$

$$
\begin{aligned}
x &= 3u-u^3 + 3v-v^3 \\
y &= i(3u+u^3)-i(3v+v^3) \\
z &= 3u^2 + 3v^2
\end{aligned}\;\Bigg\}.
$$

The lines of curvature (§ 314, Ex. 2) are

$$du^2-dv^2=0:$$

that is, they are

$$u+v=\text{constant}, \quad u-v=\text{constant}.$$

Let

$$u=a+i\beta, \quad v=a-i\beta,$$

$a$ and $\beta$ being real; the lines of curvature are $a=\text{constant}$, $\beta=\text{constant}$. For these values of $u$ and $v$, we have

$$x=6a+6a\beta^2-2a^3, \quad y=-6\beta-6a^2\beta+2\beta^3, \quad z=6a^2-6\beta^2;$$

---

[*] *Zeitschr. f. Math. u. Physik*, t. ix (1864), p. 108. In connection with this example and the preceding example, as well as for a partial discussion of the conjugate of an initial curve on any minimal surface, see a paper by the author, *Annali di Mat.*, Ser. iii, t. xxi (1913), pp. 121—142.

the surface is unicursal and of degree nine. Also, for these values of $u$ and $v$, we have

$$\frac{\mathbf{X}}{2a} = \frac{\mathbf{Y}}{2\beta} = \frac{\mathbf{Z}}{a^2 + \beta^2 - 1} = \frac{1}{a^2 + \beta^2 - 1}.$$

Consider the surface in relation to a line of curvature

$$u + v = 2c,$$

so that the equation

$$a = c$$

defines the line: then $\beta$ may be taken as a current parameter along the line. We proceed to find the conjugate of that line of curvature on the Enneper surface.

Along the line, we have

$$x\,(\beta) = 6c - 2c^3 + 6c\beta^2, \quad y\,(\beta) = -6\beta - 6c^2\beta + 2\beta^3, \quad z\,(\beta) = 6c^2 - 6\beta^2;$$

and therefore

$$dx = 12c\beta\,d\beta, \quad dy = -(6 + 6c^2 - 6\beta^2)\,d\beta, \quad dz = -12\beta\,d\beta.$$

The value of $ds$, which will be required later, is

$$ds = 6\,(1 + c^2 + \beta^2)\,d\beta.$$

To obtain the expressions of $x, y, z$, shewing the line of curvature as an initial curve, we have

$$\mathbf{Y}\,dz - \mathbf{Z}\,dy = 6\,(c^2 - 1 - \beta^2)\,d\beta,$$
$$\mathbf{Z}\,dx - \mathbf{X}\,dz = 12c\beta\,d\beta,$$
$$\mathbf{X}\,dy - \mathbf{Y}\,dx = -12c\,d\beta,$$

on reduction; hence, by § 304, we have

$$x = \tfrac{1}{2}x\,(\rho) + \tfrac{1}{2}x\,(\sigma) - \tfrac{1}{2}i\int_\sigma^\rho (\mathbf{Y}\,dz - \mathbf{Z}\,dy)$$
$$= 6c - 2c^3 + 3c\,(\rho^2 + \sigma^2) - 3i\,(c^2 - 1)\,(\rho - \sigma) + i\,(\rho^3 - \sigma^3),$$

$$y = \tfrac{1}{2}y\,(\rho) + \tfrac{1}{2}y\,(\sigma) - \tfrac{1}{2}i\int_\sigma^\rho (\mathbf{Z}\,dx - \mathbf{X}\,dz)$$
$$= -3\,(c^2 + 1)\,(\rho + \sigma) - 3ic\,(\rho^2 - \sigma^2) + \rho^3 + \sigma^3,$$

$$z = \tfrac{1}{2}z\,(\rho) + \tfrac{1}{2}z\,(\sigma) - \tfrac{1}{2}i\int_\sigma^\rho (\mathbf{X}\,dy - \mathbf{Y}\,dx)$$
$$= 6c^2 + 6ic\,(\rho - \sigma) - 3\,(\rho^2 + \sigma^2).$$

To compare $\rho$ and $\sigma$ with $u$ and $v$, we have

$$-u\,\frac{\partial z}{\partial \rho} = \frac{\partial x}{\partial \rho} + i\,\frac{\partial y}{\partial \rho}, \quad -v\,\frac{\partial z}{\partial \sigma} = \frac{\partial x}{\partial \sigma} - i\,\frac{\partial y}{\partial \sigma};$$

and therefore

$$u = c + i\rho, \quad v = c - i\sigma.$$

Again, we need the quantities $\xi, \eta, \zeta$. To determine them, we have

$$\int \mathbf{X}\,ds = \int^\mu 12c\,d\beta = 12c\mu,$$

$$\int \mathbf{Y}\,ds = \int^\mu 12\beta\,d\beta = 6\mu^2,$$

$$\int \mathbf{Z}\,ds = 6\int^\mu (c^2 - 1 + \beta^2)\,d\beta = 6\,(c^2 - 1)\,\mu + 2\mu^3;$$

and therefore, by the equations which determine a consecutive characteristic

$$2\kappa\xi = -12i\epsilon c\,(\rho - \sigma),$$
$$2\kappa\eta = -6i\epsilon\,(\rho^2 - \sigma^2),$$
$$2\kappa\zeta = -i\epsilon\,\{6\,(c^2 - 1)\,(\rho - \sigma) + 2\,(\rho^3 - \sigma^3)\}.$$

Changing from $\rho$ and $\sigma$ to $u$ and $v$, these give

$$\kappa\xi = -6\epsilon c\,(u+v-2c),$$
$$\kappa\eta = -3\epsilon\,(u+v-2c)\,(v-u)\,i,$$
$$\kappa\zeta = -\epsilon\,(u+v-2c)\,\{2c^2-1+c\,(u+v)+uv-u^2-v^2\}.$$

Consequently,

$$-\frac{\kappa}{\epsilon}\,\frac{\mathbf{X}\xi+\mathbf{Y}\eta+\mathbf{Z}\zeta}{u+v-2c}\,(1+uv)$$
$$= 6c\,(u+v)-3\,(v-u)^2+\{2c^2-1+c\,(u+v)+uv-u^2-v^2\}\,(uv-1)$$
$$= T\,(u,\,v).$$

Thus the conjugate of the initial line of curvature

$$u+v-2c=0$$

is the nearest curve on the surface given by the equation

$$T\,(u,\,v)=0,$$

which therefore constitutes the conjugate of the initial curve.

The two curves would meet where

$$3uv = -4 \pm \sqrt{13}.$$

The right-hand side is negative, for both signs; as $u$ and $v$ are conjugate complex variables, $uv$ is positive; and so the equation is impossible. Thus the two curves do not meet, and there is an unrestricted length of the initial line of curvature, while its conjugate is definite and unrestricted.

# CHAPTER X.

*Strong variations of double integrals: constituent element
for surface variations.*

**317.** Now that the tests for the weak (or regular) variations have been constructed, it is necessary to consider the possibility of strong (or non-regular) variations. When plane curves or twisted curves were under consideration, there was the necessity (in even the simplest instances, where derivatives of only the first order occurred) of considering only the simplest type of strong variations. The same necessity arises now, when we deal with surfaces; and the simplest type of strong variation is an even more restricted selection from among all the possibilities than it is in the case of curves.

In the first place, we take any curve $C$ upon the characteristic surface. But we make a provisional limitation, by choosing only a regular curve: that is, a curve without any discontinuities in direction on the surface and without any discontinuities in its main curvatures of flexion and torsion. Through this curve we draw a regular surface $\Sigma$, cutting the characteristic surface at any angle which can vary from point to point along the curve. Through the initial curve determining the characteristic surface, we draw a consecutive characteristic; that is, a new characteristic surface such that its tangent planes along the initial curve belong to a developable slightly varied from that which has determined the central characteristic. This consecutive characteristic will cut, in a curve $C'$, the regular surface $\Sigma$ drawn through $C$. We thus shall have

(i)    a part $A$ of the central characteristic surface, bounded by the initial curve and extending to the curve $C$;

(ii)    a part $A'$ of the consecutive characteristic surface, bounded by the same initial curve and extending to the curve $C'$; and

(iii)    a part $\Delta$ of the regular surface $\Sigma$, lying between the two curves $C$ and $C'$. These curves $C$ and $C'$ will be supposed not to meet: that is, we suppose our curve $C$ on the central characteristic not to be chosen near the conjugate of the original initial curve.

Then we shall consider the region $A'$ extending up to $C'$ on the consecutive characteristic, combined with the region $\Delta$ extending on the surface $\Sigma$ from the curve $C'$ to the curve $C$, as a variation of the region $A$ extending up to $C$ on the original characteristic*. Denoting the general integral by $I$, where

$$I = \iint \Omega \left(x, y, z, x_1, y_1, z_1, x_2, y_2, z_2\right) du\, dv,$$

we proceed to obtain an expression for

$$I_{A'} + I_{\Delta} - I_{A},$$

which represents the increment of the integral for the whole variation indicated.

*Strong variation of integral: the characteristic component.*

**318.** Take any elementary arc $PQ$ on the curve $C$. Let $P'Q'$ be the corresponding elementary arc on the curve $C'$, so that the point $P$ is to be regarded as displaced to $P'$; and similarly for all the points on $PQ$ in its dis-

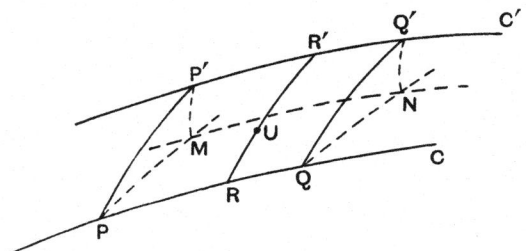

placement to $P'Q'$. Let $P'M$ and $Q'N$ be the normals from $P'$ and $Q'$ to the central characteristic; and let $PM$, $QN$, be the sections of that surface by the planes $PP'M$, $QQ'N$, respectively. We thus have $PMNQ$ as an element of the area of the central characteristic, where the curve $C$ limits the part $A$. We have $PP'Q'Q$ as an element of the strip of area of the regular surface $\Sigma$. The whole part $A'$ of the consecutive characteristic (beginning with the initial curve, and not otherwise outlined in the figure) is limited by the curve $C'$.

We denote, by $s$, the arc-length along the curve $C$ represented by $\ldots PQ\ldots$; and, by $\kappa\xi$, $\kappa\eta$, $\kappa\zeta$, the coordinates of $P'$ relative to $P$. Let $D$ be the length of the small arc $PP'$ (which, to the first order, may be taken straight); and let $\lambda$, $\mu$, $\nu$, be the direction-cosines of $P'P$ in the direction from $P'$ to $P$ (so that $\lambda$, $\mu$, $\nu$, represent an arbitrary direction at $P$ off the characteristic surface). Then

$$\kappa\xi = -\lambda D, \quad \kappa\eta = -\mu D, \quad \kappa\zeta = -\nu D.$$

We regard the whole part $A'$ of the consecutive characteristic, between $C'$ and the initial curve, as a variation of the whole part $A$ of the central characteristic, between $C$ and the initial curve. This variation is obtained by

---

* This, in essence, is the small (strong) variation considered by Kobb, *Acta Math.*, t. xvi (1892), in particular, pp. 116—140, in the memoir already quoted.

means of the weak variations $\kappa X$, $\kappa Y$, $\kappa Z$; thus $X = 0$, $Y = 0$, $Z = 0$, along the initial curve, and $X = \xi$, $Y = \eta$, $Z = \zeta$, along $C$. Then, as in § 293, we have

$$I_{A'} - I_A = \left[ \kappa \int \left\{ \left( X \frac{\partial \Omega}{\partial x_1} + Y \frac{\partial \Omega}{\partial y_1} + Z \frac{\partial \Omega}{\partial z_1} \right) dv - \left( X \frac{\partial \Omega}{\partial x_2} + Y \frac{\partial \Omega}{\partial y_2} + Z \frac{\partial \Omega}{\partial z_2} \right) du \right\} \right],$$

taken between the limits, which are the curve $C$ as the upper limit and the initial curve as the lower limit, while the line-integral is taken positively along these curves; the double integral, which occurs in $I_{A'} - I_A$, vanishes because the characteristic equations* are satisfied along $A$. Along the initial curve, the portion contributed is zero, because $X = 0$, $Y = 0$, $Z = 0$, everywhere on that curve. Along the curve $C$, we have $X = \xi$, $Y = \eta$, $Z = \zeta$. Also, along $C$, we take

$$dv = \frac{dv}{ds} ds, \quad du = \frac{du}{ds} ds,$$

$ds$ being measured positively; hence

$$I_{A'} - I_A = - \int D \left\{ \left( \lambda \frac{\partial \Omega}{\partial x_1} + \mu \frac{\partial \Omega}{\partial y_1} + \nu \frac{\partial \Omega}{\partial z_1} \right) \frac{dv}{ds} - \left( \lambda \frac{\partial \Omega}{\partial x_2} + \mu \frac{\partial \Omega}{\partial y_2} + \nu \frac{\partial \Omega}{\partial z_2} \right) \frac{du}{ds} \right\} ds,$$

where $D$ is not assumed to be constant, and is never zero, because the curves $C$ and $C'$ do not meet.

### Component from arbitrary surface.

**319.** We now require the integral $I_\Delta$, extended over the band of the surface $\Sigma$ between the curves $C$ and $C'$. We shall consider $I_\Delta$ as resolved into the sum of the components, contributed by elements of the band typically represented in the parallelogram $PQQ'P'$. (The quadrilateral is taken to be a parallelogram; the difference between the angles $Q'QC$ and $P'PC$ is negligible, for our purpose: likewise, the difference between the lengths of $P'P$ and $Q'Q$ is negligible.)

The value of the integral over this elementary parallelogram $PQQ'P'$ we shall denote by $I''$, so that

$$I'' = \iint \Omega \, du \, dv,$$

the double integration being taken over the parallelogram. Now any element of area is

$$V \, du \, dv,$$

where $V$, $= (EG - F^2)^{\frac{1}{2}}$, is expressible in terms of the derivatives of $x, y, z$, by the expressions in § 290. Denoting this element of area by $dA$, we have

$$I'' = \iint \frac{\Omega}{V} \, dA.$$

---

* The consecutive surface is made a characteristic, in order to build up (as in § 209) the general strong variation from the constituent elements described in § 322.

When the variables are changed arbitrarily to $u'$ and $v'$ (they will be made specific, almost immediately) as in § 291, and if

$$du' = \alpha \, du + \beta \, dv, \quad dv' = \gamma \, du + \delta \, dv,$$

we have

$$x_1 = \alpha x_1' + \gamma x_2', \quad y_1 = \alpha y_1' + \gamma y_2', \quad z_1 = \alpha z_1' + \gamma z_2',$$
$$x_2 = \beta x_1' + \delta x_2', \quad y_2 = \beta y_1' + \delta y_2', \quad z_2 = \beta z_1' + \delta z_2',$$

and

$$J\left(\frac{u', \, v'}{u, \, v}\right) = \alpha\delta - \beta\gamma.$$

If $\Omega'$ be the new value of $\Omega$, we have

$$\Omega = J\Omega'.$$

Also

$$V^2 = (x_1^2 + y_1^2 + z_1^2)(x_2^2 + y_2^2 + z_2^2) - (x_1 x_2 + y_1 y_2 + z_1 z_2)^2$$
$$= (y_1 z_2 - z_1 y_2)^2 + (z_1 x_2 - x_1 z_2)^2 + (x_1 y_2 - y_1 x_2)^2$$
$$= J^2\left\{(y_1' z_2' - z_1' y_2')^2 + (z_1' x_2' - x_1' z_2')^2 + (x_1' y_2' - y_1' x_2')^2\right\}$$
$$= J^2 V'^2,$$

if $V'$ be the new value of $V$. Now $V$ is always positive when it is real. Hence

$$\frac{\Omega}{V} = \frac{\Omega'}{V'}:$$

so that the subject of integration in $I''$, when $I''$ is expressed in the form $\iint \frac{\Omega}{V} dA$, is an absolute covariant.

Accordingly, let $U$ be any point in the parallelogram: and draw $R'UR$ parallel to $P'P$. Then, if

$$PP' = D, \quad R'U = T, \quad PR = S, \quad PQ = \sigma, \quad P'PQ = \chi;$$

if $L, M, N \, (= x', y', z')$ are the direction-cosines of $PRQ$; if $\lambda, \mu, \nu$, are the direction-cosines of $P'P$ (that is, of $R'UR$); and if $X, Y, Z$, are the coordinates of $U$ in space relative to $P$ in space; we have

$$X = LS - \lambda(D - T),$$
$$Y = MS - \mu(D - T),$$
$$Z = NS - \nu(D - T).$$

For the integral $I''$, we take $T$ and $S$ to be the parametric variables over the area. Moreover, we integrate completely for $T$ from 0 up to $D$; and the integration for $S$ is from 0 to $\sigma$, where $\sigma$ is a selected element $ds$ along the boundary which is not completely represented in the figure, so that (under the convention of § 275) we integrate first with regard to $T$: that is, corresponding to $I'' = \int \Omega \, du \, dv$, we take

$$I'' = \iint \frac{\Omega'}{V'} dA = \iint \frac{\Omega' \sin \chi}{V'} dT . dS,$$

where $T$ ranges from $0$ to $D$, and, in the first instance, $S$ ranges from $0$ to $\sigma$. With these variables,

$$X_1 = \lambda, \quad Y_1 = \mu, \quad Z_1 = \nu,$$
$$X_2 = L, \quad Y_2 = M, \quad Z_2 = N.$$

Thus, at any point in the parallelogram, the value of $\Omega'$ is

$$= \Omega\,(x + X,\, y + Y,\, z + Z,\, \lambda,\, \mu,\, \nu,\, L,\, M,\, N)$$
$$= \Omega\,(x,\, y,\, z,\, \lambda,\, \mu,\, \nu,\, L,\, M,\, N) + [D]$$
$$= \overline{\Omega} + [D],$$

where $[D]$ represents an aggregate of terms of the first and higher degrees in $D$, while the area of the rudimentary parallelogram is of the order $D^2$. Again

$$V' = \{(X_1^2 + Y_1^2 + Z_1^2)(X_2^2 + Y_2^2 + Z_2^2) - (X_1 X_2 + Y_1 Y_2 + Z_1 Z_2)^2\}^{\frac{1}{2}}$$
$$= \{1 - (\lambda L + \mu M + \nu N)^2\}^{\frac{1}{2}}$$
$$= \sin \chi,$$

where $\chi$ is the (non-vanishing) angle between the directions $\lambda$, $\mu$, $\nu$, and $L$, $M$, $N$. Thus

$$I'' = \iint \frac{\overline{\Omega} + [D]}{\sin \chi} \cdot dT \cdot dS \cdot \sin \chi$$
$$= \iint (\overline{\Omega} + [D])\, dT\, dS$$
$$= \overline{\Omega} D\sigma + [D^3],$$

$\sigma$ being assumed to be of the same order as $D$. As $[D^3]$ is negligible compared with $\sigma D$, we retain the important part of $I''$ by taking

$$I'' = \overline{\Omega} D\sigma.$$

The integral $I_\Delta$ is the sum of all the integrals $I''$ for the aggregate of arcs $PQ$ along the curve $C$. Consequently, we now write $\sigma = ds$, and therefore

$$I_\Delta = \int \overline{\Omega} D\, ds,$$

taken along the curve $C$. We thus have the value of the integral over the whole band of the arbitrary surface $\Sigma$ lying between $C$ and $C'$.

### Total strong variation: the $E_\Sigma$-function, and the Weierstrass test.

**320.** Consequently, for the whole variation of the original integral, we have

$$I_{\Delta'} - I_\Delta + I_\Delta = \int D \left\{ \Omega - \left( \lambda \frac{\partial \Omega}{\partial x_1} + \mu \frac{\partial \Omega}{\partial y_1} + \nu \frac{\partial \Omega}{\partial z_1} \right) \frac{dv}{ds} \right.$$
$$\left. + \left( \lambda \frac{\partial \Omega}{\partial x_2} + \mu \frac{\partial \Omega}{\partial y_2} + \nu \frac{\partial \Omega}{\partial z_2} \right) \frac{du}{ds} \right\} ds$$
$$= \int E_\Sigma D\, ds,$$

where $E_\Sigma$, the *E-function for double integrals*, is given by

$$E_\Sigma = \overline{\Omega} - \left(\lambda \frac{\partial \Omega}{\partial x_1} + \mu \frac{\partial \Omega}{\partial y_1} + \nu \frac{\partial \Omega}{\partial z_1}\right) \frac{dv}{ds} + \left(\lambda \frac{\partial \Omega}{\partial x_2} + \mu \frac{\partial \Omega}{\partial y_2} + \nu \frac{\partial \Omega}{\partial z_2}\right) \frac{du}{ds}.$$

In this expression, $x$, $y$, $z$, are the coordinates of any point on the characteristic surface; $L$, $M$, $N$, are the direction-cosines of any direction which is tangential to the surface, being the values of $x'$, $y'$, $z'$, at $P$ for the curve $PQ$; and $\lambda$, $\mu$, $\nu$, are the direction-cosines of any direction through $P$ off the surface, while

$$\overline{\Omega} = \Omega\,(x,\, y,\, z,\, \lambda,\, \mu,\, \nu,\, x',\, y',\, z').$$

We have taken a variation of the characteristic surface constituted by a complete band between two curves $C$ and $C'$, on a regular surface $\Sigma$ through the curve $C$ on the central characteristic. If we choose, we can take a more fragmentary variation by selecting only a portion $PQQ'P'P$ of that band. Then the variation, represented by taking the portion $AP'Q'BA$ of the consecutive

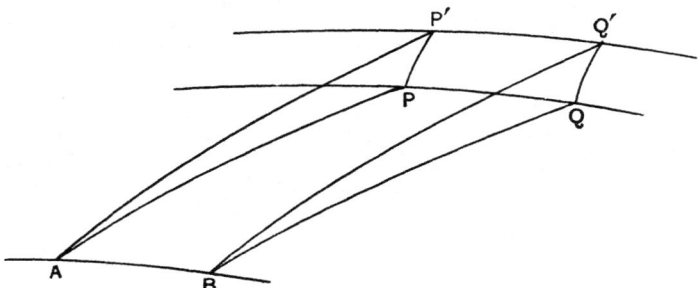

characteristic (where $AB$ is an element of arc of the initial curve) and $P'PQQ'P'$ of the surface $\Sigma$ through $PQ$, in place of the portion $APQBA$ of the central characteristic, is a possible small variation. For this small variation of the characteristic area, the variation of the integral is

$$E_\Sigma \,.\, D \,.\, ds.$$

If the original integral is to be a minimum, then its increment: alike for all variations, however comprehensive; and for each variation, however slight provided it be a definite variation: must be positive. Thus taking the slightest variation, as represented by the variation just described, we must have $E_\Sigma$ positive; and when $E_\Sigma$ is positive for each such slight variation, the aggregate for all such variations along the band is also positive, because it is equal to

$$\int E_\Sigma \,.\, D \,.\, ds,$$

and $D$ and $ds$ are positive. Thus we secure a minimum if $E_\Sigma$ is positive, for every position on the surface, for every superficial direction through a position, and for every direction through that position off the surface.

Similarly, if the original integral is to be a maximum, then its increment for the whole range of variations, the most comprehensive and the slightest

alike, must always be negative. Thus, taking the slightest variation, we must
have $E_\Sigma$ negative for each and every such variation; and then an aggregate
for such variations, representing the integral

$$\int E_\Sigma \,.\, D \,.\, ds,$$

also is negative, because $D$ and $ds$ are positive. Thus we secure a maximum if
$E_\Sigma$ is negative for every position on the surface, for every superficial direction
through the position, and for every direction through the position which lies
off the surface.

We thus have the extension of the *Weierstrass test*, as originally obtained
in connection with the simplest strong variation for an integral involving a
single independent variable, with one dependent variable and its first derivative.
This extension of the Weierstrass test is:

*The integral*

$$\iint \Omega \,(x,\, y,\, z,\, x_1,\, y_1,\, z_1,\, x_2,\, y_2,\, z_2)\, du\, dv,$$

*where $\Omega$ satisfies the four general identities, admits a maximum or admits a
minimum for a strong variation of the directional type, if, at every point $(x, y, z)$
on the characteristic surface, for every direction $(x', y', z')$ on that surface
through the point, and for every direction $(\lambda, \mu, \nu)$ off that surface through the
point, the function $E_\Sigma$ is negative or the function $E_\Sigma$ is positive, where*

$$E_\Sigma = \overline{\Omega} - \left( \lambda \frac{\partial \Omega}{\partial x_1} + \mu \frac{\partial \Omega}{\partial y_1} + \nu \frac{\partial \Omega}{\partial z_1} \right) \frac{dv}{ds} + \left( \lambda \frac{\partial \Omega}{\partial x_2} + \mu \frac{\partial \Omega}{\partial y_2} + \nu \frac{\partial \Omega}{\partial z_2} \right) \frac{du}{ds},$$

$\overline{\Omega}$ *denotes the function $\Omega \,(x, y, z, \lambda, \mu, \nu, x', y', z')$, and the s-variation is along
the arc specified at a point $(x, y, z)$ by the superficial direction.*

### Alternative forms of the $E_\Sigma$-function.

**321.** In the variation, which has been imposed on the characteristic
surface along the curve $C$, the important constituent is the element of area
$PP'Q'QP$. In order to have a non-zero variation, this element must not lie
in the tangent plane to the characteristic surface at $P$. Within the area of
the element, any direction through $P$ (such as $PP'$) may be taken. In the
formation of the $E_\Sigma$-function, any direction within the characteristic surface
is taken at $P$, quite arbitrarily. Hence any change in the direction of $P$ in
the element $PP'Q'QP$, without affecting the range for the change of position
$P$, adds nothing to the generality of the superficial variation, represented by
the element as constructed. Now the plane of this superficial variation can
be determined by the arc $PQ$ and by a direction in the plane, $PK$, perpendicular to $PQ$. We therefore proceed to express the change of the original
integral by reference to the plane of the small variation, thus determined by
the arc $PQ$, the direction-cosines of which are $x', y', z'$, and by the line $KP$,
the direction-cosines of which will be denoted by $\cos \alpha$, $\cos \beta$, $\cos \gamma$.

The total variation of the integral, taken over the band of the surface $\Sigma$, has been expressed in the form

$$\int E_\Sigma \,.\, ds \,.\, D,$$

where $D$ is the length of $PP'$ drawn from $P$ to a point $P'$ in the curve $C'$. The small area is now to be referred to $PQ$ and $PK$; so we take

$$KP = l,$$

and therefore, as $l = D \sin \chi$, where $\chi$ is the angle $P'PQ$, the increment of the integral is

$$\int \frac{E_\Sigma}{\sin \chi} \, l \, ds.$$

The angle $\chi$ cannot vanish for a non-zero variation. Then if

$$\mathbf{E} = \frac{E_\Sigma}{\sin \chi},$$

the increment becomes $\int \mathbf{E} \, l ds$: the quantity $l ds$ is always positive, and does not affect the sign of the increment; also, $\sin \chi$ does not vanish; and so $\mathbf{E}$ comes to be the critical function for the determination of the increment. Now we have

$$\lambda x' + \mu y' + \nu z' = \cos \chi, \quad \lambda \cos \alpha + \mu \cos \beta + \nu \cos \gamma = \sin \chi,$$
$$x' \cos \alpha + y' \cos \beta + z' \cos \gamma = 0;$$

and therefore

$$\lambda = x' \cos \chi + \cos \alpha \sin \chi, \quad \mu = y' \cos \chi + \cos \beta \sin \chi, \quad \nu = z' \cos \chi + \cos \gamma \sin \chi.$$

Also

$$x' \frac{\partial \Omega}{\partial x_1} + y' \frac{\partial \Omega}{\partial y_1} + z' \frac{\partial \Omega}{\partial z_1} = \Sigma \left( x_1 \frac{du}{ds} + x_2 \frac{dv}{ds} \right) \frac{\partial \Omega}{\partial x_1}$$

$$= \frac{du}{ds} \Sigma x_1 \frac{\partial \Omega}{\partial x_1} + \frac{dv}{ds} \Sigma x_2 \frac{\partial \Omega}{\partial x_1}$$

$$= \Omega \frac{du}{ds},$$

because of the identities satisfied by $\Omega$; and, similarly,.

$$x' \frac{\partial \Omega}{\partial x_2} + y' \frac{\partial \Omega}{\partial y_2} + z' \frac{\partial \Omega}{\partial z_2} = \Omega \frac{dv}{ds}.$$

Thus

$$\left( \Sigma \lambda \frac{\partial \Omega}{\partial x_1} \right) \frac{dv}{ds} - \left( \Sigma \lambda \frac{\partial \Omega}{\partial x_2} \right) \frac{du}{ds}$$

$$= \left( \Sigma x' \frac{\partial \Omega}{\partial x_1} \right) \frac{dv}{ds} \cos \chi - \left( \Sigma x' \frac{\partial \Omega}{\partial x_2} \right) \frac{du}{ds} \cos \chi$$

$$+ \left\{ \left( \Sigma \frac{\partial \Omega}{\partial x_1} \cos \alpha \right) \frac{dv}{ds} - \left( \Sigma \frac{\partial \Omega}{\partial x_2} \cos \alpha \right) \frac{du}{ds} \right\} \sin \chi$$

$$= \left\{ \left( \Sigma \frac{\partial \Omega}{\partial x_1} \cos \alpha \right) \frac{dv}{ds} - \left( \Sigma \frac{\partial \Omega}{\partial x_2} \cos \alpha \right) \frac{du}{ds} \right\} \sin \chi.$$

Consequently,

$$\mathbf{E} = \frac{E_\Sigma}{\sin \chi}$$

$$= \frac{\overline{\Omega}}{\sin \chi} - \left( \Sigma \frac{\partial \Omega}{\partial x_1} \cos \alpha \right) \frac{dv}{ds} + \left( \Sigma \frac{\partial \Omega}{\partial x_2} \cos \alpha \right) \frac{du}{ds},$$

which is Kobb's first form* for the $E$-function; where, in this form, as for $E_\Sigma$,

$$\overline{\Omega} = \Omega \left( x, y, z, \lambda, \mu, \nu, x', y', z' \right),$$

$x'$, $y'$, $z'$, being any direction in the surface at $x$, $y$, $z$; while $\cos \alpha$, $\cos \beta$, $\cos \gamma$, are the direction-cosines of a direction perpendicular to $x'$, $y'$, $z'$, taken off the surface; $\lambda$, $\mu$, $\nu$, are the direction-cosines of any line making an angle $\chi$ with $x'$, $y'$, $z'$, in the plane, which passes through the directions $x'$, $y'$, $z'$, and $\cos \alpha$, $\cos \beta$, $\cos \gamma$.

 *Ex.* On any characteristic surface, an arbitrary direction $x'$, $y'$, $z'$, is taken lying in the tangent plane at $x, y, z$; and a direction, which is perpendicular to $x'$, $y'$, $z'$, and lies in that plane, is given by means of direction-cosines $\cos \theta$, $\cos \phi$, $\cos \psi$. Prove that

$$\frac{\Omega}{V} = \left( \Sigma \frac{\partial \Omega}{\partial x_1} \cos \theta \right) \frac{dv}{ds} - \left( \Sigma \frac{\partial \Omega}{\partial x_2} \cos \theta \right) \frac{du}{ds},$$

the summation being for the three derivatives with regard to $x, y, z$.

 Thus, if the function $\Omega$ be taken for the surface $\Sigma$ along the curve $C$, with the same variables of reference,

$$\frac{\overline{\Omega}}{\sin \chi} = \left( \Sigma \frac{\partial \overline{\Omega}}{\partial \overline{x}_1} \cos a \right) \frac{dv}{ds} - \left( \Sigma \frac{\partial \overline{\Omega}}{\partial \overline{x}_2} \cos a \right) \frac{du}{ds}.$$

(Kobb.)

 Further analytical developments are possible. Thus, by the use of the result given in the preceding exercise, it can be proved that

$$\mathbf{E} = \left\{ \Sigma \left( \frac{\partial \overline{\Omega}}{\partial \overline{x}_1} - \frac{\partial \Omega}{\partial x_1} \right) \cos \alpha \right\} \frac{dv}{ds} - \left\{ \Sigma \left( \frac{\partial \overline{\Omega}}{\partial \overline{x}_2} - \frac{\partial \Omega}{\partial x_2} \right) \cos \alpha \right\} \frac{du}{ds},$$

which is Kobb's second form for the $\mathbf{E}$-function, being the extension of the form given (§ 213) for the Weierstrass $E$-function for single integrals with one dependent variable and its first derivative. From this expression, Kobb proceeds to obtain an integral expression leading to mean values $P_\mu$, $R_\mu$, $Q_\mu$, of the functions $(P, R, Q)(x, y, z, x_1{}^{(\mu)}, y_1{}^{(\mu)}, z_1{}^{(\mu)}, x_2{}^{(\mu)}, y_2{}^{(\mu)}, z_2{}^{(\mu)})$, for values of the derivatives of $x, y, z$, intermediate between those for the characteristic

---

 * *Acta Math.*, t. xvi (1892), p. 123. Kobb does not introduce the quantities $x'$, $y'$, $z'$, $\lambda$, $\mu$, $\nu$, into the expression for $\dfrac{\overline{\Omega}}{\sin \chi}$ as given: he retains $u$ and $v$ as parametric variables for the consecutive surface.

 It has been proved that $\dfrac{\Omega'}{V'}$ is invariantive, whatever variables be chosen; so that, if $\Omega'$ be $\Omega \left( x, y, z, \overline{x}_1, \overline{y}_1, \overline{z}_1, \overline{x}_2, \overline{y}_2, \overline{z}_2 \right)$ for the surface $\Sigma$, and if $V'$ denotes

$$\{ (\overline{x}_1{}^2 + \overline{y}_1{}^2 + \overline{z}_1{}^2)(\overline{x}_2{}^2 + \overline{y}_2{}^2 + \overline{z}_2{}^2) - (\overline{x}_1 \overline{x}_2 + \overline{y}_1 \overline{y}_2 + \overline{z}_1 \overline{z}_2)^2 \}^{\frac{1}{2}},$$

we have

$$\frac{\Omega'}{V'} = \frac{\overline{\Omega}}{\sin \chi}.$$

With this change, the form given for $\mathbf{E}$ accords exactly with Kobb's form.

surface and those for the surface $\Sigma$; and for the requirements of the **E**-function, there can be substituted the requirement that the inequality

$$P_\mu Q_\mu > R_\mu{}^2$$

shall be satisfied, whatever surface $\Sigma$ be chosen.

Such requirements are additional to the extended Legendre test. They belong to the ranges of directions which are definitely *off* the characteristic surface: the Legendre conditions belong to directions which are definitely *along* the characteristic surface.

The form of the $E$-function, as given in § 320, seems the simplest of all the forms. It exhibits the direction $(x', y', z')$ at every point $x, y, z$, all such directions lying in the characteristic surface; it exhibits the arbitrary direction $(\lambda, \mu, \nu)$ anywhere off that surface, each such direction lying in the varied plane, which passes through the direction $x', y', z'$, and through any arbitrary direction $\cos \alpha, \cos \beta, \cos \gamma$, at once perpendicular to the direction $x', y', z'$, and lying off the characteristic surface.

### Types of strong variation that can be constructed from the elementary type.

**322.** At the beginning of the preceding discussion, the variation allowed divergence from a weak variation solely in the property that the surface $\Sigma$, represented by the band between the curves $C$ and $C'$, cut the central characteristic surface at any arbitrary angle along the length of the curve $C$, that angle being susceptible of continuous variation when the whole band was brought into account. But when the $E_\Sigma$-function was used finally, it was taken for an arbitrarily selected element $P'PQQ'P'$ of the band: and, when the required property was exacted for each such element, its cumulative effect secured the property for the whole band. Thus the $E_\Sigma$-function has come to be associated with an elementary area, one side of which is an arbitrary elementary arc of the curve $C$ lying in the central surface, and a contiguous side of which is an arbitrary elementary arc $PP'$ lying off the central surface.

Now it has been proved (§ 297) that the quantities

$$\frac{\partial \Omega}{\partial x_1}\frac{dv}{ds} - \frac{\partial \Omega}{\partial x_2}\frac{du}{ds}, \quad \frac{\partial \Omega}{\partial y_1}\frac{dv}{ds} - \frac{\partial \Omega}{\partial y_2}\frac{du}{ds}, \quad \frac{\partial \Omega}{\partial z_1}\frac{dv}{ds} - \frac{\partial \Omega}{\partial z_2}\frac{du}{ds},$$

are continuous in transit over a curve where free discontinuities in the derivatives of $x, y, z$, may occur. Hence the $E_\Sigma$-function remains unaltered in value, even if the curve $C$ should (in whole or in part) be a line of discontinuity as regards tangential directions; in the integrals under consideration, the superficial curvatures do not affect the matter.

We thus can substitute, for the curve $C$ which is continuous in direction, a broken line which, continuous in its length on the surface, is discontinuous in its direction. The broken line can be made up of consecutive small arcs

$PQ$, the final point of any such arc being also a final point of its contiguous neighbour. Also, through each such arc $PQ$ we can draw a plane cutting the tangent plane of the surface through $PQ$ at any angle; and the angle of intersection can be taken arbitrarily for each such elementary arc. For each such elementary area, the $E_\Sigma$-function is required to possess its property.

By the composition of a number of elementary areas, with the appropriate continuous change of the angle of intersection for the consecutive arcs $PQ$ so as to secure a continuous surface composed of the successive elementary areas, we can construct a narrow fillet (in whole or in part) which shall be a deformation of the central surface to form a strong variation. As the effect of the $E_\Sigma$-function acts in the same sense for each component of the fillet, the cumulative effect for the fillet acts in that same sense.

We thus can have a strong variation which is more general than the strong variation, originally postulated by the requirement that the whole surface $\Sigma$ shall be regular. All that is necessary, is that the $E_\Sigma$-function shall act for each rudiment of area at every place (whatever orientation be adopted at the place for the rudiment) in the one sense, either persistently as an increase or persistently as a decrease.

It was pointed out, in connection with strong variations of characteristic plane curves, that the type adopted (however extended within its range) was only the simplest type of strong variation that could be conceived. So here also, the foregoing type adopted (however wide its extension may be made, under the foregoing explanation) is only the simplest type of strong variation that can be conceived. Thus we could imagine a continuous surface, bounded by the closed contour $PP'Q'QP$, and having any arbitrary continuous curvatures over its range; such a surface, for integrals that involved second derivatives of the dependent variables, would provide a strong variation while, for integrals of that character, the preceding rudimentary plane area $PP'Q'QP$ would not provide an appropriate strong variation.

### The four tests for a double integral.

**323**. Summarising the results, we thus have the four tests for a double integral, if it is to possess a maximum or a minimum:

($a$)  the Euler test, the Legendre test, and the Jacobi test, for increments that arise out of weak variations; and

($b$)  the Weierstrass test, for increments that arise out of strong variations of the indicated types constructed from the typical elementary strong variation.

Of these, the Euler test and the Weierstrass test range over 'first' variations of the integral: the Euler test making it vanish for weak variations, so that a 'second' variation has to be considered; the Weierstrass test precluding it from becoming zero for strong variations. The Legendre test and the Jacobi test range around the 'second' variation, which demands consideration only in the case of weak variations.

As for single integrals (§ 236), the Euler test, the Legendre test, and the Weierstrass test, are qualitative in their character; the Jacobi test is quantitative.

**324.** We proceed to the application of the Weierstrass test to an important class of double integrals of the first order.

*Ex.* 1. *The extended Weierstrass test is satisfied, so as to admit a minimum, for all integrals*

$$\int\int f(x, y, z)\, dS, \quad = \int\int f(x, y, z)\, V du\, dv\,;$$

the simplest example being given by $f(x, y, z) = 1$, *so that the Weierstrass test is satisfied for all minimal surfaces.*

In this case, we have

$$\Omega(x, y, z, x_1, y_1, z_1, x_2, y_2, z_2) = f(x, y, z)\, V,$$

so that

$$\frac{\overline{\Omega}}{V'} = f(x, y, z).$$

Also

$$V' = \{(x'^2 + y'^2 + z'^2)(\lambda^2 + \mu^2 + \nu^2) - (\lambda x' + \mu y' + \nu z')^2\}^{\frac{1}{2}} = \sin \chi,$$

and therefore

$$\frac{\overline{\Omega}}{\sin \chi} = f(x, y, z).$$

(I)  Take a unit sphere, centre $P$ (not shewn in the figure). Let $PC$ be the curve $C$, $PU$ be the curve $u = constant$, $PV$ be the curve $v = constant$; all in the characteristic surface. Let $PK$ be the perpendicular to $PC$ in the plane $KPC$ of the surface $\Sigma$; thus $KC = \frac{1}{2}\pi$. Let

$$KCV = \vartheta,$$

so that $\vartheta$ is the inclination between the tangent plane to the characteristic surface and the tangent plane to $\Sigma$. The direction-cosines of $PU$ are

$$x_2 G^{-\frac{1}{2}}, \quad y_2 G^{-\frac{1}{2}}, \quad z_2 G^{-\frac{1}{2}};$$

those of $PV$ are

$$x_1 E^{-\frac{1}{2}}, \quad y_1 E^{-\frac{1}{2}}, \quad z_1 E^{-\frac{1}{2}};$$

those of $PK$ are $\cos a$, $\cos \beta$, $\cos \gamma$; and those of $PC$ are $x'$, $y'$, $z'$. Then

$$\cos KV = \Sigma \frac{x_1}{E^{\frac{1}{2}}} \cos a \quad, \quad \cos KU = \Sigma \frac{x_2}{G^{\frac{1}{2}}} \cos a\,;$$

and

$$\cos KV = \cos \vartheta \sin CV, \quad \cos KU = -\cos \vartheta \sin CU,$$

because $KC = \frac{1}{2}\pi$; and if $\omega$ denote the angle $UPV$, the angle at $P$ between the curves of reference,

$$\cos \omega = \Sigma \frac{x_1}{E^{\frac{1}{2}}} \frac{x_2}{G^{\frac{1}{2}}} = \frac{F}{E^{\frac{1}{2}} G^{\frac{1}{2}}}, \quad \sin \omega = \frac{V}{E^{\frac{1}{2}} G^{\frac{1}{2}}}.$$

Finally,

$$\cos CU = \frac{x_2}{G^{\frac{1}{2}}}\, x' + \frac{y_2}{G^{\frac{1}{2}}}\, y' + \frac{z_2}{G^{\frac{1}{2}}}\, z'$$

$$= \frac{1}{G^{\frac{1}{2}}}\left(F\frac{du}{ds} + G\frac{dv}{ds}\right),$$

because $x' = x_1 \dfrac{du}{ds} + x_2 \dfrac{dv}{ds}$, and so for $y'$ and $z'$; and

$$\cos CV = \frac{x_1}{E^{\frac{1}{2}}}\, x' + \frac{y_1}{E^{\frac{1}{2}}}\, y' + \frac{z_1}{E^{\frac{1}{2}}}\, z'$$

$$= \frac{1}{E^{\frac{1}{2}}}\left(E\frac{du}{ds} + F\frac{dv}{ds}\right).$$

Because

$$\Omega = f(x, y, z) \quad V = (EG - F^2)^{\frac{1}{2}} f,$$

we have

$$\frac{\partial \Omega}{\partial x_1} = \frac{f}{V}(Gx_1 - Fx_2), \quad \frac{\partial \Omega}{\partial x_2} = \frac{f}{V}(Ex_2 - Fx_1),$$

and so for the other derivatives.  Thus

$$\left(\Sigma \frac{\partial \Omega}{\partial x_1}\cos a\right)\frac{dv}{ds} - \left(\Sigma \frac{\partial \Omega}{\partial x_2}\cos a\right)\frac{du}{ds}$$

$$= \frac{f}{V}\left\{\left(G\frac{dv}{ds} + F\frac{du}{ds}\right)(\Sigma x_1 \cos a) - \left(E\frac{du}{ds} + F\frac{dv}{ds}\right)(\Sigma x_2 \cos a)\right\}$$

$$= \frac{f}{V}E^{\frac{1}{2}}G^{\frac{1}{2}}(\cos CU \cos KV - \cos CV \cos KU)$$

$$= \frac{f}{V}E^{\frac{1}{2}}G^{\frac{1}{2}}\cos \vartheta\,(\cos CU \sin CV + \cos CV \sin CU)$$

$$= \frac{f}{V}E^{\frac{1}{2}}G^{\frac{1}{2}}\cos \vartheta \sin \omega$$

$$= f\cos \vartheta.$$

Consequently,

$$\mathbf{E} = f(x, y, z)(1 - \cos \vartheta),$$

where $\vartheta$ (being $KCV$, the angle between the characteristic surface and the varied surface $\Sigma$) is never zero for a non-zero variation.

Thus $\mathbf{E}$ is always positive for these strong variations; and therefore the surface admits a minimum.

(II)  The analysis, leading to the preceding result, can be arranged in a different manner.  We have a strong variation in the direction $PL$, taken at $P$ off the tangent plane at $P$ containing the directions $PU$, $PC$, $PV$.  The direction $PL$ makes an angle $\chi$ with $PC$.  Let $PD = D$, $PD' = D'$: the small element of area in the surface $\Sigma$ is equal to $SD$, that is, to

$$SD' \sin \chi.$$

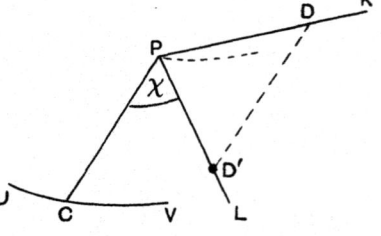

(a)  Hence the portion of the strong variation of the integral, due to the part taken over this element of area, is equal to

$$f(x, y, z)\, SD' \sin \chi,$$

to the first order of small quantities.

(b) The portion of the strong variation of the integral, due to the characteristic variation from the characteristic surface to the consecutive surface, is

$$-\int D' f(x, y, z) \left\{ \left( \lambda \frac{x_1 G - x_2 F}{V} + \mu \frac{y_1 G - y_2 F}{V} + \nu \frac{z_1 G - z_2 F}{V} \right) \frac{dv}{ds} \right.$$
$$\left. - \left( \lambda \frac{x_2 E - x_1 F}{V} + \mu \frac{y_2 E - y_1 F}{V} + \nu \frac{z_2 E - z_1 F}{V} \right) \frac{du}{ds} \right\} ds$$
$$= -f(x, y, z) SD' \left\{ \left( \Sigma \lambda \frac{x_1 G - x_2 F}{V} \right) \frac{dv}{ds} - \left( \Sigma \lambda \frac{x_2 E - x_1 F}{V} \right) \frac{du}{ds} \right\}$$

for the small length $S$, small quantities of the second and higher orders being neglected. We denote the right-hand side by

$$-f(x, y, z) SD' \square.$$

In the figure (p. 539), let

$$CU = \theta, \quad CV = \theta', \quad LU = \phi, \quad LV = \phi', \quad LC = \chi, \quad KC = \tfrac{1}{2}\pi, \quad KCV = \vartheta.$$

Then

$$\lambda x_1 + \mu y_1 + \nu z_1 = E^{\frac{1}{2}} \cos \phi', \quad \lambda x_2 + \mu y_2 + \nu z_2 = G^{\frac{1}{2}} \cos \phi;$$

and therefore

$$\square = \frac{E^{\frac{1}{2}} \cos \phi'}{V} \left( G \frac{dv}{ds} + F \frac{du}{ds} \right) - \frac{G^{\frac{1}{2}} \cos \phi}{V} \left( E \frac{du}{ds} + F \frac{dv}{ds} \right)$$
$$= \frac{E^{\frac{1}{2}} G^{\frac{1}{2}}}{V} (\cos \phi' \cos \theta - \cos \phi \cos \theta').$$

Now

$$\cos \phi' = \cos LV = \cos LC \cos VC + \sin LC \sin VC \cos LCV$$
$$= \cos \chi \cos \theta' + \sin \chi \sin \theta' \cos \vartheta;$$

and

$$\cos \phi = \cos LU = \cos LC \cos UC + \sin LC \sin UC \cos LCU$$
$$= \cos \chi \cos \theta - \sin \chi \sin \theta \cos \vartheta.$$

Hence

$$\square = \frac{E^{\frac{1}{2}} G^{\frac{1}{2}}}{V} \sin \chi \cos \vartheta \sin (\theta' + \theta)$$
$$= \frac{E^{\frac{1}{2}} G^{\frac{1}{2}}}{V} \sin \omega \sin \chi \cos \vartheta$$
$$= \sin \chi \cos \vartheta.$$

Thus the part of the integral, due to the constituent arising through the strong variation, is

$$-f(x, y, z) SD' \sin \chi \cos \vartheta.$$

(c) Consequently, the whole strong variation of the integral is

$$= f(x, y, z) SD' \sin \chi (1 - \cos \vartheta);$$

and therefore the **E**-function is

$$= f(x, y, z) (1 - \cos \vartheta),$$

as before.

*Ex.* 2. Shew that the Weierstrass test is satisfied for integrals $\iint f(x, y, z)\, dS$ under strong variations when, instead of a surface $\Sigma$ constituted by plane elements, any surface of finite (non-zero) continuous curvature is used, to join the curve $C$ on the characteristic surface to the curve $C'$ on the consecutive characteristic surface.

*The integral $\iint \Omega \, du \, dv$ cannot have a maximum or a minimum*

*when $\Omega$ is rational in the derivatives.*

**325.** It was proved (§ 218) that a simple integral $\int \Omega \, dt$, involving only a single dependent variable and its derivative of the first order, could not have a true maximum or a true minimum, if the subject of integration $\Omega$ is rational in the derived arguments. The same property holds of a double integral: *an integral*

$$\iint \Omega \left( x, y, z, x_1, y_1, z_1, x_2, y_2, z_2 \right) du \, dv$$

*cannot have a maximum or a minimum, if $\Omega$ is rational in all the arguments $x_1, y_1, z_1, x_2, y_2, z_2$.*

We have seen (§ 291) that the function $\Omega$ satisfies the relations

$$- \Omega \left( x, y, z, x_1, y_1, z_1, x_2, y_2, z_2 \right) = \Omega \left( x, y, z, -x_1, -y_1, -z_1, x_2, y_2, z_2 \right)$$
$$= \Omega \left( x, y, z, x_1, y_1, z_1, -x_2, -y_2, -z_2 \right).$$

In establishing the property, the implication is made that we are dealing with a uniform function $\Omega$: or, if with a branch of a non-uniform function, that the branch remains unaffected by any subsequent arbitrary interference, so that it is allowed to change continuously as the variables change to their opposite values. This implication, however, is not justified whenever the deliberate choice of another branch is made in case the variables change to their opposite values. For example, the element of area of a surface is $V \, du \, dv$, where

$$V = \left\{ (x_1 y_2 - y_1 x_2)^2 + (y_1 z_2 - z_1 y_2)^2 + (z_1 x_2 - z_2 x_1)^2 \right\}^{\frac{1}{2}}$$
$$= g \left( x_1, y_1, z_1, x_2, y_2, z_2 \right);$$

but we deliberately and definitely choose the positive sign as the unchanging sign of $V$ whatever be its arguments, and so the relation

$$g \left( -x_1, -y_1, -z_1, x_2, y_2, z_2 \right) = - g \left( x_1, y_1, z_1, x_2, y_2, z_2 \right)$$

is not allowed for $V$. (Of course, $V$ always satisfies the relations

$$V = x_1 \frac{\partial V}{\partial x_1} + y_1 \frac{\partial V}{\partial y_1} + z_1 \frac{\partial V}{\partial z_1}, \quad 0 = x_1 \frac{\partial V}{\partial x_2} + y_1 \frac{\partial V}{\partial y_2} + z_1 \frac{\partial V}{\partial z_2},$$

$$0 = x_2 \frac{\partial V}{\partial x_1} + y_2 \frac{\partial V}{\partial y_1} + z_2 \frac{\partial V}{\partial z_1}, \quad V = x_2 \frac{\partial V}{\partial x_2} + y_2 \frac{\partial V}{\partial y_2} + z_2 \frac{\partial V}{\partial z_2},$$

which arise out of an infinitesimal variation (§ 291)

$$du' - du = \epsilon \, du + \zeta \, dv, \quad dv' - dv = \theta \, du + \eta \, dv,$$

where $\epsilon, \zeta, \theta, \eta$, are independent small quantities. The explanation is that the full analytical change in $V$, as $x_1, y_1, z_1$, change to $-x_1, -y_1, -z_1$, is excluded by the deliberate selection of a persistent positive sign.)

Assuming, now, that $\Omega$ is rational in all its derivative arguments, we have

$$\Omega\left(x,\, y,\, z,\, -\lambda,\, -\mu,\, -\nu,\, x',\, y',\, z'\right) = -\Omega\left(x,\, y,\, z,\, \lambda,\, \mu,\, \nu,\, x',\, y',\, z'\right).$$

Owing to the deliberate choice of the persistent positive sign for $V$, we have

$$V\left(-\lambda,\, -\mu,\, -\nu,\, x',\, y',\, z'\right) = V\left(\lambda,\, \mu,\, \nu,\, x',\, y',\, z'\right).$$

Hence

$$\frac{\Omega\left(x,\, y,\, z,\, -\lambda,\, -\mu,\, -\nu,\, x',\, y',\, z'\right)}{V\left(-\lambda,\, -\mu,\, -\nu,\, x',\, y',\, z'\right)} = -\frac{\Omega\left(x,\, y,\, z,\, \lambda,\, \mu,\, \nu,\, x',\, y',\, z'\right)}{V\left(\lambda,\, \mu,\, \nu,\, x',\, y',\, z'\right)}.$$

The quantities $\dfrac{\partial\Omega}{\partial x_1},\ \dfrac{\partial\Omega}{\partial y_1},\ \dfrac{\partial\Omega}{\partial z_1},\ \dfrac{\partial\Omega}{\partial x_2},\ \dfrac{\partial\Omega}{\partial y_2},\ \dfrac{\partial\Omega}{\partial z_2},\ \dfrac{dv}{ds},\ \dfrac{du}{ds}$ are not concerned with $\lambda,\, \mu,\, \nu$, the direction-cosines of an arbitrary direction through $x,\, y,\, z$, off the surface. Hence

$$E_\Sigma\left(-\lambda,\, -\mu,\, -\nu\right) = -E_\Sigma\left(\lambda,\, \mu,\, \nu\right).$$

In other words, a reversal of the arbitrary direction chosen—such reversed direction being equally imposable, as all the variations are arbitrary—leads to a change of sign on the part of $E_\Sigma$: while $E_\Sigma$ should remain of the same sign for all directions chosen, if a true maximum or a true minimum is to be possessed.

Thus the Weierstrass test shews that no true maximum and no true minimum can exist for an integral $\displaystyle\iint\Omega\left(x,\, y,\, z,\, x_1,\, y_1,\, z_1,\, x_2,\, y_2,\, z_2\right)du\,dv$, when $\Omega$ is rational in the arguments $x_1,\, y_1,\, z_1,\, x_2,\, y_2,\, z_2$.

*Ex.* Consider the possible *minimum of the integral*

$$\iint\left(p^2+q^2\right)dx\,dy,$$

with the customary significance for $p$ and $q$.

(A) The characteristic equation is

$$-\frac{\partial}{\partial x}\left(2\,\frac{\partial z}{\partial x}\right) - \frac{\partial}{\partial y}\left(2\,\frac{\partial z}{\partial y}\right) = 0,$$

that is,

$$\frac{\partial^2 z}{\partial x^2} + \frac{\partial^2 z}{\partial y^2} = 0,$$

the potential-equation in two dimensions. The primitive is

$$z = \phi\left(x+iy\right) + \psi\left(x-iy\right),$$

where $\phi$ and $\psi$ are arbitrary functions.

The Legendre test is satisfied, because

$$\frac{\partial^2 f}{\partial p^2} = 2, \qquad \frac{\partial^2 f}{\partial q^2} = 2, \qquad \frac{\partial^2 f}{\partial p\,\partial q} = 0.$$

It admits a minimum for the integral.

Before the Jacobi test can be applied, the functions $\phi$ and $\psi$ must be determined by assigned initial conditions, *e.g.*, such that the surface, given by the characteristic equation, must pass through an assigned curve and touch a given developable along that curve. Let the curve be

$$x^2 - y^2 = b^2, \quad z = a\;;$$

and let the developable be the cone

$$b^2 (x^2 - y^2) = \{c (z - a) + b^2\}^2,$$

which manifestly passes through the curve. A characteristic surface, satisfying these conditions, is

$$2c (z - a) = x^2 - y^2 - b^2.$$

Another surface, satisfying the characteristic equation, passing through the same curve, and touching the cone

$$b^2 (x^2 - y^2) = \{c' (z - a) + b^2\}^2,$$

is

$$2c' (z - a) = x^2 - y^2 - b^2.$$

If therefore $c'$ differs slightly from $c$, we obtain a consecutive characteristic. When

$$c' = c (1 + \epsilon),$$

where $\epsilon$ is small, it appears that the angle, at which the two surfaces intersect along the curve $z = a$, $x^2 - y^2 = b^2$, is

$$\chi = \epsilon \frac{cr}{r^2 + c^2},$$

where $r = (x^2 + y^2)^{\frac{1}{2}}$.

The two characteristic surfaces do not intersect in any other curve at a finite distance. Hence the Jacobi test imposes no conjugate of the initial curve as a limit to the range of integration.

It therefore appears that the Euler test, the Legendre test, and the Jacobi test, are satisfied. A minimum exists for weak variations.

(B) When surface-variables $u$ and $v$ are introduced, the integral becomes

$$\iint \frac{1}{x_1 y_2 - y_1 x_2} \{(y_1 z_2 - z_1 y_2)^2 + (x_1 z_2 - z_1 x_2)^2\} \, du \, dv.$$

The subject of integration is rational in all the derived arguments; hence it does not satisfy the Weierstrass test. The integral does not admit a maximum or a minimum for even the simplest type of strong variation. It therefore does not possess a real maximum or a real minimum.

(C) In applying the **E**-test, only one particular kind of variation (a complete reversal of direction of variation from a point $P$) is used, in order to shew that a variation certainly does exist under which the $E$-function changes its sign and therefore prevents the qualification for the possession of a maximum or minimum. But such a variation need not be, and usually is not, the only variation which shews that a minimum (or a maximum) under a weak variation cannot stand the test of a strong variation.

In the particular illustration in (A), the characteristic surface is a hyperbolic paraboloid. An entirely different surface, not continuous in tangential character or in curvature, can be constructed along which the integral acquires a value smaller than

$$\frac{1}{c^2} \iint (x^2 + y^2) \, dx \, dy,$$

its value for the characteristic surface, being proportional to the moment of inertia, round the axis of $z$, of the projection of the surface on the plane of $z = 0$. Let an arc $PQ$ be taken

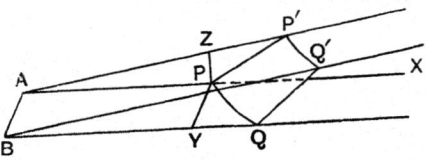

in a plane $ABQP$ parallel to $z=0$, where $PA$ and $QB$ are parallel to the axis of $x$, so that $QB\ (=l)$ is large compared to $PQ$: through $AB$ draw a plane $ABQ'P'A$ at a small inclination $\epsilon$ to $ABQPA$, the planes $Q'BQ$ and $P'AP$ being perpendicular to the axis of $y$, to which $AB$ is parallel. (The arcs $PQ$ and $P'Q'$ are hyperbolic: $PP'$ and $QQ'$ are parabolic.) Now take an element of a surface, having the planes $QBAPQ$ and $Q'BAP'Q'$ for its boundary, instead of the element of the characteristic surface $QPP'Q'Q$; let $PY\ (=dy)$, and $PZ\ (=dz)$, be parallel to the axes of $y$ and of $z$ respectively. Then

$$dS = \text{area } PQQ'P'P = \frac{x}{(x^2+y^2+c^2)^{\frac{1}{2}}}\, dy\, dz,$$

and

$$dx\, dy = \frac{c}{(x^2+y^2+c^2)^{\frac{1}{2}}}\, dS,$$

so that

$$(p^2+q^2)\, dx\, dy = \frac{x\,(x^2+y^2)}{c\,(x^2+y^2+c^2)}\, dy\, dz,$$

along the characteristic surface.

Along the (plane) surface $PABQP$, we have $p=0$, $q=0$; that element of surface contributes zero to the corresponding integral. The plane $P'ABQ'P'$ is

$$z - z_0 = (x - x_0) \tan \epsilon,$$

so that

$$p = \tan \epsilon, \quad q = 0,$$

over that element, while the area of the projection of the element is

$$l\, dy\,;$$

thus the part contributed to the integral is

$$l\, dy \tan^2 \epsilon.$$

But $l \tan \epsilon = dz$; so the contribution is

$$dy\, dz \tan \epsilon.$$

Therefore the whole contribution to the corresponding integral, for the element of the new surface, is

$$dy\, dz \tan \epsilon.$$

As, for a selected finite portion of the surface, $x$ and $y$ are finite and different from zero, the element of the new integral is small compared with the element of the old integral. Because this property can be made to hold for each corresponding pair of elements by taking a non-regular surface $PQBQ'P'AP$ instead of $PQQ'P'P$, it follows that the new integral can be made small (as small as we please), while the old integral (a minimum under weak variations) has a finite non-zero value.

The finite non-zero value is manifestly not a true minimum: its minimum quality is possessed only under the imposition of weak variations.

*Note.* The discussion shews that Dirichlet's Principle does not provide a true minimum in double integrals.

## *Isoperimetrical problems, for double integrals.*

**326.** Isoperimetrical problems arise for double integrals as naturally as for simple integrals. In the simplest case, we can be required to determine a dependent variable $z$, so that an integral

$$I = \iint g\,(x,\, y,\, z,\, p,\, q)\, dx\, dy$$

shall have a maximum or a minimum, subject to the condition that another integral

$$J = \iint \gamma\,(x,\,y,\,z,\,p,\,q)\,dx\,dy$$

shall have a constant assigned value, the range of integration for the two integrals being the same in magnitude and sense. It will be evident, from the analysis, that corresponding problems can be propounded and corresponding results can be obtained, should the subjects of integration involve derivatives of $z$ of order higher than the first.

The method adopted for the resolution of the problem will be the same as the earlier method of Chapter VIII. The increment of $I$ will be constructed for the most general variations that are possible. The integral $J$ will be subjected to these variations; and then, because $J$ is constant and its increment consequently is zero, the necessary limitations upon the complete generality of the variations will thus be indicated. The integral of $I$ need then be considered only for these limited variations; and the resulting conditions for a maximum or a minimum will emerge.

We shall discuss, not the above forms suited for special variations but the more general forms of the integrals, in which superficial parameters $u$ and $v$ are made the variables of integration. Thus we consider two integrals

$$I = \iint G\,(x,\,y,\,z,\,x_1,\,y_1,\,z_1,\,x_2,\,y_2,\,z_2)\,du\,dv,$$

$$J = \iint \Gamma\,(x,\,y,\,z,\,x_1,\,y_1,\,z_1,\,x_2,\,y_2,\,z_2)\,du\,dv,$$

where the functions $G$ and $\Gamma$ satisfy the respective sets of identities

$$\left.\begin{aligned} G &= x_1 \frac{\partial G}{\partial x_1} + y_1 \frac{\partial G}{\partial y_1} + z_1 \frac{\partial G}{\partial z_1} \\ 0 &= x_2 \frac{\partial G}{\partial x_1} + y_2 \frac{\partial G}{\partial y_1} + z_2 \frac{\partial G}{\partial z_1} \\ 0 &= x_1 \frac{\partial G}{\partial x_2} + y_1 \frac{\partial G}{\partial y_2} + z_1 \frac{\partial G}{\partial z_2} \\ G &= x_2 \frac{\partial G}{\partial x_2} + y_2 \frac{\partial G}{\partial y_2} + z_2 \frac{\partial G}{\partial z_2} \end{aligned}\right\}, \qquad \left.\begin{aligned} \Gamma &= x_1 \frac{\partial \Gamma}{\partial x_1} + y_1 \frac{\partial \Gamma}{\partial y_1} + z_1 \frac{\partial \Gamma}{\partial z_1} \\ 0 &= x_2 \frac{\partial \Gamma}{\partial x_1} + y_2 \frac{\partial \Gamma}{\partial y_1} + z_2 \frac{\partial \Gamma}{\partial z_1} \\ 0 &= x_1 \frac{\partial \Gamma}{\partial x_2} + y_1 \frac{\partial \Gamma}{\partial y_2} + z_1 \frac{\partial \Gamma}{\partial z_2} \\ \Gamma &= x_2 \frac{\partial \Gamma}{\partial x_2} + y_2 \frac{\partial \Gamma}{\partial y_2} + z_2 \frac{\partial \Gamma}{\partial z_2} \end{aligned}\right\},$$

and are regular for continuously varying arguments.

### First variation of both integrals.

**327.** As in the corresponding problem for single integrals, we impose upon $x$, $y$, $z$, a composite general weak variation, the constituents of which are simultaneous small arbitrary variations $\kappa_r X_r$, $\kappa_r Y_r$, $\kappa_r Z_r$, for $r = 1, 2, \ldots$; the constants $\kappa_1$, $\kappa_2$, ... are small, as usual; $X_r$, $Y_r$, $Z_r$, are arbitrary regular functions of $u$ and $v$, independent of $\kappa$, and also independent of one another. The otherwise arbitrary quality of all the elements in the composite variation is limited solely (i), by the possible assignment of values of $x$, $y$, $z$, along the boundary of the field of integration, and (ii), by the necessity of conserving the constant value of $J$.

Let **I** denote the consequent varied value of the integral $I$. By analysis, similar to that (§ 293) in the case of unrestricted weak variations, we find

$$\mathbf{I} - I = \Sigma \kappa_r \iint (X_r G_x + Y_r G_y + Z_r G_z)\, du\, dv$$
$$+ \Sigma \kappa_r \int \left\{ \left( X_r \frac{\partial G}{\partial x_1} + Y_r \frac{\partial G}{\partial y_1} + Z_r \frac{\partial G}{\partial z_1} \right) \frac{dv}{ds} \right.$$
$$\left. - \left( X_r \frac{\partial G}{\partial x_2} + Y_r \frac{\partial G}{\partial y_2} + Z_r \frac{\partial G}{\partial z_2} \right) \frac{du}{ds} \right\} ds + [\kappa^2]_G,$$

where $[\kappa^2]_G$ denotes the aggregate of terms of the second and higher degrees in the small quantities $\kappa$; the single integral is taken positively round the whole boundary of the field of integration in the double integral; and the definition of the quantities $G_x$, $G_y$, $G_z$, is

$$G_\mu = \frac{\partial G}{\partial \mu} - \frac{\partial}{\partial u}\left( \frac{\partial G}{\partial \mu_1} \right) - \frac{\partial}{\partial v}\left( \frac{\partial G}{\partial \mu_2} \right),$$

for $\mu = x$, $y$, $z$, in turn.

In the same way, and because the integral $J$ is to remain constant in value under all variations, we have

$$0 = \Sigma \kappa_r \iint (X_r \Gamma_x + Y_r \Gamma_y + Z_r \Gamma_z)\, du\, dv$$
$$+ \Sigma \kappa_r \int \left\{ \left( X_r \frac{\partial \Gamma}{\partial x_1} + Y_r \frac{\partial \Gamma}{\partial y_1} + Z_r \frac{\partial \Gamma}{\partial z_1} \right) \frac{dv}{ds} \right.$$
$$\left. - \left( X_r \frac{\partial \Gamma}{\partial x_2} + Y_r \frac{\partial \Gamma}{\partial y_2} + Z_r \frac{\partial \Gamma}{\partial z_2} \right) \frac{du}{ds} \right\} ds + [\kappa^2]_\Gamma ;$$

and this relation has to be satisfied by the arbitrary quantities $\kappa_r$, $X_r$, $Y_r$, $Z_r$, which otherwise are independent. For all these arbitrary quantities, thus limited, we have to make the original integral $I$ a maximum or a minimum.

*Variations leaving $J$ constant: characteristic equations.*

**328.** Consider, first of all, the variations which are zero everywhere along the whole boundary of the field and necessarily keep $J$ constant. Let

$$J_r = \iint (X_r \Gamma_x + Y_r \Gamma_y + Z_r \Gamma_z)\, du\, dv,$$

for any such variation $\kappa_r X_r$, $\kappa_r Y_r$, $\kappa_r Z_r$; then the limiting relation is

$$0 = \kappa_1 J_1 + \kappa_2 J_2 + \ldots + [\kappa^2]_\Gamma.$$

The integrals $J_1$, $J_2$, ... are taken over a surface, which now is limited by a fixed boundary; thus each integral $J_r$ is a constant.

Moreover, no one of these integrals $J_r$ is zero; if, *e.g.*, we had $J_1$ zero, while $X_r$, $Y_r$, $Z_r$, are arbitrary and independent, we should have

$$\Gamma_x = 0, \quad \Gamma_y = 0, \quad \Gamma_z = 0.$$

On the one hand, all the integrals $J_r$ would then become zero; the constancy of $J$ would be certainly maintained. On the other hand, the equations $\Gamma_x = 0$, $\Gamma_y = 0$, $\Gamma_z = 0$, would define $x$, $y$, $z$, as functions of $u$ and $v$; at the utmost, the primitive of these equations would possess only two functional forms (of precise arguments) by means of which to secure a maximum or minimum of $I$, as relatively inadequate for our general purpose as the possession of two mere constants in the case (§ 239) of single integrals of the first order. We reject, or set aside, this too restricted possibility; and so we assume that $J_1$ is not zero. Thus the limiting relation gives

$$\kappa_1 = - \sum_{r=2} \frac{J_r}{J_1} \kappa_r + K_2',$$

where $K_2'$ is of the second and higher degrees in $\kappa_2$, $\kappa_3$, ....

Similarly, let

$$I_r = \iint (X_r G_x + Y_r G_y + Z_r G_z)\, du\, dv,$$

so that, for the variations which vanish everywhere along the boundary,

$$\mathbf{I} - I = \kappa_1 I_1 + \kappa_2 I_2 + \dots + [\kappa^2]_G,$$

where the integrals $I_1$, $I_2$, ... are taken over a field limited by a fixed boundary, and thus are constants. The only limitation among the constants $\kappa_1$, $\kappa_2$, ... is that $\kappa_1$ is expressible in terms of the rest, by the relation

$$\kappa_1 = - \sum_{r=2} \frac{J_r}{J_1} \kappa_r + K_2'.$$

Hence

$$\mathbf{I} - I = \sum_{r=2} \left( I_r - \frac{J_r}{J_1} I_1 \right) \kappa_r + [\kappa^2]_G + I_1 K_2'$$

$$= \sum_{r=2} \left( I_r - \frac{J_r}{J_1} I_1 \right) \kappa_r + K_2,$$

where $K_2$ is an aggregate of terms of the second and higher degrees in $\kappa_2$, $\kappa_3$, ..., when substitution is made for $\kappa_1$.

As the integral $I$ is to be a maximum or a minimum, the quantity $\mathbf{I} - I$ must have a constant sign for all variations, the negative sign when $I$ is a maximum, the positive sign when $I$ is a minimum. All these variations depend upon the small constants $\kappa$, which can be positive or negative at choice, subject to the single foregoing relation; and account has been taken of the requirement that $J$ shall be constant, so that $\kappa_1$ no longer occurs in $\mathbf{I} - I$, and $\kappa_2$, $\kappa_3$, ... are unrestricted. Hence the terms of the first degree in $\kappa_2$, $\kappa_3$, ..., which occur in $\mathbf{I} - I$, must vanish; if not, as $K_2$ is insignificant compared with terms of the first degree, $\mathbf{I} - I$ could be made positive or negative at will. Further, the quantities $\kappa_2$, $\kappa_3$, ... are arbitrary; and they are independent of one another. Consequently, the terms of the first order in $\mathbf{I} - I$ can vanish, only if the coefficient of each of these quantities $\kappa$ vanishes, that is, if

$$\frac{I_2}{J_2} = \frac{I_3}{J_3} = \dots = \frac{I_1}{J_1}.$$

As $I_r$ and $J_r$, integrals taken round the fixed boundary, are constants, each of these fractions is a constant. Moreover, as $X_r$, $Y_r$, $Z_r$, are arbitrary in the integrals denoted by $I_r$ and $J_r$ respectively, each of these constants is arbitrary; and therefore their quotient may be taken to be an arbitrary constant, undeterminable (that is) by the analysis at this stage. Thus, denoting this arbitrary constant by $\lambda$, we have

$$I_r - \lambda J_r = 0,$$

for each of the values of $r$; and therefore

$$\iint \{X_r (G_x - \lambda \Gamma_x) + Y_r (G_y - \lambda \Gamma_y) + Z_r (G_z - \lambda \Gamma_z)\} \, du \, dv = 0,$$

for all arbitrary quantities $X_r$, $Y_r$, $Z_r$, being functions of $u$ and $v$, independent of one another. By the customary argument, this equation can be satisfied for such an arbitrary range of functions $X_r$, $Y_r$, $Z_r$, only if

$$G_x - \lambda \Gamma_x = 0, \quad G_y - \lambda \Gamma_y = 0, \quad G_z - \lambda \Gamma_z = 0,$$

which therefore are necessary equations.

Moreover, when these three equations are satisfied (as now will be assumed), each of the equations

$$I_s - \lambda J_s = 0,$$

for $s = 1, 2, \dots$, is satisfied.

### Variations leaving J constant: conditions at mobile boundary.

**329.** Consider, next, variations which are not zero along the boundary, an assumption which does not affect the foregoing three critical equations; and return to the general expression for $\mathbf{I} - I$. Let

$$\Omega = G - \lambda \Gamma,$$

so that $\Omega$ is precisely the same type of function as $G$ and $\Gamma$. We have

$$\mathbf{I} - I = \Sigma \kappa_r \iint (X_r G_x + Y_r G_y + Z_r G_z) \, du \, dv + \int L_G ds + [\kappa^2]_G,$$

while

$$0 = \Sigma \kappa_r \iint (X_r \Gamma_x + Y_r \Gamma_y + Z_r \Gamma_z) \, du \, dv + \int L_\Gamma ds + [\kappa^2]_\Gamma,$$

where $L_G$ and $L_\Gamma$ are linear in the derivatives of $G$ and of $\Gamma$, as likewise are $[\kappa^2]_G$ and $[\kappa^2]_\Gamma$.

Multiply the latter result by the foregoing constant $\lambda$, and subtract from the former. The terms involving the double integral are

$$\Sigma \kappa_r \iint \{X_r (G_x - \lambda \Gamma_x) + Y_r (G_y - \lambda \Gamma_y) + Z_r (G_z - \lambda \Gamma_z)\} \, du \, dv;$$

and they vanish because the three critical equations are satisfied.

The boundary terms, of the first degree in the quantities $\kappa$, are

$$\int (L_G - \lambda L_\Gamma) \, ds.$$

As the whole variation of the first order is to vanish, and as these are the only surviving terms, they also must vanish. Now let

$$\Sigma \kappa_r X_r = \kappa X, \quad \Sigma \kappa_r Y_r = \kappa Y, \quad \Sigma \kappa_r Z_r = \kappa Z,$$

where $\kappa$ is small of the first order; as there was only a single relation among the constants $\kappa$, and as the quantities $X_r$, $Y_r$, $Z_r$, were arbitrary functions of $u$ and $v$, each independent of the rest, the new quantities $X$, $Y$, $Z$, also are arbitrary independent functions of $u$ and $v$. The condition now becomes, on dropping the factor $\kappa$,

$$\int \left\{ \left( X \frac{\partial \Omega}{\partial x_1} + Y \frac{\partial \Omega}{\partial y_1} + Z \frac{\partial \Omega}{\partial z_1} \right) \frac{dv}{ds} - \left( X \frac{\partial \Omega}{\partial x_2} + Y \frac{\partial \Omega}{\partial y_2} + Z \frac{\partial \Omega}{\partial z_2} \right) \frac{du}{ds} \right\} ds = 0,$$

the integral being taken along the whole boundary.

As before, the condition must be satisfied for the whole of each distinct and separate portion of the boundary, because we can take a set of variations zero at each other portion: and the condition, now leaving a result only for that portion, must still be satisfied.

Similarly, as before, the condition, which is satisfied for the whole of each distinct portion of the boundary, must be satisfied everywhere on a portion ; because, among the arbitrary variations, we can take some which are arbitrarily zero, and some which are arbitrarily not zero, the choice being at our free disposal.

### Summary of requirements, from 'first' variation.

**330.** Summing up the results as regards the vanishing of the first variation, which is an essential condition for the existence of a maximum or a minimum for $I$, we have the following requirements :

We take a new *quantity $\Omega$, defined as*

$$\Omega = G - \lambda \Gamma,$$

*where* $\lambda$ *is an arbitrary constant; the equations*

$$
\begin{rcases}
\dfrac{\partial \Omega}{\partial x} - \dfrac{\partial}{\partial u}\left(\dfrac{\partial \Omega}{\partial x_1}\right) - \dfrac{\partial}{\partial v}\left(\dfrac{\partial \Omega}{\partial x_2}\right) = 0 \\[2ex]
\dfrac{\partial \Omega}{\partial y} - \dfrac{\partial}{\partial u}\left(\dfrac{\partial \Omega}{\partial y_1}\right) - \dfrac{\partial}{\partial v}\left(\dfrac{\partial \Omega}{\partial y_2}\right) = 0 \\[2ex]
\dfrac{\partial \Omega}{\partial z} - \dfrac{\partial}{\partial u}\left(\dfrac{\partial \Omega}{\partial z_1}\right) - \dfrac{\partial}{\partial v}\left(\dfrac{\partial \Omega}{\partial z_2}\right) = 0
\end{rcases}
$$

*must be satisfied everywhere in the field of integration; and the relation*

$$
\left(X\,\dfrac{\partial \Omega}{\partial x_1} + Y\,\dfrac{\partial \Omega}{\partial y_1} + Z\,\dfrac{\partial \Omega}{\partial z_1}\right)\dfrac{dv}{ds} - \left(X\,\dfrac{\partial \Omega}{\partial x_2} + Y\,\dfrac{\partial \Omega}{\partial y_2} + Z\,\dfrac{\partial \Omega}{\partial z_2}\right)\dfrac{du}{ds} = 0,
$$

*must be satisfied everywhere on each portion of the boundary of the field.*

The three equations are the *characteristic equations*, as before; and there is the same type of boundary condition, as before, for the absolute problem.

The type of characteristic equations is unaltered. But, in their primitive, there is a new arbitrary constant $\lambda$, which did not occur in the former problem. This constant is determinable by the new requirement that the integral $J$ shall conserve its assigned value.

COROLLARY. It at once follows that, *if an integral*

$$
\iint g\,(x,\,y,\,z,\,p,\,q)\,dx\,dy
$$

*is to be a maximum or a minimum, subject to the conservation of a constant value by an integral*

$$
\iint \gamma\,(x,\,y,\,z,\,p,\,q)\,dx\,dy,
$$

*and if* $f(x,\,y,\,z,\,p,\,q)$ *denote* $g(x,\,y,\,z,\,p,\,q) - \lambda\gamma(x,\,y,\,z,\,p,\,q)$, *where* $\lambda$ *is an arbitrary constant, then* $z$ *must satisfy the characteristic equation*

$$
\dfrac{\partial f}{\partial z} - \dfrac{\partial}{\partial x}\left(\dfrac{\partial f}{\partial p}\right) - \dfrac{\partial}{\partial y}\left(\dfrac{\partial f}{\partial q}\right) = 0;
$$

*and the relation*

$$
\left\{(\zeta - p\xi - q\eta)\dfrac{\partial f}{\partial p} + \xi f\right\}\dfrac{dy}{ds} - \left\{(\zeta - p\xi - q\eta)\dfrac{\partial f}{\partial q} + \eta f\right\}\dfrac{dx}{ds} = 0,
$$

*must be satisfied for all arbitrary small variations* $\kappa\xi,\ \kappa\eta,\ \kappa\zeta,$ *at the boundary which are in accord with the assigned conditions.*

*Continuity of certain derivatives through an isolated free line of discontinuity.*

**331.** It is possible—and it happens less infrequently for problems of relative maxima and minima than for problems of free maxima and minima— that curves of discontinuity (in the direction of tangent planes, or of curvature) may occur on the complete characteristic surface.

Should the contingency occur, analysis precisely similar to that, which is given in § 297 and is used in connection with a free line-discontinuity on a surface, leads to the property that *the quantities*

$$\frac{\partial \Omega}{\partial x_1}\frac{dv}{ds} - \frac{\partial \Omega}{\partial x_2}\frac{du}{ds}, \quad \frac{\partial \Omega}{\partial y_1}\frac{dv}{ds} - \frac{\partial \Omega}{\partial y_2}\frac{du}{ds}, \quad \frac{\partial \Omega}{\partial z_1}\frac{dv}{ds} - \frac{\partial \Omega}{\partial z_2}\frac{du}{ds},$$

*remain unchanged in value in passing from one region of the surface to another region on a different side of the line of discontinuity*, the function $\Omega$ denoting $G - \lambda\Gamma$ as in § 330. The march of the analysis differs from that in § 297, in much the same way as the corresponding calculations in the free problem and the relative problem for single integrals (§ 47, § 244).

## EXAMPLES.

**332.** In illustration of the tests, some problems will now be considered: in particular, the problem of determining a surface, enclosing a maximum volume with an assigned superficial area.

*Ex.* 1. A surface of given area is to have its centre of gravity as low as possible; shew that the mean measure of curvature is the reciprocal of the intercept of the normal to the surface between the surface and a fixed plane.

*Ex.* 2. *Required to find the surface which, under assigned boundary conditions, shall enclose the greatest volume within a superficial area of given extent.*

(A) Using Cartesian coordinates as the variables in the integral, we have to make the integral

$$I = \iint z \, dx \, dy$$

a maximum, subject to the condition

$$A = \iint (1 + p^2 + q^2)^{\frac{1}{2}} \, dx \, dy,$$

where $A$ is the constant magnitude of the given superficial area.

After the preceding result (§ 330, *Corollary*), we take a function $f(x, y, z, p, q)$ in the form

$$f = z - \lambda (1 + p^2 + q^2)^{\frac{1}{2}}.$$

The characteristic equation is

$$1 + \frac{\partial}{\partial x}\left\{\frac{\lambda p}{(1 + p^2 + q^2)^{\frac{1}{2}}}\right\} + \frac{\partial}{\partial y}\left\{\frac{\lambda q}{(1 + p^2 + q^2)^{\frac{1}{2}}}\right\} = 0.$$

When expanded, this equation becomes

$$\lambda \{(1 + q^2) r - 2pqs + (1 + p^2) t\} + (1 + p^2 + q^2)^{\frac{3}{2}} = 0.$$

The equation, giving the principal radii of curvature of any surface at any point, is

$$\rho^2 (rt - s^2) - \{(1 + q^2) r - 2pqs + (1 + p^2) t\} (1 + p^2 + q^2)^{\frac{1}{2}} \rho + (1 + p^2 + q^2)^2 = 0.$$

Hence, for the foregoing surface, we have

$$\frac{1}{\rho_1} + \frac{1}{\rho_2} = -\frac{1}{\lambda},$$

shewing that *the mean measure of curvature of the surface is constant.*

The constant $\lambda$ is seen to be not zero and to be not infinite, when regard is paid to its general origin. As it is the quotient of the two integrals, denoted by $I_r$ and $J_r$ in the text and involving the same arbitrary quantities $X_r$, $Y_r$, $Z_r$, we have $\lambda$ as a positive quantity. Thus the mean measure of curvature is negative. When $\rho_1$ and $\rho_2$ have the same sign, the sign must be negative; and so the surface is concave to the plane $z=0$.

The primitive of the foregoing differential equation of the second order has not been obtained in the general case when $\lambda$ is any assigned constant, the value of $\lambda$ being ultimately dependent on the initially given superficial area $A$. We set aside the case when $\lambda=\infty$; otherwise the analysis would ignore the requirement as to volume, and would give a minimal surface.

The most obvious particular case is that of the spherical surface, when

$$\rho_1 = \rho_2 = -2\lambda \; ;$$

and the value of $\lambda$ is derivable from that of $A$, limited by the boundary curve.

(B) The Legendre test gives

$$\frac{\partial^2 f}{\partial p^2} = -\lambda \, \frac{1+q^2}{(1+p^2+q^2)^{\frac{3}{2}}}, \quad \frac{\partial^2 f}{\partial q^2} = -\lambda \, \frac{1+p^2}{(1+p^2+q^2)^{\frac{3}{2}}},$$

$$\frac{\partial^2 f}{\partial p \, \partial q} = \lambda \, \frac{pq}{(1+p^2)^{\frac{3}{2}}} .$$

The requirements, that $\frac{\partial^2 f}{\partial p^2}$ and $\frac{\partial^2 f}{\partial q^2}$ do not vanish and must possess the same sign, and that they satisfy the inequality

$$\frac{\partial^2 f}{\partial p^2} \frac{\partial^2 f}{\partial q^2} > \left( \frac{\partial^2 f}{\partial p \, \partial q} \right)^2 ,$$

are satisfied. Also, as $\lambda$ is a positive constant, the common sign of $\frac{\partial^2 f}{\partial p^2}$ and $\frac{\partial^2 f}{\partial q^2}$ is negative; thus, other conditions being fulfilled, the solution provides a maximum volume.

(C) The problem, as stated, is lacking in the details of precise definition as regards limiting or boundary conditions.

For fuller precision, let it (e.g.) be required to have the surface of mean constant curvature—for simplicity, let it be taken to be a sphere—so that, its area being given, it shall contain the greatest possible volume when the rest of this volume is bound by the three plane faces of a trihedral angle.

We first have to find the conterminous relation between the spherical surface, and the three planes which bound the portion of the surface.

Along any plane giving a part of the boundary, the condition

$$\left\{ (\zeta - p\xi - q\eta) \frac{\partial f}{\partial p} + \xi f \right\} dy - \left\{ (\zeta - p\xi - q\eta) \frac{\partial f}{\partial q} + \eta f \right\} dx = 0$$

must always be satisfied. To find the conterminous relation between the spherical surface and the planes, take one of them temporarily for the plane $z=0$. Thus a part of the boundary curve lies in this plane; and for all variations, subject to the conservation of the boundary, we have $\zeta=0$. The condition becomes

$$\left\{ \xi f - (p\xi + q\eta) \frac{\partial f}{\partial p} \right\} dy - \left\{ \eta f - (p\xi + q\eta) \frac{\partial f}{\partial q} \right\} dx = 0.$$

Here $dx$, $dy$, $0$, is an elementary arc of the surface, while $p$, $q$, $-1$, are proportional to the direction-cosines of the normal. We are dealing temporarily with the plane $z=0$; and so it will be convenient to take the equation of the spherical surface in the form

$$F(x, y, z) = 0,$$

so that

$$p= -\frac{1}{\frac{\partial F}{\partial z}}\frac{\partial F}{\partial x}, \quad q= -\frac{1}{\frac{\partial F}{\partial z}}\frac{\partial F}{\partial y}.$$

As $\xi$ and $\eta$ are now arbitrary quantities, the requirement of lying in the plane being given by $\zeta=0$, the foregoing relation can be satisfied for all such arbitrary quantities, only if the coefficients of $\xi$ and of $\eta$ vanish simultaneously, that is, only if

$$\left(f-p\frac{\partial f}{\partial p}\right)dy+p\frac{\partial f}{\partial q}\,dx=0,$$

$$q\frac{\partial f}{\partial p}\,dy+\left(f-q\frac{\partial f}{\partial q}\right)dx=0;$$

hence

$$\left(f-p\frac{\partial f}{\partial p}-q\frac{\partial f}{\partial q}\right)f=0.$$

Now $f$ is not zero : so we must have

$$f-p\frac{\partial f}{\partial p}-q\frac{\partial f}{\partial q}=0.$$

But

$$f=z-\lambda\,(1+p^2+q^2)^{\frac{1}{2}},$$

and therefore

$$\frac{\lambda}{(1+p^2+q^2)^{\frac{1}{2}}}=0,$$

or

$$\lambda\frac{\frac{\partial F}{\partial z}}{\left\{\left(\frac{\partial F}{\partial x}\right)^2+\left(\frac{\partial F}{\partial y}\right)^2+\left(\frac{\partial F}{\partial z}\right)^2\right\}^{\frac{1}{2}}}=0;$$

hence

$$\frac{\partial F}{\partial z}=0,$$

and therefore the normal to the sphere lies in the plane $z=0$ : that is, the centre of the sphere lies in the boundary plane (temporarily taken to be the plane $z=0$).

Similarly, the centre of the sphere lies in each of the other two boundary planes. Therefore it is the point of intersection of the three planes.

Consequently, when we select the spherical surface, the solution of the problem is obtained by taking a sphere, with its centre at the point of the trihedral angle, and of a radius $\lambda$, such that

$$\lambda\,(A+B+C-\pi)=S,$$

where $S$ is the given superficial area, and $A$, $B$, $C$, are the angles between the pairs of the planes.

*Ex.* 3. Let the same problem, of obtaining *a maximum volume with a given superficial area*, be considered, with more general variables for integration admitting of simultaneous variations of $x$, $y$, $z$.

(A) The former notation, used (§ 301) specially for minimal surfaces, will be adopted. Thus we are able to make the volume

$$\iint zn\,du\,dv$$

a maximum, while the integral

$$\iint\{(y_1z_2-z_1y_2)^2+(z_1x_2-x_1z_2)^2+(x_1y_2-y_1x_2)^2\}^{\frac{1}{2}}\,du\,dv$$

is to be constant. We take, as variables of reference, the parameters of the nul-lines on the surface, so that

$$E = x_1{}^2 + y_1{}^2 + z_1{}^2 = 0, \quad G = x_2{}^2 + y_2{}^2 + z_2{}^2 = 0, \quad F = x_1 x_2 + y_1 y_2 + z_1 z_2,$$

$F$ not being zero. Then

$$V = \{(y_1 z_2 - z_1 y_2)^2 + (z_1 x_2 - x_1 z_2)^2 + (x_1 y_2 - y_1 x_2)^2\}^{\frac{1}{2}} = iF \; ;$$

so that the integrals are $\iint z(x_1 y_2 - y_1 x_2) \, du \, dv$, which is to be maximised ; and $i \iint F \, du \, dv$, which is to be constant. We therefore form the function

$$\Omega = z(x_1 y_2 - y_1 x_2) - i\lambda F,$$

where $\lambda$ is an arbitrary constant.

When the characteristic equations are framed, they become (after simple reductions)

$$\left. \begin{array}{l} 2i\lambda \, x_{12} + y_1 z_2 - z_1 y_2 = 0 \\ 2i\lambda \, y_{12} + z_1 x_2 - x_1 z_2 = 0 \\ 2i\lambda \, z_{12} + x_1 y_2 - y_1 x_2 = 0 \end{array} \right\}.$$

Because

$$x_1 x_{12} + y_1 y_{12} + z_1 z_{12} = 0, \quad x_2 x_{12} + y_2 y_{12} + z_2 z_{12} = 0,$$

from the nul-equations $E = 0$ and $G = 0$ respectively, the three characteristic equations are equivalent to a single equation.

With the notation of § 290, it follows that

$$2i\lambda \, x_{12} + l = 0, \quad 2i\lambda \, y_{12} + m = 0, \quad 2i\lambda \, z_{12} + n = 0 \; ;$$

and so, as

$$l x_{12} + m y_{12} + n z_{12} = VM,$$

it follows that

$$2i\lambda \, VM + V^2 = 0,$$

or, in the present case,

$$\frac{M}{F} = -\frac{1}{2\lambda}.$$

But the mean measure of curvature, always equal to

$$\frac{GL - 2FM + EN}{EG - F^2},$$

thus is equal to $-\dfrac{1}{\lambda}$ in the present case. We therefore have a *surface of constant mean curvature*.

(B) To consider it in greater detail, we begin with the general equations, applying to all surfaces. As the most general solution of the equations

$$E = x_1{}^2 + y_1{}^2 + z_1{}^2 = 0, \quad G = x_2{}^2 + y_2{}^2 + z_2{}^2 = 0,$$

we take

$$\left. \begin{array}{l} x_1 = P^2 - Q^2 \\ y_1 = i(P^2 + Q^2) \\ z_1 = 2PQ \end{array} \right\}, \qquad \left. \begin{array}{l} x_2 = R^2 - S^2 \\ y_2 = -i(R^2 + S^2) \\ z_2 = 2RS \end{array} \right\},$$

where $P, Q, R, S$ are functions of the conjugate complex variables $u$ and $v$. Then

$$F = x_1 x_2 + y_1 y_2 + z_1 z_2 = 2(QS + PR)^2,$$

and $F$ is not zero; otherwise, every arc on the surface would be zero, being given by

$$ds^2 = E \, du^2 + 2F \, du \, dv + G \, dv^2.$$

Also
$$V^2 = EG - F^2 = -F^2\,;$$
so we take
$$V = iF = 2i\,(QS + PR).$$
Then
$$
\left.
\begin{aligned}
l &= y_1 z_2 - z_1 y_2 = 2i\,(QS + PR)\,(QR + PS) \\
m &= z_1 x_2 - x_1 z_2 = 2\,(QS + PR)\,(QR - PS) \\
n &= x_1 y_2 - y_1 x_2 = 2i\,(QS + PR)\,(QS - PR)
\end{aligned}
\right\}\,;
$$

consequently, if $X$, $Y$, $Z$ are the direction-cosines of the normal to the surface,
$$X = \frac{QR + PS}{QS + PR}, \qquad Y = i\,\frac{PS - QR}{QS + PR}, \qquad Z = \frac{QS - PR}{QS + PR}.$$

From the foregoing values of $x_1$, $y_1$, $z_1$, $x_2$, $y_2$, $z_2$, we have
$$
\begin{aligned}
x_{12} &= 2\,(PP_2 - QQ_2) = \quad 2\,(RR_1 - SS_1), \\
y_{12} &= 2i\,(PP_2 + QQ_2) = -2i\,(RR_1 + SS_1), \\
z_{12} &= 2\,(QP_2 + PQ_2) = \quad 2\,(SR_1 + RS_1).
\end{aligned}
$$
The first and second of these, combined, lead to the relations
$$PP_2 = -SS_1, \qquad QQ_2 = -RR_1\,;$$
and then the third gives
$$\frac{R_1}{Q} + \frac{S_1}{P} = 0,$$
on removing the non-zero factor $QS + PR$. Hence there exists a quantity $\Omega$, such that
$$R_1 = Q\Omega, \qquad S_1 = -P\Omega, \qquad Q_2 = -R\Omega, \qquad P_2 = -S\Omega\,;$$
and, now,
$$
\left.
\begin{aligned}
x_{12} &= \quad 2\,(QR + SP)\,\Omega \\
y_{12} &= -2i\,(QR - SP)\,\Omega \\
z_{12} &= \quad 2\,(QS - RP)\,\Omega
\end{aligned}
\right\}\,.
$$
Again, we had (§ 290) quantities $L$ and $N$ as well as $M$, given by
$$
\begin{aligned}
VL &= lx_{11} + my_{11} + nz_{11}, \\
VN &= lx_{22} + my_{22} + nz_{22}.
\end{aligned}
$$
When we substitute the values of $x_1$, $y_1$, $z_1$, in the first; and the values of $x_2$, $y_2$, $z_2$, in the second; we find
$$
\left.
\begin{aligned}
L &= 2\,(QP_1 - PQ_1) \\
N &= 2\,(SR_2 - RS_2)
\end{aligned}
\right\}\,.
$$

All these formulæ belong to all surfaces, the parameters of the nul-lines being the variables of reference.

(C) For the special instance when the surface is one of mean curvature, the characteristic equations being those in (A), we have
$$-2\lambda\Omega = QS + PR,$$
so that, as $QS + PR$ is not zero, $\Omega$ is not zero. Then
$$
\begin{aligned}
-2\lambda\Omega_1 &= QS_1 + PR_1 + SQ_1 + RP_1 = SQ_1 + RP_1, \\
-2\lambda\Omega_2 &= QS_2 + PR_2 + SQ_2 + RP_2 = QS_2 + PR_2.
\end{aligned}
$$
From the former equation, expressing $\Omega_1$, we have
$$
\begin{aligned}
-2\lambda\Omega_{12} &= S_2 Q_1 + R_2 P_1 + SQ_{12} + RP_{12} \\
&= S_2 Q_1 + R_2 P_1 + S\,(-R\Omega_1 - \Omega R_1) + R\,(S\Omega_1 + \Omega S_1) \\
&= S_2 Q_1 + R_2 P_1 - \Omega^2\,(QS + PR) \\
&= S_2 Q_1 + R_2 P_1 + 2\lambda\Omega^3.
\end{aligned}
$$

The same result follows from the latter equation, expressing $\Omega_2$. Thus

$$-2\lambda\,(\Omega_{12}+\Omega^3)=S_2Q_1+R_2P_1.$$

(D) It appears that discrimination between possibilities is made by the quantities $L$ and $N$.

We have $L=2\,(QP_1-PQ_1)$; and therefore

$$L_2=2\,(Q_2P_1-P_2Q_1+QP_{12}-PQ_{12}).$$

But

$$P_{12}=S_1\Omega+S\Omega_1,\quad Q_{12}=-R_1\Omega-R\Omega_1;$$

and therefore

$$L_2=2\,(Q_2P_1-P_2Q_1)+2\,\{(QS_1+PR_1)\,\Omega+(QS+PR)\,\Omega_1\}$$
$$=-2\Omega\,(RP_1+Q_1S)-4\lambda\,\Omega\Omega_1=0.$$

Similarly, we find *

$$N_1=0.$$

Thus $L$ is independent of $v$; it therefore either is zero or is a function of $u$ alone. But

$$L=2\,(QP_1-PQ_1)\,;$$

and the analysis has not required the variable $u$ to be made precise, so that any function of $u$ still can be substituted for $u$, without affecting the fundamental equations. If, then, $L$ is a function of $u$, we absorb it into the derivatives $P_1$ and $Q_1$, and leave $L$ a pure constant. Hence, without loss of generality, we may take

$$L=2,\ \text{or}\ 0.$$

Similarly, $N$ is either zero or a function of $v$ alone; if the latter, it could be absorbed into the derivatives $R_2$ and $S_2$, and leave $N$ a pure constant. Hence, equally without loss of generality, we take

$$N=2,\ \text{or}\ 0.$$

Thus three cases emerge :

(i)  $L=0,\ N=0$ :

(ii)  $L=2,\ N=0$; or $L=0,\ N=2$; these being equivalent analytically, by interchange of the conjugate variables $u$ and $v$ :

(iii)  $L=2,\ N=2$.

It might seem as if all cases could be included in one, by taking $L=a,\ N=c$, where $a$ and $c$ are constants. But the central differential equation for $\Omega$ changes its form, according as $ac$ is not zero or is zero : when $ac$ is zero, the primitive is known; when $ac$ is not zero, the primitive is not known. The cases are analytically distinct on present knowledge, though there seems no organic geometrical cause for discontinuity in the continuous passage of the constant $ac$ through a zero value, the surface for this zero value (when $a$ and $c$ are zero together) being a sphere.

(E)  *Case* (i), $L=0,\ N=0$.

(a) We have

$$QP_1-PQ_1=0,\quad SR_2-RS_2=0\,;$$

so that

$$\frac{P_1}{P}=\frac{Q_1}{Q}=\rho,\quad \frac{R_2}{R}=\frac{S_2}{S}=\sigma.$$

Also

$$-2\lambda\Omega_1=SQ_1+RP_1,\quad -2\lambda\Omega_2=QS_2+PR_2,$$

so that

$$\rho=\frac{\Omega_1}{\Omega},\quad \sigma=\frac{\Omega_2}{\Omega}.$$

Thus

$$P=\Omega f\,(v),\quad Q=\Omega g\,(v),\quad R=\Omega\theta\,(u),\quad S=\Omega\phi\,(u),$$

---

* These equations $L_2=0$ and $N_1=0$ are the Mainardi-Codazzi relations for a surface of constant mean curvature ; see my *Lectures on Differential Geometry*, § 57.

where $f(v)$ and $g(v)$ are functions of $v$ alone, $\theta(u)$ and $\phi(u)$ are functions of $u$ alone, unrestricted by these equations.

Also

$$S_2 Q_1 + R_2 P_1 = \rho\sigma\,(QS + PR) = -2\lambda\,\frac{\Omega_1 \Omega_2}{\Omega}\;;$$

hence the equation, satisfied by $\Omega$, becomes

$$\Omega_{12} - \frac{\Omega_1 \Omega_2}{\Omega} = -\Omega^3.$$

The primitive of this equation is

$$\Omega^2 = \frac{-U'V'}{(U+V)^2},$$

where $U$ is an arbitrary function of $u$, and $V$ an arbitrary function of $v$. Moreover, the coexistence of the two sets of equations

$$\left.\begin{array}{ll} R_2 = \sigma R, & P_1 = \rho P \\ S_2 = \sigma S, & Q_1 = \rho Q \end{array}\right\},\quad \left.\begin{array}{ll} R_1 = & Q\Omega,\ \ Q_2 = -R\Omega \\ S_1 = -P\Omega,\ \ P_2 = & S\Omega \end{array}\right\},$$

where $\rho$ and $\sigma$ have their foregoing values, is secured through the equation satisfied by $\Omega$.

(b)  Thus

$$x_1 = P^2 - Q^2 = -\frac{U'V'}{(U+V)^2}(f^2 - g^2),\quad x_2 = R^2 - S^2 = -\frac{U'V'}{(U+V)^2}(\theta^2 - \phi^2),$$

$$y_1 = i(P^2 + Q^2) = -i\frac{U'V'}{(U+V)^2}(f^2 + g^2),\quad y_2 = -i(R^2 + S^2) = i\frac{U'V'}{(U+V)^2}(\theta^2 + \phi^2),$$

$$z_1 = 2PQ = -\frac{U'V'}{(U+V)^2}2fg,\quad z_2 = 2RS = -\frac{U'V'}{(U+V)^2}2\theta\phi,$$

where $f$ and $g$ involve $v$ alone, while $\theta$ and $\phi$ involve $u$ alone; and therefore

$$a + x = \frac{V'}{U+V}(f^2 - g^2) + A_1(v) = \frac{U'}{U+V}(\theta^2 - \phi^2) + B_1(u),$$

$$\beta + y = \frac{iV'}{U+V}(f^2 + g^2) + iA_2(v) = -\frac{iU'}{U+V}(\theta^2 + \phi^2) - i\,B_2(u),$$

$$\gamma + z = \frac{V'}{U+V}2fg + A_3(v) = \frac{U'}{U+V}2\theta\phi + B_3(u),$$

where $a$, $\beta$, $\gamma$ are constants; $A_1(v)$ and $B_1(u)$ are functions of $v$ and of $u$ respectively, to be chosen so as to harmonise the two expressions for $x + a$; similarly for $A_2(v)$ and $B_2(u)$ as regards $y + \beta$, and for $A_3(v)$ and $B_3(u)$ as regards $z + \gamma$: it being remembered that $U$ and $V$ are the arbitrary functions of the $\Omega$-primitive, to be retained for a general solution.

(c)  When the two values of $a + x$ agree,

$$V'(f^2 - g^2) + VA_1(v) + UA_1(v) = U'(\theta^2 - \phi^2) + UB_1(u) + VB_1(u).$$

These are the same for all values of $U$ and $V$, only if

$$A_1(v) = C_1 V + D_1,\quad B_1(u) = C_1 U + D_1',$$

where $C_1$, $D_1$, and $D_1'$ are constants; and then

$$V'(f^2 - g^2) + C_1 V^2 + D_1 V + C_1 UV + D_1 U$$
$$= U'(\theta^2 - \phi^2) + C_1 U^2 + D_1' U + C_1 UV + D_1' V.$$

Thus

$$V'(f^2 - g^2) + C_1 V^2 + (D_1 - D_1')\,V = U'(\theta^2 - \phi^2) + C_1 U^2 + (D_1' - D_1)\,U$$
$$= I_1,$$

where $I_1$ is a constant, for one side involves $v$ only and the other involves $u$ only.

In order that the two values of $\beta + y$ may agree, we similarly find that

$$A_2(v) = C_2 V + D_2, \quad B_2(u) = -C_2 U - D_2',$$

where $C_2$, $D_2$, $D_2'$ are constants, and

$$V'(f^2 + g^2) + C_2 V^2 + (D_2 - D_2') V = -U'(\theta^2 - \phi^2) - C_2 U^2 + (D_2' - D_2) U$$
$$= I_2,$$

where $I_2$ is a constant.

In order that the two values of $\gamma + z$ may agree, we similarly find that

$$A_3(v) = C_3 V + D_3, \quad B_3(u) = C_3 U + D_3',$$

where $C_3$, $D_3$, $D_3'$ are constants, and

$$V' 2fg + C_3 V^2 + (D_3 - D_3') V = U' 2\theta\phi + C_3 U^2 + (D_3' - D_3) U$$
$$= I_3,$$

where $I_3$ is a constant.

From these relations, we have

$$\{I_1 - (D_1 - D_1') V - C_1 V^2\}^2 - \{I_2 - (D_2 - D_2') V - C_2 V^2\}^2 + \{I_3 - (D_3 - D_3') V - C_3 V^2\}^2$$
$$= \{V'(f^2 - g^2)\}^2 - \{V'(f^2 + g^2)\}^2 + (V' 2fg)^2 = 0,$$

and $V$ is an arbitrary function: hence

$$I_1{}^2 - I_2{}^2 + I_3{}^2 = 0,$$
$$I_1(D_1 - D_1') - I_2(D_2 - D_2') + I_3(D_3 - D_3') = 0,$$
$$(D_1 - D_1')^2 - (D_2 - D_2')^2 + (D_3 - D_3')^2 = 2(I_1 C_1 - I_2 C_2 + I_3 C_3)$$
$$= r^2,$$

where $r$ is a constant,

$$C_1(D_1 - D_1') - C_2(D_2 - D_2') + C_3(D_3 - D_3') = 0,$$
$$C_1{}^2 - C_2{}^2 + C_3{}^2 = 0.$$

The same consequences follow from the use of the relations which involve the arbitrary function $U$.

Further, the values of $x + a$, $y + \beta$, $z + \gamma$, are given by the expressions

$$a + x = \frac{1}{U + V}[I_1 - C_1 V^2 - (D_1 - D_1') V] + C_1 V + D_1,$$

so that

$$\left. \begin{aligned} x + a - D_1 &= \frac{I_1 + (D_1' - D_1) V + C_1 U V}{U + V} \\ x + a - D_1' &= \frac{I_1 + (D_1 - D_1') U + C_1 U V}{U + V} \end{aligned} \right\},$$

and similarly,

$$\left. \begin{aligned} y + \beta - i D_2 &= i\, \frac{I_2 + (D_2' - D_2) V + C_2 U V}{U + V} \\ y + \beta - i D_2' &= i\, \frac{I_2 + (D_2 - D_2') U + C_2 U V}{U + V} \end{aligned} \right\},$$

$$\left. \begin{aligned} z + \gamma - D_3 &= \frac{I_3 + (D_3' - D_3) V + C_3 U V}{U + V} \\ z + \gamma - D_3' &= \frac{I_3 + (D_3 - D_3') U + C_3 U V}{U + V} \end{aligned} \right\}.$$

Hence

$$(x + a - D_1)^2 + (y + \beta - i D_2)^2 + (z + \gamma - D_3)^2 = \frac{r^2 V^2 + r^2 U V}{(U + V)^2},$$

$$(x + a - D_1')^2 + (y + \beta - i D_2')^2 + (z + \gamma - D_3')^2 = \frac{r^2 U^2 + r^2 U V}{(U + V)^2},$$

and so

$$(x+a-D_1)^2+(x+a-D_1')^2+(y+\beta-iD_2)^2+(y+\beta-iD_2')^2+(z+\gamma-D_3)^2+(z+\gamma-D_3')^2=r^2,$$

shewing that the surface is a sphere.

(*d*) It may be noted that a simple identification of the surface with a sphere is obtained by taking

$$\theta=\frac{1}{U'^{\frac{1}{2}}}, \quad \phi=\frac{U}{U'^{\frac{1}{2}}}, \quad g=\frac{i}{V'^{\frac{1}{2}}}, \quad f=i\,\frac{V}{V'^{\frac{1}{2}}} \, ;$$

$$D_1=0=D_2; \quad D_1'=0=D_2'; \quad D_3=\tfrac{1}{2}r=-D_3' \, ;$$

$$C_1=I_1=\tfrac{1}{2}r\,; \quad C_2=-I_2=\tfrac{1}{2}r\,;$$

and then

$$a+x=\tfrac{1}{2}r\,\frac{1+UV}{U+V}, \quad \beta+y=\frac{i}{2}r\,\frac{UV-1}{U+V}, \quad \gamma+z=\tfrac{1}{2}r\,\frac{U-V}{U+V},$$

giving a known parametric representation of the sphere

$$(x+a)^2+(y+\beta)^2+(z+\gamma)^2=\tfrac{1}{4}r^2.$$

(F) *Case* (ii), $L=2$, $N=0$; *or* $L=0$, $N=2$.

The equation for $\Omega$ is the same as in Case (i), so that

$$\Omega^2=-\frac{U'V'}{(U+V)^2}.$$

The development of the analysis is left as an exercise.

(G) *Case* (iii), $L=2$, $N=2$. We have

$$QP_1-PQ_1=1, \quad SR_2-RS_2=1,$$

for this case: and, always,

$$RP_1+SQ_1=-2\lambda\Omega_1, \quad PR_2+QS_2=-2\lambda\Omega_2,$$
$$QS+RP=-2\lambda\Omega.$$

Hence

$$\Omega P_1=P\Omega_1-\frac{S}{2\lambda} \left.\right\} \quad\quad \Omega R_2=R\Omega_2-\frac{Q}{2\lambda} \left.\right\}$$
$$\Omega Q_1=Q\Omega_1+\frac{R}{2\lambda} \left.\right\}, \quad\quad \Omega S_2=S\Omega_2+\frac{P}{2\lambda} \left.\right\};$$

and so

$$S_2Q_1+R_2P_1=-2\lambda\,\frac{\Omega_1\Omega_2}{\Omega}-\frac{1}{2\lambda\Omega}.$$

Thus the equation for $\Omega$ is

$$\Omega_{12}-\frac{\Omega_1\Omega_2}{\Omega}=\frac{1}{4\lambda^2\Omega}-\Omega^3.$$

Let

$$\Omega^2=-\frac{1}{2\lambda}e^{\theta i}\,;$$

then the equation for $\theta$ is

$$\tfrac{1}{2}\lambda\theta_{12}=\sin\theta,$$

the customary equation* associated with surfaces, having a constant mean curvature and having the quantities $L$ and $N$ simultaneously different from zero.

The primitive of this partial differential equation of the second order is not known.

---

* *Lectures on Differential Geometry*, § 57. It may be added that the equation is not, by itself, sufficient to specify such surfaces. Thus it is also the equation determining the angle $\theta$ between the asymptotic lines on a pseudo-spherical surface (that is, a surface with the Gauss measure of curvature constant and negative): *l.c.*, § 54.

*Second variation.*

**333.** The second variation can be treated very briefly; for it can be transformed, almost at once, into the second variation of the integral

$$\iint (G - \lambda \Gamma)\, du\, dv,$$

that is, into a second variation which already (§§ 305–314) has been discussed at length.

The transformation is effected as follows. The total change in $I$, for an arbitrary general weak variation, can be represented by the equation

$$\mathbf{I} - I = \kappa I_1 + \tfrac{1}{2}\kappa^2 I_2 + [\kappa^3],$$

where $\kappa I_1$ is the whole first variation, $\tfrac{1}{2}\kappa^2 I_2$ is the whole second variation, and $[\kappa^3]$ represents the aggregate of terms of the third and higher degrees in the small quantities. Also, because $J$ is constant under all variations, we have

$$0 = \kappa J_1 + \tfrac{1}{2}\kappa^2 J_2 + [\kappa^3],$$

where $\kappa J_1$ is the whole first variation for $J$, $\tfrac{1}{2}\kappa^2 J_2$ is the whole second variation, and $[\kappa^3]$ represents a corresponding aggregate of terms of the third and higher degrees in the small quantities. And, in each of these expressions for total variations, the small weak variations are of the composite character already considered (§§ 73, 239, 327).

All these composite variations, which allow $J$ to maintain its constant value and which make the first variation of $I$ vanish, are such that, for the composite terms $\kappa I_1$ and $\kappa J_1$, we have the relation

$$\kappa I_1 - \lambda \kappa J_1 = 0.$$

Hence, modifying the form $\mathbf{I} - I$, so as to limit its expression to the effect of those variations under which $J$ remains constant,

$$\mathbf{I} - I = \tfrac{1}{2}\kappa^2 (I_2 - \lambda J_2) + K_3,$$

where $K_3$ is now the modified aggregate of terms of the third and higher degrees in the small quantities. Thus the increment $\mathbf{I} - I$ is governed by the quantity

$$\tfrac{1}{2}\kappa^2 (I_2 - \lambda J_2).$$

But, if we take the function

$$F = G - \lambda \Gamma,$$

where $I = \iint G\, du\, dv$ and $J = \iint \Gamma\, du\, dv$, this dominant magnitude is the second variation of the integral

$$\iint F\, du\, dv,$$

of which the first variation has been made to vanish.

The tests, giving a definite positive value or a definite negative value for this second variation, have been established in the preceding chapter. All

F. C. V.                                                                 36 therefore that remains, for the discussion of the second variation of the general isoperimetrical problem, is to apply those tests to the function $F$, equal to the linear combination $G - \lambda\Gamma$ of the subjects of integration of the integral $I$, which is to be maximised or minimised, and of the integral $J$, the constant value of which is to be conserved through all variations.

### Strong variations in isoperimetrical problems.

**334.** We still have to consider the effect of strong variations on integrals arising in isoperimetrical problems, and to construct the corresponding form of the necessary Weierstrass test.

The method, used in § 256, will be used again. For its process, strong variations are imposed upon the integral $I$, which is to be maximised or minimised, and upon the integral $J$, which has to remain constant in value. From among these strong variations, we select those which allow the constant value of $J$ to be maintained; and then we assign the due criteria which, under the selected variations, make the increment of $I$ always positive (for the possession of a minimum) or make the increment of $I$ always negative (for the possession of a maximum). To this end, it will be sufficient (§ 322) to consider the arbitrary single strong variation of elementary type, already used for the discussion of unlimited maxima and minima for double integrals.

We take any point $P$, $(x, y, z)$, on the characteristic surface; and through $P$ we draw any arbitrary regular curve $C$ which lies in that characteristic surface, denoting by $x'$, $y'$, $z'$, the direction-cosines of the tangent to $C$ at the point $P$. Through this curve $C$, and everywhere cutting the characteristic surface at a finite angle which can vary continuously along the course of $C$, we draw a regular surface $\Sigma$. Through $P$ and lying in the surface $\Sigma$, an arbitrary direction off the curve $C$ is taken, its direction-cosines being denoted by $l$, $m$, $n$. A strong variation at $P$ is thus begun. To complete the form of this strong variation, we draw a consecutive characteristic surface through the initial curve $A$ on the central characteristic surface; and we denote, by $C'$, its curve of intersection with the regular non-characteristic surface $\Sigma$. The arbitrary direction $l$, $m$, $n$, through $P$ in this surface meets the curve $C'$ in some point, say $Q$, necessarily contiguous to $P$.

A weak characteristic variation, from the central surface (through $A$ and $C$) to the consecutive surface (through $A$ and $C'$), may be denoted by $\kappa X$, $\kappa Y$, $\kappa Z$. If $\kappa\xi$, $\kappa\eta$, $\kappa\zeta$, are the coordinates of $Q$ relative to $P$; and if $l$, $m$, $n$, are now the direction-cosines of $QP$ from $Q$ towards $P$ along the straight line $QP$, the length of which is denoted by $D$; we have

$$\kappa\xi = -lD, \quad \kappa\eta = -mD, \quad \kappa\zeta = -nD.$$

Also $X$, $Y$, $Z$, $= 0, 0, 0$, everywhere along the initial curve $A$, because the consecutive characteristic passes through $A$; while, at the arbitrarily selected point $P$ on the arbitrary curve $C$, we have $X$, $Y$, $Z$, $= \xi$, $\eta$, $\zeta$. As before (in § 318), we complete a parallelogram $PQQ'P'P$, with adjacent sides $D$ along $PP'$ and $ds$ along the arc of $C$. This elementary parallelogram on the surface $\Sigma$ is the rudimentary strong variation of the characteristic surface.

Exactly as in § 317, but now dealing with the two integrals, we construct the quantities

$$I_{A'} + I_\Delta - I_A, \quad J_{A'} + J_\Delta - J_A,$$

representing the strong variations of the integral $I$, $= \iint G\,du\,dv$, and of the integral $J$, $= \iint \Gamma\,du\,dv$. We write

$$\bar{G} = G(x, y, z, l, m, n, x', y', z'), \quad \bar{\Gamma} = \Gamma(x, y, z, l, m, n, x', y', z').$$

Then we have

$$I_{A'} + I_\Delta - I_A = \kappa \iint (X G_x + Y G_y + Z G_z)\,du\,dv$$

$$+ \kappa \int \left\{ \left( \xi \frac{\partial G}{\partial x_1} + \eta \frac{\partial G}{\partial y_1} + \zeta \frac{\partial G}{\partial z_1} \right) \frac{dv}{ds} - \left( \xi \frac{\partial G}{\partial x_2} + \eta \frac{\partial G}{\partial y_2} + \zeta \frac{\partial G}{\partial z_2} \right) \frac{du}{ds} \right\}\,ds$$

$$+ \int \bar{G} D\,ds + [\kappa^2],$$

$$J_{A'} + J_\Delta - J_A = \kappa \iint (X \Gamma_x + Y \Gamma_y + Z \Gamma_z)\,du\,dv$$

$$+ \kappa \int \left\{ \left( \xi \frac{\partial \Gamma}{\partial x_1} + \eta \frac{\partial \Gamma}{\partial y_1} + \zeta \frac{\partial \Gamma}{\partial z_1} \right) \frac{dv}{ds} - \left( \xi \frac{\partial \Gamma}{\partial x_2} + \eta \frac{\partial \Gamma}{\partial y_2} + \zeta \frac{\partial \Gamma}{\partial z_2} \right) \frac{du}{ds} \right\}\,ds$$

$$+ \int \bar{\Gamma} D\,ds + [\kappa^2],$$

where the double integrals are taken over the stretch of the characteristic surface between $A$ and $C$, the single integrals are taken positively round $C$, and $[\kappa^2]$ in each expression denotes the aggregate of terms of the second and higher degrees in $\kappa$.

The integral $J$ is to remain unchanged for all possible variations, individual and elementary like the parallelogram $PQQ'P'P$ on the surface $\Sigma$, or cumulative into a fillet represented by the integrals $\int \bar{G} D\,ds$ and $\int \bar{\Gamma} D\,ds$. Accordingly, whatever variation be adopted, the relation

$$J_{A'} + J_\Delta - J_A = 0$$

must always be satisfied.

Now the characteristic equations are

$$G_x - \lambda \Gamma_x = 0, \quad G_y - \lambda \Gamma_y = 0, \quad G_z - \lambda \Gamma_z = 0,$$

where $\lambda$ is the arbitrary constant in the general theory. By the use of these three equations, and because of the equation that represents the constancy of

$J$ under the small variations, the double integral can be eliminated from the expression for $I_{A'} + I_\Delta - I_A$. The result is a new expression for $I$ given by

$$I_{A'} + I_\Delta - I_A = \kappa \int \Bigg[ \left\{ \xi \left( \frac{\partial G}{\partial x_1} - \lambda \frac{\partial \Gamma}{\partial x_1} \right) + \eta \left( \frac{\partial G}{\partial y_1} - \lambda \frac{\partial \Gamma}{\partial y_1} \right) + \zeta \left( \frac{\partial G}{\partial z_1} - \lambda \frac{\partial \Gamma}{\partial z_1} \right) \right\} \frac{dv}{ds}$$

$$- \left\{ \xi \left( \frac{\partial G}{\partial x_2} - \lambda \frac{\partial \Gamma}{\partial x_2} \right) + \eta \left( \frac{\partial G}{\partial y_2} - \lambda \frac{\partial \Gamma}{\partial y_2} \right) + \zeta \left( \frac{\partial G}{\partial z_2} - \lambda \frac{\partial \Gamma}{\partial z_2} \right) \right\} \frac{du}{ds} \Bigg] ds$$

$$+ \int (\bar{G} - \lambda \bar{\Gamma}) D \, ds + K_2,$$

where $K_2$ is the whole aggregate of terms of the second and higher degrees in $\kappa$. But

$$\bar{G} - \bar{\Gamma} \lambda = \bar{F} = F(x, y, z, l, m, n, x', y', z');$$

and

$$\kappa \xi = - lD, \quad \kappa \eta = - mD, \quad \kappa \zeta = - nD;$$

hence

$$I_{A'} + I_\Delta - I_A = \int E_\Sigma D \, ds + K_2,$$

where

$$E_\Sigma = \bar{F} - \left( l \frac{\partial F}{\partial x_1} + m \frac{\partial F}{\partial y_1} + n \frac{\partial F}{\partial z_1} \right) \frac{dv}{ds} - \left( l \frac{\partial F}{\partial x_2} + m \frac{\partial F}{\partial y_2} + n \frac{\partial F}{\partial z_2} \right) \frac{du}{ds}.$$

The quantities $\kappa$ and $D$ are of the same order of small magnitude; and therefore $K_2$ is negligible (being of the second order) compared with the integral $\int E_\Sigma D \, ds$, which is of the first order. Thus we may take

$$I_{A'} + I_\Delta - I_A = \int E_\Sigma D \, ds,$$

the expression for the variation of the integral, due to the surface variation caused by the band of the surface $\Sigma$.

The band of the surface is as arbitrary as can be taken, alike in range, in breadth at every point $P$, and in orientation to the central surface. We have considered it composed of small parallelograms such as $PQQ'P'P$, at any

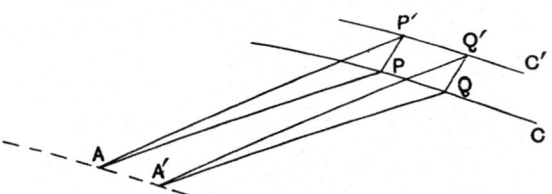

place $P$ of any curve $C$ on the characteristic surface. Thus we can regard the portion $AP'Q'A'A$ of the consecutive characteristic, together with the portion $PQQ'P'P$ of the surface $\Sigma$, as the selected arbitrary strong variation of the portion $APQA'A$ of the central characteristic: the magnitude of $D$ and the direction $PP'$ of $D$ at $P$ being arbitrary. Hence the foregoing relation must be satisfied for this variation, which now is the element $D\,ds$ of the line-integral $\int D\,ds$ constituting the band of $\Sigma$.

Consequently, if $E_\Sigma$ is always positive, for every such elementary area, for all points $P$ and for all curves $C$ on the characteristic, the increment of $I$ is always positive for all such elementary variations. When this necessary condition is satisfied for each element, and we take a band of the surface $\Sigma$, the increment of $I$ is cumulatively positive. Thus, when the condition as to $E_\Sigma$ being always positive is satisfied, $I$ has a minimum value for the characteristic surface.

Similarly, if $E_\Sigma$ is always negative within the same range of circumstances, the increment of $I$ is negative. In that event, $I$ has a maximum value for the characteristic surface.

Combining the results, we have the Weierstrass test for the integral $I$, subject to the requirement that the integral $J$ shall maintain an assigned constant value, as follows:

*Let $x$, $y$, $z$, be any point on the characteristic surface, $(x'$, $y'$, $z')$ be any direction $ds$ on the surface at that point, and $l$, $m$, $n$, be any direction off the surface at that point. Let $E_\Sigma$ denote the function*

$$E_\Sigma = F\,(x,\ y,\ z,\ l,\ m,\ n,\ x',\ y',\ z')$$
$$-\left(l\frac{\partial F}{\partial x_1}+m\frac{\partial F}{\partial y_1}+n\frac{\partial F}{\partial z_1}\right)\frac{dv}{ds}-\left(l\frac{\partial F}{\partial x_2}+m\frac{\partial F}{\partial y_2}+n\frac{\partial F}{\partial z_2}\right)\frac{du}{ds},$$

*where, if $I$ denotes $\iint G\,du\,dv$, the integral to be minimised, and $J$ denotes $\iint\Gamma\,du\,dv$, the integral to be kept constant, $F$ is equal to $G-\lambda\Gamma$. Then if, for all places $x$, $y$, $z$, on the surface; for all directions $x'$, $y'$, $z'$, on the surface; for all directions $l$, $m$, $n$, off the surface; $E_\Sigma$ has an unchanging sign: the integral $I$, taken over the characteristic surface, has a minimum value, as compared with its value over any other surface, when the unchanging sign of $E_\Sigma$ is positive; and the integral $I$ has, similarly, a maximum value when the unchanging sign of $E_\Sigma$ is negative.*

*Ex. The Weierstrass test shews that a maximum is admissible for the* problem in examples 2 and 3 of § 332 demanding the *surface which, within a given superficial area, shall enclose the greatest volume.*

The volume-integral $I$ is $\iint z\,(x_1y_2-y_1x_2)\,du\,dv$, so that
$$G=z\,(x_1y_2-y_1x_2).$$
The surface-integral $J$ is $\iint V\,du\,dv$, so that
$$\Gamma=V.$$

We therefore take

$$F = G - \lambda \Gamma = z (x_1 y_2 - y_1 x_2) - \lambda V.$$

The magnitude $E_\Sigma$ is linear in $F$ and its derivatives. We therefore may construct the portions of $E_\Sigma$ depending upon $G$ and upon $\Gamma$ separately, and then make the due linear combination.

The portion of $F$ which depends upon $G$ is

$$= \bar{G} - \frac{dv}{ds} \left( l \frac{\partial G}{\partial x_1} + m \frac{\partial G}{\partial y_1} + n \frac{\partial G}{\partial z_1} \right) + \frac{du}{ds} \left( l \frac{\partial G}{\partial x_2} + m \frac{\partial G}{\partial y_2} + n \frac{\partial G}{\partial z_2} \right)$$

$$= z (ly' - mx') - \frac{dv}{ds} (ly_2 - mx_2) z + \frac{du}{ds} \{l(-y_1) + mx_1\} z$$

$$= z (ly' - mx') - zl \left( y_2 \frac{dv}{ds} + y_1 \frac{du}{ds} \right) + zm \left( x_2 \frac{dv}{ds} + x_1 \frac{du}{ds} \right)$$

$$= z (ly' - mx') - zly' + zmx' = 0.$$

The $E_\Sigma$ for the integral $\iint \Gamma \, du \, dv$ has already been obtained in § 324: it is

$$1 - \cos \vartheta,$$

where $\vartheta$ denotes the angle at $P$ between the surface $\Sigma$ and the characteristic surface.

Hence the function $E_\Sigma$ for the problem is

$$E_\Sigma = 0 - \lambda (1 - \cos \vartheta)$$
$$= - \lambda (1 - \cos \vartheta).$$

As the surface $\Sigma$ lies off the characteristic surface, $\vartheta$ is neither 0 nor $2\pi$. Hence $E_\Sigma$ is always negative.

This test for the possession of a maximum is satisfied.

# CHAPTER XI.

DOUBLE INTEGRALS, WITH DERIVATIVES OF THE SECOND ORDER:
WEAK VARIATIONS.

*Double integral involving second derivatives: special weak variations:
'first' variation.*

**335.** Owing to the elaborate analysis necessary for a complete discussion, and owing to the fact that the characteristic equations usually are partial differential equations of the fourth order—a range of the subject in which only the rudiments of knowledge have been attained—we shall deal briefly with the criteria for a maximum or a minimum of an integral $I$, where

$$I = \iint f(x, y, z, p, q, r, s, t) \, dx \, dy.$$

As in preceding instances, we begin with a special weak variation represented by $z + \kappa Z$ in place of $z$, where $Z$ is any arbitrary regular function of $x$ and $y$, and is independent of $\kappa$. The derivatives of $Z$ are denoted by $P, Q, R, S, T$; and the whole variation of the integral under the specified small variation is

$$\kappa \iint U \, dx \, dy + K_2,$$

where

$$U = \frac{\partial f}{\partial z} Z + \frac{\partial f}{\partial p} P + \frac{\partial f}{\partial q} Q + \frac{\partial f}{\partial r} R + \frac{\partial f}{\partial s} S + \frac{\partial f}{\partial t} T,$$

while $K_2$ denotes the aggregate of terms of the second and higher degrees in $\kappa$.

To transform the first variation of $I$, we need some further inferences from the lemma of § 275. There, under stated conditions, it is proved that

$$\iint \frac{\partial \Phi}{\partial x} \, dx \, dy = \int \Phi \, dy, \quad \iint \frac{\partial \Psi}{\partial y} \, dx \, dy = -\int \Psi \, dx.$$

In the first of these results, let $\Phi = A \dfrac{\partial Z}{\partial x} - Z \dfrac{\partial A}{\partial x}$, for any function $A$; then

$$\iint \left( A \frac{\partial^2 Z}{\partial x^2} - Z \frac{\partial^2 A}{\partial x^2} \right) dx \, dy = \int \left( A \frac{\partial Z}{\partial x} - Z \frac{\partial A}{\partial x} \right) dy.$$

Hence

$$\iint \frac{\partial f}{\partial r} R \, dx \, dy = \iint Z \frac{\partial^2}{\partial x^2} \left( \frac{\partial f}{\partial r} \right) dx \, dy + \int \left\{ \frac{\partial f}{\partial r} \frac{\partial Z}{\partial x} - Z \frac{\partial}{\partial x} \left( \frac{\partial f}{\partial r} \right) \right\} dy.$$

In the second result, let $\Psi = C\dfrac{\partial Z}{\partial y} - Z\dfrac{\partial C}{\partial y}$, for any function $C$; then

$$\iint\left(C\frac{\partial^2 Z}{\partial y^2} - Z\frac{\partial^2 C}{\partial y^2}\right)dx\,dy = -\int\left(C\frac{\partial Z}{\partial y} - Z\frac{\partial C}{\partial y}\right)dx.$$

Hence

$$\iint\frac{\partial f}{\partial t}\,T\,dx\,dy = \iint Z\frac{\partial^2}{\partial y^2}\left(\frac{\partial f}{\partial t}\right)dx\,dy - \int\left\{\frac{\partial f}{\partial t}\frac{\partial Z}{\partial y} - Z\frac{\partial}{\partial y}\left(\frac{\partial f}{\partial t}\right)\right\}dx.$$

Next, for any function $B$, we have

$$\frac{\partial}{\partial x}\left(B\frac{\partial Z}{\partial y} - Z\frac{\partial B}{\partial y}\right) = B\frac{\partial^2 Z}{\partial x\,\partial y} - Z\frac{\partial^2 B}{\partial x\,\partial y} + \frac{\partial B}{\partial x}\frac{\partial Z}{\partial y} - \frac{\partial Z}{\partial x}\frac{\partial B}{\partial y},$$

$$\frac{\partial}{\partial y}\left(B\frac{\partial Z}{\partial x} - Z\frac{\partial B}{\partial x}\right) = B\frac{\partial^2 Z}{\partial x\,\partial y} - Z\frac{\partial^2 B}{\partial x\,\partial y} + \frac{\partial B}{\partial y}\frac{\partial Z}{\partial x} - \frac{\partial Z}{\partial y}\frac{\partial B}{\partial x};$$

and therefore

$$B\frac{\partial^2 Z}{\partial x\,\partial y} - Z\frac{\partial^2 B}{\partial x\,\partial y} = \tfrac{1}{2}\frac{\partial}{\partial x}\left(B\frac{\partial Z}{\partial y} - Z\frac{\partial B}{\partial y}\right) + \tfrac{1}{2}\frac{\partial}{\partial y}\left(B\frac{\partial Z}{\partial x} - Z\frac{\partial B}{\partial x}\right).$$

Consequently,

$$\iint\frac{\partial f}{\partial s}\,S\,dx\,dy = \iint Z\frac{\partial^2}{\partial x\,\partial y}\left(\frac{\partial f}{\partial s}\right)dx\,dy$$
$$+\tfrac{1}{2}\int\left\{\frac{\partial f}{\partial s}\frac{\partial Z}{\partial y} - Z\frac{\partial}{\partial y}\left(\frac{\partial f}{\partial s}\right)\right\}dy - \tfrac{1}{2}\int\left\{\frac{\partial f}{\partial s}\frac{\partial Z}{\partial x} - Z\frac{\partial}{\partial x}\left(\frac{\partial f}{\partial s}\right)\right\}dx.$$

Again,

$$\iint\frac{\partial}{\partial x}\left(Z\frac{\partial f}{\partial p}\right)dx\,dy = \int Z\frac{\partial f}{\partial p}\,dy,$$

so that

$$\iint\frac{\partial f}{\partial p}\,P\,dx\,dy = -\iint Z\frac{\partial}{\partial x}\left(\frac{\partial f}{\partial p}\right)dx\,dy + \int Z\frac{\partial f}{\partial p}\,dy;$$

and, similarly,

$$\iint\frac{\partial f}{\partial q}\,Q\,dx\,dy = -\iint Z\frac{\partial}{\partial y}\left(\frac{\partial f}{\partial q}\right)dx\,dy - \int Z\frac{\partial f}{\partial q}\,dx,$$

the line-integrals, in each instance, being taken in the positive direction for each portion of the boundary.

When these values are substituted for the respective six terms of the integral $\iint U\,dx\,dy$, it becomes

$$\iint Z\left\{\frac{\partial f}{\partial z} - \frac{\partial}{\partial x}\left(\frac{\partial f}{\partial p}\right) - \frac{\partial}{\partial y}\left(\frac{\partial f}{\partial q}\right) + \frac{\partial^2}{\partial x^2}\left(\frac{\partial f}{\partial r}\right) + \frac{\partial^2}{\partial x\,\partial y}\left(\frac{\partial f}{\partial s}\right) + \frac{\partial^2}{\partial y^2}\left(\frac{\partial f}{\partial t}\right)\right\}dx\,dy$$

$$+\int\left[\frac{\partial f}{\partial r}\frac{\partial Z}{\partial x} + \tfrac{1}{2}\frac{\partial f}{\partial s}\frac{\partial Z}{\partial y} + Z\left\{\frac{\partial f}{\partial p} - \frac{\partial}{\partial x}\left(\frac{\partial f}{\partial r}\right) - \tfrac{1}{2}\frac{\partial}{\partial y}\left(\frac{\partial f}{\partial s}\right)\right\}\right]dy$$

$$-\int\left[\tfrac{1}{2}\frac{\partial f}{\partial s}\frac{\partial Z}{\partial x} + \frac{\partial f}{\partial t}\frac{\partial Z}{\partial y} + Z\left\{\frac{\partial f}{\partial q} - \tfrac{1}{2}\frac{\partial}{\partial x}\left(\frac{\partial f}{\partial s}\right) - \frac{\partial}{\partial y}\left(\frac{\partial f}{\partial t}\right)\right\}\right]dx,$$

where the double integral is taken positively over the whole bounded field, and the single integrals are taken positively round the whole boundary of that field.

*The special variation, mobile at the boundary.*

**336.** But it is to be noted that, in the foregoing discussion, regard is had to the weak variation as a modification of the ordinate of the surface; only incidentally, and indirectly even then, is the character of the weak variation considered at the boundary. If the boundaries are definitely fixed, no hesitation need arise; along a boundary, $Z$ would then vanish, and the continuous values of $Z$ within the field of variation would have to be such as to vanish at the boundary. But the boundaries may happen not to be definitely fixed by assigned data. Thus, to take only one type of boundary, its portions may be required to lie upon given fixed surfaces, but are allowed to be mobile on those surfaces. In such an event, the variation of the surface for the integral must, at the boundary, be brought into relation with the variations that are admissible for the mobile boundary. Accordingly, we denote by $\kappa\xi, \kappa\eta, \kappa\zeta$, the small arbitrary variation of the point $(x, y, z)$ on the boundary, where the simultaneous values of $\xi, \eta, \zeta$, must be consistent with the equation (or equations) of the surface (or surfaces) upon which the mobile boundary is required to lie.

These quantities affect the first variation of the integral, in two different ways. The quantities $\kappa\xi$ and $\kappa\eta$, variations of a point on a portion of the boundary, are small variations of the limits of the integral belonging to that portion of the boundary; and they give rise, just as in § 278, to terms

$$\kappa \int (\xi f \, dy - \eta f \, dx),$$

taken positively along that portion of the boundary and, for the full first variation of the integral, aggregated for all the portions that are mobile in their several ways.

The quantity $Z$, together with the quantities $\dfrac{\partial Z}{\partial x}$ and $\dfrac{\partial Z}{\partial y}$, belonging to the rim of the bounded area when they occur in the single integral, though they have originated in the current variation within the area, must be brought into relation with $\zeta$ and its derivatives.

Anticipating the explanation of the characteristic surface which will ensue immediately, we note that $\kappa Z$ is the variation of the ordinate of that surface for all purposes after the immediately introductory analysis, because it is the surface which will provide the maximum or the minimum (if any). On the other hand, $\kappa\zeta$ is the variation in the value of the ordinate owing to the displacement of the terminal point on the boundary. Then, as in § 280, we have

$$Z = \zeta - p\xi - q\eta.$$

To bring the values of $\dfrac{\partial Z}{\partial x}$ and $\dfrac{\partial Z}{\partial y}$, at the rim, into relation with the derivatives of $\zeta$, imagine an associated surface having $\dfrac{\partial z}{\partial x} (= p)$, $x$, $y$ for the coordinates of a point corresponding to a point $z$, $x$, $y$, on the characteristic surface. The small variation of the ordinate of a point on this ancillary surface will be $\kappa \dfrac{\partial Z}{\partial x}$, while at the rim the small variations will be $\kappa\xi$, $\kappa\eta$, $\kappa \dfrac{\partial \zeta}{\partial x}$. Hence

$$\frac{\partial Z}{\partial x} = \frac{\partial \zeta}{\partial x} - \xi \frac{\partial}{\partial x}\left(\frac{\partial z}{\partial x}\right) - \eta \frac{\partial}{\partial y}\left(\frac{\partial z}{\partial x}\right)$$

$$= \frac{\partial \zeta}{\partial x} - r\xi - s\eta.$$

In the same manner, we find

$$\frac{\partial Z}{\partial y} = \frac{\partial \zeta}{\partial y} - s\xi - t\eta.$$

When these values are inserted, and the whole form is re-arranged, the complete expression for the integral $\iint U\,dx\,dy$, which is multiplied by $\kappa$ and then gives the first variation of the original integral $I$, is

$$\iint Z\left\{\frac{\partial f}{\partial z} - \frac{\partial}{\partial x}\left(\frac{\partial f}{\partial p}\right) - \frac{\partial}{\partial y}\left(\frac{\partial f}{\partial q}\right) + \frac{\partial^2}{\partial x^2}\left(\frac{\partial f}{\partial r}\right) + \frac{\partial^2}{\partial x\partial y}\left(\frac{\partial f}{\partial s}\right) + \frac{\partial^2}{\partial y^2}\left(\frac{\partial f}{\partial t}\right)\right\}dx\,dy$$

$$+ \int\left\{\left(\frac{\partial f}{\partial r}\frac{\partial \zeta}{\partial x} + \tfrac{1}{2}\frac{\partial f}{\partial s}\frac{\partial \zeta}{\partial y}\right)dy - \left(\tfrac{1}{2}\frac{\partial f}{\partial s}\frac{\partial \zeta}{\partial x} + \frac{\partial f}{\partial t}\frac{\partial \zeta}{\partial y}\right)dx\right\}$$

$$+ \int\left[f - p\left\{\frac{\partial f}{\partial p} - \frac{\partial}{\partial x}\left(\frac{\partial f}{\partial r}\right) - \tfrac{1}{2}\frac{\partial}{\partial y}\left(\frac{\partial f}{\partial s}\right)\right\} - \left(r\frac{\partial f}{\partial r} + \tfrac{1}{2}s\frac{\partial f}{\partial s}\right)\right]\xi\,dy$$

$$+ \int\left[p\left\{\frac{\partial f}{\partial q} - \tfrac{1}{2}\frac{\partial}{\partial x}\left(\frac{\partial f}{\partial s}\right) - \frac{\partial}{\partial y}\left(\frac{\partial f}{\partial t}\right)\right\} + \left(\tfrac{1}{2}r\frac{\partial f}{\partial s} + s\frac{\partial f}{\partial t}\right)\right]\xi\,dx$$

$$- \int\left[q\left\{\frac{\partial f}{\partial p} - \frac{\partial}{\partial x}\left(\frac{\partial f}{\partial r}\right) - \tfrac{1}{2}\frac{\partial}{\partial y}\left(\frac{\partial f}{\partial s}\right)\right\} + \left(s\frac{\partial f}{\partial r} + \tfrac{1}{2}t\frac{\partial f}{\partial s}\right)\right]\eta\,dy$$

$$- \int\left[f - q\left\{\frac{\partial f}{\partial q} - \tfrac{1}{2}\frac{\partial}{\partial x}\left(\frac{\partial f}{\partial s}\right) - \frac{\partial}{\partial y}\left(\frac{\partial f}{\partial t}\right)\right\} - \left(\tfrac{1}{2}s\frac{\partial f}{\partial s} + t\frac{\partial f}{\partial t}\right)\right]\eta\,dx$$

$$+ \int\zeta\left[\left\{\frac{\partial f}{\partial p} - \frac{\partial}{\partial x}\left(\frac{\partial f}{\partial r}\right) - \tfrac{1}{2}\frac{\partial}{\partial y}\left(\frac{\partial f}{\partial s}\right)\right\}dy - \left\{\frac{\partial f}{\partial q} - \tfrac{1}{2}\frac{\partial}{\partial x}\left(\frac{\partial f}{\partial s}\right) - \frac{\partial}{\partial y}\left(\frac{\partial f}{\partial t}\right)\right\}dx\right].$$

The double integral is taken over the whole of the bounded area; and each of the single integrals is taken positively along each portion (if there be more than one separate portion) of the boundary of that area. The aggregate of line-integrals in this expression will be denoted briefly by

$$\int(\Delta\,dy - \Gamma\,dx).$$

### Characteristic equation: boundary condition.

**337.** In order that the integral may possess a maximum or a minimum, its first variation must vanish for all the non-zero weak variations that are admissible; whether they (i) are such as to vanish everywhere on the boundary; or (ii) are such as to vanish, (*a*) only along a portion of a boundary, or (*b*) partly along a portion of the boundary at choice, or (*c*) nowhere on such a portion, should this last alternative be necessary for consideration. When we take the non-zero variations which vanish everywhere on the boundary, the double integral must still vanish. We infer that the relation

$$\frac{\partial f}{\partial z} - \frac{\partial}{\partial x}\left(\frac{\partial f}{\partial p}\right) - \frac{\partial}{\partial y}\left(\frac{\partial f}{\partial q}\right) + \frac{\partial^2}{\partial x^2}\left(\frac{\partial f}{\partial r}\right) + \frac{\partial^2}{\partial x\partial y}\left(\frac{\partial f}{\partial s}\right) + \frac{\partial^2}{\partial y^2}\left(\frac{\partial f}{\partial t}\right) = 0$$

must be satisfied everywhere within the region of the area of integration.

This relation is the *characteristic equation*, to be satisfied by $z$ as a function of $x$ and $y$. Manifestly, it is a partial differential equation of the fourth order, in which the terms involving derivatives of $z$ of that fourth order are

$$\frac{\partial^2 f}{\partial r^2}\frac{\partial^4 z}{\partial x^4} + 2\frac{\partial^2 f}{\partial r\partial s}\frac{\partial^4 z}{\partial x^3\partial y} + \left(\frac{\partial^2 f}{\partial s^2} + 2\frac{\partial^2 f}{\partial r\partial t}\right)\frac{\partial^4 z}{\partial x^2\partial y^2} + 2\frac{\partial^2 f}{\partial s\partial t}\frac{\partial^4 z}{\partial x\partial y^3} + \frac{\partial^2 f}{\partial t^2}\frac{\partial^4 z}{\partial y^4}.$$

The primitive of the equation contains four arbitrary functions. Whether the arguments of the four functions are distinct from one another (as generally is the fact), or may be the same for two of them, or for three of them, or even for all four of them, each such argument $u$ satisfies the critical equation

$$\frac{\partial^2 f}{\partial r^2}\left(\frac{\partial u}{\partial x}\right)^4 + 2\frac{\partial^2 f}{\partial r\partial s}\left(\frac{\partial u}{\partial x}\right)^3\frac{\partial u}{\partial y} + \left(\frac{\partial^2 f}{\partial s^2} + 2\frac{\partial^2 f}{\partial r\partial t}\right)\left(\frac{\partial u}{\partial x}\right)^2\left(\frac{\partial u}{\partial y}\right)^2$$
$$+ 2\frac{\partial^2 f}{\partial s\partial t}\frac{\partial u}{\partial x}\left(\frac{\partial u}{\partial y}\right)^3 + \frac{\partial^2 f}{\partial t^2}\left(\frac{\partial u}{\partial y}\right)^4 = 0.$$

The surface, represented by the primitive of the characteristic equation, is the *characteristic surface*.

**338.** Next, consider non-zero variations which do not vanish at the rim. For these (as for all) variations, the double integral in the condition now disappears, because the characteristic equation is satisfied everywhere in the region of variation. Therefore the surviving conditions at the boundary must be satisfied by themselves. Moreover, the arbitrary variations along the boundary are limited, if at all, only by the circumstances of the boundary; the variations for each separate portion of the boundary are independent of the variations for every other portion, and they are arbitrary along each such portion. Hence, along each separate portion, the relation

$$\int(\Delta\, dy - \Gamma\, dx) = 0$$

must be satisfied. Consequently, as the variations still possess their arbitrary quality along each portion so that, within restrictions that leave arbitrary variation possible, $\xi$, $\eta$, and $\zeta$, can be made zero or be taken different from zero at will, *the relation*

$$\Delta\, dy - \Gamma\, dx = 0$$

*must be satisfied everywhere on each portion of the boundary*, for arbitrary quantities $\kappa\xi$, $\kappa\eta$, $\kappa\zeta$, representing the displacement of a point, consistent with the observance of the general conditions of the boundary.

*Second variation: initial transformation.*

**339.** Now that the first variation of the integral has been made to vanish, both throughout the field by means of the characteristic equation, and along the whole boundary of the field by means of terminal conditions (whether in the form of assigned data, or through deduced relations), we pass to the second variation of the original integral.

The whole of the remainder of the variation of the integral can be expressed in the form

$$\tfrac{1}{2}\kappa^2 \iint \Theta \, dx \, dy + K_3,$$

where $K_3$ denotes the aggregate of terms of the third and higher degrees in $\kappa$, and where

$$\Theta = f_{55}T^2 + 2f_{45}ST + 2f_{35}RT + f_{44}S^2 + 2f_{34}RS + f_{33}R^2$$
$$+ 2\left( f_{25}QT + f_{15}PT + f_{05}ZT + f_{24}QS + f_{14}PS + f_{04}ZS + f_{23}QR + f_{13}PR + f_{03}ZR \right)$$
$$+ f_{22}Q^2 + 2f_{12}QP + 2f_{02}QZ + f_{11}P^2 + 2f_{01}PZ + f_{00}Z^2,$$

the subscripts $0, 1, 2, 3, 4, 5$, denoting derivativation with regard to $z, p, q, r, s, t$, respectively. As usual, the whole variation is now dominated by the term $\tfrac{1}{2}\kappa^2 \iint \Theta \, dx \, dy$, when $\kappa$ is sufficiently small. If either a maximum or a minimum is to occur, this quantity must maintain a persistent sign for all possible special variations. We therefore proceed to the construction of a normal form for the integral $\iint \Theta \, dx \, dy$, chosen so as to allow necessary and sufficient criteria to be framed.

For this purpose, we use the method adopted in previous instances. We add to $\Theta$ a quantity

$$\frac{\partial}{\partial x}\left( AP^2 + BQ^2 + CZ^2 + 2aQZ + 2bPZ + 2cPQ \right)$$
$$+ \frac{\partial}{\partial y}\left( KP^2 + LQ^2 + MZ^2 + 2kQZ + 2lPZ + 2mPQ \right);$$

and we use the relation

$$\iint \left( \Theta + \frac{\partial U}{\partial x} + \frac{\partial V}{\partial y} \right) dx \, dy = \iint \Theta \, dx \, dy + \int (U \, dy - V \, dx).$$

The limiting conditions fix the values of $z, p, q$, along the boundary and so make $Z, P, Q$, vanish there. Consequently the line-integral vanishes, and so allows the second variation to be determined by the double integral on the left-hand side.

To deduce the modified form, the coefficients $A, \ldots, c, K, \ldots, m$, are chosen so as to establish a relation

$$\Theta + \frac{\partial U}{\partial x} + \frac{\partial V}{\partial y} = f_{55}w^2 + 2f_{45}wv + 2f_{35}wu + f_{44}v^2 + 2f_{34}vu + f_{33}u^2,$$

where

$$\left.\begin{array}{l} u = R + \alpha P + \beta Q + \gamma Z \\ v = S + \lambda P + \mu Q + \nu Z \\ w = T + \rho P + \sigma Q + \tau Z \end{array}\right\},$$

while the newly introduced quantities $\alpha$, $\beta$, ..., $\sigma$, $\tau$, will also have to be determined.

*Equations of relation to modify the expression for the second variation.*

**340.** Manifestly the terms in $(R, S, T)^2$ on the two sides agree, without any imposed or inferred condition.

In order that the terms in $(R, S, T \chi Z, P, Q)$ may agree, the nine relations —three sets, with three in each set—must be satisfied:

$$\left.\begin{array}{l} f_{25} + L = \sigma f_{55} + \mu f_{45} + \beta f_{35} \\ f_{15} + m = \rho f_{55} + \lambda f_{45} + \alpha f_{35} \\ f_{05} + k = \tau f_{55} + \nu f_{45} + \gamma f_{35} \end{array}\right\},$$

$$\left.\begin{array}{l} f_{24} + B + m = \sigma f_{45} + \mu f_{44} + \beta f_{34} \\ f_{14} + K + c = \rho f_{45} + \lambda f_{44} + \alpha f_{34} \\ f_{04} + a + l = \tau f_{45} + \nu f_{44} + \gamma f_{34} \end{array}\right\},$$

$$\left.\begin{array}{l} f_{23} + c = \sigma f_{35} + \mu f_{34} + \beta f_{33} \\ f_{13} + A = \rho f_{35} + \lambda f_{34} + \alpha f_{33} \\ f_{03} + b = \tau f_{35} + \nu f_{34} + \gamma f_{33} \end{array}\right\}.$$

In order that the terms in $(Z, P, Q)^2$ may agree, the following six relations must be satisfied:

$$\left.\begin{array}{l} f_{22} + \dfrac{\partial L}{\partial y} + \dfrac{\partial B}{\partial x} + 2k = f_{55}\sigma^2 + 2f_{45}\sigma\mu + 2f_{35}\sigma\beta \\ \qquad\qquad\qquad\qquad + f_{44}\mu^2 + 2f_{34}\mu\beta + f_{33}\beta^2 \\[4pt] f_{12} + \dfrac{\partial c}{\partial x} + \dfrac{\partial m}{\partial y} + a + l = f_{55}\sigma\rho + f_{45}(\sigma\lambda + \mu\rho) + f_{35}(\sigma\alpha + \rho\beta) \\ \qquad\qquad\qquad\qquad + f_{44}\mu\lambda + f_{34}(\mu\alpha + \lambda\beta) + f_{33}\alpha\beta \\[4pt] f_{11} + \dfrac{\partial A}{\partial x} + \dfrac{\partial K}{\partial y} + 2b = f_{55}\rho^2 + 2f_{45}\rho\lambda + 2f_{35}\rho\alpha \\ \qquad\qquad\qquad\qquad + f_{44}\lambda^2 + 2f_{34}\lambda\alpha + f_{33}\alpha^2 \\[4pt] f_{02} + \dfrac{\partial a}{\partial x} + \dfrac{\partial k}{\partial y} + M = f_{55}\sigma\tau + f_{45}(\sigma\nu + \mu\tau) + f_{35}(\sigma\gamma + \beta\tau) \\ \qquad\qquad\qquad\qquad + f_{44}\mu\nu + f_{34}(\mu\gamma + \beta\nu) + f_{33}\beta\gamma \\[4pt] f_{01} + \dfrac{\partial b}{\partial x} + \dfrac{\partial l}{\partial y} + C = f_{55}\rho\tau + f_{45}(\rho\nu + \lambda\tau) + f_{35}(\rho\gamma + \alpha\tau) \\ \qquad\qquad\qquad\qquad + f_{44}\lambda\nu + f_{34}(\alpha\nu + \lambda\gamma) + f_{33}\alpha\gamma \\[4pt] f_{00} + \dfrac{\partial C}{\partial x} + \dfrac{\partial M}{\partial y} = f_{55}\tau^2 + 2f_{45}\tau\nu + 2f_{35}\tau\gamma \\ \qquad\qquad\qquad\qquad + f_{44}\nu^2 + 2f_{34}\nu\gamma + f_{33}\gamma^2 \end{array}\right\}.$$

When regard is paid to the results of earlier analogous reductions, it is natural to anticipate that the coefficients $\alpha, \ldots, \tau$, in the expressions for $u$, $v$, $w$, will be such as to make those three quantities vanish when $\kappa Z$, the small arbitrary variation of the characteristic ordinate, is made actually equal to $\kappa \mathbf{z}$, the small variation moving a point on to a contiguous characteristic surface; that is, we may anticipate that the relations

$$\left.\begin{aligned}
0 &= \mathbf{r} + \alpha \mathbf{p} + \beta \mathbf{q} + \gamma \mathbf{z} \\
0 &= \mathbf{s} + \lambda \mathbf{p} + \mu \mathbf{q} + \nu \mathbf{z} \\
0 &= \mathbf{t} + \rho \mathbf{p} + \sigma \mathbf{q} + \tau \mathbf{z}
\end{aligned}\right\}$$

will be satisfied. But this expectation requires the formation of the subsidiary characteristic equation, which must be satisfied by $\mathbf{z}$ because it arises out of the small characteristic variation $\kappa \mathbf{z}$.

### Subsidiary characteristic equation.

**341.** Accordingly, let $2\Omega$ denote the same function of $\mathbf{z}, \mathbf{p}, \mathbf{q}, \mathbf{r}, \mathbf{s}, \mathbf{t}$, as $\Theta$ is of $Z, P, Q, R, S, T$. When a small arbitrary variation is made on the general elements (that is, the arbitrary functions) in the primitive of the characteristic equation, the latter is still satisfied; and therefore the formal increment in this equation must be zero. The consequent changes in the members are, from $\dfrac{\partial f}{\partial z}, \dfrac{\partial f}{\partial p}, \dfrac{\partial f}{\partial q}, \dfrac{\partial f}{\partial r}, \dfrac{\partial f}{\partial s}, \dfrac{\partial f}{\partial t}$, to

$$\frac{\partial f}{\partial z} + \kappa \frac{\partial \Omega}{\partial \mathbf{z}}, \quad \frac{\partial f}{\partial p} + \kappa \frac{\partial \Omega}{\partial \mathbf{p}}, \quad \frac{\partial f}{\partial q} + \kappa \frac{\partial \Omega}{\partial \mathbf{q}}, \quad \frac{\partial f}{\partial r} + \kappa \frac{\partial \Omega}{\partial \mathbf{r}}, \quad \frac{\partial f}{\partial s} + \kappa \frac{\partial \Omega}{\partial \mathbf{s}}, \quad \frac{\partial f}{\partial t} + \kappa \frac{\partial \Omega}{\partial \mathbf{t}},$$

respectively. Therefore $\mathbf{z}$ satisfies the equation

$$\frac{\partial \Omega}{\partial \mathbf{z}} - \frac{\partial}{\partial x}\left(\frac{\partial \Omega}{\partial \mathbf{p}}\right) - \frac{\partial}{\partial y}\left(\frac{\partial \Omega}{\partial \mathbf{q}}\right) + \frac{\partial^2}{\partial x^2}\left(\frac{\partial \Omega}{\partial \mathbf{r}}\right) + \frac{\partial^2}{\partial x \partial y}\left(\frac{\partial \Omega}{\partial \mathbf{s}}\right) + \frac{\partial^2}{\partial y^2}\left(\frac{\partial \Omega}{\partial \mathbf{t}}\right) = 0,$$

which is the *subsidiary characteristic equation*.

Manifestly it is a linear partial differential equation of the fourth order in $\mathbf{z}$. The coefficients of $\mathbf{z}$ and of the derivatives of $\mathbf{z}$ are the values of the quantities $f_{mn}$ and their derivatives, when the value of $z$ given by the primitive of the central characteristic equation is substituted in them. And the primitive of the subsidiary equation is derivable through the value of $\kappa \mathbf{z}$, resulting from the most general arbitrary small variation made in the arbitrary functions that occur in the primitive of the characteristic equation. Moreover, the terms in the fourth derivatives of $\mathbf{z}$ are

$$\frac{\partial^2 f}{\partial r^2}\frac{\partial^4 \mathbf{z}}{\partial x^4} + 2\frac{\partial^2 f}{\partial r \partial s}\frac{\partial^4 \mathbf{z}}{\partial x^3 \partial y} + \left(\frac{\partial^2 f}{\partial s^2} + 2\frac{\partial^2 f}{\partial r \partial t}\right)\frac{\partial^4 \mathbf{z}}{\partial x^2 \partial y^2} + 2\frac{\partial^2 f}{\partial s \partial t}\frac{\partial^4 \mathbf{z}}{\partial x \partial y^3} + \frac{\partial^2 f}{\partial t^2}\frac{\partial^4 \mathbf{z}}{\partial y^4}.$$

These terms agree (as is to be expected) with the corresponding terms of highest order in the characteristic equation itself (§ 337). Their form exacts the same arguments, for the four arbitrary functions in the primitive of the subsidiary equation, as for the four arbitrary functions in the primitive of the characteristic.

### Resolution of the equations of § 340 by means of integrals of the subsidiary equation.

**342.** We now are in a position to construct a normal form for the second variation.

Let there be three linearly independent integrals $z_1$, $z_2$, $z_3$, of the subsidiary characteristic equation such that the determinant $\Delta$, given by

$$\Delta = \begin{vmatrix} p_1, & q_1, & z_1 \\ p_2, & q_2, & z_2 \\ p_3, & q_3, & z_3 \end{vmatrix} = |\, p_1, q_2, z_3 \,|,$$

with the notation already (§ 188) used for determinants, does not vanish.

First of all, we introduce nine quantities $\alpha$, $\beta$, $\gamma$; $\lambda$, $\mu$, $\nu$; $\rho$, $\sigma$, $\tau$; defined by the nine equations

$$r + \alpha p + \beta q + \gamma z = 0,$$

$$s + \lambda p + \mu q + \nu z = 0,$$

$$t + \rho p + \sigma q + \tau z = 0,$$

which are to hold for $z = z_1$, $z_2$, $z_3$, in succession. Then

$$\left. \begin{array}{l} \alpha\Delta + |\, r_1, q_2, z_3 \,| = 0 \\ \lambda\Delta + |\, s_1, q_2, z_3 \,| = 0 \\ \rho\Delta + |\, t_1, q_2, z_3 \,| = 0 \end{array} \right\}, \quad \left. \begin{array}{l} \beta\Delta - |\, r_1, p_2, z_3 \,| = 0 \\ \mu\Delta - |\, s_1, p_2, z_3 \,| = 0 \\ \sigma\Delta - |\, t_1, p_2, z_3 \,| = 0 \end{array} \right\}, \quad \left. \begin{array}{l} \gamma\Delta + |\, r_1, p_2, q_3 \,| = 0 \\ \nu\Delta + |\, s_1, p_2, q_3 \,| = 0 \\ \tau\Delta + |\, t_1, p_2, q_3 \,| = 0 \end{array} \right\}.$$

Next, we introduce nine quantities $A, c, b$; $K + c, B + m, a + l$; $m, L, k$; defined by other nine equations

$$\frac{\partial \Omega}{\partial r} = - A p - c q - b z,$$

$$\frac{\partial \Omega}{\partial s} = - (K + c) p - (B + m) q - (a + l) z,$$

$$\frac{\partial \Omega}{\partial t} = - m p - L q - k z,$$

which also are to hold for $z = z_1$, $z_2$, $z_3$, in succession; and we make quantities $\Omega_1$, $\Omega_2$, $\Omega_3$, successively take the place of $\Omega$ in the three substitutions respectively. In effect, the three second equations define $B$ and $K$, for $c$ is defined by the three first equations, and $m$ is defined by the three third equations. But the three second equations only define $a + l$, not $a$ and $l$ separately. Thus we have

$$A\Delta + \left| \frac{\partial \Omega_1}{\partial r_1}, \mathbf{q}_2, \mathbf{z}_3 \right| = 0$$

$$(K + c)\Delta + \left| \frac{\partial \Omega_1}{\partial s_1}, \mathbf{q}_2, \mathbf{z}_3 \right| = 0 \Biggr\},$$

$$m\Delta + \left| \frac{\partial \Omega_1}{\partial t_1}, \mathbf{q}_2, \mathbf{z}_3 \right| = 0$$

$$c\Delta - \left| \frac{\partial \Omega_1}{\partial r_1}, \mathbf{p}_2, \mathbf{z}_3 \right| = 0$$

$$(B + m)\Delta - \left| \frac{\partial \Omega_1}{\partial s_1}, \mathbf{p}_2, \mathbf{z}_3 \right| = 0 \Biggr\},$$

$$L\Delta - \left| \frac{\partial \Omega_1}{\partial t_1}, \mathbf{p}_2, \mathbf{z}_3 \right| = 0$$

$$b\Delta + \left| \frac{\partial \Omega_1}{\partial r_1}, \mathbf{p}_2, \mathbf{q}_3 \right| = 0$$

$$(a + l)\Delta + \left| \frac{\partial \Omega_1}{\partial s_1}, \mathbf{p}_2, \mathbf{q}_3 \right| = 0 \Biggr\}.$$

$$k\Delta + \left| \frac{\partial \Omega_1}{\partial t_1}, \mathbf{p}_2, \mathbf{q}_3 \right| = 0$$

Lastly, we take, subject to verifications that will be effected in §§ 344—346, the three equations

$$\frac{\partial \Omega}{\partial \mathbf{p}} = - A\mathbf{r} - (K + c)\mathbf{s} - m\mathbf{t} - \left( \frac{\partial A}{\partial x} + \frac{\partial K}{\partial y} + 2b \right)\mathbf{p}$$
$$- \left( \frac{\partial c}{\partial x} + \frac{\partial m}{\partial y} + a + l \right)\mathbf{q} - \left( \frac{\partial b}{\partial x} + \frac{\partial l}{\partial y} + C \right)\mathbf{z},$$

$$\frac{\partial \Omega}{\partial \mathbf{q}} = - c\mathbf{r} - (B + m)\mathbf{s} - L\mathbf{t} - \left( \frac{\partial c}{\partial x} + \frac{\partial m}{\partial y} + a + l \right)\mathbf{p}$$
$$- \left( \frac{\partial L}{\partial y} + \frac{\partial B}{\partial x} + 2k \right)\mathbf{q} - \left( \frac{\partial a}{\partial x} + \frac{\partial k}{\partial y} + M \right)\mathbf{z},$$

$$\frac{\partial \Omega}{\partial \mathbf{z}} = - b\mathbf{r} - (a + l)\mathbf{s} - k\mathbf{t} - \left( \frac{\partial b}{\partial x} + \frac{\partial l}{\partial y} + C \right)\mathbf{p}$$
$$- \left( \frac{\partial a}{\partial x} + \frac{\partial k}{\partial y} + M \right)\mathbf{q} - \left( \frac{\partial C}{\partial x} + \frac{\partial M}{\partial y} \right)\mathbf{z},$$

holding for $\mathbf{z} = \mathbf{z}_1, \mathbf{z}_2, \mathbf{z}_3$, in succession. We proceed to shew that these are not independent equations.

**343.** Take the last three equations of definition, and also the three preceding equations of definition arising with $\frac{\partial \Omega}{\partial r}$, $\frac{\partial \Omega}{\partial s}$, $\frac{\partial \Omega}{\partial t}$; all of them in connection with any unspecified integral of the subsidiary characteristic equation. It can be verified, by direct differentiation, that the equation

$$\frac{\partial \Omega}{\partial z} - \frac{\partial}{\partial x}\left(\frac{\partial \Omega}{\partial p}\right) - \frac{\partial}{\partial y}\left(\frac{\partial \Omega}{\partial q}\right) + \frac{\partial^2}{\partial x^2}\left(\frac{\partial \Omega}{\partial r}\right) + \frac{\partial^2}{\partial x \partial y}\left(\frac{\partial \Omega}{\partial s}\right) + \frac{\partial^2}{\partial y^2}\left(\frac{\partial \Omega}{\partial t}\right) = 0$$

is satisfied unconditionally and identically. Thus the last of the three equations of definition is a consequence of all the preceding equations of definition for any value of $z$, taken in conjunction with the characteristic equation; and this holds for $z = z_1, z_2, z_3$.

Again, the quantity $\Omega$ is a homogeneous quadratic expression in $z$, $p$, $q$, $r$, $s$, $t$. Hence we have

$$z_1 \frac{\partial \Omega_2}{\partial z_2} + p_1 \frac{\partial \Omega_2}{\partial p_2} + q_1 \frac{\partial \Omega_2}{\partial q_2} + r_1 \frac{\partial \Omega_2}{\partial r_2} + s_1 \frac{\partial \Omega_2}{\partial s_2} + t_1 \frac{\partial \Omega_2}{\partial t_2}$$
$$= z_2 \frac{\partial \Omega_1}{\partial z_1} + p_2 \frac{\partial \Omega_1}{\partial p_1} + q_2 \frac{\partial \Omega_1}{\partial q_1} + r_2 \frac{\partial \Omega_1}{\partial r_1} + s_2 \frac{\partial \Omega_1}{\partial s_1} + t_2 \frac{\partial \Omega_1}{\partial t_1},$$

$$z_2 \frac{\partial \Omega_3}{\partial z_3} + p_2 \frac{\partial \Omega_3}{\partial p_3} + q_2 \frac{\partial \Omega_3}{\partial q_3} + r_2 \frac{\partial \Omega_3}{\partial r_3} + s_2 \frac{\partial \Omega_3}{\partial s_3} + t_2 \frac{\partial \Omega_3}{\partial t_3}$$
$$= z_3 \frac{\partial \Omega_2}{\partial z_2} + p_3 \frac{\partial \Omega_2}{\partial p_2} + q_3 \frac{\partial \Omega_2}{\partial q_2} + r_3 \frac{\partial \Omega_2}{\partial r_2} + s_3 \frac{\partial \Omega_2}{\partial s_2} + t_3 \frac{\partial \Omega_2}{\partial t_2},$$

$$z_3 \frac{\partial \Omega_1}{\partial z_1} + p_3 \frac{\partial \Omega_1}{\partial p_1} + q_3 \frac{\partial \Omega_1}{\partial q_1} + r_3 \frac{\partial \Omega_1}{\partial r_1} + s_3 \frac{\partial \Omega_1}{\partial s_1} + t_3 \frac{\partial \Omega_1}{\partial t_1}$$
$$= z_1 \frac{\partial \Omega_3}{\partial z_3} + p_1 \frac{\partial \Omega_3}{\partial p_3} + q_1 \frac{\partial \Omega_3}{\partial q_3} + r_1 \frac{\partial \Omega_3}{\partial r_3} + s_1 \frac{\partial \Omega_3}{\partial s_3} + t_1 \frac{\partial \Omega_3}{\partial t_3}.$$

Multiplying these three relations by $z_3, z_1, z_2$, and adding, we have

$$\left| \frac{\partial \Omega_1}{\partial p_1}, p_2, z_3 \right| + \left| \frac{\partial \Omega_1}{\partial q_1}, q_2, z_3 \right| + \left| \frac{\partial \Omega_1}{\partial r_1}, r_2, z_3 \right| + \left| \frac{\partial \Omega_1}{\partial s_1}, s_2, z_3 \right| + \left| \frac{\partial \Omega_1}{\partial t_1}, t_2, z_3 \right| = 0;$$

and, similarly,

$$\left| \frac{\partial \Omega_1}{\partial z_1}, z_2, q_3 \right| + \left| \frac{\partial \Omega_1}{\partial p_1}, p_2, q_3 \right| + \left| \frac{\partial \Omega_1}{\partial r_1}, r_2, q_3 \right| + \left| \frac{\partial \Omega_1}{\partial s_1}, s_2, q_3 \right| + \left| \frac{\partial \Omega_1}{\partial t_1}, t_2, q_3 \right| = 0,$$

$$\left| \frac{\partial \Omega_1}{\partial z_1}, z_2, p_3 \right| + \left| \frac{\partial \Omega_1}{\partial q_1}, q_2, p_3 \right| + \left| \frac{\partial \Omega_1}{\partial r_1}, r_2, p_3 \right| + \left| \frac{\partial \Omega_1}{\partial s_1}, s_2, p_3 \right| + \left| \frac{\partial \Omega_1}{\partial t_1}, t_2, p_3 \right| = 0.$$

Now from the expressions for $\frac{\partial \Omega}{\partial p}$ and $\frac{\partial \Omega}{\partial q}$, for $z = z_1, z_2, z_3$, we have

$$\left| \frac{\partial \Omega_1}{\partial p_1}, p_2, z_3 \right| = -A \left| r_1, p_2, z_3 \right| - (K+c) \left| s_1, p_2, z_3 \right| - m \left| t_1, p_2, z_3 \right|$$
$$+ \Delta \left( \frac{\partial c}{\partial x} + \frac{\partial m}{\partial y} + a + l \right),$$

$$\left| \frac{\partial \Omega_1}{\partial q_1}, q_2, z_3 \right| = -c \left| r_1, q_2, z_3 \right| - (B+m) \left| s_1, q_2, z_3 \right| - L \left| t_1, q_2, z_3 \right|$$
$$- \Delta \left( \frac{\partial c}{\partial x} + \frac{\partial m}{\partial y} + a + l \right);$$

and from the like expressions for $\dfrac{\partial \Omega}{\partial \mathbf{r}}$, $\dfrac{\partial \Omega}{\partial \mathbf{s}}$, $\dfrac{\partial \Omega}{\partial \mathbf{t}}$, we have

$$\left|\frac{\partial \Omega_1}{\partial \mathbf{r}_1}, \mathbf{r}_2, \mathbf{z}_3\right| = A \mid \mathbf{r}_1, \mathbf{p}_2, \mathbf{z}_3 \mid + c \mid \mathbf{r}_1, \mathbf{q}_2, \mathbf{z}_3 \mid,$$

$$\left|\frac{\partial \Omega_1}{\partial \mathbf{s}_1}, \mathbf{s}_2, \mathbf{z}_3\right| = (K+c) \mid \mathbf{s}_1, \mathbf{p}_2, \mathbf{z}_3 \mid + (B+m) \mid \mathbf{s}_1, \mathbf{q}_2, \mathbf{z}_3 \mid,$$

$$\left|\frac{\partial \Omega_1}{\partial \mathbf{t}_1}, \mathbf{t}_2, \mathbf{z}_3\right| = m \mid \mathbf{t}_1, \mathbf{p}_2, \mathbf{z}_3 \mid + L \mid \mathbf{t}_1, \mathbf{q}_2, \mathbf{z}_3 \mid.$$

The first of the three deduced relations now shews that the two expressions for $\dfrac{\partial c}{\partial x} + \dfrac{\partial m}{\partial y} + a + l$ are equivalent to one another. Similarly, the postulated relations for $\dfrac{\partial \Omega}{\partial \mathbf{p}}$ and $\dfrac{\partial \Omega}{\partial \mathbf{z}}$ provide two expressions for $\dfrac{\partial b}{\partial x} + \dfrac{\partial l}{\partial y} + C$, equivalent in virtue of the second of the three relations; and the postulated relations for $\dfrac{\partial \Omega}{\partial \mathbf{q}}$ and $\dfrac{\partial \Omega}{\partial \mathbf{z}}$ provide two equivalent expressions for

$$\frac{\partial a}{\partial x} + \frac{\partial k}{\partial y} + M.$$

**344.** Returning to the equations of definition, we use the results in § 342 deduced from them; and we proceed to express the quantities $A, \ldots, k$, in terms of the coefficients $f_{r,n}$ and of the quantities $\alpha, \ldots, \tau$. For each of the three variables $\mathbf{z} = \mathbf{z}_1, \mathbf{z}_2, \mathbf{z}_3$, we have

$$\frac{\partial \Omega}{\partial \mathbf{r}} = f_{03}\mathbf{z} + f_{13}\mathbf{p} + f_{23}\mathbf{q} + f_{33}\mathbf{r} + f_{34}\mathbf{s} + f_{35}\mathbf{t}.$$

Therefore

$$\left|\frac{\partial \Omega_1}{\partial \mathbf{r}_1}, \mathbf{q}_2, \mathbf{z}_3\right| = f_{13}\Delta + f_{33} \mid \mathbf{r}_1, \mathbf{q}_2, \mathbf{z}_3 \mid + f_{34} \mid \mathbf{s}_1, \mathbf{q}_2, \mathbf{z}_3 \mid + f_{35} \mid \mathbf{t}_1, \mathbf{q}_2, \mathbf{z}_3 \mid$$

$$= f_{13}\Delta - \Delta\,(f_{33}\alpha + f_{34}\lambda + f_{35}\rho):$$

that is,

$$f_{13} + A = \alpha f_{33} + \lambda f_{34} + \rho f_{35}.$$

Similarly, from $\left|\dfrac{\partial \Omega_1}{\partial \mathbf{r}_1}, \mathbf{p}_2, \mathbf{z}_3\right|$ and from $\left|\dfrac{\partial \Omega_1}{\partial \mathbf{r}_1}, \mathbf{p}_2, \mathbf{q}_3\right|$, we have

$$f_{23} + c = \beta f_{33} + \mu f_{34} + \sigma f_{35},$$

$$f_{03} + b = \gamma f_{33} + \nu f_{34} + \tau f_{35}.$$

Again, we have

$$\frac{\partial \Omega}{\partial \mathbf{s}} = f_{04}\mathbf{z} + f_{14}\mathbf{p} + f_{24}\mathbf{q} + f_{34}\mathbf{r} + f_{44}\mathbf{s} + f_{45}\mathbf{t},$$

$$\frac{\partial \Omega}{\partial \mathbf{t}} = f_{05}\mathbf{z} + f_{15}\mathbf{p} + f_{25}\mathbf{q} + f_{35}\mathbf{r} + f_{45}\mathbf{s} + f_{55}\mathbf{t}.$$

Proceeding in the same way, we find

$$f_{14} + K + c = \alpha f_{34} + \lambda f_{44} + \rho f_{45}$$
$$f_{24} + B + m = \beta f_{34} + \mu f_{44} + \sigma f_{45}$$
$$f_{04} + a + l = \gamma f_{34} + \nu f_{44} + \tau f_{45}$$

and

$$f_{15} + m = \alpha f_{35} + \lambda f_{45} + \rho f_{55}$$
$$f_{25} + L = \beta f_{35} + \mu f_{45} + \sigma f_{55}$$
$$f_{05} + k = \gamma f_{35} + \nu f_{45} + \tau f_{55}$$

**345.** Expressions for all the coefficients $A$, $B$, $C$, $a$, $b$, $c$, $K$, $L$, $M$, $k$, $l$, $m$, have been obtained; but $a$ and $l$ occur in the combination $a + l$, while $C$ and $M$ occur in the respective combinations $\frac{\partial b}{\partial x} + \frac{\partial l}{\partial y} + C$ and $\frac{\partial a}{\partial x} + \frac{\partial k}{\partial y} + M$. Combinations, simpler than the latter, can be obtained. From the equations already given, we easily find

$$\frac{\partial \Omega}{\partial \mathbf{p}} - \frac{\partial}{\partial x}\left(\frac{\partial \Omega}{\partial \mathbf{r}}\right) = -m\mathbf{t} - k\mathbf{s} - \left(\frac{\partial K}{\partial y} + b\right)\mathbf{p} - \left(\frac{\partial m}{\partial y} + a + l\right)\mathbf{q} - \left(\frac{\partial l}{\partial y} + C\right)\mathbf{z},$$

$$\frac{\partial \Omega}{\partial \mathbf{q}} - \frac{\partial}{\partial y}\left(\frac{\partial \Omega}{\partial \mathbf{t}}\right) = -c\mathbf{r} - B\mathbf{s} - \left(\frac{\partial c}{\partial x} + a + l\right)\mathbf{p} - \left(\frac{\partial B}{\partial x} + k\right)\mathbf{q} - \left(\frac{\partial a}{\partial x} + M\right)\mathbf{z},$$

so that

$$\left| \left(\frac{\partial \Omega_1}{\partial \mathbf{p}_1} - \frac{\partial}{\partial x}\left(\frac{\partial \Omega_1}{\partial \mathbf{r}_1}\right)\right),\ \mathbf{p}_2,\ \mathbf{q}_3 \right| = -m\left|\ \mathbf{t}_1,\ \mathbf{p}_2,\ \mathbf{q}_3\ \right| - K\left|\ \mathbf{s}_1,\ \mathbf{p}_2,\ \mathbf{q}_3\ \right| - \Delta\left(\frac{\partial l}{\partial y} + C\right)$$

$$= -\tau\left|\ \frac{\partial \Omega_1}{\partial \mathbf{t}_1},\ \mathbf{q}_2,\ \mathbf{z}_3\ \right| - \nu\left|\ \frac{\partial \Omega_1}{\partial \mathbf{s}_1},\ \mathbf{q}_2,\ \mathbf{z}_3\ \right| - \nu\left|\ \frac{\partial \Omega_1}{\partial \mathbf{r}_1},\ \mathbf{p}_2,\ \mathbf{z}_3\ \right| - \Delta\left(\frac{\partial l}{\partial y} + C\right),$$

and

$$\left| \left(\frac{\partial \Omega_1}{\partial \mathbf{q}_1} - \frac{\partial}{\partial y}\left(\frac{\partial \Omega_1}{\partial \mathbf{t}_1}\right)\right),\ \mathbf{p}_2,\ \mathbf{q}_3 \right| = -c\left|\ \mathbf{r}_1,\ \mathbf{p}_2,\ \mathbf{q}_3\ \right| - B\left|\ \mathbf{s}_1,\ \mathbf{p}_2,\ \mathbf{q}_3\ \right| - \Delta\left(\frac{\partial a}{\partial x} + M\right)$$

$$= \gamma\left|\ \frac{\partial \Omega_1}{\partial \mathbf{r}_1},\ \mathbf{p}_2,\ \mathbf{z}_3\ \right| + \nu\left|\ \frac{\partial \Omega_1}{\partial \mathbf{s}_1},\ \mathbf{p}_2,\ \mathbf{z}_3\ \right| + \nu\left|\ \frac{\partial \Omega_1}{\partial \mathbf{t}_1},\ \mathbf{q}_2,\ \mathbf{z}_3\ \right| - \Delta\left(\frac{\partial a}{\partial x} + M\right).$$

We thus have expressions for

$$\frac{\partial l}{\partial y} + C, \quad \frac{\partial a}{\partial x} + M, \quad a + l.$$

We have, as well, an expression for $\frac{\partial C}{\partial x} + \frac{\partial M}{\partial y}$, which must be consistent with the expressions for the combination

$$\frac{\partial}{\partial x}\left(\frac{\partial l}{\partial y} + C\right) + \frac{\partial}{\partial y}\left(\frac{\partial a}{\partial x} + M\right) - \frac{\partial^2}{\partial x \partial y}(a + l):$$

it proves to be consistent with these expressions, through the subsidiary characteristic equation satisfied separately by $\mathbf{z}_1$, $\mathbf{z}_2$, $\mathbf{z}_3$.

Also, the expression added (§ 339) to $\Theta$, in order to obtain the modified normal form, is

$$\frac{\partial}{\partial x}\left(AP^2 + 2bPZ + 2cPQ + BQ^2\right) + \frac{\partial}{\partial y}\left(KP^2 + 2kQZ + 2mPQ + LQ^2\right)$$

$$+ Z^2\left(\frac{\partial C}{\partial x} + \frac{\partial M}{\partial y}\right) + 2\left(C + \frac{\partial l}{\partial y}\right)PZ + 2\left(M + \frac{\partial a}{\partial x}\right)QZ + 2\left(a + l\right)(ZS + PQ):$$

that is, the coefficients arise in the combinations which have been determined. Nothing has arisen to impose further restrictions on the choice of quantities satisfying the values of

$$\frac{\partial l}{\partial y} + C, \quad \frac{\partial a}{\partial x} + M, \quad a + l,$$

which involve $C$, $M$, $a$, $l$. One simple choice, in the absence of any other requirement, is to take $a = l$—the choice $a + l = 0$ being already excluded.

*Ex.* Relations involving $f_{12}$ can be derived from $\dfrac{\partial\Omega}{\partial\mathbf{p}}$ and $\dfrac{\partial\Omega}{\partial\mathbf{q}}$, in the forms

$$\left|\frac{\partial\Omega_1}{\partial\mathbf{p}_1}, \mathbf{p}_2, \mathbf{z}_3\right| = -f_{12}\Delta + f_{13}\left|\mathbf{r}_1, \mathbf{p}_2, \mathbf{z}_3\right| + f_{14}\left|\mathbf{s}_1, \mathbf{p}_2, \mathbf{z}_3\right| + f_{15}\left|\mathbf{t}_1, \mathbf{p}_2, \mathbf{z}_3\right|,$$

$$\left|\frac{\partial\Omega_1}{\partial\mathbf{q}_1}, \mathbf{q}_2, \mathbf{z}_3\right| = \phantom{-}f_{12}\Delta + f_{23}\left|\mathbf{r}_1, \mathbf{q}_2, \mathbf{z}_3\right| + f_{24}\left|\mathbf{s}_1, \mathbf{q}_2, \mathbf{z}_3\right| + f_{25}\left|\mathbf{t}_1, \mathbf{q}_2, \mathbf{z}_3\right|.$$

Verify that these two relations are equivalent to one another.

Obtain, similarly, two relations involving $f_{02}$, from $\dfrac{\partial\Omega}{\partial\mathbf{q}}$ and $\dfrac{\partial\Omega}{\partial\mathbf{z}}$; and two relations, involving $f_{01}$, from $\dfrac{\partial\Omega}{\partial\mathbf{z}}$ and $\dfrac{\partial\Omega}{\partial\mathbf{p}}$. Verify the necessary equivalence of the respective relations.

**346.** Lastly, we have, for each of the values of $\mathbf{z} = \mathbf{z}_1, \mathbf{z}_2, \mathbf{z}_3$,

$$\frac{\partial\Omega}{\partial\mathbf{p}} = f_{01}\mathbf{z} + f_{11}\mathbf{p} + f_{12}\mathbf{q} + f_{13}\mathbf{r} + f_{14}\mathbf{s} + f_{15}\mathbf{t},$$

$$\frac{\partial\Omega}{\partial\mathbf{q}} = f_{02}\mathbf{z} + f_{12}\mathbf{p} + f_{22}\mathbf{q} + f_{23}\mathbf{r} + f_{24}\mathbf{s} + f_{25}\mathbf{t},$$

$$\frac{\partial\Omega}{\partial\mathbf{z}} = f_{00}\mathbf{z} + f_{01}\mathbf{p} + f_{02}\mathbf{q} + f_{03}\mathbf{r} + f_{04}\mathbf{s} + f_{05}\mathbf{t}.$$

Then

$$\left|\frac{\partial\Omega_1}{\partial\mathbf{p}_1}, \mathbf{p}_2, \mathbf{q}_3\right| = f_{01}\Delta + f_{13}\left|\mathbf{r}_1, \mathbf{p}_2, \mathbf{q}_3\right| + f_{14}\left|\mathbf{s}_1, \mathbf{p}_2, \mathbf{q}_3\right| + f_{15}\left|\mathbf{t}_1, \mathbf{p}_2, \mathbf{q}_3\right|$$

$$= f_{01}\Delta - f_{13}\gamma\Delta - f_{14}\nu\Delta - f_{15}\tau\Delta.$$

From the relations at the end of § 342, we have

$$\left|\frac{\partial\Omega_1}{\partial\mathbf{p}_1}, \mathbf{p}_2, \mathbf{q}_3\right| = -A\left|\mathbf{r}_1, \mathbf{p}_2, \mathbf{q}_3\right| - (K+c)\left|\mathbf{s}_1, \mathbf{p}_2, \mathbf{q}_3\right| - m\left|\mathbf{t}_1, \mathbf{p}_2, \mathbf{q}_3\right| - \Delta\left(\frac{\partial b}{\partial x} + \frac{\partial l}{\partial y} + C\right)$$

$$= A\gamma\Delta + (K+c)\nu\Delta + m\tau\Delta - \Delta\left(\frac{\partial b}{\partial x} + \frac{\partial l}{\partial y} + C\right).$$

Equating these two values of $\left|\dfrac{\partial \Omega_1}{\partial \mathbf{p}_1}, \mathbf{p}_2, \mathbf{q}_3\right|$, and using the results already obtained (§ 344) for the expressions of $A$, $K+c$, and $m$, we have

$$f_{01} + \frac{\partial b}{\partial x} + \frac{\partial l}{\partial y} + C = \; \gamma\,(\alpha f_{33} + \lambda f_{34} + \rho f_{35})$$
$$+\, \nu\,(\alpha f_{34} + \lambda f_{44} + \rho f_{45})$$
$$+\, \tau\,(\alpha f_{35} + \lambda f_{45} + \rho f_{55}).$$

Similarly, we obtain

$$f_{02} + \frac{\partial a}{\partial x} + \frac{\partial k}{\partial y} + M = \; \gamma\,(\beta f_{33} + \mu f_{34} + \sigma f_{35})$$
$$+\, \nu\,(\beta f_{34} + \mu f_{44} + \sigma f_{45})$$
$$+\, \tau\,(\beta f_{35} + \mu f_{45} + \sigma f_{55});$$

$$f_{00} + \frac{\partial C}{\partial x} + \frac{\partial M}{\partial y} = \gamma^2 f_{33} + 2\gamma\nu f_{34} + 2\gamma\tau f_{35} + \nu^2 f_{44} + 2\nu\tau f_{45} + \tau^2 f_{55};$$

$$f_{11} + \frac{\partial A}{\partial x} + \frac{\partial K}{\partial y} + 2b = \alpha^2 f_{33} + 2\alpha\lambda f_{34} + 2\alpha\rho f_{35} + \lambda^2 f_{44} + 2\lambda\rho f_{45} + \rho^2 f_{55};$$

$$f_{22} + \frac{\partial B}{\partial x} + \frac{\partial L}{\partial y} + 2k = \beta^2 f_{33} + 2\beta\mu f_{34} + 2\beta\sigma f_{35} + \mu^2 f_{44} + 2\mu\sigma f_{45} + \sigma^2 f_{55};$$

and

$$f_{12} + \frac{\partial c}{\partial x} + \frac{\partial m}{\partial y} + a + l = \; \beta\,(\alpha f_{33} + \lambda f_{34} + \rho f_{35})$$
$$+\, \mu\,(\alpha f_{34} + \lambda f_{44} + \rho f_{45})$$
$$+\, \sigma\,(\alpha f_{35} + \lambda f_{45} + \rho f_{55});$$

agreeing with the former aggregate of expressions (§ 340) which would have to be established for the reduction of the second variation to its normal canonical form.

*Note.* It will be noted that some of these relations occur for any one solution of the subsidiary characteristic equation, and that some of them involve three linearly independent solutions. In the latter event, various verifications, as to the consistency of the deduced results among one another and with the general defining equations have been given; and other verifications can be made.

*Ex.* Prove that

$$\frac{\partial \Delta}{\partial x} + (a + \mu)\,\Delta = 0, \quad \frac{\partial \Delta}{\partial y} + (\lambda + \sigma)\,\Delta = 0;$$

and verify that the relation

$$\left|\frac{\partial \Omega_1}{\partial \mathbf{p}_1}, \mathbf{p}_2, \mathbf{z}_3\right| = -A\,|\,\mathbf{r}_1, \mathbf{p}_2, \mathbf{z}_3\,| - (K+c)\,|\,\mathbf{s}_1, \mathbf{p}_2, \mathbf{z}_3| - m\,|\,\mathbf{t}_1, \mathbf{p}_2, \mathbf{z}_3| + \Delta\left(\frac{\partial c}{\partial x} + \frac{\partial m}{\partial y} + a + l\right)$$

is satisfied by the values of $A$, $K+c$, $m$, $c$, $a+l$, which have been obtained.

*Relation between two integrals of the subsidiary equation.*

**347.** But, thus far, relations involving only two linearly independent integrals of the subsidiary characteristic equation (which have arisen frequently in earlier investigations, §§ 91, 173) have not been used, save in the construction of the expressions which have involved three such integrals taken in pairs.

A different use of a pair of integrals can be made as follows. By direct substitution, we find

$$\frac{\partial\Omega}{\partial\mathbf{p}} - \frac{\partial}{\partial x}\left(\frac{\partial\Omega}{\partial\mathbf{r}}\right) - \frac{\partial}{\partial y}\left(\frac{\partial\Omega}{\partial\mathbf{s}}\right) = B\mathbf{t} + c\mathbf{s} + \frac{\partial B}{\partial y}\mathbf{q} + \left(\frac{\partial c}{\partial y} - b\right)\mathbf{p} + \left(\frac{\partial a}{\partial x} - C\right)\mathbf{z}$$

for any integral; and therefore, for two independent integrals $\mathbf{z}_1$ and $\mathbf{z}_2$,

$$\mathbf{z}_1\left\{\frac{\partial\Omega_2}{\partial\mathbf{p}_2} - \frac{\partial}{\partial x}\left(\frac{\partial\Omega_2}{\partial\mathbf{r}_2}\right) - \frac{\partial}{\partial y}\left(\frac{\partial\Omega_2}{\partial\mathbf{s}_2}\right)\right\} + \mathbf{p}_1\frac{\partial\Omega_2}{\partial\mathbf{r}_2}$$

$$- \mathbf{z}_2\left\{\frac{\partial\Omega_1}{\partial\mathbf{p}_1} - \frac{\partial}{\partial x}\left(\frac{\partial\Omega_1}{\partial\mathbf{r}_1}\right) - \frac{\partial}{\partial y}\left(\frac{\partial\Omega_1}{\partial\mathbf{s}_1}\right)\right\} - \mathbf{p}_2\frac{\partial\Omega_1}{\partial\mathbf{r}_1}$$

$$= \frac{\partial}{\partial y}\left\{B\left(\mathbf{z}_1\mathbf{q}_2 - \mathbf{z}_2\mathbf{q}_1\right) + c\left(\mathbf{z}_1\mathbf{p}_2 - \mathbf{z}_2\mathbf{p}_1\right)\right\}.$$

Similarly,

$$\mathbf{z}_1\left\{\frac{\partial\Omega_2}{\partial\mathbf{q}_2} - \frac{\partial}{\partial x}\left(\frac{\partial\Omega_2}{\partial\mathbf{s}_2}\right) - \frac{\partial}{\partial y}\left(\frac{\partial\Omega_2}{\partial\mathbf{t}_2}\right)\right\} + \mathbf{q}_1\frac{\partial\Omega_2}{\partial\mathbf{t}_2}$$

$$- \mathbf{z}_2\left\{\frac{\partial\Omega_1}{\partial\mathbf{q}_1} - \frac{\partial}{\partial x}\left(\frac{\partial\Omega_1}{\partial\mathbf{s}_1}\right) - \frac{\partial}{\partial y}\left(\frac{\partial\Omega_1}{\partial\mathbf{t}_1}\right)\right\} - \mathbf{q}_2\frac{\partial\Omega_1}{\partial\mathbf{t}_1}$$

$$= \frac{\partial}{\partial x}\left\{K\left(\mathbf{z}_1\mathbf{p}_2 - \mathbf{z}_2\mathbf{p}_1\right) + m\left(\mathbf{z}_1\mathbf{q}_2 - \mathbf{z}_2\mathbf{q}_1\right)\right\}.$$

Also

$$\mathbf{z}_1\frac{\partial\Omega_2}{\partial\mathbf{s}_2} - \mathbf{z}_2\frac{\partial\Omega_1}{\partial\mathbf{s}_1} = -\left(K + c\right)\left(\mathbf{z}_1\mathbf{p}_2 - \mathbf{z}_2\mathbf{p}_1\right) - \left(B + m\right)\left(\mathbf{z}_1\mathbf{q}_2 - \mathbf{z}_2\mathbf{q}_1\right).$$

On combining these results, we find

$$\frac{\partial}{\partial x}\left[\mathbf{z}_1\left\{\frac{\partial\Omega_2}{\partial\mathbf{p}_2} - \frac{\partial}{\partial x}\left(\frac{\partial\Omega_2}{\partial\mathbf{r}_2}\right) - \frac{\partial}{\partial y}\left(\frac{\partial\Omega_2}{\partial\mathbf{s}_2}\right)\right\} + \mathbf{p}_1\frac{\partial\Omega_2}{\partial\mathbf{r}_2}\right.$$

$$\left. - \mathbf{z}_2\left\{\frac{\partial\Omega_1}{\partial\mathbf{p}_1} - \frac{\partial}{\partial x}\left(\frac{\partial\Omega_1}{\partial\mathbf{r}_1}\right) - \frac{\partial}{\partial y}\left(\frac{\partial\Omega_1}{\partial\mathbf{s}_1}\right)\right\} - \mathbf{p}_2\frac{\partial\Omega_1}{\partial\mathbf{r}_1}\right]$$

$$+ \frac{\partial}{\partial y}\left[\mathbf{z}_1\left\{\frac{\partial\Omega_2}{\partial\mathbf{q}_2} - \frac{\partial}{\partial x}\left(\frac{\partial\Omega_2}{\partial\mathbf{s}_2}\right) - \frac{\partial}{\partial y}\left(\frac{\partial\Omega_2}{\partial\mathbf{t}_2}\right)\right\} + \mathbf{q}_1\frac{\partial\Omega_2}{\partial\mathbf{t}_2}\right.$$

$$\left. - \mathbf{z}_2\left\{\frac{\partial\Omega_1}{\partial\mathbf{q}_1} - \frac{\partial}{\partial x}\left(\frac{\partial\Omega_1}{\partial\mathbf{s}_1}\right) - \frac{\partial}{\partial y}\left(\frac{\partial\Omega_1}{\partial\mathbf{t}_1}\right)\right\} - \mathbf{q}_2\frac{\partial\Omega_1}{\partial\mathbf{t}_1}\right]$$

$$+ \frac{\partial^2}{\partial x\partial y}\left(\mathbf{z}_1\frac{\partial\Omega_2}{\partial\mathbf{s}_2} - \mathbf{z}_2\frac{\partial\Omega_1}{\partial\mathbf{s}_1}\right) = 0,$$

a relation between derivatives of $\Omega_1$ and $\Omega_2$, free from their derivatives with regard to $\mathbf{z}$.

This last relation can be obtained as follows, directly from the subsidiary characteristic equation. Because $\Omega_1$ is a quadratic homogeneous form in $z_1$, $p_1$, $q_1$, $r_1$, $s_1$, $t_1$, and $\Omega_2$ is the same quadratic homogeneous form in $z_2$, $p_2$, $q_2$, $r_2$, $s_2$, $t_2$, we have the identity

$$z_1 \frac{\partial\Omega_2}{\partial z_2} + p_1 \frac{\partial\Omega_2}{\partial p_2} + q_1 \frac{\partial\Omega_2}{\partial q_2} + r_1 \frac{\partial\Omega_2}{\partial r_2} + s_1 \frac{\partial\Omega_2}{\partial s_2} + t_1 \frac{\partial\Omega_2}{\partial t_2}$$

$$= z_2 \frac{\partial\Omega_1}{\partial z_1} + p_2 \frac{\partial\Omega_1}{\partial p_1} + q_2 \frac{\partial\Omega_1}{\partial q_1} + r_2 \frac{\partial\Omega_1}{\partial r_1} + s_2 \frac{\partial\Omega_1}{\partial s_1} + t_2 \frac{\partial\Omega_1}{\partial t_1}.$$

Because $\Omega_1$ and $\Omega_2$ satisfy the subsidiary characteristic equation, we have

$$\frac{\partial\Omega_2}{\partial z_2} - \frac{\partial}{\partial x}\left(\frac{\partial\Omega_2}{\partial p_2}\right) - \frac{\partial}{\partial y}\left(\frac{\partial\Omega_2}{\partial q_2}\right) + \frac{\partial^2}{\partial x^2}\left(\frac{\partial\Omega_2}{\partial r_2}\right) + \frac{\partial^2}{\partial x\partial y}\left(\frac{\partial\Omega_2}{\partial s_2}\right) + \frac{\partial^2}{\partial y^2}\left(\frac{\partial\Omega_2}{\partial t_2}\right) = 0;$$

and similarly for $\Omega_1$. When the values of $\dfrac{\partial\Omega_2}{\partial z_2}$ and $\dfrac{\partial\Omega_1}{\partial z_1}$, thus given, are substituted, the left-hand side of the preceding identity becomes

$$z_1 \frac{\partial}{\partial x}\left(\frac{\partial\Omega_2}{\partial p_2}\right) + p_1 \frac{\partial\Omega_2}{\partial p_2} - z_1 \frac{\partial^2}{\partial x^2}\left(\frac{\partial\Omega_2}{\partial r_2}\right) + r_1 \frac{\partial\Omega_2}{\partial r_2}$$

$$+ z_1 \frac{\partial}{\partial y}\left(\frac{\partial\Omega_2}{\partial q_2}\right) + q_1 \frac{\partial\Omega_2}{\partial q_2} - z_1 \frac{\partial^2}{\partial y^2}\left(\frac{\partial\Omega_2}{\partial t_2}\right) + t_1 \frac{\partial\Omega_2}{\partial t_2}$$

$$- z_1 \frac{\partial^2}{\partial x\partial y}\left(\frac{\partial\Omega_2}{\partial s_2}\right) + s_1 \frac{\partial\Omega_2}{\partial s_2}.$$

There is a similar expression for the right-hand side. Now

$$z_1 \frac{\partial}{\partial x}\left(\frac{\partial\Omega_2}{\partial p_2}\right) + p_1 \frac{\partial\Omega_2}{\partial p_2} = \frac{\partial}{\partial x}\left(z_1 \frac{\partial\Omega_2}{\partial p_2}\right),$$

$$z_1 \frac{\partial^2}{\partial x^2}\left(\frac{\partial\Omega_2}{\partial r_2}\right) - r_1 \frac{\partial\Omega_2}{\partial r_2} = \frac{\partial}{\partial x}\left\{z_1 \frac{\partial}{\partial x}\left(\frac{\partial\Omega_2}{\partial r_2}\right) - p_1 \frac{\partial\Omega_2}{\partial r_2}\right\},$$

$$z_1 \frac{\partial}{\partial y}\left(\frac{\partial\Omega_2}{\partial q_2}\right) + q_1 \frac{\partial\Omega_2}{\partial q_2} = \frac{\partial}{\partial y}\left(z_1 \frac{\partial\Omega_2}{\partial q_2}\right),$$

$$z_1 \frac{\partial^2}{\partial y^2}\left(\frac{\partial\Omega_2}{\partial t_2}\right) - t_1 \frac{\partial\Omega_2}{\partial t_2} = \frac{\partial}{\partial y}\left\{z_1 \frac{\partial}{\partial y}\left(\frac{\partial\Omega_2}{\partial t_2}\right) - q_1 \frac{\partial\Omega_2}{\partial t_2}\right\},$$

$$z_1 \frac{\partial^2}{\partial x\partial y}\left(\frac{\partial\Omega_2}{\partial s_2}\right) - s_1 \frac{\partial\Omega_2}{\partial s_2} = \frac{\partial}{\partial x}\left\{z_1 \frac{\partial}{\partial y}\left(\frac{\partial\Omega_2}{\partial s_2}\right)\right\} + \frac{\partial}{\partial y}\left\{z_1 \frac{\partial}{\partial x}\left(\frac{\partial\Omega_2}{\partial s_2}\right)\right\} - \frac{\partial^2}{\partial x\partial y}\left(z_1 \frac{\partial\Omega_2}{\partial s_2}\right).$$

Therefore the left-hand side of the transformed identity becomes

$$\frac{\partial}{\partial x}\left[z_1\left\{\frac{\partial\Omega_2}{\partial p_2} - \frac{\partial}{\partial x}\left(\frac{\partial\Omega_2}{\partial r_2}\right) - \frac{\partial}{\partial y}\left(\frac{\partial\Omega_2}{\partial s_2}\right)\right\} + p_1 \frac{\partial\Omega_2}{\partial r_2}\right]$$

$$+ \frac{\partial}{\partial y}\left[z_1\left\{\frac{\partial\Omega_2}{\partial q_2} - \frac{\partial}{\partial x}\left(\frac{\partial\Omega_2}{\partial s_2}\right) - \frac{\partial}{\partial y}\left(\frac{\partial\Omega_2}{\partial t_2}\right)\right\} + q_1 \frac{\partial\Omega_2}{\partial t_2}\right] + \frac{\partial^2}{\partial x\partial y}\left(z_1 \frac{\partial\Omega_2}{\partial s_2}\right).$$

The right-hand side of that transformed identity becomes the corresponding expression derived by interchanging $\Omega_2$ and $\Omega_1$, and $z_1$ and $z_2$, and the derivatives of $z_1$ and $z_2$. Equating the two new expressions, we have the required relation.

*Normal form of second variation : the Legendre test.*

**348.** The normal form for $\frac{1}{2}\kappa^2 \iint \Theta \, dx \, dy$ follows from the use of these relations. When we take

$$\Theta + \frac{\partial}{\partial x}(AP^2 + BQ^2 + CZ^2 + 2aQZ + 2bPZ + 2cPQ)$$

$$+ \frac{\partial}{\partial y}(KP^2 + LQ^2 + MZ^2 + 2kQZ + 2lPZ + 2mPQ),$$

and use the values of the quantities $A$, $B$, $C$, $a$, $b$, $c$, $K$, $L$, $M$, $k$, $l$, $m$, as expressed in terms of the coefficient $f_{rn}$ in $\Theta$ and of the three independent integrals of the subsidiary characteristic equation, the whole expression becomes

$$f_{33}u^2 + 2f_{34}uv + 2f_{35}uw + f_{44}v^2 + 2f_{45}vw + f_{55}w^2, = \Upsilon,$$

where

$$u = R + \alpha P + \beta Q + \gamma Z = \frac{1}{\Delta}\begin{vmatrix} R, & P, & Q, & Z \\ r_1, & p_1, & q_1, & z_1 \\ r_2, & p_2, & q_2, & z_2 \\ r_3, & p_3, & q_3, & z_3 \end{vmatrix},$$

$$v = S + \lambda P + \mu Q + \nu Z = \frac{1}{\Delta}\begin{vmatrix} S, & P, & Q, & Z \\ s_1, & p_1, & q_1, & z_1 \\ s_2, & p_2, & q_2, & z_2 \\ s_3, & p_3, & q_3, & z_3 \end{vmatrix},$$

$$w = T + \rho P + \sigma Q + \tau Z = \frac{1}{\Delta}\begin{vmatrix} T, & P, & Q, & Z \\ t_1, & p_1, & q_1, & z_1 \\ t_2, & p_2, & q_2, & z_2 \\ t_3, & p_3, & q_3, & z_3 \end{vmatrix}.$$

When we form the double integral $\iint \Upsilon \, dx \, dy$, there is the term $\iint \Theta \, dx \, dy$. In addition, there are the terms

$$\int (AP^2 + BQ^2 + CZ^2 + 2aQZ + 2bPZ + 2cPQ)\, dy$$

$$- \int (KP^2 + LQ^2 + MZ^2 + 2kQZ + 2lPZ + 2mPQ)dx,$$

taken positively round the whole boundary. Now the vanishing of the first variation has settled all the variations at the boundary; and the only variations now admissible are those which vary in the field but are completely fixed at the boundary. Thus the foregoing additional terms vanish everywhere on the boundary; and so the line-integral is zero. Hence *the second variation of the original integral can be expressed* in the form $\iint \Upsilon \, dx \, dy$, that is, *in the form*

$$\iint (f_{33} u^2 + 2f_{34} uv + 2f_{35} uw + f_{44} v^2 + 2f_{45} vw + f_{55} w^2) \, dx \, dy.$$

Tests for the existence of a maximum or a minimum can now be stated at once. The first variation has been made to vanish. The second variation must preserve an unchanging sign for all non-zero variations. To meet the latter requirement, the quantity under the sign of integration must be always positive or always negative; and non-zero variations must be excluded which would make $u$, $v$, $w$, zero simultaneously.

One set of conditions, therefore, is that *the relations*

$$f_{33} f_{44} - f_{34}^2 > 0, \quad f_{44} f_{55} - f_{45}^2 > 0, \quad f_{55} f_{33} - f_{35}^2 > 0,$$

*must be satisfied everywhere throughout the field of integration.* When these conditions are satisfied, *the quantities*

$$f_{33}, \; f_{44}, \; f_{55},$$

*have a common sign* throughout that field, no one of the quantities vanishing. If that common sign be positive, a minimum is admissible for the integral; if the common sign be negative, a maximum is admissible. The conditions are the extended form of the *Legendre* test.

*Note.* No one of the greater inequalities can be turned into a lesser inequality, without the certain loss of a maximum or a minimum as the case may be. It is not impossible to have an inequality changed to an equality, such as $f_{33} f_{44} - f_{34}^2 = 0$; but then for variations, such that $w = 0$ and $f_{33} u + f_{34} v = 0$ while $Z$ is not zero, the second variation would vanish. For a maximum or a minimum, the third variation would have to vanish for those variations—a necessary additional requirement, though not sufficient to secure the result. As in previous instances, we omit the general discussion.

### Characteristic variations.

**349.** The other set of conditions arises out of the requirement that non-zero variations, which can make the arguments $u$, $v$, $w$ (in § 348) vanish simultaneously, are to be excluded—an exclusion which, after the earlier precedents, will be expected to depend upon the range of the field of integration.

These quantities $u$, $v$, $w$, involve three linearly independent integrals $z_1$, $z_2$, $z_3$, of the subsidiary characteristic equation, such that $\Delta$, where

$$\Delta = \begin{vmatrix} p_1, & q_1, & z_1 \\ p_2, & q_2, & z_2 \\ p_3, & q_3, & z_3 \end{vmatrix},$$

is not zero. Now we can always choose three quantities $\xi$, $\eta$, $\zeta$ such that

$$Z = \xi z_1 + \eta z_2 + \zeta z_3,$$
$$P = \xi p_1 + \eta p_2 + \zeta p_3,$$
$$Q = \xi q_1 + \eta q_2 + \zeta q_3,$$

because $\Delta$ is not zero. Because of the relations $P = \dfrac{\partial Z}{\partial x}$, $Q = \dfrac{\partial Z}{\partial y}$, we must have

$$0 = z_1 \frac{\partial \xi}{\partial x} + z_2 \frac{\partial \eta}{\partial x} + z_3 \frac{\partial \zeta}{\partial x},$$

$$0 = z_1 \frac{\partial \xi}{\partial y} + z_2 \frac{\partial \eta}{\partial y} + z_3 \frac{\partial \zeta}{\partial y}.$$

If then values $u = 0$, $v = 0$, $w = 0$, could exist simultaneously through the range, we should have

$$R = \xi r_1 + \eta r_2 + \zeta r_3,$$
$$S = \xi s_1 + \eta s_2 + \zeta s_3,$$
$$T = \xi t_1 + \eta t_2 + \zeta t_3.$$

Because of the relations

$$R = \frac{\partial P}{\partial x}, \quad S = \frac{\partial Q}{\partial x} = \frac{\partial P}{\partial y}, \quad T = \frac{\partial Q}{\partial y},$$

we must therefore have

$$0 = p_1 \frac{\partial \xi}{\partial x} + p_2 \frac{\partial \eta}{\partial x} + p_3 \frac{\partial \zeta}{\partial x},$$

$$0 = q_1 \frac{\partial \xi}{\partial x} + q_2 \frac{\partial \eta}{\partial x} + q_3 \frac{\partial \zeta}{\partial x},$$

$$0 = p_1 \frac{\partial \xi}{\partial y} + p_2 \frac{\partial \eta}{\partial y} + p_3 \frac{\partial \zeta}{\partial y},$$

$$0 = q_1 \frac{\partial \xi}{\partial y} + q_2 \frac{\partial \eta}{\partial y} + q_3 \frac{\partial \zeta}{\partial y}.$$

As $\Delta$ is not zero, these six conditions require that

$$\frac{\partial \xi}{\partial x} = 0, \quad \frac{\partial \eta}{\partial x} = 0, \quad \frac{\partial \zeta}{\partial x} = 0; \quad \frac{\partial \xi}{\partial y} = 0, \quad \frac{\partial \eta}{\partial y} = 0, \quad \frac{\partial \zeta}{\partial y} = 0;$$

so that $\xi = a$, $\eta = b$, $\zeta = c$, where $a$, $b$, $c$ are constant quantities. Thus, for our purpose, it is necessary and sufficient that there shall be no non-zero variation valid throughout the field of integration such as to allow

$$Z = a z_1 + b z_2 + c z_3,$$

where $a$, $b$, $c$, are constants; $z_1$, $z_2$, $z_3$, are linearly independent integrals of the subsidiary characteristic equation; and $\kappa Z$ is a special variation of the characteristic surface, the relation of which to the possibility of a maximum or a minimum is under discussion.

Now each variation $\kappa z_1$, $\kappa z_2$, $\kappa z_3$, represents a small variation from the characteristic surface to a contiguous surface, say a characteristic variation; and therefore such a variation $\kappa Z$ also represents a characteristic variation. We must therefore consider what are the possibilities of such a variation.

### Consecutive characteristic surface.

**350.** The differential equation of the characteristic surface is of the fourth order. Its Cauchy primitive is made determinate and unique by the assignment of arbitrary functions of $x$ and $y$ as the values of $z$, $\dfrac{\partial z}{\partial x}$, $\dfrac{\partial^2 z}{\partial x^2}$, $\dfrac{\partial^3 z}{\partial x^3}$, along any given curve $\theta(x, y) = 0$, provided that curve does not satisfy the critical equation* in § 337,

$$\left(\frac{\partial^2 f}{\partial s^2} + 2\frac{\partial^2 f}{\partial r\partial t}\right)\left(\frac{\partial\theta}{\partial x}\right)^2\left(\frac{\partial\theta}{\partial y}\right)^2$$

$$+ \frac{\partial^2 f}{\partial r^2}\left(\frac{\partial\theta}{\partial x}\right)^4 + 2\frac{\partial^2 f}{\partial r\partial s}\left(\frac{\partial\theta}{\partial x}\right)^3\frac{\partial\theta}{\partial y} + 2\frac{\partial^2 f}{\partial s\partial t}\frac{\partial\theta}{\partial x}\left(\frac{\partial\theta}{\partial y}\right)^3 + \frac{\partial^2 f}{\partial t^2}\left(\frac{\partial\theta}{\partial y}\right)^4 = 0:$$

for otherwise the arguments of the arbitrary functions would be constant, and not variable, along the given curve. The assigned values of $z$ and $\dfrac{\partial z}{\partial x}$ along $\theta(x, y) = 0$ give the value of $\dfrac{\partial z}{\partial y}$ along that curve. From these, and from the value of $\dfrac{\partial^2 z}{\partial x^2}$ along the curve, we deduce the values of $\dfrac{\partial^2 z}{\partial x\partial y}$ and $\dfrac{\partial^2 z}{\partial y^2}$ along the curve; and then, with the value of $\dfrac{\partial^3 z}{\partial x^3}$ given along the curve, we deduce the values of $\dfrac{\partial^3 z}{\partial x^2\partial y}$, $\dfrac{\partial^3 z}{\partial x\partial y^2}$, $\dfrac{\partial^3 z}{\partial y^3}$, along the curve. In turn, from these values of the third derivatives but now by means of the single differential equation also, we deduce the values of all the fourth derivatives of $z$ along the curve. And so on in succession: but always provided that, at the first stage and at every succeeding stage, the function $\theta$ does not satisfy the critical equation. We are thus led to Cauchy's establishment of the existence of the primitive of the characteristic equation, obtained in the form of a converging series.

Now the assignment of a slightly different value to $\dfrac{\partial^3 z}{\partial x^3}$, along the initial curve $\theta = 0$, leads to a slightly different primitive: that is, a small variation in the assignment of values to derivatives of the third order, without affecting those of the derivatives of the first order and the second order, leads to a slightly varied primitive. A small variation of this type is the only deviation from the conditions which lead to the determination of the unique central

---

* The equation is merely the extension, to the fourth order, of the corresponding equation that occurs in the second order. It is the equation of the 'characteristics,' in the sense of the term (due to Monge) as used in Cauchy's method of establishing the existence of the primitive: see my *Theory of Differential Equations*, vol. vi, chaps. xii, xx.

characteristic; and it leads to the construction of another characteristic, which is closest to that central characteristic in intrinsic properties. Accordingly, it is called a *consecutive characteristic*.

Now take three *contiguous* characteristics through the initial curve, differing from the central characteristic under discussion by the assignment of added increments $\kappa\alpha_1$, $\kappa\alpha_2$, $\kappa\alpha_3$, to the values of $p_1$, $p_2$, $p_3$; by the assignment of added increments $\kappa\beta_1$, $\kappa\beta_2$, $\kappa\beta_3$, to the values of $r_1$, $r_2$, $r_3$; and by the assignment of added increments $\kappa\gamma_1$, $\kappa\gamma_2$, $\kappa\gamma_3$, to the values of $\frac{\partial^3 z_1}{\partial x^3}$, $\frac{\partial^3 z_2}{\partial x^3}$, $\frac{\partial^3 z_3}{\partial x^3}$. Each of these added increments might be (if we wished) an assigned continuous function along the initial curve, but we shall assume them to be constants; and all of them are at our disposal to satisfy conditions that may be desired. We shall assume that the determinant

$$\begin{vmatrix} \alpha_1, & \alpha_2, & \alpha_3 \\ \beta_1, & \beta_2, & \beta_3 \\ \gamma_1, & \gamma_2, & \gamma_3 \end{vmatrix}$$

is, in general, distinct from zero along the initial curve. Each set of values $\kappa\alpha_\mu$, $\kappa\beta_\mu$, $\kappa\gamma_\mu$ (for sets given by $\mu = 1, 2, 3$) determines a contiguous characteristic. Each of these three contiguous characteristics differs from the central characteristic, because of the differences in the initial values assigned to $\frac{\partial z}{\partial x}$, $\frac{\partial^2 z}{\partial x^2}$, $\frac{\partial^3 z}{\partial x^3}$, along the initial curve.

Consider the contiguous characteristic represented by $\kappa Z$, where

$$Z = a z_1 + b z_2 + c z_3,$$

this contiguous characteristic also passing through the initial curve. Along that common initial curve, we have $z_1 = 0$, $z_2 = 0$, $z_3 = 0$; and $Z = 0$, because the new contiguous surface passes through the curve. Consequently, out of this source, no relation attaches to $a$, $b$, $c$.

We shall have $P = 0$ along the initial curve, if the condition

$$a\alpha_1 + b\alpha_2 + c\alpha_3 = 0$$

be assigned; this condition will be postulated. Then, as $a z_1 + b z_2 + c z_3 = 0$ along the initial curve, we have also

$$a q_1 + b q_2 + c q_3 = 0,$$

along the curve (a condition not affecting imposed boundary relations).

We shall have $R = 0$ along the initial curve, if the condition

$$a\beta_1 + b\beta_2 + c\beta_3 = 0$$

be assigned; this condition also will be postulated. Then we have also

$$a s_1 + b s_2 + c s_3 = 0, \quad a t_1 + b t_2 + c t_3 = 0,$$

along the curve (conditions which do not affect imposed boundary relations).

Lastly, we shall have $\dfrac{\partial^3 z}{\partial x^3}$ distinct from zero along the initial curve, if the condition

$$a\gamma_1 + b\gamma_2 + c\gamma_3 \gtrless 0$$

be assigned. This condition is possible, because the determinant

$$\begin{vmatrix} \alpha_1, & \alpha_2, & \alpha_3 \\ \beta_1, & \beta_2, & \beta_3 \\ \gamma_1, & \gamma_2, & \gamma_3 \end{vmatrix} \gtrless 0.$$

Thus the contiguous characteristic, determined by $z + \kappa Z$ with this value of $Z$, has the same values of $z, \dfrac{\partial z}{\partial x}, \dfrac{\partial^2 z}{\partial x^2}$, along the initial curve as the central characteristic; but the value of $\dfrac{\partial^3 z}{\partial x^3}$ along that curve for the contiguous characteristic differs from the value of $\dfrac{\partial^3 z}{\partial x^3}$ for the central surface. Hence the new contiguous characteristic thus determined, distinct from the central surface, is a *consecutive characteristic*. The constants $a, b, c$, satisfy the two independent equations

$$a\alpha_1 + b\alpha_2 + c\alpha_3 = 0, \quad a\beta_1 + b\beta_2 + c\beta_3 = 0,$$

so that we may take

$$a = \alpha_2\beta_3 - \alpha_3\beta_2, \quad b = \alpha_3\beta_1 - \alpha_1\beta_3, \quad c = \alpha_1\beta_2 - \alpha_2\beta_1;$$

and the contiguous characteristic is given by

$$Z = az_1 + bz_2 + cz_3.$$

### *Conjugate of the initial curve: the Jacobi test.*

**351.** The consecutive characteristic passes through the initial curve. Along that curve, it has complete contact, of the first order and of the second order, with the central surface. Being distinct from that surface, it may or may not meet the central surface again; and if intersection does take place, there may be several curves of intersection. The curve of intersection, if there be only one, or the earliest of such curves of intersection, if there be more than one, as we pass over the field from the initial curve, we shall call the *conjugate* of the initial curve.

If then the field of integration, bounded on one side by an initial curve, should extend as far as the conjugate of that initial curve, we should be able to have a non-zero variation from the central surface given by

$$Z = az_1 + bz_2 + cz_3,$$

because such a variation $\kappa Z$ satisfies the requirement of being zero at the initial boundary and at the final boundary. In that event, the quantities $u, v, w$, could vanish everywhere in the field of integration; for all small variations of this type, the second variation of the original integral would vanish. We could not then assert the existence of a maximum or a minimum, unless and until the third variation of the integral for variations of the selected type is proved to vanish and its fourth variation has been examined. As in earlier instances, such an investigation is set aside from the general discussion, as belonging at the utmost to particular instances. But, anyhow, it can be excluded even from special consideration, by a definite restriction of the range so as to make it fall short of the conjugate curve. Accordingly, with the definition of a consecutive characteristic as a surface, which is determined by the same initial values of $z$, $\dfrac{\partial z}{\partial x}$, $\dfrac{\partial^2 z}{\partial x^2}$, as the characteristic surface, and by a different initial value of $\dfrac{\partial^3 z}{\partial x^3}$, and with the definition of the conjugate, we have the *Jacobi test*:

*The field of integration, bounded by an initial curve, must not extend as far as the conjugate of that initial curve on the characteristic surface.*

## General weak variations: transformation of the integral.

**352.** A discussion of the double integral involving derivatives of the second order, so as to allow full variations to all the variables that occur, variations moreover which should include the strong types as well as the weak types, demands an extensive amount of purely algebraical calculations in the first instance. For adequate treatment, there is the need of a more comprehensive knowledge, than at present is possessed, concerning the primitives of partial differential equations of order four. The discussion of the effect of strong variations upon such integrals may not unreasonably be deferred until the topic has been sufficiently treated for the much simpler corresponding investigation (§§ 222, 223) for single integrals containing ordinary derivatives of the second order.

Accordingly, we shall consider here the theory, in its association with general weak variations, only so far as to obtain the characteristic equations and the boundary conditions. We shall shew that, in essence, they correspond to the characteristic equation under special variations, and to the boundary conditions for special variations, amplified as usual by the necessary extension of those special variations in their significance at a mobile boundary curve or curves.

As before, in order to impose weak variations on $x$, $y$, $z$, simultaneously throughout the field of variation, we make $x$, $y$, $z$, functions of two new independent variables $u$ and $v$. The first and the second derivatives will be denoted by

$$\theta,\ \theta_1,\ \theta_2,\ \theta_{11},\ \theta_{12},\ \theta_{22},$$

for $\theta = x,\ y,\ z$, separately and successively. The notation of § 290 will be repeated. We write

$$l,\ m,\ n = \left\| \begin{array}{ccc} x_1, & y_1, & z_1 \\ x_2, & y_2, & z_2 \end{array} \right\|$$

The values of $p$ and $q$, in terms of the new derivatives of $x,\ y,\ z$, are

$$\frac{p}{z_1 y_2 - y_1 z_2} = \frac{q}{x_1 z_2 - z_1 x_2} = \frac{1}{x_1 y_2 - y_1 x_2},$$

so that

$$np = -l, \quad nq = -m.$$

The values of $r,\ s,\ t$, have been stated in § 291.

When the transformation to the new independent variables $u$ and $v$ has been effected upon the double integral

$$\iint f(x,\ y,\ z,\ p,\ q,\ r,\ s,\ t)\, dx\, dy,$$

with the due substitutions for $p,\ q,\ r,\ s,\ t$, it becomes

$$\iint \Omega\, du\, dv,$$

where

$$\Omega = (x_1 y_2 - y_1 x_2) f = nf.$$

Now $\Omega$ involves eighteen arguments, viz. the three variables $x,\ y,\ z$, their six first derivatives ($\theta_1$ and $\theta_2$, for $\theta = x,\ y,\ z$), and their nine second derivatives ($\theta_{11},\ \theta_{12},\ \theta_{22}$, for $\theta = x,\ y,\ z$). The function $f$ involves only eight arguments. It may therefore be expected that $\Omega$, the subject of integration in the modified integral, will satisfy ten identities. These ten identities can be obtained as follows.

### Identities satisfied by $\Omega$, the integrand in the modified integral.

**353.** The integral $\iint \Omega\, du\, dv$ is an absolute covariant, whatever independent variables $u'$ and $v'$ be chosen: it is always equal to $\iint f\, dx\, dy$, and the independent variables are not intrinsic to the problem. Hence

$$\iint \Omega'\, du'\, dv' = \iint \Omega\, du\, dv = \iint \Omega J\left(\frac{u,\ v}{u',\ v'}\right) du'\, dv',$$

where $\Omega'$ is the same function of the derivatives of $x,\ y,\ z$, with respect to $u'$ and $v'$, as $\Omega$ is of their derivatives with respect to $u$ and $v$. Let

$$du' = \alpha\, du + \beta\, dv = (1 + \epsilon)\, du + \beta\, dv,$$
$$dv' = \gamma\, du + \delta\, dv = \qquad \gamma\, du + (1 + \eta)\, dv,$$

where, for our purpose, $\epsilon$, $\beta$, $\gamma$, $\eta$, are small arbitrary independent variable functions of $u$ and $v$. Then, as

$$\frac{\partial u'}{\partial u} = 1 + \epsilon, \quad \frac{\partial u'}{\partial v} = \beta, \quad \frac{\partial v'}{\partial u} = \gamma, \quad \frac{\partial v'}{\partial v} = 1 + \eta,$$

we have

$$\epsilon_2 = \beta_1 = \lambda, \quad \gamma_2 = \eta_1 = \mu;$$

so that ten small quantities $\epsilon$, $\eta$, $\beta$, $\gamma$, $\epsilon_1$, $\gamma_1$, $\eta_2$, $\beta_2$, $\lambda$, $\mu$, occur. The changes in the derivatives of $x$, $y$, $z$, up to the first degree of small quantities, are:

$$\theta_1 = \theta_1' + \epsilon\ \theta_1' + \gamma\ \theta_2',$$
$$\theta_2 = \theta_2' + \beta\ \theta_1' + \eta\ \theta_2',$$
$$\theta_{11} = \theta_{11}' + \epsilon_1\ \theta_1' + \gamma_1\ \theta_2' + 2\epsilon\theta_{11}' + 2\gamma\quad\theta_{12}',$$
$$\theta_{12} = \theta_{12}' + \lambda\ \theta_1' + \mu\ \theta_2' + \beta\theta_{11}' + (\epsilon + \eta)\ \theta_{12}' + \gamma\ \theta_{22}',$$
$$\theta_{22} = \theta_{22}' + \beta_2\theta_1' + \eta_2\theta_2' \qquad\qquad + 2\beta\quad\theta_{12}' + 2\eta\ \theta_{22}',$$

for $\theta = x, y, z$. From the covariance of the integral, we have

$$\Omega' = \Omega J\left(\frac{u', v'}{u, v}\right)$$
$$= (1 + \epsilon + \eta)\ \Omega,$$

to the first degree of small quantities. In $\Omega$, we substitute the foregoing values of the derivatives with respect to $u$ and $v$ in terms of the derivatives with respect to $u'$ and $v'$; we expand in powers of the small quantities; we equate the several coefficients of their first powers on the right-hand side to zero\*, as no such terms occur on the left-hand side consisting of $\Omega'$ alone; and then we omit accents. We thus obtain the set of ten identities in question, viz.

$$\Sigma x_1 \frac{\partial \Omega}{\partial x_1} + 2\Sigma x_{11}\frac{\partial \Omega}{\partial x_{11}} + \cdot\Sigma x_{12}\frac{\partial \Omega}{\partial x_{12}} = \Omega$$
$$\Sigma x_2 \frac{\partial \Omega}{\partial x_1} + 2\Sigma x_{12}\frac{\partial \Omega}{\partial x_{11}} + \Sigma x_{22}\frac{\partial \Omega}{\partial x_{12}} = 0$$
$$\Sigma x_1 \frac{\partial \Omega}{\partial x_2} + \Sigma x_{11}\frac{\partial \Omega}{\partial x_{12}} + 2\Sigma x_{12}\frac{\partial \Omega}{\partial x_{22}} = 0$$
$$\Sigma x_2 \frac{\partial \Omega}{\partial x_2} + \Sigma x_{12}\frac{\partial \Omega}{\partial x_{12}} + 2\Sigma x_{22}\frac{\partial \Omega}{\partial x_{22}} = \Omega$$

from the coefficients of $\epsilon$, $\gamma$, $\beta$, $\eta$, respectively; and

$$\Sigma x_1 \frac{\partial \Omega}{\partial x_{11}} = 0, \quad \Sigma x_2 \frac{\partial \Omega}{\partial x_{11}} = 0$$
$$\Sigma x_1 \frac{\partial \Omega}{\partial x_{12}} = 0, \quad \Sigma x_2 \frac{\partial \Omega}{\partial x_{12}} = 0$$
$$\Sigma x_1 \frac{\partial \Omega}{\partial x_{22}} = 0, \quad \Sigma x_2 \frac{\partial \Omega}{\partial x_{22}} = 0$$

---

\* The process is the customary process of forming the infinitesimal transformations belonging to the continuous group, which makes $u'$ and $v'$ any quite general functions of $u$ and $v$.

from the coefficients of $\epsilon_1$, $\gamma_1$; $\lambda$, $\mu$; $\beta_2$, $\eta_2$; respectively: the summation in each symbol being over the variables $x$, $y$, $z$.

For the reduction of the full expression of the second variation of $\iint \Omega\, du\, dv$ from its initial form, we should need the relations between the second derivatives of $\Omega$ with regard to its arguments. Many of these relations are identities, arising out of the ten identities satisfied by $\Omega$. Others of them are consequences of the characteristic equations hereafter (§ 354) obtained. Among the relations derived from the identities, the following may be noted:

$$\frac{1}{l^2}\frac{\partial^2\Omega}{\partial x_{11}^2} = \frac{1}{lm}\frac{\partial^2\Omega}{\partial x_{11}\partial y_{11}} = \frac{1}{ln}\frac{\partial^2\Omega}{\partial x_{11}\partial z_{11}} = \frac{1}{m^2}\frac{\partial^2\Omega}{\partial y_{11}^2} = \frac{1}{mn}\frac{\partial^2\Omega}{\partial y_{11}\partial z_{11}} = \frac{1}{n^2}\frac{\partial^2\Omega}{\partial z_{11}^2};$$

$$\frac{1}{l^2}\frac{\partial^2\Omega}{\partial x_{12}^2} = \frac{1}{lm}\frac{\partial^2\Omega}{\partial x_{12}\partial y_{12}} = \frac{1}{ln}\frac{\partial^2\Omega}{\partial x_{12}\partial z_{12}} = \frac{1}{m^2}\frac{\partial^2\Omega}{\partial y_{12}^2} = \frac{1}{mn}\frac{\partial^2\Omega}{\partial y_{12}\partial z_{12}} = \frac{1}{n^2}\frac{\partial^2\Omega}{\partial z_{12}^2};$$

$$\frac{1}{l^2}\frac{\partial^2\Omega}{\partial x_{22}^2} = \frac{1}{lm}\frac{\partial^2\Omega}{\partial x_{22}\partial y_{22}} = \frac{1}{ln}\frac{\partial^2\Omega}{\partial x_{22}\partial z_{22}} = \frac{1}{m^2}\frac{\partial^2\Omega}{\partial y_{22}^2} = \frac{1}{mn}\frac{\partial^2\Omega}{\partial y_{22}\partial z_{22}} = \frac{1}{n^2}\frac{\partial^2\Omega}{\partial z_{22}^2};$$

$$\frac{1}{l^2}\frac{\partial^2\Omega}{\partial x_{12}\partial x_{22}} = \frac{1}{lm}\frac{\partial^2\Omega}{\partial x_{12}\partial y_{22}} = \frac{1}{ln}\frac{\partial^2\Omega}{\partial x_{12}\partial z_{22}} = \frac{1}{m^2}\frac{\partial^2\Omega}{\partial y_{12}\partial y_{22}} = \frac{1}{mn}\frac{\partial^2\Omega}{\partial y_{12}\partial z_{22}}$$
$$= \frac{1}{lm}\frac{\partial^2\Omega}{\partial x_{22}\partial y_{12}} = \frac{1}{ln}\frac{\partial^2\Omega}{\partial x_{22}\partial z_{12}} = \frac{1}{mn}\frac{\partial^2\Omega}{\partial z_{12}\partial y_{22}} = \frac{1}{n^2}\frac{\partial^2\Omega}{\partial z_{12}\partial z_{22}};$$

$$\frac{1}{l^2}\frac{\partial^2\Omega}{\partial x_{11}\partial x_{22}} = \frac{1}{lm}\frac{\partial^2\Omega}{\partial x_{11}\partial y_{22}} = \frac{l}{ln}\frac{\partial^2\Omega}{\partial x_{11}\partial z_{22}} = \frac{1}{m^2}\frac{\partial^2\Omega}{\partial y_{11}\partial y_{22}} = \frac{1}{mn}\frac{\partial^2\Omega}{\partial y_{11}\partial z_{22}}$$
$$= \frac{1}{lm}\frac{\partial^2\Omega}{\partial x_{22}\partial y_{11}} = \frac{1}{ln}\frac{\partial^2\Omega}{\partial x_{22}\partial z_{11}} = \frac{1}{mn}\frac{\partial^2\Omega}{\partial y_{22}\partial z_{11}} = \frac{1}{n^2}\frac{\partial^2\Omega}{\partial z_{11}\partial z_{22}};$$

$$\frac{1}{l^2}\frac{\partial^2\Omega}{\partial x_{11}\partial x_{12}} = \frac{1}{lm}\frac{\partial^2\Omega}{\partial x_{11}\partial y_{12}} = \frac{1}{ln}\frac{\partial^2\Omega}{\partial x_{11}\partial z_{12}} = \frac{1}{m^2}\frac{\partial^2\Omega}{\partial y_{11}\partial y_{12}} = \frac{1}{mn}\frac{\partial^2\Omega}{\partial y_{11}\partial z_{12}}$$
$$= \frac{1}{lm}\frac{\partial^2\Omega}{\partial x_{12}\partial y_{11}} = \frac{1}{ln}\frac{\partial^2\Omega}{\partial x_{12}\partial z_{11}} = \frac{1}{mn}\frac{\partial^2\Omega}{\partial y_{12}\partial z_{11}} = \frac{1}{n^2}\frac{\partial^2\Omega}{\partial z_{11}\partial z_{12}}.$$

All of these arise through the equivalence of the operators

$$\frac{1}{l}\frac{\partial}{\partial x_{11}} = \frac{1}{m}\frac{\partial}{\partial y_{11}} = \frac{1}{n}\frac{\partial}{\partial z_{11}} = \frac{1}{n^3}\left\{y_2^2\frac{\partial}{\partial r} - x_2 y_2\frac{\partial}{\partial s} + x_2^2\frac{\partial}{\partial t}\right\},$$

$$\frac{1}{l}\frac{\partial}{\partial x_{12}} = \frac{1}{m}\frac{\partial}{\partial y_{12}} = \frac{1}{n}\frac{\partial}{\partial z_{12}} = -\frac{1}{n^3}\left\{2y_1 y_2\frac{\partial}{\partial r} - (x_1 y_2 + y_1 x_2)\frac{\partial}{\partial s} + 2x_1 x_2\frac{\partial}{\partial t}\right\},$$

$$\frac{1}{l}\frac{\partial}{\partial x_{22}} = \frac{1}{m}\frac{\partial}{\partial y_{22}} = \frac{1}{n}\frac{\partial}{\partial z_{22}} = \frac{1}{n^3}\left\{y_1^2\frac{\partial}{\partial r} - x_1 y_1\frac{\partial}{\partial s} + x_1^2\frac{\partial}{\partial t}\right\};$$

and they shew that forty-five of the second derivatives of $\Omega$ are expressible in terms of the six quantities

$$\frac{\partial^2 f}{\partial r^2},\quad \frac{\partial^2 f}{\partial r\partial s},\quad \frac{\partial^2 f}{\partial r\partial t},\quad \frac{\partial^2 f}{\partial s^2},\quad \frac{\partial^2 f}{\partial s\partial t},\quad \frac{\partial^2 f}{\partial t^2}.$$

*Ex.* Prove that the fifty-four second derivatives of $\Omega$, of the type $\dfrac{\partial^2\Omega}{\partial z_1\partial z_{11}}$, ..., are expressible in terms of nine additional quantities

$$\left(\frac{\partial}{\partial p},\ \frac{\partial}{\partial q}\right)\!\!\left(\frac{\partial}{\partial r},\ \frac{\partial}{\partial s},\ \frac{\partial}{\partial t}\right)f,\quad \left(\frac{\partial}{\partial r},\ \frac{\partial}{\partial s},\ \frac{\partial}{\partial t}\right)f\,;$$

that the twenty-one second derivatives of $\Omega$, of the type $\dfrac{\partial^2\Omega}{\partial z_1{}^2},\ \dfrac{\partial^2\Omega}{\partial z_1\partial z_2}$, ..., are expressible in terms of six additional quantities

$$\left(\frac{\partial}{\partial p},\ \frac{\partial}{\partial q}\right)^{\!2}f,\quad \left(\frac{\partial}{\partial p},\ \frac{\partial}{\partial q}\right)f\,;$$

that the twenty-seven second derivatives of $\Omega$, of the type $\dfrac{\partial^2\Omega}{\partial z\partial z_{11}}$, ..., are expressible in terms of nine additional quantities

$$\left(\frac{\partial}{\partial x},\ \frac{\partial}{\partial y},\ \frac{\partial}{\partial z}\right)\!\!\left(\frac{\partial}{\partial r},\ \frac{\partial}{\partial s},\ \frac{\partial}{\partial t}\right)f\,;$$

that the eighteen second derivatives of $\Omega$, of the type $\dfrac{\partial^2\Omega}{\partial z\partial z_1}$, ..., are expressible in terms of nine additional quantities

$$\left(\frac{\partial}{\partial x},\ \frac{\partial}{\partial y},\ \frac{\partial}{\partial z}\right)\!\!\left(\frac{\partial}{\partial p},\ \frac{\partial}{\partial q}\right)f,\quad \left(\frac{\partial}{\partial x},\ \frac{\partial}{\partial y},\ \frac{\partial}{\partial z}\right)f,$$

and that the six second derivatives of $\Omega$, of the type $\dfrac{\partial^2\Omega}{\partial z^2}$, are expressible in terms of the six quantities

$$\left(\frac{\partial}{\partial x},\ \frac{\partial}{\partial y},\ \frac{\partial}{\partial z}\right)^{\!2}f.$$

*First variation of the integral: three characteristic equations.*

**354.** When the integral $I,=\iint\Omega\,du\,dv$, is submitted to a general weak variation, whereby $x,\ y,\ z$, become $x+\kappa X,\ y+\kappa Y,\ z+\kappa Z$, while $X,\ Y,\ Z$, are arbitrary regular functions of $u$ and $v$, independent of one another and not involving $\kappa$, and when the new value of $I$ is denoted by $\mathbf{I}$, the whole variation $\mathbf{I}-I$ is equal to

$$\kappa I_1+[\kappa^2],$$

where $\kappa I_1$ is the aggregate of terms involving the first power of $\kappa$ only, and where $[\kappa^2]$ is the aggregate of terms involving the second and higher powers of $\kappa$. Also, when $\theta$ stands for $x,\ y,\ z$, in turn, while $\Theta$ stands for $X,\ Y,\ Z$, in corresponding turn, we have

$$I_1=\Sigma\iint\!\left(\frac{\partial\Omega}{\partial\theta}\Theta+\frac{\partial\Omega}{\partial\theta_1}\Theta_1+\frac{\partial\Omega}{\partial\theta_2}\Theta_2+\frac{\partial\Omega}{\partial\theta_{11}}\Theta_{11}+\frac{\partial\Omega}{\partial\theta_{12}}\Theta_{12}+\frac{\partial\Omega}{\partial\theta_{22}}\Theta_{22}\right)du\,dv,$$

the summation being for the values $x,\ y,\ z$, of $\theta$ and $X,\ Y,\ Z$, of $\Theta$. Exactly as for the special weak variation (§ 335), we have

$$\iint \frac{\partial \Omega}{\partial \theta_{11}} \Theta_{11} du\, dv = \iint \Theta \frac{\partial^2}{\partial u^2} \left(\frac{\partial \Omega}{\partial \theta_{11}}\right)\ du\, dv + \int \left\{ \frac{\partial \Omega}{\partial \theta_{11}} \Theta_1 - \Theta \frac{\partial}{\partial u} \left(\frac{\partial \Omega}{\partial \theta_{11}}\right) \right\} dv,$$

$$\iint \frac{\partial \Omega}{\partial \theta_{12}} \Theta_{12} du\, dv = \iint \Theta \frac{\partial^2}{\partial u\, \partial v} \left(\frac{\partial \Omega}{\partial \theta_{12}}\right) du\, dv + \tfrac{1}{2} \int \left\{ \frac{\partial \Omega}{\partial \theta_{12}} \Theta_2 - \Theta \frac{\partial}{\partial v} \left(\frac{\partial \Omega}{\partial \theta_{12}}\right) \right\} dv$$

$$- \tfrac{1}{2} \int \left\{ \frac{\partial \Omega}{\partial \theta_{12}} \Theta_1 - \Theta \frac{\partial}{\partial u} \left(\frac{\partial \Omega}{\partial \theta_{12}}\right) \right\} du,$$

$$\iint \frac{\partial \Omega}{\partial \theta_{22}} \Theta_{22} du\, dv = \iint \Theta \frac{\partial^2}{\partial v^2} \left(\frac{\partial \Omega}{\partial \theta_{22}}\right) du\, dv - \int \left\{ \frac{\partial \Omega}{\partial \theta_{22}} \Theta_2 - \Theta \frac{\partial}{\partial v} \left(\frac{\partial \Omega}{\partial \theta_{22}}\right) \right\} du,$$

$$\iint \frac{\partial \Omega}{\partial \theta_1} \Theta_1 du\, dv = - \iint \Theta \frac{\partial}{\partial u} \left(\frac{\partial \Omega}{\partial \theta_1}\right) du\, dv + \int \Theta \frac{\partial \Omega}{\partial \theta_1} dv,$$

$$\iint \frac{\partial \Omega}{\partial \theta_2} \Theta_2 du\, dv = - \iint \Theta \frac{\partial}{\partial v} \left(\frac{\partial \Omega}{\partial \theta_2}\right) du\, dv - \int \Theta \frac{\partial \Omega}{\partial \theta_2} du.$$

When $I_1$ is transformed by these relations, it is changed into the sum of a double integral $\overline{\overline{I}}_1$ and a single integral $\overline{I}_1$, where

$$\overline{\overline{I}}_1 = \Sigma \iint \Theta \left\{ \frac{\partial \Omega}{\partial \theta} - \frac{\partial}{\partial u} \left(\frac{\partial \Omega}{\partial \theta_1}\right) - \frac{\partial}{\partial v} \left(\frac{\partial \Omega}{\partial \theta_2}\right) + \frac{\partial^2}{\partial u^2} \left(\frac{\partial \Omega}{\partial \theta_{11}}\right) + \frac{\partial^2}{\partial u\, \partial v} \left(\frac{\partial \Omega}{\partial \theta_{12}}\right) + \frac{\partial^2}{\partial v^2} \left(\frac{\partial \Omega}{\partial \theta_{22}}\right) \right\} du\, dv,$$

taken over the whole bounded field of variation positively, while

$$\overline{I}_1 = \Sigma \int \left[ \frac{\partial \Omega}{\partial \theta_{11}} \Theta_1 + \tfrac{1}{2} \frac{\partial \Omega}{\partial \theta_{12}} \Theta_2 + \Theta \left\{ \frac{\partial \Omega}{\partial \theta_1} - \frac{\partial}{\partial u} \left(\frac{\partial \Omega}{\partial \theta_{11}}\right) - \tfrac{1}{2} \frac{\partial}{\partial v} \left(\frac{\partial \Omega}{\partial \theta_{12}}\right) \right\} \right] dv$$

$$- \int \left[ \frac{\partial \Omega}{\partial \theta_{22}} \Theta_2 + \tfrac{1}{2} \frac{\partial \Omega}{\partial \theta_{12}} \Theta_1 + \Theta \left\{ \frac{\partial \Omega}{\partial \theta_2} - \tfrac{1}{2} \frac{\partial}{\partial u} \left(\frac{\partial \Omega}{\partial \theta_{12}}\right) - \frac{\partial}{\partial v} \left(\frac{\partial \Omega}{\partial \theta_{22}}\right) \right\} \right] du,$$

each of these integrals being taken in the positive direction of each portion of the boundary, and along the whole of each portion of the boundary, of the whole field of variation.

In order that the integral $I$ may have a maximum or a minimum, its first variation must vanish whatever small variation be imposed upon its variables; and therefore

$$\overline{\overline{I}}_1 + \overline{I}_1 = 0,$$

for all admissible variations whatever.

When those variations are selected which, being distinct from zero throughout the general field, are zero everywhere along the boundary, they make $\overline{I}_1$ vanish unconditionally; and the condition then requires that

$$\overline{\overline{I}}_1 = 0.$$

In this double integral, the quantities $X$, $Y$, $Z$, being the three quantities for which $\Theta$ stands in the summation, are arbitrary regular functions of $u$ and $v$, independent of one another, and otherwise completely unrestricted. Thus, by the usual argument, the condition can be satisfied only if the coefficient of $X$, that of $Y$, and that of $Z$, vanish separately. Accordingly, we have

$$
\begin{aligned}
E_x &= \frac{\partial \Omega}{\partial x} - \frac{\partial}{\partial u}\left(\frac{\partial \Omega}{\partial x_1}\right) - \frac{\partial}{\partial v}\left(\frac{\partial \Omega}{\partial x_2}\right) + \frac{\partial^2}{\partial u^2}\left(\frac{\partial \Omega}{\partial x_{11}}\right) + \frac{\partial^2}{\partial u \partial v}\left(\frac{\partial \Omega}{\partial x_{12}}\right) + \frac{\partial^2}{\partial v^2}\left(\frac{\partial \Omega}{\partial x_{22}}\right) = 0 \\
E_y &= \frac{\partial \Omega}{\partial y} - \frac{\partial}{\partial u}\left(\frac{\partial \Omega}{\partial y_1}\right) - \frac{\partial}{\partial v}\left(\frac{\partial \Omega}{\partial y_2}\right) + \frac{\partial^2}{\partial u^2}\left(\frac{\partial \Omega}{\partial y_{11}}\right) + \frac{\partial^2}{\partial u \partial v}\left(\frac{\partial \Omega}{\partial y_{12}}\right) + \frac{\partial^2}{\partial v^2}\left(\frac{\partial \Omega}{\partial y_{22}}\right) = 0 \\
E_z &= \frac{\partial \Omega}{\partial z} - \frac{\partial}{\partial u}\left(\frac{\partial \Omega}{\partial z_1}\right) - \frac{\partial}{\partial v}\left(\frac{\partial \Omega}{\partial z_2}\right) + \frac{\partial^2}{\partial u^2}\left(\frac{\partial \Omega}{\partial z_{11}}\right) + \frac{\partial^2}{\partial u \partial v}\left(\frac{\partial \Omega}{\partial z_{12}}\right) + \frac{\partial^2}{\partial v^2}\left(\frac{\partial \Omega}{\partial z_{22}}\right) = 0
\end{aligned} \Big\} ;
$$

and these equations must be satisfied everywhere in the field.

### General boundary condition.

**355.** Moreover, when these equations are thus satisfied, we have

$$\overline{\overline{I}}_1 = 0$$

for all small variations, whether they vanish on the boundary or not. Hence, when we take the variations which do not vanish everywhere along the boundary, the requirement as to the vanishing of the first variation demands, for all of them, the relation

$$\overline{I}_1 = 0.$$

The only limitation upon the variations $\kappa X$, $\kappa Y$, $\kappa Z$, is that they are regular (and therefore continuous) while they are independent, subject to the conditions as to the character of the boundary. This boundary might be a fixed curve or fixed curves: no condition would emerge. Some portion might be a fixed curve, while another portion might be required to be a mobile curve on a given surface, mobile (that is) to meet requirements; or each portion might be such a mobile curve, with a different surface for each different portion This last alternative manifestly is the most general.

In such an event, all the admissible variations being independent of one another, we must have $\overline{I}_1 = 0$ for each separate and distinct portion of the boundary; because there are variations, zero for every other portion and not zero for that portion.

Also along each separate portion, the variations are arbitrary subject to continuity. Hence, in order that $\overline{I}_1$ may vanish for all these arbitrary variations, the element contributed by each rudiment of arc in the portion must vanish; for otherwise we could select only positive elements at will, and only negative elements at will, keeping all the other elements zero at each such selection. Hence we must have, everywhere along each separate portion of the whole boundary of the field,

$$
\begin{aligned}
B = \Sigma &\left\{ \left(\frac{\partial \Omega}{\partial \theta_{11}} \Theta_1 + \tfrac{1}{2}\frac{\partial \Omega}{\partial \theta_{12}} \Theta_2\right) dv - \left(\frac{\partial \Omega}{\partial \theta_{22}} \Theta_2 + \tfrac{1}{2}\frac{\partial \Omega}{\partial \theta_{12}} \Theta_1\right) du \right\} \\
&+ \Sigma\Theta \left[ \left\{\frac{\partial \Omega}{\partial \theta_1} - \frac{\partial}{\partial u}\left(\frac{\partial \Theta}{\partial \theta_{11}}\right) - \tfrac{1}{2}\frac{\partial}{\partial v}\left(\frac{\partial \Theta}{\partial \theta_{12}}\right)\right\} dv - \left\{\frac{\partial \Omega}{\partial \theta_2} - \tfrac{1}{2}\frac{\partial}{\partial u}\left(\frac{\partial \Theta}{\partial \theta_{12}}\right) - \frac{\partial}{\partial v}\left(\frac{\partial \Theta}{\partial \theta_{22}}\right)\right\} du \right] = 0.
\end{aligned}
$$

Thus the necessary requirement, that the first variation of the integral shall vanish for all small weak variations, is met when two sets of conditions are satisfied:

(i), the equations

$$E_x = 0, \quad E_y = 0, \quad E_z = 0,$$

must be satisfied everywhere in the field of variation: and

(ii), the equation

$$B = 0$$

must be satisfied everywhere on each portion of the boundary of the field.

The three equations, in (i), are the *characteristic equations*; determining $x, y, z$, as functions of $u$ and $v$, they potentially determine the *characteristic surface*. The condition $B = 0$, universal along the boundary, is to be satisfied at the limits of the region of the characteristic surface selected for the integral.

*The three characteristic equations are equivalent to a single equation.*

**356.** With the special weak variation, we found (§ 337) a single characteristic equation. With the general weak variation, we have found three such equations satisfied by $\Omega$, the subject of integration; and $\Omega$ satisfies certain identities. We now proceed to shew that, when account is taken of these identities, the three equations $E_x = 0$, $E_y = 0$, $E_z = 0$, are satisfied in virtue of a single equation.

Differentiating the identity $\Sigma x_1 \dfrac{\partial \Omega}{\partial x_{11}} = 0$ twice with respect to $u$, the identity $\Sigma x_1 \dfrac{\partial \Omega}{\partial x_{12}} = 0$ with respect to $u$ and $v$ in succession, and the identity $\Sigma x_1 \dfrac{\partial \Omega}{\partial x_{22}} = 0$ twice with respect to $v$, we find respectively,

$$\Sigma x_1 \frac{\partial^2}{\partial u^2} \left( \frac{\partial \Omega}{\partial x_{11}} \right) = - 2\Sigma x_{11} \frac{\partial}{\partial u} \left( \frac{\partial \Omega}{\partial x_{11}} \right) - \Sigma x_{111} \frac{\partial \Omega}{\partial x_{11}},$$

$$\Sigma x_1 \frac{\partial^2}{\partial u \partial v} \left( \frac{\partial \Omega}{\partial x_{12}} \right) = - \Sigma x_{12} \frac{\partial}{\partial u} \left( \frac{\partial \Omega}{\partial x_{12}} \right) - \Sigma x_{11} \frac{\partial}{\partial v} \left( \frac{\partial \Omega}{\partial x_{12}} \right) - \Sigma x_{112} \frac{\partial \Omega}{\partial x_{12}},$$

$$\Sigma x_1 \frac{\partial^2}{\partial v^2} \left( \frac{\partial \Omega}{\partial x_{22}} \right) = - 2\Sigma x_{12} \frac{\partial}{\partial v} \left( \frac{\partial \Omega}{\partial x_{22}} \right) - \Sigma x_{122} \frac{\partial \Omega}{\partial x_{22}}.$$

Again, differentiating the identity

$$\Omega = \Sigma x_1 \frac{\partial \Omega}{\partial x_1} + 2\Sigma x_{11} \frac{\partial \Omega}{\partial x_{11}} + \Sigma x_{12} \frac{\partial \Omega}{\partial x_{12}},$$

with respect to $u$, we have

$$\Sigma x_1 \frac{\partial \Omega}{\partial x} + \Sigma x_{11} \frac{\partial \Omega}{\partial x_1} + \Sigma x_{12} \frac{\partial \Omega}{\partial x_2} + \Sigma x_{111} \frac{\partial \Omega}{\partial x_{11}} + \Sigma x_{112} \frac{\partial \Omega}{\partial x_{12}} + \Sigma x_{122} \frac{\partial \Omega}{\partial x_{22}}$$

$$= \Sigma x_{11} \frac{\partial \Omega}{\partial x_1} + \Sigma x_1 \frac{\partial}{\partial u} \left( \frac{\partial \Omega}{\partial x_1} \right) + 2\Sigma x_{111} \frac{\partial \Omega}{\partial x_{11}} + 2\Sigma x_{11} \frac{\partial}{\partial u} \left( \frac{\partial \Omega}{\partial x_{11}} \right) + \Sigma x_{112} \frac{\partial \Omega}{\partial x_{12}} + \Sigma x_{12} \frac{\partial}{\partial u} \left( \frac{\partial \Omega}{\partial x_{12}} \right),$$

so that

$$\Sigma x_1 \frac{\partial \Omega}{\partial x} - \Sigma x_1 \frac{\partial}{\partial u} \left( \frac{\partial \Omega}{\partial x_1} \right)$$

$$= - \Sigma x_{12} \frac{\partial \Omega}{\partial x_2} + \Sigma x_{111} \frac{\partial \Omega}{\partial x_{11}} - \Sigma x_{122} \frac{\partial \Omega}{\partial x_{22}} + 2\Sigma x_{11} \frac{\partial}{\partial u} \left( \frac{\partial \Omega}{\partial x_{11}} \right) + \Sigma x_{12} \frac{\partial}{\partial u} \left( \frac{\partial \Omega}{\partial x_{12}} \right).$$

Further, differentiating the identity

$$\Sigma x_1 \frac{\partial \Omega}{\partial x_2} + \Sigma x_{11} \frac{\partial \Omega}{\partial x_{12}} + 2\Sigma x_{12} \frac{\partial \Omega}{\partial x_{22}} = 0$$

with respect to $v$, we have

$$- \Sigma x_1 \frac{\partial}{\partial v} \left( \frac{\partial \Omega}{\partial x_2} \right) = \Sigma x_{12} \frac{\partial \Omega}{\partial x_2} + \Sigma x_{112} \frac{\partial \Omega}{\partial x_{12}} + \Sigma x_{11} \frac{\partial}{\partial v} \left( \frac{\partial \Omega}{\partial x_{12}} \right) + 2\Sigma x_{122} \frac{\partial \Omega}{\partial x_{22}} + 2\Sigma x_{12} \frac{\partial}{\partial v} \left( \frac{\partial \Omega}{\partial x_{22}} \right).$$

Combining these results, we find

$$\Sigma x_1 \frac{\partial \Omega}{\partial x} - \Sigma x_1 \frac{\partial}{\partial u} \left( \frac{\partial \Omega}{\partial x_1} \right) - \Sigma x_1 \frac{\partial}{\partial v} \left( \frac{\partial \Omega}{\partial x_2} \right)$$

$$+ \Sigma x_1 \frac{\partial^2}{\partial u^2} \left( \frac{\partial \Omega}{\partial x_{11}} \right) + \Sigma x_1 \frac{\partial^2}{\partial u \partial v} \left( \frac{\partial \Omega}{\partial x_{12}} \right) + \Sigma x_1 \frac{\partial^2}{\partial v^2} \left( \frac{\partial \Omega}{\partial x_{22}} \right) = 0 :$$

that is, having regard to the expressions for $E_x$, $E_y$, $E_z$, we have

$$x_1 E_x + y_1 E_y + z_1 E_z = 0,$$

a result obtained by using five out of the ten identities satisfied by $\Omega$.

Using the other five identities in a precisely similar manner with interchange of the differential operations with regard to $u$ and $v$, we have

$$x_2 E_x + y_2 E_y + z_2 E_z = 0.$$

Hence there is a quantity $E$ such that

$$E_x = lE, \quad E_y = mE, \quad E_z = nE :$$

that is, *when account is taken of the identities satisfied by* $\Omega$, *the three characteristic equations* $E_x = 0$, $E_y = 0$, $E_z = 0$, *are equivalent to a single characteristic equation*

$$E = 0.$$

### *Identification with the characteristic equation of* § 337.

**357.** This equation $E = 0$ is the same as the former single characteristic equation, which emerged under the special weak variation.

Selecting the equation $E_z = 0$ for the purpose of this identification, we begin with the relation

$$\Omega = nf.$$

Noting the relations (§ 291) between $p$, $q$, $r$, $s$, $t$, and the first and second derivatives of $x$, $y$, $z$, with regard to $u$ and $v$, we have

$$\frac{\partial \Omega}{\partial z_1} = y_2 \frac{\partial f}{\partial p} - x_2 \frac{\partial f}{\partial q}$$

$$+ \frac{1}{n^2} \frac{\partial f}{\partial r} \{-y_2(x_{11}y_2^2 - 2x_{12}y_1y_2 + x_{22}y_1^2) + x_2(y_{11}y_2^2 - 2y_{12}y_1y_2 + y_{22}y_1^2)\}$$

$$+ \frac{1}{n^2} \frac{\partial f}{\partial s} [\ y_2\{x_{11}x_2y_2 - x_{12}(x_1y_2 + y_1x_2) + x_{22}x_1y_1\}$$

$$- x_2\{y_{11}x_2y_2 - y_{12}(x_1y_2 + y_1x_2) + y_{22}x_1y_1\}]$$

$$+ \frac{1}{n^2} \frac{\partial f}{\partial t} \{-y_2(x_{11}x_2^2 - 2x_{12}x_1x_2 + x_{22}x_1^2) + x_2(y_{11}x_2^2 - 2y_{12}x_1x_2 + y_{22}x_1^2)\},$$

$$\frac{\partial \Omega}{\partial z_2} = -y_1 \frac{\partial f}{\partial p} + x_1 \frac{\partial f}{\partial q}$$

$$+ \frac{1}{n^2} \frac{\partial f}{\partial r} \{\ y_1(x_{11}y_2^2 - 2x_{12}y_1y_2 + x_{22}y_1^2) - x_1(y_{11}y_2^2 - 2y_{12}y_1y_2 + y_{22}y_1^2)\}$$

$$+ \frac{1}{n^2} \frac{\partial f}{\partial s} [-y_1\{x_{11}x_2y_2 - x_{12}(x_1y_2 + y_1x_2) + x_{22}x_1y_1\}$$

$$+ x_1\{y_{11}x_2y_2 - y_{12}(x_1y_2 + y_1x_2) + y_{22}x_1y_1\}]$$

$$+ \frac{1}{n^2} \frac{\partial f}{\partial t} \{\ y_1(x_{11}x_2^2 - 2x_{12}x_1x_2 + x_{22}x_1^2) - x_1(y_{11}x_2^2 - 2y_{12}x_1x_2 + y_{22}x_1^2)\},$$

$$\frac{\partial \Omega}{\partial z_{11}} = \frac{1}{n}\left(y_2^2 \frac{\partial f}{\partial r} - x_2y_2 \frac{\partial f}{\partial s} + x_2^2 \frac{\partial f}{\partial t}\right),$$

$$-\tfrac{1}{2}\frac{\partial \Omega}{\partial z_{12}} = \frac{1}{n}\left\{y_1y_2 \frac{\partial f}{\partial r} - \tfrac{1}{2}(x_1y_2 + y_1x_2)\frac{\partial f}{\partial s} + x_1x_2 \frac{\partial f}{\partial t}\right\},$$

$$\frac{\partial \Omega}{\partial z_{22}} = \frac{1}{n}\left(y_1^2 \frac{\partial f}{\partial r} - x_1y_1 \frac{\partial f}{\partial s} + x_1^2 \frac{\partial f}{\partial t}\right).$$

Also

$$\frac{\partial}{\partial u} = x_1 \frac{\partial}{\partial x} + y_1 \frac{\partial}{\partial y}, \qquad \frac{\partial}{\partial v} = x_2 \frac{\partial}{\partial x} + y_2 \frac{\partial}{\partial y},$$

so that

$$n\frac{\partial}{\partial x} = y_2 \frac{\partial}{\partial u} - y_1 \frac{\partial}{\partial v}, \qquad n\frac{\partial}{\partial y} = -x_2 \frac{\partial}{\partial u} + x_1 \frac{\partial}{\partial v}.$$

Then, by direct substitution, we find

$$\frac{\partial \Omega}{\partial z_1} - \frac{\partial}{\partial u}\left(\frac{\partial \Omega}{\partial z_{11}}\right) - \tfrac{1}{2}\frac{\partial}{\partial v}\left(\frac{\partial \Omega}{\partial z_{12}}\right)$$

$$= y_2\left\{\frac{\partial f}{\partial p} - \frac{\partial}{\partial x}\left(\frac{\partial f}{\partial r}\right) - \tfrac{1}{2}\frac{\partial}{\partial y}\left(\frac{\partial f}{\partial s}\right)\right\} - x_2\left\{\frac{\partial f}{\partial q} - \tfrac{1}{2}\frac{\partial}{\partial x}\left(\frac{\partial f}{\partial s}\right) - \frac{\partial}{\partial y}\left(\frac{\partial f}{\partial t}\right)\right\},$$

$$\frac{\partial \Omega}{\partial z_2} - \tfrac{1}{2}\frac{\partial}{\partial u}\left(\frac{\partial \Omega}{\partial z_{12}}\right) - \frac{\partial}{\partial v}\left(\frac{\partial \Omega}{\partial z_{22}}\right)$$

$$= -y_1\left\{\frac{\partial f}{\partial p} - \frac{\partial}{\partial x}\left(\frac{\partial f}{\partial r}\right) - \tfrac{1}{2}\frac{\partial}{\partial y}\left(\frac{\partial f}{\partial s}\right)\right\} + x_1\left\{\frac{\partial f}{\partial q} - \tfrac{1}{2}\frac{\partial}{\partial x}\left(\frac{\partial f}{\partial s}\right) - \frac{\partial}{\partial y}\left(\frac{\partial f}{\partial t}\right)\right\}.$$

Consequently,

$$\frac{\partial}{\partial u}\left\{\frac{\partial\Omega}{\partial z_1} - \frac{\partial}{\partial u}\left(\frac{\partial\Omega}{\partial z_{11}}\right) - \frac{1}{2}\frac{\partial}{\partial v}\left(\frac{\partial\Omega}{\partial z_{12}}\right)\right\} + \frac{\partial}{\partial v}\left\{\frac{\partial\Omega}{\partial z_2} - \frac{1}{2}\frac{\partial}{\partial u}\left(\frac{\partial\Omega}{\partial z_{12}}\right) - \frac{\partial}{\partial v}\left(\frac{\partial\Omega}{\partial z_{22}}\right)\right\}$$

$$= \left(y_2\frac{\partial}{\partial u} - y_1\frac{\partial}{\partial v}\right)\left\{\frac{\partial f}{\partial p} - \frac{\partial}{\partial x}\left(\frac{\partial f}{\partial r}\right) - \frac{1}{2}\frac{\partial}{\partial y}\left(\frac{\partial f}{\partial s}\right)\right\}$$

$$+ \left(x_1\frac{\partial}{\partial v} - x_2\frac{\partial}{\partial u}\right)\left\{\frac{\partial f}{\partial q} - \frac{1}{2}\frac{\partial}{\partial x}\left(\frac{\partial f}{\partial s}\right) - \frac{\partial}{\partial y}\left(\frac{\partial f}{\partial t}\right)\right\}$$

$$= n\frac{\partial}{\partial x}\left\{\frac{\partial f}{\partial p} - \frac{\partial}{\partial x}\left(\frac{\partial f}{\partial r}\right) - \frac{1}{2}\frac{\partial}{\partial y}\left(\frac{\partial f}{\partial s}\right)\right\} + n\frac{\partial}{\partial y}\left\{\frac{\partial f}{\partial q} - \frac{1}{2}\frac{\partial}{\partial x}\left(\frac{\partial f}{\partial s}\right) - \frac{\partial}{\partial y}\left(\frac{\partial f}{\partial t}\right)\right\}.$$

Also

$$\frac{\partial\Omega}{\partial z} = n\frac{\partial f}{\partial z}.$$

Hence

$$E_z = \frac{\partial\Omega}{\partial z} - \frac{\partial}{\partial u}\left(\frac{\partial\Omega}{\partial z_1}\right) - \frac{\partial}{\partial v}\left(\frac{\partial\Omega}{\partial z_2}\right) + \frac{\partial^2}{\partial u^2}\left(\frac{\partial\Omega}{\partial z_{11}}\right) + \frac{\partial^2}{\partial u\,\partial v}\left(\frac{\partial\Omega}{\partial z_{12}}\right) + \frac{\partial^2}{\partial v^2}\left(\frac{\partial\Omega}{\partial z_{22}}\right)$$

$$= n\left\{\frac{\partial f}{\partial z} - \frac{\partial}{\partial x}\left(\frac{\partial f}{\partial p}\right) - \frac{\partial}{\partial y}\left(\frac{\partial f}{\partial q}\right) + \frac{\partial^2}{\partial x^2}\left(\frac{\partial f}{\partial r}\right) + \frac{\partial^2}{\partial x\,\partial y}\left(\frac{\partial f}{\partial s}\right) + \frac{\partial^2}{\partial y^2}\left(\frac{\partial f}{\partial t}\right)\right\}$$

$$= nE:$$

that is, the characteristic equation deduced from the special weak variation, and the single dominant characteristic equation derived from the general weak variation (account being taken of the identities satisfied by $\Omega$), are identical.

*Ex.* Verify the relations

$$y_2\frac{\partial f}{\partial r} - \frac{1}{2}x_2\frac{\partial f}{\partial s} = x_1\frac{\partial\Omega}{\partial z_{11}} + \frac{1}{2}x_2\frac{\partial\Omega}{\partial z_{12}},$$

$$-y_1\frac{\partial f}{\partial r} + \frac{1}{2}x_1\frac{\partial f}{\partial s} = x_2\frac{\partial\Omega}{\partial z_{22}} + \frac{1}{2}x_1\frac{\partial\Omega}{\partial z_{12}},$$

$$\frac{1}{2}y_2\frac{\partial f}{\partial s} - x_2\frac{\partial f}{\partial t} = y_1\frac{\partial\Omega}{\partial z_{11}} + \frac{1}{2}y_2\frac{\partial\Omega}{\partial z_{12}},$$

$$-\frac{1}{2}y_1\frac{\partial f}{\partial s} + x_1\frac{\partial f}{\partial t} = y_2\frac{\partial\Omega}{\partial z_{22}} + \frac{1}{2}y_1\frac{\partial\Omega}{\partial z_{12}};$$

deduce that, along the boundary,

$$\left(\frac{\partial\Omega}{\partial z_{11}}\zeta_1 + \frac{1}{2}\frac{\partial\Omega}{\partial z_{12}}\zeta_2\right)dv - \left(\frac{\partial\Omega}{\partial z_{22}}\zeta_2 + \frac{1}{2}\frac{\partial\Omega}{\partial z_{12}}\zeta_1\right)du$$

$$= \left(\frac{\partial f}{\partial r}\frac{\partial\zeta}{\partial x} + \frac{1}{2}\frac{\partial f}{\partial s}\frac{\partial\zeta}{\partial y}\right)dy - \left(\frac{1}{2}\frac{\partial f}{\partial s}\frac{\partial\zeta}{\partial x} + \frac{\partial f}{\partial t}\frac{\partial\zeta}{\partial y}\right)dx;$$

and hence verify the agreement of the boundary terms in $\zeta$, arising out of the general weak variations (§ 355), with the corresponding terms arising out of the special weak variations (§ 338).

These results may suffice to shew that, before the calculations for the problem are carried further when general weak variations are imposed, it is desirable that the problem shall be more completely advanced towards its solution when only special weak variations are imposed.

# CHAPTER XII.

### Lemma in triple integration.

**358.** We shall deal only briefly with maxima and minima of triple integrals, which are volume-integrals when geometrical interpretations are used: one principal reason being that, in even the simplest instances, the characteristic partial differential equations involved are of the second order, while the number of independent variables is three. The general theory of such equations has received scant attention: though one particular exception, that of the equation of the potential in free space, is notable.

To shorten the analysis, one preliminary lemma\* in integration must be established, as follows:

*Let $P, Q, R$, denote any three functions of three independent variables $x, y, z$, regular everywhere throughout a volume bounded by surfaces that do not intersect; then*

$$\iiint \left( \frac{\partial P}{\partial x} + \frac{\partial Q}{\partial y} + \frac{\partial R}{\partial z} \right) dx\, dy\, dz = \iint (lP + mQ + nR)\, dS,$$

*where the triple integral† is to be taken positively for increases of the variables throughout the volume; the double integral extends over the whole superficial boundary of the volume; and $l, m, n$, are the direction-cosines of the normal at the boundary surface drawn outwards from the enclosed volume.*

We have

$$\iiint \frac{\partial P}{\partial x}\, dx\, dy\, dz = \iint [P]\, dy\, dz,$$

where $[P]$ denotes that the value of $P$ has to be taken at the various places at which the increasing coordinate $x$ enters or leaves the bounded volume. Thus, if the value of $P$ at the elements of surface $a'b'c'd'$, $A'B'C'D'$, $abcd$, $ABCD$, be $P_1, P_2, P_3, P_4$, respectively, we have

$$[P] = - P_1 + P_2 - P_3 + P_4,$$

(assuming, merely for illustration, that the four elements constitute the whole of the boundary met by the line parallel to the axis of $x$). If the correspond-

---

\* The lemma is of frequent occurrence in mathematical physics: it is included here, because of its immediate use.

† We assume, as a convention for triple integrals similar to the convention assumed (§ 275) for double integrals, that the integrations are to be performed, firstly with regard to the differential element $dx$ that occurs first, secondly with regard to the differential element that occurs second, and so on, in the sequence from left to right.

ing elements of area be $dS_1$, $dS_2$, $dS_3$, $dS_4$; and if $l_1$, $l_2$, $l_3$, $l_4$, be the direction-cosines (towards the positive direction of the axis of $x$) of the respective normals, drawn outwards from the volume, that is, away from the volume

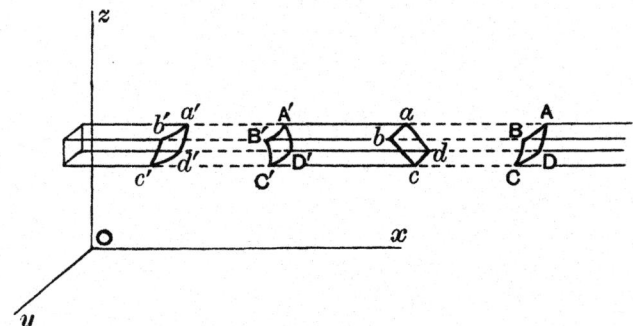

between $ABCD$ and $abcd$, and away from the volume between $A'B'C'D'$ and $a'b'c'd'$, then

$$- l_1 dS_1 = l_2 dS_2 = - l_3 dS_3 = l_4 dS_4 = dy\,dz.$$

Hence

$$[P]\,dy\,dz = P_4 l_4 dS_4 + P_3 l_3 dS_3 + P_2 l_2 dS_2 + P_1 l_1 dS_1 :$$

that is, the contribution of the whole surface, contributed in connection with $dy\,dz$ is $Pl\,dS$, where $l$ is the cosine of the angle between the positive direction of the axis of $x$ and the normal drawn outwards from the enclosed volume.  Thus

$$\iiint \frac{\partial P}{\partial x}\,dx\,dy\,dz = \iint lP\,dS.$$

Again, by the customary theorem for the change of variables in an integral, we have

$$\iiint \frac{\partial Q}{\partial y}\,dx\,dy\,dz = \iiint \frac{\partial Q}{\partial y}\,J\left(\frac{x,\,y,\,z}{y,\,z,\,x}\right)\,dy\,dz\,dx = \iiint \frac{\partial Q}{\partial y}\,dy\,dz\,dx.$$

In the same manner as for the preceding integral, we have

$$\iiint \frac{\partial Q}{\partial y}\,dy\,dz\,dx = \iint mQ\,dS,$$

where $m$ is the cosine of the inclination between the positive direction of the axis of $y$ and the normal to the superficial boundary drawn outwards from the enclosed volume wherever it is met by the axis of $y$ : the surface-integral being taken over the whole of the boundary.

Similarly,

$$\iiint \frac{\partial R}{\partial z}\,dx\,dy\,dz = \iiint \frac{\partial R}{\partial z}\,J\left(\frac{x,\,y,\,z}{z,\,x,\,y}\right)\,dz\,dx\,dy$$

$$= \iiint \frac{\partial R}{\partial z}\,dz\,dx\,dy$$

$$= \iiint nR\,dS,$$

where $n$ is the cosine of the inclination of the positive direction of the axis of $z$ and the normal to the superficial boundary drawn outwards from the enclosed volume wherever it is met by the axis of $z$, the surface-integral being taken over the whole of the boundary.

Combining the results, we have

$$\iiint \left(\frac{\partial P}{\partial x} + \frac{\partial Q}{\partial y} + \frac{\partial R}{\partial z}\right) dx\,dy\,dz = \iint (lP + mQ + nR)\,dS,$$

where the double integral is taken over the whole boundary of the volume, and $l$, $m$, $n$, are the direction-cosines of the normal along each portion of that whole boundary drawn outwards from the enclosed volume.

### 'First' variation of triple integral, under special variation.

**359.** We now proceed to the consideration of an integral

$$\iiint f(x, y, z, v, v_1, v_2, v_3)\,dx\,dy\,dz,$$

where $v$ is a dependent variable; $v_1, v_2, v_3$, denote $\dfrac{\partial v}{\partial x}$, $\dfrac{\partial v}{\partial y}$, $\dfrac{\partial v}{\partial z}$, respectively; and the function $f$ is regular within the space of integration which has definite boundaries. These boundaries may be expressed by taking the $x$-range from $x_0$ to $x_1$, where $x_0$ and $x_1$ are defined functions of $y$ and $z$; the $y$-range, after the $x$-integration is achieved, from $y_0$ to $y_1$, where $y_0$ and $y_1$ are defined functions of $z$; and the $z$-range from $z_0$ to $z_1$: or they may be expressed in any equivalent suitable mode. It is required to make the integral a maximum or a minimum, so far as this can be secured, (i) by the determination of $v$ as a function of $x$, $y$, $z$, throughout the volume, and (ii) by ancillary conditions at the boundary.

As usual, we begin with a special variation imposed upon $v$ alone through the space of integration; and we denote it by a change from $v$ to $v + \kappa V$. But in order to allow some freedom at the boundary so as to satisfy conditions, which may be descriptive rather than particularised in complete detail, we shall assume a variation at the upper limits represented by changes of $x_1$, $y_1$, $z_1$, into $x_1 + \kappa X_1$, $y_1 + \kappa Y_1$, $z_1 + \kappa Z_1$, and at the lower limits represented by changes of $x_0, y_0, z_0$ into $x_0 + \kappa X_0, y_0 + \kappa Y_0, z_0 + \kappa Z_0$. (The consequences of the last variations upon the expression of $V$ at the boundary will be considered later.) Then the new integral is

$$\int_{x_0+\kappa X_0}^{x_1+\kappa X_1} \int_{y_0+\kappa Y_0}^{y_1+\kappa Y_1} \int_{z_0+\kappa Z_0}^{z_1+\kappa Z_1} f(x, y, z, v + \kappa V, v_1 + \kappa V_1, v_2 + \kappa V_2, v_3 + \kappa V_3)\,dx\,dy\,dz,$$

which, for the moment, is denoted by

$$\int_{x_0+\kappa X_1}^{x_1+\kappa X_1} \int_{y_0+\kappa Y_0}^{y_1+\kappa Y_1} \int_{z_0+\kappa Z_0}^{z_1+\kappa Z_1} F\,dx\,dy\,dz.$$

Now

$$\int_{x_0+\kappa X_0}^{x_1+\kappa X_1} \Theta\, dx = \left( \int_{x_1}^{x_1+\kappa X_1} + \int_{x_0}^{x_1} - \int_{x_0}^{x_0+\kappa X_0} \right) \Theta\, dx,$$

for any function $\Theta$. Also

$$\int_{x_1}^{x_1+\kappa X_1} \Theta\, dx = \kappa X_1 \Theta_1 + [\kappa^2],$$

$$\int_{x_0}^{x_0+\kappa X_0} \Theta\, dx = \kappa X_0 \Theta_0 + [\kappa^2],$$

where $\Theta_1$ is the value of $\iint_{y_0 z_0}^{y_1 z_1} f\, dy\, dz$ for $x = x_1$, and $\Theta_0$ is its value for $x = x_0$ while $[\kappa^2]$ represents an aggregate of terms of the second and higher degrees in $\kappa$. Also, taking account of the fact that $dy\, dz = l\, dS$ at an upper limit and $= - l\, dS$ at a lower limit, we have our varied integral

$$= \kappa \iint l f X\, dS + \int_{x_0}^{x_1} \int_{y_0+\kappa Y_0}^{y_1+\kappa Y_1} \int_{z_0+\kappa Z_0}^{z_1+\kappa Z_1} F\, dx\, dy\, dz + [\kappa^2].$$

Again,

$$\int_{x_0}^{x_1} \int_{y_0+\kappa Y_0}^{y_1+\kappa Y_1} \int_{z_0+\kappa Z_0}^{z_1+\kappa Z_1} F\, dx\, dy\, dz = \int_{y_0+\kappa Y_0}^{y_1+\kappa Y_1} \int_{z_0+\kappa Z_0}^{z_1+\kappa Z_1} \int_{x_0}^{x_1} F\, dy\, dz\, dx.$$

As before, using the property

$$\int_{y_0+\kappa Y_0}^{y_1+\kappa Y_1} \Theta\, dy = \left( \int_{y_1}^{y_1+\kappa Y_1} + \int_{y_0}^{y_1} - \int_{y_0}^{y_0+\kappa Y_0} \right) \Theta\, dy,$$

and writing $\Theta_3$ and $\Theta_2$ as the values of $\int_{z_0}^{z_1} \int_{x_0}^{x_1} f\, dz\, dx$ when $y = y_1$ and when $y = y_0$ respectively, we have

$$\int_{y_1}^{y_1+\kappa Y_1} \Theta\, dx = \kappa Y_1 \Theta_3 + [\kappa^2], \qquad \int_{y_0}^{y_0+\kappa Y_0} \Theta\, dx = \kappa Y_0 \Theta_2 + [\kappa^2].$$

Thus

$$\int_{y_0+\kappa Y_0}^{y_1+\kappa Y_1} \int_{z_0+\kappa Z_0}^{z_1+\kappa Z_1} \int_{x_0}^{x_1} F\, dy\, dz\, dx$$

$$= \kappa \iint m f Y\, dS + \int_{y_0}^{y_1} \int_{z_0+\kappa Z_0}^{z_1+\kappa Z_1} \int_{x_0}^{x_1} F\, dy\, dz\, dx + [\kappa^2].$$

Lastly,

$$\int_{y_0}^{y_1} \int_{z_0+\kappa Z_0}^{z_1+\kappa Z_1} \int_{x_0}^{x_1} F\, dy\, dz\, dx = \int_{z_0+\kappa Z_0}^{z_1+\kappa Z_1} \int_{x_0}^{x_1} \int_{y_0}^{y_1} F\, dz\, dx\, dy.$$

The right-hand side, in the same way, is equal to

$$\kappa \iint n f Z\, dS + \int_{z_0}^{z_1} \int_{x_0}^{x_1} \int_{y_0}^{y_1} F\, dz\, dx\, dy + [\kappa^2]:$$

that is, to

$$\kappa \iint n f Z\, dS + \int_{x_0}^{x_1} \int_{y_0}^{y_1} \int_{z_0}^{z_1} F\, dx\, dy\, dz + [\kappa^2].$$

Hence, when the results are combined, our original integral is equal to

$$\kappa \iint (lX + mY + nZ) f \, dS + \int_{x_0}^{x_1} \int_{y_0}^{y_1} \int_{z_0}^{z_1} F \, dx \, dy \, dz + [\kappa^2].$$

Thus the variation of the integral, consequent upon the whole of the special variation that has been imposed, is equal to

$$\int_{x_0}^{x_1} \int_{y_0}^{y_1} \int_{z_0}^{z_1} (F - f) \, dx \, dy \, dz + \kappa \iint (lX + mY + nZ) f \, dS + [\kappa^2].$$

Again,

$$F - f = \kappa \left( V \frac{\partial f}{\partial v} + V_1 \frac{\partial f}{\partial v_1} + V_2 \frac{\partial f}{\partial v_2} + V_3 \frac{\partial f}{\partial v_3} \right) + [\kappa^2];$$

and therefore the whole variation of the original integral is

$$\kappa \int_{x_0}^{x_1} \int_{y_0}^{y_1} \int_{z_0}^{z_1} \left( V \frac{\partial f}{\partial v} + V_1 \frac{\partial f}{\partial v_1} + V_2 \frac{\partial f}{\partial v_2} + V_3 \frac{\partial f}{\partial v_3} \right) dx \, dy \, dz$$

$$+ \kappa \iint (lX + mY + nZ) f \, dS + [\kappa^2].$$

In order that the original integral may have a maximum or a minimum, its whole variation must maintain a uniform sign for all admissible variations. To secure this result, the first necessity is that the portion of the whole variation involving the first power of $\kappa$—called, as usual, the first variation of the integral—must vanish; if it did not vanish, its value would dominate the whole variation, and would change in sign with a change of sign in $\kappa$.

The first variation, which has to vanish, consists of two portions: (i) a surface-integral; and (ii) a volume-integral, which will be transformed before the conditions for the vanishing of the first variation are inferred. By our lemma,

$$\iiint \left\{ V_1 \frac{\partial f}{\partial v_1} + V \frac{\partial}{\partial x} \left( \frac{\partial f}{\partial v_1} \right) \right\} dx \, dy \, dz = \iiint \frac{\partial}{\partial x} \left( V \frac{\partial f}{\partial v_1} \right) dx \, dy \, dz = \iint V l \, \frac{\partial f}{\partial v_1} \, dS,$$

$$\iiint \left\{ V_2 \frac{\partial f}{\partial v_2} + V \frac{\partial}{\partial y} \left( \frac{\partial f}{\partial v_2} \right) \right\} dx \, dy \, dz = \iiint \frac{\partial}{\partial y} \left( V \frac{\partial f}{\partial v_2} \right) dx \, dy \, dz = \iint V m \, \frac{\partial f}{\partial v_2} \, dS,$$

$$\iiint \left\{ V_3 \frac{\partial f}{\partial v_3} + V \frac{\partial}{\partial z} \left( \frac{\partial f}{\partial v_3} \right) \right\} dx \, dy \, dz = \iiint \frac{\partial}{\partial z} \left( V \frac{\partial f}{\partial v_3} \right) dx \, dy \, dz = \iint V n \, \frac{\partial f}{\partial v_3} \, dS :$$

and therefore

$$\iiint \left( V \frac{\partial f}{\partial v} + V_1 \frac{\partial f}{\partial v_1} + V_2 \frac{\partial f}{\partial v_2} + V_3 \frac{\partial f}{\partial v_3} \right) dx \, dy \, dz$$

$$= \iiint V \left\{ \frac{\partial f}{\partial v} - \frac{\partial}{\partial x} \left( \frac{\partial f}{\partial v_1} \right) - \frac{\partial}{\partial y} \left( \frac{\partial f}{\partial v_2} \right) - \frac{\partial}{\partial z} \left( \frac{\partial f}{\partial v_3} \right) \right\} dx \, dy \, dz$$

$$+ \iint V \left( l \frac{\partial f}{\partial v_1} + m \frac{\partial f}{\partial v_2} + n \frac{\partial f}{\partial v_3} \right) dS.$$

Thus the first variation takes the form

$$\kappa \iiint V \left\{ \frac{\partial f}{\partial v} - \frac{\partial}{\partial x} \left( \frac{\partial f}{\partial v_1} \right) - \frac{\partial}{\partial y} \left( \frac{\partial f}{\partial v_2} \right) - \frac{\partial}{\partial z} \left( \frac{\partial f}{\partial v_3} \right) \right\} dx\, dy\, dz$$

$$+ \kappa \iint \left\{ V \left( l \frac{\partial f}{\partial v_1} + m \frac{\partial f}{\partial v_2} + n \frac{\partial f}{\partial v_3} \right) + (lX + mY + nZ) f \right\} dS.$$

This quantity must vanish for all admissible variations $\kappa X$, $\kappa Y$, $\kappa Z$.

### Characteristic equation: boundary condition.

**360.** Consider, first, those variations, which differ from zero throughout the bounded space and are zero everywhere on the boundary. The magnitudes $V$, $X$, $Y$, $Z$, are zero everywhere in the surface-integral for such variations: the volume-integral then must vanish by itself. If the quantity multiplied by $V$ in the subject of integration does not vanish everywhere, we can make the integral positive by choosing $V$ of the same sign as the quantity, and we can make it negative by choosing $V$ of the opposite sign—a possibility of value in the integral which must be excluded, because the integral could then be made positive or negative. Moreover, the sign of $\kappa$ is arbitrary. Consequently, we must have

$$\frac{\partial f}{\partial v} - \frac{\partial}{\partial x} \left( \frac{\partial f}{\partial v_1} \right) - \frac{\partial}{\partial y} \left( \frac{\partial f}{\partial v_2} \right) - \frac{\partial}{\partial z} \left( \frac{\partial f}{\partial v_3} \right) = 0,$$

everywhere through the bounded space. This equation is called the *characteristic equation*; and the geometrical representation of $v$ as a function of $x$, $y$, $z$, may be called the *characteristic volume*.

Consider, next, other variations not required to be zero on the boundary. The volume-integral now vanishes because of the condition just established; and therefore the surface-integral must vanish by itself: that is, we must have

$$\iint \left\{ V \left( l \frac{\partial f}{\partial v_1} + m \frac{\partial f}{\partial v_2} + n \frac{\partial f}{\partial v_3} \right) + (lX + mY + nZ) f \right\} dS = 0.$$

Now $\kappa X$, $\kappa Y$, $\kappa Z$, are variations of $x$, $y$, $z$, while $\kappa V$ is the change of $v$ on the supposition that the variation is special, viz., that it is an arbitrary increment of $v$ while $x$, $y$, $z$, do not suffer change. Accordingly, let $\kappa U$ be the small variation of $v$ on the boundary consequent on the variations $\kappa X$, $\kappa Y$, $\kappa Z$, in the independent variables. The relation between quantities $\kappa U$ and $\kappa V$ must be obtained, and it can be derived simply through an extension of the notion of geometric tangency. We imagine a characteristic volume given by

$$v = v(x, y, z),$$

and another characteristic volume through $x + \kappa X$, $y + \kappa Y$, $z + \kappa Z$, given by

$$v + \kappa U = v(x + \kappa X, y + \kappa Y, z + \kappa Z);$$

the flat triple amplitude, tangential to the latter, has for its equation

$$\frac{\partial v}{\partial x}(\bar{x}-\kappa X)+\frac{\partial v}{\partial y}(\bar{y}-\kappa Y)+\frac{\partial v}{\partial z}(\bar{z}-\kappa Z)-(\bar{v}-\kappa U)=0,$$

when referred to $v$, $x$, $y$, $z$, as origin, $\bar{v}$, $\bar{x}$, $\bar{y}$, $\bar{z}$ being current coordinates in that amplitude. When we take a special variation for the characteristic surface, we increase $v$ while leaving $x$, $y$, $z$ unaltered: that is, $\bar{v}=\kappa V$ when $\bar{x}=0$, $\bar{y}=0$, $\bar{z}=0$; consequently,

$$V=U-X\frac{\partial v}{\partial x}-Y\frac{\partial v}{\partial y}-Z\frac{\partial v}{\partial z}=U-Xv_1-Yv_2-Zv_3.$$

Thus the vanishing surface-integral becomes

$$\iint\Big[\Big\{(U-Xv_1-Yv_2-Zv_3)\frac{\partial f}{\partial v_1}+Xf\Big\}l$$
$$+\Big\{(U-Xv_1-Yv_2-Zv_3)\frac{\partial f}{\partial v_2}+Yf\Big\}m$$
$$+\Big\{(U-Xv_1-Yv_2-Zv_3)\frac{\partial f}{\partial v_3}+Zf\Big\}n\Big]dS.$$

If the boundary consists of detached and separate surfaces, this integral must vanish over each such separate surface: for we can have variations, which are non-zero over any one separate surface and are zero over all the others, and then the condition can only be satisfied by the vanishing of the one surviving integral. Again, for each separate surface, the subject of integration must vanish everywhere: otherwise, we can select a region on the boundary where it is positive, fringing to zero in extent, and outside that region it can be made zero, so that the integral would be positive; and we can select a region on the boundary where it is negative, fringing to zero in extent, and outside that region it can be made zero, so that the integral would be negative; contrary to the requirement that the integral shall vanish. Thus the subject of integration of the surface-integral must vanish everywhere on the boundary.

We thus have two distinct requirements, necessary and sufficient, to make the first variation of our integral vanish for all the admissible special variations:

(i), *the characteristic equation*

$$\frac{\partial f}{\partial v}-\frac{\partial}{\partial x}\Big(\frac{\partial f}{\partial v_1}\Big)-\frac{\partial}{\partial y}\Big(\frac{\partial f}{\partial v_2}\Big)-\frac{\partial}{\partial z}\Big(\frac{\partial f}{\partial v_3}\Big)=0$$

*must be satisfied everywhere throughout the volume of integration: and* (ii), *the condition*

$$l\Big(V\frac{\partial f}{\partial v_1}+Xf\Big)+m\Big(V\frac{\partial f}{\partial v_2}+Yf\Big)+n\Big(V\frac{\partial f}{\partial v_3}+Zf\Big)=0,$$

*where* $V=U-Xv_1-Yv_2-Zv_3$, *and* $\kappa U$, $\kappa X$, $\kappa Y$, $\kappa Z$, *are small variations consistent with boundary relations, must be satisfied everywhere on the whole boundary of the included volume, and l, m, n, are the direction-cosines of the outward-drawn normal to the surface limiting the characteristic volume.*

### *Primitive of the characteristic equation: Cauchy's existence-theorem.*

**361.** The characteristic equation is a partial differential equation of the second order in $v$, there being three independent variables $x$, $y$, $z$. Cauchy's existence-theorem[*] shews that a unique integral exists under certain conditions. This integral is a regular function of the variables, which acquires an assigned arbitrary value (being a regular function of position everywhere on an assigned surface), and one first derivative of which acquires another assigned arbitrary value (being a regular function also of position over that surface). But this surface—it may be called the initial surface—must not belong to one or other of particular sets of surfaces generically associated with the differential equation. (From the data, we can then infer the values of all the other derivatives everywhere on the surface.) The reason, why these surfaces are precluded from constituting initial surfaces, appears as follows.

Let an initial surface be

$$\theta\,(x,\,y,\,z) = 0,$$

or equal to a parameter, constant over the surface: along the surface,

$$\frac{\partial\theta}{\partial x}\,dx + \frac{\partial\theta}{\partial y}\,dy + \frac{\partial\theta}{\partial z}\,dz = 0.$$

Let the arbitrary value of position assigned to $v$ over the surface be $\phi\,(x,\,y,\,z)$; and let the arbitrary value of position, which is assigned to $v_1$ over the surface, be $\psi\,(x,\,y,\,z)$. Then we have

$$dv = v_1 dx + v_2 dy + v_3 dz,$$

always; and therefore we have

$$\frac{\partial\phi}{\partial x}\,dx + \frac{\partial\phi}{\partial y}\,dy + \frac{\partial\phi}{\partial z}\,dz = dv$$
$$= v_1 dx + v_2 dy + v_3 dz$$

along the surface, that is, provided

$$\frac{\partial\theta}{\partial x}\,dx + \frac{\partial\theta}{\partial y}\,dy + \frac{\partial\theta}{\partial z}\,dz = 0.$$

Hence along the surface, we have

$$P = v_1 = \frac{\partial\phi}{\partial x} + \lambda\,\frac{\partial\theta}{\partial x}, \quad Q = v_2 = \frac{\partial\phi}{\partial y} + \lambda\,\frac{\partial\theta}{\partial y}, \quad R = v_3 = \frac{\partial\phi}{\partial z} + \lambda\,\frac{\partial\theta}{\partial z}.$$

Now $v_1$ is given as equal to $\psi\,(x,\,y,\,z)$ along the surface. Thus the first equation gives the value of $\lambda$; and then the second and the third give the value of $v_2$ and the value of $v_3$. Hence the data imply that the values of all the first derivatives of $v$ can be regarded as known functions of position on the initial surface.

[*] See my *Theory of Differential Equations*, vol. vi, chap. xxiv; and an earlier memoir, *Phil. Trans.* (1898), pp. 1—86.

Next, we have

$$v_{11}dx + v_{12}dy + v_{13}dz = dv_1,$$
$$v_{12}dx + v_{22}dy + v_{23}dz = dv_2,$$
$$v_{13}dx + v_{23}dy + v_{33}dz = dv_3,$$

always. Therefore

$$v_{11}dx + v_{12}dy + v_{13}dz = \frac{\partial P}{\partial x}\,dx + \frac{\partial P}{\partial y}\,dy + \frac{\partial P}{\partial z}\,dz,$$

$$v_{12}dx + v_{22}dy + v_{23}dz = \frac{\partial Q}{\partial x}\,dx + \frac{\partial Q}{\partial y}\,dy + \frac{\partial Q}{\partial z}\,dz,$$

$$v_{13}dx + v_{23}dy + v_{33}dz = \frac{\partial R}{\partial x}\,dx + \frac{\partial R}{\partial y}\,dy + \frac{\partial R}{\partial z}\,dz,$$

along the surface in each instance, that is, provided

$$\frac{\partial \theta}{\partial x}\,dx + \frac{\partial \theta}{\partial y}\,dy + \frac{\partial \theta}{\partial z}\,dz = 0.$$

Hence there are quantities $\rho$, $\sigma$, $\tau$, such that

$$\left.\begin{aligned}
v_{11} &= \frac{\partial P}{\partial x} + \rho \frac{\partial \theta}{\partial x} \\[4pt]
v_{12} &= \frac{\partial P}{\partial y} + \rho \frac{\partial \theta}{\partial y} = \frac{\partial Q}{\partial x} + \sigma \frac{\partial \theta}{\partial x} \\[4pt]
v_{13} &= \frac{\partial P}{\partial z} + \rho \frac{\partial \theta}{\partial z} \qquad\quad = \frac{\partial R}{\partial x} + \tau \frac{\partial \theta}{\partial x} \\[4pt]
v_{22} &= \qquad\qquad\quad \frac{\partial Q}{\partial y} + \sigma \frac{\partial \theta}{\partial y} \\[4pt]
v_{23} &= \qquad\qquad\quad \frac{\partial Q}{\partial z} + \sigma \frac{\partial \theta}{\partial z} = \frac{\partial R}{\partial y} + \tau \frac{\partial \theta}{\partial y} \\[4pt]
v_{33} &= \qquad\qquad\qquad\qquad\quad \frac{\partial R}{\partial z} + \tau \frac{\partial \theta}{\partial z}
\end{aligned}\right\}.$$

Substituting the values of $P$ and $Q$ in the equal expressions for $v_{12}$, we have

$$\frac{\partial \lambda}{\partial y}\frac{\partial \theta}{\partial x} - \frac{\partial \lambda}{\partial x}\frac{\partial \theta}{\partial y} = \sigma \frac{\partial \theta}{\partial x} - \rho \frac{\partial \theta}{\partial y};$$

and, similarly, from the equal values of $v_{13}$, and from the equal values of $v_{23}$,

$$\frac{\partial \lambda}{\partial x}\frac{\partial \theta}{\partial z} - \frac{\partial \lambda}{\partial z}\frac{\partial \theta}{\partial x} = \rho \frac{\partial \theta}{\partial z} - \tau \frac{\partial \theta}{\partial x},$$

$$\frac{\partial \lambda}{\partial z}\frac{\partial \theta}{\partial y} - \frac{\partial \lambda}{\partial y}\frac{\partial \theta}{\partial z} = \tau \frac{\partial \theta}{\partial y} - \sigma \frac{\partial \theta}{\partial z}.$$

Consequently, we have

$$\frac{\rho - \dfrac{\partial \lambda}{\partial x}}{\dfrac{\partial \theta}{\partial x}} = \frac{\sigma - \dfrac{\partial \lambda}{\partial y}}{\dfrac{\partial \theta}{\partial y}} = \frac{\tau - \dfrac{\partial \lambda}{\partial z}}{\dfrac{\partial \theta}{\partial z}},$$

in which $\lambda$ is a known quantity. Thus we can express the three quantities $\rho$, $\sigma$, $\tau$, in terms of a single unknown quantity $M$, taken to be the common value of the last three fractions. It follows that

$$v_{11} = \frac{\partial P}{\partial x} + \frac{\partial \theta}{\partial x} \left( \frac{\partial \lambda}{\partial x} + M \frac{\partial \theta}{\partial x} \right) = \frac{\partial^2 \phi}{\partial x^2} + 2 \frac{\partial \theta}{\partial x} \frac{\partial \lambda}{\partial x} + \lambda \frac{\partial^2 \theta}{\partial x^2} + M \left( \frac{\partial \theta}{\partial x} \right)^2,$$

along the surface. But $\theta = 0$ along the surface; hence

$$v_{11} = \frac{\partial^2}{\partial x^2} (\phi + \lambda \theta) + M \left( \frac{\partial \theta}{\partial x} \right)^2,$$

along the surface. Similarly for $v_{12}$, $v_{13}$, $v_{22}$, $v_{23}$, $v_{33}$, along the surface, the aggregate of values being

$$v_{11} = \frac{\partial^2}{\partial x^2} (\phi + \lambda \theta) + M \left( \frac{\partial \theta}{\partial x} \right)^2, \quad v_{23} = \frac{\partial^2}{\partial y \partial z} (\phi + \lambda \theta) + M \frac{\partial \theta}{\partial y} \frac{\partial \theta}{\partial z},$$

$$v_{22} = \frac{\partial^2}{\partial y^2} (\phi + \lambda \theta) + M \left( \frac{\partial \theta}{\partial y} \right)^2, \quad v_{31} = \frac{\partial^2}{\partial z \partial x} (\phi + \lambda \theta) + M \frac{\partial \theta}{\partial z} \frac{\partial \theta}{\partial x},$$

$$v_{33} = \frac{\partial^2}{\partial z^2} (\phi + \lambda \theta) + M \left( \frac{\partial \theta}{\partial z} \right)^2, \quad v_{12} = \frac{\partial^2}{\partial x \partial y} (\phi + \lambda \theta) + M \frac{\partial \theta}{\partial x} \frac{\partial \theta}{\partial y},$$

in each case along the surface, all of them expressed in terms of the single unknown quantity $M$. The characteristic equation has to be satisfied. Let

$$c_{mn} = \frac{\partial^2 f}{\partial v_m \partial v_n},$$

for the combinations $m, n, = 1, 2, 3$; and substitute these values in the equation. The quantity $M$ will be unique and determinate, unless

$$c_{11} \left( \frac{\partial \theta}{\partial x} \right)^2 + 2 c_{12} \frac{\partial \theta}{\partial x} \frac{\partial \theta}{\partial y} + 2 c_{13} \frac{\partial \theta}{\partial x} \frac{\partial \theta}{\partial z} + c_{22} \left( \frac{\partial \theta}{\partial y} \right)^2 + 2 c_{23} \frac{\partial \theta}{\partial y} \frac{\partial \theta}{\partial z} + c_{33} \left( \frac{\partial \theta}{\partial z} \right)^2 = 0.$$

When this equation is not satisfied, $M$ is known; and then the values of all the derivatives $v_{mn}$ are known.

Similarly for derivatives of the third order, and derivatives of all higher orders in succession. The value of $v$ then can be regarded as known, these derivatives being the coefficients in a power-series. The convergence of the series is established, as completing the proof of the existence of an integral from the given data, given (that is) everywhere on a surface which does not satisfy the foregoing critical equation.

*Special forms of boundary condition.*

**362.** There remains the condition

$$l \left( V \frac{\partial f}{\partial v_1} + Xf \right) + m \left( V \frac{\partial f}{\partial v_2} + Yf \right) + n \left( V \frac{\partial f}{\partial v_3} + Zf \right) = 0,$$

which must be satisfied everywhere on the superficial boundary; $\kappa U$, $\kappa X$, $\kappa Y$, $\kappa Z$, are simultaneous variations connected with the boundary; and $V$ denotes $U - X v_1 - Y v_2 - Z v_3$.

(i) If $v$ be given as a constant over a portion of the boundary, being a complete surface, then over that portion

$$V = 0;$$

and therefore

$$lX + mY + nZ = 0,$$

where now $\kappa X$, $\kappa Y$, $\kappa Z$, are any variation on that surface. Thus the direction $\kappa X$, $\kappa Y$, $\kappa Z$, is perpendicular to the direction $l$, $m$, $n$; and therefore the portion of the boundary becomes merely a part of the surface limiting the characteristic volume.

(ii) If $v$ be given as a function $G(x, y, z)$ along a part of the boundary or a complete portion of the boundary, then

$$V = \frac{\partial G}{\partial x} X + \frac{\partial G}{\partial y} Y + \frac{\partial G}{\partial z} Z;$$

and thus the condition becomes

$$X \left( lf + v_1 \frac{\partial G}{\partial x} \Sigma l \frac{\partial f}{\partial v_1} \right) + Y \left( mf + v_2 \frac{\partial G}{\partial y} \Sigma l \frac{\partial f}{\partial v_1} \right) + Z \left( nf + v_3 \frac{\partial G}{\partial z} \Sigma l \frac{\partial f}{\partial v_1} \right) = 0.$$

If the part of the boundary be

$$\theta(x, y, z) = 0,$$

so that

$$\frac{\partial \theta}{\partial x} X + \frac{\partial \theta}{\partial y} Y + \frac{\partial \theta}{\partial z} Z = 0,$$

we must have

$$\frac{1}{\frac{\partial \theta}{\partial x}} \left( lf + v_1 \frac{\partial G}{\partial x} \Sigma l \frac{\partial f}{\partial v_1} \right) = \frac{1}{\frac{\partial \theta}{\partial y}} \left( mf + v_2 \frac{\partial G}{\partial y} \Sigma l \frac{\partial f}{\partial v_1} \right) = \frac{1}{\frac{\partial \theta}{\partial z}} \left( nf + v_3 \frac{\partial G}{\partial z} \Sigma l \frac{\partial f}{\partial v_1} \right)$$

along that part of the boundary.

(iii) Similarly, for data in other forms.

When once the boundary conditions have been satisfied, we can regard the circumstances at the boundary as fixed, admitting of no subsequent variation; and thus, when the second variation is considered, we need consider only those space-variations which are obliged to vanish at the boundary.

### Second variation.

**363.** Now that the first variation of the original integral $I$ has been made to vanish, by means of the characteristic equation and the deduced conditions at the surface boundaries, the whole variation of the integral is dominated by the aggregate of terms involving the power $\kappa^2$—that is, by the 'second' variation. Let

$$\frac{\partial^2 f}{\partial v^2} = c_{00}, \quad \frac{\partial^2 f}{\partial v \partial v_r} = c_{0r}, \quad \frac{\partial^2 f}{\partial v_m \partial v_r} = c_{mr},$$

for $m, r, = 1, 2, 3$; and write

$$\Theta = c_{11}V_1^2 + 2c_{12}V_1V_2 + 2c_{13}V_1V_3 + c_{22}V_2^2 + 2c_{23}V_2V_3 + c_{33}V_3^2$$
$$+ 2c_{01}VV_1 + 2c_{02}VV_2 + 2c_{03}VV_3 + c_{00}V^2,$$

where the value of $v$, provided by the primitive of the characteristic equation, is supposed to have been substituted in the coefficients $c_{mr}$. The whole variation of $I$ is equal to

$$\tfrac{1}{2}\kappa^2 \iiint \Theta \, dx \, dy \, dz + K_3,$$

where $K_3$ is the aggregate of terms involving the third and higher powers of $\kappa$. Thus, for the discussion of the second variation, it is desirable that

$$\iiint \Theta \, dx \, dy \, dz$$

should be expressed in a convenient normal form.

We take three assignable functions of $x, y, z$, denoted by $A, B, C$; and we determine them, so that the quantity

$$\Theta + \frac{\partial}{\partial x}(Av^2) + \frac{\partial}{\partial y}(Bv^2) + \frac{\partial}{\partial z}(Cv^2)$$

shall be expressible in the form $\Upsilon$, where

$$\Upsilon = c_{11}u_1^2 + 2c_{12}u_1u_2 + 2c_{13}u_1u_3 + c_{22}u_2^2 + 2c_{23}u_2u_3 + c_{33}u_3^2,$$

the arguments in $\Upsilon$ being

$$u_1 = V_1 - \alpha V, \quad u_2 = V_2 - \beta V, \quad u_3 = V_3 - \gamma V.$$

The necessary and sufficient conditions are

$$A + c_{01} = -c_{11}\alpha - c_{12}\beta - c_{13}\gamma,$$
$$B + c_{02} = -c_{12}\alpha - c_{22}\beta - c_{23}\gamma,$$
$$C + c_{03} = -c_{13}\alpha - c_{23}\beta - c_{33}\gamma,$$
$$\frac{\partial A}{\partial x} + \frac{\partial B}{\partial y} + \frac{\partial C}{\partial z} + c_{00} = c_{11}\alpha^2 + 2c_{12}\alpha\beta + 2c_{13}\alpha\gamma + c_{22}\beta^2 + 2c_{23}\beta\gamma + c_{33}\gamma^2.$$

The elimination of $A, B, C$, among these conditions can obviously be effected at once, and leaves only a single condition to be satisfied by the three quantities. We may therefore expect that $\alpha, \beta, \gamma$, can be represented in terms of some one magnitude satisfying an equation, which is the equivalent of the surviving single condition.

Assuming this algebraic representation possible, we have

$$\iiint \Upsilon \, dx \, dy \, dz = \iiint \Theta \, dx \, dy \, dz + \iiint \left\{ \frac{\partial}{\partial x}(AV^2) + \frac{\partial}{\partial y}(BV^2) + \frac{\partial}{\partial z}(CV^2) \right\} dx \, dy \, dz$$

$$= \iiint \Theta \, dx \, dy \, dz + \iint V^2 (lA + mB + nC) \, dS,$$

with the former notation. Circumstances have been settled at the boundary surfaces, so that no variation is now admissible on any part of the boundary. Thus $V$ is everywhere zero in the surface-integral; consequently, the whole of the second variation is equal to

$$\tfrac{1}{2}\kappa^2 \iiint \Upsilon \, dx \, dy \, dz.$$

Thus $\Upsilon$ is the quantity to be considered, for the significance of the second variation.

In former instances (§§ 89, 93, 110, 168, 311) at this stage, it appeared that the arguments in the modified quadratic form could be made to vanish, were it possible to make the small arbitrary variation a constant multiple of the small variation in passing from one characteristic (curve or surface, as the instances came) to a consecutive characteristic. We are therefore led to consider a small characteristic variation in the present instance.

### Subsidiary characteristic equation.

**364.** The central characteristic equation is

$$\frac{\partial f}{\partial v} - \frac{\partial}{\partial x}\left(\frac{\partial f}{\partial v_1}\right) - \frac{\partial}{\partial y}\left(\frac{\partial f}{\partial v_2}\right) - \frac{\partial}{\partial z}\left(\frac{\partial f}{\partial v_3}\right) = 0.$$

The primitive involves arbitrary elements, whatever be their form; and the equation is satisfied when the provided value of $v$ is substituted. Accordingly, let small arbitrary variations be imposed on the arbitrary elements; and let the consequent value of $v$ be denoted by $v + \kappa U$, where $U$ can be derived from the primitive when the latter is known. The characteristic equation remains unchanged: but the change of $v$ into $v + \kappa U$, with the changes in the derivatives, causes an increment in the left-hand side of the equation, and the increment must therefore vanish by itself.

To express this requirement, we proceed as before. Let

$$2\Omega = c_{11}U_1{}^2 + 2c_{12}U_1U_2 + 2c_{13}U_1U_3 + c_{22}U_2{}^2 + 2c_{23}U_2U_3 + c_{33}U_3{}^2$$
$$+ 2c_{01}UU_1 + 2c_{02}UU_2 + 2c_{03}UU_3 + c_{00}U^2,$$

so that $2\Omega$ is the same function of $U$ and its first derivatives, as $\Theta$ is of $V$ and its first derivatives. The effect of the change of $v$ into $v + \kappa U$ is to change

$$\frac{\partial f}{\partial v} \text{ into } \frac{\partial f}{\partial v} + \kappa \frac{\partial \Omega}{\partial U},$$

$$\frac{\partial f}{\partial v_r} \text{ into } \frac{\partial f}{\partial v_r} + \kappa \frac{\partial \Omega}{\partial U_r},$$

for $r = 1, 2, 3$. As the increment of the left-hand side of the characteristic equation is to vanish because the equation remains unchanged, we have

$$\frac{\partial \Omega}{\partial U} - \frac{\partial}{\partial x}\left(\frac{\partial \Omega}{\partial U_1}\right) - \frac{\partial}{\partial y}\left(\frac{\partial \Omega}{\partial U_2}\right) - \frac{\partial}{\partial z}\left(\frac{\partial \Omega}{\partial U_3}\right) = 0.$$

This is a linear partial equation in $U$, of the second order, the coefficients of the derivatives being functions of $x$, $y$, $z$, after substitution is made for $v$. It is the *subsidiary characteristic equation*. Its primitive can be derived from that of the central characteristic equation, if the latter be known. But the primitive must be constructed by integrating the subsidiary equation, if only a special integral of the central equation is known.

### Normal form : the Legendre test.

**365.** The quantities $\alpha$, $\beta$, $\gamma$, have to satisfy the single equation which results from the elimination of $A$, $B$, $C$, between the four relations in § 363. All these relations are satisfied by taking

$$\alpha = \frac{U_1}{U}, \quad \beta = \frac{U_2}{U}, \quad \gamma = \frac{U_3}{U},$$

where $U$ is any integral of the subsidiary equation. For then

$$\frac{\partial \Omega}{\partial U_1} = c_{01} U + c_{11} U_1 + c_{12} U_2 + c_{13} U_3 = -A U,$$

$$\frac{\partial \Omega}{\partial U_2} = c_{02} U + c_{12} U_1 + c_{22} U_2 + c_{23} U_3 = -B U,$$

$$\frac{\partial \Omega}{\partial U_3} = c_{03} U + c_{13} U_1 + c_{23} U_2 + c_{33} U_3 = -C U;$$

and

$$\begin{aligned}
\frac{\partial \Omega}{\partial U} &= c_{00} U + c_{01} U_1 + c_{02} U_2 + c_{03} U_3 \\
&= U (c_{00} + c_{01}\alpha + c_{02}\beta + c_{03}\gamma) \\
&= -U (A\alpha + B\beta + C\gamma) - U \left( \frac{\partial A}{\partial x} + \frac{\partial B}{\partial y} + \frac{\partial C}{\partial z} \right) \\
&= -A U_1 - U \frac{\partial A}{\partial x} - B U_2 - U \frac{\partial B}{\partial y} - C U_3 - U \frac{\partial C}{\partial z} \\
&= -\frac{\partial}{\partial x} (A U) - \frac{\partial}{\partial y} (B U) - \frac{\partial}{\partial z} (C U) ;
\end{aligned}$$

and the necessary condition for the coexistence of these relations, being

$$\frac{\partial \Omega}{\partial U} = \frac{\partial}{\partial x} \left( \frac{\partial \Omega}{\partial U_1} \right) + \frac{\partial}{\partial y} \left( \frac{\partial \Omega}{\partial U_2} \right) + \frac{\partial}{\partial z} \left( \frac{\partial \Omega}{\partial U_3} \right),$$

is satisfied. Accordingly, the arguments $u_1$, $u_2$, $u_3$, are

$$u_1 = V_1 - \frac{U_1}{U} V, \quad u_2 = V_2 - \frac{U_2}{U} V, \quad u_3 = V_3 - \frac{U_3}{U} V,$$

where $\kappa V$ is an arbitrary variation from the characteristic volume, and $\kappa U$ is a variation (a *characteristic variation*) that leads to a consecutive characteristic volume.

The second variation has been expressed in the form

$$\tfrac{1}{2}\kappa^2 \iiint \Upsilon\, dx\, dy\, dz\,;$$

and it must have a persistent sign whatever arbitrary variations are adopted. This sign depends upon the quantity $\Upsilon$; if $\Upsilon$ could be sometimes positive, and sometimes negative, we could make the integral positive by choosing $V = 0$ where $\Upsilon$ otherwise would be negative, and we could make the integral negative by choosing $V = 0$ where $\Upsilon$ otherwise would be positive. Such alternatives are such as to exclude a uniform sign; and therefore $\Upsilon$ itself must have a uniform sign.

When we take non-zero arguments, the necessary and sufficient conditions * are that

$$c_{22}c_{33} > c_{23}{}^2, \quad c_{33}c_{11} > c_{31}{}^2, \quad c_{11}c_{22} > c_{12}{}^2\,;$$

and when these are satisfied, then $c_{11}$, $c_{22}$, $c_{33}$, have the same sign. Should this common sign be positive, $\Upsilon$ and the second variation are positive; and a minimum is possible. Should the common sign be negative, $\Upsilon$ and the second variation are negative; and a maximum is possible.

We exclude a condition of equality represented (*e.g.*) by

$$c_{11}c_{22} - c_{12}{}^2 = 0\,;$$

otherwise, for non-zero variations such that $V = Ug\,(x,\,y)$, where

$$c_{11}{}^{\tfrac{1}{2}}\frac{\partial g}{\partial x} \pm c_{22}{}^{\tfrac{1}{2}}\frac{\partial g}{\partial y} = 0,$$

the quantity $\Upsilon$ can vanish. It would become necessary to examine the third variation (which would have to vanish) and the fourth variation of the original integral, for variations of this type. Such an investigation belongs to particular instances; as before, it can be made when they actually occur.

We thus require *the conditions*

$$c_{22}c_{33} - c_{23}{}^2 > 0, \quad c_{33}c_{11} > c_{31}{}^2, \quad c_{11}c_{22} > c_{12}{}^2,$$

*with the implied requirement that* $c_{11}$, $c_{22}$, $c_{33}$ *have the same sign;* and these are to be satisfied *throughout the characteristic volume.* These conditions are the extension of the Legendre test.

### The Jacobi test.

**366.** Further, in order that the second variation may conserve a non-zero value of uniform sign, it is necessary that its arguments shall not vanish simultaneously; the concern being, of course, with non-zero variations. Such an excluded possibility would arise, if we could have

$$V = hU,$$

---

* Other forms of the conditions can be stated; those, which are stated, seem the most convenient.

where $h$ is a pure constant. At the lower limit, we are to have $V = 0$; and therefore, in such an event, we must have $U = 0$. Now $\kappa U$ is the small variation from one volume to another, the volumes being bounded; consequently, the two volumes must have the same initial superficial boundary. If a new ordinate be introduced to represent $v$, so that the vocabulary of four-dimensional geometry would be used, we thus have two separate volumes, one the original characteristic volume represented by the ordinate $v$, the other the consecutive characteristic volume represented by the ordinate $v + \kappa U$; and these two volumes (triple manifolds) have the same initial surface (double manifold). If then the range of integration extended so far as to allow these two distinct volumes, meeting in a common initial surface, to meet again in a further second surface, then $U$ would be zero all over this second surface; and we now could have $V = hU$, still satisfying the terminal conditions as to $V$. The circumstance, that allows this possibility, must be prevented. Consequently, the range of integration must not extend so far as this second surface.

Analytically, the restrictive condition is imposed as follows. The characteristic volume is such that, all over an initial surface, $v$ and $v_1$ (and therefore, by § 361, $v_2$ and $v_3$) are given. For the consecutive characteristic volume, the value of $v$ is unchanged over that initial surface, so that $U = 0$ initially, while small changes are imposed upon $v_1$ (and consequently are incidental to $v_2$ and $v_3$); and $v + \kappa U$ represents the consecutive characteristic volume, for which $U$ vanishes over the initial surface. *The range of integration must not extend so far as the nearest surface where $U$, zero initially, next vanishes.*

This condition is the extension of the *Jacobi* test.

*Ex. Required to determine $v$ as a function of $x$, $y$, $z$, such as to minimise the integral*

$$\iiint \left\{ \left(\frac{\partial v}{\partial x}\right)^2 + \left(\frac{\partial v}{\partial y}\right)^2 + \left(\frac{\partial v}{\partial z}\right)^2 \right\} dx \, dy \, dz,$$

*extended throughout any region of variation.* (The so-called 'Dirichlet's Principle' asserts the existence of a minimum.)

(I) The characteristic equation is

$$\frac{\partial}{\partial x}\left(2\frac{\partial v}{\partial x}\right) + \frac{\partial}{\partial y}\left(2\frac{\partial v}{\partial y}\right) + \frac{\partial}{\partial z}\left(2\frac{\partial v}{\partial z}\right) = 0,$$

that is,

$$\frac{\partial^2 v}{\partial x^2} + \frac{\partial^2 v}{\partial y^2} + \frac{\partial^2 v}{\partial z^2} = 0,$$

a persistently recurring equation in mathematical physics. Thus it is the equation of the potential of gravitational attraction (and of other attractions, with the same law) in free space.

Of this partial differential equation of the second order, there are many particular solutions, specifically expressed in finite terms, such as the functions known as Solid Harmonics. There are other solutions of a more general type, in the form of infinite series, satisfying the Cauchy 'initial' conditions at a given boundary surface. Again, there are other solutions, analogous in character, expressible in the form of definite integrals, sometimes single, sometimes double. But it still lacks a primitive which, in form, at once obviously satisfies initial conditions at an arbitrarily assigned boundary and at the same time is conveniently suitable for general analytical calculations.

The Legendre test for the triple integral is satisfied; it makes a minimum value admissible. For we have

$$c_{11}=2,\ c_{22}=2,\ c_{33}=2\ ;\quad c_{23}=0,\ c_{31}=0,\ c_{12}=0\ ;$$

manifestly meeting the requirements of the test.

As a suitable primitive of the characteristic equation is lacking, it is not possible to construct the analytical expression of the Jacobi test in general. If we take any particular solution of the characteristic equation, it is not more easily possible to integrate the subsidiary characteristic equation than it is to integrate the central characteristic equation. The two equations are one and the same in form, because the subsidiary equation, obtained by varying $v$ into $v + \kappa U$, is

$$\frac{\partial^2 U}{\partial x^2} + \frac{\partial^2 U}{\partial y^2} + \frac{\partial^2 U}{\partial z^2} = 0.$$

Consequently, the general analytical determination of the limitation (if any) upon the range in space, within which even any particular solution of the central characteristic equation provides a minimum, has not yet been effected.

(II) There is, of course, a vast range of achieved investigation, often called Harmonic Analysis, for the expression of an integral of the central characteristic equation satisfying certain assigned conditions. For example, a solution of that equation, expressible in polar coordinates $r$, $\theta'$, $\phi'$, for space, has been obtained in various ways, satisfying the conditions :

(i) of assuming an assigned value $f(\theta, \phi)$ of position on the surface of a sphere, centre the origin and radius $r = a$ : and

(ii) of tending to the value zero, at distances from the origin increasing beyond any finite limit.

An expression* for it is

$$v = \frac{1}{4\pi a} \int_0^{2\pi} \int_0^\pi \frac{r^2 (r^2 - a^2) f(\theta', \phi') \sin \theta'\, d\theta'\, d\phi'}{[r^2 - 2ar \{\cos \theta \cos \theta' + \sin \theta \sin \theta' \cos (\phi - \phi')\} + a^2]^{\frac{3}{2}}}.$$

The customary detailed applications of this solution deal with the use of the expansion of the subject of integration in powers of $\frac{a}{r}$; even so, they are mainly concerned with the terms involving the lowest powers of $\frac{a}{r}$ in this expansion.

(III) It is to be noted that, so far as the three tests are concerned—the characteristic equation (or Euler-Lagrange test), the Legendre test, and the Jacobi test—only weak variations have been considered. Indeed, only special weak variations have been considered. The experience of earlier investigations, however, shews that, when general weak varia-

---

* See (e.g.) Thomson and Tait, *Natural Philosophy*, vol. i (1890), at p. 199 of the section devoted to Spherical Harmonic Analysis (*l.c.*, pp. 171—218). See also Ferrers, *Spherical Harmonics*; and Heine, *Handbuch der Kugelfunctionen*; among many treatises.

tions are imposed, the resulting essential tests are of the same character as for the special variation : and that no new essential test emerges, so far as weak variations are concerned.

The test, emerging from the imposition of strong variations, remains for consideration (§ 385).

*General weak variations.*

**367.** In order to consider the effect, on triple integrals, of general weak variations and also of strong variations, imposed simultaneously upon $x, y, z, v$, we take three new independent variables $\rho, \sigma, \tau$; and we make the four variables $x, y, z, v$, functions of $\rho, \sigma, \tau$. Any relation, which expresses $v$ explicitly as a function of $x, y, z$, or which connects the four variables without explicit expression of $v$, is obtainable by the elimination of $\rho, \sigma, \tau$, from the four equations defining $x, y, z, v$. We write*

$$\frac{\partial \theta}{\partial \rho} = \theta_1, \quad \frac{\partial \theta}{\partial \sigma} = \theta_2, \quad \frac{\partial \theta}{\partial \tau} = \theta_3,$$

for $\theta = x, y, z, v$, in turn, with the corresponding signification for $\theta_{11}, \theta_{12}, \theta_{13}, \theta_{22}, \theta_{23}, \theta_{33}$: and, as $x, y, z$, are independent variables, there is the manifest requirement that the quantity

$$\begin{vmatrix} x_1, & y_1, & z_1 \\ x_2, & y_2, & z_2 \\ x_3, & y_3, & z_3 \end{vmatrix}$$

shall be distinct from zero. Let

$$J_x = \begin{vmatrix} y_1, & z_1, & v_1 \\ y_2, & z_2, & v_2 \\ y_3, & z_3, & v_3 \end{vmatrix}, \quad J_y = \begin{vmatrix} z_1, & x_1, & v_1 \\ z_2, & x_2, & v_2 \\ z_3, & x_3, & v_3 \end{vmatrix}, \quad J_z = \begin{vmatrix} x_1, & y_1, & v_1 \\ x_2, & y_2, & v_2 \\ x_3, & y_3, & v_3 \end{vmatrix}, \quad J_v = \begin{vmatrix} x_1, & y_1, & z_1 \\ x_2, & y_2, & z_2 \\ x_3, & y_3, & z_3 \end{vmatrix}$$

Because

$$v_r = \frac{\partial v}{\partial x} x_r + \frac{\partial v}{\partial y} y_r + \frac{\partial v}{\partial z} z_r,$$

for $r = 1, 2, 3$, in turn, we have

$$\frac{\partial v}{\partial x} = \frac{J_x}{J_v}, \quad \frac{\partial v}{\partial y} = \frac{J_y}{J_v}, \quad \frac{\partial v}{\partial z} = \frac{J_z}{J_v}.$$

When the triple integral

$$\iiint f\left(x, y, z, v, \frac{\partial v}{\partial x}, \frac{\partial v}{\partial y}, \frac{\partial v}{\partial z}\right) dx\, dy\, dz$$

is subjected to the transformation, it becomes

$$\iiint f J \left(\frac{x, y, z}{\rho, \sigma, \tau}\right) d\rho\, d\sigma\, d\tau,$$

---

* Changes are made here in the meanings of $v_1, v_2, v_3$, which are assigned in § 359.

which will be expressed in the form

$$\iiint F \, d\rho \, d\sigma \, d\tau.$$

Then $F$ is a function of $\theta, \theta_1, \theta_2, \theta_3$, where $\theta$ denotes $x, y, z, v$, in turn; and

$$F = J_v f = \begin{vmatrix} x_1, & y_1, & z_1 \\ x_2, & y_2, & z_2 \\ x_3, & y_3, & z_3 \end{vmatrix} f\left(x, y, z, v, \frac{J_x}{J_v}, \frac{J_y}{J_v}, \frac{J_z}{J_v}\right).$$

### Identities satisfied by the transformed integrand.

**368.** The function $F$ is a function of sixteen arguments, whereas $f$ is a function of only seven arguments; we therefore expect that $F$ shall satisfy nine identities. To obtain these, let $\rho', \sigma', \tau'$, denote slightly varied values of $\rho, \sigma, \tau$, the widest variation being admitted; and let the consequent differential elements be

$$d\rho' = (1 + \alpha_1) \, d\rho + \quad \alpha_2 \, d\sigma + \quad \alpha_3 \, d\tau,$$
$$d\sigma' = \quad \beta_1 \, d\rho + (1 + \beta_2) \, d\sigma + \quad \beta_3 \, d\tau,$$
$$d\tau' = \quad \gamma_1 \, d\rho + \quad \gamma_2 \, d\sigma + (1 + \gamma_3) \, d\tau,$$

where $\alpha_\mu, \beta_\mu, \gamma_\mu$ (for $\mu = 1, 2, 3$) are small independent arbitrary quantities. For each of the variables $v, x, y, z$, we have

$$\theta_1' d\rho' + \theta_2' d\sigma' + \theta_3' d\tau' = d\theta = \theta_1 d\rho + \theta_2 d\sigma + \theta_3 d\tau;$$

and therefore

$$\theta_1 - \theta_1' = \alpha_1 \theta_1' + \beta_1 \theta_2' + \gamma_1 \theta_3',$$
$$\theta_2 - \theta_2' = \alpha_2 \theta_1' + \beta_2 \theta_2' + \gamma_2 \theta_3',$$
$$\theta_3 - \theta_3' = \alpha_3 \theta_1' + \beta_3 \theta_2' + \gamma_3 \theta_3'.$$

When $F'$ denotes the new value of $F$, so that

$$\iiint F' \, d\rho' \, d\sigma' \, d\tau' = \iiint F \, d\rho \, d\sigma \, d\tau,$$

we have

$$F' J\left(\frac{\rho', \sigma', \tau'}{\rho, \sigma, \tau}\right) = F,$$

that is,

$$(1 + \alpha_1 + \beta_2 + \gamma_3) F' = F,$$

the small quantities of order higher than the first being omitted, as is usual for all infinitesimal transformations of a continuous group. When we substitute on the right-hand side for $\theta_1, \theta_2, \theta_3$, expand both sides in powers of the nine small quantities $\alpha_\mu, \beta_\mu, \gamma_\mu$, which are independent of one another and are arbitrary, and equate coefficients, reverting from the derivatives of

$F'$ to those of $F$, we obtain the nine identities

$$F = \Sigma\theta_1 \frac{\partial F}{\partial\theta_1}, \qquad 0 = \Sigma\theta_1 \frac{\partial F}{\partial\theta_2}, \qquad 0 = \Sigma\theta_1 \frac{\partial F}{\partial\theta_3}$$

$$0 = \Sigma\theta_2 \frac{\partial F}{\partial\theta_1}, \qquad F = \Sigma\theta_2 \frac{\partial F}{\partial\theta_2}, \qquad 0 = \Sigma\theta_2 \frac{\partial F}{\partial\theta_3} \Bigg\},$$

$$0 = \Sigma\theta_3 \frac{\partial F}{\partial\theta_1}, \qquad 0 = \Sigma\theta_3 \frac{\partial F}{\partial\theta_2}, \qquad F = \Sigma\theta_3 \frac{\partial F}{\partial\theta_3}$$

where each summation is for four terms given by taking $\theta = x, y, z, v$, in turn.

The relations between the first derivatives of $F$ and the first derivatives of $f$, with regard to their respective arguments, are needed later. We change the notation of § 359 for $\dfrac{\partial v}{\partial x}, \dfrac{\partial v}{\partial y}, \dfrac{\partial v}{\partial z}$, because $v_1, v_2, v_3$ are now used in a new significance, which agrees with the former when $\rho = x, \sigma = y, \tau = z$; and we write

$$\frac{\partial v}{\partial x} = p, \qquad \frac{\partial v}{\partial y} = q, \qquad \frac{\partial v}{\partial z} = r,$$

so that

$$p = \frac{J_x}{J_v}, \qquad q = \frac{J_y}{J_v}, \qquad r = \frac{J_z}{J_v}.$$

The relation between $F$ and $f$ is

$$F = J_v f.$$

The following relations are derivable :

$$\frac{\partial F}{\partial x} = J_v \frac{\partial f}{\partial x}, \quad \frac{\partial F}{\partial y} = J_v \frac{\partial f}{\partial y}, \quad \frac{\partial F}{\partial z} = J_v \frac{\partial f}{\partial z}, \quad \frac{\partial F}{\partial v} = J_v \frac{\partial f}{\partial v};$$

$$\frac{\partial F}{\partial x_1} = (y_2 z_3 - z_2 y_3)\, T \qquad\qquad\quad + (v_2 z_3 - v_3 z_2)\frac{\partial f}{\partial q} + (v_3 y_2 - v_2 y_3)\frac{\partial f}{\partial r}$$

$$\frac{\partial F}{\partial x_2} = (y_3 z_1 - z_3 y_1)\, T \qquad\qquad\quad + (v_3 z_1 - v_1 z_3)\frac{\partial f}{\partial q} + (v_1 y_3 - v_3 y_1)\frac{\partial f}{\partial r} \Bigg\};$$

$$\frac{\partial F}{\partial x_3} = (y_1 z_2 - z_1 y_2)\, T \qquad\qquad\quad + (v_1 z_2 - v_2 z_1)\frac{\partial f}{\partial q} + (v_2 y_1 - v_1 y_2)\frac{\partial f}{\partial r}$$

$$\frac{\partial F}{\partial y_1} = (z_2 x_3 - x_2 z_3)\, T + (v_3 z_2 - v_2 z_3)\frac{\partial f}{\partial p} \qquad\qquad + (v_2 x_3 - v_3 x_2)\frac{\partial f}{\partial r}$$

$$\frac{\partial F}{\partial y_2} = (z_3 x_1 - x_3 z_1)\, T + (v_1 z_3 - v_3 z_1)\frac{\partial f}{\partial p} \qquad\qquad + (v_3 x_1 - v_1 x_3)\frac{\partial f}{\partial r} \Bigg\};$$

$$\frac{\partial F}{\partial y_3} = (z_1 x_2 - x_1 z_2)\, T + (v_2 z_1 - v_1 z_2)\frac{\partial f}{\partial p} \qquad\qquad + (v_1 x_2 - v_2 x_1)\frac{\partial f}{\partial r}$$

$$\frac{\partial F}{\partial z_1} = (x_2 y_3 - y_2 x_3)\, T + (v_2 y_3 - v_3 y_2)\frac{\partial f}{\partial p} + (v_3 x_2 - v_2 x_3)\frac{\partial f}{\partial q}$$

$$\frac{\partial F}{\partial z_2} = (x_3 y_1 - y_3 x_1)\, T + (v_3 y_1 - v_1 y_3)\frac{\partial f}{\partial p} + (v_1 x_3 - v_3 x_1)\frac{\partial f}{\partial q} \Bigg\};$$

$$\frac{\partial F}{\partial z_3} = (x_1 y_2 - y_1 x_2)\, T + (v_1 y_2 - v_2 y_1)\frac{\partial f}{\partial p} + (v_2 x_1 - v_1 x_2)\frac{\partial f}{\partial q}$$

$$\frac{\partial F}{\partial v_1} = (y_2 z_3 - z_2 y_3) \frac{\partial f}{\partial p} + (z_2 x_3 - x_2 z_3) \frac{\partial f}{\partial q} + (x_2 y_3 - y_2 x_3) \frac{\partial f}{\partial r}$$

$$\frac{\partial F}{\partial v_2} = (y_3 z_1 - z_3 y_1) \frac{\partial f}{\partial p} + (z_3 x_1 - x_3 z_1) \frac{\partial f}{\partial q} + (x_3 y_1 - y_3 x_1) \frac{\partial f}{\partial r} \Bigg\} ;$$

$$\frac{\partial F}{\partial v_3} = (y_1 z_2 - z_1 y_2) \frac{\partial f}{\partial p} + (z_1 x_2 - x_1 z_2) \frac{\partial f}{\partial q} + (x_1 y_2 - y_1 x_2) \frac{\partial f}{\partial r}$$

where

$$T = f - p \frac{\partial f}{\partial p} - q \frac{\partial f}{\partial q} - r \frac{\partial f}{\partial r}.$$

*Ex.* Verify that

$$\sum_{\mu=1}^{3} x_\mu \frac{\partial F}{\partial x_\mu} = \left( f - p \frac{\partial f}{\partial p} \right) J_v,$$

$$\sum_{\mu=1}^{3} y_\mu \frac{\partial F}{\partial y_\mu} = \left( f - q \frac{\partial f}{\partial q} \right) J_v$$

$$\sum_{\mu=1}^{3} z_\mu \frac{\partial F}{\partial z_\mu} = \left( f - r \frac{\partial f}{\partial r} \right) J_v,$$

$$\sum_{\mu=1}^{3} v_\mu \frac{\partial F}{\partial v_\mu} = \left( p \frac{\partial f}{\partial p} + q \frac{\partial f}{\partial q} + r \frac{\partial f}{\partial r} \right) J_v.$$

*General weak variation of the integral: the four characteristic equations: boundary surface condition.*

**369.** Now let a general weak variation of $x + \kappa X$, $y + \kappa Y$, $z + \kappa Z$, $v + \kappa U$, be imposed upon the variables $x$, $y$, $z$, $v$, so as to affect the integral

$$I = \iiint F \, d\rho \, d\sigma \, d\tau.$$

The first variation of the integral is

$$\kappa \iiint \Sigma \left( \frac{\partial F}{\partial \theta} \Theta + \frac{\partial F}{\partial \theta_1} \Theta_1 + \frac{\partial F}{\partial \theta_2} \Theta_2 + \frac{\partial F}{\partial \theta_3} \Theta_3 \right) d\rho \, d\sigma \, d\tau.$$

When integration by parts is effected, this first variation becomes

$$= \kappa \iiint \Sigma \Theta \left\{ \frac{\partial F}{\partial \theta} - \frac{\partial}{\partial \rho} \left( \frac{\partial F}{\partial \theta_1} \right) - \frac{\partial}{\partial \sigma} \left( \frac{\partial F}{\partial \theta_2} \right) - \frac{\partial}{\partial \tau} \left( \frac{\partial F}{\partial \theta_3} \right) \right\} d\rho \, d\sigma \, d\tau$$

$$+ \kappa \iint \Sigma \Theta \left( \frac{\partial F}{\partial \theta_1} d\sigma \, d\tau + \frac{\partial F}{\partial \theta_2} d\tau \, d\rho + \frac{\partial F}{\partial \theta_3} d\rho \, d\sigma \right),$$

where the summation is for $\theta$, $\Theta$, $= x$, $X$; $y$, $Y$; $z$, $Z$; $v$, $U$; where the triple integral is taken throughout the bounded volume in the $(\rho, \sigma, \tau)$ space; and where the double integrals are taken over the whole of the surface which bounds the included volume.

The variations $\Theta$ (that is, $X$, $Y$, $Z$, $U$) within the volume are arbitrary, provided they are regular, and are independent of one another and of $\kappa$. We assume that the derivatives of $F$ within the region, and up to its boundary, suffer no discontinuities. Then the usual argument requires, as the necessary condition for the possession of a maximum or a minimum by the integral $I$, that the first variation of $I$ shall vanish under all admissible small weak variations. In order that the first variation may vanish, it is necessary (i), that the *four characteristic equations*

$$E_x = \frac{\partial F}{\partial x} - \frac{\partial}{\partial \rho}\left(\frac{\partial F}{\partial x_1}\right) - \frac{\partial}{\partial \sigma}\left(\frac{\partial F}{\partial x_2}\right) - \frac{\partial}{\partial \tau}\left(\frac{\partial F}{\partial x_3}\right) = 0,$$

$$E_y = \frac{\partial F}{\partial y} - \frac{\partial}{\partial \rho}\left(\frac{\partial F}{\partial y_1}\right) - \frac{\partial}{\partial \sigma}\left(\frac{\partial F}{\partial y_2}\right) - \frac{\partial}{\partial \tau}\left(\frac{\partial F}{\partial y_3}\right) = 0,$$

$$E_z = \frac{\partial F}{\partial z} - \frac{\partial}{\partial \rho}\left(\frac{\partial F}{\partial z_1}\right) - \frac{\partial}{\partial \sigma}\left(\frac{\partial F}{\partial z_2}\right) - \frac{\partial}{\partial \tau}\left(\frac{\partial F}{\partial z_3}\right) = 0,$$

$$E_v = \frac{\partial F}{\partial v} - \frac{\partial}{\partial \rho}\left(\frac{\partial F}{\partial v_1}\right) - \frac{\partial}{\partial \sigma}\left(\frac{\partial F}{\partial v_2}\right) - \frac{\partial}{\partial \tau}\left(\frac{\partial F}{\partial v_3}\right) = 0,$$

shall be satisfied everywhere within the volume : and (ii), that the condition

$$\iint \Sigma \Theta \left(\frac{\partial F}{\partial \theta_1}\, d\sigma\, d\tau + \frac{\partial F}{\partial \theta_2}\, d\tau\, d\rho + \frac{\partial F}{\partial \theta_3}\, d\rho\, d\sigma\right) = 0$$

shall be satisfied along the superficial boundary of that volume.

*The four characteristic equations are equivalent to the single characteristic equation in § 360.*

**370.** In the first place, we find that, when account is taken of the nine identities satisfied by $F$, the four characteristic equations are, in fact, satisfied by means of a single equation alone.

We have

$$x_1 \frac{\partial}{\partial \rho}\left(\frac{\partial F}{\partial x_1}\right) + y_1 \frac{\partial}{\partial \rho}\left(\frac{\partial F}{\partial y_1}\right) + z_1 \frac{\partial}{\partial \rho}\left(\frac{\partial F}{\partial z_1}\right) + v_1 \frac{\partial}{\partial \rho}\left(\frac{\partial F}{\partial v_1}\right)$$

$$= \frac{\partial}{\partial \rho}\left(\Sigma \theta_1 \frac{\partial F}{\partial \theta_1}\right) - \left(x_{11}\frac{\partial F}{\partial x_1} + y_{11}\frac{\partial F}{\partial y_1} + z_{11}\frac{\partial F}{\partial z_1} + v_{11}\frac{\partial F}{\partial v_1}\right)$$

$$= \frac{\partial F}{\partial \rho} - \Sigma \theta_{11}\frac{\partial F}{\partial \theta_1},$$

by using the first identity in the first line of identities contained in § 368 (p. 620). Similarly,

$$x_1 \frac{\partial}{\partial \sigma}\left(\frac{\partial F}{\partial x_2}\right) + y_1 \frac{\partial}{\partial \sigma}\left(\frac{\partial F}{\partial y_2}\right) + z_1 \frac{\partial}{\partial \sigma}\left(\frac{\partial F}{\partial z_2}\right) + v_1 \frac{\partial}{\partial \sigma}\left(\frac{\partial F}{\partial v_2}\right)$$

$$= \frac{\partial}{\partial \sigma}\left(\Sigma \theta_1 \frac{\partial F}{\partial \theta_2}\right) - \left(x_{12}\frac{\partial F}{\partial x_2} + y_{12}\frac{\partial F}{\partial y_2} + z_{12}\frac{\partial F}{\partial z_2} + v_{12}\frac{\partial F}{\partial v_2}\right)$$

$$= - \Sigma \theta_{12}\frac{\partial F}{\partial \theta_2},$$

by using the second identity in that line; and, by using the third identity in the line,

$$x_1 \frac{\partial}{\partial \tau}\left(\frac{\partial F}{\partial x_3}\right) + y_1 \frac{\partial}{\partial \tau}\left(\frac{\partial F}{\partial y_3}\right) + z_1 \frac{\partial}{\partial \tau}\left(\frac{\partial F}{\partial z_3}\right) + v_1 \frac{\partial}{\partial \tau}\left(\frac{\partial F}{\partial v_3}\right)$$

$$= -\Sigma \theta_{13} \frac{\partial F}{\partial \theta_3}.$$

Also,

$$\frac{\partial F}{\partial \rho} = \Sigma \theta_1 \frac{\partial F}{\partial \theta} + \Sigma \theta_{11} \frac{\partial F}{\partial \theta_1} + \Sigma \theta_{12} \frac{\partial F}{\partial \theta_2} + \Sigma \theta_{13} \frac{\partial F}{\partial \theta_3},$$

on taking the full derivative of $F$ with regard to the independent variable $\rho$. Hence

$$x_1 E_x + y_1 E_y + z_1 E_z + v_1 E_v$$

$$= \Sigma \theta_1 \left\{ \frac{\partial F}{\partial \theta} - \frac{\partial}{\partial \rho}\left(\frac{\partial F}{\partial \theta_1}\right) - \frac{\partial}{\partial \sigma}\left(\frac{\partial F}{\partial \theta_2}\right) - \frac{\partial}{\partial \tau}\left(\frac{\partial F}{\partial \theta_3}\right) \right\}$$

$$= \Sigma \theta_1 \frac{\partial F}{\partial \theta} - \frac{\partial F}{\partial \rho} + \Sigma \theta_{11} \frac{\partial F}{\partial \theta_1} + \Sigma \theta_{12} \frac{\partial F}{\partial \theta_2} + \Sigma \theta_{13} \frac{\partial F}{\partial \theta_3}$$

$$= 0.$$

Similarly, it is proved that

$$x_2 E_x + y_2 E_y + z_2 E_z + v_2 E_v = 0,$$

$$x_3 E_x + y_3 E_y + z_3 E_z + v_3 E_v = 0.$$

Consequently,

$$\frac{E_x}{J_x} = \frac{E_y}{J_y} = \frac{E_z}{J_z} = -\frac{E_v}{J_v} = -E.$$

Hence the four characteristic equations are satisfied in virtue of one single equation

$$E = 0.$$

**371.** This single equation we can identify with the former characteristic equation (§ 360) by using the relations in § 368. We choose $E_v$, to make the identification. We have

$$\frac{\partial}{\partial \rho}\left\{(y_2 z_3 - z_2 y_3)\frac{\partial f}{\partial p}\right\} + \frac{\partial}{\partial \sigma}\left\{(y_3 z_1 - z_3 y_1)\frac{\partial f}{\partial p}\right\} + \frac{\partial}{\partial \tau}\left\{(y_1 z_2 - z_1 y_2)\frac{\partial f}{\partial p}\right\}$$

$$= \left\{(y_2 z_3 - z_2 y_3)\frac{\partial}{\partial \rho} + (y_3 z_1 - z_3 y_1)\frac{\partial}{\partial \sigma} + (y_1 z_2 - z_1 y_2)\frac{\partial}{\partial \tau}\right\}\frac{\partial f}{\partial p}$$

$$+ \frac{\partial f}{\partial p}\left\{\frac{\partial}{\partial \rho}(y_2 z_3 - z_2 y_3) + \frac{\partial}{\partial \sigma}(y_3 z_1 - z_3 y_1) + \frac{\partial}{\partial \tau}(y_1 z_2 - z_1 y_2)\right\}.$$

In the last line, the coefficient of $\frac{\partial f}{\partial p}$ vanishes identically. For any function

$W$, we have

$$\frac{\partial W}{\partial \rho} = x_1 \frac{\partial W}{\partial x} + y_1 \frac{\partial W}{\partial y} + z_1 \frac{\partial W}{\partial z},$$

$$\frac{\partial W}{\partial \sigma} = x_2 \frac{\partial W}{\partial x} + y_2 \frac{\partial W}{\partial y} + z_2 \frac{\partial W}{\partial z},$$

$$\frac{\partial W}{\partial \tau} = x_3 \frac{\partial W}{\partial x} + y_3 \frac{\partial W}{\partial y} + z_3 \frac{\partial W}{\partial z};$$

and therefore

$$J_v \frac{\partial W}{\partial x} = (y_2 z_3 - z_2 y_3) \frac{\partial W}{\partial \rho} + (y_3 z_1 - z_3 y_1) \frac{\partial W}{\partial \sigma} + (y_1 z_2 - z_1 y_2) \frac{\partial W}{\partial \tau}.$$

Consequently,

$$\frac{\partial}{\partial \rho} \left\{ (y_2 z_3 - z_2 y_3) \frac{\partial f}{\partial p} \right\} + \frac{\partial}{\partial \sigma} \left\{ (y_3 z_1 - z_3 y_1) \frac{\partial f}{\partial p} \right\} + \frac{\partial}{\partial \tau} \left\{ (y_1 z_2 - z_1 y_2) \frac{\partial f}{\partial p} \right\} = J_v \frac{\partial}{\partial x} \left( \frac{\partial f}{\partial p} \right).$$

Similarly,

$$\frac{\partial}{\partial \rho} \left\{ (z_2 x_3 - x_2 z_3) \frac{\partial f}{\partial q} \right\} + \frac{\partial}{\partial \sigma} \left\{ (z_3 x_1 - x_3 z_1) \frac{\partial f}{\partial q} \right\} + \frac{\partial}{\partial \tau} \left\{ (z_1 x_2 - x_1 z_2) \frac{\partial f}{\partial q} \right\} = J_v \frac{\partial}{\partial y} \left( \frac{\partial f}{\partial q} \right).$$

and

$$\frac{\partial}{\partial \rho} \left\{ (x_2 y_3 - y_2 x_3) \frac{\partial f}{\partial r} \right\} + \frac{\partial}{\partial \sigma} \left\{ (x_3 y_1 - y_3 x_1) \frac{\partial f}{\partial r} \right\} + \frac{\partial}{\partial \tau} \left\{ (x_1 y_2 - y_1 x_2) \frac{\partial f}{\partial r} \right\} = J_v \frac{\partial}{\partial z} \left( \frac{\partial f}{\partial r} \right).$$

Also, as $F = J_v f$, we have

$$\frac{\partial F}{\partial v} = J_v \frac{\partial f}{\partial v}.$$

Hence, having regard to the expressions for $\dfrac{\partial F}{\partial v_1}$, $\dfrac{\partial F}{\partial v_2}$, $\dfrac{\partial F}{\partial v_3}$, in § 368, we find

$$E_v = \frac{\partial F}{\partial v} - \frac{\partial}{\partial \rho} \left( \frac{\partial F}{\partial v_1} \right) - \frac{\partial}{\partial \sigma} \left( \frac{\partial F}{\partial v_2} \right) - \frac{\partial}{\partial \tau} \left( \frac{\partial F}{\partial v_3} \right)$$

$$= J_v \left\{ \frac{\partial f}{\partial v} - \frac{\partial}{\partial x} \left( \frac{\partial f}{\partial p} \right) - \frac{\partial}{\partial y} \left( \frac{\partial f}{\partial q} \right) - \frac{\partial}{\partial z} \left( \frac{\partial f}{\partial r} \right) \right\} = J_v E,$$

which establishes the significance of the single equation $E = 0$. It is the former characteristic equation obtained in § 360.

*Identification of the boundary conditions in §§ 360 and 369.*

**372.** Next, the condition to be satisfied along the boundary surface is

$$\iint \left\{ \left( X \frac{\partial F}{\partial x_1} + Y \frac{\partial F}{\partial y_1} + Z \frac{\partial F}{\partial z_1} + U \frac{\partial F}{\partial v_1} \right) d\sigma\, d\tau \right.$$

$$+ \left( X \frac{\partial F}{\partial x_2} + Y \frac{\partial F}{\partial y_2} + Z \frac{\partial F}{\partial z_2} + U \frac{\partial F}{\partial v_2} \right) d\tau\, d\rho$$

$$\left. + \left( X \frac{\partial F}{\partial x_3} + Y \frac{\partial F}{\partial y_3} + Z \frac{\partial F}{\partial z_3} + U \frac{\partial F}{\partial v_3} \right) d\rho\, d\sigma \right\} = 0,$$

for all admissible variations. The former full condition (§ 360), in the present notation, is

$$\iint \left\{ (U - pX - qY - rZ) \left( \frac{\partial f}{\partial p} \, dy\, dz + \frac{\partial f}{\partial q} \, dz\, dx + \frac{\partial f}{\partial r} \, dx\, dy \right) \right.$$
$$\left. + (X\, dy\, dz + Y\, dz\, dx + Z\, dx\, dy) f \right\} = 0.$$

These two conditions should be equivalent; the equivalence can be established as follows.

In the double integral in the first form of the condition, the coefficient of $U$ is

$$\frac{\partial F}{\partial v_1} \, d\sigma\, d\tau + \frac{\partial F}{\partial v_2} \, d\tau\, d\rho + \frac{\partial F}{\partial v_3} \, d\rho\, d\sigma.$$

On the surface, let $\phi$ and $\psi$ be independent parameters in terms of which $\rho, \sigma, \tau$, for the later condition, and $x, y, z$, for the earlier condition are expressed. Then the transformed elements of the integral are

$$d\sigma\, d\tau = J\left(\frac{\sigma, \tau}{\phi, \psi}\right) d\phi\, d\psi, \quad d\tau\, d\rho = J\left(\frac{\tau, \rho}{\phi, \psi}\right) d\phi\, d\psi, \quad d\rho\, d\sigma = J\left(\frac{\rho, \sigma}{\phi, \psi}\right) d\phi\, d\psi.$$

Now

$$\frac{\partial F}{\partial v_1} \, d\sigma\, d\tau + \frac{\partial F}{\partial v_2} \, d\tau\, d\rho + \frac{\partial F}{\partial v_3} \, d\rho\, d\sigma$$

$$= \frac{\partial f}{\partial p} \left\{ (y_2 z_3 - z_2 y_3) J\left(\frac{\sigma, \tau}{\phi, \psi}\right) + (y_3 z_1 - z_3 y_1) J\left(\frac{\tau, \rho}{\phi, \psi}\right) \right.$$
$$\left. + (y_1 z_2 - z_1 y_2) J\left(\frac{\rho, \sigma}{\phi, \psi}\right) \right\} d\phi\, d\psi$$

$$+ \frac{\partial f}{\partial q} \left\{ (z_2 x_3 - x_2 z_3) J\left(\frac{\sigma, \tau}{\phi, \psi}\right) + (z_3 x_1 - x_3 z_1) J\left(\frac{\tau, \rho}{\phi, \psi}\right) \right.$$
$$\left. + (z_1 x_2 - x_1 z_2) J\left(\frac{\rho, \sigma}{\phi, \psi}\right) \right\} d\phi\, d\psi$$

$$+ \frac{\partial f}{\partial r} \left\{ (x_2 y_3 - y_2 x_3) J\left(\frac{\sigma, \tau}{\phi, \psi}\right) + (x_3 y_1 - y_3 x_1) J\left(\frac{\tau, \rho}{\phi, \psi}\right) \right.$$
$$\left. + (x_1 y_2 - y_1 x_2) J\left(\frac{\rho, \sigma}{\phi, \psi}\right) \right\} d\phi\, d\psi.$$

In the double integral in the second form of the condition,

$$l\, dS = dy\, dz = \left( \frac{\partial y}{\partial \phi} \frac{\partial z}{\partial \psi} - \frac{\partial z}{\partial \phi} \frac{\partial y}{\partial \psi} \right) d\phi\, d\psi.$$

Also,

$$\frac{\partial y}{\partial \phi} = y_1 \frac{\partial \rho}{\partial \phi} + y_2 \frac{\partial \sigma}{\partial \phi} + y_3 \frac{\partial \tau}{\partial \phi}, \quad \frac{\partial y}{\partial \psi} = y_1 \frac{\partial \rho}{\partial \psi} + y_2 \frac{\partial \sigma}{\partial \psi} + y_3 \frac{\partial \tau}{\partial \psi};$$

and likewise for the derivatives of $z$. Hence, on substitution,

$$dy\, dz = \left\{ (y_2 z_3 - z_2 y_3) J\left(\frac{\sigma, \tau}{\phi, \psi}\right) + (y_3 z_1 - z_3 y_1) J\left(\frac{\tau, \rho}{\phi, \psi}\right) \right.$$
$$\left. + (y_1 z_2 - z_1 y_2) J\left(\frac{\rho, \sigma}{\phi, \psi}\right) \right\} d\phi\, d\psi.$$

Similarly for $dz\,dx$, and for $dx\,dy$. Consequently,

$$\frac{\partial F}{\partial v_1}\,d\sigma\,d\tau + \frac{\partial F}{\partial v_2}\,d\tau\,d\rho + \frac{\partial F}{\partial v_3}\,d\rho\,d\sigma = \frac{\partial f}{\partial p}\,dy\,dz + \frac{\partial f}{\partial q}\,dz\,dx + \frac{\partial f}{\partial r}\,dx\,dy,$$

so that the coefficients of $U$ in the integrals in the two forms of condition are the same.

Again, in the second form of condition, the coefficient of $X$ under the sign of the double integral is

$$\frac{\partial F}{\partial x_1}\,d\sigma\,d\tau + \frac{\partial F}{\partial x_2}\,d\tau\,d\rho + \frac{\partial F}{\partial x_3}\,d\rho\,d\sigma.$$

When the values of $\dfrac{\partial F}{\partial x_1}$, $\dfrac{\partial F}{\partial x_2}$, $\dfrac{\partial F}{\partial x_3}$, as given in § 368, are substituted, we find that this coefficient

$$= T\{(y_2z_3 - z_2y_3)\,d\sigma\,d\tau + (y_3z_1 - z_3y_1)\,d\tau\,d\rho + (y_1z_2 - z_1y_2)\,d\rho\,d\sigma\}$$
$$+ \frac{\partial f}{\partial q}\{(v_2z_3 - z_2v_3)\,d\sigma\,d\tau + (v_3z_1 - z_3v_1)\,d\tau\,d\rho + (v_1z_2 - z_1v_2)\,d\rho\,d\sigma\}$$
$$+ \frac{\partial f}{\partial r}\{(y_2v_3 - v_2y_3)\,d\sigma\,d\tau + (y_3v_1 - v_3y_1)\,d\tau\,d\rho + (y_1v_2 - v_1y_2)\,d\rho\,d\sigma\}.$$

By the preceding analysis, the first line

$$= T\,dy\,dz$$
$$= \left(f - p\frac{\partial f}{\partial p} - q\frac{\partial f}{\partial q} - r\frac{\partial f}{\partial r}\right)dy\,dz.$$

It is easy to verify the identical relation

$$(v_2z_3 - z_2v_3)\,J_v + J_x\,(z_2x_3 - z_3x_2) - J_y\,(y_2z_3 - z_2y_3) = 0,$$

that is,

$$v_2z_3 - z_2v_3 - q\,(y_2z_3 - z_2y_3) = -p\,(z_2x_3 - z_3x_2);$$

and similarly,

$$v_3z_1 - z_3v_1 - q\,(y_3z_1 - z_3y_1) = -p\,(z_3x_1 - z_1x_3),$$
$$v_1z_2 - z_1v_2 - q\,(y_1z_2 - z_1y_2) = -p\,(z_1x_2 - z_2x_1).$$

Hence the whole quantity multiplying $\dfrac{\partial f}{\partial q}$ in the coefficient is

$$= -q\,dy\,dz$$
$$+ q\{(y_2z_3 - z_2y_3)\,d\sigma\,d\tau + (y_3z_1 - z_3y_1)\,d\tau\,d\rho + (y_1z_2 - z_1y_2)\,d\rho\,d\sigma\}$$
$$- p\{(z_2x_3 - z_3x_2)\,d\sigma\,d\tau + (z_3x_1 - z_1x_3)\,d\tau\,d\rho + (z_1x_2 - z_2x_1)\,d\rho\,d\sigma\}$$
$$= -q\,dy\,dz + q\,dy\,dz - p\,dz\,dx;$$

or the part involving $\dfrac{\partial f}{\partial q}$ is

$$-p\frac{\partial f}{\partial q}\,dz\,dx.$$

Similarly, the part involving $\frac{\partial f}{\partial r}$ is

$$-p \frac{\partial f}{\partial r} dx dy.$$

Consequently, the coefficient of $X$ is equal to

$$\left(f - p \frac{\partial f}{\partial p}\right) dy\, dz - p \frac{\partial f}{\partial q}\, dz\, dx - p \frac{\partial f}{\partial r}\, dx\, dy$$

$$= f dy\, dz - p \left(\frac{\partial f}{\partial p}\, dy\, dz + \frac{\partial f}{\partial q}\, dz\, dx + \frac{\partial f}{\partial r}\, dx\, dy\right);$$

and this is the coefficient of $X$ in the double integral in the earlier expression (in § 360) of the terminal condition.

Similarly, the coefficients of $Y$ and of $Z$, under the sign of the respective double integrals, agree.

Hence the two forms of terminal conditions are equivalent to one another.

### Summary of conditions, under general weak variations.

**373.** Finally, the mobile variations along a part of the boundary are subject only to the relations (if any) that connect $X$, $Y$, $Z$, $U$, through some fixed surface at which the part of the boundary shall terminate. Hence, by the usual argument, we can state the terminal conditions to be that the relation

$$\left(\Sigma\Theta \frac{\partial F}{\partial \theta_1}\right) d\sigma\, d\tau + \left(\Sigma\Theta \frac{\partial F}{\partial \theta_2}\right) d\tau\, d\rho + \left(\Sigma\Theta \frac{\partial F}{\partial \theta_3}\right) d\rho\, d\sigma = 0$$

must be satisfied at every place on the whole of the surfaces constituting the complete boundary of the volume through which the triple integral extends.

Summarising, we see that, in order to have a vanishing first variation under general weak variations, the *four characteristic equations must be satisfied throughout the volume*, and the *terminal condition must be satisfied everywhere on the boundary of the volume*.

### The Legendre test.

**374.** We shall not deal with the normal form of the 'second' variation when the integral is subjected to general weak variations. The experience, derived from the preceding cases which have progressively differed from one another only in the detailed range of the analysis, but not in its general character nor in the nature of the issues it has raised, shews that, when once the characteristic equations and the terminal conditions have been established, we should expect no tests, other than the extended form of the Legendre test and the extended form of the Jacobi test, would emerge from the general weak variation. In each case, the set of characteristic equations has been brought into agreement with the single characteristic equation for the special weak variation. Also, the terminal conditions for the general weak variation have been identified with the terminal conditions for the special weak variation, and thus have been used to confirm the complete expression of those earlier conditions by the particular considerations required at the limits, where formerly it was necessary to introduce a non-special terminal variation accordant with the terminal condition.

For the triple integral, in which only first derivatives occur in a form

$$\iiint F\, d\rho\, d\sigma\, d\tau,$$

the Legendre test is as follows:

Either singly, or in pair-combinations, the second derivatives of $F$ with respect to the twelve quantities $x_1$, $x_2$, $x_3$; $y_1$, $y_2$, $y_3$; $z_1$, $z_2$, $z_3$; $v_1$, $v_2$, $v_3$; are expressible, in terms of six magnitudes $c_{11}$, $c_{12}$, $c_{13}$, $c_{22}$, $c_{23}$, $c_{33}$, by relations

$$\frac{1}{J_x^2}\frac{\partial^2 F}{\partial x_\mu^2} = \frac{1}{J_x J_y}\frac{\partial^2 F}{\partial x_\mu \partial y_\mu} = \frac{1}{J_x J_z}\frac{\partial^2 F}{\partial x_\mu \partial z_\mu} = \frac{1}{J_y^2}\frac{\partial^2 F}{\partial y_\mu^2} = \frac{1}{J_y J_z}\frac{\partial^2 F}{\partial y_\mu \partial z_\mu} = \frac{1}{J_z^2}\frac{\partial^2 F}{\partial z_\mu^2}$$

$$= \frac{-1}{J_x J_v}\frac{\partial^2 F}{\partial x_\mu \partial v_\mu} = \frac{-1}{J_y J_v}\frac{\partial^2 F}{\partial y_\mu \partial v_\mu} = \frac{-1}{J_z J_v}\frac{\partial^2 F}{\partial z_\mu \partial v_\mu} = \frac{1}{J_v^2}\frac{\partial^2 F}{\partial v_\mu^2} = c_{\mu\mu},$$

for $\mu = 1, 2, 3$: and

$$\frac{2}{J_x^2}\frac{\partial^2 F}{\partial x_m \partial x_n} = \frac{2}{J_y^2}\frac{\partial^2 F}{\partial y_m \partial y_n} = \frac{2}{J_z^2}\frac{\partial^2 F}{\partial z_m \partial z_n}$$

$$= \frac{1}{J_x J_y}\left(\frac{\partial^2 F}{\partial x_m \partial y_n} + \frac{\partial^2 F}{\partial x_n \partial y_m}\right) = \frac{1}{J_x J_z}\left(\frac{\partial^2 F}{\partial x_m \partial z_n} + \frac{\partial^2 F}{\partial x_n \partial z_m}\right) = \frac{1}{J_y J_z}\left(\frac{\partial^2 F}{\partial y_m \partial z_n} + \frac{\partial^2 F}{\partial y_n \partial z_m}\right)$$

$$= \frac{-1}{J_x J_v}\left(\frac{\partial^2 F}{\partial x_m \partial v_n} + \frac{\partial^2 F}{\partial x_n \partial v_m}\right) = \frac{-1}{J_y J_v}\left(\frac{\partial^2 F}{\partial y_m \partial v_n} + \frac{\partial^2 F}{\partial y_n \partial v_m}\right) = \frac{-1}{J_z J_v}\left(\frac{\partial^2 F}{\partial z_m \partial v_n} + \frac{\partial^2 F}{\partial z_n \partial v_m}\right)$$

$$= \frac{2}{J_v^2}\frac{\partial^2 F}{\partial v_m \partial v_n} = 2c_{mn},$$

for* $m, n, = 1, 2, 3$. The Legendre test requires that the relations

$$c_{11}c_{22} > c_{12}^2, \quad c_{22}c_{33} > c_{23}^2, \quad c_{33}c_{11} > c_{13}^2,$$

shall be satisfied throughout the characteristic volume; and these can be satisfied, only if $c_{11}$, $c_{22}$, $c_{33}$, vanish nowhere in the volume and have the same sign. When this common sign is positive, a minimum is admissible; when it is negative, a maximum is admissible.

*The Jacobi test.*

**375.** The extended form of the Jacobi test cannot be stated, at present, in similar precise analytical form. It remains in the merely descriptive shape, of limiting the range of the triple integral so that it shall not extend as far as the conjugate of the initial boundary, this conjugate boundary being the

---

* We need only take $m$ unequal to $n$; for $m = n$, the forms are included in the preceding expressions $c_{\mu\mu}$.

superficial area next common to two consecutive characteristic volumes which have the initial boundary in common. In order to construct the corresponding analytical expression of the Jacobi test, the primitive of the characteristic equations would have to be known in explicit form; and the primitive of the associated subsidiary equations would have to be deduced. Even the first of these requirements is not met in the present processes of analysis. In establishing the analytical expression, incidental properties of surfaces and volumes would be required: but no critical test, additional in character, need be expected to emerge.

### Strong variations: method of geometrical representation.

**376.** We pass to a brief consideration of strong variations. In the circumstances, only the simplest kind of strong variation can be treated.

We therefore seek the extended form of the Weierstrass test, which really is the 'first' variation (in the customary sense of that term) under the simplest type of strong variation. For the purpose, it will be convenient to utilise the rudiments of four-dimensional geometry (mainly as regards vocabulary) in connection with the four variables $x$, $y$, $z$, $v$. The problem will only require the association of imagined configurations in abstract geometry, with configurations in the customary three-dimensional geometry for the independent variables $\rho$, $\sigma$, $\tau$.

(i) An element of arc at a place $x$, $y$, $z$, $v$, in the quadruple manifold is given by

$$ds^2 = dx^2 + dy^2 + dz^2 + dv^2.$$

(Should the fourth variable $v$ be different in nature from the other three, it can be multiplied by a constant of the dimensions which are appropriate to make the product of the same nature as those three: the product can then still be denoted by $v$.) In order to deal with triple integrals, we have represented $x$, $y$, $z$, $v$, as functions (§ 367) of three independent variables $\rho$, $\sigma$, $\tau$. These functions are initially unknown; and their analytical determination constitutes the problem. We write, as usual,

$$d\theta = \theta_1 d\rho + \theta_2 d\sigma + \theta_3 d\tau,$$

for $\theta = x$, $y$, $z$, $v$, in turn. Then

$$ds^2 = A\,d\rho^2 + B\,d\sigma^2 + C\,d\tau^2 + 2F\,d\sigma\,d\tau + 2G\,d\tau\,d\rho + 2H\,d\rho\,d\sigma,$$

where

$$
\begin{aligned}
A &= x_1^2 + y_1^2 + z_1^2 + v_1^2, & F &= x_2 x_3 + y_2 y_3 + z_2 z_3 + v_2 v_3 \\
B &= x_2^2 + y_2^2 + z_2^2 + v_2^2, & G &= x_3 x_1 + y_3 y_1 + z_3 z_1 + v_3 v_1 \\
C &= x_3^2 + y_3^2 + z_3^2 + v_3^2, & H &= x_1 x_2 + y_1 y_2 + z_1 z_2 + v_1 v_2
\end{aligned}.
$$

The quantities $A$, $B$, $C$, $F$, $G$, $H$, are fundamental for any volume in which $\rho$, $\sigma$, $\tau$, are variables, similar to the quantities $E$, $F$, $G$, for any surface (§ 290) in ordinary space represented by two parameters.

(ii) At a place $x$, $y$, $z$, $v$, there are three prime directions, constituted there by the combinations

$$\sigma = b,\ \tau = c\ ;\quad \tau = c,\ \rho = a\ ;\quad \rho = a,\ \sigma = b\ ;$$

severally : where $a$, $b$, $c$, are three parametric constants. For these three directions at $x$, $y$, $z$, $v$, we have three sets of direction-cosines

$$x_1 A^{-\frac{1}{2}},\quad y_1 A^{-\frac{1}{2}},\quad z_1 A^{-\frac{1}{2}},\quad v_1 A^{-\frac{1}{2}},$$
$$x_2 B^{-\frac{1}{2}},\quad y_2 B^{-\frac{1}{2}},\quad z_2 B^{-\frac{1}{2}},\quad v_2 B^{-\frac{1}{2}},$$
$$x_3 C^{-\frac{1}{2}},\quad y_3 C^{-\frac{1}{2}},\quad z_3 C^{-\frac{1}{2}},\quad v_3 C^{-\frac{1}{2}},$$

respectively. If $l_1$, $m_1$, $n_1$, $k_1$, denote the first set ; $l_2$, $m_2$, $n_2$, $k_2$, and $l_3$, $m_3$, $n_3$, $k_3$, the other two sets ; we have

$$l_1^2 + m_1^2 + n_1^2 + k_1^2 = 1,\quad l_2^2 + m_2^2 + n_2^2 + k_2^2 = 1,\quad l_3^2 + m_3^2 + n_3^2 + k_3^2 = 1.$$

A line-direction at a place $x$, $y$, $z$, $v$, in the quadruple manifold can be represented by a mere place in an associated descriptive triple manifold—which can be any general three-dimensional space, but is not the three-dimensional space of the variables $\rho$, $\sigma$, $\tau$. Accordingly, we represent these three prime line-directions by three points $a$, $b$, $c$.

(iii) Two line-directions at a place $x$, $y$, $z$, $v$, in the quadruple manifold determine a two-way space in that manifold. Thus the two-way space $\tau = c$ at the place is the two-way space through the line-directions, represented by the points $a$ and $b$ in the associated descriptive triple manifold. It can be represented in that triple manifold by the line $ab$.

If $\gamma$ be the angle between the prime directions at $x$, $y$, $z$, $v$, represented by $a$ and $b$, we have

$$\cos \gamma = \frac{x_1}{A^{\frac{1}{2}}} \frac{x_2}{B^{\frac{1}{2}}} + \frac{y_1}{A^{\frac{1}{2}}} \frac{y_2}{B^{\frac{1}{2}}} + \frac{z_1}{A^{\frac{1}{2}}} \frac{z_2}{B^{\frac{1}{2}}} + \frac{v_1}{A^{\frac{1}{2}}} \frac{v_2}{B^{\frac{1}{2}}} = \frac{H}{(AB)^{\frac{1}{2}}},$$

so that

$$\frac{\cos \gamma}{H} = \frac{\sin \gamma}{(AB - H^2)^{\frac{1}{2}}} = \frac{1}{(AB)^{\frac{1}{2}}}.$$

We thus can measure the magnitude of the line $ab$ by $\gamma$. If, similarly, the lines $ca$ and $bc$ are measured by $\beta$ and by $\alpha$ respectively, we have

$$\frac{\cos \alpha}{F} = \frac{\sin \alpha}{(BC - F^2)^{\frac{1}{2}}} = \frac{1}{(BC)^{\frac{1}{2}}},$$

$$\frac{\cos \beta}{G} = \frac{\sin \beta}{(CA - G^2)^{\frac{1}{2}}} = \frac{1}{(CA)^{\frac{1}{2}}}.$$

If we please, we can take coordinates of a two-way space, normal at the place $x$, $y$, $z$, $v$, to the two-way space represented by the line $ab$, in the form

$$\frac{1}{(AB - H^2)^{\frac{1}{2}}} \left\| \begin{array}{cccc} x_1, & y_1, & z_1, & v_1 \\ x_2, & y_2, & z_2, & v_2 \end{array} \right\| ,$$

these six coordinates $l_{12}$, $m_{12}$, $n_{12}$, $\lambda_{12}$, $\mu_{12}$, $\nu_{12}$ $(= y_1 z_2 - z_1 y_2,\ z_1 x_2 - x_1 z_2,\ x_1 y_2 - y_1 x_2,$ $x_1 v_2 - v_1 x_2,\ y_1 v_2 - v_1 y_2,\ z_1 v_2 - v_1 z_2)$ being connected by the relations

$$l_{12}\lambda_{12} + m_{12}\mu_{12} + n_{12}\nu_{12} = 0,$$
$$l_{12}^2 + \lambda_{12}^2 + m_{12}^2 + \mu_{12}^2 + n_{12}^2 + \nu_{12}^2 = 1.$$

(iv) Three line-directions at a place $x$, $y$, $z$, $v$, in the quadruple manifold determine a three-way space in that manifold. Thus the three-way space, limited at $x$, $y$, $z$, $v$, by the three two-way spaces represented in the associated descriptive triple manifold, that is, by the three two-way spaces, $\rho = a$, $\sigma = b$, $\tau = c$, can be represented by the triangle which has the lines $bc$, $ca$, $ab$, representing those spaces.

If $l$, $m$, $n$, $k$, be direction-cosines for a direction perpendicular to each of the three directions, represented in the descriptive triple manifold by the points $a$, $b$, $c$, we have

$$lx_1 + my_1 + nz_1 + kv_1 = 0,$$
$$lx_2 + my_2 + nz_2 + kv_2 = 0,$$
$$lx_3 + my_3 + nz_3 + kv_3 = 0.$$

Hence

$$\frac{l}{J_x} = \frac{m}{J_y} = \frac{n}{J_z} = \frac{-k}{J_v}.$$

Let $\Delta$ denote the common value of these fractions; then, as

$$l^2 + m^2 + n^2 + k^2 = 1,$$

we have

$$\Delta^2 = J_x^2 + J_y^2 + J_z^2 + J_v^2$$
$$= \begin{vmatrix} A, & H, & G \\ H, & B, & F \\ G, & F, & C \end{vmatrix},$$

as may easily be verified. Thus

$$l = \frac{J_x}{\Delta}, \quad m = \frac{J_y}{\Delta}, \quad n = \frac{J_z}{\Delta}, \quad -k = \frac{J_v}{\Delta}.$$

(v) At the place $x$, $y$, $z$, $v$, in the quadruple manifold, take three arbitrary directions, distinct from the three directions represented by $a$, $b$, $c$, and different from one another, but all three within the characteristic volume for any problem. Let their representative positions in the plane triangle $abc$ in the descriptive triple manifold be $\delta'$, $\delta''$, $\delta'''$, respectively; and let $ds'$, $ds''$, $ds'''$, be the respective three arc-elements at $x$, $y$, $z$, $v$. The three sets of direction-cosines are such that

$$\left.\begin{array}{l} x'^2 + y'^2 + z'^2 + v'^2 = 1 \\ x''^2 + y''^2 + z''^2 + v''^2 = 1 \\ x'''^2 + y'''^2 + z'''^2 + v'''^2 = 1 \end{array}\right\}.$$

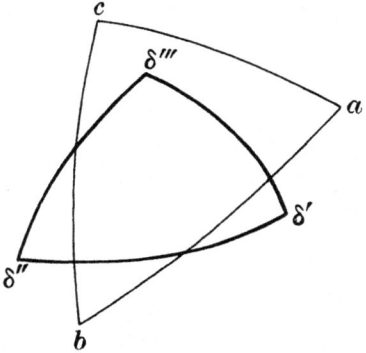

Also,

$$\left.\begin{array}{l} x' = x_1\rho' + x_2\sigma' + x_3\tau' \\ y' = y_1\rho' + y_2\sigma' + y_3\tau' \\ z' = z_1\rho' + z_2\sigma' + z_3\tau' \\ v' = v_1\rho' + v_2\sigma' + v_3\tau' \end{array}\right\};$$

with corresponding sets of expressions for $x''$, $y''$, $z''$, $v''$, in terms of $\rho''$, $\sigma''$, $\tau''$; and for $x'''$, $y'''$, $z'''$, $v'''$, in terms of $\rho'''$, $\sigma'''$, $\tau'''$.

If the inclinations $\theta'$, $\phi'$, $\psi'$, of the direction $\delta'$ to $a$, $b$, $c$, are desired, we have

$$\cos\theta' = x'\frac{x_1}{A^{\frac{1}{2}}} + y'\frac{y_1}{A^{\frac{1}{2}}} + z'\frac{z_1}{A^{\frac{1}{2}}} + v'\frac{v_1}{A^{\frac{1}{2}}},$$

and so for $\phi'$, $\psi'$; that is,

$$\left.\begin{array}{l} A^{\frac{1}{2}}\cos\theta' = A\dfrac{d\rho}{ds'} + H\dfrac{d\sigma}{ds'} + G\dfrac{d\tau}{ds'} \\[2mm] B^{\frac{1}{2}}\cos\phi' = H\dfrac{d\rho}{ds'} + B\dfrac{d\sigma}{ds'} + F\dfrac{d\tau}{ds'} \\[2mm] C^{\frac{1}{2}}\cos\psi' = G\dfrac{d\rho}{ds'} + F\dfrac{d\sigma}{ds'} + C\dfrac{d\tau}{ds'} \end{array}\right\}.$$

Similarly for the inclinations $\theta''$, $\phi''$, $\psi''$, of the direction $\delta''$ to $a$, $b$, $c$, in terms of $\dfrac{d\rho}{ds''}$, $\dfrac{d\sigma}{ds''}$, $\dfrac{d\tau}{ds''}$; and for the inclinations $\theta'''$; $\phi'''$, $\psi'''$, of the direction $\delta'''$ to $a$, $b$, $c$, in terms of $\dfrac{d\rho}{ds'''}$, $\dfrac{d\sigma}{ds'''}$, $\dfrac{d\tau}{ds'''}$.

*Representation of volumes:* (i) *characteristic* ; (ii) *rudimentary strong variation.*

**377.** We have to consider a characteristic volume, that is, a volume determined by the functional forms of $x$, $y$, $z$, $v$, in terms of $\rho$, $\sigma$, $\tau$, the functions being so chosen as to satisfy the characteristic equations in any problem. The volume has to be bounded at some lower range: such boundary may be called the initial boundary. We take any portion of this volume having the initial boundary; the portion will have a boundary at its own upper range. In that upper boundary which itself can be chosen arbitrarily, let any place be denoted by $x$, $y$, $z$, $v$; and let $ds'$, $ds''$, be two arbitrary arc-elements of reference at $x$, $y$, $z$, $v$, chosen in that boundary. In the portion of the characteristic volume with this upper boundary, any direction at $x$, $y$, $z$, $v$, belonging to the characteristic volume, may be denoted by an arc-element $ds'''$. For the portion of the characteristic volume with the selected upper surface-boundary, the three directions $ds'$, $ds''$, $ds'''$, are represented in the descriptive triple manifold by the points $\delta'$, $\delta''$, $\delta'''$, respectively. In the triangle $\delta'\delta''\delta'''$, the line $\delta'\delta''$ is taken as a line of reference.

**378.** We are to effect a strong variation upon this characteristic volume.

In the case of single integrals, the strong variations (§ 208) were composed of (or resolved into) ultimate elementary strong variations. Each of these elementary variations is compounded of (i) a characteristic weak variation of the curve, zero at the initial boundary (point) and non-zero at the typical general place $P$ selected for the construction of the test: and (ii) a direct non-characteristic return, from the varied position of $P$ off the curve, back to $P$ on the curve.

In the case of double integrals, the strong variations (§ 318) were similarly composed of ultimate elementary strong variations. Each of these elementary variations is compounded of (i) a characteristic weak variation of the surface, zero at the initial boundary (curve) and non-zero at the typical general curve $C$ selected for the construction of the test: and (ii) a direct non-characteristic return, from the varied position of $C$ off the surface, back to $C$ on the surface. For the construction of the test, only an elementary arc at a point $P$ of $C$ is retained ultimately. The test is required to apply in connection with each such elementary arc: and the element of area for the variation is made, by combining this arc with the length of the small displacement of $P$, off the curve $C$ and away from the characteristic surface.

In a similar manner for triple integrals, the strong variations of the type that are considered, are similarly composed of ultimate elementary strong variations. Each of these elementary variations is compounded of (i) a characteristic weak variation of the volume, zero at the initial boundary (two-way space) and non-zero at the typical general two-way space $\delta'\delta''$ selected for the construction of the test: and (ii) a direct non-characteristic return, from the displaced boundary of the volume away out of the portion of characteristic volume, back to the original boundary of the portion. For the construction of the test, only an elementary two-way space of the boundary through $P$ is retained ultimately, this elementary space being determined by the two arbitrary arc-elements, $ds'$ and $ds''$, at $P$. The test is required to apply in connection with each such elementary space: and the element of volume (three-way space) is made, by combining this elementary space with the length of the small displacement of $P$, away from the boundary (two-way space) at $P$, and away out of the portion of the characteristic volume. This non-characteristic displacement of $P$ is taken in any arbitrary direction $ds$, represented by a point $\delta$; and $\delta$ lies outside the continuous configuration $\delta'\delta''\delta'''$ in the descriptive triple manifold, so that now all the three dimensions of the descriptive manifold are used.

When, to the triple integral, we apply an ultimate elementary strong variation of the type thus constituted, we consider separately the contributions to the variation of the integral made by the two constituent portions of the strong variation.

*Component of the variation of the integral, due to the characteristic
constituent of the strong variation.*

**379.** The contribution from the constituent weak characteristic variation
is obtained by means of the limiting terms in the expression for the first
variation of the integral. (The requirement, that the variation shall be
characteristic, is used in connection with the general resolution and the
general composition of strong variations, in their relation to the ultimate
elementary variations: it does not affect the expression of the contribution
at $P$.) To have such a variation, we take a small continuous variation
$x + \kappa\xi$, $y + \kappa\eta$, $z + \kappa\zeta$, $v + \kappa v$, of $x$, $y$, $z$, $v$, where $\xi$, $\eta$, $\zeta$, $v$, satisfy the equations
subsidiary to the characteristic equations and are required to vanish at the
initial boundary of the characteristic volume (three-way space) concerned.
At $P$, their values may be denoted by $X$, $Y$, $Z$, $U$; so that, if $\bar{P}$ be the
displaced position of $P$, its coordinates are $x + \kappa X$, $y + \kappa Y$, $z + \kappa Z$, $v + \kappa U$.
Also, let the length of $\bar{P}P$ be $D$; and, as we have to make the abrupt non-
characteristic return from $\bar{P}$ to $P$, we denote the direction-cosines of $\bar{P}P$ by
$l$, $m$, $n$, $k$. Thus

$$\kappa X = -lD, \quad \kappa Y = -mD, \quad \kappa Z = -nD, \quad \kappa U = -kD.$$

Moreover, $-l$, $-m$, $-n$, $-k$, constitute any arbitrary direction at $P$.

The integral, which is to be subjected to the ultimate elementary strong
variation, is (§ 367)

$$I = \iiint F \, d\rho \, d\sigma \, d\tau,$$

where $F$ is a function of $x$, $y$, $z$, $v$, and of their first derivatives with
regard to $\rho$, $\sigma$, $\tau$. If $I_C$ denote its value for the characteristic volume (three-
way space), and $I_\Gamma$ its value for the contiguous characteristic volume, we
have (up to first powers of $\kappa$)

$$I_C - I_\Gamma = \kappa \iint \left( X \frac{\partial F}{\partial x_1} + Y \frac{\partial F}{\partial y_1} + Z \frac{\partial F}{\partial z_1} + U \frac{\partial F}{\partial v_1} \right) d\sigma \, d\tau$$

$$+ \kappa \iint \left( X \frac{\partial F}{\partial x_2} + Y \frac{\partial F}{\partial y_2} + Z \frac{\partial F}{\partial z_2} + U \frac{\partial F}{\partial v_2} \right) d\tau \, d\rho$$

$$+ \kappa \iint \left( X \frac{\partial F}{\partial x_3} + Y \frac{\partial F}{\partial y_3} + Z \frac{\partial F}{\partial z_3} + U \frac{\partial F}{\partial v_3} \right) d\rho \, d\sigma,$$

the double integrals being taken along the boundary (two-way space) of the
volume concerned. At a place $P$, the directions in this boundary are
indicated by $ds'$ and $ds''$. Now, for the integral-elements,

$$d\sigma \, d\tau = J \left( \frac{\sigma, \tau}{s', s''} \right) ds' ds'', \quad d\tau \, d\rho = J \left( \frac{\tau, \rho}{s', s''} \right) ds' ds'', \quad d\rho \, d\sigma = J \left( \frac{\rho, \sigma}{s', s''} \right) ds' ds''.$$

Thus we obtain an integral

$$\iint \mathbf{I}\,ds'\,ds''$$

taken over this boundary, where

$$\mathbf{I} = \kappa \left( X\frac{\partial F}{\partial x_1} + Y\frac{\partial F}{\partial y_1} + Z\frac{\partial F}{\partial z_1} + U\frac{\partial F}{\partial v_1} \right) J\left(\frac{\sigma,\ \tau}{s',\ s''}\right)$$

$$+ \kappa \left( X\frac{\partial F}{\partial x_2} + Y\frac{\partial F}{\partial y_2} + Z\frac{\partial F}{\partial z_2} + U\frac{\partial F}{\partial v_2} \right) J\left(\frac{\tau,\ \rho}{s',\ s''}\right)$$

$$+ \kappa \left( X\frac{\partial F}{\partial x_3} + Y\frac{\partial F}{\partial y_3} + Z\frac{\partial F}{\partial z_3} + U\frac{\partial F}{\partial v_3} \right) J\left(\frac{\rho,\ \sigma}{s',\ s''}\right).$$

As the variations are entirely arbitrary upon the surface, and are at our disposal, we can consider the double integral for any rudimentary two-way space $ds'\,ds''$ alone. We, also, can consider the double integral separately, for each such rudimentary element as part of its whole range. For small lengths $S'$ and $S''$ along $ds'$ and $ds''$ respectively, we have, for the rudimentary element,

$$\iint \mathbf{I}\,ds'\,ds'' = \{\mathbf{I} + [\kappa]\}\,S'\,S'' = \mathbf{I}\,S'\,S'',$$

up to the first power of $\kappa$ inclusive, where $[\kappa]$ denotes an aggregate of terms of the first and higher degrees in $\kappa$, and is therefore negligible compared with $\mathbf{I}$.

Thus the contribution of the rudimentary element to $I_C - I_\Gamma$ is $\mathbf{I}\,S'\,S''$. Let the values of $\kappa X,\ \kappa Y,\ \kappa Z,\ \kappa U$, in terms of $D$, be substituted. Then the contribution is equal to

$$-\,\square\,S'\,S''\,D,$$

where

$$\square = \left( l\frac{\partial F}{\partial x_1} + m\frac{\partial F}{\partial y_1} + n\frac{\partial F}{\partial z_1} + k\frac{\partial F}{\partial v_1} \right) J\left(\frac{\sigma,\ \tau}{s',\ s''}\right)$$

$$+ \left( l\frac{\partial F}{\partial x_2} + m\frac{\partial F}{\partial y_2} + n\frac{\partial F}{\partial z_2} + k\frac{\partial F}{\partial v_2} \right) J\left(\frac{\tau,\ \rho}{s',\ s''}\right)$$

$$+ \left( l\frac{\partial F}{\partial x_3} + m\frac{\partial F}{\partial y_3} + n\frac{\partial F}{\partial z_3} + k\frac{\partial F}{\partial v_3} \right) J\left(\frac{\rho,\ \sigma}{s',\ s''}\right).$$

**380.** This expression for $\square$ may be modified. In the integral $I$, where $\rho,\ \sigma,\ \tau$, are the current variables of integration, let there be a change of variables to lengths $s',\ s'',\ s'''$, along representative directions at any point. Then we have

$$x' = x_1\rho' + x_2\sigma' + x_3\tau',\quad x'' = x_1\rho'' + x_2\sigma'' + x_3\tau'',\quad x''' = x_1\rho''' + x_2\sigma''' + x_3\tau''',$$

expressing $x_1,\ x_2,\ x_3$, in terms of $x',\ x'',\ x'''$; and similarly for $y_1,\ y_2,\ y_3$, in terms of $y',\ y'',\ y'''$; for $z_1,\ z_2,\ z_3$, in terms of $z',\ z'',\ z'''$; and for $v_1,\ v_2,\ v_3$, in terms of $v',\ v'',\ v'''$; always with derivatives of $\rho,\ \sigma,\ \tau$, as linear coefficients in the expressions. Now

$$I = \iiint F \, d\rho \, d\sigma \, d\tau$$

$$= \iiint F J \left( \frac{\rho, \sigma, \tau}{s', s'', s'''} \right) ds' \, ds'' \, ds'''$$

$$= \iiint \Omega \, ds' \, ds'' \, ds''',$$

where

$$\Omega = F J \left( \frac{\rho, \sigma, \tau}{s', s'', s'''} \right) = F \bar{J},$$

substitution for the derivatives $x_\mu, y_\mu, z_\mu, v_\mu$ ($\mu = 1, 2, 3$), which occur in $F$ being made in terms of derivatives of $x, y, z, v$, with regard to $s', s'', s'''$. Thus

$$\bar{J} \frac{\partial F}{\partial x_1} = \frac{\partial \Omega}{\partial x'} \rho' + \frac{\partial \Omega}{\partial x''} \rho'' + \frac{\partial \Omega}{\partial x'''} \rho''',$$

$$\bar{J} \frac{\partial F}{\partial x_2} = \frac{\partial \Omega}{\partial x'} \sigma' + \frac{\partial \Omega}{\partial x''} \sigma'' + \frac{\partial \Omega}{\partial x'''} \sigma''',$$

$$\bar{J} \frac{\partial F}{\partial x_3} = \frac{\partial \Omega}{\partial x'} \tau' + \frac{\partial \Omega}{\partial x''} \tau'' + \frac{\partial \Omega}{\partial x'''} \tau''';$$

with similar expressions for the derivatives of $F$ with regard to the derivatives of $y_1, y_2, y_3$; $z_1, z_2, z_3$; $v_1, v_2, v_3$.

In $\square$, the coefficient of $l$

$$= \frac{\partial F}{\partial x_1} J \left( \frac{\sigma, \tau}{s', s''} \right) + \frac{\partial F}{\partial x_2} J \left( \frac{\tau, \rho}{s', s''} \right) + \frac{\partial F}{\partial x_3} J \left( \frac{\rho, \sigma}{s', s''} \right)$$

$$= \frac{1}{\bar{J}} \left\{ \frac{\partial \Omega}{\partial x'} J \left( \frac{\rho, \sigma, \tau}{s', s', s''} \right) + \frac{\partial \Omega}{\partial x''} J \left( \frac{\rho, \sigma, \tau}{s'', s', s''} \right) + \frac{\partial \Omega}{\partial x'''} J \left( \frac{\rho, \sigma, \tau}{s''', s', s''} \right) \right\}$$

$$= \frac{\partial \Omega}{\partial x'''}.$$

Similarly, the coefficients of $m$, of $n$, and of $k$, respectively, are

$$\frac{\partial \Omega}{\partial y'''}, \quad \frac{\partial \Omega}{\partial z'''}, \quad \frac{\partial \Omega}{\partial v'''}.$$

Hence

$$\square = l \frac{\partial \Omega}{\partial x'''} + m \frac{\partial \Omega}{\partial y'''} + n \frac{\partial \Omega}{\partial z'''} + k \frac{\partial \Omega}{\partial v'''}.$$

Consequently, the contribution of the rudimentary element $S'S''$ to $I_C - I_\Gamma$, that is, to $\iint \mathbf{I} \, ds' \, ds''$, is equal to

$$- \left( l \frac{\partial \Omega}{\partial x'''} + m \frac{\partial \Omega}{\partial y'''} + n \frac{\partial \Omega}{\partial z'''} + k \frac{\partial \Omega}{\partial v'''} \right) S'S''D.$$

**381.** The integral $I$ has been transformed to the shape

$$\iiint \Omega \, ds' \, ds'' \, ds'''.$$

The intermediate form $\iiint F\,d\rho\,d\sigma\,d\tau$ has been used for the transformation: but the transformation can be effected directly from

$$\iiint f\left(x,\,y,\,z,\,v,\,\frac{\partial v}{\partial x},\,\frac{\partial v}{\partial y},\,\frac{\partial v}{\partial z}\right)dx\,dy\,dz.$$

For we have

$$v' = \frac{\partial v}{\partial x}\,x' + \frac{\partial v}{\partial y}\,y' + \frac{\partial v}{\partial z}\,z',\quad v'' = \frac{\partial v}{\partial x}\,x'' + \frac{\partial v}{\partial y}\,y'' + \frac{\partial v}{\partial z}\,z'',$$

$$v''' = \frac{\partial v}{\partial x}\,x''' + \frac{\partial v}{\partial y}\,y''' + \frac{\partial v}{\partial z}\,z''';$$

so that, if

$$\bar{J}_x = |\ y'\ ,\ z''\ ,\ v'''\ |,\ \bar{J}_y = |\ z'\ ,\ x''\ ,\ v'''\ |,\ \bar{J}_z = |\ x'\ ,\ y''\ ,\ v'''\ |,\ \bar{J}_v = |\ x'\ ,\ y''\ ,\ z'''\ |,$$

$$= \begin{vmatrix} y'\ ,\ z'\ ,\ v' \\ y''\ ,\ z''\ ,\ v'' \\ y'''\ ,\ z'''\ ,\ v''' \end{vmatrix},\quad = \begin{vmatrix} z'\ ,\ x'\ ,\ v' \\ z''\ ,\ x''\ ,\ v'' \\ z'''\ ,\ x'''\ ,\ v''' \end{vmatrix},\quad = \begin{vmatrix} x'\ ,\ y'\ ,\ v' \\ x''\ ,\ y''\ ,\ v'' \\ x'''\ ,\ y'''\ ,\ v''' \end{vmatrix},\quad = \begin{vmatrix} x'\ ,\ y'\ ,\ z' \\ x''\ ,\ y''\ ,\ z'' \\ x'''\ ,\ y'''\ ,\ z''' \end{vmatrix},$$

we have

$$\frac{1}{\bar{J}_x}\frac{\partial v}{\partial x} = \frac{1}{\bar{J}_y}\frac{\partial v}{\partial y} = \frac{1}{\bar{J}_z}\frac{\partial v}{\partial z} = \frac{1}{\bar{J}_v}.$$

Hence the relation between the subject of integration $\Omega$ and the original subject of integration $f$ is

$$\Omega = \bar{J}_v f\left(x,\,y,\,z,\,v,\,\frac{\bar{J}_x}{\bar{J}_v},\,\frac{\bar{J}_y}{\bar{J}_v},\,\frac{\bar{J}_z}{\bar{J}_v}\right).$$

For a more complete expression of the contribution in question, we write

$$p = \frac{\partial v}{\partial x} = \frac{\bar{J}_x}{\bar{J}_v},\quad q = \frac{\partial v}{\partial y} = \frac{\bar{J}_y}{\bar{J}_v},\quad r = \frac{\partial v}{\partial z} = \frac{\bar{J}_z}{\bar{J}_v}.$$

Now, on writing

$$\mathbf{J}_x = \begin{vmatrix} y'\ ,\ z'\ ,\ v' \\ y''\ ,\ z''\ ,\ v'' \\ m,\ n,\ k \end{vmatrix},\ \mathbf{J}_y = \begin{vmatrix} z'\ ,\ x'\ ,\ v' \\ z''\ ,\ x''\ ,\ v'' \\ n,\ l,\ k \end{vmatrix},\ \mathbf{J}_z = \begin{vmatrix} x'\ ,\ y'\ ,\ v' \\ x''\ ,\ y''\ ,\ v'' \\ l,\ m,\ k \end{vmatrix},\ \mathbf{J}_v = \begin{vmatrix} x'\ ,\ y'\ ,\ z' \\ x''\ ,\ y''\ ,\ z'' \\ l,\ m,\ n \end{vmatrix},$$

we have the relations

$$\left(l\frac{\partial}{\partial x'''} + m\frac{\partial}{\partial y'''} + n\frac{\partial}{\partial z'''} + k\frac{\partial}{\partial v'''}\right)(\bar{J}_x,\,\bar{J}_y,\,\bar{J}_z,\,\bar{J}_v) = (\mathbf{J}_x,\,\mathbf{J}_y,\,\mathbf{J}_z,\,\mathbf{J}_v).$$

Hence

$$\square = l\frac{\partial \Omega}{\partial x'''} + m\frac{\partial \Omega}{\partial y'''} + n\frac{\partial \Omega}{\partial z'''} + k\frac{\partial \Omega}{\partial v'''}$$

$$= \mathbf{J}_x\frac{\partial \Omega}{\partial \bar{J}_x} + \mathbf{J}_y\frac{\partial \Omega}{\partial \bar{J}_y} + \mathbf{J}_z\frac{\partial \Omega}{\partial \bar{J}_z} + \mathbf{J}_v\frac{\partial \Omega}{\partial \bar{J}_v}$$

$$= \mathbf{J}_v\left(f - p\frac{\partial f}{\partial p} - q\frac{\partial f}{\partial q} - r\frac{\partial f}{\partial r}\right) + \mathbf{J}_x\frac{\partial f}{\partial p} + \mathbf{J}_y\frac{\partial f}{\partial q} + \mathbf{J}_z\frac{\partial f}{\partial r}.$$

Consequently, when the original integral has the form

$$\iiint f\,(x,\,y,\,z,\,v,\,p,\,q,\,r)\,dx\,dy\,dz,$$

with the significance $p,\,q,\,r,=\dfrac{\partial v}{\partial x},\,\dfrac{\partial v}{\partial y},\,\dfrac{\partial v}{\partial z}$, the contribution of the rudimentary element $S'S''$ to the constituent $I_C-I_\Gamma$ of the variation of the integral $I$ under the strong variation is

$$-\left\{\mathbf{J}_x\frac{\partial f}{\partial p}+\mathbf{J}_y\frac{\partial f}{\partial q}+\mathbf{J}_z\frac{\partial f}{\partial r}+\mathbf{J}_v\left(f-p\frac{\partial f}{\partial p}-q\frac{\partial f}{\partial q}-r\frac{\partial f}{\partial r}\right)\right\}S'S''D.$$

### Component of the variation of the integral, from the non-characteristic constituent of the strong variation.

**382.** For the contribution to the variation of the integral, from the constituent of the strong variation which is the non-characteristic return from the upper boundary of the contiguous characteristic volume to the upper boundary of the central characteristic volume, we need the integral taken through the volume between these two upper boundaries. Moreover, as the two-way upper boundary is divided into elements for the double integral

$$\iint \Upsilon\,ds'\,ds''$$

in the former constituent of the strong variation, so for the present constituent we resolve the whole of the volume into elements for the triple integral, with lengths $S'$ and $S''$ corresponding to the unchanged boundary base of the volume and the length $D$, in the arbitrary direction $P\bar{P}$ distinct from the direction $ds'''$ and drawn outwards from the characteristic volume.

Any place $\bar{x},\,\bar{y},\,\bar{z},\,\bar{v}$, within this region of integration, is given by the equations

$$\bar{x}=x+x's'+x''s''-l\ (D-t),$$
$$\bar{y}=y+y's'+y''s''-m\,(D-t),$$
$$\bar{z}=z+z's'+z''s''-n\,(D-t),$$
$$\bar{v}=v+v's'+v''s''-k\,(D-t),$$

when $s',\,s'',\,t$, are the three variables of the region, $s'$ ranging from 0 to $S'$, $s''$ ranging from 0 to $S''$, and $t$ from 0 (that is, from $\bar{P}$ the displaced position of $P$) to $D$ (returning, that is, to the original place $P$). Thus, with these variables,

$$\left.\begin{aligned}
\bar{x}_1&=x', & \bar{x}_2&=x'', & \bar{x}_3&=l\\
\bar{y}_1&=y', & \bar{y}_2&=y'', & \bar{y}_3&=m\\
\bar{z}_1&=z', & \bar{z}_2&=z'', & \bar{z}_3&=n\\
\bar{v}_1&=v', & \bar{v}_2&=v'', & \bar{v}_3&=k
\end{aligned}\right\}.$$

The contribution of the constituent to the variation of the integral $I$ under the strong variation is

$$\iiint \overline{\Omega} \, ds' \, ds'' \, dt,$$

where

$$\overline{\Omega} = \Omega \left( \overline{x}, \overline{y}, \overline{z}, \overline{v} \; ; \; x', x'', l \; ; \; y', y'', m \; ; \; z', z'', n \; ; \; v', v'', k \right)$$

$$= \Omega \left( x, y, z, v \; ; \; x', x'', l \; ; \; y', y'', m \; ; \; z', z'', n \; ; \; v', v'', k \right) + [1] = \mathbf{F} + [1],$$

in which $[1]$ represents an aggregate of terms of the first and higher orders of magnitudes in $s'$, $s''$, $t$, that is, in $S'$, $S''$, $D$. Hence, as we are retaining only the important terms in the constituent, the contribution of the triple integral is

$$\mathbf{F} S' S'' D.$$

Now

$$\Omega = \overline{J}_v f \left( x, y, z, v, \frac{\overline{J}_x}{\overline{J}_v}, \frac{\overline{J}_y}{\overline{J}_v}, \frac{\overline{J}_z}{\overline{J}_v} \right);$$

and therefore

$$\mathbf{F} = \mathbf{J}_v f \left( x, y, z, v, \frac{\mathbf{J}_x}{\mathbf{J}_v}, \frac{\mathbf{J}_y}{\mathbf{J}_v}, \frac{\mathbf{J}_z}{\mathbf{J}_v} \right).$$

### *Total strong variation of the integral: the Weierstrass function.*

**383.** Having obtained the expressions for the two constituents, we combine them. The whole variation of the integral, as contributed by the element $S'S''$ of the boundary through the strong variation based on that element with a displacement $D$ out of the characteristic volume, is

$$S'S'' D \,.\, M,$$

where

$$M = \mathbf{J}_v f \left( x, y, z, v, \frac{\mathbf{J}_x}{\mathbf{J}_v}, \frac{\mathbf{J}_y}{\mathbf{J}_v}, \frac{\mathbf{J}_z}{\mathbf{J}_v} \right)$$

$$- \mathbf{J}_x \frac{\partial f}{\partial p} - \mathbf{J}_y \frac{\partial f}{\partial q} - \mathbf{J}_z \frac{\partial f}{\partial r} - \mathbf{J}_v \left( f - p \frac{\partial f}{\partial p} - q \frac{\partial f}{\partial q} - r \frac{\partial f}{\partial r} \right),$$

in the second line of which $f$ denotes

$$f(x, y, z, v, p, q, r).$$

The quantity $\mathbf{J}_v S'S'' D$ admits of simple expression. Consider the elementary solid, with a length $S'$ in the direction $ds'$, that is, $x', y', z', v'$: a length $S''$ in the direction $ds''$, that is, $x'', y'', z'', v''$; and a length $D$ in the direction $dt$, that is, $l, m, n, k$: and imagine the parallelepiped completed, with its corner at $P$ and its three edges in these three directions. Let $\omega$ be the angle between the directions $PS'$ and $PS''$, and $\mu$ the angle between the direction $PD$ and the plane $S'PS''$. Then

$$\sin \mu \sin \omega = (1 - \cos^2 \omega_1 - \cos^2 \omega_2 - \cos^2 \omega + 2 \cos \omega \cos \omega_1 \cos \omega_2)^{\frac{1}{2}},$$

where

$$\cos \omega = x'x'' + y'y'' + z'z'' + v'v'',$$
$$\cos \omega_1 = x'l + y'm + z'n + v'k,$$
$$\cos \omega_2 = x''l + y''m + z''n + v''k,$$

with the conditions

$$\Sigma x'^2 = 1, \quad \Sigma x''^2 = 1, \quad \Sigma l^2 = 1.$$

Now

$$\mathbf{J}_x{}^2 + \mathbf{J}_y{}^2 + \mathbf{J}_z{}^2 + \mathbf{J}_v{}^2 = \begin{vmatrix} \Sigma x'^2 , & \Sigma x'x'', & \Sigma x'l \\ \Sigma x'x'', & \Sigma x''^2 , & \Sigma x''l \\ \Sigma x'l , & \Sigma x''l , & \Sigma l^2 \end{vmatrix} = \begin{vmatrix} 1 , & \cos \omega , & \cos \omega_1 \\ \cos \omega , & 1 , & \cos \omega_2 \\ \cos \omega_1 , & \cos \omega_2 , & 1 \end{vmatrix}$$

If then $\overline{V}$ denote the volume of the parallelepiped $S'S''D$, we have

$$\overline{V} = S'S'' \sin \omega \,.\, D \sin \mu \,;$$

and therefore

$$S'S''D = \overline{V}(\mathbf{J}_x{}^2 + \mathbf{J}_y{}^2 + \mathbf{J}_z{}^2 + \mathbf{J}_v{}^2)^{-\frac{1}{2}},$$

where the radical is to be taken with the positive sign.

Finally, we take

$$S'S''DM = \mathbf{E}\,\overline{V},$$

so that $\mathbf{E}\,\overline{V}$ is the whole of the strong variation, composed of the element $S'S''$ and the displacement $D$; and now

$$\mathbf{E} = \frac{M}{(\mathbf{J}_x{}^2 + \mathbf{J}_y{}^2 + \mathbf{J}_z{}^2 + \mathbf{J}_v{}^2)^{\frac{1}{2}}}$$

$$= \frac{\mathbf{J}_v}{(\mathbf{J}_x{}^2 + \mathbf{J}_y{}^2 + \mathbf{J}_z{}^2 + \mathbf{J}_v{}^2)^{\frac{1}{2}}} f\left(x, y, z, v, \frac{\mathbf{J}_x}{\mathbf{J}_v}, \frac{\mathbf{J}_y}{\mathbf{J}_v}, \frac{\mathbf{J}_z}{\mathbf{J}_v}\right)$$

$$- \frac{1}{(\mathbf{J}_x{}^2 + \mathbf{J}_y{}^2 + \mathbf{J}_z{}^2 + \mathbf{J}_v{}^2)^{\frac{1}{2}}} \left\{ \mathbf{J}_x \frac{\partial f}{\partial p} + \mathbf{J}_y \frac{\partial f}{\partial q} + \mathbf{J}_z \frac{\partial f}{\partial r} + \mathbf{J}_v \left( f - p\frac{\partial f}{\partial p} - q\frac{\partial f}{\partial q} - r\frac{\partial f}{\partial r} \right) \right\}.$$

We call $\mathbf{E}$ the *Weierstrass function* for triple integrals.

### The Weierstrass test for triple integrals.

**384.** In accordance with the definitions of a maximum and a minimum, every increment of the integral due to any small variation, weak or strong, must be a positive increment (an increase) for a minimum and must be a negative increment (a decrease) for a maximum. Moreover, this requirement must be satisfied for every specimen of the small variation, taken individually, and for any aggregate of such specimens taken collectively.

The strong variation, which has just been considered, has been applied to the most rudimentary element of the integral $I$ extended throughout a characteristic volume. In that volume, any place $P$, the variables at which are $x, y, z, v$, has been selected arbitrarily. At that place, and still in the volume, two directions denoted by $ds'$ and $ds''$, have been selected, also arbitrarily, and the selection of each was independent of the selection of the other. Still at that place $P$, but now out of the characteristic volume in the quadruple manifold, another arbitrary direction was selected, denoted by $dt$. In the three directions, denoted by $ds'$, $ds''$, $dt$, three short lengths $S'$, $S''$, $D$, were taken; with these three lengths in the specified directions, two of which belong to the characteristic volume and the third of which does not so belong, a rudimentary volume $\overline{V}$ has been determined. This rudimentary volume provides a rudimentary strong variation of the characteristic volume; and the increment of the integral $I$ for this rudimentary strong variation is

$$\mathbf{E}\,\overline{V}.$$

This rudimentary variation may be imposed alone. In that event, $\mathbf{E}$ must be positive for a minimum, and negative for a maximum. The place $P$ has been chosen unconditionally within the characteristic volume; and, at $P$, two independent directions $ds'$ and $ds''$ have been chosen within the volume, also unconditionally. If, then, we observe the necessary requirement, that $\mathbf{E}$ shall be always positive, or that $\mathbf{E}$ shall be always negative, for each rudimentary variation: any accumulation of small rudimentary variations will provide an increment that is definitely positive in the former alternative, and definitely negative in the latter alternative. There thus will be admitted the possibility of a minimum in the former of the alternatives, and the possibility of a maximum in the latter. The provision for any variation, which is less restricted in extent, is secured by the persistence of the provision for that typical variation, which is the most restricted in extent: rudimentary variations, all in the same sense, accumulate in that sense. Provision has to be made for all possible small variations, within the whole range of their extent. It is therefore initially necessary to have the requirement met for every rudimentary small variation. When the requirement is thus met, it is satisfied cumulatively for variations less restricted in extent. The volume $\overline{V}$ is arbitrary, and is necessarily a positive magnitude.

Accordingly, the $E$-test, for strong variations of the simplest type, can be enunciated in the form:

*The* **E**-*function, constructed for an arbitrary place in a characteristic volume, for any two arbitrary directions in the characteristic volume at that place, and for any arbitrary direction out of the characteristic volume at that place, must be always positive if the integral is to have a minimum value; and it must be always negative if the integral is to have a maximum value.*

This is the extended form of the Weierstrass test.

**385.** We consider some applications of the Weierstrass test: the first is concerned with Dirichlet's Principle.

*Ex.* 1. Consider *the integral*

$$\iiint \left\{ \left(\frac{\partial v}{\partial x}\right)^2 + \left(\frac{\partial v}{\partial y}\right)^2 + \left(\frac{\partial v}{\partial z}\right)^2 \right\} dx\, dy\, dz,$$

*which has been proved* (§ 366) *to satisfy the conditions for a minimum under weak variations, the variable v being determined by the characteristic equation*

$$\frac{\partial^2 v}{\partial x^2} + \frac{\partial^2 v}{\partial y^2} + \frac{\partial^2 v}{\partial z^2} = 0.$$

The subject of integration is given by

$$f = \left(\frac{\partial v}{\partial x}\right)^2 + \left(\frac{\partial v}{\partial y}\right)^2 + \left(\frac{\partial v}{\partial z}\right)^2$$

$$= \frac{1}{J_v^2}(J_x^2 + J_y^2 + J_z^2).$$

Thus the **E**-function is

$$\mathbf{E} = (\mathbf{J}_x^2 + \mathbf{J}_y^2 + \mathbf{J}_z^2 + \mathbf{J}_v^2)^{-\frac{1}{2}} \left\{ \frac{\mathbf{J}_x^2 + \mathbf{J}_y^2 + \mathbf{J}_z^2}{\mathbf{J}_v} - 2\left(\mathbf{J}_x\frac{J_x}{J_v} + \mathbf{J}_y\frac{J_y}{J_v} + \mathbf{J}_z\frac{J_z}{J_v}\right) + \mathbf{J}_v\left(\frac{J_x^2 + J_y^2 + J_z^2}{J_v^2}\right) \right\}$$

$$= \frac{(\mathbf{J}_x J_v - \mathbf{J}_v J_x)^2 + (\mathbf{J}_y J_v - \mathbf{J}_v J_y)^2 + (\mathbf{J}_z J_v - \mathbf{J}_v J_z)^2}{\mathbf{J}_v J_v^2 (\mathbf{J}_x^2 + \mathbf{J}_y^2 + \mathbf{J}_z^2 + \mathbf{J}_v^2)^{\frac{1}{2}}}.$$

Hence the sign of $E$ depends upon the sign of $\mathbf{J}_v$, that is, of

$$\begin{vmatrix} x', & y', & z' \\ x'', & y'', & z'' \\ l, & m, & n \end{vmatrix},$$

where $l, m, n$, are three of the direction-cosines of the arbitrary direction $D$ at $P$.

Whenever the sign of $E$ is positive for a direction $l, m, n, k$, it is negative for the directions $-l, -m, -n, k$, and $-l, -m, -n, -k$. (We need not consider the case when $\mathbf{J}_v$ is zero.) Thus for some directions of a strong variation at any place in the characteristic volume, **E** is positive; for directions in the same line but the opposite sense, **E** is negative.

Hence **E** does not maintain a uniform sign for all strong variations. *The integral does not possess a minimum value for all small variations: it has no true minimum.*

It thus appears that while the integral acquires, for the selected value of $v$, a value which is a minimum for weak variations, that minimum value can be decreased by strong variations.

The so-called "Dirichlet's Principle"—that the triple integral has a minimum—is not admissible: a conclusion to which Weierstrass* appears to have first drawn attention. Upon an assumption of this principle, Lejeune-Dirichlet, Riemann, Thomson and Tait†, and others, based a proof that the potential in space, of a mass of matter, distributed over a surface and attracting according to the Newtonian law, is uniquely determinate. The particular proof is therefore not valid.

The latter theorem by itself, viz. that the potential in question is uniquely determinate with the assigned distribution, must be established otherwise. For the distribution, the value of $v$ is determinate over the surface, and the components of force $-\dfrac{\partial v}{\partial x}, -\dfrac{\partial v}{\partial y}, -\dfrac{\partial v}{\partial z}$, are determinate over the surface; so that we can regard $v$ and its derivatives as definitely given over the surface. Then Cauchy's existence-theorem‡ shews that an integral of the equation

$$\frac{\partial^2 v}{\partial x^2} + \frac{\partial^2 v}{\partial y^2} + \frac{\partial^2 v}{\partial z^2} = 0$$

---

* *Ges. Werke*, vol. ii, pp. 49—53; the date attached to this short paper is 14 July, 1870.

† See Thomson and Tait, *Natural Philosophy*, vol. i, p. 169, vol. ii, p. 52. This proof takes account only of special weak variations of the triple integral.

‡ *Theory of Differential Equations*, vol. vi, chap. xx.

exists, uniquely determinate under the assignment of values of $v$ and of its derivatives along a surface. Consequently, a value of $v$, satisfying the equation and satisfying the assigned surface requirements, and obtained in any manner, is the unique solution of the problem.

*Ex. 2. Find the minimum value of the integral*

$$\iiint \left\{ 1 + \left(\frac{\partial v}{\partial x}\right)^2 + \left(\frac{\partial v}{\partial y}\right)^2 + \left(\frac{\partial v}{\partial z}\right)^2 \right\}^{\frac{1}{2}} dx\, dy\, dz,$$

*v being the dependent variable.*

This problem is the extension, to four-dimensional space with $x$, $y$, $z$, $v$, as coordinates, of the problem of determining the minimal surface in ordinary space, for which

$$\iint \left\{ 1 + \left(\frac{\partial z}{\partial x}\right)^2 + \left(\frac{\partial z}{\partial y}\right)^2 \right\}^{\frac{1}{2}} dx\, dy$$

is a minimum.

The integral, as stated, expresses the element of volume in the quadruple manifold in terms of the projection of that volume into the manifold $v = \text{constant}$. Other forms are

$$\iiint \left\{ 1 + \left(\frac{\partial x}{\partial y}\right)^2 + \left(\frac{\partial x}{\partial z}\right)^2 + \left(\frac{\partial x}{\partial v}\right)^2 \right\}^{\frac{1}{2}} dy\, dz\, dv,$$

$$\iiint \left\{ 1 + \left(\frac{\partial y}{\partial z}\right)^2 + \left(\frac{\partial y}{\partial x}\right)^2 + \left(\frac{\partial y}{\partial v}\right)^2 \right\}^{\frac{1}{2}} dz\, dx\, dv,$$

$$\iiint \left\{ 1 + \left(\frac{\partial z}{\partial x}\right)^2 + \left(\frac{\partial z}{\partial y}\right)^2 + \left(\frac{\partial z}{\partial v}\right)^2 \right\}^{\frac{1}{2}} dx\, dy\, dv,$$

which are equivalent to one another. The single comprehensive form is

$$\iiint \{ J_x^2 + J_y^2 + J_z^2 + J_v^2 \}^{\frac{1}{2}}\, d\rho\, d\sigma\, d\tau,$$

when $x$, $y$, $z$, $v$, are made functions of three new independent variables $\rho$, $\sigma$, $\tau$.

(I) Taking the integral in the form initially stated, we have

$$f = (1 + v_x^2 + v_y^2 + v_z^2)^{\frac{1}{2}} = \Delta^{\frac{1}{2}}.$$

The characteristic equation is

$$\frac{\partial}{\partial x}\left(\frac{v_x}{\Delta}\right) + \frac{\partial}{\partial y}\left(\frac{v_y}{\Delta}\right) + \frac{\partial}{\partial z}\left(\frac{v_z}{\Delta}\right) = 0,$$

which, in full, is *

$$(1 + v_y^2 + v_z^2)\, v_{xx} + (1 + v_z^2 + v_x^2)\, v_{yy} + (1 + v_x^2 + v_y^2)\, v_{zz} - 2\,(v_x v_y v_{xy} + v_y v_z v_{yz} + v_z v_x v_{zx}) = 0.$$

(II) The quantities connected with the Legendre test are

$$\frac{\partial^2 f}{\partial v_x^2} = \Delta^{-\frac{3}{2}}(1 + v_y^2 + v_z^2), \quad \frac{\partial^2 f}{\partial v_y^2} = \Delta^{-\frac{3}{2}}(1 + v_z^2 + v_x^2), \quad \frac{\partial^2 f}{\partial v_z^2} = \Delta^{-\frac{3}{2}}(1 + v_x^2 + v_y^2),$$

$$\frac{\partial^2 f}{\partial v_y \partial v_z} = -\Delta^{-\frac{3}{2}} v_y v_z \quad , \quad \frac{\partial^2 f}{\partial v_z \partial v_x} = -\Delta^{-\frac{3}{2}} v_z v_x \quad , \quad \frac{\partial^2 f}{\partial v_x \partial v_y} = -\Delta^{-\frac{3}{2}} v_x v_y$$

* The equation is an expression of the property that zero is the sum of the three principal curvatures at any point in the characteristic volume, the analogue of the zero mean curvature property (§ 289, Ex. 1) of minimal surfaces in ordinary space.

Then

$$\frac{\partial^2 f}{\partial v_x{}^2}\frac{\partial^2 f}{\partial v_y{}^2} - \left(\frac{\partial^2 f}{\partial v_x \partial v_y}\right)^2 = (1+v_z{}^2)\,\Delta^{-2},$$

$$\frac{\partial^2 f}{\partial v_y{}^2}\frac{\partial^2 f}{\partial v_z{}^2} - \left(\frac{\partial^2 f}{\partial v_y \partial v_z}\right)^2 = (1+v_x{}^2)\,\Delta^{-2},$$

$$\frac{\partial^2 f}{\partial v_z{}^2}\frac{\partial^2 f}{\partial v_x{}^2} - \left(\frac{\partial^2 f}{\partial v_z \partial v_x}\right)^2 = (1+v_y{}^2)\,\Delta^{-2},$$

all of which are positive. Also $\dfrac{\partial^2 f}{\partial v_x{}^2}$, $\dfrac{\partial^2 f}{\partial v_y{}^2}$, $\dfrac{\partial^2 f}{\partial v_z{}^2}$, are positive. Thus the Legendre test is satisfied, and renders a minimum admissible.

(III) In the absence of the primitive of the characteristic equation, the analytical expression connected with the Jacobi test cannot be constructed ; and therefore a precise statement cannot yet be made as to the limitation (if any) upon the range of integration for the triple integral within which a minimum value can be ensured.

(IV) For the Weierstrass test, we have
$$f = (1 + p^2 + q^2 + r^2)^{\frac{1}{2}},$$
in the notation of § 368 (p. 620). We write
$$J_x{}^2 + J_y{}^2 + J_z{}^2 + J_v{}^2 = \Delta^2, \quad \mathbf{J}_x{}^2 + \mathbf{J}_y{}^2 + \mathbf{J}_z{}^2 + \mathbf{J}_v{}^2 = \Delta^2;$$
so that, as
$$p = v_x = \frac{J_x}{J_v}, \quad q = v_y = \frac{J_y}{J_v}, \quad r = v_z = \frac{J_z}{J_v},$$
we have
$$f\left(x,\,y,\,z,\,v,\,\frac{J_x}{J_v},\,\frac{J_y}{J_v},\,\frac{J_z}{J_v}\right) = \frac{\Delta}{J_v},$$
$$f\left(x,\,y,\,z,\,v,\,\frac{\mathbf{J}_x}{\mathbf{J}_v},\,\frac{\mathbf{J}_y}{\mathbf{J}_v},\,\frac{\mathbf{J}_z}{\mathbf{J}_v}\right) = \frac{\Delta}{\mathbf{J}_v}.$$
Also
$$f - p\frac{\partial f}{\partial p} - q\frac{\partial f}{\partial q} - r\frac{\partial f}{\partial r} = \frac{1}{f} = \frac{J_v}{\Delta},$$
$$\frac{\partial f}{\partial p} = \frac{J_x}{\Delta}, \quad \frac{\partial f}{\partial q} = \frac{J_y}{\Delta}, \quad \frac{\partial f}{\partial r} = \frac{J_z}{\Delta}.$$

Hence the $E$-function is given by
$$\mathbf{E} = \frac{\mathbf{J}_v}{\Delta}\cdot\frac{\Delta}{\mathbf{J}_v} - \frac{1}{\Delta}\left(\mathbf{J}_x\frac{J_x}{\Delta} + \mathbf{J}_y\frac{J_y}{\Delta} + \mathbf{J}_z\frac{J_z}{\Delta} + \mathbf{J}_v\frac{J_v}{\Delta}\right)$$
$$= 1 - \frac{1}{\Delta\Delta}\,(J_x\mathbf{J}_x + J_y\mathbf{J}_y + J_z\mathbf{J}_z + J_v\mathbf{J}_v).$$

Now, having regard to the values of $J_x$, $J_y$, $J_z$, $J_v$; $\mathbf{J}_x$, $\mathbf{J}_y$, $\mathbf{J}_z$, $\mathbf{J}_v$, we note that $\dfrac{J_x}{\Delta}$, $\dfrac{J_y}{\Delta}$, $\dfrac{J_z}{\Delta}$, $\dfrac{-J_v}{\Delta}$, are (§ 376, p. 631) the direction-cosines $L, M, N, K$, of a direction in the quadruple manifold, perpendicular to each of the three directions $x'$, $y'$, $z'$, $v'$ ; $x''$, $y''$, $z''$, $v''$ ; $x'''$, $y'''$, $z'''$, $v'''$, in the characteristic volume : while $\dfrac{\mathbf{J}_x}{\Delta}$, $\dfrac{\mathbf{J}_y}{\Delta}$, $\dfrac{\mathbf{J}_z}{\Delta}$, $\dfrac{-\mathbf{J}_v}{\Delta}$, are the direction-cosines $\mathbf{L}, \mathbf{M}, \mathbf{N}, \mathbf{K}$, of a direction in the quadruple manifold, perpendicular to each of the three directions $x'$, $y'$, $z'$, $v'$ ; $x''$, $y''$, $z''$, $v''$ ; $l, m, n, k$ ; the first two of which directions lie in the characteristic volume and the third of which passes out of that volume. If, then, $\vartheta$ be the inclination of these two directions, we have
$$\mathbf{E} = 1 - (L\mathbf{L} + M\mathbf{M} + N\mathbf{N} + K\mathbf{K})$$
$$= 1 - \cos\vartheta.$$

The $E$-function is therefore positive*, unless $\mathbf{L}=L$, $\mathbf{M}=M$, $\mathbf{N}=N$, $\mathbf{K}=K$, relations which imply that the strong variation does not march out of the volume. The integral therefore is a minimum for strong variations, as well as for weak variations.

The *solution of the characteristic equation therefore provides a true minimum* within the limitations (if any, in a particular case) due to the Jacobi test.

*Ex.* 3. *Any integral*

$$\iiint \phi\,(x,\,y,\,z,\,v) \left\{ 1 + \left(\frac{\partial v}{\partial x}\right)^2 + \left(\frac{\partial v}{\partial y}\right)^2 + \left(\frac{\partial v}{\partial z}\right)^2 \right\}^{\frac{1}{2}} dx\,dy\,dz$$

*satisfies the Weierstrass test for the possession of a minimum.*

In the construction of the $E$-function for this integral, the analysis differs, solely by the persistence of a factor $\phi\,(x,\,y,\,z,\,v)$ in $f$ and in each derivative of $f$, from the analysis leading to the construction of the $E$-function in the preceding example.

Accordingly, for the integral, we find

$$\mathbf{E} = (1 - \cos\vartheta)\,\phi\,(x,\,y,\,z,\,v),$$

which is always positive unless $\vartheta = 0$: in which case there is no strong variation of the characteristic volume.

Because $\mathbf{E}$ is always positive, the integral admits a minimum for strong variations.

## *Conclusion.*

**386.** On the attainment of this stage, the present account of the Calculus of Variations will end. Several matters of enquiry have been omitted during its course, even within the selected range; four of these may be mentioned at the close.

For single integrals of the first order, which involve one dependent variable, the relation between an initial limit and its conjugate is reciprocal; the critical function (§ 66) shews that the initial point is the rear conjugate of its own forward conjugate. But this reciprocal quality is not proved to be possessed for single integrals of the second order, involving one dependent variable; the critical function (§ 117) is not symmetrical between an initial point $P$ and its conjugate $Q$, the limiting position of the intersection of the characteristic curve by a consecutive characteristic. With the initial point $P$,

---

* The positive value can be deduced also from the property

$$\Delta^2\Delta^2 - (J_x\mathbf{J}_x + J_y\mathbf{J}_y + J_z\mathbf{J}_z + J_v\mathbf{J}_v)^2$$

$$= \begin{vmatrix} x', & y', & z', & v' \\ x'', & y'', & z'', & v'' \\ x''', & y''', & z''', & v''' \\ l, & m, & n, & k \end{vmatrix}^2 \{(x'^2 + y'^2 + z'^2 + v'^2)(x''^2 + y''^2 + z''^2 + v''^2) - (x'x'' + y'y'' + z'z'' + v'v'')^2\}$$

$$= \begin{vmatrix} x', & y', & z', & v' \\ x'', & y'', & z'', & v'' \\ x''', & y''', & z''', & v''' \\ l, & m, & n, & k \end{vmatrix}^2 \sin^2\omega,$$

where $\omega$ is the angle between the directions $x'$, $y'$, $z'$, $v'$, and $x''$, $y''$, $z''$, $v''$.

a conjugate $R$ can be differently chosen, as the limiting position of the intersection of the characteristic curve by a subconsecutive characteristic; the critical function (§ 121) shews that the relation between $P$ and $R$ is reciprocal. Investigation, as to the relative positions of $Q$ and $R$, and as to the relative positions of $P$ and the rear conjugate of $Q$, is desirable.

In the class of isoperimetrical problems (§ 262), where explicit relations bind variables in an integral to be maximised or minimised, parametric multipliers have been introduced (§§ 266, 273). They occur in the characteristic equations. Investigation of their influence, upon the expression of the normal form of the second variation and upon the derived tests, is suggested.

Not infrequently, it is declared that the characteristic surface for a double integral of the first order is determinate under the condition of passing through two curves in space. The Cauchy primitive of the corresponding characteristic equation is determinate (§ 282) under the condition of touching an assigned developable along an assigned curve. Are the two determinations equivalent? If so, is there any limitation on the form or on the position of the final curve, similar to the limitation (§§ 70—72) in the determination of the characteristic curve for a single integral of the first order?

The formation of the Weierstrass function for strong variations of single integrals of the second order has received little discussion; the way to fuller development is free. General weak variations of double integrals of the second order have been considered only to a slight extent; strong variations of such integrals, requiring the consideration of the curvature of characteristic surfaces, await investigation.

# INDEX.

*(Arabic figures refer to pages.)*

Printed by Printforce, United Kingdom